Lecture Notes in Computer Science

Lecture Notes in Artificial Intelligence 14179

Founding Editor

Jörg Siekmann

Series Editors

Randy Goebel, *University of Alberta, Edmonton, Canada*
Wolfgang Wahlster, *DFKI, Berlin, Germany*
Zhi-Hua Zhou, *Nanjing University, Nanjing, China*

The series Lecture Notes in Artificial Intelligence (LNAI) was established in 1988 as a topical subseries of LNCS devoted to artificial intelligence.

The series publishes state-of-the-art research results at a high level. As with the LNCS mother series, the mission of the series is to serve the international R & D community by providing an invaluable service, mainly focused on the publication of conference and workshop proceedings and postproceedings.

Xiaochun Yang · Heru Suhartanto ·
Guoren Wang · Bin Wang · Jing Jiang · Bing Li ·
Huaijie Zhu · Ningning Cui
Editors

Advanced Data Mining and Applications

19th International Conference, ADMA 2023
Shenyang, China, August 21–23, 2023
Proceedings, Part IV

 Springer

Editors
Xiaochun Yang
Northeastern University
Shenyang, China

Guoren Wang
Beijing Institute of Technology
Beijing, China

Jing Jiang
University of Technology Sydney
Sydney, NSW, Australia

Huaijie Zhu
Sun Yat-sen University
Guangzhou, China

Heru Suhartanto
The University of Indonesia
Depok, Indonesia

Bin Wang
Northeastern University
Shenyang, China

Bing Li
Agency for Science, Technology
and Research (A*STAR)
Singapore, Singapore

Ningning Cui
Anhui University
Hefei, China

ISSN 0302-9743   ISSN 1611-3349 (electronic)
Lecture Notes in Artificial Intelligence
ISBN 978-3-031-46673-1   ISBN 978-3-031-46674-8 (eBook)
https://doi.org/10.1007/978-3-031-46674-8

LNCS Sublibrary: SL7 – Artificial Intelligence

This Springer imprint is published by the registered company Springer Nature Switzerland AG
The registered company address is: Gewerbestrasse 11, 6330 Cham, Switzerland

Paper in this product is recyclable.

Preface

The 19th International Conference on Advanced Data Mining and Applications (ADMA 2023) was held in Shenyang, China, during August 21–23, 2023. Researchers and practitioners from around the world came together at this leading international forum to share innovative ideas, original research findings, case study results, and experienced insights into advanced data mining and its applications. With the ever-growing importance of appropriate methods in these data-rich times, ADMA has become a flagship conference in this field. ADMA 2023 received a total of 503 submissions from 22 countries across five continents. After a rigorous double-blind review process involving 318 reviewers, 216 regular papers were accepted to be published in the proceedings, 123 were selected to be delivered as oral presentations at the conference, 85 were selected as poster presentations, and 8 were selected as industry papers. This corresponds to a full oral paper acceptance rate of 24.4%. The Program Committee (PC), composed of international experts in relevant fields, did a thorough and professional job of reviewing the papers submitted to ADMA 2023, and each paper was reviewed by an average of 2.97 PC members. With the growing importance of data in this digital age, papers accepted at ADMA 2023 covered a wide range of research topics in the field of data mining, including pattern mining, graph mining, classification, clustering and recommendation, multi-objective, optimization, augmentation, and database, data mining theory, image, multimedia and time series data mining, text mining, web and IoT applications, finance and healthcare. It is worth mentioning that ADMA 2023 was organized as a physical-only event, allowing for in-person gatherings and networking. We thank the PC members for completing the review process and providing valuable comments within tight schedules. The high-quality program would not have been possible without the expertise and dedication of our PC members. Moreover, we would like to take this valuable opportunity to thank all authors who submitted technical papers and contributed to the tradition of excellence at ADMA. We firmly believe that many colleagues will find the papers in these proceedings exciting and beneficial for advancing their research. We would like to thank Microsoft for providing the CMT system, which is free to use for conference organization, Springer for their long-term support, the host institution, Northeastern University, for their hospitality and support, Niu Translation and Shuangzhi Bo for their sponsorship. We are grateful for the guidance of the steering committee members, Osmar R. Zaiane, Chengqi Zhang, Michael Sheng, Guodong Long, Xue Li, Jianxin Li, and Weitong Chen. With their leadership and support, the conference ran smoothly. We also would like to acknowledge the support of the other members of the organizing committee. All of them helped to make ADMA 2023 a success. We appreciate the local arrangements, registration and finance management from the local arrangement chairs, registration management chairs and finance chairs Kui Di, Baoyan Song, Junchang Xin, Donghong Han, Guoqiang Ma, Yuanguo Bi, and Baiyou Qiao, the time and effort of the proceedings chairs, Bing Li, Huaijie Zhu, and Ningning Cui, the effort in advertising the conference by the publicity chairs and social network and social media coordination chairs, Xin Wang, Yongxin

Tong, Lina Wang, and Sen Wang, and the effort of managing the Tutorial sessions by the tutorial chairs, Zheng Zhang and Shuihua Wang, We would like to give very special thanks to the web chair, industry chairs, and PhD school chairs Faming Li, Chi Man Pun, Sen Wang, Linlin Ding, M. Emre Celebi, and Zheng Zhang, for creating a successful and memorable event. We also thank sponsorship chair Hua Shao for his sponsorship. Finally, we would like to thank all the other co-chairs who have contributed to the conference.

August 2023   Xiaochun Yang
                                                                         Bin Wang
                                                                       Jing Jiang

Organization

Chair of the Steering Committee

Xue Li   University of Queensland, Australia

Steering Committee

Osmar R. Zaiane   University of Alberta, Canada
Chengqi Zhang   Sydney University of Technology, Australia
Michael Sheng   Macquarie University, Australia
Guodong Long   Sydney University of Technology, Australia
Xue Li   University of Queensland, Australia
Jianxin Li   Deakin University, Australia
Weitong Chen   Adelaide University, Australia

Honor Chairs

Xingwei Wang   Northeastern University, China
Xuemin Lin   Shanghai Jiao Tong University, China
Ge Yu   Northeastern University, China

General Chairs

Xiaochun Yang   Northeastern University, China
Heru Suhartanto   The University of Indonesia, Indonesia
Guoren Wang   Beijing Institute of Technology, China

Program Chairs

Bin Wang   Northeastern University, China
Jing Jiang   University of Technology Sydney, Australia

Local Arrangement Chairs

Kui Di   Northeastern University, China
Baoyan Song   Liaoning University, China
Junchang Xin   Northeastern University, China

Registration Management Chairs

Donghong Han   Northeastern University, China
Guoqiang Ma   Northeastern University, China
Yuanguo Bi   Northeastern University, China

Finance Chair

Baiyou Qiao   Northeastern University, China

Sponsorship Chair

Hua Shao   Shenyang Huaruibo Information Technology Co.,
                            Ltd., China

Publicity Chairs

Xin Wang   Tianjin University, China
Yongxin Tong   Beihang University, China
Lina Wang   Wuhan University, China

Social Network and Social Media Coordination Chair

Sen Wang   University of Queensland, Australia

Proceeding Chairs

Bing Li   Agency for Science, Technology and Research
                            (A*STAR), Singapore
Huaijie Zhu   Sun Yat-sen University, China
Ningning Cui   Anhui University, China

Tutorial Chairs

Zheng Zhang   Harbin Institute of Technology, Shenzhen, China
Shuihua Wang   University of Leicester, UK

Web Chair

Faming Li   Northeastern University, China

Industry Chairs

Chi Man Pun   University of Macau, China
Sen Wang   University of Queensland, Australia
Linlin Ding   Liaoning University, China

PhD School Chairs

M. Emre Celebi   University of Central Arkansas, USA
Zheng Zhang   Harbin Institute of Technology, Shenzhen, China

Program Committee

Meta Reviewers

Bohan Li   Nanjing University of Aeronautics and
                                     Astronautics, China
Can Wang   Griffith University, Australia
Chaokun Wang   Tsinghua University, China
Cheqing Jin   East China Normal University, China
Guodong Long   University of Technology Sydney, Australia
Hongzhi Wang   Harbin Institute of Technology, China
Huaijie Zhu   Sun Yat-sen University, China
Jianxin Li   Deakin University, Australia
Jun Gao   Peking University, China
Lianhua Chi   La Trobe University, Australia
Lin Yue   University of Newcastle, Australia
Tao Shen   University of Technology Sydney, Australia

Wei Emma Zhang	University of Adelaide, Australia
Weitong Chen	Adelaide University, Australia
Xiang Lian	Kent State University, USA
Xiaoling Wang	East China Normal University, China
Xueping Peng	University of Technology Sydney, Australia
Xuyun Zhang	Macquarie University, Australia
Yanjun Zhang	Deakin University, Australia
Zheng Zhang	Harbin Institute of Technology, Shenzhen, China

Reviewers

Abdulwahab Aljubairy	Macquarie University, Australia
Adita Kulkarni	SUNY Brockport, USA
Ahoud Alhazmi	Macquarie University, Australia
Akshay Peshave	GE Research, USA
Alex Delis	Univ. of Athens, Greece
Alexander Zhou	Hong Kong University of Science and Technology, China
Baoling Ning	Heilongjiang University, China
Bin Zhao	Nanjing Normal University, China
Bing Li	Institute of High Performance Computing, A*STAR, Singapore
Bo Tang	Southern University of Science and Technology, China
Carson Leung	University of Manitoba, Canada
Changdong Wang	Sun Yat-sen University, China
Chao Zhang	Tsinghua University, China
Chaokun Wang	Tsinghua University, China
Chaoran Huang	University of New South Wales, Australia
Chen Wang	Chongqing University, China
Chengcheng Yang	East China Normal University, China
Chenhao Zhang	University of Queensland, Australia
Cheqing Jin	East China Normal University, China
Chuan Ma	Zhejiang Lab, China
Chuan Xiao	Osaka University and Nagoya University, Japan
Chuanyu Zong	Shenyang Aerospace University, China
Congbo Ma	University of Adelaide, Australia
Dan He	University of Queensland, Australia
David Broneske	German Centre for Higher Education Research and Science Studies, Germany

Dechang Pi	Nanjing University of Aeronautics and Astronautics, China
Derong Shen	Northeastern University, China
Dima Alhadidi	University of New Brunswick, Canada
Dimitris Kotzinos	ETIS, France
Dong Huang	South China Agricultural University, China
Dong Li	Liaoning University, China
Dong Wen	University of New South Wales, Australia
Dongxiang Zhang	Zhejiang University, China
Dongyuan Tian	Jilin University, China
Dunlu Peng	University of Shanghai for Science and Technology, China
Eiji Uchino	Yamaguchi University, Japan
Ellouze Mourad	University of Sfax, Tunisia
Elsa Negre	LAMSADE, Paris-Dauphine University, France
Faming Li	Northeastern University, China
Farid Nouioua	Université Mohamed El Bachir El Ibrahimi de Bordj Bou Arréridj, Algeria
Genoveva Vargas-Solar	CNRS, France
Gong Cheng	Nanjing University, China
Guanfeng Liu	Macquarie University, Australia
Guangquan Lu	Guangxi Normal University, China
Guangyan Huang	Deakin University, Australia
Guannan Dong	University of Macau, China
Guillaume Guerard	ESILV, France
Guodong Long	University of Technology Sydney, Australia
Haïfa Nakouri	ISG Tunis, Tunisia
Hailong Liu	Northwestern Polytechnical University, China
Haojie Zhuang	University of Adelaide, Australia
Haoran Yang	University of Technology Sydney, Australia
Haoyang Luo	Harbin Institute of Technology (Shenzhen), China
Hongzhi Wang	Harbin Institute of Technology, China
Huaijie Zhu	Sun Yat-sen University, China
Hui Yin	Deakin University, Australia
Indika Priyantha Kumara Dewage	Tilburg University, The Netherlands
Ioannis Konstantinou	University of Thessaly, Greece
Jagat Challa	BITS Pilani, India
Jerry Chun-Wei Lin	Western Norway University of Applied Sciences, Norway
Jiabao Han	NUDT, China
Jiajie Xu	Soochow University, China
Jiali Mao	East China Normal University, China

Jianbin Qin	Shenzhen University, China
Jianhua Lu	Southeast University, China
Jianqiu Xu	Nanjing University of Aeronautics and Astronautics, China
Jianxin Li	Deakin University, Australia
Jianxing Yu	Sun Yat-sen University, China
Jiaxin Jiang	National University of Singapore, Singapore
Jiazun Chen	Peking University, China
Jie Shao	University of Electronic Science and Technology of China, China
Jie Wang	Indiana University, USA
Jilian Zhang	Jinan University, China, China
Jingang Yu	Shenyang Institute of Computing Technology, Chinese Academy of Sciences
Jing Du	University of New South Wales, Australia
Jules-Raymond Tapamo	University of KwaZulu-Natal, South Africa
Jun Gao	Peking University, China
Junchang Xin	Northeastern University, China
Junhu Wang	Griffith University, Australia
Junshuai Song	Peking University, China
Kai Wang	Shanghai Jiao Tong University, China
Ke Deng	RMIT University, Australia
Kun Han	University of Queensland, Australia
Kun Yue	Yunnan University, China
Ladjel Bellatreche	ISAE-ENSMA, France
Lei Duan	Sichuan University, China
Lei Guo	Shandong Normal University, China
Lei Li	Hong Kong University of Science and Technology (Guangzhou), China
Li Li	Southwest University, China
Lin Guo	Changchun University of Science and Technology, China
Lin Mu	Anhui University, China
Linlin Ding	Liaoning University, China
Lizhen Cui	Shandong University, China
Long Yuan	Nanjing University of Science and Technology, China
Lu Chen	Swinburne University of Technology, Australia
Lu Jiang	Northeast Normal University, China
Lukui Shi	Hebei University of Technology, China
Maneet Singh	IIT Ropar, India
Manqing Dong	Macquarie University, Australia

Mariusz Bajger	Flinders University, Australia
Markus Endres	University of Applied Sciences Munich, Germany
Mehmet Ali Kaygusuz	Middle East Technical University, Turkey
Meng-Fen Chiang	University of Auckland, New Zealand
Ming Zhong	Wuhan University, China
Minghe Yu	Northeastern University, China
Mingzhe Zhang	University of Queensland, Australia
Mirco Nanni	CNR-ISTI Pisa, Italy
Misuk Kim	Sejong University, South Korea
Mo Li	Liaoning University, China
Mohammad Alipour Vaezi	Virginia Tech, USA
Mourad Nouioua	Mohamed El Bachir El Ibrahimi University, Bordj Bou Arreridj, Algeria
Munazza Zaib	Macquarie University, Australia
Nabil Neggaz	Université des Sciences et de la Technologie d'Oran Mohamed Boudiaf, Algeria
Nicolas Travers	Léonard de Vinci Pôle Universitaire, Research Center, France
Ningning Cui	Anhui University, China
Paul Grant	Charles Sturt University, Australia
Peiquan Jin	University of Science and Technology of China, China
Peng Cheng	East China Normal University, China
Peng Peng	Hunan University, China
Peng Wang	Fudan University, China
Pengpeng Zhao	Soochow University, China
Philippe Fournier-Viger	Shenzhen University, China
Ping Lu	Beihang University, China
Pinghui Wang	Xi'an Jiaotong University, China
Qiang Yin	Shanghai Jiao Tong University, China
Qing Liao	Harbin Institute of Technology (Shenzhen), China
Qing Liu	Data61, CSIRO, Australia
Qing Xie	Wuhan University of Technology, China
Quan Chen	Guangdong University of Technology, China
Quan Z. Sheng	Macquarie University, Australia
Quoc Viet Hung Nguyen	Griffith University, Australia
Rania Boukhriss	University of Sfax, Tunisia
Riccardo Cantini	University of Calabria, Italy
Rogério Luís Costa	Polytechnic of Leiria, Portugal
Rong Zhu	Alibaba Group, China
Ronghua Li	Beijing Institute of Technology, China
Rui Zhou	Swinburne University of Technology, Australia

Rui Zhu	Shenyang Aerospace University, China
Sadeq Darrab	Otto von Guericke University Magdeburg, Germany
Saiful Islam	Griffith University, Australia
Sayan Unankard	Maejo University, Thailand
Senzhang Wang	Central South University, China
Shan Xue	University of Wollongong, Australia
Shaofei Shen	University of Queensland, Australia
Shi Feng	Northeastern University, China
Shiting Wen	Zhejiang University, China
Shiyu Yang	Guangzhou University, China
Shouhong Wan	University of Science and Technology of China, China
Shuhao Zhang	Singapore University of Technology and Design, Singapore
Shuiqiao Yang	UNSW, Australia
Shuyuan Li	Beihang University, China
Silvestro Roberto Poccia	University of Turin, Italy
Sonia Djebali	Léonard de Vinci Pôle Universitaire, Research Center, France
Suman Banerjee	IIT Jammu, India
Tao Qiu	Shenyang Aerospace University, China
Tao Zhao	National University of Defense Technology, China
Tarique Anwar	University of York, UK
Thanh Tam Nguyen	Griffith University, Australia
Theodoros Chondrogiannis	University of Konstanz, Germany
Tianrui Li	Southwest Jiaotong University, China
Tianyi Chen	Peking University, China
Tieke He	Nanjing University, China
Tiexin Wang	Nanjing University of Aeronautics and Astronautics, China
Tiezheng Nie	Northeastern University, China
Uno Fang	Deakin University, Australia
Wei Chen	University of Auckland, New Zealand
Wei Deng	Southwestern University of Finance and Economics, China
Wei Hu	Nanjing University, China
Wei Li	Harbin Engineering University, China
Wei Liu	University of Macau, Sun Yat-sen University, China
Wei Shen	Nankai University, China
Wei Song	Wuhan University, China

Weijia Zhang	University of Newcastle, Australia
Weiwei Ni	Southeast University, China
Weixiong Rao	Tongji University, China
Wen Zhang	Wuhan University, China
Wentao Li	Hong Kong University of Science and Technology (Guangzhou), China
Wenyun Li	Harbin Institute of Technology (Shenzhen), China
Xi Guo	University of Science and Technology Beijing, China
Xiang Lian	Kent State University, USA
Xiangguo Sun	Chinese University of Hong Kong, China
Xiangmin Zhou	RMIT University, Australia
Xiangyu Song	Swinburne University of Technology, Australia
Xianmin Liu	Harbin Institute of Technology, China
Xianzhi Wang	University of Technology Sydney, Australia
Xiao Pan	Shijiazhuang Tiedao University, China
Xiaocong Chen	University of New South Wales, Australia
Xiaofeng Gao	Shanghai Jiaotong University, China
Xiaoguo Li	Singapore Management University, Singapore
Xiaohui (Daniel) Tao	University of Southern Queensland, Australia
Xiaoling Wang	East China Normal University, China
Xiaowang Zhang	Tianjin University, China
Xiaoyang Wang	University of New South Wales, Australia
Xiaojun Xie	Nanjing Agricultural University, China
Xin Cao	University of New South Wales, Australia
Xin Wang	Southwest Petroleum University, China
Xinqiang Xie	Neusoft, China
Xiuhua Li	Chongqing University, China
Xiujuan Xu	Dalian University of Technology, China
Xu Yuan	Harbin Institute of Technology, Shenzhen, China
Xu Zhou	Hunan University, China
Xupeng Miao	Carnegie Mellon University, USA
Xuyun Zhang	Macquarie University, Australia
Yajun Yang	Tianjin University, China
Yanda Wang	Nanjing University of Aeronautics and Astronautics, China
Yanfeng Zhang	Northeastern University, China
Yang Cao	Hokkaido University, China
Yang-Sae Moon	Kangwon National University, South Korea
Yanhui Gu	Nanjing Normal University, China
Yanjun Shu	Harbin Institute of Technology, China
Yanlong Wen	Nankai University, China

Yanmei Hu	Chengdu University of Technology, China
Yao Liu	University of New South Wales, Australia
Yawen Zhao	University of Queensland, Australia
Ye Zhu	Deakin University, Australia
Yexuan Shi	Beihang University, China
Yicong Li	University of Technology Sydney, Australia
Yijia Zhang	Jilin University, China
Ying Zhang	Nankai University, China
Yingjian Li	Harbin Institute of Technology, Shenzhen, China
Yingxia Shao	BUPT, China
Yishu Liu	Harbin Institute of Technology, Shenzhen, China
Yishu Wang	Northeastern University, China
Yixiang Fang	Chinese University of Hong Kong, Shenzhen, China
Yixuan Qiu	The University of Queensland, Australia
Yong Zhang	Tsinghua University, China
Yongchao Liu	Ant Group, China
Yongpan Sheng	Southwest University, China
Yongqing Zhang	Chengdu University of Information Technology, China
Youwen Zhu	Nanjing University of Aeronautics and Astronautics, China
Yu Gu	Northeastern University, China
Yu Liu	Huazhong University of Science and Technology, China
Yu Yang	Hong Kong Polytechnic University, China
Yuanbo Xu	Jilin University, China
Yucheng Zhou	University of Technology Sydney, Australia
Yue Tan	University of Technology Sydney, Australia
Yunjun Gao	Zhejiang University, China
Yunzhang Huo	Hong Kong Polytechnic University, China
Yurong Cheng	Beijing Institute of Technology, China
Yutong Han	Dalian Minzu University, China
Yutong Qu	University of Adelaide, Australia
Yuwei Peng	Wuhan University, China
Yuxiang Zeng	Hong Kong University of Science and Technology, China
Zesheng Ye	University of New South Wales, Sydney, Australia
Zhang Anzhen	Shenyang Aerospace University, China
Zhaojing Luo	National University of Singapore, Singapore
Zhaonian Zou	Harbin Institute of Technology, China
Zheng Liu	Nanjing University of Posts and Telecommunications, China

Zhengyi Yang	University of New South Wales, Australia
Zhenying He	Fudan University, China
Zhihui Wang	Fudan University, China
Zhiwei Zhang	Beijing Institute of Technology, China
Zhixin Li	Guangxi Normal University, China
Zhongnan Zhang	Xiamen University, China
Ziyang Liu	Tsinghua University, China

Contents – Part IV

Deep Learning

SL-TeaE: An Efficient Method for Improving the Precision of Teaching Evaluation

Xianzhi Huang[1], Lina Chen[2(✉)], Yuzhou Zheng[3], Hongjie Guo[2], Fangyao Shen[2], and Hong Gao[2]

[1] College of Physics and Electronic Information Engineering, Zhejiang Normal University, Jinhua, China
`huangxianzhi@zjnu.edu.cn`
[2] School of Computer Science and Technology, Zhejiang Normal University, Jinhua, China
`chenlina@zjnu.cn`, `{guohongjie,fyshen,honggao}@zjnu.edu.cn`
[3] School of Mechanical Engineering, Jiangnan University, Wuxi, China
`1985592749@qq.com`

Abstract. This study introduces an effective method for analyzing emotional states in evaluation data with enhanced precision by constructing a sentiment lexicon for teaching evaluation (SL-TeaE). We expand a general basic sentiment lexicon based on teaching evaluation data from our university's academic system by creating a list of adverbs of degree and negative words. We use the TextRank algorithm to select sentiment seed words from user data and the SO-PMI algorithm to generate a user-based domain sentiment vocabulary in the teaching domain. Finally, we merge the user-based domain sentiment vocabulary and the expanded general foundation sentiment lexicon to construct the SL-TeaE. The experimental results indicate that the proposed method has excellent performance in both sentiment classification and quantitative evaluation score. Specifically, the F1 values for positive and negative teaching comments have been improved by 27.3% and 16.7%, respectively, compared to the general basic sentiment lexicon. Furthermore, the sentiment classification performance of our method surpasses that of commonly used sentiment classification algorithms. In addition, compared to the general basic sentiment lexicon and the expanded general basic sentiment lexicon, the proposed method in this paper achieves the lowest mean absolute error and root mean square error for course comprehensive evaluation scores (MAE = 2.57, RMSE = 2.62).

Keywords: Sentiment Lexicon · Teaching Evaluation · SO-PMI Algorithm · Sentiment Classification

1 Introduction

With the advancement of educational digitization, Student Evaluations of Teaching (SET) have become an increasingly important method in education reform and have attracted extensive attention. Despite the widespread implementation of online SET activities in many universities, the lack of domain-specific sentiment lexicons has resulted in large discrepancies between evaluation results and actual values, which is not conducive to the development and promotion of SET activities.

X. Yang et al. (Eds.): ADMA 2023, LNAI 14179, pp. 3–17, 2023.
https://doi.org/10.1007/978-3-031-46674-8_1

Student Evaluations of Teaching (SET) are a valuable tool for collecting feedback from students about teaching quality in courses. Analyzing the underlying emotional states in SET can identify existing problems in university evaluations and provide solutions to improve teaching quality and students' learning interests. However, despite the widespread implementation of online SET activities in many universities, a lack of domain-specific sentiment lexicons has resulted in significant discrepancies between evaluation results and actual values. This situation is not conducive to the development and promotion of SET activities. The construction of a sentiment lexicon for teaching evaluation offers a new research perspective to address this issue.

Sentiment analysis (SA) is a method that analyzes the emotional tendencies expressed in texts with subjective descriptions. Due to its ability to handle unstructured data, such as texts, many researchers have applied SA to the field of education. For example, F. Balahadia et al. [2] used the sentiments expressed in SET as a performance evaluation standard for teachers and developed a performance evaluation system based on sentiment analysis and opinion mining. Lin et al. [3] utilized multiple machine learning methods to automatically extract sentiment information from SET and applied it to the student evaluation section of the college teaching management system. Wang et al. [4] used a sentiment lexicon approach to analyze the emotional tendencies of educational news texts, combining the sentiment lexicon to calculate the emotional weights of different categories of news texts to obtain sentiment classification results. Taboada et al. [5] utilized a sentiment lexicon and grammar rules to calculate the sentiment score of text and determine the sentiment of the text. However, these studies relied on general sentiment lexicons, which restricted the exploration of deeper information embedded in data and may have impacted the objectivity and accuracy of sentiment analysis.

To address the above issues, this study introduces SL-TeaE (sentiment lexicon for teaching evaluation), a professional sentiment lexicon specifically designed for the field of teaching evaluation. Additionally, we propose a sentiment analysis method that utilizes SL-TeaE to analyze student evaluations of teaching. The main contributions of this method are as follows:

1. User-based domain sentiment vocabulary mining. The TextRank algorithm is used to select sentiment seed words, and the SO-PMI algorithm generates a user-based domain sentiment vocabulary in the field of education, improving the model's generalization ability and the accuracy of sentiment classification.
2. Expansion of a general foundation sentiment lexicon. By assigning different weights to adverbs of degree using the gradient descent formula and constructing a negative word list based on negation judgment, a precise analysis of changes in emotions expressed in evaluation comments can be achieved.
3. Generation of a sentiment lexicon for evaluating teaching. The user-based domain sentiment vocabulary is merged into the expanded general foundation sentiment lexicon, resulting in a more effective sentiment lexicon for evaluating teaching.

2 Related Work

Cai et al. [6] proposed that using different sentiment lexicons for different fields, which include domain-specific professional words in the lexicon, can achieve better performance. There are three prevalent methods for constructing sentiment lexicons: manual annotation, knowledge-based methods, and corpus-based methods. Although manually generated lexicons being universally applicable, their huge labor costs make it difficult to cover emotional words in different fields and result in poor domain adaptation.

Knowledge-based methods utilize resources such as Wordnet and HowNet to determine the polarity of sentimental words by analyzing conceptual relationships in the knowledge base (e.g., synonymy, antonymy, hypernymy/hyponymy, etc.). Liu et al. [7] expanded HowNet by choosing a set of frequently used emotional words to create a basic emotional lexicon, calculating each word's semantic distance score to obtain its emotional orientation value. Yang et al. [8] employed both HowNet and NTUSD while computing similarity or frequency statistics to determine the emotional tendencies of words. Zhou et al. [9] adopted a cross-lingual technique to extract English semantic elements from HowNet that relate to SentiWordNet, computing the average emotional intensity of these factors to establish the relevant Chinese emotional intensity. Knowledge-based methods possess an advantage in that they do not rely on corpus data and promptly generate a sentiment lexicon based on a comprehensive semantic knowledge base, which is considerably versatile and practical where less accuracy is required. However, the disadvantage of this method is that the sentiment lexicon usually only covers general domains and may lack domain specificity.

Corpus-based methods for constructing sentiment lexicons can be divided into two types: conjunction relations and co-occurrence rules. Conjunction relations use conjunction words in the sentence to determine the polarity relationship between words before and after the conjunction. Hatzivassiloglou et al. [10] summarized the language rules and connection patterns in English text and proved through extensive experimental data that the polarity relationship between words before and after the conjunction is reliable. They ascertained the emotional orientations of adjectives by utilizing a corpus and an emotional seed word set. Huang et al. [11] used conjunctions to determine the polarity relationship between words and established an emotional polarity constraint matrix based on the negation form of words. They then used the PMI algorithm to predict the sentiment polarity of words. Co-occurrence rules assume that emotionally similar words are likely to appear in the same sentence, and usually require a small number of sentiment seed words to be manually annotated before estimating the sentiment polarity of other words based on the co-occurrence frequency with the seed words in the corpus. Turney et al. [12] proposed a statistical approach based on the hypothesis that emotionally similar words are likely to have the same sentiment polarity. They selected the sentiment seed words "excellent" and "poor" to determine the sentiment polarity of other words. Wawer [13] used an automatic approach by searching for patterns in a corpus using a search engine to obtain polar seed words for SO-PMI calculation. Bollegala et al. [14] first annotated the polarity of all words co-occurring with candidate words as the overall sentiment polarity of the comments, replaced the words with their parts of speech to form features, and then used PMI and calculated the correlation between the candidate word features and known emotional words to determine their sentiment polarity to construct the sentiment lexicon. Yang et al. [15] used Baidu search results for seed words and other words to calculate SO-PMI to determine the sentiment polarity of words. Gao et al. [16] constructed a domain-specific sentiment lexicon by incorporating special lexicons such as phrase lexicons, negation lexicons, and adverb lexicons into a general sentiment lexicon and combining them with domain-specific professional vocabulary.

This paper uses the co-occurrence method in the corpus-based sentiment lexicon construction approach to build a sentiment lexicon for teaching evaluation.

3   Sentiment Analysis Model for Teaching Evaluation

3.1   Sentiment Analysis Model

The model diagram for sentiment analysis of student evaluations of teaching (SET) is shown in Fig. 1.

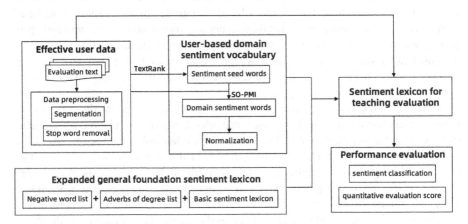

Fig. 1. The model diagram for sentiment analysis of SET

The model diagram for sentiment analysis of student evaluations of teaching (SET) can be divided into five parts:

1. Effective user data generation. The evaluation text is preprocessed, including word segmentation and stop word removal. The jieba library is utilized to perform Chinese word segmentation, the stop word dictionary is invoked, and stop words appearing in the text are removed by iterating through the segmented text.
2. Expansion of a general foundation sentiment lexicon. This includes selecting a basic sentiment lexicon, constructing a negative word list, constructing a adverbs of degree list, and constructing grammar rules.
3. User-based domain sentiment vocabulary generation.
   a. Sentiment seed word generation. Using the TextRank algorithm, sentiment seed words are selected from effective user data to generate domain-specific sentiment words.
   b. Domain sentiment word generation. With the generated sentiment seed words, domain-specific sentiment words that need to be expanded are selected from effective user data using the SO-PMI algorithm, and the sentiment polarity and inclination values of domain-specific sentiment words are obtained.

  c. Normalization of sentiment inclination values. Make the sentiment intensity of domain-specific sentiment words conform to the sentiment intensity of the general foundation sentiment lexicon.
4. Generation of a sentiment lexicon for evaluating teaching. The normalized domain-specific sentiment words and the expanded general foundation sentiment lexicon are merged to generate a sentiment lexicon for evaluating teaching.
5. Performance evaluation. Performance evaluations are divided into classification and quantitative evaluation scores. By comparing the performance of the general foundation sentiment lexicon, the expanded general foundation sentiment lexicon, and the sentiment lexicon for evaluating teaching on evaluation data in terms of sentiment classification and calculation of course comprehensive evaluation scores, the effectiveness of expanding the general foundation sentiment lexicon and expanding domain-specific sentiment vocabulary for teaching is verified.

3.2 Expansion of the General Foundation Sentiment Lexicon

Selection of Basic Sentiment Lexicon. This paper uses the Dalian University of Technology's Ontology of Chinese Sentiment Words [17], which describes Chinese vocabulary from different perspectives such as part of speech, emotional intensity, and emotional polarity. Emotional intensity is divided into five levels: 1, 3, 5, 7, and 9, with 9 representing the strongest emotional intensity and 1 representing the weakest. Compared with other sentiment lexicons, the emotional intensity in this one is subdivided more finely. Emotional polarity includes neutral, positive, and negative, corresponding to values of 0, 1, and 2 respectively. Table 1 shows some data from Dalian University of Technology's Ontology of Chinese Sentiment Words.

Table 1. Example of Dalian University of Technology's Ontology of Chinese Sentiment Words

Words	Part-of-speech tags	Sense numbers	Sentiment categories	Emotion intensity	Polarity
Exaggeration	idiom	1	NN	5	2
Tight finances	idiom	1	NE	7	0
Beautiful	noun	1	NN	5	1
Thoughtful	adj	1	PH	5	1

  To facilitate sentiment calculation, this paper modifies the negative sentiment polarity value 2 in the basic sentiment lexicon to -1 [18]. The sentiment value formula for the sentiment word t is shown in Eq. (1):

$$s(t) = w(t)g(t) \qquad (1)$$

  In Eq. (1), $s(t)$ represents the sentiment value of the sentiment word t, $w(t)$ represents the sentiment intensity of the sentiment word t, and $g(t)$ represents the sentiment polarity of the sentiment word t.

Construction of Negative Word List. The appearance of negative words often causes a reversal of the sentiment polarity in teaching comments. For cases where negative words appear before sentiment words, the sentiment intensity of the sentiment word needs to be multiplied by -1 in the algorithm design for sentiment calculation. Table 2 shows a list of common negative words.

Table 2. Negative word list

Weight values	Negative words
-1	not, is not, non, cannot, not quite, do not want, without, not necessarily, never, must not, etc

Construction of Adverbs of Degree List. The appearance of adverbs of degree often induces a certain degree of change in the sentiment intensity of sentiment words, such as in teaching comments like "very careful explanation," where the adverb of degree "very" enhances the sentiment intensity of the sentiment word "careful" to a certain extent. Referring to the degree-level words in the Hownet sentiment lexicon, this paper divides adverbs of degree into six levels representing different strengths of sentiment inclination. Gradient descent Equations are used to assign corresponding weight values to each level of adverbs of degree. Among them, the gradient descent formula is shown in Eq. (2):

$$W_{i+1} = W_1 \left(\frac{\sqrt{2}}{2} \right)^i, i = 1, 2, 3, 4, 5 \tag{2}$$

In Eq. (2), W_1 represents the weight value of the first level "extremely/most" in the adverbs of degree list, and the constant $\frac{\sqrt{2}}{2}$ represents the gradient descent rate. W_{i+1} represents the weight value of the $(i + 1)$-th level adverb of degree. Table 3 shows the constructed adverbs of degree list.

Table 3. Adverbs of degree list

Levels	Weight values	Adverbs of degree(partial)	Quantities
Extremely/most	3	extremely, very, absolute, overly	69
Super	2.1	excessively, biased, partial	30
Very	1.5	particularly, especially, more, greatly	42
Quite	1.06	even more, relatively, compared to, further	37
A little	0.75	slightly, somewhat, slightly, somewhat	29
Lacking	0.53	a little bit, not very, not great, not too much	12

3.3 User-Based Domain Sentiment Vocabulary Generation

To select domain-specific sentiment words that need to be expanded from actual student teaching evaluations, this paper uses the Semantic Orientation Pointwise Mutual Information (SO-PMI) algorithm. SO-PMI uses PMI to evaluate the semantic orientation of words and measures the similarity between two words using PMI. The definition is shown in Eq. (3):

$$PMI(term_1, term_2) = \log_2 \frac{p(term_1 \& term_2)}{p(term_1)p(term_2)} \tag{3}$$

In Eq. (3), $p(term_1 \& term_2)$ represents the probability of co-occurrence of the words $term_1$ and $term_2$, and $p(term_1)$ and $p(term_2)$ represent the probabilities of word $term_1$ and word $term_2$ appearing alone, respectively.

When $PMI(term_1, term_2)$ is larger, it indicates that the correlation between word $term_1$ and $term_2$ is closer, and their semantic orientations are more consistent. There are three cases of $PMI(term_1, term_2)$, as shown in Eq. (4):

$$PMI(term_1, term_2) \begin{cases} > 0, term_1 \text{ and } term_2 \text{ are related} \\ = 0, term_1 \text{ and } term_2 \text{ are independent} \\ < 0, term_1 \text{ and } term_2 \text{ are mutually exclusive} \end{cases} \tag{4}$$

Sentiment Seed Word Generation. To obtain sentiment seed words with clear emotional tendencies, we employ TextRank algorithm to sort sentiment words in the teaching evaluation corpus according to their importance from high to low and ultimately selects the top 10 positive and negative sentiment seed words.

TextRank is a graph-based ranking algorithm commonly used in keyword extraction. TextRank segments a text into a set of nodes (words), constructs a word-node connected graph, and computes the similarities between words as edge weights. Through iterative calculation of word weights, TextRank selects the top K words with the highest TextRank values. The algorithm is based on a graph representation of the text, and it can efficiently extract meaningful keywords from large amounts of text data.

Domain-Specific Sentiment Word Generation. The formula for calculating the sentiment orientation value of a word term in teaching comments is shown in Eq. (5):

$$SO - PMI(term) = \sum_{i=1}^{n} [PMI(term, Pterm_i) - PMI(term, Nterm_i)] \tag{5}$$

In Eq. (5), $Pterm_i$ represents the i-th positive sentiment seed word, $Nterm_i$ represents the i-th negative sentiment seed word, and $SO - PMI(term)$ has three cases, as shown in Eq. (6):

$$SO - PMI(term) \begin{cases} > 0, term \text{ has a positive sentiment tendency} \\ = 0, term \text{ has no sentiment tendency} \\ < 0, term \text{ has a negative sentiment tendency} \end{cases} \tag{6}$$

Among the 148 positive sentiment words and 79 negative sentiment words obtained through the SO-PMI algorithm have been normalized and expanded into a generic basic sentiment lexicon to generate a new domain-specific sentiment lexicon.

Normalization of Sentiment Values. To make the emotion intensity of the newly expanded domain-specific sentiment words suitable for that of the basic sentiment lexicon, normalization procedures are performed on the sentiment orientation values of the domain-specific sentiment words. The normalization Equation is shown in Eq. (7):

$$y = (SP - SP_{max})/(SP_{max} - SP_{min}) \tag{7}$$

In Eq. (7), y represents the normalized sentiment polarity value of domain sentiment words, SP represents the sentiment polarity value of candidate words, SP_{max} represents the maximum sentiment polarity value among the domain sentiment words, and SP_{min} represents the minimum sentiment polarity value among the domain sentiment words. In order to make the sentiment strength of domain sentiment words compatible with that of a general basic sentiment lexicon, the normalized value of y is divided into five intervals: [0, 0.2), [0.2, 0.4), [0.4, 0.6), [0.6, 0.8), and [0.8, 1]. These intervals are assigned corresponding sentiment strengths of 1, 3, 5, 7, and 9, respectively, in order to complete the normalization process of sentiment polarity values for domain sentiment words.

3.4 Sentiment Lexicon for Teaching Evaluation Generation

The normalized domain-specific sentiment words, corresponding emotion intensities, and polarities are added to the basic sentiment lexicon to form a sentiment lexicon for teaching evaluation. Similar to the basic sentiment lexicon, the sentiment polarity of the sentiment word can be neutral, positive, or negative, corresponding to values of 0, 1, and 2, respectively. Table 4 shows selected entries of the sentiment lexicon for teaching evaluation.

Table 4. Sentiment lexicon for teaching evaluation(partial)

Sentiment words	Emotion intensity	Polarity	Sentiment words	Emotion intensity	Polarity
Meticulous	7	1	Vivid	3	1
Patient	5	1	Gain	1	1
Caring	5	1	Boring	5	2
Clear thinking	3	1	Inaudible	3	2
Clear	3	1	Unable to catch up	3	2

3.5 3.5. Performance Evaluation.

Sentiment Classification. Sentiment classification is carried out by the sentiment lexicon for teaching evaluation, which includes the following four steps:

1. input effective teaching evaluations.
2. read the sentiment lexicon for teaching evaluation, negative word list, and adverbs of degree list.
3. traverse the negative words and adverbs of degree between each sentiment word in the teaching evaluation and calculate the corresponding weight value to obtain the sentiment value of every sentiment class in the teaching evaluation according to Eq. (8):

$$y(t) = n(t)a(t)p(t)s(t) \tag{8}$$

In Eq. (8), $y(t)$ represents the emotional value of emotional words, $s(t)$ represents the emotional value of sentiment words, $n(t)$ represents the weight value of negative words, $a(t)$ represents the sum of weights of all adverbs of degree before the sentiment word, and $p(t)$ represents the relative positional relation between adverbs of degree and negative words before the sentiment word. The specific formulas for $n(t)$, $a(t)$, and $p(t)$ are shown in Eqs. (9), (10), and (11), respectively.

$$n(t) = (-1)^n \tag{9}$$

$$a(t) = \sum_{i=1}^{m} a_i \tag{10}$$

$$p(t) = \begin{cases} 0.5, \text{ the negative word appears before the degree adverb} \\ 1.0, \text{ the degree adverb appears before the negative word} \end{cases} \tag{11}$$

In Eq. (9), n represents the number of negative words before the sentiment word. In Eq. (10), m represents the number of adverbs of degree before the sentiment word, and a_i represents the weight value of the i-th adverb of degree. In Eq. (11), if there is a negative word before the adverb of degree, $p(t) = 0.5$; otherwise, $p(t) = 1.0$.
4. The overall sentiment value of each teaching comment can be obtained by Eq. (12):

$$Y(r) = \sum_{t \in r} y(t) \tag{12}$$

In Eq. (12), r represents the set of emotional words in that comment, $Y(r)$ represents the overall sentiment value, and $Y(r) \geq 0$ indicates a positive sentiment tendency, whereas $Y(r) < 0$ indicates a negative sentiment tendency.

Quantitative Evaluation Score. We propose a sentiment-based quantitative scoring mechanism for teaching evaluation. The mechanism calculates the sentiment values using a sentiment lexicon and derives the course comprehensive evaluation scores by employing sentiment normalization and conversion to hundred-mark system. By comparing the comprehensive evaluation scores with the actual evaluation scores in the academic system, it reflects the performance of the sentiment lexicon in quantitative evaluation score. The steps of the proposed scoring mechanism are as follows:

1. Quantification based on emotional values. The overall emotional value of each teaching comment in the course is calculated, and each teaching comment is normalized

to a 1–5 score (five-point system) according to its emotional value from low to high. The five-point system score for the course is calculated as shown in Eq. (13):

$$y = \sum_{i=1}^{5} i \frac{num_i}{N} \qquad (13)$$

In Eq. (13), i represents the score in the five-point system, N is the total number of teaching comments in the course, num_i represents the number of teaching comments in the course that received a score of i in the five-point system, and y represents the five-point system score for the corresponding course taught by the teacher.

2. Conversion to hundred-mark system. Since the comprehensive evaluation score in our university's education system is represented in a hundred-mark system, a comparison experiment requires conversion of the five-point system score to the hundred-mark system score, which is calculated as shown in Eq. (14):

$$Y = 20y \qquad (14)$$

In Eq. (14), y represents the five-point system score for the course, and Y represents the hundred-mark system score for the course.

4   Experimental Results Analysis

4.1   Experimental Data

The experimental data consist of 508 valid comments from the SET data of four teachers in our university's academic system after data preprocessing.

4.2   Experimental Evaluation Metrics

In sentiment classification, we employ common evaluation metrics used in sentiment analysis models: precision (P), recall (R), and F1 score (F1). The confusion matrix is defined as shown in Table 5.

Table 5. Confusion matrix

	Judged as positive class	Judged as negative class
positive class	TP	FN
negative class	FP	TN

Precision, recall, and F1 score are represented by Eq. (15), Eq. (16), and Eq. (17), respectively:

$$P = TP/(TP + FP) \qquad (15)$$

$$R = TP/(TP + FN) \tag{16}$$

$$F1 = 2 \times P \times R/(P + R) \tag{17}$$

For the quantification of teaching evaluation scores, we use mean absolute error (MAE) and root mean squared error (RMSE) as the evaluation metrics in experiments, as shown in Eq. (18) and Eq. (19), respectively.

$$MAE = \frac{1}{N} \sum_{i=1}^{N} |\mu_i - \hat{y}_i| \tag{18}$$

$$RMSE = \sqrt{\frac{1}{N} \sum_{i=1}^{N} (\mu_i - \hat{y}_i)^2} \tag{19}$$

In Eq. (18) and Eq. (19), N represents the number of experiments, μ_i represents the true value of the i-th group of data, and $\hat{y}_i$ represents the predicted value of the i-th group of data.

4.3 Performance Comparison

Comparison of Sentiment Classification Performance. The teaching comments encompass both positive and negative evaluations. By comparing with the general basic sentiment lexicon (Dalian University of Technology's Ontology of Chinese Sentiment Words) and the expanded general basic sentiment lexicon, the effectiveness of constructing a sentiment lexicon for teaching evaluation proposed in this study is validated. Additionally, comparative experiments were conducted with common sentiment classification algorithms including K-Nearest Neighbors (KNN), Naïve Bayes, Maximum Entropy, and Support Vector Machine (SVM) to verify the sentiment classification performance of the sentiment lexicon for teaching evaluation(SL-TeaE). The experimental results are shown in Table 6.

From Table 6, it can be observed that: 1) The performance of sentiment classification with only the general basic sentiment lexicon is poor. However, after the addition of the negative word list and the adverb of degree list, the recall for negative teaching comments significantly increased by 21.8%, from 49.1% to 70.9%, and the F1 score increased from 53.1% to 63.7%, a 10.6% increase in performance. The precision saw a slight improvement. The expanded general basic sentiment lexicon showed a slight improvement in both the precision and F1 score for positive teaching comments, thereby proving the effectiveness of expanding the general basic sentiment lexicon. 2) The SL-TeaE demonstrates a significant improvement in sentiment classification performance compared to the expanded general basic sentiment lexicon. This improvement can be attributed to the expansion of the vocabulary of teaching domain-specific sentiment words on top of the expanded general basic sentiment lexicon. For positive teaching comments, the sentiment classification of the SL-TeaE showed good results, with varying degrees of improvement in precision, recall, and F1 score. The precision, recall, and F1

Table 6. Comparison of the performance of three sentiment lexicons in terms of sentiment classification

Sentiment classification methods	Comment polarity	P(%)	R(%)	F1(%)
General basic sentiment lexicon	positive teaching comments	87.1	90.5	88.8
	negative teaching comments	57.7	49.1	53.1
Expanded general basic sentiment lexicon	positive teaching comments	92.6	87.6	90.0
	negative teaching comments	57.9	70.9	63.7
KNN	positive teaching comments	86.4	87.9	87.2
	negative teaching comments	64.2	61.0	62.6
Naïve Bayes	positive teaching comments	89.7	91.8	90.8
	negative teaching comments	75.8	71.1	73.4
Maximum Entropy	positive teaching comments	90.5	89.7	90.1
	negative teaching comments	71.8	73.6	72.7
SVM	positive teaching comments	91.7	90.6	91.2
	negative teaching comments	74.2	76.6	75.4
SL-TeaE	positive teaching comments	**95.2**	**96.1**	**95.7**
	negative teaching comments	**82.2**	**78.7**	**80.4**

score increased from 92.6%, 87.6%, and 90% to 95.2%, 96.1%, and 95.7%, respectively, resulting in increases of 2.6%, 8.5%, and 5.7%. For negative teaching comments, the precision saw a significant increase from 57.9% to 82.2%, an increase of 24.3%, and there were also significant increases in recall and F1 score, from 70.9% and 63.7% to 78.7% and 80.4%, respectively, with increases of 7.8% and 16.7%. This proves the effectiveness of expanding the vocabulary of teaching domain-specific sentiment words. 3) Compared to common sentiment classification algorithms such as KNN, Naïve Bayes, Maximum Entropy, and SVM, the SL-TeaE exhibits superior sentiment classification performance in terms of precision, recall, and F1 score. Specifically, for positive teaching comments, it achieves a maximum precision improvement of 8.8%, a minimum precision improvement of 3.5%, and an average precision improvement of 5.3%. The recall shows a maximum improvement of 8.2%, a minimum improvement of 4.3%, and an average improvement of 5.9%. The F1 score demonstrates a maximum improvement of 8.5%, a minimum improvement of 4.5%, and an average improvement of 5.6%. For negative teaching comments, the precision shows a maximum improvement of 18%, a minimum improvement of 6.4%, and an average improvement of 10.7%. The recall exhibits a maximum improvement of 17.7%, a minimum improvement of 2.1%, and an average improvement of 8.1%. The F1 score presents a maximum improvement of 17.8%, a minimum improvement of 5%, and an average improvement of 9.4%.

Based on the comparative experiments, constructing a sentiment lexicon for teaching evaluation has good sentiment classification performance in the SET field and has practical significance for SET domain sentiment analysis.

Comparison of Quantitative Evaluation Scores. Our school's academic system not only includes the textual data of students' evaluations but also includes the overall evaluation scores of students for each course. In this paper, we conducted a comparative experiment on four courses taught by different teachers to verify the accuracy of our model in quantitative evaluation scores. We compared the course comprehensive evaluation scores obtained by three sentiment lexicons through scoring mechanisms to the actual evaluation scores of each course. The comparative results of the quantitative scores and error analysis for the evaluation are presented in Table 7 and Table 8, respectively.

Table 7. Comparison of the quantitative scores of the three sentiment lexicons

Teacher	General basic sentiment lexicon	Expanded general basic sentiment lexicon	SL-TeaE	Actual evaluation scores
Teacher1	73.54	73.96	**94.49**	97.54
Teacher2	82.70	83.92	**96.56**	98.77
Teacher3	81.73	82.09	**93.38**	96.45
Teacher4	83.59	85.45	**96.15**	98.10

Table 8. Comparison of the error of three sentiment lexicons

Dictionary type	MAE	RMSE
General basic sentiment lexicon	17.32	17.76
Expanded general basic sentiment lexicon	16.36	16.90
SL-TeaE	**2.57**	**2.62**

Three sentiment lexicons were used to calculate course comprehensive evaluation scores for each course and compared to the actual evaluation scores, with the results shown in Table 7 and Table 8. The course comprehensive evaluation scores calculated by the general basic sentiment lexicon differed greatly from the actual evaluation scores, with the highest mean absolute error and root mean squared error. The expanded general basic sentiment lexicon had smaller errors, while the evaluation scores calculated by the SL-TeaE were closest to the actual evaluation scores, with the smallest mean absolute error and root mean squared error (2.57 and 2.62, respectively). Additionally, the ranking of the course comprehensive evaluation scores calculated by the SL-TeaE was consistent with the ranking of the actual evaluation scores.

Through comparative experiments, we found that building a sentiment lexicon for teaching evaluation yielded the smallest mean absolute error and root mean squared error, which was closer to the actual evaluation scores, demonstrating its strong sentiment analysis performance in the teaching evaluation domain.

5 Conclusion

In this study, we constructed a list of adverbs of degree and negative words, selected sentiment seed words using TextRank algorithm, and implemented the SO-PMI algorithm to mine a professional sentiment lexicon based on user-generated content to enhance the generalization of our model and significantly improve its sentiment analysis performance. Our approach outperformed the general basic sentiment lexicon in terms of F1 score for positive and negative teaching comments, with improvements of 27.3% and 16.7%, respectively. Furthermore, the sentiment classification performance of our method surpasses that of commonly used sentiment classification algorithms. Moreover, the resulting domain-specific sentiment lexicon displayed higher accuracy in quantitative evaluation scores for courses, showcasing the smallest mean absolute error and root mean squared error and closer correlation to actual course comprehensive evaluation scores, while maintaining consistent ranking with them. These results validate the effectiveness of our sentiment lexicon construction method.

Acknowledgements. This study was supported by the Key Project of Regional Innovation and Development Joint Fund of National Natural Science Foundation of China (Grant No. U22A2025).

References

1. Wang, B.: Research on Sentiment Analysis of Student Evaluations of Teaching Based on Deep Learning. Chongqing University of Posts and Telecommunications (2021)
2. Balahadia, F.F., Fernando, M., Juanatas I.C.: Teachers performance evaluation tool using opinion mining with sentiment analysis. In: 2016 IEEE Region 10 Symposium (TENSYMP), pp. 95–98. IEEE (2016)
3. Lin, Q., Zhu, Y., Zang, S., et al.: Lexical based automated teaching evaluation via students short reviews. Comput. Appl. Eng. Educ. **27**(1), 194–205 (2019)
4. Wang, B., Gao, L., An, T., et al: A method of educational news classification based on emotional dictionary. In: 2018 Chinese Control and Decision Conference. Shenyang, China, pp. 3547–3551. IEEE (2018)
5. Taboada, M., Brooke, J., Tofiloski, M., et al.: Lexicon based methods for sentiment analysis. Comput. Linguist. **37**(2), 267–307 (2011)
6. Cai, Y., Yang, K., Zhou, Z., et al.: A hybrid model for opinion mining based on domain sentiment dictionary. Int. J. Mach. Learn. Cybern. **10**(8), 2131–2142 (2019)
7. Liu, W., Zhu, Y., Li, C., et al.: Research on building Chinese basic semantic lexicon. J. Comput. Appl. **29**(10), 2875–2877 (2009)
8. Yang, C., Feng, S., Wang, D., et al.: Analysis on web public opinion orientation based on extending sentiment lexicon. J. Chin. Comput. Syst. **31**(4), 691–695 (2010)
9. Zhou, Y., Yang, J., Yang, A.: A method on building Chinese sentiment lexicon for text sentiment analysis. J. Shandong Univ. (Eng. Sci.) 6, 27–33 (2013)
10. Hatzivassiloglou, V., Mckeown, K.R.: Predicting the semantic orientation of adjectives. Proc. Acl (2002)
11. Huang, S., Niu, Z., Shi, C.: Automatic construction of domain specific sentiment lexicon based on constrained label propagation. Knowl.-Based Syst. 56, 191–200 (2014)
12. Turney, P.D.: Thumbs up or thumbs down? semantic orientation applied to unsupervised classification of reviews. In: Proceedings of the 40th Annual Meeting of the Association for Computational Linguistics, July 6–12, 2002, Philadelphia, PA, USA, pp. 417–424 (2002)

13. Wawer, A.: Mining co-occurrence matrices for SO-PMI paradigm word candidates. In: Proceedings of the Student Research Workshop at the 13th Conference of the European Chapter of the Association for Computational Linguistics, pp.74–80 (2012)
14. Bollegala, D., Weir, D., Carroll, J.A.: Using multiple sources to construct a sentiment sensitive thesaurus for cross domain sentiment classification. In: Proceedings of the 49th Annual Meeting of the Association for Computational Linguistics: Human Language Technologies, vol. 1, pp. 132–141 (2011)
15. Yang, A., Lin, J., Zhou, Y.: Method on building chinese text sentiment lexicon. J. Front. Comput. Sci. Technol. 11, 1033–1039 (2013)
16. Gao, H., Zhang, J.: Sentiment analysis and visualization of hotel reviews based on sentiment dictionary. Comput. Eng. Softw. 42(01), 45–47 (2021)
17. Xu, L., Lin, H., Pan, Y., et al.: Constructing the affective lexicon ontology. J. China Soc. Sci. Tech. Inf. 27(2), 180–185 (2017)
18. Dun, X., Zhang, Y., Yang, K.: Fine grained sentiment analysis based on weibo. Data Anal. Knowl. Discovery 7, 61–72 (2017)
19. Liu, Q., Shen, W.: Research of keyword extraction of political news based on Word2Vec and TextRank. Inf. Res. 6, 22–27 (2018)
20. Chi, Y., Zhao, S., Luo, Y. , et al.: Text data preprocessing based on term frequency statistics rules. Comput. Sci. 44(10), 276–282+288 (2017)

Graph Fusion Multimodal Named Entity Recognition Based on Auxiliary Relation Enhancement

Guohui Ding, Wenjing Tang[✉], Zhaoyi Yuan, Lulu Sun, and Chunlong Fan

Shenyang Aerospace University, Shenyang, China
{dingguohui,FanCHL}@sau.edu.cn, tangwj1999@163.com

Abstract. Multimodal Named Entity Recognition (MNER) aims to use images to locate and classify named entities in a given free text. The mainstream MNER method based on a pre-trained model ignores the syntactic relations in the text and associations between different data; however, these relations can provide crucial missing auxiliary information for the MNER task. Therefore, we propose an auxiliary and syntactic relation enhancement graph fusion (ASGF) method for MNER based on the cross-modal information between similar texts and long-distance inter-word syntactic dependencies in the text. First, for each text image pair (training sample), we search for a sample that is most similar to its text because similar samples may contain similar entity information. We then exploit a multimodal relation graph to model the association between different modal data of the two similar samples; that is, we use the similar sample to supplement the entity information of the text to be recognized. Second, we parse the syntax of the text to capture the syntactic dependencies between different words and integrate them into the relation graph to further enhance its semantic information. Finally, the relation graph is input into the graph neural network, multimodal information is interactively fused through the attention and gating mechanisms, and final MNER label sequence is predicted through CRF decoding. Extensive experimental results show that compared to mainstream methods, the proposed model achieved competitive recognition accuracy on public datasets.

Keywords: Named entity recognition · Multimodal learning · Graph neural network

1 Introduction

Named entity recognition (NER) is an important task in the field of natural language processing (NLP) that aims to recognize text spans as specific entity types, such as people, locations, and organizations. The extracted named entities can support various NLP tasks, including question answering [9] and relation extraction [25], etc.

X. Yang et al. (Eds.): ADMA 2023, LNAI 14179, pp. 18–32, 2023.
https://doi.org/10.1007/978-3-031-46674-8_2

Recently, with the surge in the number of posts published on social media, platforms such as Twitter and Weibo have been increasingly incorporating multimodal information, such as combinations of text and images. As shown in Fig. 1, when the meaning behind "Argentina" is interpreted from the text alone, it is likely to be classified as "Location", but when given the information provided by the paired image, "Argentina" can be easily classified as "Organization", indicating that "Argentina" refers to the national soccer team and not the country. Therefore, multimodal information, such as images, can improve the accuracy of NER. Consequently, multimodal named entity recognition (MNER) has attracted considerable attention. In this task, text is considered the primary source of information, and image information serves as an auxiliary source that can help improve the overall accuracy of NER. This task can be applied to many scenarios, such as multimodal relation extraction [25] and multimodal retrieval [24], etc.

Argentina[ORG] is the best !

Fig. 1. An example where visual information helps NER, we can easily tell that "Argentina" is tagged as "ORG" and not "LOC" by matching images.

In recent years, many studies have been conducted on MNER and achieved good accuracies. However, existing methods that model a single image-text pair ignore connections with other relational pairs in the training data. Through observation and research, we discovered that in social media, when the meanings and themes expressed by two pieces of text are similar, the paired images are also similar or complementary. As shown in Fig. 2, the matched text in both images have similar themes and content. In the text shown in Fig. 2(b), "DannyWelbeck" is an entity of type "PER", but there is no visual object of type "Person" in its paired image. However, in Fig. 2(a), the visible object of type "Person" in the matched image of the similar text can assist in determining the entity type of "Danny Welbeck". These findings clearly revealed that the joint modeling of two relational pairs with similar semantics could provide effective missing supplementary information for MNER, improving its recognition accuracy.

Mainstream MNER methods extract text features using pretrained language models, ignoring the syntactic information such as syntactic dependencies con-

Happy birthday to
former Gunner David Platt !

(a)

Happy birthday to **Danny Welbeck** !

(b)

Fig. 2. An example of the auxiliary effect of text with similar semantics on MNER tasks in a dataset.

tained in the text. Research on traditional NER tasks has obtained useful discrete features from dependent structures [2] or structural constraints [5] to help complete NER tasks. The example sentence in Fig. 3 illustrates the relationship between syntactic dependencies and named entities.

The dependency from "near" to "premises" indicates that there is a direct relation "pobj" (pre-object) between them, which can help the model judge that "premises" is a named entity of type LOC (location). However, existing MNER models do not consider such syntactic dependencies, thereby missing the interactions and important dependencies between the distant words in sentences. The utilization of the complete syntactic-dependent structure and its effective integration into the MNER task is a research problem that remains unsolved.

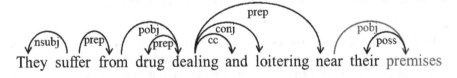

Fig. 3. Example of the auxiliary effect of texts with similar semantics on the MNER task in the datase.

To solve the problems, we propose an **A**uxiliary and **S**yntactic relation enhancement **G**raph **F**usion (ASGF) method for MNER tasks. First, for each image-text pair (training sample), the sample most like its text is determined using the similarity retrieval method. Images paired with similar texts in social media posts potentially contain similar entity information. Therefore, we extracted the visual region related to the entity in an image and inputted the relevant visual region and two similar texts into the pretrained language model to encode the image and text respectively. Then, we constructed a multimodal relation graph for the encoded text and image features. In this graph, each node represents a semantic unit, that is, a word or visual object. We introduced inter-modal and intra-modal edges to transmit information between the different nodes. Using this graph, we modeled the association between different data and used similar samples to supplement the entity information of the text to be recognized. Second, to address the problem of the insufficient use of syntactic information in previous methods and the omission of important semantic relationships between words in sentences, we used a syntactic parser to parse the syntax of the text, captured the syntactic dependencies between different words, introduced a syntactic relation matrix modeling the semantic connections between different words in the text, and incorporated them into a multimodal relation graph to further enhance their semantic information. A graph neural network (GNN) can process graph-structured data and propagate information through the edges, achieving fusion between different modalities. Therefore, in the final stage, we used a GNN that incorporated attention and gating mechanisms to model the relationships between both similar and different modalities, achieving the integration of multimodal information interaction. Conditional random field (CRF) decoding was used to predict the final label sequence for conditional sequence labeling. We conducted extensive experiments on public datasets and the results demonstrated the effectiveness of the proposed method. Overall, the main contributions of our work are as follows:

(1) We designed a method for MNER called ASGF to fully utilize the syntactic information of texts. To the best of our knowledge, our method is the first to apply syntactic dependencies to MNER and incorporate them into GNNs.
(2) We jointly modeled two text-image pairs with similar content or topics in the training data as a multimodal relation graph to fully utilize the effective auxiliary or supplementary information between different data.
(3) Extensive experiments on public datasets demonstrate the effectiveness of our proposed method.

2   Related Work

Named entity recognition (NER) has received extensive attention from scholars as a key part of information extraction. Traditional methods rely on rule-based feature extractors and sequence tagging techniques. However, with the development of social media, text data often appears together with multimodal information such as images, and studies show that incorporating image information can improve NER performance [1], leading to the emergence of MNER. In our work, we use GNN to jointly model the relationship between images and text in MNER tasks. Therefore, in the following sections, we provide a brief overview of both MNER and GNN.

2.1   MNER

For MNER, Zhang et al. [22], Moon, Neves, Carvalho and Lu et al. [14] explored the task during the same period. Early research attempted to encode text via

RNN and image regions via CNN [14,22]. With the advent of pre-trained models and their successful application in the field of natural language processing, Yu et al. [20] used BERT [3] as the sentence encoder, ResNet [4] as the image encoder, and a multimodal interaction module to capture the relationship between words and images. Zhang et al. [21] used a text-guided object detection model to obtain objects relevant to the text, and introduced graph modeling in MNER task to eliminate interference from irrelevant objects. However, none of the above methods considered the relationship between different text-image pairs in the training data, and the syntactic dependency information in the text, which are essential for the MNER task.

2.2   GNN

Graph Neural Network (GNN) can handle graph structures that neural networks such as CNN cannot handle. Among them, graph convolutional network (GCN) [7] and graph attention network (GAT) [18] have been widely used in many tasks and achieved remarkable results. Zhang et al. [21] first proposed to integrate GNN into the MNER task. Treat words in text and visual objects identified by noun phrases as nodes, and use edges to transfer useful information between different modalities. Zhao et al. [23] utilized the external matching relations of the dataset to build a graph model on the whole dataset.

Different from the above work, our model fully mines the text information by using the relation and syntactic dependence between different text image pairs in the training data, and further improves the MNER performance by using the relation between similar text image pairs.

3   Methodology

As GNN can transmit information through edges and achieve information interaction between different modalities, in this work, we proposed a novel MNER method based on GNN, called ASGF. Figure 4 shows the overall architecture of ASGF, which consists of four main modules: 1) feature extraction module; 2) multimodal relation graph construction module; 3) multimodal interaction module; 4) CRF decoding module. Below, we first introduce the task definition, and then illustrate the four main components of the ASGF model separately.

3.1   Task Definition

In this work, MNER task is formulated as a sequence labeling problem. Given a text sequence $X = \{x_1, x_2 \ldots x_n\}$ and its associated image O. MNER aims to identify entity boundaries from text using BIO tags and classify identified entities into predefined categories, including Person(PER), Organization(ORG), Location(LOC), and Other(MISC). The output of the MNER model is a series of labels with input text $Y = \{y_1, y_2 \ldots y_i\}$. In this work, $y_i \in \{O, B - PER, I - PER, B - ORG, I - ORG, B - LOC, I - LOC, B - OTHER, I - OTHER\}$ [16].

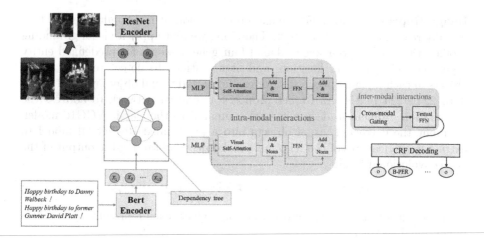

Fig. 4. The overall architecture of the model.

3.2   Feature Extraction Module

According to our observation, when two texts have similar meanings, their paired images are also similar or complementary. When the named entity in a text does not appear in its paired image, the similar text and its paired image can provide effective supplementary information for the current text and improve the recognition performance of MNER. Therefore, we first use cosine similarity to retrieve the text in the training data that is most similar to the input text, and jointly input the two pairs of relationships into the feature extraction module along with their paired images.

Text Representation. Traditional word vector representation methods, such as GloVe [15] or Word2Vec [13], generate word representations that cannot jointly consider contextual information, and cannot solve polysemy problems. In contrast, pre-trained language models can address these issues. With the widespread use of the pre-trained language model BERT in the NLP field, this paper will jointly input two similar sentences after similarity retrieval into the BERT-based model [3]. BERT input vector $E_n(n \in N)$ consists of word embedding vector, segment embedding vector and position encoding vector. In our model, we encode segment embedding vectors with 0 and 1 to distinguish two different input texts, and finally the input text is decoded to obtain the embedded contextual representation as shown in Eq. (1):

$$H_x = BERT(x) \tag{1}$$

where $H_x \in R^{n \times d}$ denotes the text representations, d is the hidden dimension, and n is the length of the sentence.

Image Representation. For image representation, it is difficult to extract all visually relevant objects from text. Therefore, we first apply a visual grounding toolkit [19] and use prior knowledge of four general words for predefined entity types (i.e. misc, person, location, and organization) to detect relevant visual objects in each image. Specifically, for each detected visual object, we first resize it to 224×224 pixels. As Resnet proposes a residual network to address the degradation problem in deep networks compared to the classic VGG16 model, we input the extracted visual objects into a pre-trained ResNet [4] model to extract image features of the relevant visual objects, obtaining the output of the last convolutional layer:

$$ResNet(I) = \{r_j | r_j \in R^{2048}, j = 1, 2, \ldots, 49\} \tag{2}$$

It divides the original image into $7 \times 7 = 49$ regions, each represented by a 2048 dimensional vector r_j.

Finally, we separately input the text representation and the image representation into two independent multi-layer perceptrons (MLP) with ReLU activation functions, forming a dimension of d. This is done to embed different features of the two modalities into the same feature space.

3.3 Multimodal Relation Graph Construction

In our work, we use a multimodal relation graph to model cross-modal relations. Formally, our graph is an undirected graph, which can be expressed as G=(V,E). In the following, we will describe in detail how we construct the multimodal relational graph.

Node. The multimodal relation graph contains two types of nodes: text nodes and image nodes. We use the text representation of the words in the input text after being encoded by BERT as a text node, and the image representation of the detected visual object processed by the ResNet model as an image node. For example, in Fig. 4, the two texts contain a total of 12 words, so there are 12 text nodes, and there are two image nodes because two visual regions were detected. Due to space limitations, we only show a schematic diagram of the multimodal relational graph here.

Edge. We introduce two types of edges in the multimodal relation graph: intra-modal edges and inter-modal edges. Intra-modal edges connect nodes within the same modals, while inter-modal edges connect nodes between different modals. To fully exploit the syntactic information, capture the long-distance dependencies between words in a sentence, and alleviate the information loss caused by the excessively long text, we integrate the syntactic dependencies into the multimodal relation graph. Many GNN-based methods have demonstrated that edge weights are crucial for graph information aggregation [10], Therefore, we use SpaCy syntax parser to perform syntactic parsing on each input text and introduce a syntactic dependency matrix to model the semantic relationships between different words in the text. If there is a syntactic dependency relation between two nodes, the edge weight between them is increased by 1, otherwise the weight is 0. As shown in Eq. (3):

$$A^T_{i,j} = \begin{cases} 1, & if \ D(w_i, w_j) \ and \ i < n, j < n \\ 0, & otherwise \end{cases} \tag{3}$$

where n is the length of the input sentence. $D_{i,j}$ represents the syntactic dependency relationship between two nodes. In the model, we model the multimodal relation graph as an adjacency matrix. We add the adjacency matrix and the syntactic weight matrix to integrate syntactic dependencies into the multimodal graph, so as to facilitate the learning of distant words and contextual dependencies. We will discuss edge weights in detail in the experimental section. Finally, a fully connected multimodal relation graph with weighted edges is constructed.

3.4 Multimodal Interaction Module

Inspired by [21], we divide the multimodal interaction module into two parts: intra-modal interaction and inter-modal interaction.

Intra-modal Interactions. After the construction of the multimodal relation graph, we need to model the relation between different modalities based on the graph. Since self-attention model tends to overly focus on the current position when encoding information, the multi-head self-attention mechanism can solve this problem. Therefore, for nodes in the same modality (text or image), we use the multi-head self-attention mechanism to gather the neighbor information of each node. Through the multi-head self-attention mechanism, the model can focus on more useful information for entity recognition between the same modalities, thus achieving interaction between different nodes of the same modality. Formally, the context representation M_x of all text nodes is shown in Eq. (4):

$$M_x = MHatt(H_x, H_x, H_x) \tag{4}$$

where MHatt(Q, K, V) is a multi-head self-attention function, which takes query matrix Q, key matrix K and value matrix V as input. H_x is the state representation of the current text node. Similarly, we generate the contextual representation M_o of all visual nodes as shown in Eq. (5):

$$M_o = MHatt(H_o, H_o, H_o) \tag{5}$$

Due to limited space, we omit the description of residual connection and normalization.

Inter-modal Interactions. The input for inter-modal interactions is the representation of text and image nodes that have undergone intra-modal interaction. Since our model jointly models two pairs of similar text image pairs and inputs them into GNN for training, in reality, images and text may not be entirely related, and noisy images may be introduced, interfering with the final recognition performance. Therefore, in order to exclude the negative interference of irrelevant visual regions on the final recognition, for nodes between different modalities, we use a gating mechanism to aggregate cross-modal neighbor information for each node between different modalities. Specifically, we set up a modulation factor β_o, which can dynamically adjust Rx according to the degree of contribution of the visual object to the text. Thus more incorporation of visuals that are more relevant to the text modality. After obtaining β_o we generate the representation R_x of each text node:

$$\beta_o = sigmoid(W_1 M_x + W_2 M_o) \tag{6}$$

$$R_x = M_x + \beta_o M_o \tag{7}$$

where W_1 and W_2 are parameter matrices. Since the text should dominant in the MNER task, and the image information is only used as an auxiliary. So different from [21], in the final stage, we only integrate the useful information from the image modality into the text modality through the gating mechanism, omitting the process of integrating text modality information into the image modality, simplifying the model architecture while ensuring recognition performance and making the model more concise. Finally, we use the position feedforward network to generate the final text node representation H_x:

$$H_x = FFN(R_x) \tag{8}$$

3.5 CRF Dcoder

As conditional random field (CRF) can consider the correlation between labels in the neighborhood, and the mainstream methods of MNER and NER currently use CRF for model decoding, in order to make a fair comparison in the final decoding stage, we apply CRF [8,21,23] to perform conditional sequence labeling. The input of the CRF is the representation of the text node that incorporates the relevant image information generated by the multimodal interaction module. Let Y_0 denote the set of all possible label sequences of the input sentence X, the probability of the label sequence Y can be calculated as:

$$p(y|H_x) = \frac{\prod_{i=1}^{N} F_i(y_{i-1}, y_i, H_x)}{\sum_{y' \in Y} \prod_{i=1}^{N} F_i(y'_{i-1}, y'_i, H_x)} \tag{9}$$

where $F_i(y_{i-1}, y_i, H_x)$ and $F_i(y'_{i-1}, y'_i, H_x)$ are potential function. We use the label with the highest probability score as the final prediction result. Finally, we apply maximum conditional likelihood estimation as the loss function of the model. The loss function is shown in Eq. (11):

$$L(p(y|H_x)) = \sum_i logp(y|H_x) \tag{10}$$

4 Experiment

4.1 Datasets and Evaluation Metrics

To evaluate the performance of our proposed model, we use the publicly available Twitter dataset, Twitter-2015, constructed by Zhang et al. [22] for MNER. Table 1 shows the number of entities of each type and the number of multimodal tweets in the training, development and test sets of the dataset.

We used standard precision (P), recall (R), and F1 score (F1) to evaluate the overall performance on the Twitter MNER dataset and reported the F1 score for each individual type of entity.

Table 1. The statistics summary of Twitter datasets.

Entity Type	Twitter2015		
	Train	Dev	Test
Person	2217	552	1816
Location	2091	522	1697
Organization	928	247	839
Misc	940	225	726
Total	6176	1546	5078
Tweets	4000	1000	3257

4.2 Implement Details

In our model, word representations are initialized with a pretrained cased BERTbase [3] model and visual representations are initialized with a pretrained ResNet-152 model [4], both are fine-tuned during training. We set the maximum length of input sentence to 128, the mini-batch size to 8, and the number of attention heads to 8.

We choose the model with the best F1 score on the dev set and evaluate its performance on the test set. We use the Adam optimizer [6] to minimize the loss function, with the learning rate as 1e-4, the dropout rate as 0.4, and the tradeoff parameter as 0.5.

4.3 Compared Methods

To fully evaluate the performance of our model, we mainly compare it with two groups of baselines.

The first group consists of representative text-based methods for NER: (1) CNN-BiLSTM-CRF [12] and HBiLSTM-CRF [8] both use word-level information based on BiLSTM. (2) BERT [3] and BERT-CRF both use multilayer bidirectional transformer encoder.

The second group consists of several competitive multimodal methods for MNER: (3) VG-ATT [11]: propose visual attention to combine HBiLSTM-CRF with visual context. (4) Ada-Co-ATT [22]: A multi-modal approach based on CNNBiLSTM-CRF, fusing word-guided visual representation and image-guided textual representation through filter gates with adaptive co-attention networks. (5) RpBERT [17] is a multi-task method for MNER and text-image relation classification, where an external Twitter annotation dataset with text-image relations is used for text-image relation classification.

Table 2. Performance comparison of different competing unimodal and multimodal methods for NER.

Modality	Approaches	Twitter-2015						
		Single Type($F1$)				Overall		
		PER	LOC	ORG	MISC	Pre	Rec	F1
Text	CNN-BiLSTM-CRF	80.86	75.39	47.77	32.61	66.24	68.09	67.15
	HBiLSTM-CRF	82.34	6.83	51.59	32.52	70.32	68.05	69.17
	BERT	84.72	79.91	58.26	38.81	68.30	**74.61**	71.32
	BERT-CRF	84.74	80.51	60.27	37.29	69.22	74.59	71.81
Text+Image	VG-ATT	82.66	77.21	55.06	35.25	73.96	67.90	70.80
	Ada-Co-ATT	81.98	78.95	53.07	34.02	72.75	68.74	70.69
	RpBERT	**85.18**	81.19	58.68	37.88	71.15	74.30	72.69
	ASGF(Ours)	84.40	**82.03**	**59.07**	**38.82**	**73.98**	73.20	**73.15**

4.4   Experimental Results

We conduct experiments on the Twitter-2015 datasets. As shown in Table 2, similar to previous works, we mainly focus on the overall F1 score. Based on these results, we can make some observations:

1): For the single-modal entity recognition model, the BERT-CRF method achieves 2.4%, 3.68%, 8.68% and 4.77% improvements in the single type F1 of the Twitter-2015 dataset compared to the HBiLSTM-CRF method. This indicates that the pre-trained language model BERT provides rich syntactic and semantic features that are significantly better than traditional neural network methods. In addition, through comparison, it is found that BERT-CRF is slightly better than the BERT model. Therefore, we can conclude that CRF decoding considers the relationship between neighboring words, can improve the performance of entity recognition.

2): In multimodal methods, VG-ATT is 0.34%, 0.38%, 3.47% and 2.73% higher than its corresponding unimodal method HBiLSTM-CRF in the single-type F1 of Twitter-2015, respectively. Similarly, Ada-Co-ATT is 1.12%, 3.56%, 5.3%, 1.41% higher than its corresponding unimodal method CNN-BiLSTM-CRF in the single-type F1 of Twitter-2015, respectively. From this we can conclude that incorporating image information as a supplement in traditional uni-modal entity recognition can improve the performance of entity recognition.

3): For multimodal methods, RpBERT is an extension of the BERT model in the multimodal direction. It uses the attention mechanism combined with an external Twitter annotation dataset with text-image relations to classify text-image relations, eliminating noise interference from visual objects and effectively improving the performance of multimodal entity recognition. The accuracy RpBERT is 1.89% and 2% higher than VG-ATT and Ada-Co-ATT in the overall F1 of Twitter-2015, respectively.

4): Compared with RpBERT, our proposed ASGF achieves competitive results on the dataset, outperforming RpBERT by 0.46% on overall F1. Furthermore, for a single type, our model outperforms RpBERT by up to 0.94% on the dataset. These results further reveal the effectiveness of our model.

Table 3. Ablation study of our ASGF.

Twitter-2015			
Approaches	Overall		
	Pre	Rec	F1
baseline+dep+sim(**ASGF**)	73.10	73.20	75.15
baseline+dep	71.85	73.69	72.76
baseline	70.62	72.94	71.76

4.5 Ablation Study

In order to study the impact of different components in our proposed method, we conducted an ablation study on the two modules in ASGF. The results are shown in Table 3, where "baseline" means the baseline method after removing the two modules, "dep" denotes the syntactic dependency part of our method, and "sim" denotes the joint similarity semantic text-image relationship part of our method. We add each module in turn. From the table 3, we can intuitively see that after adding each module separately, the overall precision, recall rate, and F1 of our proposed method have improved. Among them, the model that only adds syntactic dependencies is better than the model that only adds syntactic dependencies outperforms the baseline method by 1% on F1, while our proposed method outperforms the baseline method by 1.99% on F1 after adding two modules. Therefore, we conclude that the syntactic dependencies of the text itself and the connections between different texts are crucial for the MNER task.

The Choice of Similarity Threshold. In our model, we first need to use similarity retrieval to find the text in the training data that is most similar to the text to be predicted. However, the training data is limited, not all of the

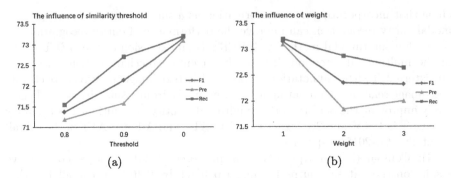

Fig. 5. The influence of the threshold value in similarity retrieval and the selection of edge weights in the multimodal relation graph on the final recognition performance.

most similar texts have high similarity between them. So in order to compare the impact of the similarity threshold on the final MNER recognition performance, we set the threshold to 0.8, 0.9 and no threshold for comparison. We use overall recall, precision and F1 score as evaluation metrics. We plot the comparison results as a line graph, as shown in Fig. 5(a). From the line chart, we can clearly see that when the threshold is increased from 0.8 to 0.9, the overall recall rate, precision and F1 score are all improve. That is, when the similarity between two pieces of text is higher, the paired images for each of them will provide greater assistance to the final recognition performance of our model. This further proves that in our model, when two texts express similar meaning and themes, their paired images are also similar or complementary. When we do not set thresholds, we get the best overall performance. This indicates the effectiveness of considering the connections between different text-image pairs in the dataset and jointly modeling them in our method. It also illustrates the positive effect of the gating mechanism introduced by our model in the multimodal interaction module to exclude the interference of irrelevant image modalities. Through the above analysis, we do not set a threshold when performing text similarity retrieval.

The Choice of Edge Weights. In the process of building a multimodal relation graph, we incorporate syntactic dependencies relation into the graph. If there is a dependency relation between two words in the text, we assign weights to the edges between them. To compare the impact of different edge weights on the final entity recognition performance, we assign weights of 1, 2, and 3 to the edges respectively for comparison. We use overall recall, precision and F1 score as evaluation metrics. The results are shown in the line chart in Fig. 5(b). From the figure, we can clearly see that the model achieves the best performance when the edge weight is 1. Therefore, in the process of building a multimodal relation graph, if two words there is a dependency between them, we assign a weight of 1 to the edge between them.

5   Conclusion

In this paper, we propose ASGF, an auxiliary relation and syntactic relation augmentation graph fusion method for MNER. This method is based on the graph neural network and integrates the text syntax dependency relation to fully exploit the text syntax information. By combining text similarity retrieval with similar text-image pairs in the training data, the accuracy of MNER recognition is further improved. Finally, extensive experimental results show that our model achieves competitive recognition performance compared to other state-of-the-art methods.

In future work, we hope to utilize more external resources beyond the training data to further improve the recognition performance of MNER.

Acknowledgement. This research work is supported by the Sci-Tech Innovation 2030 "- New Generation Artificial Intelligence" Major Project (2018AAA0102100); the Natural Science Foundation of Liaoning Province (2021-MS-261);the Natural Science Foundation Key Project of Zhejiang Province(LZ22F020014); Key R&D Sub- Project of the Ministry of Science and Technology(2021YFF0307505).

References

1. Chen, S., Aguilar, G., Neves, L., Solorio, T.: Can images help recognize entities? A study of the role of images for multimodal NER. arXiv preprint arXiv:2010.12712 (2020)
2. Cucchiarelli, A., Velardi, P.: Unsupervised named entity recognition using syntactic and semantic contextual evidence. Comput. Linguist. **27**(1), 123–131 (2001)
3. Devlin, J., Chang, M.W., Lee, K., Toutanova, K.: BERT: pre-training of deep bidirectional transformers for language understanding. arXiv preprint arXiv:1810.04805 (2018)
4. He, K., Zhang, X., Ren, S., Sun, J.: Deep residual learning for image recognition. In: Proceedings of the IEEE Conference on Computer Vision and Pattern Recognition, pp. 770–778 (2016)
5. Jie, Z., Muis, A., Lu, W.: Efficient dependency-guided named entity recognition. In: Proceedings of the AAAI Conference on Artificial Intelligence, vol. 31 (2017)
6. Kingma, D.P., Ba, J.: Adam: a method for stochastic optimization. arXiv preprint arXiv:1412.6980 (2014)
7. Kipf, T.N., Welling, M.: Semi-supervised classification with graph convolutional networks. arXiv preprint arXiv:1609.02907 (2016)
8. Lample, G., Ballesteros, M., Subramanian, S., Kawakami, K., Dyer, C.: Neural architectures for named entity recognition. arXiv preprint arXiv:1603.01360 (2016)
9. Lao, N., Cohen, W.W.: Relational retrieval using a combination of path-constrained random walks. Mach. Learn. **81**, 53–67 (2010)
10. Lou, C., Liang, B., Gui, L., He, Y., Dang, Y., Xu, R.: Affective dependency graph for sarcasm detection. In: Proceedings of the 44th International ACM SIGIR Conference on Research and Development in Information Retrieval, pp. 1844–1849 (2021)

11. Lu, D., Neves, L., Carvalho, V., Zhang, N., Ji, H.: Visual attention model for name tagging in multimodal social media. In: Proceedings of the 56th Annual Meeting of the Association for Computational Linguistics (Volume 1: Long Papers), pp. 1990–1999 (2018)

12. Ma, X., Hovy, E.: End-to-end sequence labeling via bi-directional LSTM-CNNs-CRF. arXiv preprint arXiv:1603.01354 (2016)

13. Mikolov, T., Sutskever, I., Chen, K., Corrado, G.S., Dean, J.: Distributed representations of words and phrases and their compositionality. In: Advances in Neural Information Processing Systems 26 (2013)

14. Moon, S., Neves, L., Carvalho, V.: Multimodal named entity recognition for short social media posts. arXiv preprint arXiv:1802.07862 (2018)

15. Pennington, J., Socher, R., Manning, C.D.: Glove: Global vectors for word representation. In: Proceedings of the 2014 Conference on Empirical Methods in Natural Language Processing (EMNLP), pp. 1532–1543 (2014)

16. Sang, E.F., Veenstra, J.: Representing text chunks. arXiv preprint cs/9907006 (1999)

17. Sun, L., Wang, J., Zhang, K., Su, Y., Weng, F.: RpBERT: a text-image relation propagation-based BERT model for multimodal NER. In: Proceedings of the AAAI Conference on Artificial Intelligence, vol. 35, pp. 13860–13868 (2021)

18. Veličković, P., Cucurull, G., Casanova, A., Romero, A., Lio, P., Bengio, Y.: Graph attention networks. arXiv preprint arXiv:1710.10903 (2017)

19. Yang, Z., Gong, B., Wang, L., Huang, W., Yu, D., Luo, J.: A fast and accurate one-stage approach to visual grounding. In: Proceedings of the IEEE/CVF International Conference on Computer Vision, pp. 4683–4693 (2019)

20. Yu, J., Jiang, J., Yang, L., Xia, R.: Improving multimodal named entity recognition via entity span detection with unified multimodal transformer. In: Association for Computational Linguistics (2020)

21. Zhang, D., Wei, S., Li, S., Wu, H., Zhu, Q., Zhou, G.: Multi-modal graph fusion for named entity recognition with targeted visual guidance. In: Proceedings of the AAAI Conference on Artificial Intelligence, vol. 35, pp. 14347–14355 (2021)

22. Zhang, Q., Fu, J., Liu, X., Huang, X.: Adaptive co-attention network for named entity recognition in tweets. In: Proceedings of the AAAI Conference on Artificial Intelligence, vol. 32 (2018)

23. Zhao, F., Li, C., Wu, Z., Xing, S., Dai, X.: Learning from different text-image pairs: a relation-enhanced graph convolutional network for multimodal NER. In: Proceedings of the 30th ACM International Conference on Multimedia, pp. 3983–3992 (2022)

24. Zhao, Y., Wang, W., Zhang, H., Hu, B.: Learning homogeneous and heterogeneous co-occurrences for unsupervised cross-modal retrieval. In: 2021 IEEE International Conference on Multimedia and Expo (ICME), pp. 1–6. IEEE (2021)

25. Zheng, C., Wu, Z., Feng, J., Fu, Z., Cai, Y.: MNRE: a challenge multimodal dataset for neural relation extraction with visual evidence in social media posts. In: 2021 IEEE International Conference on Multimedia and Expo (ICME), pp. 1–6. IEEE (2021)

Sentence-Level Event Detection Without Triggers via Prompt Learning and Machine Reading Comprehension

Tongtao Ling, Lei Chen$^{(\boxtimes)}$ ⓘ, Huangxu Sheng, Zicheng Cai, and Hai-Lin Liu ⓘ

Guangdong University of Technology, Guangzhou, China
chenlei3@gdut.edu.cn

Abstract. Sentence-level event detection has traditionally been carried out in two key steps: trigger identification and trigger classification. The trigger words first are identified from sentences and then utilized to categorize event types. However, this classification hugely relies on a substantial amount of annotated trigger words along with the accuracy of the trigger identification process. This annotation of trigger words is labor-intensive and time-consuming in real-world environments. As a solution to this, we propose a model that does not require any triggers for event detection. This model reformulates event detection into a two-tower model that uses machine learning comprehension and prompt learning. Compared to the existing methods, which are either trigger-based or trigger-free, experimental studies on two benchmark event detection datasets (ACE2005 and MAVEN) reveal that our proposed method can achieve competitive performance.

Keywords: Event detection · Prompt learning · Machine reading comprehension

1 Introduction

Information extraction (IE) is an important application of Natural Language Processing (NLP). Event detection (ED) is a fundamental part of IE, aiming at identifying trigger words and classifying event types, which could be divided into two sub-tasks: trigger identification and trigger classification [1]. For example, consider the following sentence *"To **assist** in managing the vessel traffic, Chodkiewicz **hired** a few sailors, mainly Livonian"*. The trigger words are "assist" and "hired", the trigger-based event detection model is used to locate the position of the trigger words and classify them into the corresponding event types, *Assistance* and *Employment* respectively.

Contemporary mainstream studies on ED concentrate on trigger-based methods. These methods involve initially identifying the triggers and then categorizing the types of events [2–4]. This approach changes the ED task into a multi-stage classification issue, with the outcome of trigger identification also impacting the categorization of triggers. Therefore, it is crucial to identify trigger words correctly, which requires datasets containing multiple annotated trigger words and event types [5]. However, it is time-consuming to annotate trigger words in a real scenario, especially in a long sentence. Due to the expensive annotation of the corpus, the application of existing ED approaches is greatly limited. It should be noted that trigger words are considered an extra supplement for trigger classification, but event triggers may not be essential for ED [6].

From a problem-solving perspective, ED aims to categorize the type of events and therefore triggers can be seen as an intermediate result of this task [6]. To alleviate manual effort, we aim to explore how to detect events without triggers. Event detection can be considered a text classification problem if the event triggers are missing. But three challenges should be solved: (1) **Multi-label problem**: since a sentence can contain multiple events or no events at all, which is called a multi-label text classification problem in NLP. (2) **Insufficient event information**: triggers are important and helpful for ED [2,7]. Without trigger words, the ED model may lack sufficient information to detect the event type, and we need to find other ways to enrich the sentence semantic information and learn the correlation between the input sentence and the corresponding event type. (3) **Imbalance Data Distribution**: the data distribution in the real world is long-tail, which means that most event types have only a small number of instances and many sentences may not have events occurring. The goal of ED is also to evaluate its ability in the long-tail scenario.

To detect events without triggers and solve these problems, we propose a two-tower model via machine reading comprehension (MRC) [8] and prompt learning [9]. Figure 1 illustrates the structure of our proposed model with two parts: reading comprehension encoder (RCE) and event type classifier (ETC). In the first-tower, we employ BERT [10] as backbone, and the input sentence concatenates with all event tokens are fed into BERT simultaneously[1]. Such a way is inspired by the MRC task, extracting event types is formalized as extracting answer position for the given sequence of event type tokens. In other words, the input sentences are deemed as "Question" and the sequence of event type tokens deemed as "Answer". This way allows BERT to automatically learn semantic relations between the input sentences and event tokens through self-attention mechanism [11]. In the second-tower, we use the same backbone of RCE and utilize prompt learning methods to predict event types. Specifically, when adding the prompt "This sentence describes a [MASK] event" after the original sentence, this prompt can be viewed as a cloze-style question and the answer is related to the target event type. Therefore, ETC aims to fill the [MASK] token and can output the scores for each vocabulary token. We only use event type tokens in vocabulary and predict event types that score higher than the ⟨none⟩ event type. In the inference time, only when these two-tower models predict results are correct can they be used as the final correct answer. In our

---

[1] For example, we convert event token *employment* to "⟨*employment*⟩" and add it to vocabulary. All events operate like this. In addition, we add a special token "⟨*none*⟩" that no events have occurred.

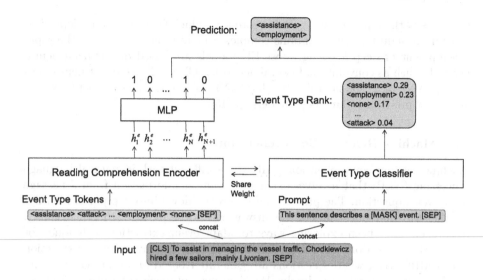

Fig. 1. Overview of our proposed EDPRC. It consists of two modules: reading comprehension encoder (RCE) and event type classifier (ETC).

example from Fig. 1, RCE can predict the answer tokens are ⟨*assistance*⟩ and ⟨*employment*⟩ respectively. In addition, since ⟨*assistance*⟩ and ⟨*employment*⟩ both have higher values than ⟨*none*⟩, we predict *Assistance* and *Employment* as the event type in this sentence.

In summary, we propose a two-tower model to solve the ED task without triggers and call our model **EDPRC: E**vent **D**etection via **P**rompt learning and machine **R**eading **C**omprehension. The main contributions of our work are: (1) We propose a trigger-free event detection method based on prompt learning and machine reading comprehension that does not require triggers. The machine reading comprehension method can capture the semantic relations between sentence and event tokens. The prompt learning method can evaluate the scores of all event tokens in vocabulary; (2) Our experiments can achieve competitive results compared with other trigger-based methods and outperform other trigger-free baselines on ACE2005 and MAVEN; (3) Further analysis of attention weight also indicates that our trigger-free model can identify the relation between input sentences and events, and appropriate prompts in a specific topic can guide pre-trained language models to predict correct events.

2   Related Work

2.1   Sentence-Level Event Detection

Conventional sentence-level event detection models based on pattern matching methods mainly utilize syntax trees or regular expressions [12]. These pattern-matching methods largely rely on the expression form of text to recognize triggers and classify them into event types in sentences, which fails to learn in-depth features from plain text that contains complex semantic relations. With the rapid development of deep learning, most ED models are based on artificial neural networks such as convolutional neural networks (CNN) [2], recurrent neural network (RNN) [3], graph neural network (GNN) [13] and transformer network [14], and other pre-trained language models [10,15].

2.2   Machine Reading Comprehension

Machine reading comprehension (MRC) is a difficult task in natural language processing (NLP) that involves extracting relevant information from a passage to answer a question. The process can be broken down into two parts: identifying the start and end points of the answer within the passage [16,17]. Recently, researchers have been exploring ways to adapt event extraction techniques for use in MRC question answering. One approach is to convert event extraction into a MRC task, where questions are generated based on event schemas and answers are retrieved accordingly [18]. Another approach is to utilize a mechanism like DRC, which employs self-attention to understand the relationships between context and events, allowing for more accurate answer retrieval [19].

2.3   Prompt Learning

In recent years, there has been significant progress in natural language processing (NLP) tasks using prompt-based methods [9]. Unlike traditional model fine-tuning, prompt-tuning involves adding prompts to the raw input to extract knowledge from pre-trained language models like BERT [10] and GPT3 [20]. This new approach allows for the creation of tailored prompts for specific downstream tasks such as text classification, relation extraction, and text generation. By doing so, it bridges the gap between pre-trained tasks and downstream tasks, reducing training time significantly [21]. Additionally, prompt-based learning enables pre-trained language models to gain prior knowledge of a particular downstream task, ultimately improving performance [22].

3   Methodology

In this section, we present the proposed EDPRC in detail for sentence-level event detection without triggers.

3.1   Problem Description

Formally, denote $\mathcal{X}$, $\mathcal{Y}$ as the sentence set and the event type set, respectively. $\mathcal{X} = \{x_i | i \in [1, M]\}$ contains M sentences, and each sentence x_i in $\mathcal{S}$ is a token sequence $x_i = (w_1, w_2, ..., w_L)$ with maximum length L. In sentence-level event detection, given a sentence x_i and its ground-truth $y_i \in \mathcal{Y}$, $\mathcal{Y} = \{e_1, e_2, ..., e_N\}$, we need to detect the corresponding event types for each instance. For sentences where no event occurred, we add a special token "⟨*None*⟩" as their event type. This problem can be reformulated as a multi-label classification task with $N+1$ event types.

3.2 Reading Comprehension Encoder

Inspired by the MRC task, we employ BERT as backbone to design a reading comprehension encoder due to its capability in learning contextual representations of the input sequence. We describe it as follows:

$$Input = [\text{CLS}] \textbf{ Sentence } [\text{SEP}] \textbf{ Events} \tag{1}$$

where **Sentence** is the input sentence and **Events** is the event type set (also including "⟨*None*⟩"). [CLS] and [SEP] stand for the start token and separator token in BERT, respectively. For some event types such as "Business:Lay off" fails to map to a single token according to the vocabulary. In this case, we employ an angle bracket around each event type and remove the prefix, e.g., the event type of "Business:Lay off" is converted to a lower-case "⟨*lay_off*⟩". Then, we add $N+1$ event tokens to the vocabulary and randomly initialize its embeddings. Our objective is to utilize BERT for understanding the correlation between the event types and input sentence, producing accurate representations of event tokens.

After that, we get the token representations by using BERT:

$$h_{[CLS]}, h_1^w, ..., h_L^w, h_{[SEP]}, h_1^e, ..., h_N^e, h_{N+1}^e = BERT(Input) \tag{2}$$

where h_i^w is the hidden state of the i-th input token. This setup is close to MRC that chooses the correct option to answer question "What happened in the sentence?". Unlike traditional fine-tuning methods that utilize the [CLS] token to complete classification, we use the hidden states of event tokens to predict the probability of each token being the correct answer. The representation of event tokens:

$$E = h_1^e, ..., h_N^e, h_{N+1}^e \tag{3}$$

where $E \in \mathbb{R}^{N \times D}$, D is the dimension of token representation. The probability of each event token as follows:

$$P = softmax(E \cdot W) \in \mathbb{R}^{N \times 2} \tag{4}$$

where $W \in \mathbb{R}^{D \times 2}$ is a trainable weight matrix. During training time, we therefore have the following loss for predictions:

$$\mathcal{L}_{RCE} = CE(P, Y) \tag{5}$$

where Y is the ground-truth label of each event token e_i being the correct answer.

3.3   Event Type Classifier

We describe the implementation of ETC in this subsection. Inspired by the cloze-style prompt learning paradigm for text classification with pre-trained language models, event type classification can be realized by filling the [MASK] answer using a prompt function.

First, the prompt function wraps the input sentence by inserting pieces of natural language text. For prompt function f_p, as illustrated in Fig. 1, we use "[SENTENCE] This sentence describes a [MASK] event" as a prompt function for our model. Let $\mathcal{M}$ be pre-trained language model (i.e., BERT), and $\mathbf{x}$ be the input sentence. The prediction score of each token v in vocabulary being filled in [MASK] token can be computed as:

$$p_v = \mathcal{M}([\text{MASK}] = v | f_p(x)) \qquad (6)$$

After that, the other key of prompt learning is answer engineering. We aim to construct a mapping function from event token space to event type space. In the first tower (RCE), it learns the relation between the input sentence and event tokens. RCE and ETC share the same weights of BERT. Then, we only select tokens in $\mathcal{Y} = \{e_1, e_2, ..., e_N\}$ and compute the scores of event tokens:

$$p_e = \sigma(p_v | v \in \mathcal{Y}) \qquad (7)$$

where $\sigma(\cdot)$ determines which function to transform the scores into the probability of event tokens, such as *softmax*.

Finally, as shown in Fig. 1, we predict all event tokens that score higher than the "$\langle None \rangle$" token as the predicted result. In our example, since both "$\langle assistance \rangle$" and "$\langle employment \rangle$" have higher scores than "$\langle None \rangle$", we predict *Assistance* and *Employment* as target event types.

In the process of training, we calculate two losses due to the problem of imbalance data distribution. The first loss is defined as:

$$\mathcal{L}_1 = \frac{1}{|T|} \sum_{t \in T} \log \frac{\exp(\mathcal{M}([\text{MASK}] = t | f_p(x)))}{\sum_{t' \in \{t, \langle none \rangle\}} \exp(\mathcal{M}([\text{MASK}] = t' | f_p(x)))} \qquad (8)$$

where T is the set of event tokens that score higher than "$\langle None \rangle$" in the sentence. The second loss is defined as follows:

$$\mathcal{L}_2 = \log \frac{\exp(\mathcal{M}([\text{MASK}] = \langle none \rangle | f_p(x)))}{\sum_{t' \in \{\langle none \rangle\} \cup \overline{T}} \exp(\mathcal{M}([\text{MASK}] = t' | f_p(x)))} \qquad (9)$$

where $\overline{T}$ is the set of event tokens that score lower than "$\langle None \rangle$" in the sentence. Note that in Eq. 8, we only compare the prediction scores that higher than the "$\langle None \rangle$" event token. The reason is that we aim to improve the score of each event token that is higher than "$\langle None \rangle$". In Eq. 9, we compare to event tokens that lower than the "$\langle None \rangle$", which can decrease the score of them. The training loss of ETC is defined as:

$$\mathcal{L}_{ETC} = \frac{1}{M} \sum_{x \in \mathcal{S}} (\mathcal{L}_1 + \mathcal{L}_2) \qquad (10)$$

In the training time, the total loss of our model is defined as:

$$\mathcal{L} = \mathcal{L}_{RCE} + \mathcal{L}_{ETC} \tag{11}$$

4   Experiments

In this section, we introduce the experimental datasets, evaluation metrics, implementation details, and experimental results.

4.1   Dataset and Evaluation

To evaluate the potential of **EDPRC** under different size datasets, we conducted our experiments on two benchmark datasets, ACE2005 [23] and MAVEN [24]. Details of statistics are available in Table 1.

- The **ACE2005** is globally recognized as the primary multilingual dataset applied for event extraction. Our use focuses on the English version that includes 599 documents and 33 types of events. We engage two versions in line with prior data split pre-processing: ACE05-E [25] and ACE05-E$^+$ [26]. In contrast with ACE05-E, ACE05-E$^+$ incorporates roles for pronouns and multi-token event triggers.
- **MAVEN**, constructed from Wikipedia[2] and FrameNet [27], is a vast event detection dataset encompassing 4,480 documents and 168 different types of events.

For data split and preprocessing, following previous work [24–26], we split 599 documents of ACE2005 into 529/30/40 for train/dev/test set, respectively. Then, we use the same processing that splits 4480 documents of MAVEN into 2913/710/857 for train/dev/test set respectively.

To assess the performance of our event detection model, we employ three commonly used evaluation metrics: precision (P), recall (R), and micro F1-score (F1) [2]. These metrics provide a comprehensive picture of our model's accuracy and effectiveness.

4.2   Baseline

We compare our method to baselines with trigger-based and trigger-free methods. For trigger-based methods, we compare with: (1)**DMCNN** [2], which utilizes a convolutional neural network (CNN) and a dynamic multi-pooling mechanism to learn sentence-level features; (2) **BiLSTM** [28], which uses bi-directional long short-term memory network (LSTM) to capture the hidden states of triggers and classify them into corresponding event types; (3)**MOGANDED** [29], which proposes multi-order syntactic relations in dependency trees to improve event detection; (4)**BERT** [10], fine-tuning BERT on the ED task via a sequence labeling manner; (5)**DMBERT** [4], which adopts BERT as backbone and utilizes a

---

[2] https://www.wikipedia.org/.

Table 1. Dataset statistics of ACE05-E, ACE05-E$^+$ and MAVEN.

Dataset	Split	#Sentences	#Events	#Documents
ACE05-E	Train	17,172	4,202	529
	Dev	923	450	30
	Test	832	403	40
ACE05-E$^+$	Train	19216	4419	529
	Dev	901	468	30
	Test	676	424	40
MAVEN	Train	32431	73496	2913
	Dev	8042	17726	710
	Test	9400	20389	857

dynamic multi-pooling mechanism to aggregate textual features. For trigger-free methods, we compare with: (6)**TBNNAM** [6], the first work on detecting events without triggers, which uses LSTM and attention mechanisms to detect events; (7)**TEXT2EVENT** [30], proposing a sequence-to-sequence model and extracting events from the text in an end-to-end manner; (8)**DEGREE** [31], formulating event detection as a conditional generation problem and extracting final predictions from the generated sentence with a deterministic algorithm.

We re-implemented some trigger-based baselines for comparison, including DMCNN, BiLSTM, MOGANDED, BERT and DMBERT. The other baseline results are from the original paper.

4.3   Implementation Details

We utilize the Transformers toolkit [32] and PyTorch to implement our proposed model. Specifically, we employ the *bert-base-uncased*[3] model as the backbone and optimize it with AdamW optimizer, setting the learning rate to 2e-5, maximum gradient norm to 1.0, and weight decay to 5e-5. We limit the maximum sequence length to 128 for ACE2005 and 256 for MAVEN, and apply a dropout rate of 0.3. Our model is trained on a single Nvidia RTX 3090 GPU for 10 epochs, selecting the checkpoint with the highest validation performance on the development set. Our code is publicly available at https://github.com/rickltt/event_detection.

4.4   Main Results

Table 2 reports main results. Compared with trigger-free methods, we can find out that our method achieves a much better performance than other trigger-free baselines (TBNNAM, TEXT2EVENT and DEGREE). Obviously, ED_PRC can achieve improvements of 0.4% (73.3% v.s. 73.7%) F1 score of the best trigger-free baseline (DEGREE) in ACE05-E, and 2.1% (71.8% v.s. 73.9%) F1 score of

---

[3] https://huggingface.co/bert-base-uncased.

Table 2. Event detection results on both trigger-based and trigger-free methods of the ACE2005 corpora. "-" means not reported in original paper. ∗ indicates results cited from the original paper.

Category	Models	ACE05-E			ACE05-E$^+$		
		P	R	F-1	P	R	F-1
Trigger-based	DMCNN	74.3	66.8	70.3	67.0	73.5	70.1
	BiLSTM	73.6	72.3	72.9	73.5	71.3	72.4
	MOGANED	74.6	71.1	72.8	74.2	72.2	73.2
	BERT	72.5	74.2	73.3	75.2	72.4	73.8
	DMBERT	76.4	71.9	74.1	74.9	73.5	74.2
Trigger-free	TBNNAM∗	76.2	64.5	69.9	-	-	-
	TEXT2EVENT∗	69.6	74.4	71.9	71.2	72.5	71.8
	DEGREE∗	-	-	73.3	-	-	70.9
	ED_PRC (Ours)	76.1	71.5	73.7	74.6	73.2	73.9

TEXT2EVENT in ACE05-E$^+$. It proves the overall superiority and effectiveness of our model in the absence of triggers. Compared to trigger-based methods, despite the absence of trigger annotations, ED_PRC can achieve competitive results with other trigger-based baselines, which is only 0.4% (73.7% vs. 74.1%) in ACE05-E and 0.3% (73.9% vs. 74.2%) in ACE05-E$^+$ less than the best trigger-based baseline (DMBERT). The result shows that prompt-based method can greatly utilize pre-trained language models to adapt ED task and our MRC module is capable of learning relations between the input text and the target event tokens under low trigger clues scenario.

To further evaluate the effectiveness of our model on large-scale corpora, we show the result of MAVEN on various trigger-based baselines and our model in Table 3. We can see that our model also can achieve competitive performance on various trigger-based baselines, reaching 69.1% F1 score. Compared with CNN-based (DMCNN), RNN-based (BiLSTM) and GNN-based (MOGANED) method, BERT-based methods (BERT, DMBERT and ED_PRC) can outperform high improvements, which indicates pre-trained language models can greatly capture contextual representation of input text. However, ED_PRC can achieve only improvements of 0.1% (67.2% v.s. 67.3%) F1 score on BERT and is 0.8% (67.3% v.s. 68.1%) less than DMBERT. This can be attributed to more triggers and events on MAVEN than that on ACE2005. We conjecture that trigger-based event detection models can greatly outperform trigger-free models when sufficient event information is available. All in all, our ED_PRC is proven competitive in both ACE2005 dataset and MAVEN dataset.

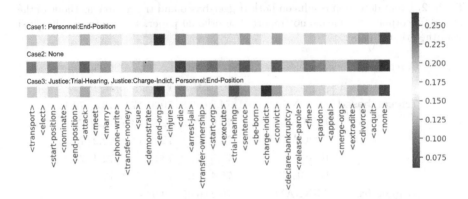

Fig. 2. The ACE2005 examples visualization of attention weight in event tokens. We show three cases, the first with only one event, the second with no events and the third with multiple events.

5   Analysis

In this section, we demonstrate further analysis and give an insight into the effectiveness of our method.

5.1   Effective of Reading Comprehension Encoder

Figure 2 shows a few examples with different target event types and their attention weight visualizations learned by the reading comprehension encoder. In the first case, the target event type is "Personnel:End-Position" and our reading comprehension encoder successfully captures this feature by giving "⟨end − org⟩" a high attention score. In addition, in the second case, it is a negative sample that no event happened in this sentence and our reading comprehension encoder can correctly give a high attention score for "⟨none⟩" and give low attention scores for other event tokens. Moreover, three events occur in the third case, "Justice:Trial-Hearing", "Justice:Charge-Indict" and "Personnel:End-Position", respectively. Our approach can also give high attention scores to "⟨trial − hearing⟩", "⟨charge − indict⟩" and "⟨end − org⟩". We argue that, although triggers are absent, our model can learn the relations between input text and event tokens and assign the ground-truth event tokens with high attention scores.

5.2   Effective of Different Prompts

Generally, as the key factor in prompt learning, the prompt can be divided into two categories: hard prompt and soft prompt. The hard prompt is also called a discrete template, which inserts tokens into the original input sentence. Soft prompt is also called continuous template, which is a learnable prompt that does not need any textual templates. To further analyze the influence of prompts, we

Table 3. Event detection results on MAVEN corpus.

Models	P	R	F-1
DMCNN	66.5	58.4	62.2
BiLSTM	64.7	68.2	62.4
MOGANED	65.9	65.1	65.5
BERT	64.3	70.5	67.2
DMBERT	68.9	67.4	68.1
ED_RRC (Ours)	66.0	68.7	67.3

Table 4. Results on ACE2005 datasets with different prompts.

Models	P	R	F-1
Prompt_1	74.2	72.9	73.5
Prompt_2	75.6	71.4	73.4
Prompt_3	74.7	71.2	72.9
Prompt_4	74.1	73.5	73.8
Soft	73.5	72.7	73.1

design four different textual templates (hard prompt) to predict event types: (1) What happened? [SENTENCE] This sentence describes a [MASK] event; (2) [SENTENCE] What event does the previous sentence describe? It was a [MASK] event; (3) [SENTENCE] It was [MASK]; (4) A [MASK] event: [SENTENCE]. For soft prompt, we insert four trainable tokens into the original sentence, such as "[TOKEN] [TOKEN] [SENTENCE] [TOKEN] [TOKEN] [MASK]". The results of our method on ACE2005 are shown in Table 4.

Prompt_1 and Prompt_2 perform similarly, and both of them work better than Prompt_3. The reason for this may be that Prompt_3 provides less information and less topic-specific. And both Prompt_1 and Prompt_2 add a common phrase "sentence describe" and a question to prompt the model to focus on the previous sentence. Unlike previous prompts, Prompt_4 puts [MASK] at the beginning of a sentence, and the result indicates that it might be slightly better to put the [MASK] at the end of the sentence. Compared with hard prompt, soft prompt eliminate the need for manual human design and construct trainable tokens that be optimized during training time. The result of soft prompt achieve performance that was fairly close to the hard prompt.

6   Conclusion

In this paper, we transform sentence-level event detection to a two-tower model via prompt learning and machine reading comprehension, which can detect events without trigger words. By using machine reading comprehension framework to formulate a reading comprehension encoder, we can learn the relation between input text and event tokens. Besides, we utilize prompt-based learning methods to construct an event type classifier and final predictions are based on two towers. To make effective use of prompts, we design four manual hard prompts and compare with soft prompt. Experiments and analyses show that ED_PRC can even achieves competitive performance compared to mainstream approaches using annotated triggers. In the future, we are interested in exploring more event detection methods without triggers by using prompt learning or other techniques.

Acknowledgements. This work was supported in part by the National Natural Science Foundation of China (62006044, 62172110). Additionally, support was also partly provided by the Natural Science Foundation of Guangdong Province (2022A1515010130), and the Programme of Science and Technology of Guangdong Province (2021A0505110004) contributed in part to this work.

References

1. Li, Q., et al.: A survey on deep learning event extraction: approaches and applications. IEEE Trans. Neural Netw. Learn. Syst. **PP**, 1–21 (2022)
2. Chen, Y., Xu, L., Liu, K., Zeng, D., Zhao, J.: Event extraction via dynamic multi-pooling convolutional neural networks. In: Proceedings of the 53rd Annual Meeting of the Association for Computational Linguistics and the 7th International Joint Conference on Natural Language Processing (ACL-IJCNLP), pp. 167–176 (2015)
3. Sha, L., Qian, F., Chang, B., Sui, Z.: Jointly extracting event triggers and arguments by dependency-bridge RNN and tensor-based argument interaction. In: Proceedings of the AAAI Conference on Artificial Intelligence, vol. 32 (2018)
4. Wang, X., Han, X., Liu, Z., Sun, M., Li, P.: Adversarial training for weakly supervised event detection. In: Proceedings of the 2019 Conference of the North American Chapter of the Association for Computational Linguistics: Human Language Technologies (NAACL-HLT), pp. 998–1008 (2019)
5. Lai, V.D., Nguyen, T.H., Dernoncourt, F.: Extensively matching for few-shot learning event detection. In: Proceedings of the First Joint Workshop on Narrative Understanding, Storylines, and Events, pp. 38–45 (2020)
6. Liu, S., Li, Y., Zhang, F., Yang, T., Zhou, X.: Event detection without triggers. In: Proceedings of the 2019 Conference of the North American Chapter of the Association for Computational Linguistics: Human Language Technologies (NAACL-HLT) (2019)
7. Zhang, Z., Kong, X., Liu, Z., Ma, X., Hovy, E.: A two-step approach for implicit event argument detection. In: Proceedings of the 58th Annual Meeting of the Association for Computational Linguistics (ACL), pp. 7479–7485 (2020)
8. Li, X., Feng, J., Meng, Y., Han, Q., Wu, F., Li, J.: A unified MRC framework for named entity recognition. In: Proceedings of the 58th Annual Meeting of the Association for Computational Linguistics (ACL), pp. 5849–5859 (2020)
9. Schick, T., Schütze, H.: Exploiting cloze-questions for few-shot text classification and natural language inference. In: Proceedings of the 16th Conference of the European Chapter of the Association for Computational Linguistics (EACL), pp. 255–269 (2021)
10. Devlin, J., Chang, M.W., Lee, K., Toutanova, K.: BERT: pre-training of deep bidirectional transformers for language understanding. In: Proceedings of the 2019 Conference of the North American Chapter of the Association for Computational Linguistics: Human Language Technologies (NAACL-HLT), pp. 4171–4186 (2019)
11. Vaswani, A., et al.: Attention is all you need. In: Advances in Neural Information Processing Systems 30 (2017)
12. Ahn, D.: The stages of event extraction. In: Proceedings of the Workshop on Annotating and Reasoning about Time and Events, pp. 1–8 (2006)
13. Cui, S., Yu, B., Liu, T., Zhang, Z., Wang, X., Shi, J.: Edge-enhanced graph convolution networks for event detection with syntactic relation. In: Findings of the Association for Computational Linguistics: EMNLP 2020, pp. 2329–2339 (2020)

14. Yang, S., Feng, D., Qiao, L., Kan, Z., Li, D.: Exploring pre-trained language models for event extraction and generation. In: Proceedings of the 57th Annual Meeting of the Association for Computational Linguistics (ACL), pp. 5284–5294 (2019)
15. Wei, Y., et al.: DESED: Dialogue-based explanation for sentence-level event detection. In: Proceedings of the 29th International Conference on Computational Linguistics (COLING), pp. 2483–2493 (2022)
16. Seo, M., Kembhavi, A., Farhadi, A., Hajishirzi, H.: Bidirectional attention flow for machine comprehension. arXiv preprint arXiv:1611.01603 (2016)
17. Shen, Y., Huang, P.S., Gao, J., Chen, W.: ReasoNet: learning to stop reading in machine comprehension. In: Proceedings of the 23rd ACM SIGKDD International Conference on Knowledge Discovery and Data Mining, pp. 1047–1055 (2017)
18. Liu, J., Chen, Y., Liu, K., Bi, W., Liu, X.: Event extraction as machine reading comprehension. In: Proceedings of the 2020 Conference on Empirical Methods in Natural Language Processing (EMNLP), pp. 1641–1651 (2020)
19. Zhao, J., Yang, H.: Trigger-free event detection via derangement reading comprehension. arXiv preprint arXiv:2208.09659 (2022)
20. Brown, T., et al.: Language models are few-shot learners. Adv. Neural. Inf. Process. Syst. **33**, 1877–1901 (2020)
21. Liu, P., Yuan, W., Fu, J., Jiang, Z., Hayashi, H., Neubig, G.: Pre-train, prompt, and predict: a systematic survey of prompting methods in natural language processing. ACM Comput. Surv. **55**(9), 1–35 (2023)
22. Wei, Y., Mo, T., Jiang, Y., Li, W., Zhao, W.: Eliciting knowledge from pretrained language models for prototypical prompt verbalizer. In: Artificial Neural Networks and Machine Learning - ICANN 2022, pp. 222–233 (2022)
23. Doddington, G., Mitchell, A., Przybocki, M., Ramshaw, L., Strassel, S., Weischedel, R.: The automatic content extraction (ACE) program - tasks, data, and evaluation. In: Proceedings of the Fourth International Conference on Language Resources and Evaluation (LREC) (2004)
24. Wang, X., et al.: MAVEN: a massive general domain event detection dataset. In: Proceedings of the 2020 Conference on Empirical Methods in Natural Language Processing (EMNLP), pp. 1652–1671 (2020)
25. Wadden, D., Wennberg, U., Luan, Y., Hajishirzi, H.: Entity, relation, and event extraction with contextualized span representations. In: Proceedings of the 2019 Conference on Empirical Methods in Natural Language Processing and the 9th International Joint Conference on Natural Language Processing (EMNLP-IJCNLP), pp. 5784–5789 (2019)
26. Lin, Y., Ji, H., Huang, F., Wu, L.: A joint neural model for information extraction with global features. In: Proceedings of the 58th Annual Meeting of the Association for Computational Linguistics (ACL), pp. 7999–8009 (2020)
27. Baker, C.F., Fillmore, C.J., Lowe, J.B.: The Berkeley Framenet project. In: COLING 1998 Volume 1: The 17th International Conference on Computational Linguistics (1998)
28. Hochreiter, S., Schmidhuber, J.: Long short-term memory. Neural Comput. **9**(8), 1735–1780 (1997)
29. Yan, H., Jin, X., Meng, X., Guo, J., Cheng, X.: Event detection with multi-order graph convolution and aggregated attention. In: Proceedings of the 2019 Conference on Empirical Methods in Natural Language Processing and the 9th International Joint Conference on Natural Language Processing (EMNLP-IJCNLP), pp. 5766–5770 (2019)

30. Lu, Y., et al.: Text2Event: Controllable sequence-to-structure generation for end-to-end event extraction. In: Proceedings of the 59th Annual Meeting of the Association for Computational Linguistics and the 11th International Joint Conference on Natural Language Processing (ACL-IJCNLP), pp. 2795–2806 (2021)
31. Hsu, I.H., et al.: DEGREE: a data-efficient generation-based event extraction model. In: Proceedings of the 2022 Conference of the North American Chapter of the Association for Computational Linguistics: Human Language Technologies (NAACL-HLT), pp. 1890–1908 (2022)
32. Wolf, T., et al.: Transformers: State-of-the-art natural language processing. In: Proceedings of the 2020 Conference on Empirical Methods in Natural Language Processing (EMNLP), pp. 38–45 (2020)

Multi-grained Logical Graph Network for Reasoning-Based Machine Reading Comprehension

Jiaqi Wang[1], Jia Zhong[1]([✉]) [iD], Hong Yin[1], Chen Wang[1], Qizhu Dai[1], Yang Xia[2], and Rongzhen Li[1]

[1] Chongqing University, Chongqing, China
{wangjiaqi,zhongjiang,yinhong,chenwang,daiqizhu,lirongzhen}@cqu.edu.cn
[2] Chongqing Academy of Big Data Co., Ltd., Chongqing, China

Abstract. Machine reading comprehension (MRC) is a crucial and challenging task in natural language processing (NLP). In order to equip machines with logical reasoning abilities, the challenging logical reasoning tasks are proposed. Existing approaches use graph-based neural models based on either sentence-level or entity-level graph construction methods which designed to capture a logical structure and enable inference over it. However, sentence-level methods result in a loss of fine-grained information and difficulty in capturing implicit relationships, while entity-level methods fail to capture the overall logical structure of the text. To address these issues, we propose a multi-grained graph-based mechanism for solving logical reasoning MRC. To combine the advantages of sentence-level and entity-level information, we mine elementary discourse units (EDUs) and entities from texts to construct graph, and learn the logical-aware features through a graph network for subsequent answer prediction. Furthermore, we implement a positional embedding mechanism to enforce the positional dependence, which facilitates logical reasoning. Our experimental results demonstrate that our approach provides significant and consistent improvements via multi-grained graphs, outperforming competitive baselines on both ReClor and LogiQA benchmarks.

Keywords: Machine Reading comprehension · Logical Reasoning · Multi-grained Graph

1 Introduction

Machine Reading Comprehension (MRC) is a fundamental task in Natural Language Processing (NLP) that seeks to teach machines to comprehend the meaning of human text and answer questions [50]. With the advancement of unsupervised learning and pre-trained language models (LM), many neural methods have achieved remarkable success on inchoate and simple datasets. For instance, BERT [7] has outperformed human performance in SQuAD [27]. Recently, MRC

X. Yang et al. (Eds.): ADMA 2023, LNAI 14179, pp. 47–62, 2023.
https://doi.org/10.1007/978-3-031-46674-8_4

tasks have become more challenging by raising the difficulty of contexts and questions, which aim to better evaluate model capabilities, such as multi-hop reasoning [17,23,39,43], numerical reasoning [8,52] and commonsense reasoning [13,35].

In addition to the above capabilities, logical reasoning is also a crucial aspect of human intelligence that plays a significant role in cognition and judgment [21]. It was also a primary research topic in the early days of AI [12,22]. However, most existing MRC models struggle to capture the logical structure of contexts due to the lack of logical reasoning ability. This limitation often leads to poor performance in logical reasoning MRC questions. To drive the development of logical reasoning, ReClor [48] and LogiQA [21] were proposed. These two datasets are multiple-choice MRC datasets, constructed by selecting logical reasoning questions from standardized exams. A logical reasoning problem example from LogiQA dataset is shown in Fig. 1. It contains a context, a question, and four answer options, among which only one option is correct.

Context

Left-handed people suffer from immune disorders, such as allergies, more often than right-handed people. However, left-handers often have an advantage over right-handers in accomplishing tasks controlled by the right hemisphere of the brain, and most people's mathematical reasoning ability is strongly affected by the right hemisphere of the brain.

Question

If the above information is true, which one of the following assumptions can it best support?

Option

A. Most people with allergies or other immune disorders are left-handed rather than right-handed.
B. Most left-handed mathematicians have some kind of allergy.
C. The proportion of left-handed people who have stronger mathematical reasoning ability than the average is higher than the proportion of left-handed people who have weaker mathematical reasoning ability than the average.
D. The proportion of people with immune dysfunction such as allergies is higher than that of left-handed people or people with unusually good mathematical reasoning skills.

Fig. 1. A logical reasoning based MRC example from LogiQA dataset.

To solve this task, previous research has employed methods that involves mining logical units and constructing a logical structure to facilitate reasoning over the context and question, ultimately predicting the correct answer. The two main granularities of information used are sentence-level and entity-level. For instance, at the sentence-level, AdaLoGN [18] mines a set of elementary discourse units (EDUs) from texts, converts discourse relations to logical relations to construct graphs, and then extends the graphs based on some inference rules. Finally, it uses a graph neural network (GNN) [32] to predict the answer. For another instance, at the entity-level, FocalReasoner [25] mines "Entity-Predicate-Entity" triplets as fact units from each sentence in the context, and finds the co-reference relations among entities in fact units. It then constructs a supergraph on the top of fact units and enhances graph presentation by GNN to predict the answer.

Although previous works have made significant advancements in logical reasoning, they also have their limitations. The sentence-level method is at a coarse-grained level, simply averaging the EDUs vector as the node representation, which could cause the loss of fine-grained information [1,9], particularly for the keyword in texts. Moreover, there are instances where there are no explicit logical conjunctions like "because" and "if" in the texts, making it challenging for prior works to mine explicit logical relations, as shown in the example in Fig. 1. In the LogiQA dataset [21], 3092 data points out of 8678 data points could not mine any explicit logical relations. If the logical relations cannot be extracted, AdaLoGN can only rely on the adjacent relations between EDUs to construct graphs and can't derive implicit logical relation via logical reasoning, resulting in sparse graph construction, significantly affecting graph information interaction and answer prediction. Additionally, the entity-level method is at a fine-grained level, ignoring the overall logical structure of texts. It only focuses on entity-level information, neglecting sentence-level interaction.

Inspired by previous work [9,41,51], we found that combining sentence-level and entity-level information can be more comprehensive and effective to make full use of the information in the text, which can address the above problems. In this paper, we present a new approach, MLGNet, for logical reasoning-based MRC with a multi-grained graph, as the overall model architecture depicted in Fig. 2. The aim is to combine the advantages of sentence-level and entity-level information in texts to create a better logical reasoning-based MRC model. We propose to construct graphs that contain EDUs and entities and present their relations, and then use graph network to learn the logical-aware features for subsequent answer prediction. With such multi-grained graphs, we can (i) not only mine logical structure via sentence-level information but also focus on local perception via entity-level information; and (ii) capture the implicit relations of EDUs through entities simultaneously. Furthermore, nodes in graphs are not arranged in a sequence, which may lead to a loss of order information for EDUs mined from the text, especially with the introduction of entity nodes and relations between EDUs and entities. Therefore, we propose a spatial encoding mechanism to strengthen the positional dependency of EDUs.

The contributions of this paper are three-fold:

- We introduce a heterogeneous multi-grained logical graph (MLG) with a graph-based neural network to model the logical relations of texts and offer logical-aware features.
- We present a positional embedding mechanism to reinforce the positional information to facilitate logical reasoning.
- Our experiment results demonstrate that MLGNet can boost the performance compared with strong baselines on two datasets ReClor and LogiQA.

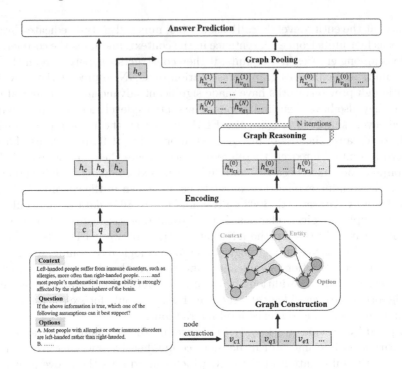

Fig. 2. Overall architecture of MLGNet.

2   Related Work

2.1   Machine Reading Comprehension

In recent years, there has been a surge of interest in complex machine reading comprehension (MRC) that evaluates model capabilities from various angles. For instance, HotpotQA [43], WikiHop [39], OpenBookQA [23], and MultiRC [17] require models to possess multi-hop reasoning capabilities, while DROP [8] and MA-TACO [52] require models to perform numerical reasoning. Besides, WIQA [35] and CosmosQA [13] test models' commonsense reasoning abilities. In addition to these abilities, logical reasoning is a crucial component of human intelligence and is receiving significant attention from researchers. Several MRC datasets that require logical reasoning have been proposed, including ReClor [48] and LogiQA [21]. ReClor is based on standardized exams such as GMAT and LSAT in the United States, while LogiQA is derived from the National Civil Servants Examination of China. These datasets contain 6138 and 8678 data points respectively, providing ample data to support logical reasoning in MRC.

2.2   Reasoning-Based MRC

Previous work has attempted to use semantic information extracted from text to construct logic graphs, which are then used to pass messages and update graph representations for answer prediction. These methods rely on two levels of information to construct graphs, sentence-level, and entity-level. For sentence-level, DAGN [14] constructs logic graphs using elementary discourse units split by discourse relations, but the chain-type graph is too sparse to facilitate effective node interaction. On this basis, AdaLoGN [18] extends the graph using symbolic logical reasoning to make it more densely connected, and Logiformer [42] constructs causal and syntax graphs simultaneously to capture logical and co-occurrence relations. However, these methods simply average the sentence vector as node representation, resulting in a loss of fine-grained information [1,9]. For entity-level, FocalReasoner [25] extracts fact units in the form of entity-predicate-entity triplets to construct supergraphs and updates nodes using GNNs, but it overlooks the logical relationship between sentences, which does not fully emphasize the logical structure of the text. Therefore, we propose to combine the advantages of sentence-level and entity-level information to build a multi-grained graph.

3 Methodology

In this work, we consider the multiple-choice MRC task, which can be described as a triplet $\langle c, q, O \rangle$, where c is a context, q is a question over c and O is a set of options. Our goal is to find only one correct option in O. Our framework is shown in Fig. 2. We first construct a multi-grained graph via texts, then conduct encoding and graph reasoning to make information fully interactive, and finally aggregate graph information for answer prediction.

3.1 Graph Construction

Multi-grained Logical Graph Definition. To model the logical information from text, a Multi-grained Logical Graph (MLG) is constructed. MLG is a directed graph, which can be represented as $G = \langle V, E \rangle$, where V is a set of nodes and E is a set of edges. MLG has two different kinds of nodes: EDUs V_E and entities V_e, where $V_E \cup V_e = V$. And the edges E present the relationship between EDUs and entities, which correspond to three situations.

- **Logic Edge:** Two EDUs having logical relation are connected with a logic edge. Similar to AdaLoGN [18], we set five types of common logical relations between EUDs as logical edge $L = \{conj, disj, impl, neg, rev\}$, where conjunction (*conj*), disjunction (*disj*), implication (*impl*), and negation (*neg*) are standard logical connectives in propositional logic and reversed implication (*rev*) is introduced to represent the inverse relation of *impl*.
- **Context Edge:** EDUs adjacent in the text, including the last EDU of c and the first EDU of o, are connected with a context edge. In this way, the context relations among the text could be modeled.
- **Containment Edge:** A EDU node is connected with a entity node if EDU containing the entity. With such connections, entity can enhance the representation of EDUs, at the same time EDUs containing the same entity can interact through the entity directly.

Multi-grained Logical Graph Construction. For each sample, we construct graph based on context and option, since the question is usually doesn't carry logical units in existing datasets [14].

For EDUs nodes, we follow the method of AdaLoGN [18], using Graphene [2] to mine EDUs from context and options, and mine the rhetorical relations between them. Rhetorical relations are mapping to logical relations via Table 1. For the relation $\langle v_i, impl, v_j \rangle$, we set the relation $\langle v_j, rev, v_i \rangle$. We also use syntactic rules based on part-of-speech tags and dependencies to find the EDUs which are negate each others, and connect them using *neg* relations.

Table 1. Mapping of logical relation with rhetorical relation.

Rhetorical relation	Logical relation
LIST, CONTRAST	*conj*
DISJUNCTION	*disj*
RESULT	*impl*
CAUSE, PURPOSE, CONDITION	*rev*

And for entity nodes, we employ an entity extractor based on part-of-speech tagging of each EDU, as nouns generally contain the richest semantic information in a sentence. We then select the top k nouns with the most occurrences as entity nodes, where k is a predefined hyper-parameter. For entities and EDUs containing this entity, we establish an *in* relation to indicate the EDU-entity containment relation.

To construct graph, we convert the relations extracted before to the edges in MLG. The logical relations between EDUs are converted to logic edges, and the *in* relations are converted into containment edge. Apart from that, for each EDUs that adjacent in text but are not connected with logic edge, we connect them with context edge. The last EDU of c and the first EDU of o are also connected with context edge.

3.2 Multi-grained Logical Graph Network

We propose the Multi-grained Logical Graph Network with the constructed graph to leverage the logical structure and multi-grained information of text for subsequent answer prediction. It consists of three module: graph encoding, graph filtering and graph pooling.

Graph Encoding. First of all, it's necessary to initialize representation for each node. Similar to previous works, we use RoBERTa [20] encoder to model graph nodes. It takes graph nodes as input and computes a context-aware representation for each token. Specifically, given V_c, V_o denoting EDUs nodes mining from context and option and V_e denoting entities node where $V_c \cup V_o = V_E$ and

$V_E \cup V_e = V$, we pack these nodes in to a single sequence and separate V_c, V_o, V_e with special tokens:

$$[h_{\langle s \rangle}, h_{v_{1_1}}, \cdots, h_{|}, \cdots, h_{\langle/s \rangle}, h_{v_{|V_c|+1_1}}, \cdots, h_{\langle/s \rangle}, h_{v_{|V_E|+1_1}}, \cdots, h_{\langle/s \rangle}]$$
$$= \text{RoBERTa}(\langle s \rangle v_{1_1} \cdots | \cdots \langle/s \rangle v_{|V_c|+1_1} \cdots \langle/s \rangle v_{|V_E|+1_1} \cdots \langle/s \rangle) \quad (1)$$

where $\langle s \rangle$ and $\langle/s \rangle$ is the special tokens for RoBERTa and $|$ is a special token to separate nodes inside V_c, V_o, V_e. For the representation of each node $v_i \in V$, we use the average hidden state.

$$h_{v_i} = \frac{1}{|v_i|} \sum_{j=1}^{|v_i|} h_{v_{i_j}} \quad (2)$$

For graphs always lack of sentence original position information, which aggravated with introduction of the entities nodes, we use the spatial encoding mechanism proposed in [42, 46] to keep the original order information of EDUs in the text. Concretely, for each node $v_i \in V_E$ we compute the positional embeddings of v_i:

$$h_{v_i}^{(0)} = h_{v_i} + \text{PosEmbed}(idx(v_i)) \quad (3)$$

where $idx(v_i)$ returns the index of v_i, and PosEmbed() provides a $|V_E|$-dimensional embedding for each EDUs node. We take the result $h_{v_i}^{(0)}$ as initial representation of each node.

We also use the same pre-train language model RoBERTa as graph node encoding to model the texts in context, question and options for subsequent operation. Given context $c = \{c_i\}_{i=1}^{|c|}$, question $q = \{q_j\}_{j=1}^{|q|}$ and option $o = \{o_k\}_{k=1}^{|o|}$, we calculate for each token a context-aware representation:

$$[h_{\langle s \rangle}, h_{c_1}, \cdots, h_{\langle/s \rangle}, h_{q_1}, \cdots, h_{\langle/s \rangle}, h_{o_1}, \cdots, h_{\langle/s \rangle}]$$
$$= \text{RoBERTa}(\langle s \rangle c_1 \cdots \langle/s \rangle q_1 \cdots \langle/s \rangle o_1 \cdots \langle/s \rangle) \quad (4)$$

We take the average embedding as output of the representations of c, q, o:

$$h_c = \frac{1}{|c|} \sum_{i=1}^{|c|} h_{c_i}, h_q = \frac{1}{|q|} \sum_{i=1}^{|q|} h_{q_i}, h_o = \frac{1}{|o|} \sum_{i=1}^{|o|} h_{o_i} \quad (5)$$

Graph Reasoning. We utilize an iterative neural reasoning method, proposed in [18], to extend previous constructed graph, as well as update nodes representation. We construct the graph described in Sect. 3.1 and then initialize the nodes representation described in Sect. 3.2. Since the entities nodes may introduce irrelevant information, we implement a entity selection strategy to filter the entities nodes in the graph to avoid too much noise. For each candidate entity node $v_i \in V_e$, we calculate the matching score between entity and the text to judge whether relevant to answering the question:

$$rel_e = sigmoid(linear(v_i \,||\, h_o)) \quad (6)$$

where $\|$ represents vector concatenation. We set a predefined threshold τ_e to judge which entity we choose to construct graph. If $rel_e > \tau_e$, we select the entity.

Then, we feed the filtered graph into the iterative reasoning mechanism. In the $(n+1)$-th iteration, we start graph reasoning with the node representation $h_{v_i}^{(n)}$ from the n-th iteration. Since some logical relations are implicit in text which is hard to mine from text, we perform logical inference over the extracted explicit logical relations to derive implicit logical relation according to inference rules. Here we apply three logical equivalence rules:

- Transposition:

$$v_i \rightarrow v_j \Rightarrow \neg v_i \rightarrow \neg v_j \tag{7}$$

- Hypothetical Syllogism:

$$(v_i \rightarrow v_j) \wedge (v_j \rightarrow v_k) \Rightarrow v_i \rightarrow v_k \tag{8}$$

- Adjacency-Transmission:

$$(v_i \sim v_j) \wedge (v_j \,|\, v_k) \Rightarrow v_i \sim v_k \tag{9}$$

where $\sim \in \{\wedge, \vee, \rightarrow\}$ and $|$ represents the context edge, which is adjacency in text.

While these rules may cause misleading, we introduce a mechanism to judge whether the candidate extension is relevant to answering the question. For each candidate extension ϵ applied inference rule over a set of nods $V_\epsilon \in V$, we calculate the relevance score of ϵ:

$$rel_\epsilon = sigmoid(linear(h_\epsilon \,\|\, h_o)),$$
$$h_\epsilon = \frac{1}{V_\epsilon} \sum_{v_i \in V_\epsilon} h_{v_i}^{(n)} \tag{10}$$

where $\|$ represents vector concatenation. We set a predefined threshold τ_ϵ to judge which extension can be admitted to extend graph. If $rel_\epsilon > \tau_\epsilon$, we accept this extension.

After graph extension, we performs to fuse the multi-grained information by interaction of nodes and update node representation from $h_{v_i}^{(n)}$ to $h_{v_i}^{(n+1)}$. Let $\mathcal{N}^i$ indicate the neighbors of node v_i, and $\mathcal{N}_r^i \subseteq \mathcal{N}^i$ indicate the subset under relation $r \in R$. The node representations are updated with message propagation mechanism in R-GCN [?]:

$$h_{u_i}^{(n+1)} = ReLU(\sum_{r \in R} \sum_{v_j \in \mathcal{N}_r^i} \frac{\alpha_{i,j}}{\mathcal{N}_r^i} W_r^{(n)} h_{v_j}^{(n)} + W_0^n h_{v_j}^{(n)}), \text{ where}$$
$$\alpha_{i,j} = softmax_{idx(a_{i,j})}([\cdots, a_{i,j}, \cdots]^T), \text{ for all } u_j \in N_i, \tag{11}$$
$$a_{i,j} = LeakyReLU(linear(h_{v_i}^{(n)} \,\|\, h_{v_j}^{(n)})),$$

where $W_r^{(n)}, W_0^{(n)}$ are matrices and $idx(a_{i,j})$ returns the index of $a_{i,j}$.

Graph Pooling. After N iterations, for each node v_i we fuse the representation over all iterations:

$$h_{v_i}^{fus} = h_{v_i}^{(0)} + linear(h_{v_i}^{(1)}||\cdots||h_{v_i}^{(N)}) \tag{12}$$

In order to avoid the influence of entities on the text sequence, we only consider the representation of the node $v_i \in V_E$ and feed it into a bidirectional residual GRU layer [4], ignoring the representation of nodes $v_i \in V_e$.

$$[h_{v_i}^{fnl}, \cdots, h_{v_{|V_E|}}^{fnl}] = \text{Res-BiGRU}([h_{v_i}^{fnl}, \cdots, h_{v_{|V_E|}}^{fnl}]) \tag{13}$$

We aggregate the node representations by computing an o-attended weighted sum:

$$h_{V_E} = \sum_{v_i \in V_E} \alpha_i h_{v_i}^{fnl}, \text{ where}$$

$$\alpha_i = softmax_i([a_1, \cdots, a_{|V_E|}]^T), \tag{14}$$

$$a_i = LeakyReLU(linear(h_o || h_{v_i}^{fnl}))$$

We concatenate h_{V_E} and the relevance scores to form the representation of G:

$$h_G = (h_{V_E} || rel_{\mathcal{E}^{(1)}} || \cdots || rel_{\mathcal{E}^{(N)}}), \text{ where}$$

$$rel_{\mathcal{E}^{(n)}} = \frac{1}{\mathcal{E}^{(n)}} \sum_{\epsilon \in \mathcal{E}^{(n)}} rel_\epsilon \tag{15}$$

where $\mathcal{E}^{(n)}$ is the set of candidate extensions in the n-th iteration.

3.3 Answer Prediction

To predict the correct answer, we concatenate the representation of text from backbone pre-train model and the representation of our Multi-grained Logical Graph.

$$score_o = linear(tanh(linear(h_c || h_q || h_o || h_G))) \tag{16}$$

where h_c, h_q, h_o is the results of Eq. 5 and $score_o$ is the final score of each option in one example. Finally we choose the option with the highest score as the predicted answer.

3.4 Loss Function

Let $o_t \in O$ be the ground truth of the sample. We use cross-entropy loss with label smoothing optimizing.

$$\mathcal{L} = -(1 - \gamma)score'_{o_t} - \gamma \frac{1}{|O|} \sum_{o_i \in O} score'_{o_i}, \text{ where}$$

$$score'_{o_i} = \log \frac{\exp(score_{o_i})}{\sum_{o_j \in O} \exp(score_{o_j})} \tag{17}$$

where γ is a predefined smoothing factor.

4   Experiment

4.1   Datasets

We evaluate the performance on two logical reasoning based MRC datasets: ReClor [48] and LogiQA [21].

- **ReClor:** The Reading Comprehension dataset requiring logical reasoning for reasoning-based MRC. It consists of 6138 four-option multi-choice questions sourced from actual exams of GMAT and LSAT, which are split into 4638 for training, 500 for validation and 1000 for testing. In order to fully assess the logical reasoning ability, the dataset divided into EASY set and HARD set according to the performance of pre-trained language models.
- **LogiQA:** It consists of 8678 four-option multi-choice questions sourced from National Civil Servants Examination of China, which are split into 7376 for training, 651 for validation and 651 for testing.

4.2   Baselines

To compare our multi-grained graph-based method with prior work, we main employ several sentence-level and entity-level baselines on logical reasoning based MRC task as follow:

- **DAGN** [14]: It propose a discourse-aware graph network that reasongs relying on the extracted discourse structure of texts, which used the sentence-level information of texts, and facilitates logical reasoning via graph neural networks.
- **FocalReasoner** [25]: It defines and extracts fact units from text, which are the entity-level information of text, to construct a supergraph, and enhance the supergraph with graph attention network.
- **AdaLoGN** [18]: It extracts the discourse structure and the explicit logical relation, and further extend them to find implicit logical relation based on several logical rules via a iterative mechanism, which is realized on sentence-level information.
- **Logiformer** [42]: It utilizes two different strategies to extract logic and syntax units, and construct the logical graph and the syntax graph respectively. After that it feed the extracted node sequence to the fully connected transformer to each graph, and use a dynamic gate mechanism to fuse the features from two branches.

4.3   Overall Results

Table 2 presents the overall results of the logical reasoning-based MRC task, comparing our method with baselines such as sentence-level methods DAGN, AdaLoGN, Logiformer, and entity-level method FocalReasoner. Our multi-grained graphs approach achieve the best performance among all other graph-based method on both ReClor and LogiQA. MLGNet reaches 64.07% of test accuracy on ReClor, and reaches 43.39% of test accuracy on LogiQA. Specifically, MLGNet achieves the highest test accuracy among all models, with 64.07% on ReClor and 43.39% on LogiQA. These results confirm our hypothesis that multi-grained graphs are effective in capturing and utilizing the logical structure of texts.

Table 2. Experimental results (accuracy %) compared with baselines on ReClor and LogiQA.

Methods	ReClor				LogiQA	
	Valid	Test	Test-E	Test-H	Valid	Test
DAGN	65.80	58.30	75.91	44.46	36.87	39.32
FocalReasoner	66.80	58.90	77.05	44.64	41.01	40.25
AdaLoGN	65.20	60.20	79.32	45.18	39.94	40.71
Logiformer	68.40	63.50	79.09	51.25	42.24	42.55
MLGNet	70.02	64.07	79.32	51.60	43.08	43.39

4.4 Ablation Study

We design an ablation study to verify the feasibility of the main contributions in our method: multi-grained logical graph construction and positional embedding mechanism. The results are reported in Table 3.

Table 3. Ablation study results (accuracy %) on LogiQA.

Methods	Valid	Δ	Test	Δ
MLGNet	43.08		43.39	
multi-grained logical graph				
MLGNet w/o entities	40.70	-2.38	41.08	-2.31
MLGNet w/ all entities	41.31	-1.77	42.20	-1.19
positional embedding				
MLGNet w/o position	42.93	-0.15	42.93	-0.46

Multi-grained Logical Graph. In graph construction, we build multi-grained graph by selecting and introducing entities as nodes on the existing methods using EDUs, hence we ablate the effects of whether introducing entities nodes and whether selecting entities nodes. Using the modified graphs with introducing no entities or introducing all extracted entities, the results show that the performance all decrease whether introduce no entities or all entities. This verifies the feasibility of multi-grained logical graph construction and entity selection strategy.

Positional Embedding. We remove the positional embedding and only use the average of RoBERTa outputs as node initial representation. The accuracy results decrease 0.15% in dev set and 0.46% in test set, which indicates positional embedding beneficial for subsequent graph reasoning and graph pooling.

4.5 Effect of Entities Nodes Introduction

To evaluate the effectiveness of entity nodes introduction, we compare MLGNet with other MRC models. We suspected that our method would be more effective for data points where explicit logical relations could not be extracted. To verify this, we split the original dev set and test set of LogiQA into four subsets based on the number of extracted explicit logical relations, as the statistics shown in Table 4. We display the accuracy of AdaLoGN and our proposed MLGNet in Fig. 3, and find that our model outperforms the baseline models on all divided subsets, demonstrating the effectiveness of our model for different extracted explicit logical relations. While MLGNet performs better when the number of extracted explicit logical relations is in the range of $[0, 3)$ and $[3, 6)$, the reason for this could be that our method effectively supplements information when the available information is less.

Table 4. Distribution of explicit logical relations on dev set and test set of LogiQA.

Dataset	$[0, 3)$	$[3, 6)$	$[6, 9)$	$[9, \infty)$
LogiQA-dev	55.4%	20.0%	12.1%	12.5%
LogiQA-test	52.8%	25.0%	11.8%	10.4%

4.6 Case Study

This section provides a case study, using the example described in Fig. 1 which is fail with previous works but successful with our method, to vividly show the effectiveness of our method. The case is shown in Fig. 4. We totally extract six nodes based on Graphene and part-of-speech tags, including five EDUs nodes and one entity node. Among them four pairs of context edges (U1-U2, U2-U3, U3-U4, U4-U5) and two pairs of containment edges (U1-U6, U5-U6) are detected. We can see that MLGNet can build a bridge for context and option to interact with each other, i.e. the path U1-U6-U5, especially in the graph without logic edge. In the same time, the entity "allergies" is the key word of sentence that it appears, so the entity node can also enhance vital information to the sentences.

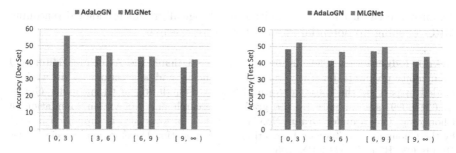

Fig. 3. Accuracy of models on number of extracted explicit logical relations on dev set (left) and test set (right) of LogiQA.

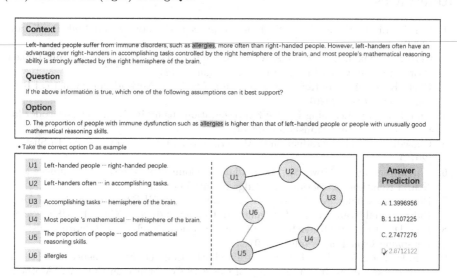

Fig. 4. A successful example in our method.

5   Conclusion

This paper presents MLGNet, a novel approach for logical reasoning based machine reading comprehension (MRC) that leverages both sentence-level and entity-level information. The approach involves the extraction of elementary discourse units (EDUs) and entity nodes, and the construction of multi-grained logical graphs containing three types of relations between nodes. An entity selection process is applied to filter the entity nodes, and the resulting graphs are used to facilitate information interaction and prediction. The proposed multi-grained graph-based mechanism effectively captures the logical structure of texts, and a positional embedding mechanism is employed to intensify the positional dependency of EDUs. The results show that MLGNet outperforms baseline models on two datasets ReClor and LogiQA. This study represents the first exploration of multi-grained graph-based methods for logical reasoning, which investigate and demonstrate the feasibility of multi-grained logical graphs for logical reasoning MRC, opening up potential avenues for future research.

Acknowledgements. The authors would like to thank the Associate Editor and anonymous reviewers for their valuable comments and suggestions. This work is funded in part by the National Natural Science Foundation of China under Grants No.62176029, and in part by the graduate research and innovation foundation of Chongqing, China under Grants No.CYB21063. This work also is supported in part by the Chongqing Technology Innovation and Application Development Special under Grants CSTB2022TIAD-KPX0206.

References

1. Cai, L., Li, J., Wang, J., Ji, S.: Line Graph Neural Networks for Link Prediction (2020)
2. Cetto, M., Niklaus, C., Freitas, A., Handschuh, S.: Graphene: Semantically-Linked Propositions in Open Information Extraction (2018)
3. Chen, K., et al.: Question Directed Graph Attention Network for Numerical Reasoning over Text (2020)
4. Cho, K., et al.: Learning Phrase Representations using RNN Encoder-Decoder for Statistical Machine Translation (2014)
5. Chowdhury, G.G.: Introduction to Modern Information Retrieval. Facet Publishing (2010)
6. Christopoulou, F., Miwa, M., Ananiadou, S.: Connecting the Dots: Document-level Neural Relation Extraction with Edge-oriented Graphs (2019)
7. Devlin, J., Chang, M.-W., Lee, K., Toutanova, K.: BERT: Pre-training of Deep Bidirectional Transformers for Language Understanding. arXiv:1810.04805 (2019)
8. Dua, D., et al.: DROP: A Reading Comprehension Benchmark Requiring Discrete Reasoning Over Paragraphs (2019)
9. Fan, F., Feng, Y., Zhao, D.: Multi-grained attention network for aspect-level sentiment classification. In: Proceedings of the 2018 Conference on Empirical Methods in Natural Language Processing. Brussels, Belgium, pp. 3433–3442. Association for Computational Linguistics (2018)
10. Hermann, K.M., et al.: Teaching machines to read and comprehend. In: Advances in Neural Information Processing Systems, vol. 28. Curran Associates Inc. (2015)
11. Hirschman, L., Light, M., Breck, E., Burger, J.D.: Deep Read: a reading comprehension system. In: Proceedings of the 37th Annual Meeting of the Association for Computational Linguistics. College Park, Maryland, USA, pp. 325–332. Association for Computational Linguistics (1999)
12. "History of programming languages–II — ACM Other Books. https://dl.acm.org/doi/abs/10.1145/234286
13. Huang, L., Bras, R.L., Bhagavatula, C., Choi, Y.: Cosmos QA: Machine Reading Comprehension with Contextual Commonsense Reasoning (2019)
14. Huang, Y., Fang, M., Cao, Y., Wang, L., Liang, X.: DAGN: Discourse-Aware Graph Network for Logical Reasoning (2021)
15. Jia, R., Cao, Y., Tang, H., Fang, F., Cao, C., Wang, S.: Neural extractive summarization with hierarchical attentive heterogeneous graph network. In: Proceedings of the 2020 Conference on Empirical Methods in Natural Language Processing (EMNLP), pp. 3622–3631. Online: Association for Computational Linguistics (2020)

16. Jiao, F., Guo, Y., Song, X., Nie, L.: MERIt: Meta-Path Guided Contrastive Learning for Logical Reasoning (2022)
17. Khashabi, D., Chaturvedi, S., Roth, M., Upadhyay, S., Roth, D.: Looking beyond the surface: a challenge set for reading comprehension over multiple sentences. In: Proceedings of the 2018 Conference of the North American Chapter of the Association for Computational Linguistics: Human Language Technologies, Volume 1 (Long Papers), pp. 252–262. Association for Computational Linguistics, New Orleans, Louisiana (2018)
18. Li, X., Cheng, G., Chen, Z., Sun, Y., Qu, Y.: AdaLoGN: Adaptive Logic Graph Network for Reasoning-Based Machine Reading Comprehension (2022)
19. Liu, S., Zhang, X., Zhang, S., Wang, H., Zhang, W.: Neural machine reading comprehension: methods and trends. Appl. Surf. Sci. **471**, 18–22 (2019)
20. Liu, Y., et al.: RoBERTa: a robustly optimized BERT pretraining approach (2019)
21. Liu, J., Cui, L., Liu, H., Huang, D., Wang, Y., Zhang, Y.: LogiQA: a challenge dataset for machine reading comprehension with logical reasoning (2020)
22. McCarthy, J.: Artificial intelligence, logic and formalizing common sense. In: Thomason, R.H. (ed.) Philosophical Logic and Artificial Intelligence, pp. 161–190. Springer, Netherlands, Dordrecht (1989). https://doi.org/10.1007/978-94-009-2448-2_6
23. Mihaylov, T., Clark, P., Khot, T., Sabharwal, A.: Can a suit of armor conduct electricity? A new dataset for open book question answering (2018)
24. Nguyen, T. et al.: MS MARCO: A Human Generated MAchine Reading COmprehension Dataset (2016)
25. Ouyang, S., Zhang, Z., Zhao, H.: Fact-driven Logical Reasoning (2021)
26. Paszke, A., et al.: Automatic differentiation in PyTorch (2017)
27. Rajpurkar, P., Zhang, J., Lopyrev, K., Liang, P.: SQuAD: 100,000+ questions for machine comprehension of text (2016)
28. Rajpurkar, P., Jia, R., Liang, P.: Know what you don't know: unanswerable questions for SQuAD (2018)
29. Ran, Q., Lin, Y., Li, P., Zhou, J., Liu, Z.: NumNet: machine reading comprehension with numerical reasoning (2019)
30. Riloff, E., Thelen, M.: A rule-based question answering system for reading comprehension tests. In: ANLP-NAACL 2000 Workshop: Reading Comprehension Tests as Evaluation for Computer-Based Language Understanding Systems (2000)
31. Sahu, S.K., Christopoulou, F., Miwa, M., Ananiadou, S.: Inter-sentence relation extraction with document-level graph convolutional neural network (2019)
32. Scarselli, F., Gori, M., Tsoi, A.C., Hagenbuchner, M., Monfardini, G.: The graph neural network model. IEEE Trans. Neural Netw. **20**(1), 61–80 (2009)
33. Seo, M., Kembhavi, A., Farhadi, A., Hajishirzi, H.: Bidirectional Attention Flow for Machine Comprehension (2018)
34. Sugawara, S., Aizawa, A.: An analysis of prerequisite skills for reading comprehension. In: Proceedings of the Workshop on Uphill Battles in Language Processing: Scaling Early Achievements to Robust Methods, pp. 1–5. Association for Computational Linguistics, Austin, TX (2016)
35. Tandon, N., Mishra, B.D., Sakaguchi, K., Bosselut, A., Clark, P.: WIQA: a dataset for "What if..." reasoning over procedural text (2019)
36. Wang, S., Jiang, J.: Machine Comprehension Using Match-LSTM and Answer Pointer (2016)
37. Wang, S., et al.: Logic-driven context extension and data augmentation for logical reasoning of text (2021)

38. Wang, S., et al.: From LSAT: the progress and challenges of complex reasoning (2021)
39. Welbl, J., Stenetorp, P., Riedel, S.: Constructing datasets for multi-hop reading comprehension across documents. Trans. Assoc. Comput. Linguist. **6**, 287–302 (2018)
40. Xu, K., Wu, L., Wang, Z., Yu, M., Chen, L., Sheinin, V.: Exploiting rich syntactic information for semantic parsing with graph-to-sequence model (2018)
41. Xu, M., Li, L., Wong, D.F., Liu, Q., Chao, L.S.: Document graph for neural machine translation (2021)
42. Xu, F., Liu, J., Lin, Q., Pan, Y., Zhang, L.: Logiformer: a two-branch graph transformer network for interpretable logical reasoning. In: Proceedings of the 45th International ACM SIGIR Conference on Research and Development in Information Retrieval, pp. 1055–1065 (2022)
43. Yang, Z., et al.: HotpotQA: a dataset for diverse, explainable multi-hop question answering (2018)
44. Yang, Z., Dai, Z., Yang, Y., Carbonell, J., Salakhutdinov, R.R., Le, Q.V.: XLNet: generalized autoregressive pretraining for language understanding
45. Yao, L., Mao, C., Luo, Y.: Graph convolutional networks for text classification. Proc. AAAI Conf. Artif. Intell. **33**(01), 7370–7377 (2019)
46. Ying, C.T., et al.: Transformers really perform bad for graph representation?"
47. Yu, A.W., et al.: QANet: combining local convolution with global self-attention for reading comprehension (2018)
48. Yu, W., Jiang, Z., Dong, Y., Feng, J.: ReClor: a reading comprehension dataset requiring logical reasoning (2020)
49. Zeng, S., Xu, R., Chang, B., Li, L.: Double graph based reasoning for document-level relation extraction (2020)
50. Zhang, Z., Zhao, H., Wang, R.: Machine reading comprehension: the role of contextualized language models and beyond. arXiv:2005.06249 (2020)
51. Zheng, B., et al.: Document Modeling with Graph Attention Networks for Multi-grained Machine Reading Comprehension (2020)
52. Zhou, B., Khashabi, D., Ning, Q., Roth, D.: Going on a vacation takes longer than "Going for a walk": a study of temporal commonsense understanding (2019)

Adaptive Prototype Network with Common and Discriminative Representation Learning for Few-Shot Relation Extraction

Wenyue Hu[1], Jiang Zhong[1(✉)] [iD], Yang Xia[2], Yangmei Zhou[3], and Rongzhen Li[1]

[1] Chongqing University,Chongqing, China
{huwenyue,zhongjiang,lirongzhen}@cqu.edu.cn
[2] Chongqing Academy of Big Data Co., Ltd.,Chongqing, China
[3] Chongqing Academy of Science and Technology,Chongqing, China

Abstract. The task of few-shot relation extraction presents a significant challenge as it requires predicting the potential relationship between two entities based on textual data using only a limited number of labeled examples for training. Recently, quite a few studies have proposed to handle this task with task-agnostic and task-specific weights, among which prototype networks have proven to achieve the best performance. However, these methods often suffer from overfitting novel relations because every task is treated equally. In this paper, we propose a novel methodology for prototype representation learning in task-adaptive scenarios, which builds on two interactive features: 1) common features are used to rectify the biased representation and obtain the relative class-centered prototype as much as possible, and 2) discriminative features help the model better distinguish similar relations by the representation learning of the entity pairs and instances. We obtain the hybrid prototype representation by combining common and discriminative features to enhance the adaptability and recognizability of few-shot relation extraction. Experimental results on FewRel dataset, under various few-shot settings, showcase the improved accuracy and generalization capabilities of our model.

Keywords: Relation extraction · Few-shot learning · Adaptive prototype network

1 Introduction

Relation extraction plays a vital role within the domain of natural language processing by identifying and extracting the relationships between different entities mentioned in textual data. It has significant applications in various downstream tasks, including text mining [25], knowledge graph [18] and social networking [21]. This problem is commonly approached as a supervised classification

X. Yang et al. (Eds.): ADMA 2023, LNAI 14179, pp. 63–77, 2023.
https://doi.org/10.1007/978-3-031-46674-8_5

task. However, the process of labeling large datasets of sentences for relation extraction is not only time-consuming but also expensive. This limitation often leads to a scarcity of labeled data available for training relation extraction models, hindering the progress and evolution of this task. To address this challenge, researchers have delved into a new field of study called few-shot relation extraction (FSRE). FSRE is viewed a model aligned with human learning and is considered the most promising method for tackling this task. Existing studies [6,19,22,23] have achieved remarkable results and surpassed human levels in general tests. Some researchers [2,11], however, observe that most proposed models are ineffective in real applications and are looking into specific reasons such as task difficulty. Most task-agnostic and task-specific models [16] actually struggle to handle the FSRE task adaptively with varying difficulties. Figure 1 depicts a process of FSRE augmented by the relation instances. Due to the high similarity of relationships and the limited number of training instances, the existing methods are hard to identify the relation *mother* between entity pair (*Isabella Jagiellon, Bona Sforza*) for the query instance.

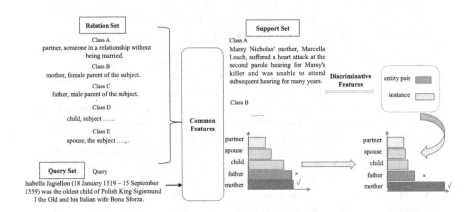

Fig. 1. A novel 5-way 1-shot relation extraction pipeline. First, use the relation set (left) and then enhance the relation representation with the support set (right). Finally, self-adapt to obtain true value for similar relationships in difficult tasks.

The FSRE task is commonly tackled through two prominent approaches: meta-learning based methods and metric-learning based methods [15,32]. Meta-learning, with its emphasis on the acquisition of learning abilities, focuses on training a task-agnostic model with various subtasks, enabling it to learn how to predict scores for novel classes. In contrast, metric-based approaches aim to discover an improved metric for quantifying distribution discrepancies and effectively discerning different categories. Among them, the prototype network [11,27] is widely used as a metric-based approach and has demonstrated remarkable performance in FSRE scenarios. Compared to meta-learning based methods, metric-based methods in FSRE task usually focus on training task-specific models that have poorer adaptability to new tasks. Conversely, the meta-learning based methods may struggle with the discriminative features of relations between entity pairs during the training process, leading to underwhelming performance. This can be attributed primarily to the scarcity of training instances for each subtask in FSRE scenarios. As shown in Fig. 1, when only using the common features from relation instances as the metric index, it is difficult to distinguish the pair $(IsabellaJagiellon, BonaSforza)$ in the query sentence from the five similar relations. While the discriminative features of the support instances are introduced, the features have more obvious distinctions, and the score of the *mother* is higher than others. In order to address these challenges in few-shot relation extraction, we present a novel adaptive prototype network representation that incorporates both relation-meta features and various instance features, called AdaProto. Firstly, we introduce relation-meta learning to obtain the common features of the relations so that the model does not excessively deviate from its relational classes, thus making the model have better adaptability. Besides, we propose entity pair rectification and instance representation learning to gain discriminative features of the task, thus increasing the discriminability of the relations. Finally, we employ a multi-task training strategy to achieve a model of superior quality.

We can summarize our main contributions into three distinct aspects: **1)** We present a relation-meta learning approach that enables the extraction of common features from relation instances, resulting in enhanced adaptability of the model. **2)** Our model combines common features with discriminative features learned from the support set to enhance the recognizability of relational classification. **3)** We assess the performance of our proposed model on the extensively utilized datasets FewRel through experiments employing the multi-task training strategy. The experimental results confirm the model's excellent performance.

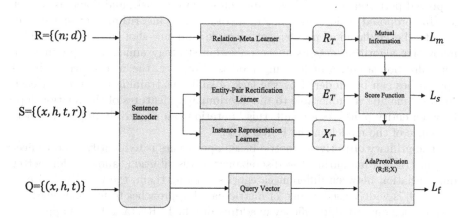

Fig. 2. The overall framework of AdaProto. The relation-meta learner exploits relation embeddings and relation meta to transfer the relative common features. To capture the discriminative features, E_t and X_t represent the entity pair rectification and instance embeddings, respectively. Finally, multi-task training adaptively fuses the above two features into different tasks.

2 Related Work

2.1 Few-Shot Relation Extraction

Relation extraction, a fundamental task within the domain of natural language processing (NLP), encompassing a wide range of applications, including but not limited to text mining [25] and knowledge graph [18]. Its primary objective is to discern the latent associations among entities explicitly mentioned within sentences. Few-shot relation extraction (FSRE), as a specialized approach, delves into the prediction of novel relations through the utilization of models trained on a restricted set of annotated examples. Han et al. [13] introduce a comprehensive benchmark called FewRel, which serves as a large-scale evaluation platform specifically designed for FSRE. Expanding upon this initiative, Gao et al. [9] presented FewRel 2.0, an extended iteration of the dataset that introduces real-world challenges, including domain adaptation and none-of-the-above detection. These augmentations render FewRel 2.0 a more veracious and efficacious milieu for the comprehensive evaluation of FSRE techniques.

Meta-learning based and metric-learning based methods are the two predominant approaches widely employed in solving the FSRE task. Meta-learning, with its emphasis on the acquisition of learning abilities, strives to educate a task-agnostic model can acquire knowledge and making predictions for novel classes through exposure to diverse subtasks. The model-agnostic meta-learning (MAML) algorithm was introduced by Finn et al. [7]. This algorithm is an adaptable technique applicable to a broad spectrum of learning scenarios by leveraging gradient descent-based training. Dong et al. [5] introduced a meta-information guided meta-learning framework that utilizes semantic notions of classes to enhance the initialization and adaptation stages of meta-learning, resulting in improved performance and robustness across various tasks and domains. Lee et al. [15] proposed a domain-agnostic meta-learning algorithm that specifically targets the challenging problem of cross-domain few-shot classification. This innovative algorithm acts as an optimization strategy aimed at improving the generalization abilities of learning models. However, the meta-learning based approaches can be hindered by the lack of sufficient training instances for each subtask, making it vulnerable to the discriminative features of the relationships between entity pairs. As a result, this limitation can negatively impact the performance of the model on the FSRE task.

The primary objective of metric-based approaches is to identify a more effective metric for measuring the discrepancy in distribution, allowing for better differentiation between different categories. Among them, the prototypical network [29] is widely used as one of metric-based approaches to few-shot learning and has demonstrated its efficacy in addressing the FSRE task. In this approach, every example is encoded as a vector by the feature extractor. The prototype representation for each relation is obtained by taking the average of all the exemplar vectors belonging to that relation. To classify a query example, its representation is compared to the prototypical representations of the candidate relations, and it is assigned a class based on the nearest neighbor rule. There have been many previous works that enhanced the model performance of different problems in the FSRE task based on prototypical networks. Ye and Ling [36] introduced a modified version of the prototypical network that incorporates multi-level matching and aggregation techniques. Gao et al. [8] proposed a novel approach called hybrid attention-based prototypical networks. This innovative approach aims to mitigate the impact of noise and improve the overall performance of the system. Yang et al. [34] proposed an enhancement to the prototypical network by incorporating relation and entity descriptions. And Yang et al. [35] introduced the inherent notion of entities to offer supplementary cues for relationship extraction, consequently augmenting the overall effectiveness of relationship prediction. Ren et al. [26] introduced a two-stage prototype network that incorporates prototype attention alignment and triple loss, aiming to enhance the performance of their model in complex classification tasks. Han et al. [11] devised an innovative method that leverages supervised contrastive learning and task adaptive focal loss, specifically focusing on hard tasks in FSRE scenario. Brody et al. [2] conducted a comprehensive analysis of the effectiveness of robust few-shot models and investigated the reliance on entity type information in FSRE models. Despite these notable progressions, the task adaptability of FSRE model is still underexplored. In this study, we propose an adaptive prototype network representation with relation-meta features and various instance features, which enhances the adaptability of the model and yields superior performance on relationships with high similarity.

2.2 Adaptability in Few-Shot Learning

Few-shot learning (FSL) is a prominent approach aimed at training models to comprehend and identify new categories using a limited amount of labeled data. However, the inherent limitations of the available information from a restricted number of samples in the N-way-K-shot setting of FSL pose a significant challenge for both task-agnostic and task-specific models to adaptively handle the FSL task. Consequently, the performance of FSL models can exhibit substantial variations across different environments, particularly when employing metrics-based approaches that typically train the task-specific model. One of the challenges in FSL is to enable models to quickly adapt to dynamic environments and previously unseen categories. Simon et al. [28] proposed a framework that incorporates dynamic classifiers constructed from a limited set of samples. This approach enhances the few-shot learning model's robustness against perturbations. Lai et al. [14] proposed an innovative meta-learning approach that captures a task-adaptive classifier-predictor, enabling the generation of customized classifier weights for few-shot classification tasks. Xiao et al. [33] presented an adaptive mixture mechanism that enhances generation of interactive class prototypes. They also introduced a loss function for joint representation learning, which seamlessly adapts the encoding process for each support instance. Their approach is also built upon the prototypical framework. Han et al. [12] proposed an adaptive instance revaluing network to tackle the biased representation problem. Their proposed method involves an enhanced bilinear instance representatio and the integration of two original structural losses, resulting in a more robust and accurate regulation of the instance revaluation process. Inspired by these works, we introduce relation-meta learning to obtain the common features that were used to rectify the biased representation, making the model generalize better and have stronger adaptability.

3   Methodology

The general architecture of the AdaProto model, proposed by us, is depicted in Fig. 2. This model is designed to tackles the task of extracting relationships between entity pairs in few-shot scenarios. The model consists of four distinct modules:

3.1   Relation-Meta Learning

To obtain relation-meta information from relation instances, we introduce a relation-meta learner, inspired by MetaR [3], that learns the relative common features, thus maintaining the task-specific weights for the relational classes. Unlike the previous method, our approach obtains the relation-meta directly from the relation set, not from the entity pairs, and transfers the common features to entity pairs for rectifying the relation representations. Firstly, we generate a template called relation set by combining the relation name and description as *"name:description"* $(n; d)$, and feed them into the encoder to produce the feature representation. To enhance the relation representation, we employ the mean value of the hidden states of the sentence tokens and concatenate it with their corresponding $[CLS]$ token to represent the relation classes $\{r_i \in R^{2d}; i = 1, \dots, N\}$. Next, the relation-meta learner compares all the N relation representations obtained from the encoder and generates the only representation specific to each relational label in the task. Finally, to represent the common features through the relation-meta learner, we design a nonlinear two-layer fully connected neural network and a mutual information mechanism to explore the feature representations by training the relation-meta learner.

$$R_m = GELU(H_{r_i}W_1 + b_1)W_2 + b_2 \tag{1}$$

where R_m is the representation of relation-meta, H_{r_i} is an output of the encoder corresponding to r_i, and W_i, b_i are the learning parameters. The hidden embeddings are further exploited by the fully-connected two layers network with $GELU(.)$ activation function. We assume that the ideal relation-meta R_m only retains the relation specific to the i-th class r_i and is orthogonal to other relations. Therefore, for all relational categories R, when $i \neq j$, the classes should satisfy their mutual information $MI(R_i, R_j)=0$, and the relative discriminative representation of each relation is independent. To achieve this goal, we design a training strategy based on mutual information, which constrains the relative common feature representation of each relation that only contains the information specific to the relational classes by minimizing the mutual information loss function.

$$L_m = \sum_{1 \leq i,j \leq N, i \neq j} MI\left(R_i, R_j\right) \qquad (2)$$

3.2 Entity-Pair Rectification Learning

To transfer the common features and obtain discriminative features, following the previous transE [1], we design score function to evaluate the interaction of entity pairs and common features learned from the relation-meta learner. The function can be formulated as the interaction between the head entity, tail entity, and relations to generate the rectified relation representation:

$$s\left(h_i, t_i\right) = \| R_m + h_i - t_i \| \qquad (3)$$

where R_m represents the relation-meta, h_i is the head entity embedding and t_i is the tail entity embedding. The L_2 norm is denoted by $\|.\|$. With regards to the relational classes in the given task, the entity-pair rectification learner evaluates the relevance of each instance in the support set. To efficiently evaluate the effectiveness of entity pair, inspired by MetaR [3], we design loss function to update the learned relational classes R_m with the score of the entity pairs and common features in the following way:

$$Ls(S_r) = \sum_{h,t \in S_r} \left(\lambda + s\left(h_i, t_i\right) - s\left(h_i, t_i'\right)\right) \qquad (4)$$

where λ is the margin hyperparameter, e.g., $\lambda=1$, t_i' is the negative example of t_i, and $s(h_i, t_i')$ is the negative instance score corresponding to current positive entity pair $s\left(h_i, t_i\right) \in S_r$. And the rectified relations E_r can be represented as follows:

$$E_r = R_m - \nabla_{R_m} Ls(S_r) \qquad (5)$$

where ∇ is the gradient.

3.3 Instance Representation Learning

Textual context is the main source of the relation classification, while the instances vary differently in the randomly sampled set, resulting in the discrepancy in relation features, especially for the task with similar relations. The keywords play a vital role in discriminating different relations in a sentence, so we design attention mechanisms to distinguish the relations from instance representation. We allocate varying weights to the tokens based on the similarity between instances with the relations. The k-th relational feature representation R_i^k is obtained by computing the similarity between each instance embedding s_k^i and the relation embedding r_i.

$$R_i^k = \sum_{n=1}^{l_r} \alpha_r^n r_i^n \qquad (6)$$

where $k = 1, ..., K$ is k-th instance, l_r is the length of r_i, and in the matrix for the instances r_i, s_k, the variable n represents the n-th row.

$$\alpha_r^n = \text{softmax}(\text{sum}(r_i(s_k^n)^T)) \tag{7}$$

The discriminative features are defined by the set of K features.

$$X_r = 1/K \sum_{k=1}^{K} R_k^i \tag{8}$$

3.4   Multi-task Joint Training

We concatenate the common and discriminative features to form the hybrid prototype representation by relation-meta R_t, entity-pair rectification E_t, and instance representation X_t.

$$C_j = [R_t^j; E_t^j; X_t^j] \tag{9}$$

where t denotes the task, R_t^j represents the j-th relations by the relation-meta learner, E_t^j represents the updated j-th relations by the entity-pair rectification learner, and X_t^j represents the contextualized j-th relations by the instance learner at the task-level learning. The model would calculate the classification probability of the query Q based on the given query and prototype representation of N relations.

$$P(y = j|Q, S) = \exp(d(C_j,\ Q))/ \sum_{k=1}^{N} \exp(d(C_k, Q)) \tag{10}$$

where variable N represents the number of classes, and the function $d(.,.)$ refers to the Euclidean distance. And the training objective in this scenario is to minimize the cross-entropy loss by employing the negative log-likelihood estimation probability with labeled instances:

$$L_c(Q, S) = -\text{log} P(y = t|Q, S) \tag{11}$$

where t denotes the relation label ground truth. There are only a few number of instances for each task and the tasks vary differently in difficulty. Training a model using only cross-entropy can be challenging to achieve satisfactory results. To tackle this issue, we employ a multi-task joint strategy to enhance the model's training process. Following previous work [4], we use mutual information loss and relax the constraints because of the complexity.

Next, to fully exploit the interaction between the relation-meta and entity pairs, the rectification loss is introduced. Furthermore, we take L_f with focal loss [17] instead of L_c to balance the different tasks. The focal loss can be expressed in the following manner:

$$L_f(Q, S) = -(1 - P(y = t|Q, S))^\gamma \text{log} P(y = t|Q, S) \tag{12}$$

where factor $\gamma \geq 0$ is utilized to regulate the rate at which easy examples are down-weighted. Finally, the training loss is constructed as follows:

$$L = L_f + \alpha L_m + \beta L_s \tag{13}$$

where hyperparameters α, β control the relative importance of L_m, L_s, respectively, e.g., $\beta=0$ means no rectified relations. Following previous similar work [30], we simply set $\alpha = 0.7$ and $\beta = 0.3$, and have obtained better performance. And other rates can be explored in the future.

4   Experiments

4.1   Dataset and Evaluation Metrics

Datasets. Our AdaProto model is assessed on the FewRel 1.0 [13] and FewRel 2.0 [9] datasets, which are publicly available and considered as large-scale datasets for FSRE task. These datasets contain 100 relations, each consisting of 700 labeled instances derived from Wikipedia. To ensure fairness, We divide the corpus of 100 relations into three distinct subsets, with a ratio of 64:16:20, correspondingly dedicated to the tasks of training, validation and testing, following the official benchmarks. The relation set is typically provided in the auxiliary dataset. While the label data for the test set is not made public, we can still assess our model's performance by submitting our prediction results and using the official test script to obtain test scores.

Evaluation. The distribution of FewRel dataset in various scenarios is frequently simulated using the N-way-K-shot (N-w-K-s) approach. Under the typical N-w-K-s setting, each evaluation episode involves sampling N relations, each of which contains K labeled instances, along with additional query instances. The objective is for the models to accurately classify the query instances into the sampled N relations, based on the provided $N{\times}K$ labeled data. Accuracy is employed as the performance metric in the N-w-K-s scenario. This metric measures the proportion of correctly predicted instances out of the total number of instances in the dataset. Building upon previous baselines, we have selected N values of 5 and 10, and K values of 1 and 5, resulting in four distinct scenarios.

4.2   Implementation Details

To effectively capture the contextual information of each entity and learn a robust representation for instances, leveraging the methodology introduced by Han et al. [11], we utilize the uncased $BERT_{base}$ as the encoder, which is a 12-layers transformer and has a 768 hidden size. To generate the statements for each instance, we merge the hidden states of the beginning tokens of both entity mentions together. To enhance the representation of each entity pair, we further incorporate their word embedding, entity id embedding, and entity type embedding. In addition, we utilize the label name and its corresponding description of each relation to generate a template. To optimize our model, we utilize the AdamW optimizer [20] and set the learning rate to 0.00002. Our model is implemented using PyTorch framework and deployed on a server equipped with two NVIDIA RTX A6000 GPUs.

4.3   Comparison to Baselines

We assess the performance of our model by comparing it to a set of strong baseline models:

Proto: Snell et al. [29] proposed the original prototype network algorithm.

GNN: Garcia et al. [10] introduced a meta-learning method based on graph neural networks.

MLMAN: Ye and Ling [36] introduced a modified version of the prototypical network that incorporates multi-level matching and aggregation.

REGRAB: Qu et al. [24] proposed a method for bayesian meta-relational learning that incorporates global relational descriptions.

BERT-PAIR: Gao et al. [9] proposed a novel approach to measure the similarity between pairs of sentences.

ConceptFERE: Yang et al. [35] put forward a methodology that incorporates the inherent concepts of entities, introducing additional cues derived from the entities involved, to enhance relation prediction and elevate the overall effectiveness of relation classification.

DRK: Wang et al. [31] proposed a innovative method for knowledge extraction based on discriminative rules.

HCRP: Han et al. [11] designed a hybrid prototype representation learning method based on contrastive learning, considering task difficulty.

Most existing models use BERT as an encoder, and we follow what is known a priori.

4.4   Main Results

Overall Results. In our evaluation, We compare our proposed AdaProto model against a set of strong baselines on the FewRel dataset. And the results of this comparison are presented in Table 1 and 2. The model's performance is demonstrated in Table 1, our model effectively improves prediction accuracy and outperforms the strong models in each setting on FewRel 1.0, demonstrating better generalization ability. Besides, the performances gains from the 5-shot settings over the second best method (i.e., HCRP) are larger than those of 1-shot scenario. This observation may suggest that the availability of only one instance per relation class restricts the extraction of discriminative features. And the semantic features of a single instance are more prone to deviating from the common features shared by its relational class. By observational analysis, the performance is mainly due to three reasons. 1) Relation-meta learning allows obtaining common features while adapting to different tasks. 2) Entity pair rectification and instance representation learning empowers prototype network with discriminative features. 3) Joint training also plays an important role.

Table 1. Main results from the validation and testing of the FewRel 1.0. The evaluation metric used in this study is accuracy(%).

Model	5-w-1-s	5-w-5-s	10-w-1-s	10-w-5-s
Proto	82.92/80.68	91.32/89.60	73.24/71.48	83.68/82.89
GNN	-/75.66	-/89.06	-/70.08	-/76.93
MLMAN(CNN)	79.01/82.98	88.86/92.66	67.37/75.59	80.07/87.29
REGRAB	87.95/90.30	92.54/94.25	80.26/84.09	81.76/87.02
BERT-PAIR	85.66/88.32	89.48/93.22	76.84/80.63	81.76/87.02
ConceptFERE	-/84.28	-/90.34	-/74.00	-/81.82
DRK	-/89.94	-/92.42	-/81.94	-/85.23
HCRP	90.90/93.76	93.22/95.66	84.11/89.95	87.79/92.10
AdaProto(ours)	**91.19/94.26**	**93.87/96.19**	**85.91/90.61**	**89.59/93.38**

Table 2. The performance on the domain adaptation test set of FewRel 2.0.

Model	5-w-1-s	5-w-5-s	10-w-1-s	10-w-5-s
Proto	40.12	51.50	26.45	36.39
BERT-PAIR	67.41	78.57	54.89	66.85
HCRP	76.34	83.03	63.77	72.94
AdaProto(ours)	77.13	84.29	65.17	73.85

Performance on FewRel 2.0 Dataset. In order to assess the adaptability of our proposed model, we performed cross-domain experiments using the FewRel 2.0 dataset. Table 2 clearly illustrates the impressive domain adaptability of our model, as evidenced by the results. In particular, our model obtain better results on 5-shot, probably because the local features of the task play a dominant role compared to the common features.

Performance on Hard Tasks in Few-Shot Scenarios. To showcase the adaptability and efficacy of our model in handling difficult tasks, we use Han's setup [11] to assess the overall capabilities of the models on the validation set of FewRel 1.0 dataset. We considered three distinct 3-way-1-shot scenarios and display the main results in Table 3. *Easy* denotes tasks with easily distinguishable

Table 3. The performance on random, easy, and hard task.

Model	Random	Easy	Hard
Proto	87.37	98.51	35.63
BERT-PAIR	91.14	99.76	38.21
HCRP	93.86	99.93	62.40
AdaProto(ours)	94.75	99.91	65.37

relations, *Random* encompass a set of 10,000 tasks randomly sampled from the validation relations, and *Hard* denotes tasks with similar relations. We choose *Mother*, *Child* and *Spouse* as the three difficult relations because they not only have similar relation descriptions but also have the same entity types. The performance degrades sharply for the difficult tasks, as shown in the Table 3, which indicates that the hard task still faces a great challenge and also illustrates the significant advantage of our AdaProto model.

4.5   Ablation Study

To assess the efficacy of the various modules in our AdaProto model, we conducted an ablation study where we disabled each module individually and assessed the impact on the model's overall performance. Table 4 illustrates the outcomes of the ablation study, showcasing the performance decline observed for each module when disabled.

Table 4. The Ablation study performance of AdaProto model on FewRel 2.0.

Model	5-w-1-s	5-w-5-s	10-w-1-s	10-w-5-s
AdaProto	77.13	84.29	65.17	73.85
-Relation-Meta	71.16	78.65	58.43	67.92
-Entity-Pair Rectification	74.98	81.17	63.23	72.19
-Instance Representation	73.11	79.86	61.34	69.25

Experiments were performed on the validation set using the FewRel 2.0 dataset for domain adaptation. Table 4 presents the detailed experimental results. The second part of the Table 4 indicates the results after removing certain modules. It is observed that every module affects performance, especially for the relation-meta learner. In the common features, we eliminate the relation-meta learner and degrade the model to the discriminative features without relation descriptions, and the performance has a large degradation, indicating that common features contribute more to the FSRE task. In the discriminate features, we remove the entity pair rectification and instance representation, respectively, and the results demonstrate that textual context is still the main source of the discriminative features. In addition, removing entity pair rectification also shows a more significant decrease in performance.

5 Conclusion

In this study, we propose AdaProto, an adaptive prototype network representation that incorporates relation-meta features and various instance features to address the challenges in FSRE tasks. The adaptability of the task heavily relies on the effectiveness of the embedding space. To enhance the model's generalization, we introduce relation-meta learning to learn common features and better capture the class-centered prototype representation. Additionally, we propose entity pair rectification and instance representation learning to identify discriminative features that differentiate similar relations. Our approach also utilizes a multi-task training strategy to ensure the development of a high-quality model. Furthermore, the model can adapt to different tasks by leveraging multiple feature extractors. Through comparative studies on FewRel in various few-shot settings, we demonstrate that AdaProto, along with each of its components, boosts the classification performance of FSRE tasks and effectively enhances the overall effectiveness and robustness.

Acknowledgements. The authors would like to thank the Associate Editor and anonymous reviewers for their valuable comments and suggestions. This work is funded in part by the National Natural Science Foundation of China under Grants No.62176029. This work also is supported in part by the Chongqing Technology Innovation and Application Development Special under Grants CSTB2022TIAD-KPX0206. Any opinions, findings, and conclusions, or recommendations expressed in this material are those of the authors and do not necessarily reflect those of the sponsor.

References

1. Bordes, A., Usunier, N., Garcia-Duran, A., Weston, J., Yakhnenko, O.: Translating embeddings for modeling multi-relational data. In: Advances in Neural Information Processing Systems, vol. 26 (2013)
2. Brody, S., Wu, S., Benton, A.: Towards realistic few-shot relation extraction. In: Proceedings of the 2021 Conference on Empirical Methods in Natural Language Processing, pp. 5338–5345 (2021)
3. Chen, M., Zhang, W., Zhang, W., Chen, Q., Chen, H.: Meta relational learning for few-shot link prediction in knowledge graphs. In: Proceedings of the 2019 Conference on Empirical Methods in Natural Language Processing and the 9th International Joint Conference on Natural Language Processing (EMNLP-IJCNLP), pp. 4217–4226 (2019)
4. Cheng, P., et al.: Improving disentangled text representation learning with information-theoretic guidance. arXiv preprint: arXiv:2006.00693 (2020)
5. Dong, B., et al.: Meta-information guided meta-learning for few-shot relation classification. In: Proceedings of the 28th International Conference on Computational Linguistics, pp. 1594–1605 (2020)

6. Dong, M., Pan, C., Luo, Z.: MapRE: an effective semantic mapping approach for low-resource relation extraction. In: Proceedings of the 2021 Conference on Empirical Methods in Natural Language Processing, pp. 2694–2704 (2021)
7. Finn, C., Abbeel, P., Levine, S.: Model-agnostic meta-learning for fast adaptation of deep networks. In: International Conference on Machine Learning, pp. 1126–1135. PMLR (2017)
8. Gao, T., Han, X., Liu, Z., Sun, M.: Hybrid attention-based prototypical networks for noisy few-shot relation classification. In: Proceedings of the AAAI Conference on Artificial Intelligence, vol. 33, pp. 6407–6414 (2019)
9. Gao, T., et al.: FewRel 2.0: towards more challenging few-shot relation classification. arXiv preprint: arXiv:1910.07124 (2019)
10. Garcia, V., Bruna, J.: Few-shot learning with graph neural networks. arXiv preprint: arXiv:1711.04043 (2017)
11. Han, J., Cheng, B., Lu, W.: Exploring task difficulty for few-shot relation extraction. In: Proceedings of the 2021 Conference on Empirical Methods in Natural Language Processing, pp. 2605–2616 (2021)
12. Han, M., et al.: Not all instances contribute equally: Instance-adaptive class representation learning for few-shot visual recognition. IEEE Trans. Neural Netw. Learn. Syst. (2022)
13. Han, X., et al.: FewRel: a large-scale supervised few-shot relation classification dataset with state-of-the-art evaluation. arXiv preprint: arXiv:1810.10147 (2018)
14. Lai, N., Kan, M., Han, C., Song, X., Shan, S.: Learning to learn adaptive classifier-predictor for few-shot learning. IEEE Trans. Neural Netw. Learn. Syst. $32(8)$, 3458–3470 (2020)
15. Lee, W.Y., Wang, J.Y., Wang, Y.C.F.: Domain-agnostic meta-learning for cross-domain few-shot classification. In: ICASSP 2022–2022 IEEE International Conference on Acoustics, Speech and Signal Processing (ICASSP), pp. 1715–1719. IEEE (2022)
16. Li, W.H., Liu, X., Bilen, H.: Cross-domain few-shot learning with task-specific adapters. In: Proceedings of the IEEE/CVF Conference on Computer Vision and Pattern Recognition, pp. 7161–7170 (2022)
17. Lin, T.Y., Goyal, P., Girshick, R., He, K., Dollár, P.: Focal loss for dense object detection. In: Proceedings of the IEEE International Conference on Computer Vision, pp. 2980–2988 (2017)
18. Lin, Y., Liu, Z., Sun, M., Liu, Y., Zhu, X.: Learning entity and relation embeddings for knowledge graph completion. In: Twenty-Ninth AAAI Conference on Artificial Intelligence (2015)
19. Liu, Y., Hu, J., Wan, X., Chang, T.H.: Learn from relation information: towards prototype representation rectification for few-shot relation extraction. In: Findings of the Association for Computational Linguistics: NAACL 2022, pp. 1822–1831 (2022)
20. Loshchilov, I., Hutter, F.: Decoupled weight decay regularization. arXiv preprint: arXiv:1711.05101 (2017)
21. Nasution, M.K.: Social network mining (SNM): a definition of relation between the resources and SNA. arXiv preprint: arXiv:2207.06234 (2022)
22. Peng, H., et al.: Learning from context or names? An empirical study on neural relation extraction. In: Proceedings of the 2020 Conference on Empirical Methods in Natural Language Processing (EMNLP), pp. 3661–3672 (2020)
23. Popovic, N., Färber, M.: Few-shot document-level relation extraction. arXiv preprint: arXiv:2205.02048 (2022)

24. Qu, M., Gao, T., Xhonneux, L.P., Tang, J.: Few-shot relation extraction via Bayesian meta-learning on relation graphs. In: International Conference on Machine Learning, pp. 7867–7876. PMLR (2020)
25. Quan, C., Wang, M., Ren, F.: An unsupervised text mining method for relation extraction from biomedical literature. PLoS ONE **9**(7), e102039 (2014)
26. Ren, H., Cai, Y., Chen, X., Wang, G., Li, Q.: A two-phase prototypical network model for incremental few-shot relation classification. In: Proceedings of the 28th International Conference on Computational Linguistics, pp. 1618–1629 (2020)
27. Ren, H., Cai, Y., Lau, R.Y.K., Leung, H.F., Li, Q.: Granularity-aware area prototypical network with bimargin loss for few shot relation classification. IEEE Trans. Knowl. Data Eng. **35**(5), 4852–4866 (2022)
28. Simon, C., Koniusz, P., Nock, R., Harandi, M.: Adaptive subspaces for few-shot learning. In: Proceedings of the IEEE/CVF Conference on Computer Vision and Pattern Recognition, pp. 4136–4145 (2020)
29. Snell, J., Swersky, K., Zemel, R.: Prototypical networks for few-shot learning. In: Advances in Neural Information Processing Systems, vol. 30 (2017)
30. Tran, V.H., Ouchi, H., Watanabe, T., Matsumoto, Y.: Improving discriminative learning for zero-shot relation extraction. In: Proceedings of the 1st Workshop on Semiparametric Methods in NLP: Decoupling Logic from Knowledge, pp. 1–6. Association for Computational Linguistics, Dublin, Ireland and Online (2022). https://doi.org/10.18653/v1/2022.spanlp-1.1, https://aclanthology.org/2022.spanlp-1.1
31. Wang, M., Zheng, J., Cai, F., Shao, T., Chen, H.: DRK: discriminative rule-based knowledge for relieving prediction confusions in few-shot relation extraction. In: Proceedings of the 29th International Conference on Computational Linguistics, pp. 2129–2140 (2022)
32. Wang, Y., Salamon, J., Bryan, N.J., Bello, J.P.: Few-shot sound event detection. In: ICASSP 2020–2020 IEEE International Conference on Acoustics, Speech and Signal Processing (ICASSP), pp. 81–85. IEEE (2020)
33. Xiao, Y., Jin, Y., Hao, K.: Adaptive prototypical networks with label words and joint representation learning for few-shot relation classification. IEEE Trans. Neural Netw. Learn. Syst. (2021)
34. Yang, K., Zheng, N., Dai, X., He, L., Huang, S., Chen, J.: Enhance prototypical network with text descriptions for few-shot relation classification. In: Proceedings of the 29th ACM International Conference on Information & Knowledge Management, pp. 2273–2276 (2020)
35. Yang, S., Zhang, Y., Niu, G., Zhao, Q., Pu, S.: Entity concept-enhanced few-shot relation extraction. arXiv preprint: arXiv:2106.02401 (2021)
36. Ye, Z.X., Ling, Z.H.: Multi-level matching and aggregation network for few-shot relation classification. arXiv preprint: arXiv:1906.06678 (2019)

Fine-Grained Knowledge Enhancement for Empathetic Dialogue Generation

Ai Chen[ID], Jiang Zhong[✉][ID], Qizhu Dai, Chen Wang, and Rongzhen Li

Chongqing University, Shapingba, Chongqing, China
ai.chen@stu.cqu.edu.cn,
{zhongjiang,daiqizhu,chenwang,lirongzhen}@cqu.edu.cn

Abstract. An engaging dialogue system is supposed to generate empathetic responses, which requires a cognitive understanding of users' situations and an affective perception of their emotions. Most of the existing work only focuses on modeling the latter, while neglecting the importance of the former. Despite some efforts to enhance chatbots' empathy in both cognition and affection, limited cognition conditions and inaccessible fine-grained information still impair the effectiveness of empathy modeling. To address this issue, we propose a novel fine-grained knowledge-enhanced empathetic dialogue generation model KEEM. We first explore strategies to filter fine-grained commonsense and emotional knowledge and leverage knowledge to construct cognitive and affective context graphs. And we learn corresponding context representations from the two knowledge-enhanced context graphs. Then we encode the raw dialogue context to learn the original cognitive and affective representations and fuse them with the knowledge-enhanced representations in cognition and affection. Finally, we feed the two fused representations into a decoder to produce empathetic replies. Extensive experiments conducted on the benchmark dataset EMPATHETICDIALOGUES verify the effectiveness of our model in comparison with several competitive models.

Keywords: Empathetic dialogue generation · Commonsense and emotional knowledge · Cognition and affection

1 Introduction

Empathy, an important aspect of engaging human conversations [14], usually refers to the ability to understand others' situations, perceive their emotions, and respond to them appropriately [5]. Research in social psychology has also shown that empathy is a vital step towards a more humanized dialogue system [20]. Consequently, we concentrate on the task of empathetic dialogue generation, which aims to empathize with users and realize a more human-like chatbot.

It is demonstrated that empathy is a complex construct involving cognition and affection [15], where cognitive empathy attends to users' situations [2] while affective one focuses on users' emotions instead [3]. But most of the existing methods [7–10,14] in empathetic dialogue generation only rely on detecting

© The Author(s), under exclusive license to Springer Nature Switzerland AG 2023
X. Yang et al. (Eds.): ADMA 2023, LNAI 14179, pp. 78–93, 2023.
https://doi.org/10.1007/978-3-031-46674-8_6

users' emotions and modeling emotional dependencies, while ignoring the importance of realizing cognitive empathy. To achieve empathy in both two aspects, Sabour et al. [15] make an attempt to use commonsense to enhance the modeling of cognitive and affective empathy and several cognition conditions would be inferred in this method. However, limited cognitive conditions and inaccessible fine-grained information still result in inaccurate recognization of users' circumstances and feelings, thereby impairing the empathetic effect of the generated responses.

Providing external knowledge to dialogue systems has been proven to operate in favor of modeling empathy in cognition and affection [15]. Intuitively, fine-grained knowledge would be beneficial for a more comprehensive understanding of user situations and a more accurate perception of user feelings, which is shown in Fig. 1. In this case, with the related cognitive concept of "walk", the chatbot understands the user's situation that he or she encountered a snake while walking so it asks what the user did. Also, the affective concepts "dazed", "demon" and "poisonous" help the robot perceive the terrified feeling of the user.

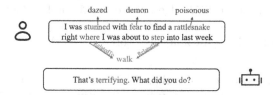

Fig. 1. An example from EMPATHETICDIALOGUES. Words related to cognition and affection are highlighted in red color, cognitive concepts and relations in green, and affective concepts in blue. (Color figure online)

In this paper, we propose a fine-grained **K**nowledge-**E**nhanced **EM**pathetic dialogue generation model (**KEEM**). We first explore novel knowledge selection strategies to filter commonsense, i.e. structural and semantic knowledge, and emotional knowledge, and use the selected fine-grained knowledge to construct cognitive and affective context graphs. We also learn the corresponding context representations from the above two knowledge-enhanced context graphs. Then we encode the original dialogue context to acquire the original information about cognition and affection, and fuse them with the two knowledge-enhanced representations. Finally, we feed the two fused cognitive and affective representations to a decoder to generate empathetic responses with coherent content and appropriate emotion. Extensive experiments are conducted on the benchmark dataset EMPATHETICDIALOGUES [14] for empathetic dialogue generation and the empirical results have verified that our model can produce more empathetic responses in comparison with several competitive models.

Our contributions can be summarized as follows:

- We propose KEEM, a novel approach that models cognitive and affective empathy by constructing and encoding corresponding context graphs.

- We explore knowledge selection strategies for cognitive and emotional knowledge to obtain more accurate and fine-grained knowledge.
- We conduct automatic and human evaluations and analyses to demonstrate the effectiveness of KEEM.

2 Preliminaries

2.1 Commonsense and Emotional Knowledge

In this work, we leverage the commonsense knowledge graph ConceptNet [17] and the emotional lexicon NRC-VAD [12] to infer the speakers' situations and their emotions, which enhances the cognition and affective empathy and leads to more empathetic responses.

ConceptNet is a large-scale knowledge graph that connects words and phrases of natural language with labeled edges. It represents the general knowledge, allowing models to better understand the meanings behind the words [11]. It contains 34 relations, over $21M$ edges, and over $8M$ nodes. The edges stored in ConceptNet can be concisely represented as the quadruples of their start node, relation label, end node, and confidence score: (h, r, t, s).

NRC-VAD is a lexicon of more than $20k$ English words and their vectors of three independent dimensions, i.e. valence (positiveness-negativeness/pleasure-displeasure), arousal (active-passive), and dominance (dominant-submissive), abbreviated as VAD. The values of VAD vectors are fine-grained real numbers in the interval from 0 (lowest) to 1 (highest).

2.2 Task Formulation

Dialogue context C is a sequence of M utterances: $C = [U_1, U_2, \cdots, U_M]$, where the i-th utterance $U_i = [w_i^1, w_i^2, \cdots, w_i^{m_i}]$ consists of m_i words. Following [9], we flat C as a token sequence, and prepend a CLS token to it, thus obtaining a new context sequence: $C = [CLS, w_1^1, \cdots, w_1^{m_1}, \cdots, w_M^1, \cdots, w_M^{m_M}]$. Given C, the task of empathetic dialogue generation is to generate an empathetic response $Y = [y_1, y_2, \cdots, y_n]$ with coherent content and appropriate emotion.

3 Methodology

3.1 Overview

Figure 2 shows an overview of our proposed model. Our model (KEEM) is composed of four stages: 1)cognitive context graph constructing and encoding; 2)affective context graph constructing and encoding; 3)cognition and affection fusion; and 4)empathetic response generation, where the first two stages can be performed simultaneously.

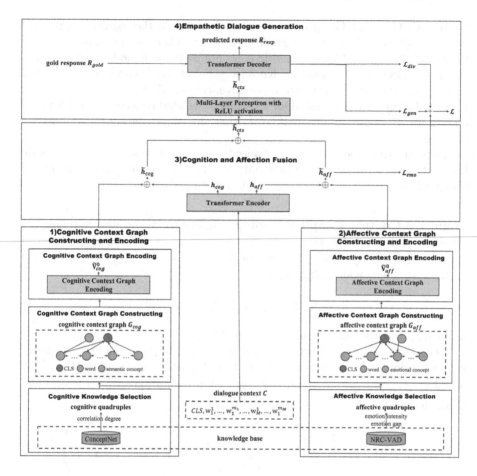

Fig. 2. Overview of our model (KEEM).

3.2 Cognitive Context Graph Constructing and Encoding

Cognitive Knowledge Selection. The structural and semantic information of ConceptNet, i.e. the relations and concepts therein, can help enhance cognition. Hence, for each non-stop word w_i^j of C, we first retrieve all of its quadruples from ConceptNet as candidates. And then we filter the cognitive knowledge by the following heuristic steps:1) We remove the quadruples that have low confidence scores (i.e. scores lower than 1.0) or inappropriate relations (i.e. relations unrelated to cognition). 2) We define and calculate the correlation degrees of the retrieved concepts. The correlation degree of a concept is the number of C's words that link to it in ConceptNet. If the concept itself appears in the dialogue context, the correlation degree is increased by one. 3) We rank the correlation degrees of the quadruples and select the top K_1 ones as the knowledge needed in cognitive graph construction.

Cognitive Context Graph Constructing. To build the cognitive graph, we first take all the tokens of C, including the CLS token, as the initial nodes of the graph. We also add the semantic concepts as the new nodes to the graph, which are selected from Cognitive Knowledge Selection. Then we connect vertex pairs of three types: 1) every two consecutive words in dialogue context; 2) every word in dialogue context and all of its semantic-related concepts; 3) CLS token and all words in dialogue context. Note that the edges connecting vertices are directed.

Thus, the dialogue context is enhanced by external structural and semantic knowledge and represented as the cognitive context graph G_{cog}, and the links of the graph are stored in the adjacency matrix A_{cog}.

Cognitive Context Graph Encoding. Similar to [8], we first initialize the cognitive vector presentation of every vertex v_{sem}^i by summing up its word embedding $\boldsymbol{E}_w\left(v_{cog}^i\right) \in \mathbb{R}^{d_{model}}$, positional embedding $\boldsymbol{E}_p\left(v_{cog}^i\right) \in \mathbb{R}^{d_{model}}$, and dialogue state embedding $\boldsymbol{E}_s\left(v_{cog}^i\right)$:

$$\mathbf{v}_{cog}^i = \boldsymbol{E}_w\left(v_{cog}^i\right) + \boldsymbol{E}_p\left(v_{cog}^i\right) + \boldsymbol{E}_s\left(v_{cog}^i\right) \tag{1}$$

where d_{model} is the dimension of the embeddings.

Then, we adopt a multi-head graph-attention mechanism, followed by a residual connection and layer normalization, thus v_{cog}^i attending to all its immediate neighbors $\left\{v_{cog}^j\right\}_{j \in A_{cog}^i}$ to update its cognitive presentation with structural and semantic knowledge:

$$\hat{\mathbf{v}}_{cog}^i = \text{LayerNorm}\left(v_{cog}^i + \|_{n=1}^H \sum_{j \in A_{cog}^i} \text{att}_{cog}^n\left(\mathbf{v}_{cog}^i, \mathbf{v}_{cog}^j\right) \mathbf{W}_{cog}^{nv} \mathbf{v}_{cog}^j\right) \tag{2}$$

where LayerNorm is layer normalization, $\|$ is the concatenation of H attention heads, A_{cog}^i are the immediate neighbors of v_{cog}^i presented in the adjacency matrix A_{cog}, $\mathbf{W}_{cog}^{nv} \in \mathbb{R}^{d_{model} \times d_h}$ is the linear transformation, $d_h = \frac{d}{H}$ is the dimension of each head, and $\text{att}_{cog}^n\left(\mathbf{v}_{cog}^i, \mathbf{v}_{cog}^j\right)$ is the self-attention mechanism for the n-th attention head:

$$\text{att}_{cog}^n\left(\mathbf{v}_{cog}^i, \mathbf{v}_{cog}^j\right) = \frac{\exp\left(\left(\mathbf{W}_{cog}^{nq} \mathbf{v}_{cog}^i\right)^{\top} \mathbf{W}_{cog}^{nk} \mathbf{v}_{cog}^j\right)}{\sum_{k \in A_{cog}^i} \exp\left(\left(\mathbf{W}_{cog}^{nq} \mathbf{v}_{cog}^i\right)^{\top} \mathbf{W}_{cog}^{nk} \mathbf{v}_{cog}^k\right)} \tag{3}$$

where $\mathbf{W}_{cog}^{nq} \in \mathbb{R}^{d_{model} \times d_h}$, $\mathbf{W}_{cog}^{nk} \in \mathbb{R}^{d_{model} \times d_h}$ are the linear transformations. And notably, when $\mathbf{v}_{cog}^i$ is the vector of a word in dialogue context and $\mathbf{v}_{cog}^j$ is the counterpart of a concept semantic-related, following [1], we update $\mathbf{v}_{cog}^j$ with the subtraction between the concept embedding and the corresponding relation embedding:

$$\mathbf{v}_{cog}^j = \mathbf{v}_{cog}^j - \boldsymbol{E}_r\left(\mathbf{r}_{cog}^{ij}\right) \tag{4}$$

where $\boldsymbol{E}_r\left(\boldsymbol{r}_{cog}^{ij}\right) \in \mathbb{R}^{d_{model}}$ is the relation embedding between $\mathbf{v}_{cog}^i$ and $\mathbf{v}_{cog}^j$.

After that, we apply Transformer layers [18] to update the vector representations of vertices, incorporating global cognitive information into all vertices in G_{cog}:

$$\tilde{\mathbf{v}}_{cog}^i = \text{TRSEnc}\left(\hat{\mathbf{v}}_{cog}^i\right) \tag{5}$$

where $\tilde{\mathbf{v}}_{cog}^i \in \mathbb{R}^{d_{model}}$ is the semantic vector representation of v_{cog}^i, and TRSEnc represents Transformer encoder layers.

Finally, we use the cognitive vertex presentation of CLS token, i.e. $\tilde{\mathbf{v}}_{cog}^0$, as the global cognitive representation of G_{cog}.

3.3 Affective Context Graph Constructing and Encoding

The procedures of Subsects. 3.2 and 3.3 are similar, with the strategies of knowledge selection and the approaches of graph encoding being different.

Affective Knowledge Selection. To inject appropriate affective knowledge into the dialogue context to detect users' emotions, we filter emotional concepts with high emotional intensity value and low emotion gap value between them and the dialogue context.

Be analogous to Cognitive Knowledge Selection., for non-stop word w_i^j of C, we first acquire its quadruples from ConceptNet and filter the affective knowledge by removing the quadruples that have low confidence scores or affection-unrelated relations.

Then we retrieve the VAD vectors (mentioned in Sect. 2) of the emotional concepts and calculate their emotion intensity values [8,21]. The formula for calculating the emotional intensity value of concept c is as follows:

$$EI(c) = \text{min} - \text{max} \left(\left\| V(c) - \frac{1}{2}, \ \frac{A(c)}{2} \right\| \right) \tag{6}$$

where $\text{min} - \text{max}$ is the min-max normalization, $\| \|$ is L_2 norm, $V(c)$ and $A(c)$ are concept c's values of the valence and arousal dimensions in the VAD vector, respectively. If c is not in NRC-VAD, $EI(c)$ will be set to 0.

Besides, we compute the values of the emotion gap between each nonstop word and its emotional concepts. The formula for computing the emotion gap value between word w and concept c is as follows:

$$EG(w, c) = \frac{abs(V(w) - V(c)) + abs(A(w) - A(c))}{2} \tag{7}$$

where abs is the operation to get the absolute value.

Eventually, we filter the quadruples whose emotion gap values between head concept and tail concept (i.e. word in dialogue context and its emotional concept) are lower than 0.5, rank the emotion intensity values of the quadruples, and select the top K_2 ones.

Affective Context Graph Constructing. We construct the affective context graph in a way similar to Cognitive Context Graph Constructing and represent it as G_{aff}, and the emotional links in the knowledge-enhanced graph are stored in the affective adjacency matrix A_{aff}.

Affective Context Graph Encoding. Taking the same steps as Cognitive Context Graph Encoding, we initialize the affective vector presentation of every vertex v^i_{aff}, and then update v^i_{aff} with affective knowledge. But notice that, when encoding the affective context graph, the relations between vertices would not be considered. We also employ Transformer layers [18] to update the vector representations of vertices, incorporating global emotional information into all vertices in G_{aff}.

The affective vertex presentation of the CLS token, that is $\tilde{v}^0_{aff}$, is also used as the global affective representation of G_{aff}.

3.4 Cognition and Affection Fusion

To full exploit the original information of the dialogue context, we encode the raw dialogue context. We use the same initialization method as Cognitive Context Graph Encoding to acquire the embeddings of context sequence, i.e. E_C, and then feed it into new Transformer encoder layers to get the hidden representations of C:

$$H = TRSEnc(C) \tag{8}$$

Then we use the hidden representation of the CLS token to represent the context sequence:

$$h = H[0] \tag{9}$$

And then we perform the cognitive and affective linear transformation on the hidden representation h to obtain the corresponding representations of the dialogue context, respectively:

$$h_{cog} = \mathbf{W}_{coc}h \tag{10}$$

$$h_{aff} = \mathbf{W}_{aoc}h \tag{11}$$

where $\mathbf{W}_{coc} \in \mathbb{R}^{d_{model} \times d_{model}}$, $\mathbf{W}_{aoc} \in \mathbb{R}^{d_{model} \times d_{model}}$ are the cognitive and affective linear transformations.

Emotion Classification. Our proposed model learns to predict the users' emotional state to guide the empathetic response generation. We concatenate $\tilde{v}^0_{aff}$ with h_{aff} to obtain a fused affective representation $\tilde{h}^0_{aff}$:

$$\tilde{h}_{aff} = \tilde{v}^0_{aff} \oplus h_{aff} \tag{12}$$

where $\oplus$ denotes concatenation and $\tilde{h}_{aff} \in \mathbb{R}^{d_{model}}$.

Hence, we pass $\tilde{h}_{aff}$ through a linear layer followed by a Softmax operation to produce the emotion category distribution $P_{emo} \in \mathbb{R}^q$, where q is the number of emotion categories:

$$P_{emo} = \text{Softmax}\left(W_{emo}\tilde{h}_{aff}\right) \tag{13}$$

where $W_{emo} \in \mathbb{R}^{2d_{model} \times q}$ is the emotional linear transformation. During training, we conduct the parameter learning by minimizing the Cross-Entropy (CE) loss between the ground truth label e^* and the predicted label e:

$$\mathcal{L}_{emo} = -\log\left(P_{emo}\left(e = e^*\right)\right) \tag{14}$$

Information Integration. To generate empathetic responses, we integrate cognitive and affective information into the dialogue context. First concatenate the global cognitive representation of the cognitive context graph, i.e. $\tilde{\mathbf{v}}_{cog}^0$, and the cognitive representation of the raw dialogue context, i.e. h_{cog}, to obtain a fused cognitive representation:

$$\tilde{h}_{cog} = \tilde{\mathbf{v}}_{cog}^0 \oplus h_{cog} \tag{15}$$

where $\tilde{h}_{cog} \in \mathbb{R}^{2d_{model}}$.

Then $\tilde{h}_{cog}$ and $\tilde{h}_{aff}$ are concatenated and the combination of them is passed through a Multi-Layer Perceptron with ReLU activation, which aims to learn a contextualized representation with adequate cognition and affective information:

$$\hat{h}_{ctx} = \tilde{h}_{cog} \oplus \tilde{h}_{aff} \tag{16}$$

$$\tilde{h}_{ctx} = \text{MLP}\left(\sigma\left(\hat{h}_{ctx}\right) \odot \hat{h}_{ctx}\right) \tag{17}$$

where $\hat{h}_{ctx} \in \mathbb{R}^{d_{model}}$, MLP denotes Multi-Layer Perceptron, and $\odot$ denotes element-wise multiplication.

3.5 Empathetic Response Generation

The contextualized representation of cognition and affection, i.e. $\tilde{h}_{ctx}$, and the word embeddings of the target response R_{gold}, i.e. $E_w\left(R_{gold}\right)$, are fed as the inputs into the Transformer decoder layers to generate a response:

$$O = \text{TRSDec}\left(E_w\left(R_{gold}\right), \tilde{h}_{ctx}\right) \tag{18}$$

$$P_{resp} = \text{softmax}\left(W_o O\right) \tag{19}$$

$$p\left(R_t \mid R_{<t}, G_{cog}, G_{aff}\right) = P_{resp}\left[t\right] \tag{20}$$

where $O \in \mathbb{R}^{l_R \times d_{model}}$, l_R is the length of the predicted response, TRSDec represents the Transformer decoder layers, $P_{resp} \in \mathbb{R}^{l_R \times |V|}$, $|V|$ is the vocabulary size, and $p\left(R_t \mid R_{<t}, G_{\text{cog}}, G_{\text{aff}}\right)$ is the distribution over the vocabulary V for the t-th word R_t.

Then a standard Negative Log-Likelihood (NLL) is used to optimize generated responses:

$$\mathcal{L}_{gen} = - \sum_{t=1}^{l_R} \log p\left(R_t \mid R_{<t}, G_{cog}, G_{aff}\right) \tag{21}$$

To avoid generating generic empathetic responses, following [15], we adopt Frequency-Aware Cross-Entropy (FACE) [4] as an additional loss to penalize high-frequency tokens. Therefore, during the training process, we first compute the relative frequency of each token $word_i$ in the training corpus:

$$RF_i = \frac{\text{freq}\left(word_i\right)}{\sum_{j=1}^{V} \text{freq}\left(word_i\right)} \tag{22}$$

where V is the vocabulary size of the training corpus. Accordingly, the frequency-based weight w_i can be calculated as follows:

$$w_i = a \times RF_i + 1 \tag{23}$$

where $a = -\left(\max_{1 \leq j \leq V}\left(RF_j\right)\right)^{-1}$ is the frequency slope, 1 is added as the bias so that w_i falls into $[0, 1]$. As done by [15], we normalize w_i to have a mean of 1. The diversity loss is finally computed as below:

$$\mathcal{L}_{div} = - \sum_{t=1}^{T} \sum_{i=1}^{V} w_i \delta_t\left(c_i\right) \log \mathrm{P}\left(c_i \mid y_{<t}, C\right) \tag{24}$$

where c_i is a candidate token in the vocabulary and $\delta_t\left(c_i\right)$ is the indicator function, which equals to 1 only if $c_i = y_t$ and 0 otherwise.

Eventually, all the parameters of our proposed model are trained and optimized by jointly minimizing the emotional loss (Eq. 14), the generation loss (Eq. 21) and the diversity loss (Eq. 24) as follows:

$$\mathcal{L} = \gamma_1 \mathcal{L}_{emo} + \gamma_2 \mathcal{L}_{gen} + \gamma_3 \mathcal{L}_{div} \tag{25}$$

where γ_1, γ_2, γ_3 are hyper-parameters to balance the above three losses.

4   Experiments

4.1   Baselines

We compare our proposed model with the following baselines:

- **MoEL** [9]: A variation of Transformer consisting of one encoder and several decoders that focus on each emotion accordingly.

- **EmpDG** [7]: A adversarial model that encodes semantic context and multi-resolution emotional context respectively and interacts with user feedback.
- **MIME** [10]: Another variation of Transformer. It does emotion grouping and applies emotion stochastic sampling and emotion mimicry.
- **KEMP** [8]: A knowledge-enriched model that uses commonsense and emotional lexical knowledge to explicitly understand and express emotions.
- **CEM** [15]: Another knowledge-enriched Transformer-based model that uses commonsense to obtain more information about users' situations.

4.2 Implementation Details

We conduct our experiments on EMPATHETICDIALOGUES [14], a large-scale dataset of $25k$ conversations, grounded in emotional situations. the dataset considers 32 emotion labels, of which the distribution is close to evenly distributed. For our experiments, we use the original 8:1:1 train/validation/test split of this dataset.

We use Pytorch[1] to implement the proposed model. The word embeddings are initialized with pre-trained Glove vectors[2] [13], and the relation embeddings are randomly initialized and fixed during training. For the positional embeddings, we follow the original paper [18]. The dimension of embeddings is set to 300 empirically. The maximum introducing numbers of external concepts per dialogue and per token are set as 10 and 1, respectively. And the loss weights γ_1, γ_2, γ_3 are all set to 1. We set the same hyper-parameters of Transformer as [8], including the hidden size, the number of attention heads, etc. When training our proposed model, we use Adam and early stopping with a batch size of 16 and an initial learning rate of $1e5$. We varied the learning rate during training following [18] and use a batch size of 1 and a maximum of 30 decoding steps during testing and inference.

4.3 Evaluations

We evaluate our models from two aspects, i.e. automatic and human evaluations.

Automatic Evaluation. To evaluate the performance of KEEM, we first adopt Emotion Accuracy, i.e. the accuracy of emotion detection. The **Perplexity** [16] is also utilized to measure the high-level general quality of the generation model. A response with higher confidence will result in a lower perplexity. Furthermore, **Distinct-1** and **Distinct-2** [6] are used to measure the proportion of the distinct unigrams and bigrams in all the generated results, which indicate the diversity of the produced responses.

---

[1] https://pytorch.org/.

[2] https://github.com/stanfordnlp/GloVe.

Table 1. Results of automatic evaluation.

Models	Accuracy (%)	Perplexity	Distinct-1	Distinct-2
MoEL	32.00	38.04	0.44	2.10
EmpDG	34.31	37.29	0.46	2.02
MIME	34.24	37.09	0.47	1.90
KEMP	39.31	36.89	0.55	2.29
CEM	39.11	**36.11**	0.66	2.99
KEEM (ours)	**40.54**	36.28	**0.72**	**3.10**
w/o Cog	40.01	36.21	0.56	2.32
w/o Aff	39.50	36.25	0.67	3.02

The results of the automatic and manual evaluations are shown in Table 1. In Table 1, we observe that KEEM achieves the highest emotion accuracy, which suggests the new strategy of selecting affective knowledge is beneficial for users' emotion detection. Although CEM [15] gets a slightly lower perplexity score than ours, our proposed model also considerably outperforms the baselines in terms of Distinct-1 and Distinct-2, which highlights the importance of the novel approaches to incorporating commonsense knowledge and constructing the cognitive context graph.

Human Evaluation. For qualitative evaluation, we take human A/B tests to compare KEEM and five baselines, following [15]. For a given dialogue context, our model's response is paired with a response from the baselines, and annotators are asked to choose the better one from the following three aspects: 1) **Empathy**: which one shows more understanding of the user's situation and feelings; 2) **Coherence**: which one is more on-topic and relevant to the context; 3) **Fluency**: which one is more fluent and natural. We randomly sample 100 dialogues and their corresponding results from our model as well as the baselines, and then assign three crowdsourcing workers to annotate each pair.

As displayed in Table 2, responses generated by KEEM are more often preferred by human judges in empathy and coherence compared to the baselines. This also demonstrates that, with the enhancement of commonsense and emotional knowledge, our model is able to produce more empathetic and relevant responses. We also notice that KEEM does not significantly outperform the baselines not enriched by external knowledge in fluency, which might imply that the knowledge incorporated has a negative influence on fluency. There is one reasonable explanation that selected knowledge contains not only useful information but also noises, which decrease the fluency of the responses.

Table 2. Results of human evaluation (%).

Comparisons	Aspects	Win	Lose	Tie
KEEM vs. MoEL	Empathy	32	9	59
	Coherence	51	8	41
	Fluency	36	33	31
KEEM vs. EmpDG	Empathy	36	10	54
	Coherence	53	10	37
	Fluency	34	35	31
KEEM vs. MIME	Empathy	50	9	41
	Coherence	39	9	52
	Fluency	24	27	49
KEEM vs. KEMP	Empathy	40	26	34
	Coherence	44	23	33
	Fluency	34	36	30
KEEM vs. CEM	Empathy	47	35	29
	Coherence	45	31	24
	Fluency	24	18	58

4.4 Ablation Study

We conduct the ablation study to verify the effect and contribution of each component of our KEEM. More specifically, we consider the following three variants of KEEM:

- **w/o Cog**: We remove the procedures in Sect. 3.2) and delete the concatenation of the global and original cognitive representation of dialogue context (Eq. 15). The fused cognitive representation is replaced with the latter in subsequent calculations.
- **w/o Aff**: We ablate the new approach to filtering affective knowledge (Eq. 7). The effects of introducing affective knowledge and building and encoding affective context graphs have been proven in [8].

The results of the above two variants are in Table 1, which show that each component contributes to KEEM from different aspects. Specifically, removing the cognitive knowledge decreases most of the performance, suggesting that incorporating extra cognitive information helps to recognize the users' situations and identify their emotions, to varying degrees. And ablating the novel strategy of selecting affective knowledge influences the emotion classification considerably and impairs the quality of the generated results slightly.

4.5 Case Study

Cases from KEEM and five baseline models are listed in Table 3. In the first case, KEEM shows the best recognition of the user's action and feeling, by incorporating the concepts of "destroy", "joy", and "great", which correspond to the word "fell", "cheer", "nice" in dialogue context. In contrast, the baseline MoEL only attends to "baby" and ignores what the user did for the baby, and the other four baselines are only able to identify the positivity of the dialogue context, producing some context-unrelated content. In the second case, KEEM generates the most context-consistent and emotion-appropriate response, which expresses a good wish for the user about his future promotion, while MoEL and EmpDG do not even detect the user's hopeful mood. Both two cases demonstrate that KEEM can generate responses of empathy.

Table 3. Responses generated by different models.

Emotion	caring
Context	Speaker:So my friend's baby <u>fell</u> down and scraped his knee yesterday and i was trying so hard to <u>cheer</u> him up.
	Listener:That is <u>nice</u> of you. What did you do?
	Speaker:I was singing nursery songs :)
Gold	**Which one?**
MoEL	Oh no, I bet he was a good kid.
EmpDG	That is a great attitude to have.
MIME	That is a good thing to do. I am glad you have a good friend!
KEMP	That is a great feeling. I am sure you will be fine.
CEM	That is awesome! I bet you were very proud of him.
Sem. Know.	(fell, RelatedTo, destroy), (scraped, RelatedTo, destroy)
Emo. Know.	(nice, great), (cheer, joy)
KEEM	**That is a great thing to do. I am sure he'll be fine.**
Emotion	hopeful
Context	Speaker:I guess I will be the <u>next</u> team <u>leader</u>.
	Lister:Really? Is that a <u>promotion</u>?
	Speaker:Yes. I was expecting that very much.
Gold	Good for you.
MoEL	I am not sure what you are going to do.
EmpDG	What kind of leader you?
MIME	That is awesome! I hope you get it!
KEMP	I am sure you will get a new one.
CEM	That is great!
Sem. Know.	(promotion, RelatedTo, leader), (leader, RelatedTo, promotion)
Emo. Know.	(next, new), (leader, hero)
KEEM	**I am sure you will do well.**

5 Related Work

With the support of the newly proposed datasets [14,19], research in empathetic dialogue generation has developed rapidly. Rashkin et al. [14] exploit adaptions of dialogue models for empathetic responding. Lin et al. [9] design listeners (i.e. decoders) in responding to different emotions and softly combining different listeners' outputs. Li et al. [7] propose a multi-resolution adversarial model to capture the nuances of user emotion and consider the potential of user feedback. Majumder et al. [10] believe that empathic responses often mimic users' emotions to varying degrees. Li et al. [8] construct an emotional context graph to perceive implicit emotions and learn emotional interactions. All of them are about how to perceive and express emotions.

Recently, some work [15] has attempted to boost both the cognitive and affection empathy of dialogue models. Sabour et al. [15] argue that the cognitive empathy of the user's situation should be considered, and they introduce commonsense to further enhance it. Nonetheless, due to some weaknesses of the knowledge base used (e.g. the limited commonsense relations, the unsatisfactory inference accuracy, and the inability to acquire fine-grained information), the effectiveness of this proposed approach would be impacted.

6 Conclusion

In this paper, we propose a novel Knowledge-Enhanced EMpathetic dialogue generation model (KEEM) to demonstrate how leveraging commonsense and emotional knowledge is beneficial to the cognition of users' situations and the detection of users' feelings, which helps produce more empathetic responses. We conduct experiments on EMPATHETICDIALOGUES dataset, and our automatic and manual evaluations have empirically proven the significance of our approach in empathetic response generation. Nevertheless, as the results demonstrate, the model still has shortcomings in terms of fluency, which is one of the future directions for us to challenge.

Acknowledgements. The authors would like to thank the Associate Editor and anonymous reviewers for their valuable comments and suggestions. This work is funded in part by the National Natural Science Foundation of China under Grants No.62176029. This work also is supported in part by the Chongqing Technology Innovation and Application Development Special under Grants CSTB2022TIAD-KPX0206. Any opinions, findings, and conclusions, or recommendations expressed in this material are those of the authors and do not necessarily reflect those of the sponsor.

References

1. Bordes, A., Usunier, N., Garcia-Duran, A., Weston, J., Yakhnenko, O.: Translating embeddings for modeling multi-relational data. In: Advances in Neural Information Processing Systems, vol. 26 (2013)
2. Cuff, B., Brown, S.J., Taylor, L., Howat, D.J.: Empathy: a review of the concept. Emot. Rev. **8**, 144–153 (2014)
3. Elliott, R., Bohart, A.C., Watson, J.C., Murphy, D.: Therapist empathy and client outcome: an updated meta-analysis. Psychotherapy **55**(4), 399 (2018)
4. Jiang, S., Ren, P., Monz, C., de Rijke, M.: Improving neural response diversity with frequency-aware cross-entropy loss. In: The World Wide Web Conference, pp. 2879–2885 (2019)
5. Keskin, S.C.: From what isn't empathy to empathic learning process. Procedia. Soc. Behav. Sci. **116**, 4932–4938 (2014)
6. Li, J., Galley, M., Brockett, C., Gao, J., Dolan, W.B.: A diversity-promoting objective function for neural conversation models. In: Proceedings of the 2016 Conference of the North American Chapter of the Association for Computational Linguistics: Human Language Technologies, pp. 110–119 (2016)
7. Li, Q., Chen, H., Ren, Z., Ren, P., Tu, Z., Chen, Z.: EmpDG: multi-resolution interactive empathetic dialogue generation. In: Proceedings of the 28th International Conference on Computational Linguistics, pp. 4454–4466 (2020)
8. Li, Q., Li, P., Ren, Z., Ren, P., Chen, Z.: Knowledge bridging for empathetic dialogue generation. In: Proceedings of the AAAI Conference on Artificial Intelligence, vol. 36, pp. 10993–11001 (2022)
9. Lin, Z., Madotto, A., Shin, J., Xu, P., Fung, P.: Moel: Mixture of empathetic listeners. In: Proceedings of the 2019 Conference on Empirical Methods in Natural Language Processing and the 9th International Joint Conference on Natural Language Processing (EMNLP-IJCNLP), pp. 121–132 (2019)
10. Majumder, N., et al.: Mime: mimicking emotions for empathetic response generation. In: Proceedings of the 2020 Conference on Empirical Methods in Natural Language Processing (EMNLP), pp. 8968–8979 (2020)
11. Mikolov, T., Chen, K., Corrado, G., Dean, J.: Efficient estimation of word representations in vector space. Computer Science (2013)
12. Mohammad, S.: Obtaining reliable human ratings of valence, arousal, and dominance for 20,000 English words. In: Proceedings of the 56th Annual Meeting of the Association for Computational Linguistics (volume 1: Long Papers), pp. 174–184 (2018)
13. Pennington, J., Socher, R., Manning, C.D.: Glove: global vectors for word representation. In: Proceedings of the 2014 Conference on Empirical Methods In Natural Language Processing (EMNLP), pp. 1532–1543 (2014)
14. Rashkin, H., Smith, E.M., Li, M., Boureau, Y.L.: Towards empathetic open-domain conversation models: a new benchmark and dataset. In: Proceedings of the 57th Annual Meeting of the Association for Computational Linguistics, pp. 5370–5381 (2019)
15. Sabour, S., Zheng, C., Huang, M.: CEM: commonsense-aware empathetic response generation. In: Proceedings of the AAAI Conference on Artificial Intelligence, vol. 36, pp. 11229–11237 (2022)
16. Serban, I.V., Sordoni, A., Bengio, Y., Courville, A., Pineau, J.: Hierarchical neural network generative models for movie dialogues. arXiv preprint arXiv:1507.04808 **7**(8), 434–441 (2015)

17. Speer, R., Chin, J., Havasi, C.: Conceptnet 5.5: an open multilingual graph of general knowledge. In: Proceedings of the AAAI Conference on Artificial Intelligence, vol. 31 (2017)
18. Vaswani, A., et al.: Attention is all you need. In: Advances in Neural Information Processing Systems, vol. 30 (2017)
19. Welivita, A., Pu, P.: A taxonomy of empathetic response intents in human social conversations. In: Proceedings of the 28th International Conference on Computational Linguistics, pp. 4886–4899 (2020)
20. Zech, E., Rimé, B.: Is talking about an emotional experience helpful? effects on emotional recovery and perceived benefits. Clin. Psychol. Psychoth. Int. J. Theory Pract. **12**(4), 270–287 (2005)
21. Zhong, P., Wang, D., Miao, C.: Knowledge-enriched transformer for emotion detection in textual conversations. In: Proceedings of the 2019 Conference on Empirical Methods in Natural Language Processing and the 9th International Joint Conference on Natural Language Processing (EMNLP-IJCNLP), pp. 165–176 (2019)

Implicit Sentiment Extraction Using Structure Generation with Sentiment Instructor Prompt Template

YuXuan Liu and Jiang Zhong[(✉)] [iD]

College of Computer Science, ChongQing University, ChongQing, China
{liuyuxuan,zhongjiang}@cqu.edu.cn

Abstract. Aspect-Category-Opinion-Sentiment quadruple extraction (ACOS) is the novel and challenging sentiment analysis task, which aims to analyze the full range of emotional causes. Existing approaches focus on solving explicit sentiment, but struggle with analyzing implicit sentiment reviews. In this paper, to address the issue, we propose SI-TS, a framework that takes implicit sentiment extraction into account. Specifically, we design target structure (TS) to capture implicit sentiment by converting sentiment elements into a structured format. Furthermore, to adaptively generate appropriate TS according to different sentiment scenarios, we design an prompt template based sentiment instructor(SI). It assists the framework in effectively extracting implicit sentiment elements from the reviews. Extensive experiments were conducted on two widely used ACOS benchmarks, and improvements in F1 values were observed. Specifically, we achieved a 1.05% and 1.28% improvement in F1 values for Laptop-ACOS and Restaurant-ACOS, respectively. Notably, significant results were achieved in extracting implicit sentiment.

Keywords: The ACOS Task · Implicit Sentiment Extraction · Target Structure (TS) · Sentiment Instructor (SI) · Prompt Template

1 Introduction

Aspect-Category-Opinion-Sentiment quadruple extraction(ACOS) [1] is a challenging task in aspect-based sentiment analysis(ABSA) [2,3], This is a typical multitask with four sub-tasks, which are (1) identifying aspect term mentions, (2) detecting aspect categories, (3) extracting opinions linked to aspects, and (4) classifying the sentiments belonging to aspect. These help us to understand the aspect-level opinions in the reviews and provide a complete story. Give an example review "The ambience is wonderful for a date or group outing." The review can be extracted as a pair of four elements of aspect-category-opinion-sentiment (a, c, o, s), which are aspect (a): "service", category (c): "food quality", opinion (o): "wonderful", and sentiment (s): "positive".

The typical studies tackle this task by using either extractive or generative methods. (1) Extractive methods require designing specific sentiment element

X. Yang et al. (Eds.): ADMA 2023, LNAI 14179, pp. 94–108, 2023.
https://doi.org/10.1007/978-3-031-46674-8_7

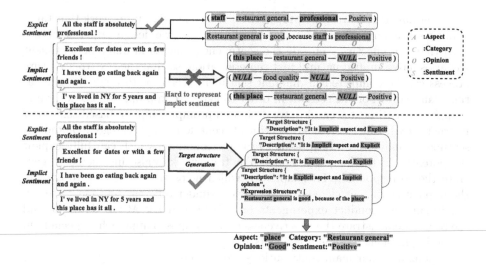

Fig. 1. Comparison between existing frameworks and our framework. Above the split dashed line is the existing framework. Below the split dashed line is our proposed framework. We believe that using target structure to represent sentiment quadruples in a unified way can comprehensively consider a variety of implicit sentiments, This approach allows better assisting the framework to mine implicit sentiments.

extraction modules to jointly model individual or multiple sentiment elements. Cai et al. [1] Perform category classification on candidate aspect-opinion pairs based on Peng et al. [4] and Wan et al. [5] for sentiment quadruplet extraction. They also extended the labeling model using a sequential labeling approach inspired by this work [6]. (2) Generative methods employ sequence generation approach to generate sentiment elements directly. For instance, Yan et al [7], proposed a generative framework. Meanwhile, models such as GAS [8], PARA-PHRASE [9] and USI [10] generate sentiment elements or paraphrases directly in the task. The generative approach is preferred over the extractive approach for generating sentiment using generic knowledge, leading to a shift in research towards generative approaches for sentiment extraction.

However, existing approaches to the ACOS task struggle with cases where reviews include implicit aspects or implicit opinions. According to recent studies [11,12], a significant portion of reviews in the corpus contain implicit expressions, with over 30% including implicit aspects or opinions, and more than 8% containing both [1]. We observe that most of the current research focuses on explicit sentiment and neglects modeling implicit sentiment(aspects or opinions).

To address the above limitation, our focus is on addressing the ACOS task, with particular emphasis on the extraction of implicit sentiment in formulations. The major difference between our work and current work is shown in Fig. 1. Firstly, we re-examine the sentiment features in reviews and categorize them into different sentiment scenarios, including three types of implicit sentiment scenarios and one type of explicit sentiment scenario. Secondly, based on the above sentiment scenarios, we propose SI-TS, a generation framework that takes implicit sentiment into account, Specifically, for one, we design target structure (TS) that can effectively encode the sentiment elements in different sentiment scenarios into a unified natural text representation, so that the text-to-structure process can be completed in different sentiment scenarios through the generation framework. For the second, in order to generate appropriate TS according to different sentiment scenarios, we design an prompt template based sentiment instructor (SI), which consists of sentiment scenario prompt, category candidate, sentiment candidate and expression prompt candidate, which can fully exploit the potential of the language model. The framework is also instructed to generate the corresponding TS adaptively, which helps the framework to explore the implicit sentiment elements from reviews under different sentiment scenarios. Finally, we conduct experiments on two benchmark ACOS datasets, and the results demonstrate the effectiveness of our proposed approach, especially in extracting implicit sentiment elements.

In summary, our main contributions are as follows:

1. We propose SI-TS, an implicit sentiment generation framework that contains an elaborate target structure (TS) that represent reviews in a variety of implicit sentiment scenarios.
2. To instruct the framework to generate suitable target structure, we propose an prompt template based sentiment instructor (SI) to help the model effectively mine implicit sentiment elements from the reviews.
3. Our experiments on two popular benchmarks show F1 improvements of 1.05% and 1.28% for Laptop-ACOS and Restaurant-ACOS, respectively. In addition, we achieved significant results in implicit sentiment extraction.

2   Problem Statement

2.1   ACOS Task Definition

Given a review sentence r the ACOS task aims to extract aspect-level quadruples $Q_i = (a_i, c_i, o_i, s_i)$. The elements a_i, c_i, o_i, s_i, corresponding to the i-th aspect, aspect category, opinion, and sentiment in the review r, these elements are collectively referred to as sentiment elements. The sentiment polarity s_i and $s_i \in \{POS, NEU, NEG\}$, where $\{POS, NEU, NEG\}$ stands for positive, neutral, negative, respectively.

2.2   Types of Sentiment Scenario

As shown in Fig. 2, we define aspect terms and opinion terms with explicit sentiment representation in review as EA (Explicit Aspect) and EO (Explicit Opinion), and if there are no aspect terms or opinion terms with explicit sentiment representation, we define them as IA (Implicit Aspect) and IO (Implicit Opinion). Reviews are divided into four sentiment scenarios, i.e. $\varepsilon = \{EA\&EO, EA\&IO, IA\&EO, IA\&IO\}$, i.e. the explicit sentiment scenario containing both EA and EO, and the three implicit sentiment scenarios containing either a single IA or IO or both IA and IO $\{(EA\&IO, IA\&EO, IA\&IO)\}$.

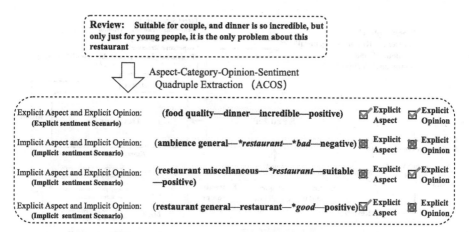

Fig. 2. The example review sentence in the figure contains four clauses, each of which contains a pair of sentiment quadruples. We define the clauses with explicit aspect and opinion as explicit sentiment scenarios (e.g. EA&EO) and the clauses without explicit aspect or opinion (with implicit aspect or opinion) as implicit sentiment scenarios (e.g. EA&IO, IA&EO, IA&IO). The cross in the figure means that it does not contain this item and the tick means that it does.

3   Methodology

In this section, we (1) introduce TS, a target structure that represents multiple implicit sentiment scenarios. And (2) propose SI, a prompt template based sentiment instructor, which controls the framework adaptively generating the corresponding target structure. The overall framework of SI-TS is shown in Fig. 3.

3.1   Sentiment Mapping

To better capture the semantics in the sentiment polarity set $\{s_i\}$, we first map the sentiment polarity s_i as follows.

$$P_s(s_i) = \begin{cases} great, & \text{if } s_i = POS \\ ok, & \text{if } s_i = NEU \\ bad, & \text{if } s_i = NEG \end{cases} \qquad (1)$$

After performing the sentiment mapping operation $P_s(.)$, The model is aware of selecting stronger semantics and more appropriate sentiments. Note that the particular mapping can be predefined using the commonsense knowledge in Eq. 1, or it can rely on the dataset, using the most common consensus terms for each sentiment polarity as sentiment expressions.

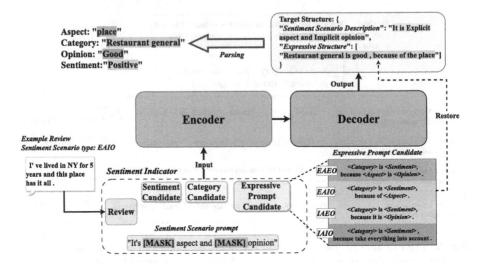

Fig. 3. The overall framework of SI-TS. Our framework first feeds the raw review into the encoder after a sentiment instructor (SI) transformation. Specifically, our sentiment instructor consists of Sentiment Scenario Prompt, Sentiment Candidate, Category Candidate, and Expression Prompt Candidate. After the transformation of the SI, decoder adaptively constructs the Target Structure (TS) based on the Sentiment Scenario Prompt, and the Expression Prompt Candidate. Finally, we parse the TS obtained from the output according to the location of the specific design to acquire the final quadruples. The expression structure we designed is shown in the bottom right part of the figure, and we design one expression prompt for each sentiment scenario (e.g., EA&EO, EA&IO, IA&EO, IA&IO).

3.2 Sentiment Instructor(SI) Components and Input Transformations

To leverage the capabilities of the language model and enable it to distinguish implicit sentiment scenarios, inspired by this paper [10], we employed a sentiment-instruction based prompting method to convert the input reviews r. The sentiment instructor we designed consist of the following components.

Sentiment Scenario Prompt. In order to allow the model to learn the discrepancy between sentiment scenarios, we designed the sentiment scenario prompt as part of the sentiment instructor. We define Sentiment Scenario prompt as I_{mask}. I_{mask} will be constructed for input in the following form:

$$I_{mask} = It's < extra\_id\_0 > aspect \ and < extra\_id\_1 > opinion \quad (2)$$

Here "$< extra\_id\_0 >$", "$< extra\_id\_1 >$" tokens are special mask tokens in the T5 model [29]. it allows better exploitation of the generic knowledge of language models by modeling downstream tasks as pre-trained targets. (i.e. mask language modeling). This approach, also called prompt learning [31], has shown powerful effectiveness on a wide range of NLP tasks.

Sentiment Candidate. We use the set of sentiment expressions $\{P_s(s_i)\}$ as candidates after sentiment mapping as described above. It enables the model to aware of which choice of sentiment expressions is required.

Category Candidate. The ACOS task also involves classifying aspect categories. Therefore, we also input all the set of categories $\{c\}$ as category candidates. It enables the model to be aware of which category choices are required.

Expressive Prompt Candidate. We designed the expression prompt T_{ep}, which makes the model aware of which expression structure may be generated. We designed four expression prompt corresponding to four sentiment scenarios, and combined together as expression prompt candidates as part of the sentiment instruction. Our expression prompt are shown in the bottom right corner of Fig. 3. To sum up, We concatenate the reviews r with all the instruction elements mentioned in the appeal, separated by separator, to obtain the input $\mathcal{X}$ refactored by the prompt template SI.

$$\mathcal{X} = [I_{mask}; [SSEP]; r; [SSEP]; \{P_s(s)\}; [SSEP]; \{c\}; [SSEP]; \{T_{ep}\}] \quad (3)$$

The input is represented by $\mathcal{X}$, with "$[SSEP]$" denoting the separator and ";" representing the connection operation.

3.3   Target Structure(TS) Generation for Implicit Sentiment

We adopt an encoder-decoder style architecture to compose our generative framework for extracting implicit sentiment. Our purposes are that:

(1) The corresponding target structures should be generated adaptively according to the implicit sentiment scenarios.
(2) In addition, to generate quadruples in one step and model multiple implicit sentiments scenarios, we adopt structures as output targets instead of sequential labels such as $< aspect >$, $< opinion >$, $< category >$, $< sentiment >$. Based on the above mentioned, inspired by this work. [32], we propose TS, a target structure that represent implicit sentiment. In the following we will describe our TS and its generation method.

Firstly, the decoder restores the special mask token (from SI) in I_{mask}, and during training, we restore the corresponding positions of the special mask token to "explicit" and "implicit". For the sentiment scenario $k/in/varepsilon$, the recovered I_{mask} we call the sentiment scenario description, which is defined as $I_{des}(k)$.

$$I_{des}(k) = \begin{cases} It's\ explicit\ aspect\ and\ explicit\ opinion. & if\ k\ is\ EA\&EO; \\ It's\ implicit\ aspect\ and\ explicit\ opinion. & if\ k\ is\ IA\&EO; \\ It's\ explicit\ aspect\ and\ implicit\ opinion. & if\ k\ is\ EA\&IO; \\ It's\ implicit\ aspect\ and\ implicit\ opinion. & if\ k\ is\ IA\&IO; \end{cases} \quad (4)$$

Second, we train to restore the special prompt tokens (e.g., $< category >$, $< emotion >$, $< view >$, $< opinion >$) which from the T_{ep} in the above SI.

During training. We employ the teacher forcing strategy to train the model to generate accurate sentiment elements corresponding to the special prompt tokens. We refer to the recovered T_{ep} as the expression structure, which is defined as $T_{et}(Q_i)$.

$$
T_{aop}(a_i, o_i, k) = \begin{cases} because \ a_i \ is \ o_i. & if \ k \ is \ EA\&EO; \\ because \ it \ is \ o_i. & if \ k \ is \ IA\&EO; \\ because \ of \ the \ a_i. & if \ k \ is \ EA\&IO; \\ taking \ everything \ into \ account. & if \ k \ is \ IA\&IO; \end{cases} \tag{5}
$$

$$
T_{et}(Q_i) = c_i \ is \ P_s(s_i) \ T_{aop}(a_i, o_i, k) \tag{6}
$$

$$
T_{output}(Q_i, k) = I_{des}(k) \ ; \ T_{et}(Q_i) \tag{7}
$$

Here, $T_{aop}(.)$ is a linear mapping function used to generate target structures for aspect-opinion pairs according to the sentiment scenario k. It maps different TS depending on the sentiment scenario. Considering that the review r probably contains more than one sentiment quadruple, we connect multiple sentiments using the special concatenation symbol "&&", and the final output TS form $\mathcal{Y}$ is:

$$
\mathcal{Y} = T_{output}(Q_1, k) \ \&\& \cdots \&\& \ T_{output}(Q_n, k) \tag{8}
$$

Notice that the target structure $T_{output}(Q_i, k)$ of our final output is reduced by $\mathcal{X}$. Also it is determined by the sentiment scenario k and the quadruplet $Q_i = a_i, c_i, o_i, s_i$. This indicates that the target structure we designed can be generated adaptively according to the sentiment scenario k, which achieves our purpose.

3.4   Model Architecture and Training

We employ the T5 [29], an encoder-decoder language model that utilizes an autoregressive generation process to model the conditional probability of generating the next token y_i, based on the input sequence $\mathcal{X}$ and previously generated tokens $y_{<i}$. The model can calculate the entire probability $p(\mathcal{Y} \mid \mathcal{X})$ of generating the output sequence $\mathcal{Y}$, given the input sequence $\mathcal{X}$, which is expressed as follows: $\mathcal{Y}$ is:

$$
p(\mathcal{Y} \mid \mathcal{X}) = \prod_{i=1}^{|\mathcal{Y}|} p(y_i \mid y_{<i}, \mathcal{X}) \tag{9}
$$

The $\mathcal{X}$ and $\mathcal{Y}$ mentioned here are the same as those mentioned in Eq. 3and Eq. 8. Therefore, we will use expressions in the form of $\mathcal{X} = (x_1, x_2, \cdots, x_m)$ for ease of understanding.

In the training phase, the model parameters θ are initialized with the pretrained weights. Our model is trained using the teacher-forcing strategy with both input $\mathcal{X}$ and ground truth target $\mathcal{Y}$. The loss function of the model is as follows:

$$
\mathcal{L}(\mathcal{D}) = - \sum_{j=1}^{|\mathcal{D}|} \sum_{i=1}^{n} \log p_\theta(y_i \mid y_1, \ldots, y_{i-1}, \mathcal{X}_j) \tag{10}
$$

θ represents all the trainable parameters, $\mathcal{D}$ represents the training set samples, $|\mathcal{D}|$ represents the number of training set samples, $\mathcal{X}_j$ denotes the input sequence from the jth sample transformed by the prompt template containing the SI, and n is the length of the output sequence. We use the Adam optimizer with weight decay [33] to update the model parameters during the training phase.

3.5 Inference and Parsing

After the training phase, we use beam search to generate predicted results in an autoregressive manner. Then, we segment the possible structures according to the special delimiter "[SSEP]" mentioned in Eq. 3, and obtain the segmented predicted TS. Finally, We employ the reverse parsing strategy of TS to obtain the predicted quadruple $Q' = (a', c', o', s')$ and evaluate it against the golden emotional quadruple Q.

4 Experiments Setup

We detail the experience setup for evaluating our techniques on the ACOS task.

4.1 Datasets

We employed two large-scale ACOS datasets, Restaurant-ACOS and Laptop-ACOS. Both were annotated by Cai et al [1] and cover various sentiment scenarios. Table 1 shows the statistics of sentiment quadruples in each dataset.

4.2 Implementation Details

We selected the T5-large model from huggingface [34] for training. The training process was executed on an NVIDIA RTX A6000 with 48Gbit memory and included a warm-up strategy for 10 epochs. We utilized the Adam optimizer with weight decay [33].

Table 1. Sentiment quadruplets statistics in different sentiment scenarios in the Restaurant-ACOS and Laptop-ACOS datasets.

Class	Restaurant-ACOS		Laptop-ACOS	
	train	test	train	test
Categories	13		121	
#Type				
EA&EO	971	366	1619	467
EA&IO	187	76	814	225
IA&EO	303	128	535	128
IA&IO	225	90	249	64
All	1686	660	3217	884

4.3   Baselines

We compared seven prevailing ACOS technologies as follows:

Double Propagation-ACOS is a representative rule-based ACOS method based on improved Double Propagation [35] by Cai et al. [1].

JET-ACOS was proposed by Cai et al. [1], which was improved from the JET model [6] for aspect-category-sentiment-opinion quadruple extraction.

TAS-BERT-ACOS is a two-step pipelined approach proposed by Cai et al. [1]. It utilizes TAS-BERTcite [22] to extract quaternions.

Extract-Classify-ACOS was proposed by Cai et al. [1]. It adopts a representative aspect-opinion co-extraction system to accommodate ACOS quadruple extraction.

GAS-ACOS was improved from GAS [8], a model for generating sentiment elements through annotated and extractive paradigms. We adapted GAS for the ACOS task and named the resulting model GAS-ACOS.

PARAPHRASE [9] is an effective aspect-category-sentiment-opinion quadratic generative model, which is a text-to-text paradigm based work.

USI was proposed by Wang et al. [10], aiming to accomplish the ACOS task with a generative approach, which is currently the state-of-the-art method in the field in terms of performance.

5   Result and Analysis

We evaluate the extraction task of the sentiment quadruplets using precision, recall, and F1 metrics, where a correct extraction requires that all components are correct.

Table 2. Main results of our model and baselines on the Restaurant-ACOS and the Laptop-ACOS. The one marked with † is the baseline we reproduce.

Model	Restaurant-ACOS			Laptop-ACOS		
	Pre	Rec	F1	Pre	Rec	F1
Double-Propagation-ACOS	34.67	15.08	21.04	13.04	5.71	8.01
JET-ACOS	59.81	28.94	39.01	44.52	16.25	23.81
TAS-BERT-ACOS	26.29	46.29	33.53	**47.15**	19.22	27.31
Extract-Classify-ACOS	38.54	52.96	44.61	45.56	29.48	35.8
GAS-ACOS †	56.01	56.01	56.01	42.04	40.91	41.47
PARAPHRASE †	58.76	59.3	59.08	45.06	41.88	43.47
USI	<u>60.07</u>	<u>61.14</u>	<u>60.61</u>	44.57	**43.91**	<u>44.24</u>
SI-TS	**62.36**	**61.41**	**61.89**	<u>46.71</u>	<u>43.58</u>	**45.29**

5.1   Main Results

Our experimental results are shown in Table 2, with some noteworthy observations. Firstly the performance of the extractive methods is far from satisfactory, likely due to error-accumulation over several sub-tasks and limited pre-training alignment. Secondly, in the generative methods, the GAS-ACOS, PARAPHRASE and USI largely outperform the previous extractive methods, which shows that sequence-sequence modeling is effective for the ACOS task. Thirdly, it can be observed that our proposed method demonstrates effective performance across all metrics in both datasets. Specifically, we achieved a 1.05% improvement in F1 value for Laptop-ACOS and a 1.28% improvement for Restaurant-ACOS.

5.2   Ablation Study

We conducted an ablation study to further quantify the contribution of each component of the proposed method. The Restaurant-ACOS dataset was chosen for this experiment. Our ablation experiment is shown in Table 3, where we explored the following different situations separately. We also eliminated all the modules we designed, and the average evaluation metric was reduced by 3.48.

Table 3. Ablation study of our method on Restaurant-ACOS dataset. The Avg.$\triangle$ represents the average difference in all evaluation metrics between removing the module and not removing it.

Model	Pre.	Rec.	F1	Avg.$\triangle$
Ours	62.36	61.41	61.89	-
-(TS)	60.33	59.87	60.11	−1.78
-(SI)	59.46	58.83	59.14	−2.74
-(All)	58.71	58.11	58.41	−3.48

Firstly, We eliminated the SI. We found that performance had a significant decrease, with an average decrease of 2.74 for all evaluation metrics. This indicates the necessity of our SI. Secondly, We also conducted experiments by excluding TS. Our results show that the overall performance of our framework decreases on average by 1.76 after excluding TS. This validates our intuition that changing the structure of our design can effectively model the sentiment elements. Also it is more adaptable to subsequent sentiment quadruplet parsing and is effective for sentiment quadruplet extraction.

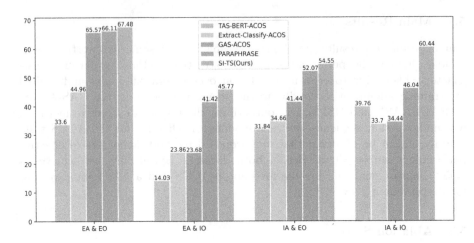

Fig. 4. The F1 performance of our model and a baseline in various sentiment scenarios, and compared the results.

5.3   Performance on Implicit Sentiment Scenarios

As we have mentioned, the ACOS task contains several different sentiment scenarios. Therefore, effective modeling of different sentiment scenarios is vital. To quantify the performance improvement of our approach under different sentiment scenarios, we explore the performance of our model. Specifically, we divided the testing set into four subsets and observed the performance on different subsets, while we compared the performance of our model with the baseline model, as shown in Fig. 4. For such situations we analyzed that previous researches such as Zhang et al [9]. and Wang et al [10]. only considered the distinction between EAEO and IAEO and only replaced the implicit aspect with it. However, we also considered the case of other implicit sentiment scenarios (e.g., EAIO and IAIO) and designed for them a suitable text structure TS, which can effectively model implicit sentiments not noticed by previous work. Meanwhile, under the guidance of SI, the model is trained to generate the corresponding TS.

5.4   The Effectiveness of Sentiment Instructor

To verify the effectiveness of SI in differentiating between four sentiment scenarios on the model, we generated t-SNE [36] visualizations of the average merged final encoder layer. Figure 5 shows the results on the Restaurant-ACOS dataset. Our findings indicate that SI is able to effectively distinguish implicit sentiment scenes. To a certain extent, the implicit sentiment scenes are distinguished secondarily. This also demonstrates that our SI can help the framework distinguish between explicit sentiment samples and implicit sentiment samples.

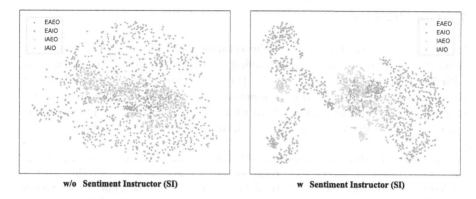

Fig. 5. T-SNE visualization of the last layer of the encoder on the Restaurant-ACOS dataset. Our SI indication model distinguishes four sentiment scenarios in the reviews.

6   Related Work

Implicit Sentiment Extraction. Implicit sentiment extraction aims to extract aspects and opinions that are not specifically described. Lazhar et al. [13] proposed a method based on dependency grammar to extract sentiment-opinion pairs obtaining implicit opinions. Fang et al. [14] developed a clustering algorithm to construct feature classes to extract the implicit opinions. Xu et al. [15] proposed the extracted aspect-specific opinion words to extract the implicit aspects. Zhang et al. [16] proposed an improved knowledge-based topic modeling KTM to extract the implicit aspects. Prasojo et al. [17] used a method based on adjective-to-aspect mapping to extract implicit aspects. Nandhini and Pradeep [18] proposed a co-occurrence and ranking-based algorithm to extract implicit aspects. However, implicit sentiment extraction remains a challenging task.

Aspect-Category-Opinion-Sentiment Quadruples. Recent researches [19–25]have shown that multiple extraction of sentiment elements can be effective in analyzing aspect-based reasons for sentiment. To fully analyze the sentiment in the review the Aspect-Category-Opinion-Sentiment quadruples(ACOS) task [1, 9] was proposed. Pre-trained transformer-based [26] models like BERT [27], BART [28], and T5 [29] are commonly used for the ACOS task. Xu et al. [6,30] used Bert for sentiment element extraction, while Cai et al. [1] improved the JET model for aspect-category-sentiment-opinion quadruple extraction. Yan et al. [7] and Zhang et al. [8] designed unified generative frameworks using Bart and T5 models, respectively. Zhang et al. [9] treated aspect-category-sentiment-opinion quadruple extraction as a paraphrase generation. Wang et al. [10] proposed a framework based on multi-task instruction tuning. However, in the ACOS task, current methods still struggle to extract implicit sentiment elements, which has motivated us to explore this direction.

7 Conclusions

In this work, we introduce TS, a modified target structure for ACOS generation. It effectively models a full range of implicit sentiment scenarios. We combine this with SI, our novel task-specific application of prompt learning that adaptively generate the corresponding target structures and assist the model in effectively mining implicit sentiment elements in reviews. Our proposed SI-TS framework demonstrate its effectiveness, especially in implicit sentiment scenarios.

Acknowledgements. The authors would like to thank the Associate Editor and anonymous reviewers for their valuable comments and suggestions. This work is funded in part by the National Natural Science Foundation of China under Grants No.62176029. This work also is supported in part by the Chongqing Technology Innovation and Application Development Special under Grants CSTB2022TIAD-KPX0206. Any opinions, findings, and conclusions, or recommendations expressed in this material are those of the authors and do not necessarily reflect those of the sponsor.

References

1. Cai, H., Xia, R., Yu, J.: Aspect-category-opinion-sentiment quadruple extraction with implicit aspects and opinions. In: Proceedings of the 59th Annual Meeting of the Association for Computational Linguistics and the 11th International Joint Conference on Natural Language Processing (Volume 1: Long Papers), pp. 340–350 (2021)
2. Bing, L.: Sentiment analysis and opinion mining (synthesis lectures on human language technologies). University of Illinois, Chicago, IL, USA (2012)
3. Pontiki, M., et al.: Semeval-2016 task 5: aspect based sentiment analysis. In: ProWorkshop on Semantic Evaluation (SemEval-2016), pp. 19–30. Association for Computational Linguistics (2016)
4. Peng, H., Xu, L., Bing, L., Huang, F., Lu, W., Si, L.: Knowing what, how and why: a near complete solution for aspect-based sentiment analysis. In: Proceedings of the AAAI Conference on Artificial Intelligence, vol. 34, pp. 8600–8607 (2020)
5. Wang, W., Pan, S.J., Dahlmeier, D., Xiao, X.: Coupled multi-layer attentions for co-extraction of aspect and opinion terms. In: Proceedings of the AAAI conference on artificial intelligence, vol. 31 (2017)
6. Xu, L., Li, H., Lu, W., Bing, L.: Position-aware tagging for aspect sentiment triplet extraction. arXiv preprint arXiv:2010.02609 (2020)
7. Yan, H., Dai, J., Qiu, X., Zhang, Z., et al.: A unified generative framework for aspect-based sentiment analysis. arXiv preprint arXiv:2106.04300 (2021)
8. Zhang, W., Li, X., Deng, Y., Bing, L., Lam, W.: Towards generative aspect-based sentiment analysis. In: Proceedings of the 59th Annual Meeting of the Association for Computational Linguistics and the 11th International Joint Conference on Natural Language Processing (Volume 2: Short Papers), pp. 504–510 (2021)
9. Zhang, W., Deng, Y., Li, X., Yuan, Y., Bing, L., Lam, W.: Aspect sentiment quad prediction as paraphrase generation. arXiv preprint arXiv:2110.00796 (2021)
10. Wang, Z., Xia, R., Yu, J.: UnifiedABSA: a unified ABSA framework based on multi-task instruction tuning. arXiv preprint arXiv:2211.10986 (2022)
11. Wang, S., et al.: Causal intervention improves implicit sentiment analysis. arXiv preprint arXiv:2208.09329 (2022)

12. Li, Z., Zou, Y., Zhang, C., Zhang, Q., Wei, Z.: Learning implicit sentiment in aspect-based sentiment analysis with supervised contrastive pre-training. arXiv preprint arXiv:2111.02194 (2021)
13. Lazhar, F., Yamina, T.G.: Mining explicit and implicit opinions from reviews. Int. J. Data Mining Model. Manag. **8**(1), 75–92 (2016)
14. Fang, Z., Zhang, Q., Tang, X., Wang, A., Baron, C.: An implicit opinion analysis model based on feature-based implicit opinion patterns. Artif. Intell. Rev. **53**, 4547–4574 (2020)
15. Xu, X., Cheng, X., Tan, S., Liu, Y., Shen, H.: Aspect-level opinion mining of online customer reviews. China Commun. **10**(3), 25–41 (2013)
16. Zhang, F., Xu, H., Wang, J., Sun, X., Deng, J.: Grasp the implicit features: hierarchical emotion classification based on topic model and SVM. In: 2016 International Joint Conference on Neural Networks (IJCNN), pp. 3592–3599. IEEE (2016)
17. Prasojo, R.E., Kacimi, M., Nutt, W.: Entity and aspect extraction for organizing news comments. In: Proceedings of the 24th ACM International on Conference on Information and Knowledge Management, pp. 233–242 (2015)
18. Devi Sri Nandhini, M., Pradeep, G.: A hybrid co-occurrence and ranking-based approach for detection of implicit aspects in aspect-based sentiment analysis. SN Comput. Sci. **1**, 1–9 (2020)
19. He, R., Lee, W.S., Ng, H.T., Dahlmeier, D.: An interactive multi-task learning network for end-to-end aspect-based sentiment analysis. arXiv preprint arXiv:1906.06906 (2019)
20. Li, X., Bing, L., Li, P., Lam, W.: A unified model for opinion target extraction and target sentiment prediction. In: Proceedings of the AAAI Conference on Artificial Intelligence, vol. 33, pp. 6714–6721 (2019)
21. Hu, M., Peng, Y., Huang, Z., Li, D., Lv, Y.: Open-domain targeted sentiment analysis via span-based extraction and classification. arXiv preprint arXiv:1906.03820 (2019)
22. Wan, H., Yang, Y., Du, J., Liu, Y., Qi, K., Pan, J.Z.: Target-aspect-sentiment joint detection for aspect-based sentiment analysis. In: Proceedings of the AAAI Conference on Artificial Intelligence, vol. 34, pp. 9122–9129 (2020)
23. Chen, S., Liu, J., Wang, Y., Zhang, W., Chi, Z.: Synchronous double-channel recurrent network for aspect-opinion pair extraction. In: Proceedings of the 58th Annual Meeting of the Association for Computational Linguistics, pp. 6515–6524 (2020)
24. Zhao, H., Huang, L., Zhang, R., Lu, Q., Xue, H.: SpanMLT: a span-based multitask learning framework for pair-wise aspect and opinion terms extraction. In: Proceedings of the 58th Annual Meeting of the Association for Computational Linguistics, pp. 3239–3248 (2020)
25. Wu, Z., Ying, C., Zhao, F., Fan, Z., Dai, X., Xia, R.: Grid tagging scheme for aspect-oriented fine-grained opinion extraction. arXiv preprint arXiv:2010.04640 (2020)
26. Vaswani, A., et al.: Attention is all you need. In: Advances in Neural Information Processing Systems, vol. 30 (2017)
27. Devlin, J., Chang, M.W., Lee, K., Toutanova, K.: BERT: pre-training of deep bidirectional transformers for language understanding. arXiv preprint arXiv:1810.04805 (2018)
28. Lewis, M., et al.: BART: denoising sequence-to-sequence pre-training for natural language generation, translation, and comprehension. arXiv preprint arXiv:1910.13461 (2019)
29. Raffel, C., et al.: Exploring the limits of transfer learning with a unified text-to-text transformer. J. Mach. Learn. Res. **21**(1), 5485–5551 (2020)

30. Xu, L., Chia, Y.K., Bing, L.: Learning span-level interactions for aspect sentiment triplet extraction. arXiv preprint arXiv:2107.12214 (2021)
31. Liu, P., et al.: Pre-train, prompt, and predict: a systematic survey of prompting methods in natural language processing. ACM Comput. Surv. **55**(9), 1–35 (2023)
32. Lu, Y., et al.: Unified structure generation for universal information extraction. arXiv preprint arXiv:2203.12277 (2022)
33. Loshchilov, I., Hutter, F.: Decoupled weight decay regularization. arXiv preprint arXiv:1711.05101 (2017)
34. Wolf, T., et al.: HuggingFace's transformers: state-of-the-art natural language processing. arXiv preprint arXiv:1910.03771 (2019)
35. Qiu, G., Liu, B., Bu, J., Chen, C.: Opinion word expansion and target extraction through double propagation. Comput. Linguist. **37**(1), 9–27 (2011)
36. Van der Maaten, L., Hinton, G.: Visualizing data using T-SNE. J. Mach. Learn. Res. **9**(11), 1–8 (2008)

SE-Prompt: Exploring Semantic Enhancement with Prompt Tuning for Relation Extraction

Cai Wang, Dongyang Li, and Xiaofeng He[✉]

School of Computer Science and Technology, East China Normal University, Shanghai, China
51255901085@stu.ecnu.edu.cn , hexf@cs.ecnu.edu.cn

Abstract. Compared to traditional supervised learning methods, utilizing prompt tuning for relation extraction tasks is a challenging endeavor in the real world. By inserting a template segment into the input, prompt tuning has proven effective for certain classification tasks. However, applying prompt tuning to relation extraction tasks, which involve mapping multiple words to a single label, poses challenges due to difficulties in precisely defining a template and mapping labels to the appropriate words. Prior approaches do not take full advantage of entities and have also overlooked the semantic connections between words in relation label. To address these limitations, we propose a semantic enhancement with prompt (SE-Prompt) which integrates entity and relation knowledge by incorporating two main contributions: semantic enhancement and subject-object relation refinement. These methods empower our model to effectively leverage relation labels and tap into the knowledge contained in pre-trained models. Our experiments on three datasets, under both fully supervised and low-resource settings demonstrate the effectiveness of our approach for relation extraction.

Keywords: Relation Extraction · Prompt Tuning · Semantic Enhancement · Entity Expansion

1 Introduction

The main goal of the relation extraction (RE) task is to predict the relations between a subject and an object given the input sentence and the subject and object present of the sentence [29]. Currently, relation extraction plays a key role in the field of natural language processing, such as information extraction and construction of knowledge graphs. In the past, numerous research scholars have devoted considerable effort to relation extraction [8,21,25].

Recently, various pre-trained models (PLMs) have made a splash in the field of natural language processing, such as: BERT, RoBERTa, GPT, etc. [4,15, 18,19]. Many state-of-the-art (SOTA) results have also been achieved by using

X. Yang et al. (Eds.): ADMA 2023, LNAI 14179, pp. 109–122, 2023.
https://doi.org/10.1007/978-3-031-46674-8_8

fine-tuning for relation extraction tasks. However, as shown in Fig. 1, the implementation of fine-tuning for various downstream tasks necessitates the inclusion of extra layers to the PLM, such as a linear classification layer. This approach may restrict the complete exploitation of the knowledge present in the PLM, owing to the disparity between the downstream tasks and the pre-training phase tasks. This problem has recently been alleviated due to the advent of prompt-tuning [20]. Taking Fig. 1 as an example, prompt-tuning transforms the original problem into a cloze-style task to predict the target words. The method consists of an original input and a template, and the goal is to predict words at [MASK] position, after which the results are mapped to the corresponding set.

However, for the task with a large number of labels such as relation extraction, manually mapping labels to labels words becomes very challenging and time-consuming. For example, for the labels "per:city_of_death" and "org:city_of_branch", it is difficult to accurately map them to the appropriate label words in the vocabulary. At the meantime, manually determining the appropriate template for the task requires expert knowledge. Although previous studies have been made to address the above situation: [22] proposed a method to automatically find templates and label words using gradient optimization, and [5] used cross-validation methods to obtain optimal templates and label words. However, the automatic prompt generation method requires a lot of time to search, while the performance of the generated prompt in most cases is not as good as those specified by human experts.

In view of the above discussion, we propose our approach: semantic enhancement with prompt tuning (SE-Prompt) which consists of two parts: entity expansion representation (EER) and relation expansion representation (RER). The EER injects the additional relevant word information of the entity and RER incorporates the semantic information of the relation label. We inject the EER into a template which is inspired by the process of human answering questions. Unlike fixed manual templates, our approach can accommodate to different types of relations. Then, we propose the relation refinement which exploits the semantic relationship between the subject-object and relation to further optimize entity and relation representations. Specifically, we insert into the template the expansion representations corresponding to the subject and object in the input text, which allows the model to learn the most relevant information from the context. It's worth noting that we do not map a relation label to a word in the vocabulary, but represent it with the knowledge of the relation triples contained in the label such as $[r_1], [r_2]$. We conduct experiments on three datasets: TACRED [30], TACREV [1], and ReTACRED [24]. The experimental results demonstrate the effectiveness and feasibility of our proposed method. The contributions of this paper can be summarized as follows:

- We propose a semantic enhancement model with prompt tuning (SE-Prompt) to perform relation extraction tasks in few-shot setting. Unlike previous works, SE-Prompt incorporates extended information about relation entities and does not rely on expert knowledge to construct template which is inspired by the human question and answer process.

- Based on the structured information contained in the relation label, we propose relation refinement to utilize the similarity between EER and RER to further optimize the representations.
- Under fully supervised and few-shot settings, we demonstrate that SE-prompt outperforms existing methods on three public datasets. The results show the superiority and robustness of SE-prompt.

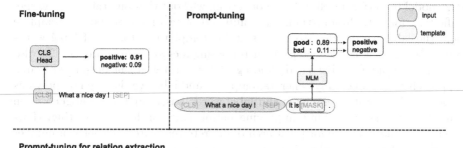

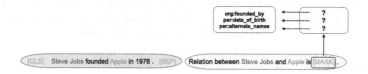

Fig. 1. The example of fine-tuning and prompt-tuning method.

2   Related Work

Relation extraction is an important task in the field of NLP, where the main goal is to predict the relation between entities given an input and two entities. Over the years, various approaches have been proposed for relation extraction including feature-based approaches [10], CNN and RNN-based approaches [28, 31], kernel-based approaches [17], and graph-based approaches [6]. Although many scholars have proposed various methods for RE in the past and achieved impressive results. However, due to the emergence of pre-trained models in recent years, a consensus in the NLP field is to use pre-trained models for fine-tuning in downstream tasks, such as BERT [4], GPT [18], RoBERTa [15], etc. Although PLMs have shown amazing performance in many areas, the gap that exists between the goals of downstream tasks and pre-trained goals cannot be ignored in reality. In other words, there is still a need for a data-specific fine-tuning process to serve downstream tasks. For example, BERT's pre-training phase tasks are masked language model (MLM) and next sentence prediction (NSP), but the downstream tasks are actually completely different from these two tasks, e.g. sentiment analysis, speech recognition. Also, a series of knowledge enhancement methods are proposed based on pre-trained models [9,16,23,27]. Specifically, these methods use external knowledge bases to enhance the PLM.

One of the major advances in prompt-tuning is the proposal of GPT3 [2], which is a major advance in the field of NLP. Prompt-tuning is the SOTA method after the pre-training model was proposed [13]. The main idea of the method is to convert the downstream task into a form that matches the pre-trained task. Specifically, the downstream task is converted into close-style task [20]. This approach enables the model to recall knowledge from the pre-training phase, allowing the model to perform well in few-shot setting without the constraints of the specific task format. Prompt-tuning is composed of two main parts: the template and the label words. How to get a suitable template and a set of label words for a specific task is a key step. After proposing a cloze-style approach, [20] also explored automatic label words finding. [22] used a gradient-optimized search approach to automatically generate templates and label words based on previous studies. [5] generates templates and label words based on T5 model generation and ranking methods, while improving performance on few-shot setting. However, these methods are for the search of discrete templates. Recently, a number of methods have emerged for continuous template generation [12,14]. Unlike discrete template, continuous template can be inserted into the input using a series of learnable embeddings, freeing them from the drawbacks of manually designed template. Unfortunately, these methods do not directly serve RE.

3   Background

Here we will give the relevant mathematical notation and some details of fine-tuning and prompt-tuning implementation. We define a relation extraction dataset as $D = \{X, Y\}$, where X is the sample set and Y is the relation label set. For each input $x \in X$, it can be represented as $x = \{t_1, t_2, ..., t_i, ..., t_{|x|}\}$, where t_i is the i-th word of x, $|x|$ is the length of x. Each input x has a label $y_x \in Y$.

3.1   Fine-Tuning for PLMs

Given a pre-trained model M, the fine-tuning method first converts the input x into a form that can be received by M, that is, $x^* = \{[CLS], x, [SEP]\}$. Then, M encodes x^* to obtain the model's output hidden vectors $\{h_{[CLS]}, h_x, h_{[SEP]}\}$, where h_x is a set of hidden vectors of x. For a specific downstream task, $h_{[CLS]}$ is usually used for the classification task. Specifically, a linear classification layer followed by a softmax function can be used to compute the probability distribution $p(\cdot|x) = Softmax(Wh_{[CLS]} + b)$ over the label set Y, where W is a randomly initialized matrix which can be optimized, b is the bias. The parameters of model M, W will be tuned to maximize $p(y_x|x)$.

3.2   Prompt-Tuning for PLMs

Prompt-tuning is proposed to better trigger the knowledge of pre-trained models. This method consists of two main parts: the template $T(\cdot)$ and the set of label words V. The key step is how to construct T and V. For each input x, the template is used to construct the corresponding prompt input $x_{prompt} = T(x)$. The template involves the place and number of custom words. It is worth noting that there is at least one [MASK] position in the template for M to predict the label words. Taking Fig. 1 as an example, with input $x =$ "*What a nice day !*", we use $T(\cdot)$ to generate $x_{prompt} =$ "*x It is* [*MASK*] *.*". After that we input x_{prompt} into M and we can get the result of [MASK], which is the hidden vector $h_{[MASK]}$. For each $v \in V$, we can calculate the probability of the distribution $p([MASK] = v | x_{prompt})$ of token v at the position of [MASK]. Meanwhile, $F : Y \rightarrow V$ is an mapping function that links the set of labels to the set of label words. Thus, using the mapping function F, we can formalize $p(y|x)$ with the probability distribution over V at the masked position, that is, $p(y|x) = p([MASK] = F(y)|x_{prompt})$. In Fig. 1, we can map the two labels: "positive, negative" to "good, bad" respectively. If the result predicted by M at the [MASK] position is "good", the prediction of sample x is "positive", otherwise it is "negative". Finally, the goal of prompt-tuning is to maximize the probability $p([MASK] = F(y_x)|T(x))$.

4   Methodology

As discussed before, previous approaches for prompt-tuning are not directly usable for RE. In this section, we propose our approach: as shown in Fig. 2, first we use the expansion representation to construct the template (Sect. 4.1 and Sect. 4.2), and then use the relation triples for relation refinement (Sect. 4.3).

4.1   Template Construction

The template construction is inspired by the human process of answering questions. When answering someone's question, it is necessary to understand the intention of the questioner. Therefore, we turn the template into a question-like pattern such as $S1 =$ "in one's opinion?". We also use phrases such as $S2 =$ "based on the text" to extend the template. As shown in Fig. 2, we construct the template as { "*Steve said,*",x, $\langle ent \rangle$, "*Alan*", $\langle /ent \rangle$, [*MASK*], $\langle ent \rangle$, "*KFC*", $\langle /ent \rangle$, $S1, S2$}, where x is the original input, $\langle ent \rangle$ and $\langle /ent \rangle$ denote an expansion representation that can be dynamically tuned by the model. We use Steve as the narrator, although other names can also be used. The [MASK] is the location where the model need to predict.

4.2   Semantic Enhancement

Entity Expansion Representation (EER). We can determine the scope of entities by utilizing the subject-object knowledge in the relation label. For example, given the label "*per : employer_of*", we can identify the two entities in the input sample as "*person*" and "*employer*". Also, we consider not only the entity itself, but also the associated words. Therefore, we inject the entity expansion

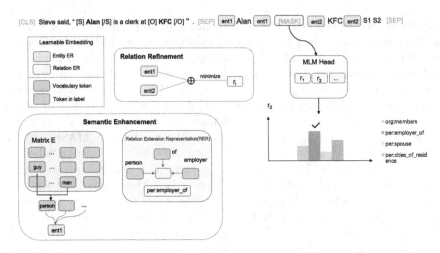

Fig. 2. The approach we propose consists of two main parts: Semantic Enhancement and Relation Refinement. The pink color represents the EER, which is capable of perceive the range of entities. The creamy yellow represents RER, which effectively utilizes the semantic knowledge within the label. To obtain a prediction result, the output from the [MASK] location is compared with RER. (Color figure online)

representation into the template around the entities. In this way, EER can aid the pre-trained model in better discerning the identities of the two entities in a given sentence. It is important to note that this expansion represent is an embedding, which allows the model optimize it. Specifically, we use Word2Vec [11] to get the expansion representation of the entities. Word2Vec is a method used to generate word embeddings. We use Word2Vec to find words related to entities to enrich the representation. For example, for the entity "person", the words associated with it are "guy", "man" and so on. The initialized EER is obtained by averaging the embedding of the first m related words and the entity itself. The initialized EER is given by:

$$ EER_{ent} = \frac{1}{1+m}\left(e_{ent} + \sum_{rel\in Rel} e_{rel}\right) \tag{1} $$

where e_{ent} is the embedding of the entity, e_{rel} is the embedding of the related words, Rel is the set of related words. Since the Word2Vec method derives words related to entities based on semantic similarity, the initial embedding of the expansion representation inserted into the template can be scope-aware and can be further optimized at a later stage.

Algorithm 1 Get m related words for entity in the label

Require:
1: Word2Vec embedding matrix E trained on the training set with vocabulary V.
2: Each Entity e_i for each relation label $y_i \in Y$ and an expansion number of entity m.
Ensure: Entity Expansion Representation set S.
3: **for** $y_i \in Y$ **do**
4:    **for** $e_i \in y_i$ **do**
5:       **if** e_i appears for the first time **then**
6:          $distance\_list = []$
7:          **for** $v \in V$ **do**
8:             $current\_dis = getDistance(v, e_i)$
9:             $distance\_list.add(current\_dis)$
10:         **end for**
11:         $relate\_m\_words = minDistance(dis\_list)$
12:         $S^i = relate\_m\_words$
13:       **end if**
14:    **end for**
15: **end for**
16: **return** S

The process of obtaining an EER is described in Algorithm 1. We first get an embedding matrix E using the training set. For each entity e_i in the label, we use E to obtain m words associated with it. Specifically, we use Euclidean distance to find the m nearest words to e_i, which shows that they are semantically related. We define the EER set $S = \{S^1, S^2, ..., S^i\}$, where S^i is the set of related words for the i-th entity.

Relation Expansion Representation (RER). Similar to EER, the initialization of RER is composed of the embedding of the words that make up the relation label. For example, the words that make up the label *"per : employer_of"* are *"person"*, *"employer"*, and *"of"*. Then, we use Word2Vec to get the related tokens that make up the label words. We take the average embedding of all related tokens as the initialization of RER. The initialization formula RER for label y_i is as follows:

$$RER_{y_i} = \frac{1}{n}\left(\sum e_m + \sum e_{rel}\right) \tag{2}$$

where n is the total number of words that make up the label and related tokens, e_m is the embedding of words that make up the label, e_{rel} is the embedding of the related tokens. In contrast to previous prompt-tuning methods that only map the label to a label word, our approach considers the relevance of words in the relation label and leverages semantic knowledge in the initialization process. Thus, the probability $p([MASK] = F(y_x)|T(x))$ can be reformalized as $p([MASK] = RER_{y_x}|T(x))$. Eventually, after getting the initialized RER, the model can be further optimized to get the best representation based on the context.

4.3   Relation Refinement

To obtain more precise representations, we compress the dimensions of both the EER and RER, as shown by following:

$$EER_{ent1} = \theta_{e1} \cdot h_{EER_{ent1}}$$
$$EER_{ent2} = \theta_{e2} \cdot h_{EER_{ent2}} \qquad (3)$$
$$RER_{rel} = \theta_{rel} \cdot h_{RER_{rel}}$$

where h is the hidden vector, θ_{e1}, θ_{e2} and $\theta_{rel} \in \mathbb{R}^{w \times d}$ are the parameters of linear layer, w is the dimension of the hidden vector after compression. Based on the above discussion, we obtain the initialized EER and RER. However, in order to further combine the subject, object and relationship in the relation triple, we propose relation refinement. Relation refinement reduce interference by allowing the model to prioritize its attention on the objects and the associations between relations. This, in turn, helps mitigate the impact of extraneous contextual information, thereby enhancing the robustness and accuracy of the model. For instance, consider the label *"per : employer_of"*, which includes the entities *"person"* and *"employer"*. We expect the EER of *"person"* and *"employer"* to closely align with the RER of the label. Therefore, we minimize the distance between them to strengthen the understanding of triples within the relation label. We calculate the similarity by using cosine similarity as it is appropriate for vector similarity calculation and eliminates the impact of length, focusing on direction. The loss function of relation refinement is defined as follows:

$$L_{RE} = -log[CosSim(EER_{ent1} + EER_{ent2}, RER_{rel})] \qquad (4)$$

where EER_{ent1}, EER_{ent2} corresponding to the expansion representation of two entities in the input sample respectively. RER_{rel} is the expansion representation of the label. The $CosSim(a, b)$ denotes the function that calculates the cosine similarity of the vectors a and b. It should be emphasized that Relation Refinement enables more effective exploration of implicit relations between entities, and allows for optimization of EER and RER to produce improved representations.

At the same time, the initializations of EER and RER we obtained are not optimal and need to be further optimized. For our experiment, we use the cross-entropy loss function. The difference with the general prompt tuning method is that instead of directly comparing the output at [MASK] position with the label, we compare it with the previously described RER to calculate the loss. According to the probability distribution $p(y|x) = p([MASK] = RER_{y_x}|T(x))$ over RER at the masked position, the loss function is shown as follows:

$$L_{[MASK]} = -\frac{1}{h} \sum y \, log \, p(y|x) \qquad (5)$$

where h is the number of samples in the dataset. Thus, the loss function of our method is

$$L = L_{[MASK]} + \alpha L_{RE} \qquad (6)$$

where α is a hyperparameter.

5 Experiment

5.1 Datasets and Experimental Settings

In our experiments, we used three datasets for evaluation: TACRED [30], TACREV [1], and ReTACRED [24]. The details of the datasets are shown in Table 1. In the relation extraction experiment, we use F1 value as the evaluation metric. We first give the definitions of precision(P) and recall(R), as follows:

$$P = \frac{TP}{TP + FP} \tag{7}$$

$$R = \frac{TP}{TP + FN} \tag{8}$$

where TP is the number of correctly identified objects, FP is the number of unrelated identified objects, and FN is the number of unidentified objects present in the dataset. The F1 is obtained by considering precision and recall together. The formula for F1 is as follows:

$$F1 = 2 * \left(\frac{P * R}{P + R} \right) \tag{9}$$

Table 1. Statistics of the datasets, where relation represents the number of labels.

Dataset	Train	Dev	Test	Relation
TACRED	68124	22631	15509	42
TACREV	68124	22631	15509	42
ReTACRED	58465	19584	13418	40

In order to compare our method with previous fine-tuning methods, we use RoBERTA_large as the pre-trained model in our relation extraction experiments. We also evaluate the effectiveness and feasibility of our method on complete data as well as on few-shot settings, respectively. We finetune the model on the full data to optimize the EER and RER proposed in our method. First we use RoBERTA_large for direct fine-tuning as our baseline for comparison. We also use MTB, SpanBERT, KnowBERT, LUKE as our baseline, which utilize external knowledge to enhance the pre-trained model. For the prompt-tuning method, we select two typical models. PTR employs logic rules to create prompts with multiple sub-prompts, enabling the encoding of prior knowledge for each class into prompt tuning. KnowPrompt inject latent knowledge contained in relation labels into prompt construction. For the few-shot setting, we randomly sampled 16 and 32 instances of each label from the training dataset to form the corresponding few-shot dataset.

5.2   Main Results

The results of the experiments conducted under full data conditions are shown in Table 2. It can be found that simply using RoBERTA_large for fine-tuning results in the worst performance, suggesting that simply using a pre-trained model for relation extraction does not effectively leverage the vast amount of knowledge learned. Also, the four models injected with external knowledge perform better than the model fine-tuned alone, which shows that using the knowledge base to enhance the models is effective. However, our proposed expansion representation and relation refinement method achieves improvements on almost all baselines. The experimental result illustrates that the approach of introducing external knowledge to enhance the model is not optimal, and our proposed approach is better able to stimulate knowledge of the pre-trained model as well as to exploit the semantic knowledge contained in the relation labels for our task.

Table 2. F1 scores (%) on the three datasets under Full Data setting. The best results are bold.

Model	TACRED	TACREV	ReTACRED
Fine-tuning methods			
RoBERT-LARGE [15]	68.7	76.0	84.9
MTB [23]	70.1	-	-
SPANBERT [9]	70.8	78.0	85.3
KNOWBERT [16]	71.5	79.3	89.1
LUKE [27]	90.3	80.6	72.7
Prompt tuning methods			
PTR [7]	72.4	81.4	90.9
KnowPrompt [3]	72.4	82.4	91.3
Ours	**73.2**	**82.5**	**91.4**

The results of our experiments under the few-shot settings are displayed in Table 3. It can be seen that our proposed method consistently outperforms the fine-tuning across all three datasets, which confirms the effectiveness of our method in the few-shot settings. Notably, when K = 16, our method exhibits the most significant improvement compared to the results of fine-tuning. As K increases from 16 to 32, the improvement of our method over fine-tuning gradually decreases. This makes sense, because as the data size increases, the model learns more from the larger datasets. Compared to the previous best prompt-based methods, SE-Prompt achieves an improvement of 0.3%–1.7% on both the ReTACRED and TACREV datasets.

Table 3. $F1$ scores (%) on the three datasets under few-shot setting. We compare the experimental results with the standard fine-tuning. The best and the second best results are in bold and underlined, respectively.

Method	TACRED		TACREV		ReTACRED	
	K = 16	K = 32	K = 16	K = 32	K = 16	K = 32
Fine-tuning	21.5	28.0	22.3	28.2	49.5	56.0
GDPNET [26]	22.5	28.8	23.8	29.1	50.0	56.5
PTR [7]	30.7	32.1	31.4	32.4	56.2	62.1
KnowPrompt [3]	**35.4**	**36.5**	<u>33.1</u>	<u>34.7</u>	<u>63.3</u>	<u>65.0</u>
Ours	<u>35.1</u>	<u>35.4</u>	**33.4**	**36.4**	**64.0**	**65.5**

Table 4. Ablation experiments performed on the dataset TACRED. We removed the EER, RER and RE separately to observe the effectiveness of our method.

	K=16	K=32	Full
-w/o EER	34.0(-1.1)	34.7(-0.7)	72.7(-0.5)
-w/o RER	31.5(-3.6)	33.8(-1.6)	72.3(-0.9)
-w/o RR	34.5(-0.6)	34.9(-0.5)	72.9(-0.3)
Ours	35.1	35.4	73.2

5.3 Ablation Study

To verify the effectiveness of our proposed semantic enhancement, we conduct ablation experiment on this, and the experimental results are shown in Table 4, where w/o EER denotes the experiment conducted using random initialized entity embedding to represent EER. It can be found that the score F1 decreases from 35.1% to 34.0% when K = 16, and from 73.2% to 72.7% under full data condition. This shows that the effect of EER gradually decreases as the size of the dataset increases. However, EER is still effective for the relation extraction task.

The ablation experiments performed for RER were similar to EER. For w/o RER, the score F1 decreases both under few-shot and full data setting, which is the same as in the ERE ablation experiment, indicating that both EER and RER are effective for the task. Surprisingly, the RER is more effective than the EER for the task. Specifically, when K = 16, the RER for score F1 improvement is 3.6%, which is higher than the EER for F1 improvement 1.1%. This suggests that RER can make better use of the structured information in relation label. On the other hand, the significance of the triple information in the relation label (RER) surpasses that of the individual entity (EER). The optimal outcomes are attained through the amalgamation of EER and RER.Please note that as per LNCS standard color tables will be printed in black/white and color will be displayed in online. Kindly modify the color specification in Table 4 amend if necessary.

Finally, we evaluated the relation refinement. Concretely, we removed it to observe the results (shown as -w/o RR in table). As we can see, although relation refinement does not improve the effect as much as sementic enhancement, it was still found to be feasible. This is because relation refinement helps to establish connections between entities and relations in the relation triple, thereby enhancing the overall understanding of the relation extraction task.

In summary, our results demonstrate that both EER and RER are important methods for relation extraction, with RER being the most effective, especially in few-shot settings. Our findings suggest that structured information in relation labels can be effectively leveraged for relation extraction.

5.4   Effect of the Number of Related Words

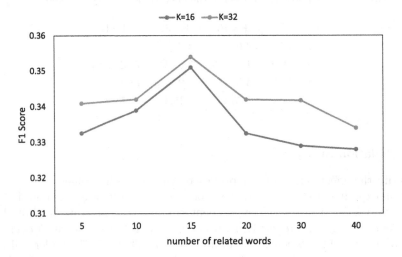

Fig. 3. Evaluating the effect of the number of related words on $F1$ value using the dataset TACRED.

We propose the EER in Sect. 4.2, which contains m: the number of words associated with the entity. The ablation experiments in Sect. 5.3 demonstrate the validity of EER. The EER inserted into the template is able to perceive the entities. We believe that a suitable number of m can well help initialize the EER and provide a good precondition for subsequent optimization. For this reason, we further explore the effect of m under the few-shot conditions of $K = 16$ and $K = 32$. The results are shown in Fig. 3. It can be found that the value of F1 first increases and then decreases in general, reaching a maximum around $m = 15$. This phenomenon shows that m is too large to perceive the scope of entity representation well, even if a large number of related words are incorporated. In any case, the change in the value of m has a non-negligible impact on the results and proves the great potential of EER. These findings indicate the great potential of EER, and we recommend further research on this approach.

6   Conclusion

In this paper, we propose the SE-Prompt, a novel approach to relation extraction that incorporates semantic enhancement and relation refinement. By explicitly incorporating EER and RER, the model can effectively leverage the semantic information present in entity and labels. An additional relation refinement method can further discover the implied relations in the labels. The contributions described above enable the effective utilization of structured information contained in relation labels, while reducing the need for manual template design. Our experiments on datasets demonstrate the effectiveness of our method, achieving significant improvements over the baseline without the need for additional knowledge, especially in few-shot settings. Moving forward, we aim to expand the application of this approach to other domains and explore its potential further.

References

1. Alt, C., Gabryszak, A., Hennig, L.: TACRED revisited: a thorough evaluation of the TACRED relation extraction task. arXiv preprint arXiv:2004.14855 (2020)
2. Brown, T., et al.: Language models are few-shot learners. Adv. Neural. Inf. Process. Syst. **33**, 1877–1901 (2020)
3. Chen, X., et al.: KnowPrompt: knowledge-aware prompt-tuning with synergistic optimization for relation extraction. In: Proceedings of the ACM Web Conference 2022, pp. 2778–2788 (2022)
4. Devlin, J., Chang, M.W., Lee, K., Toutanova, K.: BERT: pre-training of deep bidirectional transformers for language understanding. arXiv preprint arXiv:1810.04805 (2018)
5. Gao, T., Fisch, A., Chen, D.: Making pre-trained language models better few-shot learners. arXiv preprint arXiv:2012.15723 (2020)
6. Guo, Z., Zhang, Y., Lu, W.: Attention guided graph convolutional networks for relation extraction. arXiv preprint arXiv:1906.07510 (2019)
7. Han, X., Zhao, W., Ding, N., Liu, Z., Sun, M.: PTR: prompt tuning with rules for text classification. AI Open **3**, 182–192 (2022)
8. Harting, T., Mesbah, S., Lofi, C.: LOREM: language-consistent open relation extraction from unstructured text. In: Proceedings of the Web Conference 2020, pp. 1830–1838 (2020)
9. Joshi, M., Chen, D., Liu, Y., Weld, D.S., Zettlemoyer, L., Levy, O.: SpanBERT: improving pre-training by representing and predicting spans. Trans. Assoc. Comput. Linguist. **8**, 64–77 (2020)
10. Kambhatla, N.: Combining lexical, syntactic, and semantic features with maximum entropy models for information extraction. In: Proceedings of the ACL Interactive Poster and Demonstration Sessions, pp. 178–181 (2004)
11. Le, Q., Mikolov, T.: Distributed representations of sentences and documents. In: International Conference on Machine Learning, pp. 1188–1196. PMLR (2014)
12. Li, X.L., Liang, P.: Prefix-tuning: optimizing continuous prompts for generation. arXiv preprint arXiv:2101.00190 (2021)
13. Liu, P., Yuan, W., Fu, J., Jiang, Z., Hayashi, H., Neubig, G.: Pre-train, prompt, and predict: a systematic survey of prompting methods in natural language processing. ACM Comput. Surv. **55**(9), 1–35 (2023)

14. Liu, X., et al.: GPT understands, too. arXiv preprint arXiv:2103.10385 (2021)
15. Liu, Y., et al.: RoBERTa: a robustly optimized BERT pretraining approach. arXiv preprint arXiv:1907.11692 (2019)
16. Peters, M.E., et al.: Knowledge enhanced contextual word representations. arXiv preprint arXiv:1909.04164 (2019)
17. Qian, L., Zhou, G., Kong, F., Zhu, Q., Qian, P.: Exploiting constituent dependencies for tree kernel-based semantic relation extraction. In: Proceedings of the 22nd International Conference on Computational Linguistics (Coling 2008), pp. 697–704 (2008)
18. Radford, A., Narasimhan, K., Salimans, T., Sutskever, I., et al.: Improving language understanding by generative pre-training (2018)
19. Raffel, C., et al.: Exploring the limits of transfer learning with a unified text-to-text transformer. J. Mach. Learn. Res. **21**(1), 5485–5551 (2020)
20. Schick, T., Schütze, H.: Exploiting cloze questions for few shot text classification and natural language inference. arXiv preprint arXiv:2001.07676 (2020)
21. Shen, Y., Ma, X., Tang, Y., Lu, W.: A trigger-sense memory flow framework for joint entity and relation extraction. In: Proceedings of the Web Conference 2021, pp. 1704–1715 (2021)
22. Shin, T., Razeghi, Y., Logan IV, R.L., Wallace, E., Singh, S.: AutoPrompt: eliciting knowledge from language models with automatically generated prompts. arXiv preprint arXiv:2010.15980 (2020)
23. Soares, L.B., FitzGerald, N., Ling, J., Kwiatkowski, T.: Matching the blanks: distributional similarity for relation learning. arXiv preprint arXiv:1906.03158 (2019)
24. Stoica, G., Platanios, E.A., Póczos, B.: Re-TACRED: addressing shortcomings of the TACRED dataset. In: Proceedings of the AAAI Conference on Artificial Intelligence, vol. 35, pp. 13843–13850 (2021)
25. Wu, T., et al.: Curriculum-meta learning for order-robust continual relation extraction. In: Proceedings of the AAAI Conference on Artificial Intelligence, vol. 35, pp. 10363–10369 (2021)
26. Xue, F., Sun, A., Zhang, H., Chng, E.S.: GDPNet: refining latent multi-view graph for relation extraction. In: Proceedings of the AAAI Conference on Artificial Intelligence, vol. 35, pp. 14194–14202 (2021)
27. Yamada, I., Asai, A., Shindo, H., Takeda, H., Matsumoto, Y.: LUKE: deep contextualized entity representations with entity-aware self-attention. arXiv preprint arXiv:2010.01057 (2020)
28. Zeng, D., Liu, K., Chen, Y., Zhao, J.: Distant supervision for relation extraction via piecewise convolutional neural networks. In: Proceedings of the 2015 Conference on Empirical Methods in Natural Language Processing, pp. 1753–1762 (2015)
29. Zeng, D., Liu, K., Lai, S., Zhou, G., Zhao, J.: Relation classification via convolutional deep neural network. In: Proceedings of COLING 2014, the 25th International Conference on Computational Linguistics: Technical Papers, pp. 2335–2344 (2014)
30. Zhang, Y., Zhong, V., Chen, D., Angeli, G., Manning, C.D.: Position-aware attention and supervised data improve slot filling. In: Conference on Empirical Methods in Natural Language Processing (2017)
31. Zhou, P., et al.: Attention-based bidirectional long short-term memory networks for relation classification. In: Proceedings of the 54th Annual Meeting of the Association for Computational Linguistics (volume 2: Short papers), pp. 207–212 (2016)

Self-supervised Multi-view Clustering Framework with Graph Filtering and Contrast Fusion

Yongchao Lu[1], Bing Kong[1(✉)], Guowang Du[2], Congming Bao[3], Lihua Zhou[1], and Hongmei Chen[1]

[1] School of Information Science and Engineering, Yunnan University, Kunming 650504, China
`12021215117@mail.ynu.edu.cn`, {`kongbing,lhzhou,hmchen`}`@ynu.edu.cn`
[2] School of Physics and Astronomy, Yunnan University, Kunming 650091, China
[3] School of Software, Yunnan University, Kunming 650504, China
`chmbao@ynu.edu.cn`

Abstract. With the increasing prevalence of multi-view data in practical applications, multi-view clustering has become popular due to its ability to integrate complementary information from multiple views to enhance clustering accuracy and robustness. However, existing multi-view clustering methods easily suffer from the following limitations: 1) Most existing methods assume that each sample can be well represented in the original data space, but real-world data inevitably contains redundancy and noise; 2) Existing comparative methods have sampling bias when selecting negative samples; 3) The pseudo-labels generated by model iteration are underused. To address these challenges, this paper proposes a self-supervised multi-view clustering framework with graph filtering and contrast fusion named SMCGC. Specifically, SMCGC first eliminates redundancies and noise in the original data through graph filtering and then uses contrast fusion to enhance the discriminative ability of the samples. This approach avoids the potential challenges related to negative sample selection and does not require the construction of positive and negative samples. For pseudo-labels, the framework utilizes self-supervised mechanisms to guide model operation and optimize cluster representation. Extensive experimental results on six benchmark datasets demonstrate the effectiveness of the proposed SMCGC against the existing state-of-the-art methods.

Keywords: Multi-view clustering · Contrast fusion · Self-supervised learning

1 Introduction

Multi-view data [3], which refers to data represented by multiple views, has become increasingly prevalent in various fields. For example, news articles can be represented by text and image content, videos can be represented by visual

X. Yang et al. (Eds.): ADMA 2023, LNAI 14179, pp. 123–138, 2023.
https://doi.org/10.1007/978-3-031-46674-8_9

and audio content, and faces can be represented by frontal and profile views. The availability of multi-view data demands the development of multi-view learning methods, which consider multiple views of the data to achieve better performance. In contrast to single-view learning methods, such as *k-means* [10], which are based solely on a single representation of the data, it is difficult to handle complex data with multiple aspects using only one representation. On the other hand, multi-view learning methods can not only capture multiple aspects of the data to provide a more comprehensive understanding but also leverage complementary information from multiple views to improve learning performance. However, multi-view data also poses new challenges for knowledge discovery in data mining, such as data integration, feature selection, consistency, complementary learning, and view importance evaluation [21].

Multi-view clustering (MVC) is a technique that uses multiple views of data from different data sources or different feature extraction methods of the same data source for clustering. By clustering multiple views, we can more accurately describe the structure and characteristics of the data and capture more information about the data. Existing multi-view clustering methods can be roughly divided into three categories: subspace-based, graph-based, and deep learning-based. The core idea of subspace-based algorithms [5,16,26] is to find a shared representation space for multiple views while preserving the distribution information unique to each view as much as possible. Graph-based algorithms [17,25] aim to find a fusion graph that is shared by all views and can characterize the relationships among all views. Clustering results can be obtained by using graph-cut algorithms [2] or other spectral techniques on the fusion graph. Deep learning-based algorithms [7,8] use deep learning's powerful nonlinear fitting ability to perform deep feature learning from large-scale data, which can express more complex objective functions and improve the performance of learning tasks such as clustering. Although the above algorithms have achieved good performance, there are two problems that need to be solved for existing multi-view clustering methods:

(1) How to reduce noise in raw data and effectively integrate sample features and structural information? Most existing methods assume that each sample can learn a good representation through the original data space, which may result in flawed representations because real-world data inevitably contains redundancy and noise. Therefore, it is difficult to get a good representation to support subsequent clustering tasks. In addition, some existing MVC methods ignore the structural information of samples. In reality, samples are usually not isolated, and there may be correlations between different samples. Structural information usually reflects the potential similarity between data samples, including direct relationships (i.e., first-order structure) and higher-order structures. For example, in the case of second-order structure, even if two samples do not have a direct relationship, they should have similar representations if they have many common neighboring samples. Therefore, fusing neighboring samples to enrich their attribute information is helpful for clustering because a sample usually has a stronger correlation with neighboring samples than with other samples.

(2) How can we effectively supervise the operation of a model during an unsupervised clustering process? As the clustering process is unsupervised, the true labels of samples in the clusters are unknown, and only pseudo-labels are provided during each iteration. While these labels may not be entirely accurate, they still provide some valuable information. By leveraging the data provided by these labels, the model can be directed toward the correct path during the learning process. Therefore it is necessary to use the information conveyed by pseudo-labels.

To address these issues, we propose a graph filtering-based multi-view clustering framework SMCGC, which effectively integrates topological information and sample features, reduces high-frequency noise in the original features, and optimizes clustering representation in a completely unsupervised manner. Specifically, we first construct k-nearest neighbor graphs based on sample features from different datasets to reveal the underlying data structure of each view. Then, we use graph filtering technology to encode structural information into features. The feature transformation in this method can fully utilize the structural information of the data and effectively enhance the feature representation by reducing high-frequency noise. In addition, in order to improve the discrimination of samples, we reduce the redundancy between samples in the potential space by approximating the sample correlation matrix of cross-view to the identity matrix. This effectively enhances the discernment of the underlying space and alleviates the phenomenon of collapse, that is, mapping samples of different classes to similar representations. We also integrate representation learning and clustering tasks into a unified step that can be optimized together to directly obtain clustering pseudo labels without the need for additional post-processing steps. To further optimize the clustering representation, we use clustering pseudo labels to perform reverse supervised training on the model. In this way, the learned low-dimensional representations can be applied to subsequent clustering tasks to fully explore the correlation between representation learning and clustering tasks. The specific contributions of this paper are summarized as follows:

1) A multi-view clustering model (SMCGC) is proposed, which can effectively utilize sample features and topological information while eliminating data redundancy and noise through graph filtering. Additionally, self-supervised training and clustering pseudo-labels can be used for iterative optimization of clustering representations, making the generated low-dimensional representation more cluster-oriented.

2) To enhance the discriminative power of samples and mitigate sampling bias caused by selecting negative samples, a comparison fusion method was employed, which does not require the use of negative samples or the introduction of any asymmetric information in the network architecture. By making the cross-view sample correlation matrix approach the identity matrix, redundant information between vectors in the latent space is reduced, while the intrinsic attribute information of the samples was better preserved.

3) Extensive experiments are conducted on six real-world datasets, and the results demonstrate that the clustering performance of the SMCGC algorithm outperforms other state-of-the-art algorithms.

2   Related Work

2.1   Notation

Let $\chi = \{X^1, X^2, ..., X^V\}$ denote the corresponding multi-view dataset, where V denotes the number of views. The data in the v-th view is represented by the matrix $X^v \in R^{N \times d^v}$ where d^v represents the feature dimension of the v-th view, and $X^v = \{x_1^v, x_2^v, ..., x_N^v\}^T$ is a set of sample feature vectors of N sample in the v-th view.

2.2   Graph Filtering

For a given $X \in R^{N \times d}$ with N samples and d features, which can be viewed as d N-dimensional graph signals. From the underlying graph, we can know that the natural signal should exhibit similarity in close proximity to its neighboring nodes. Specifically, this means that the eigenvalues of nearby nodes are likely to be similar to each other. From an alternative viewpoint, smooth signals contain a greater proportion of low-frequency base signals as compared to high-frequency base signals. Typically, high-frequency components are categorized as noise. Consequently, smooth signals ought to be free of noise, thereby contributing to downstream analysis. Therefore, adjacent nodes can be assigned similar values through graph filtering to reduce their frequencies and make the signal smoother. In order to obtain a high-quality signal $\bar{X}$, the high-frequency components can be filtered out by a filter, while the low-frequency components are retained. This can be solved by the following formula:

$$\min_{\bar{X}} \|\bar{X} - X\|_F^2 + \mu \mathrm{Tr}\left(\bar{X}^\top L \bar{X}\right) \tag{1}$$

where $\mu > 0$ is a balance parameter. The first term is a fidelity term and the second term of the formula is graph Laplace regularization which meaning that if the samples i and j are similar in the original space, then $\overline{x}_i$ and $\overline{x}_j$ should be similar as well. By taking the derivative of Eq. (1) and set it to zero, we get $\bar{X} = (I + \mu * L)^{-1} X$. To avoid matrix transpose $\left(o\left(n^3\right)\right.$ in time complexity), we approximate it by expanding the first-order Taylor formula for $\bar{X}$, then, we get $\bar{X} = (I - \mu * L)X$, and t-order filtering we can write as

$$\bar{X} = (I - \mu * L)^t X \tag{2}$$

t is a non-negative integer. Through graph filtering, we can keep the geometric features of the graph and filter out the bad high-frequency noise, so as to obtain a smoother representation.

2.3   Multi-view Clustering

Compared to traditional methods that rely on shallow linear embedding functions to reveal the internal structure of data, there are issues such as high time complexity. Deep learning has received extensive attention due to its excellent performance in capturing complex nonlinear relationships in data, such as DCCAE [22] based on canonical correlation analysis (CCA) [9], which extracts nonlinear features from each view through two networks and top-level CCA to maximize the correlation between features from different views. In addition, given that the topological information of data can provide implicit relationships between samples, such as correlations between samples, CMEGC [23] adopts a multi-graph attention fusion encoder to encode multi-view structural information and introduces a multi-view mutual information maximization module to maintain the similarity of neighboring features in each view. Our proposed algorithm SMCGC combines structural graph information with the contrast method, and related algorithms to SMCGC include SMC [5], SMVSC [18], and CONAN [14]. Both SMCGC and CONAN algorithms use contrastive learning, but CONAN suffers from sampling bias and does not take structural information into account. The SMC and SMVSC algorithms use graph structural information, but they do not integrate representation learning and clustering in a unified framework. SMCGC simultaneously combines graph structural information with self-supervised training and does not require consideration of negative samples, effectively improving clustering accuracy.

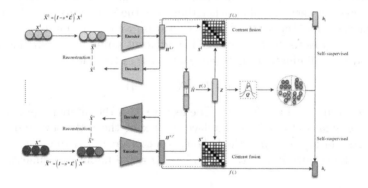

Fig. 1. SMCGC model architecture diagram

3   The Framework of SMCGC Network

In this section, we provide a detailed description of our proposed method, SMCGC, which overall framework is presented in Fig. 1. The framework comprises three main modules: the graph filtering module, the contrastive fusion module, and the self-supervised module. The first module utilizes graph filtering techniques to combine sample and neighbor information while eliminating high-frequency noise. The second module focuses on enhancing sample discrimination. Finally, the self-supervised module integrates representation learning and clustering tasks into a single step, enabling them to work in tandem and reinforce each other.

3.1   Graph Filtering Module

Construction of K-Nearest Neighbor Graph (knn). The topological structure of data can reveal the hidden relationship between data points, such as the similarity between samples. For this purpose, the topological structure on the scatter data sample is captured by the knn. Here, we first calculate the similarity matrices $M_{i,j}^v$ of all views through a certain distance measure (For the text data set, the inner product method is used. For the image or video data set the Gaussian kernel method is used) where $M_{i,j}^v$ denotes the similarity between samples x_i^v and x_j^v.

After computing the similarity matrices $\{M_{i,j}^v\}_{v=1}^V$ for all views, the top k most similar points for each sample are selected as its neighbors in each view, then put edges between the one and its neighbors to construct an undirected knn graph, and finally we can obtain V adjacency matrices $\{A^v\}_{v=1}^V$.

Graph Filtering. In this step, we employ graph filtering to fuse two types of information: sample features and knn graphs. This procedure generates an aggregated representation for each sample that summarizes the nearest neighbors centered on each sample, rather than just the sample itself. Additionally, The Laplacian smoothing filter enhances feature representation by aggregating neighbors' features to the t order, making adjacent samples have similar feature values and thus eliminating high-frequency noise in the data.

$$\bar{X}^v = (I - \mu * L^v)^t X^v \quad v \in [1, 2, \dots V] \tag{3}$$

As shown in the above Eq. (3), The feature $\bar{X}$ is the filtered representation, X is the original data, L^v the Laplace matrix of the v-th view and t is the desired order of filtering, μ is the balancing factor greater than 0. Based on the theoretical knowledge in ADAPT [6], we set $\mu = 0.6$ in our experiment.

3.2   Contrastive Fusion Module

Multi-view Private Information Learning. For depth clustering, it is crucial to learn an effective data expression method, for the sake of generality, we employ the basic auto-encoder to learn the representations of the raw data in order to accommodate different kinds of data characteristics.

We set $2 * R-1$ layers for each auto-encoder, and the R represents the number of layers. Specifically, the expression learned from the r-th layer $H^{v,r}$ of the v encoder can be obtained as follows:

$$H^{v,r} = \phi(H^{v,r-1}W^{v,r} + b^{v,r}) \tag{4}$$

where ϕ is the activation function of the fully connected layers such as the Relu or Sigmoid function. $W^{v,r}$ and $b^{v,r}$ are the weights and bias of the layer r of the automatic encoder for the v-th view. Then, we use $\bar{X}^v$ as the raw input for the v-th auto-encoder. The structure of the decoder is similar to that of the encoder,

$$H^{v,r+1} = \phi(H^{v,r}W^{v,r+1} + b^{v,r+1}) \tag{5}$$

where $W^{v,r+1}$ and $b^{v,r+1}$ are the weights and bias of the layer $r+1$ of the automatic decoder for the v-th view. The output of the decoder part is the reconstruction of the raw data $\tilde{X}^v = H^{v,2*r-1}$. The loss function of this part is as follows:

$$\mathcal{L}_{res} = \sum_{j=1}^{v} \frac{1}{2N} \| \bar{X}^j - \tilde{X}^j \|_F^2 \tag{6}$$

Fusion Reduction. To avoid the challenge of taking a negative sample, no negative sample-free loss function is used here to improve the discriminant ability of samples. To facilitate comparison, an intermediate variable is introduced to align each view. First, $\overrightarrow{H} = \mathrm{Cat}(H^{1,r}, H^{2,r}, ...H^{v,r})$ is obtained by concatenating the low-dimensional representation of each view and integrating the complementary information of multiple views, and then dimensionally shrink $\overrightarrow{H}$ through the full connection layer $p()$ to obtain intermediate variable $Z = p(\overrightarrow{H})$.

$$S_{ij}^v = \frac{(H^{v,r})_i (Z_j)^{\mathrm{T}}}{\|(H^{v,r})_i\| \|Z_j\|}, \forall i, j \in [1, N] \tag{7}$$

$$\mathcal{L}_{contrastive} = \sum_{v=1}^{V} \frac{1}{N^2} \sum (S^v - I)^2$$

$$= \sum_{v=1}^{V} \left(\frac{1}{N} \sum_{i=1}^{N} (S_{ii}^v - 1)^2 + \frac{1}{N^2 - N} \sum_{i=1}^{N} \sum_{j \neq i} (S_{ij}^v)^2 \right) \tag{8}$$

$(H^{v,r})_i$, Z_j respectively represent the i-th sample from the v-th view and the j-th sample of Z. S^v is calculated by cosine similarity, the first term in Eq. (8) ensures that the diagonal elements of S^v are equal to 1, preserving the augmentation invariance of the embedding vectors. The second term sets the non-diagonal elements of S^v to 0, allowing for the identification of the uncorrelated components of the embedding vectors, and reducing the redundancy among output units. As a result, the output units contain non-redundant information about the samples. This decorrelation operation enhances the discriminative power of the learned embeddings, reduces the redundancy in the information, and improves the discriminative power of the samples.

3.3 Self-supervised Module

Since the clustering process is unsupervised, the ground truth labels are not available during training. To alleviate this issue, inspired by Deep Embedded Clustering [24], we introduce a clustering embedding layer to integrate representation learning and clustering and perform joint training to mutually benefit each other. For the i-th sample and the j-th cluster, we use Student's t-distribution [19] as the kernel to measure the similarity between the data representation z_i and the cluster centroid vector λ_j. Here, z_i denotes the i-th row of Z, and λ_j is obtained by initializing k-$means$ with the representations learned by pre-training autoencoder. The formula for the clustering distribution is shown as follows:

$$q_{ij} = \frac{\left(1 + \|z_i - \lambda_j\|^2\right)^{-1}}{\sum_{j'} \left(1 + \|z_i - \lambda_{j'}\|^2\right)^{-1}} \tag{9}$$

q_{ij} can be considered as the probability of allocating sample i to cluster j. To further enhance clustering efficiency, the target distribution P can be used to supervise $Q = [q_{ij}]$ in order to improve clustering coherence and optimize data representation. The target distribution P is formulated as follows:

$$p_{ij} = \frac{q_{ij}^2/f_j}{\sum_{j'} q_{ij'}^2/f_{j'}} \tag{10}$$

The frequency of clustering, denoted as $f_j = \sum_i q_{ij}$, is squared normalized for each cluster in both the target distribution P and the estimated distribution Q, which enhances the reliability of the values in the distributions. The clustering embedding layer can be regarded as a self-supervised training module, which calculates the target distribution P from the estimated distribution Q and uses it to supervise the updating of Q. By minimizing the $Kullback - Leibler$ divergence (KL) [15] divergence loss between the target distribution P and the estimated distribution Q, the clustering module can learn better representations for clustering tasks.

$$\mathcal{L}_{clu} = KL(P\|Q) = \sum_i \sum_j p_{ij} log \frac{p_{ij}}{q_{ij}} \tag{11}$$

After the model has stabilized, the soft assignment from the clustering distribution Q is chosen as the final clustering result. The label assigned to sample i is:

$$y_i = \arg\max_j Q_{ij} \tag{12}$$

To further optimize the clustering representation, the reverse supervision of the clustering distribution Q is employed to operate on each view.

$$\mathcal{L}_{s-supervised} = -\sum_v^V \frac{1}{N} \sum_i \sum_c^g y_{i,c} log(h_v)_{i,c} \tag{13}$$

The low-dimensional representations $H^{v,r} \in R^{n \times m}$ learned from each view are first projected get $h_v \in R^{n \times g}$ through a fully connected layer $f(.)$, where g represents the number of clusters. The softmax function is applied to h_v to normalize it into a probability distribution. Here, $y_{i,c}$ represents 1 if the prediction category of sample x_i is c, otherwise 0. And $(h_v)_{i,c}$ represents the output probability of sample x_i in class c in the v-th view. Through the reverse supervision of pseudo-labels, each view can learn more valuable representations and better optimize the low-dimensional representation.

3.4 The Over Cost Function

The overall loss function of our proposed SMCGC is:

$$\mathcal{L} = \mathcal{L}_{clu} + \alpha \mathcal{L}_{s-supervised} + \beta \mathcal{L}_{contrastive} + \theta \mathcal{L}_{res} \qquad (14)$$

where α, β, θ are all parameters and are all greater than 0.

4 Experiments

4.1 Datasets

We evaluated the effectiveness of the proposed algorithm on six real datasets, and the corresponding statistical information is shown in Table 1.

Table 1. Dataset description.

View	BBCSport[1]	VOC [12]	Caltech101-20[2]/7[2]/all[2]	CCV[3]
1	View1(3183)	Gist(599)	Gabor(48)	Stip(5000)
2	View2(3203)	Word Frequency(319)	Wavelet Moments (40)	Sift(5000)
3			Centrist (254)	Mfcc(4000)
4			Hog (1984)	
5			Gist (512)	
6			Lbp (928)	
Feature	Text	Image	Image	Video
Date samples	544	5649	2386/1474/9144	6773
Cluster number	5	20	20/7/101	20

[1]http://mlg.ucd.ie/datasets/segment.html.
[2]http://www.vision.caltech.edu/ImageData sets/Caltech101.
[3]http://www.ee.columbia.edu/dvmm/CCV.

Table 2. Ablation comparisons on six datasets.

Dataset	Metrics	Baseline	Baseline-S	Baseline-C	Baseline-P-C
BBCSport	ACC	79.7	91.7	95.7	98.2
	NMI	80.3	86.7	91.7	92.7
VOC	ACC	62.6	63.4	66.7	67.1
	NMI	60.6	63.7	65.9	67.2
Caltech101-20	ACC	55.2	56.1	56.7	61.7
	NMI	52.2	54.3	54.9	57.0
Caltech101-7	ACC	62.4	76.0	77.8	83.9
	NMI	59.9	62.3	64.5	65.6
CCV	ACC	11.2	14.8	18.3	21.1
	NMI	13.4	13.9	14.8	16.0
Caltech101-all	ACC	17.4	23.4	27.6	28.3
	NMI	27.4	34.6	38.8	40.5

4.2 Experiment Setup

Evaluation Metrics. The clustering performance is measured using two standard evaluation matrices. Clustering Accuracy (ACC) and Normalized Mutual Information (NMI), where a higher value indicates better performance.

Implementation Details. We implemented SMCGC on PyTorch running on CentOS Linux 7 with an Nvidia 3090 GPU and 24 GB memory. Adam optimization was used to train the model, with the encoder size set at 2000-1000-500 and the decoder size at 500-1000-2000 (opposite of the encoder). For VOC, we set the learning rate to le-3, the dimension m of the low-order representation learned by the encoder is 256, and the hyperparameter is set to {1, 0.1, 1}. For Caltech101-7, we set m to 128 and the hyperparameter is {0.1, 0.1, 0.1}, the learning rate to le-2. For Caltech101-20, the hyperparameter is {1, 0.1, 0.1}, m is 128 and the learning rate is set to le-2. The m of CCV is 40, the hyperparameter is {1, 0.1, 1} and the learning rate is le-2, For Caltech101-all, the hyperparameter is {0.01, 10, 0.001}, m is 128 and the learning rate is set to le-2. For BBCSport, the hyperparameter is {0.001, 1, 0.001}, m is 128 and the learning rate is set to le-2, we trained SMCGC 30 times and reported the operation results of the average results. The comparison algorithms were downloaded from the author's home page or GitHub, and the experimental results were averaged through multiple experiments with the parameters suggested in the paper.

Comparison Algorithms. We choose spectral clustering and eleven multiview clustering algorithms as baselines. Specifically, Spectral clustering [20] is one of the most classic traditional clustering methods. Two representative contrastive methods, i.e., CONAN [14], CMEGA [23]. Three representative subspace methods, i.e., RMSL [16], LMVSC [13], MLRSSC [4], learn the relationships between samples by building a self-representation layer. Representative deep clustering methods, i.e., DCCA [1], DCCAE [9], AE2-NETS [27], learn the potential embedment by the deep way and implement clustering algorithms. SMC [18] and SMVSC [5] are two typical graph filter mtehods and AMGL [11] is a graph fusion method.

4.3   Performance Comparison

Table 3 reports the clustering performance of all compared methods on six benchmarks. Since DCCA and DCCAE can only deal with two views, we choose the best two views on all datasets according to their performance. From these results, we can conclude that 1) The SMCGC algorithm is always due to most baseline algorithms. Taking the results on VOC as an example, our SMCGC outperforms SMC by 1.8%, 2.3%, SMVSC by 2.9%, 4.1% on ACC and NMI indicators. SMC and SMVSC algorithms use graph filtering technology to effectively increase feature representation, but their training and clustering are separate from clustering. SMCGC introduces a cluster embedding layer to integrate representation learning and clustering together for collaborative training and mutual promotion, and uses clustering pseudo-tags to reverse supervise model training and guide the model to learn better low-dimensional representation; 2) It can be observed that the comparison-based clustering method CONAN and CMEGC has a slightly poor clustering effect because these methods do not consider the

Table 3. The clustering performance of different multi-view methods on 6 datasets, with bold representing the best value.

Dataset	BBCSPort		VOC		Caltech101-20		Caltech101-7		CCV		Caltech101-all	
Metrics	ACC	NMI	ACC	NMI	ACC	NMI	ACC	NMI	ACC	NMI	ACC	NMI
SC-best	96.6	90.6	43.2	33.4	53.9	55.7	64.9	54.7	9.3	7.4	11.6	20.7
DCCA(2013)	77.2	61.9	39.7	12.5	41.8	59.1	56.7	57.6	20.7	15.9	12.8	31.2
DCCAE(2015)	72.9	54.5	41.6	44.3	44.0	59.1	62.1	64.3	16.1	11.8	15.2	33.7
AMGL(2016)	97.0	89.8	56.9	63.0	30.1	40.5	45.1	42.4	15.5	10.9	17.9	32.2
MLRSSC(2018)	85.1	72.5	58.4	55.5	37.8	51.4	46.3	47.1	18.6	14.9	21.7	30.2
AE2-NETS(2019)	70.3	82.8	57.1	57.1	49.1	**65.3**	66.4	60.6	17.8	14.1	19.4	28.1
RMSL(2019)	97.6	91.7	47.7	44.8	52.5	53.5	63.9	47.4	**21.5**	15.7	23.6	29.8
LMVSC(2020)	73.9	59.0	50.4	53.0	53.0	52.7	72.6	51.9	19.4	15.3	26.4	33.9
CONAN(2021)	76.5	68.6	62.1	62.1	57.9	54.2	67.7	46.4	14.4	11.0	20.8	24.3
CMEGC(2021)	97.6	92.3	60.5	57.2	49.4	58.8	54.6	60.8	18.6	13.9	20.6	33.7
SMVSC(2021)	75.1	53.9	64.2	63.1	59.9	55.9	76.7	56.0	15.1	13.9	25.7	36.2
SMC(2022)	88.6	77.8	65.3	64.9	59.1	55.5	80.6	60.2	17.4	14.5	24.5	38.9
SMCGC	**98.2**	**92.7**	**67.1**	**67.2**	**61.7**	57.0	**83.9**	**65.6**	21.1	**16.0**	**28.3**	**40.5**

existence of pseudo-negative samples, which will cause sampling bias in the selection of negative samples and thus affect the clustering accuracy. The SMCGC algorithm does not need to consider the existence of negative samples and avoids the challenge of selecting negative samples and effectively improves the discriminative ability of samples by using a fusion reduction method; 3) Compared with subspace-based clustering methods, including MLRSSC, RMSL, LMVSC, and deep autoencoder method, AE2-NETS. All of them have strong learning ability of clustering representation for data without topology structure. However, these methods which rely only on attribute information cannot cluster effectively. Although AMGL uses graph fusion technology, the clustering quality largely depends on the quality of the constructed graph. Figure 5 demonstrates that our algorithm is not sensitive to the quality of the initial constructed graph; 4) since spectral clustering is directly performed on raw attributes, thus achieving unpromising results. Overall, the above observations have demonstrated the effectiveness of our proposed approach in addressing the limitations of the existing multi-view approach.

4.4 Hyperparametric Sensitivity and Ablation

Hyperparametric Sensitivity. To verify the sensitivity of the model to hyperparameters, the robustness of the model to each hyperparameter is verified. It can be seen from Fig. 4 that the θ parameter is relatively stable in all six data sets. In Fig. 2, since α represents the strength of self-supervision, too much α will lead to too much supervision. When α is equal to 0.01, the model performance is relatively stable, so $\alpha = 0.01$ can be the optimal choice. Figure 5 represents the influence of different k values on the model when constructing the knn diagram. It can be seen that with the increase of k value, ACC changes little, which proves that different types of data sets are not sensitive to the change of k (Fig. 3).

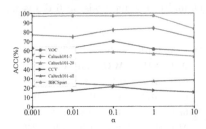

Fig. 2. α sensitivity.

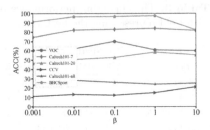

Fig. 3. β sensitivity.

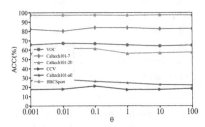

Fig. 4. θ sensitivity.

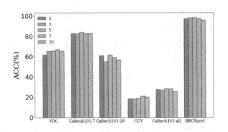

Fig. 5. k sensitivity.

Effectiveness of Graph Filtering Module. To study the effect of different filtering orders on clustering results. The nearest neighbor parameter $k = 5$ was fixed to prevent the different topological structures caused by different knn diagrams from affecting the results. Table 4 below shows the influence of different filtering order t on the experiment. The range of filtering order t was set at 0, 1, 2, 3, 4, 5, 6, 7, 8, 9, 10. $t = 0$ indicates that no graph filtering is performed. As can be seen from the Table 4, for VOC dataset, when $t = 10$, the ACC result increased by 11% and the NMI result increased by 12% compared with that without graph filtering. This is because while eliminating high-frequency noise, more neighbor information was aggregated through graph filtering, and the structural information of the data was encoded into the features which the original feature representation is effectively enhanced. The clustering effect of CCV dataset continues to increase until $t = 4$. It can also be observed that different data sets have different optimal filtering times t, so it can be concluded that for different multi-view data sets, appropriate graph filtering order t should be selected to improve the clustering performance.

Table 4. The impact of the graph filtering order t on 6 datasets.

Dataset	Metrics	BBCSport	VOC	Caltech101-7	Caltech101-20	CCV	Caltech101-all
$t = 0$	ACC	70.2	53.0	80.2	60.2	15.1	26.7
	NMI	72.2	57.0	52.4	47.0	11.2	25.9
$t = 1$	ACC	95.1	61.5	83.0	60.7	19.2	**28.3**
	NMI	88.3	62.6	63.7	50.1	13.7	**40.5**
$t = 2$	ACC	96.1	64.4	82.9	58.6	**21.1**	28.1
	NMI	88.6	64.7	62.5	52.9	**16.0**	39.8
$t = 3$	ACC	97.4	64.7	83.2	58.0	20.2	27.7
	NMI	88.7	64.3	61.4	55.8	15.0	39.3
$t = 4$	ACC	**98.2**	65.0	81.1	61.4	18.7	27.4
	NMI	**92.7**	65.3	60.7	52.5	13.2	38.4
$t = 5$	ACC	97.2	65.7	83.3	59.9	18.6	26.8
	NMI	91.3	64.8	64.7	49.6	13.6	38.6
$t = 6$	ACC	96.8	65.0	83.2	**61.7**	18.6	26.1
	NMI	90.5	66.5	64.3	**57.0**	13.0	38.1
$t = 7$	ACC	97.2	66.2	83.0	60.8	18.0	25.4
	NMI	92.3	66.3	64.5	65.7	11.8	37.5
$t = 8$	ACC	97.3	66.5	83.2	60.2	17.9	25.2
	NMI	92.4	67.1	65.3	50.5	12.5	36.8
$t = 9$	ACC	97.0	66.6	**83.9**	57.0	17.1	24.9
	NMI	92.1	67.1	**65.4**	56.9	12.5	36.6
$t = 10$	ACC	97.2	**67.1**	83.4	57.2	17.8	24.6
	NMI	91.9	**67.2**	64.9	55.7	12.6	36.1

Ablation Experiment. We conduct an ablation study to clearly verify the effectiveness of the proposed modules module and report the results in Table 2. Here we denote the clustering and refactoring loss as the Baseline. Baseline-S, Baseline-C, and Baseline-S-C denote that the baseline adopts reverse supervision, the contrast fusion, and both, From the results in Table 2, we can observe that 1) compare with baseline, Baseline-S performance improvement in all six data sets. The improvement in the BBCSport dataset is between 6% and 11%. These results show that the introduction of a Self-supervision mechanism in network training can help networks learn better clustering representation; 2) Baseline-C consistently achieves better performance than that of the baseline, It is proved that the fusion reduction method can effectively improve the discriminant ability of samples and alleviate the problem of excessive smoothing; 3) the results in the last column of Table 2 further verifies the effectiveness of both components. As seen, Baseline-S-C achieves the best results compared to other variants.

Cluster Visualization. In order to intuitively demonstrate the advantages of SMCGC model, node embedding Z learned by t-SNE [19] visualization, SMC, AMGL, LMVSC, and SMCGC algorithms on data set BBCSport was verified. The corresponding visualization results are shown in Fig. 6. As shown in Fig. 6.(a), the distribution of original data is relatively discrete, and the distance between clusters is confused. Although AMGL and SMC algorithms also achieve good results, there is still some overlap between clusters, and some points are incorrectly allocated to other clusters. SMCGC has a clearer structure, samples between clusters are more clustered, and different clusters have clear outlines, which can better reveal the internal clustering structure of data.

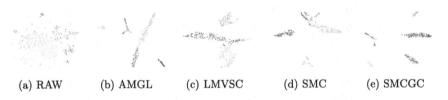

(a) RAW   (b) AMGL   (c) LMVSC   (d) SMC   (e) SMCGC

Fig. 6. Visualization of BBCSport.

5   Conclusion

In this paper, we first use graph filtering to obtain better feature representation, then use fusion reduction to enhance the discriminant ability of samples, and finally use a self-supervised approach to guide the operation of the model. The clustering task, ablation study, parameter sensitivity analysis and clustering visualization analysis on 6 data sets fully demonstrate the validity and superiority of SMCGC model in clustering performance. In addition. In the process of collecting multi-view data in the real world, there will be deficiencies. How to effectively carry out clustering task in the missing data will also be our next research focus.

Acknowledgements. This research is supported by the Program for Young and Middle-aged Academic and Technical Reserve Leaders of Yunnan Province (202205AC160033).

References

1. Andrew, G., Arora, R., Bilmes, J., Livescu, K.: Deep canonical correlation analysis. In: International Conference on Machine Learning, pp. 1247–1255. PMLR (2013)
2. Bichot, C.E., Siarry, P.: Graph Partitioning. Wiley (2013)
3. Bickel, S., Scheffer, T.: Multi-view clustering. In: ICDM, vol. 4, pp. 19–26. Citeseer (2004)
4. Brbić, M., Kopriva, I.: Multi-view low-rank sparse subspace clustering. Pattern Recogn. **73**, 247–258 (2018)

5. Chen, P., Liu, L., Ma, Z., Kang, Z.: Smoothed multi-view subspace clustering. In: Zhang, H., Yang, Z., Zhang, Z., Wu, Z., Hao, T. (eds.) NCAA 2021. CCIS, vol. 1449, pp. 128–140. Springer, Singapore (2021). https://doi.org/10.1007/978-981-16-5188-5_10

6. Cui, G., Zhou, J., Yang, C., Liu, Z.: Adaptive graph encoder for attributed graph embedding. In: Proceedings of the 26th ACM SIGKDD International Conference on Knowledge Discovery & Data Mining, pp. 976–985 (2020)

7. Du, G., Zhou, L., Li, Z., Wang, L., Lü, K.: Neighbor-aware deep multi-view clustering via graph convolutional network. Inf. Fusion 93, 330–343 (2023)

8. Du, G., Zhou, L., Yang, Y., Lü, K., Wang, L.: Deep multiple auto-encoder-based multi-view clustering. Data Sci. Eng. 6(3), 323–338 (2021)

9. Hardoon, D.R., Szedmak, S., Shawe-Taylor, J.: Canonical correlation analysis: an overview with application to learning methods. Neural Comput. 16(12), 2639–2664 (2004)

10. Hartigan, J.A., Wong, M.A.: Algorithm as 136: a K-means clustering algorithm. J. Roy. Stat. Soc. Ser. C (Appl. Stat.) 28(1), 100–108 (1979)

11. Huang, S., Kang, Z., Tsang, I.W., Xu, Z.: Auto-weighted multi-view clustering via kernelized graph learning. Pattern Recogn. 88, 174–184 (2019)

12. Hwang, S.J., Grauman, K.: Accounting for the relative importance of objects in image retrieval. In: BMVC, vol. 1, p. 5 (2010)

13. Kang, Z., Zhou, W., Zhao, Z., Shao, J., Han, M., Xu, Z.: Large-scale multi-view subspace clustering in linear time. In: Proceedings of the AAAI Conference on Artificial Intelligence, vol. 34, pp. 4412–4419 (2020)

14. Ke, G., Hong, Z., Zeng, Z., Liu, Z., Sun, Y., Xie, Y.: CONAN: contrastive fusion networks for multi-view clustering. In: 2021 IEEE International Conference on Big Data (Big Data), pp. 653–660. IEEE (2021)

15. Kullback, S., Leibler, R.A.: On information and sufficiency. Ann. Math. Stat. 22(1), 79–86 (1951)

16. Li, R., Zhang, C., Fu, H., Peng, X., Zhou, T., Hu, Q.: Reciprocal multi-layer subspace learning for multi-view clustering. In: Proceedings of the IEEE/CVF International Conference on Computer Vision, pp. 8172–8180 (2019)

17. Li, Z., Tang, C., Liu, X., Zheng, X., Zhang, W., Zhu, E.: Consensus graph learning for multi-view clustering. IEEE Trans. Multimedia 24, 2461–2472 (2021)

18. Liu, L., Chen, P., Luo, G., Kang, Z., Luo, Y., Han, S.: Scalable multi-view clustering with graph filtering. Neural Comput. Appl. 34(19), 16213–16221 (2022)

19. Van der Maaten, L., Hinton, G.: Visualizing data using t-SNE. J. Mach. Learn. Rese. 9(11) (2008)

20. Ng, A., Jordan, M., Weiss, Y.: On spectral clustering: analysis and an algorithm. In: Advances in Neural Information Processing Systems, vol. 14 (2001)

21. Sun, S.: A survey of multi-view machine learning. Neural Comput. Appl. 23, 2031–2038 (2013)

22. Wang, W., Arora, R., Livescu, K., Bilmes, J.: On deep multi-view representation learning. In: International Conference on Machine Learning, pp. 1083–1092. PMLR (2015)

23. Wang, Y., Chang, D., Fu, Z., Zhao, Y.: Consistent multiple graph embedding for multi-view clustering. IEEE Trans. Multimedia (2021)

24. Xie, J., Girshick, R., Farhadi, A.: Unsupervised deep embedding for clustering analysis. In: International Conference on Machine Learning, pp. 478–487. PMLR (2016)

25. Zhan, K., Zhang, C., Guan, J., Wang, J.: Graph learning for multiview clustering. IEEE Trans. Cybern. 48(10), 2887–2895 (2017)

26. Zhang, C., Hu, Q., Fu, H., Zhu, P., Cao, X.: Latent multi-view subspace clustering. In: Proceedings of the IEEE Conference on Computer Vision and Pattern Recognition, pp. 4279–4287 (2017)
27. Zhang, C., Liu, Y., Fu, H.: AE2-Nets: autoencoder in autoencoder networks. In: Proceedings of the IEEE/CVF Conference on Computer Vision and Pattern Recognition, pp. 2577–2585 (2019)

SSM: Semantic Selection and Multi-view Alignment for Image-Text Retrieval

Beiming Yu, Zhenfei Yang, Xiushuang Yi[(✉)], Yu Wang, and Zhangjun Bao

School of Computer Science and Engineering, Northeastern University, Shenyang, China
xsyi@mail.neu.edu.cn, {2001828,2071737}@stu.neu.edu.cn

Abstract. Image-text retrieval has been a crucial and fundamental task in multi-modal field. Benefiting from the superiority of Transformer encoder in modeling multimodal information, the Transformer-based alignment model has become the mainstream of image-text retrieval. However, current Transformer-based alignment models suffer from two major limitations: (1) The redundancy of modal features and the complexity of correlations between modalities restrict the performance of the model. (2) Current researches are typically limited to a single viewpoint during the modal alignment. To address these issues, in this paper we propose a image-text retrieval model SSM based on Semantic Selection and Multi-view alignment. Specifically, we introduce a gated attention unit to filter unnecessary information, and design an adaptive weighted similarity calculation method to dynamically adjust the importance of different features during the alignment process. On the other hand, we design a multi-view cross-modal alignment method that considers different granularity and different level of information to provide complementary benefits in representation learning. We compare SSM with other advanced image-text retrieval models in MS-COCO and Flickr30K datasets, and the results show that the SSM model has competitive performance without much interaction.

Keywords: Image-text retrieval · Multi-modal · Semantic selection · Multi-view · Contrastive learning

1 Introduction

With the growth of multimedia data on the Internet, cross-modal retrieval has been widely noticed [20]. Cross-modal retrieval aims to understand the natural semantic correlations between different modalities and hence search for semantically similar instances of different modalities. As the core task of cross-modal retrieval, the challenge of image-text retrieval is to accurately learn the semantic relatedness between image and text, and bridge the semantic gap between the two heterogeneous modalities.

Supported by organization x.

Early researches on image-text retrieval focus on alignment-based models, which encode image and text independently as feature vector representations and calculate image-text matching score via a similarity function. Faghri et al. [6] encodes the image and text as a global feature vector and aligns the features by contrastive learning. Lee et al. [11] proposes a fine-grained feature alignment method to further improve the performance of the alignment-based model. However, these works remain very inefficient for large scale image-text retrieval, limited by the weakness of CNN and RNN feature encoding capability. Chen et al. [2] proposes an interaction-based model to match image and text features by multiple iterations of neural interaction units, which fully explores the semantic association between the two modalities. But the interaction-based model, while obtaining significant gains in retrieval performance, also leads to a dramatic increase in computational cost and poses challenges for practical deployment in production environments.

In recent years, the successful deployment of Transformer models in the natural language processing [3,27] and multimodal [5,14,26] has demonstrated the superiority of Transformer modeling visual and text information. Transformer employs a multi-head attention mechanism where each part of the input representation interacts with other parts, to obtain better feature representations. Messina et al. [16] improves the alignment model using Transformer and applies it to an image-text retrieval task. Remarkably, their methods maintain the fast inference speed of the alignment-based model while achieving performance close to that of more complex interaction-based models.

Although Transformer-based alignment method has achieved acceptable performance, the current study suffers from two major drawbacks, as shown in Fig. 1: (1) The correlations between image and text are usually complex. In a mutually matching image-text pair, the text may describe only the main content of the image, and an image may require multiple sentences to be described correctly. Therefore, not all regions of image and words of text have matching relationship, especially there will be some region features in image with low contribution to retrieval. Furthermore, current researches commonly employ the Faster-RCNN model to extract image features [1,2,11,16,17], it may lead to excessive border overlap and result in the extraction of image features with redundant information. (2) Multi-layer Transformer in the process of encoding features, the vectors encoded in different layers contain different levels of information [8,22]. For example, the lower layer tend to encode basic features, and the higher layer capture complex semantic information. The previous Transformer-based alignment models [7,17,28], which commonly use the output features of the last layer, ignore the semantic differences between different layers, and these model make limited exploitation of the transformer architecture. Meanwhile, previous models [16] typically focus on local features alignment and ignore the guiding role of global features, which may lead to ambiguous representation due to local features not fully integrated with contextual information.

In this paper, we propose a novel image-text retrieval model SSM. Referring to past work, our model employs Faster-RCNN and Transformer for image feature encoding and BERT pre-trained model for text encoding. To address the redundancy of modal features and the complexity of correlations between modalities, the SSM model introduces a gated attention mechanism to filter the redundant features in image modalities. In addition, we propose an adaptive weighted similarity calculation method to dynamically attend on representative features and cast aside the interferences of uninformative features in the alignment process. In order to integrate the modal features of different views and learn the ideal modal feature representation, we propose a multi-view cross-modal alignment method to align global features and local features at the semantic level and the feature level to achieve accurate matching of image-text pairs.

Summarizing, the contributions of this paper are the following:

(1) We introduce gated attention units and adaptive weighted similarity calculation method for cross-modal semantic selection.
(2) We propose a multi-view cross-modal alignment method that captures the modal correlations of different views.
(3) We have conducted extensive experiments on two benchmark datasets to validate the effectiveness of SSM. The experimental results show that our methods can significantly improve the metrics of cross-modal retrieval.

2 Related Work

2.1 Image-Text Retrieval

Image-text retrieval is a fundamental task in the field of multimodal where the target is to find a suitable text description for an image or to find a corresponding image for a given text. Existing approaches can be divided into two main types: alignment-based and interaction-based. Notably, due to its low computational cost and fast response speed, the alignment-based method has been widely used in industry and has attracted a lot of attention in academia. The alignment-based method leverages a neural network model to encode images and text as feature representations separately and performs inter-modal alignment by contrastive learning. However, the results achieved by alignment-based method in earlier researches are not satisfactory due to encoder performance limitations [6,10,11].

Benefiting from the excellent performance of the Transformer encoder, recent studies have applied it to cross-modal alignment. Messina et al. [17] first applies Transformer as a modal encoder for image-text alignment. Qu et al. [23] enhances the feature representation capability of the model by leveraging the BERT pre-trained model and the feature summarization module. Messina et al. [16] proposes a fine-grained alignment model based on Transformer encoder to align regions of image and words of text, and achieves approximate results with the interaction model of that time. The Transformer-based alignment model has achieved promising results. However, its retrieval accuracy still has much space for improvement. In this paper, we introduce a gated attention unit and an adaptive weighted similarity calculation method to better align the image-text semantics.

2.2   Contrastive Learning

Contrastive learning is a representation learning method. It essentially aims to learn a better representation of the input by maximizing agreement between two similar data samples. The concept of contrastive learning is widely applied in cross-modal alignment. Radford et al. [24] implements the idea of contrastive learning based on large scale image-text datasets, achieving excellent performance on several multi-modal downstream tasks. Shukor et al. [25] introduces a novel triplet losses with dynamic margins that adapt to the difficulty of the task. In this work we follow the line of previous work on image-text retrieval and use triplet loss as contrastive learning loss [2,6,11,16]. Different from previous work, we design a multi-view cross-modal alignment by considering the features of different Transformer layers and the information of different granularities to obtain a high-quality modal representation.

3   Methodology

The overall framework of SSM is shown in Fig. 2, it contains three parts: image encoder, text encoder, and alignment module. In this section, we elaborate our proposed methods. Firstly, we introduce the image encoder and text encoder in Sect. 3.2 and 3.3. We then describe the adaptive weighted similarity calculation method (AWS) for local feature alignment in Sect. 3.4. Finally, we introduce the multi-view cross-modal alignment method (MVA) and objective function for image-text retrieval in Sect. 3.5.

3.1   Problem Definition

Formally, given an image-text pair, the image is represented as a visual feature of regions $I = \{r_i | i \in [1, m]\}$ and the text is represented as a text feature of words $T = \{w_i | i \in [1, n]\}$, where m and n denote the number of image regions and text words, respectively. The object of the task is to evaluate the matching score between them, thus enabling cross-modal retrieval from the database.

3.2   Text Encoder

SSM uses the pre-trained BERT as a text encoder. Considering that image features are generated by pre-trained deep neural networks, this paper uses a deeper text encoder to model the semantic relationships between words. Concretely, for the input text, each word is mapped to the embedding representation $T^e = \{T_i^e | i \in [1, n]\}$ as the input to the text encoder. We add an embedding T_0^e at the first position to aggregate the global representation of the text. The text embedding T^e consists of three parts: word embedding, position embedding, and segment embedding.

$$T^e = W + P + S \tag{1}$$

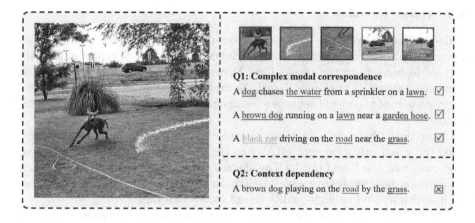

Fig. 1. Illustrations of major drawbacks for Image-Text Retrieval.

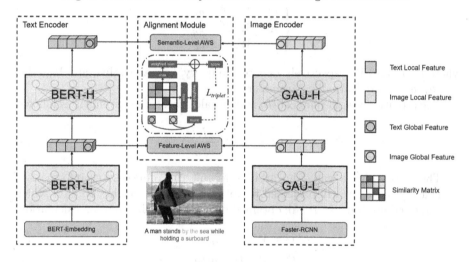

Fig. 2. Model framework of SSM.

In the Transformer architecture, different layers capture information with various semantic clues. SSM uses the first 8 layers of BERT as the low layer text encoder $(BERT_L)$ to obtain a feature-level representation of the text $T^f = \{T_i^f | i \in [0, n]\}$.

$$T^f = BERT_L(T^e) \tag{2}$$

The last 4 layers of BERT are used as the high layer text encoder $(BERT_H)$ to obtain the semantic-level features of the text $T^s = \{T_i^s | i \in [0, n]\}$.

$$T^s = BERT_H(T^f) \tag{3}$$

3.3   Image Encoder

Following recent work, we leverage Bottom-up Attention model to pre-extract features from image regions with high confidence [2,4,11]. Specifically, we pre-extract features from the image I by Faster-RCNN to obtain a set of visual sequence features $I^e = \{r_i | i \in [1, m]\}$. Notably, we add an embedding I_0^e at the first position of the visual sequence feature to aggregate the global representation.

In order to filter noise and redundant information in image features, we introduce the gated attention unit (GAU) as the image encoder. The gated attention unit, which is based on the Transformer architecture, controls the internal information flow through a gate mechanism to adaptively capture contextual information and refine high-quality image representations.

The GAU firstly projects I^e through three linear layers to obtain Q, K, and V, respectively.

$$\begin{cases} Q = W_q I^e + b_q \\ K = W_k I^e + b_k \\ V = W_v I^e + b_v \end{cases} \tag{4}$$

where Q, K and V respectively denote the query, key, and value, and W_q, W_k, W_v, b_q, b_k, b_v are learnable parameters.

The attention weight matrix $attn$ is then calculated. The formula is as follows:

$$attn = relu^2 \left(\frac{QK^T}{\sqrt{d}} \right) \tag{5}$$

The GAU adds a gated linear unit for filtering unnecessary information. Specifically, I^e is linearly projected to obtain the gating weights U, after which the intermediate features I^h are obtained by the gated self-attention calculation.

$$U = W_u I^e + b_u \tag{6}$$

$$I^h = W_h(U \otimes attnV) \tag{7}$$

where W_u, W_h, b_u are learnable parameters.

Subsequently, the image features are mapped into the same vector space as the text feature dimension by linear projection layer. The feature-level representation $I^f = \{I_i^f | i \in [0, m]\}$ of the image is calculated as follows:

$$I^f = W_f I^h + b_f \tag{8}$$

where W_f, b_f are learnable parameters.

The SSM model uses another GAU module as a high layer encoder of image features to obtain the semantic-level image features I^s.

$$I^s = GAU_H(I^f) \tag{9}$$

3.4   Adaptive Weighted Similarity Calculation

For the image-text retrieval task, different regions of the image and different words of the text make different contributions to the image-text alignment, as shown in Fig. 3. In this paper, we devise an adaptive weight similarity calculation method (AWS) to balance the importance of different features in the similarity calculation.

First, given an image I and a text T, compute the similarity matrix $M \in R^{m \times n}$ between all regions and words.

$$M_{ij} = \frac{I_i^T T_j}{||I_i||||T_j||} \qquad i \in [1, m], j \in [1, n] \tag{10}$$

where M_{ij} denotes the similarity between the i-th region feature and the j-th word feature.

We use the combination of linear layer and *softmax* function to measure the weight α of different features in similarity matching, this process is represented as:

$$\alpha = softmax(W_o I_i + b_o) \qquad i \in [1, m] \tag{11}$$

where W_o, b_o are learnable parameters.

The final image-to-text similarity score can be calculated as:

$$Sim_{i2t} = \sum_{i=1}^{m} \alpha_i max_{j=1}^{n}(M_{ij}) \qquad i \in [1, m], j \in [1, n] \tag{12}$$

For text-to-image similarity calculation, we use the same calculation as above to obtain the word-region similarity score Sim_{t2i}. The final similarity score is calculated as follows:

$$Sim = Sim_{i2t} + Sim_{t2i} \tag{13}$$

3.5   Multi-view Alignment

In this paper, we propose a multi-view cross-modal alignment method (MVA) for image-text retrieval that combines the hierarchical and granular information. We use the low-layer information for feature-level multi-grained alignment, and the high-layer information is used for multi-grained alignment at the semantic-level.

Specifically, we perform feature-level alignment using the image features I^f and text features T^f obtained from the low-layer encoder. For the local features, we use the adaptive weighted similarity calculation method proposed above to obtain the image-text similarity matrix $S^{fl} \in R^{B \times B}$, where B denotes the batch size and S_{ij}^{fl} denotes the local matching score of the i-th image and the j-th text within the same batch at the feature-level. For the global features, we calculates the cosine similarity between the feature-level global representations I_0^f and T_0^f as follows:

$$S^{fg} = \frac{I_0^{fT} T_0^f}{||I_0^f||||T_0^f||} \tag{14}$$

where $S^{fg} \in R^{B \times B}$ and S_{ij}^{fg} denotes the feature-level global matching score of the i-th image and the j-th text within the same batch.

For the semantic-level alignment, we use the last layer features as the semantic features of the image and text and calculate the global similarity matrix S^{sg} and the local similarity matrix S^{sl}. We only consider the local similarity at the semantic-level during the model validation process, the similarity of the other views is only used for the calculation of the alignment loss during model training process.

In this paper, we use a triplet contrastive loss as the optimization objective. Following Faghri et al. [6], we focus the attention on hard negatives. Our triplet contrastive loss is defined as:

$$L^* = [\lambda + S_{ij'}^* - S_{i+}^*]_+ + [\lambda + S_{i'j}^* - S_{+j}^*]_+ \qquad (15)$$

where $S^* \in \{S^{fl}, S^{fg}, S^{sl}, S^{sg}\}$, $[x]_+ = max(x, 0)$, S_{i+}^* denotes the similarity between the i-th image and the matched text, $S_{ij'}^*$ denotes the similarity between the i-th image and the hardest negative sample of text within the same batch. λ defines the minimum distance that should be maintained between a truly matched text-image positive sample pair and a negative pair. The hardest negative samples i' and j' are denoted as:

$$\begin{cases} i' = argmax(S_{*,j}) & i' \neq j \\ j' = argmax(S_{i,*}) & j' \neq i \end{cases} \qquad (16)$$

The overall training objective of our model is:

$$L = L^{fl} + L^{fg} + L^{sl} + L^{sg} \qquad (17)$$

4   Experiments

We evaluate our methods on two widely used benchmark datasets including MS-COCO [13] and Flickr30K [21], and compare the SSM model to current advanced models. We also conduct ablation studies to incrementally verify our methods.

4.1   Datasets

MS-COCO is a more general dataset for image-text retrieval, with a total of 123,287 images. Each image is given a set of 5 manual descriptions. Following the split by Karpathy and FeiFei [9] we utilize 5,000 images for validation and 5,000 images for testing and the rest for training. Flickr30K contains 31,783 images collected from social network, and each image is associated 5 captions. We use 1,000 images for validation, 1,000 images for testing and the rest for training.

4.2   Evaluation Metric

We measure the model performance with $R@k(k = 1, 5, 10)$ and $R@sum$. where $R@k$ denotes the fraction of queries for which the correct item is retrieved in the closest k points to the query, and $R@sum$ denotes the sum of recall rates of retrieval tasks.

4.3   Implementation Details

The SSM model is trained on an A5000 graphics card for 50 epochs. The batch-size is set to 30 for all experiments. The initial learning rate is set to 0.00001 and then decay to 0.1 times every 20 epochs. For the text, we utilize BERT for feature encoding, where the feature dimension is 768. For the image, we take the Faster-RCNN detector for feature pre-extraction. Each image has 36 region proposals, where the feature dimension is 2048. After feeding the region features into a GAU_L module, we add a linear layer to transform the GAU_L output to a 768-dimension vector. The layer of GAU_L and GAU_H is set to 4. The margin λ for the triplet contrastive loss is set to 0.2.

4.4   Main Results

We compare the model SSM proposed in this paper with other baseline models on two benchmark datasets, include the traditional alignment-based VSE++ [6] and SCAN [11], the interaction-based IMRAM [2] and SGRAF [4], and the Transformer-based alignment model TERN [17]. The results show that SSM significantly outperforms all other baseline models.

Table 1 shows the performance of SSM with the baseline model on the Flickr30k dataset, achieving 80.3%, 94.9%, and 98.2% for R@1, R@5 and R@10 in text retrieval, the metric on image retrieval is 62.3%, 85.9%, and 91.4% for R@1,

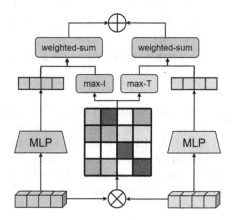

Fig. 3. Adaptive weighted similarity calculation method (AWS).

Table 1. The experimental results on Flickr30K dataset.

Models	Text Retrieval			Image Retrieval			$R@sum$
	$R@1$	$R@5$	$R@10$	$R@1$	$R@5$	$R@10$	
VSE++ [6]	52.9	80.5	87.2	39.6	70.1	79.5	409.8
SCAN [11]	67.4	90.3	95.8	48.6	77.7	85.2	465
VSRN [12]	70.4	89.2	93.7	53.0	77.9	85.7	469.9
IMRAM [2]	74.1	93.0	96.6	53.9	79.4	87.2	484.2
TERN [17]	53.2	79.4	86.9	41.1	71.9	81.2	413.7
MMCA [28]	74.2	92.8	96.4	54.8	81.4	87.8	487.4
CAMERA [23]	76.5	95.1	97.2	58.9	84.7	90.2	502.6
TERAN [16]	75.8	93.2	96.7	59.5	84.9	90.6	500.7
GASA [18]	74.9	92.7	96.8	55.3	82.5	89.3	491.5
SGRAF [4]	77.8	94.1	97.4	58.5	83.0	88.8	499.6
SSM(Ours)	**80.3**	**94.9**	**98.2**	**62.3**	**85.9**	**91.4**	**513.9**

Table 2. The experimental results on MS-COCO 1K dataset.

Models	Text Retrieval			Image Retrieval			$R@sum$
	$R@1$	$R@5$	$R@10$	$R@1$	$R@5$	$R@10$	
VSE++ [6]	64.6	90.0	95.7	52.0	84.3	92.0	478.6
SCAN [11]	72.7	94.8	98.4	58.8	88.4	94.8	507.9
VSRN [12]	76.2	94.8	98.2	62.8	89.7	95.2	516.9
IMRAM [2]	76.7	95.6	98.5	61.7	89.1	95.0	516.6
TERN [17]	63.7	90.5	96.2	51.9	85.6	93.7	481.6
MMCA [28]	74.8	95.6	97.7	61.6	89.8	95.2	514.7
CAMERA [23]	75.9	95.5	98.5	62.3	90.1	95.2	517.5
TERAN [16]	77.7	95.9	98.5	65.0	91.2	96.4	524.7
GASA [18]	77.9	96.5	98.8	63.4	90.7	96.0	523.3
SGRAF [4]	79.6	96.2	98.5	63.2	90.7	96.1	524.3
SSM(Ours)	**82.2**	**97.7**	**99.4**	**68.2**	**92.6**	**97.2**	**537.3**

R@5 and R@10. Compared to the traditional interaction method IMRAM, SSM improves retrieval speed while maintaining higher accuracy without the complex interactions. Compared with CAMERA, which also uses the BERT pre-trained model, SSM achieves a 3.4% improvement in $R@1$ for text retrieval and an even greater improvement (3.8%) for image retrieval. The SSM model also has better evaluation metrics than the Transformer-based fine-grained model TERAN [16].

Table 3. The experimental results on MS-COCO 5K dataset.

Models	Text Retrieval			Image Retrieval			$R@sum$
	$R@1$	$R@5$	$R@10$	$R@1$	$R@5$	$R@10$	
VSE++ [6]	41.3	71.1	81.2	30.3	59.4	72.4	355.7
SCAN [11]	50.4	82.2	90.0	38.6	69.3	80.4	410.9
VSRN [12]	50.3	79.6	87.9	37.9	68.5	79.4	403.6
IMRAM [2]	53.6	83.2	91.0	39.7	69.1	79.8	416.4
TERN [17]	38.4	69.5	81.3	28.7	59.7	72.7	350.3
MMCA [28]	54.0	82.5	90.7	38.7	69.7	80.8	416.4
CAMERA [23]	53.1	81.3	89.8	39.0	70.5	81.5	415.2
TERAN [16]	55.6	83.9	91.6	42.6	72.5	82.9	429.1
GASA [18]	56.7	84.8	91.8	42.3	71.2	83.1	429.9
SGRAF [4]	57.8	-	91.6	41.9	-	81.3	-
SSM(Ours)	**60.1**	**86.3**	**92.7**	**45.5**	**75.7**	**85.0**	**445.3**

Table 2, Table 3 show the bidirectional retrieval results on MS-COCO dataset with 1K and 5K test images. The results show that $R@1$ is 68.2% for image retrieval and $R@1$ is 82.2% for text retrieval on MS-COCO 1K. For MS-COCO 5K, our proposed SSM model still has a performance advantage over other models. It demonstrates that the SSM model has great generalization and robustness. Meanwhile, the performance achieved by SSM on $R@1$ verifies that the proposed methods in this paper can effectively enhance the ability of encoder.

4.5 Ablation Studies

To demonstrate the effectiveness and stability of each component in the SSM model, we carry a series of ablation experiments on the Flickr30K dataset in this section. The baseline model for comparison utilizes BERT as the text encoder and Transformer as the image encoder, and uses only the normal local alignment method during the similarity calculation. Table 4 investigates the impact of each component, where GAU denotes gated attention units, AWS denotes the adaptive weighted similarity calculation method, MVA denotes the multi-view alignment method, and w/o denotes that the current component is not used. For example, w/o AWS denotes that the adaptive weighted similarity calculation is replaced with the mainstream adopted Max-Sum fusion method, while the multi-view alignment and gated attention units are retained.

Table 4. Ablation study on Flickr30K to investigate contributions of each component.

Models	Text Retrieval			Image Retrieval			$R@sum$
	$R@1$	$R@5$	$R@10$	$R@1$	$R@5$	$R@10$	
Baseline	75.0	92.3	95.7	59.1	83.6	90.1	495.8
w/o GAU	78.7	94.5	97.5	61.4	85.3	91.1	508.5
w/o AWS	77.2	93.8	96.5	60.7	84.6	90.6	503.4
w/o MVA	78.4	94.2	96.9	61.1	84.8	91.0	506.4
SSM(Ours)	**80.3**	**94.9**	**98.2**	**62.3**	**85.9**	**91.4**	**513.9**

In Table 4 we can observe that each strategy brings an improvement on the baseline model. GAU improves 1.6% for text retrieval and 0.9% for image retrieval on R@1, demonstrating that gated attention units can filter redundant information and bring positive profits. AWS achieves a more comprehensive improvement in all metrics. It indicates that the adaptive weighted similarity calculation method, compared to the common local alignment method, is able to highlight the role played by important information in the alignment process. The results of whether or not to use MVA demonstrate that the different view information can be complementary. Finally, the final SSM model using all strategy achieves optimal result.

Table 5. Model performance with different fusion methods on Flickr30K.

Methods	Text Retrieval			Image Retrieval			$R@sum$
	$R@1$	$R@5$	$R@10$	$R@1$	$R@5$	$R@10$	
Mean-Mean	66.6	90.0	94.5	54.2	80.8	88.1	474.2
Max-Max	73.9	93.1	96.6	56.0	82.4	89.3	491.3
Max-Mean	71.7	92.5	96.5	56.6	82.3	89.3	488.9
Max-Sum	75.0	92.3	95.7	59.1	83.6	90.1	495.8
AWS(Ours)	**77.3**	**94.1**	**97.0**	**60.2**	**84.1**	**90.6**	**503.3**

Table 5 explores the effectiveness of the proposed AWS compared to the conventional Max-Mean and other variants on the baseline model without any strategy. We can observe the Mean-Mean fusion strategy is less effective, and the Max-Mean, Max-Sum and Max-Max achieve better retrieval accuracy. This may be because the Mean-Mean strategy considers the value of each feature completely equally, leading to the interference of some unnecessary features. The best performance is achieved by the AWS strategy, which indicates that our methods is able to dynamically consider the importance of different features compared to the Mean strategy. Meanwhile, compared to the Max-Max strategy which only considers the features with the highest similarity scores, our strategy is able to better utilize the information of each feature.

Table 6. Ablation study on Flickr30K to investigate contributions of multi-view alignment.

Methods				Text Retrieval			Image Retrieval			$R@sum$
sl	fl	sg	fg	$R@1$	$R@5$	$R@10$	$R@1$	$R@5$	$R@10$	
✓				78.4	94.2	97.5	61.1	84.8	91.0	506.4
✓	✓			80.1	94.6	97.9	61.6	85.4	91.2	510.8
✓	✓	✓		79.9	94.7	98.0	61.9	85.8	**91.5**	511.8
✓	✓	✓	✓	**80.3**	**94.9**	**98.2**	**62.3**	**85.9**	91.4	**513.9**

Table 6 explores the impact of different view alignment on the Flickr30K dataset, where sl denotes semantic-level local alignment, fl denotes feature-level local alignment, sg denotes semantic-level global alignment, and fg denotes feature-level global alignment. The experimental results verify that MVA can improve the retrieval accuracy of the model.

5   Conclusion

In this paper, we present a Transformer-based image-text retrieval model SSM based on semantic selection and multi-view alignment. SSM utilizes gated attention units and the adaptive weighted similarity calculation method for semantic selection and performs cross-modal alignment in multiple views. The experimental results on MS-COCO dataset and Flickr30K dataset show that SSM has excellent cross-modal retrieval performance, and the ablation experiments also demonstrate the effectiveness of each component. Our next work will explore the effectiveness in our methods on multimodal pre-trained models and investigate how to distill the knowledge from the interaction-based model to alignment-based models to achieve an overall improvement in accuracy and speed.

Acknowledgments. This work is supported by the National Key R&D Program of China under Grant No. 2021YFC3300300; the National Natural Science Foundation of China under Grant No. 62032013, 62132004.

References

1. Anderson, P., et al.: Bottom-up and top-down attention for image captioning and visual question answering. In: Proceedings of the IEEE Conference on Computer Vision and Pattern Recognition, pp. 6077–6086 (2018)
2. Chen, H., Ding, G., Liu, X., Lin, Z., Liu, J., Han, J.: IMRAM: iterative matching with recurrent attention memory for cross-modal image-text retrieval. In: Proceedings of the IEEE/CVF Conference on Computer Vision and Pattern Recognition, pp. 12655–12663 (2020)
3. Devlin, J., Chang, M.W., Lee, K., Toutanova, K.: BERT: pre-training of deep bidirectional transformers for language understanding. arXiv preprint arXiv:1810.04805 (2018)

4. Diao, H., Zhang, Y., Ma, L., Lu, H.: Similarity reasoning and filtration for image-text matching. In: Proceedings of the AAAI Conference on Artificial Intelligence, vol. 35, pp. 1218–1226 (2021)

5. Dosovitskiy, A., et al.: An image is worth 16x16 words: transformers for image recognition at scale. arXiv preprint arXiv:2010.11929 (2020)

6. Faghri, F., Fleet, D.J., Kiros, J.R., Fidler, S.: VSE++: improving visual-semantic embeddings with hard negatives. arXiv preprint arXiv:1707.05612 (2017)

7. Gabeur, V., Sun, C., Alahari, K., Schmid, C.: Multi-modal transformer for video retrieval. In: Vedaldi, A., Bischof, H., Brox, T., Frahm, J.-M. (eds.) ECCV 2020. LNCS, vol. 12349, pp. 214–229. Springer, Cham (2020). https://doi.org/10.1007/978-3-030-58548-8_13

8. Hao, Y., Dong, L., Wei, F., Xu, K.: Visualizing and understanding the effectiveness of BERT. arXiv preprint arXiv:1908.05620 (2019)

9. Karpathy, A., Fei-Fei, L.: Deep visual-semantic alignments for generating image descriptions. In: Proceedings of the IEEE Conference on Computer Vision and Pattern Recognition, pp. 3128–3137 (2015)

10. Kiros, R., Salakhutdinov, R., Zemel, R.S.: Unifying visual-semantic embeddings with multimodal neural language models. arXiv preprint arXiv:1411.2539 (2014)

11. Lee, K.-H., Chen, X., Hua, G., Hu, H., He, X.: Stacked cross attention for image-text matching. In: Ferrari, V., Hebert, M., Sminchisescu, C., Weiss, Y. (eds.) ECCV 2018. LNCS, vol. 11208, pp. 212–228. Springer, Cham (2018). https://doi.org/10.1007/978-3-030-01225-0_13

12. Li, K., Zhang, Y., Li, K., Li, Y., Fu, Y.: Visual semantic reasoning for image-text matching. In: Proceedings of the IEEE/CVF International Conference on Computer Vision, pp. 4654–4662 (2019)

13. Lin, T.-Y., et al.: Microsoft COCO: common objects in context. In: Fleet, D., Pajdla, T., Schiele, B., Tuytelaars, T. (eds.) ECCV 2014. LNCS, vol. 8693, pp. 740–755. Springer, Cham (2014). https://doi.org/10.1007/978-3-319-10602-1_48

14. Lu, J., Batra, D., Parikh, D., Lee, S.: ViLBERT: pretraining task-agnostic visiolinguistic representations for vision-and-language tasks. In: Advances in Neural Information Processing Systems, vol. 32 (2019)

15. Ma, Y., Xu, G., Sun, X., Yan, M., Zhang, J., Ji, R.: X-CLIP: end-to-end multi-grained contrastive learning for video-text retrieval. In: Proceedings of the 30th ACM International Conference on Multimedia, pp. 638–647 (2022)

16. Messina, N., Amato, G., Esuli, A., Falchi, F., Gennaro, C., Marchand-Maillet, S.: Fine-grained visual textual alignment for cross-modal retrieval using transformer encoders. ACM Trans. Multimedia Comput. Commun. Appl. (TOMM) 17(4), 1–23 (2021)

17. Messina, N., Falchi, F., Esuli, A., Amato, G.: Transformer reasoning network for image-text matching and retrieval. In: 2020 25th International Conference on Pattern Recognition (ICPR), pp. 5222–5229. IEEE (2021)

18. Miao, L., Lei, Y., Zeng, P., Li, X., Song, J.: Granularity-aware and semantic aggregation based image-text retrieval network. Comput. Sci. 49(11), 134–140 (2022)

19. Min, S., et al.: Hunyuan_tvr for text-video retrivial. arXiv preprint arXiv:2204.03382 (2022)

20. Peng, Y., Huang, X., Zhao, Y.: An overview of cross-media retrieval: concepts, methodologies, benchmarks, and challenges. IEEE Trans. Circuits Syst. Video Technol. 28(9), 2372–2385 (2017)

21. Plummer, B.A., Wang, L., Cervantes, C.M., Caicedo, J.C., Hockenmaier, J., Lazebnik, S.: Flickr30k entities: collecting region-to-phrase correspondences for richer

image-to-sentence models. In: Proceedings of the IEEE International Conference on Computer Vision, pp. 2641–2649 (2015)

22. Qiao, Y., Xiong, C., Liu, Z., Liu, Z.: Understanding the behaviors of BERT in ranking. arXiv preprint arXiv:1904.07531 (2019)

23. Qu, L., Liu, M., Cao, D., Nie, L., Tian, Q.: Context-aware multi-view summarization network for image-text matching. In: Proceedings of the 28th ACM International Conference on Multimedia, pp. 1047–1055 (2020)

24. Radford, A., et al.: Learning transferable visual models from natural language supervision. In: International Conference on Machine Learning, pp. 8748–8763. PMLR (2021)

25. Shukor, M., Couairon, G., Grechka, A., Cord, M.: Transformer decoders with multimodal regularization for cross-modal food retrieval. In: Proceedings of the IEEE/CVF Conference on Computer Vision and Pattern Recognition, pp. 4567–4578 (2022)

26. Su, W., et al.: VL-BERT: pre-training of generic visual-linguistic representations. arXiv preprint arXiv:1908.08530 (2019)

27. Vaswani, A., et al.: Attention is all you need. In: Advances in Neural Information Processing Systems, vol. 30 (2017)

28. Wei, X., Zhang, T., Li, Y., Zhang, Y., Wu, F.: Multi-modality cross attention network for image and sentence matching. In: Proceedings of the IEEE/CVF Conference on Computer Vision and Pattern Recognition, pp. 10941–10950 (2020)

Voice Conversion with Denoising Diffusion Probabilistic GAN Models

Xulong Zhang, Jianzong Wang$^{(\boxtimes)}$, Ning Cheng, and Jing Xiao

Ping An Technology (Shenzhen) Co., Ltd., Shenzhen, China
jzwang@188.com

Abstract. Voice conversion is a method that allows for the transformation of speaking style while maintaining the integrity of linguistic information. There are many researchers using deep generative models for voice conversion tasks. Generative Adversarial Networks (GANs) can quickly generate high-quality samples, but the generated samples lack diversity. The samples generated by the Denoising Diffusion Probabilistic Models (DDPMs) are better than GANs in terms of mode coverage and sample diversity. But the DDPMs have high computational costs and the inference speed is slower than GANs. In order to make GANs and DDPMs more practical we proposes DiffGAN-VC, a variant of GANs and DDPMS, to achieve non-parallel many-to-many voice conversion (VC). We use large steps to achieve denoising, and also introduce a multimodal conditional GANs to model the denoising diffusion generative adversarial network. According to both objective and subjective evaluation experiments, DiffGAN-VC has been shown to achieve high voice quality on non-parallel data sets. Compared with the CycleGAN-VC method, DiffGAN-VC achieves speaker similarity, naturalness and higher sound quality.

Keywords: Voice Conversion · DDPM · GAN

1 Introduction

Voice conversion (VC), a branch of speech signal processing also referred to as speech style transfer, entails the transformation of the linguistic attributes of the source speaker to those of the target speaker, while preserving the linguistic content of the input voice. This technology pertains to modify the linguistic style of the speech samples. VC can be viewed as a regression problem whose goal is to build a mapping function between the speech features of the target and source speaker. VC technology has important applications in many fields, including privacy and identity protection, speech imitation and camouflage, and speech enhancement [26, 31, 32, 35]. This technology is also capable of transforming standard reading speech into stylized speech, including emotional and falsetto speech [29, 30, 35]. At the same time, it can be used in music for the conversion of singers' vocal skills [2]. It also can be used to help people with language disorders voice assistance [22, 35], and can convert voice which contains noise in meeting to clear voice [13].

X. Yang et al. (Eds.): ADMA 2023, LNAI 14179, pp. 154–167, 2023.
https://doi.org/10.1007/978-3-031-46674-8_11

Traditional voice conversion technology is developed based on parallel data, which comprises recordings of both the target and source speakers speaking identical content. The current models using this data mainly include (1) statistical methods (GMMs) [34]; (2) exemplar-based methods (NMF) [36]; (3) Neural network-based methods, such as feedforward neural networks (FNNs) [21], recurrent neural networks (RNNs) [27], convolutional neural networks (CNNs) [13]. However, we often lack parallel data sets to train model. Simultaneously, collecting and producing a large parallel data set is very laborious. In addition, even if we have collected the corresponding data, we also need to carry out time alignment operation [5]. Many researchers have also explored non-parallel VC methods that don't need parallel data. Due to unfavorable training conditions, it is not as good as parallel VC methods in terms of conversion effect and speech quality. Therefore, this is a challenging and important task. This paper focuses on the development of a non-parallel VC method with the same high audio quality and conversion effects as parallel methods.

In order to implement non-parallel VC using voice data, probabilistic depth generation models have been introduced in many studies recently, mainly including RBM based methods, variational autoencoders (VAEs) [3,7,10,16,23,24,33] based methods and generative adversarial networks (GANs) [4] based methods. GANs is a powerful generative model that can learn the generation distribution. It has shown great success in the fields of image style transfer, image quality improvement and image generation. Among them, CycleGAN-VCs [9,11,12,14,15,28] have attracted extensive attention from researchers.

Recently, Denoising Diffusion Probabilistic Models (DDPM) [6,19] shows good performance in various generative tasks such as image generation [20,25], neural acoustic coding [1,17], speech enhancement [19], and speech synthesis [8,18]. Although diffusion models have been applied in several fields, but their application expensive in practice, computationally expensive, and time-consuming. Therefore, their application in the real world is limited.

DDPMs have demonstrated outstanding sample quality and variety, but their pricey sampling prevents them from being used just yet. In order to make GANs and DDPMs more practical, we propose DiffGAN-VC, a new unsupervised non-parallel many-to-many voice conversion (VC) method, which introduces multimodal conditional GANs for modeling, and reconstruct denoising diffusion probabilistic model. In the experiments, the objective is to enhance the efficiency of the denoising diffusion probabilistic model while simultaneously reducing its computational complexity. We use a large-step denoising operation, thereby reducing the total number of denoising steps and computation. Finally, let denoising diffusion GAN obtain the same sample quality and diversity as the original diffusion model, but also have the characteristics of high speed and low computational cost of the original GANs. In this paper, we use CycleGAN as our baseline model, based on which we introduce the DDPM and reconstruct the GANs, and also use an expressive multimodal distribution to parameterize the denoising distribution to achieve large step size denoising.

2   Related Work

2.1   CycleGAN-VC/VC2

CycleGAN-VC/VC2 [11,14] is a voice conversion model consisting of two generators G and two discriminators D. CycleGAN-VC/VC2 emerges as a novel approach in the domain of voice conversion, demonstrating its effectiveness in learning the transformation between different acoustic feature sequences without the need for parallel data. These advancements contribute to the ongoing progress in voice conversion research and its applications in various fields. The primary objective of CycleGAN-VC/VC2, as discussed in the research papers by Kaneko and Kameoka [11,14], is to acquire the ability to transform acoustic feature sequences belonging to source domain X, into those of target domain Y, without the reliance on parallel data. The acoustic feature sequences are represented by x $\in R^{Q \times T}$ and y $\in R^{Q \times T}$, where Q and T represents the feature dimension and the sequence length respectively.

The foundation of CycleGAN-VC/VC2 lies in the inspiration drawn from CycleGAN, an originally proposed technique for image-to-image style transfer in the field of computer vision. By leveraging the principles of CycleGAN, CycleGAN-VC/VC2 aims to learn the mapping function $G(X) \rightarrow Y$, which facilitates the conversion of an input $x \in X$ to an output $y \in Y$. To achieve this goal, CycleGAN-VC/VC2 employs several loss functions during the learning process. These include adversarial loss, cyclic consistency loss, and identity mapping loss, which collectively contribute to enhancing the quality and fidelity of the generated outputs. Furthermore, CycleGAN-VC2 [14] introduces an additional adversarial loss to further refine and improve the fine-grained details of the reconstructed features.

Generator: CycleGAN-VC aims to capture diverse temporal structures while maintaining the input structure's integrity. To accomplish this, the architecture of CycleGAN-VC comprises three essential components: a downsampling layer, a residual layer, and an upsampling layer. These components work collaboratively to effectively capture a broad range of temporal relationships present in the data. By incorporating these design choices, CycleGAN-VC can successfully preserve the original structure while accommodating a wide variety of temporal variations.

In the case of CycleGAN-VC2 [14], a 2-1-2D CNN network architecture is introduced as an extension to CycleGAN-VC. This modified architecture leverages the advantages of both 2D and 1D CNNs to enhance the performance of the voice conversion system. Specifically, 2D CNNs are employed in the upsampling and downsampling blocks, allowing the model to capture spatial and temporal dependencies simultaneously. This capability facilitates the extraction of more comprehensive and meaningful representations from the input data. On the other hand, 1D CNNs are utilized in the remaining blocks of the network, enabling the model to focus on capturing local temporal relationships and preserving the fine-grained details of the acoustic features. By combining these two types of CNNs in a carefully designed network architecture, CycleGAN-VC2 aims to improve the voice conversion performance by effectively capturing both global and local temporal structures.

The integration of a 2-1-2D CNN network architecture in CycleGAN-VC2 demonstrates a significant advancement in voice conversion research. This innovative design choice allows the model to exploit the benefits of both 2D and 1D CNNs, enabling more efficient and effective processing of temporal information in the voice conversion process. By capitalizing on the strengths of each CNN variant, CycleGAN-VC2 showcases its potential to achieve improved performance in capturing a wide range of temporal structures while preserving the input structure and maintaining the fidelity of the converted acoustic features. These advancements contribute to the ongoing progress in the field of voice conversion and hold promise for various applications that rely on accurate and high-quality voice transformation.

Discriminator: CycleGAN-VC, as introduced in the work by Kaneko [11], incorporates a discriminator structure based on a 2D convolutional neural network (CNN). This architectural choice enables the discriminator to effectively discriminate data by analyzing the 2D spectral textures present in the input. The utilization of a fully connected layer in the final layer of the discriminator further enhances its discriminative capability by considering the overall input structure. By combining these components, CycleGAN-VC successfully discriminates and distinguishes between different data samples, facilitating the voice conversion process.

However, one challenge associated with employing a 2D CNN structure in the discriminator is the potential increase in the number of parameters, which can negatively impact the model's efficiency and computational requirements. To address this issue, CycleGAN-VC2, introduced by Kaneko et al. [14], introduces a technique called Patch-GAN. Patch-GAN modifies the discriminator architecture by incorporating convolution in the final layer, the Patch-GAN approach offers a more parameter-efficient solution while still maintaining the discriminative capability of the discriminator. This modification enables CycleGAN-VC2 to mitigate the issue of a large number of parameters, thereby improving its efficiency and reducing computational complexity.

The introduction of Patch-GAN in CycleGAN-VC2 represents a significant advancement in voice conversion research. This modification not only addresses the challenge of parameter efficiency but also ensures the model's ability to discriminate and differentiate between input samples effectively. By leveraging convolutional layers instead of fully connected layers, CycleGAN-VC2 achieves a more streamlined and efficient architecture, facilitating faster training and inference while maintaining high discriminative performance.

2.2 DDPM

DDPM [6,19] has good performance in various generative tasks such as image generation [20,25], neural acoustic coding [1,17], speech enhancement [19], and speech synthesis [8,18]. DDPM utilizes Markov chains to gradually transform

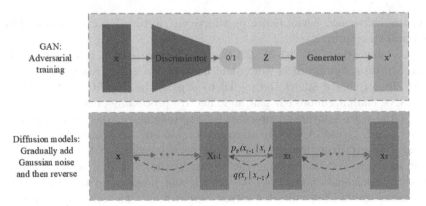

Fig. 1. The distinction between DDPM and GANs is depicted. The preceding illustration illustrates the adversarial training procedure employed by GANs, whereas the subsequent depiction showcases the gradual introduction and removal of Gaussian noise in the diffusion process.

simple distributions with Gaussian distributions into complex data distributions. In order to learn the transformation model, there are two distinct processes at play in DDPM: the forward diffusion process and the reverse generation process. The forward diffusion process involves a gradual incorporation of Gaussian noise into the data, leading to its transformation into random noise over time. This process serves to introduce controlled perturbations that explore the data distribution. On the other hand, the reverse generation process focuses on denoising, aiming to restore the original data by mitigating the impact of the noise accumulated during the diffusion process. Consequently, the diffusion process plays a crucial role in effectively denoising the data, facilitating the recovery of its inherent characteristics. Then the reverse process is a denoising process. Gradually denoising can generate a real sample, so the reverse process is also the process of generating data process. DDPM avoids the generator and discriminator in GANs being less stable during training and the "backward collapse" problem. Compared with GANs, diffusion model training is more stable and can generate more diverse high-quality data. However, since DDPM requires hundreds of iterations to generate data, their inference speed is relatively slow compared with VAEs and GANs. The comparison between diffusion model and GAN is shown in Fig. 1. The current development of DDPM in speech translation is hampered by two main challenges: (1) Although DDPM is essentially gradient-based model, the guarantee of high sample quality is usually at the expense of thousands or hundreds of denoising steps. This limits the wide application of DDPM in real life. (2) When reducing the denoising step, the diffusion model exhibits significant degradation in model convergence due to the complex data distribution, resulting in blurry and over-smoothed predictions in the mel-spectrogram.

3   Method

Generative Adversarial Networks (GANs) can quickly generate high-quality samples, but suffer from poor pattern coverage and lack of diversity in the generated samples. Although the diffusion model beats GANs in image generation, the inverse process is computationally expensive. To speed up the diffusion model and reduce computation, we use a large-step denoising operation, thereby reducing the total number and computation of denoising steps. Meanwhile, we introduce multimodal conditional GANs in the diffusion model, developing a new generative model, which we call denoising diffusion GANs. This paper adopts CycleGAN as the baseline model, and proposes a novel reconstruction method for the denoising diffusion model. At the same time, we also use an expressive multimodal distribution to parameterize the denoising distribution to achieve large stride denoising and improve model performance. Below we introduce the large-step denoising operation, the multimodal distribution design, and the parameterization of the denoising distribution, respectively.

The forward process of diffusion model [6] which gradually adds Gaussian noise to the data $x_0 \to q(x_0)$ is performed over a discrete T time step. Where $q(x_0)$ is the data distribution of a true Mel-spectrogram, and x_0 denotes a sample from input data X. x_t is sampled from a unit Gaussian independent from x_0 using the Markovian forward process as follows:

$$q(x_{1:T}|x_0) = \prod_{t \geq 1} \mathbb{N}(x_t; \sqrt{1 - \beta_t}x_{t-1}, \beta_t I) \tag{1}$$

where, $t \in [1 : T]$, and β_t is pre-defined variance schedule. The denoising process [6] is defined as:

$$p_\theta(x_{0:T}) = p(x_T) \prod_{t \geq 1} \mathbb{N}(x_{t-1}; \mu_\theta(x_t, t), \sigma_t^2 I) \tag{2}$$

where σ_t^2 is fixed according to each t. DDPMs typically presume that the denoising distribution may be roughly modeled as a Gaussian distribution. It is well-established that the assumption of Gaussianity is only applicable in the context of infinitesimally small denoising steps. This limitation results in a slow sampling problem when a large number of denoising steps are employed in the reverse process. In order to speed up the model operation and reduce the amount of computation, we optimize the model by reducing the number of denoising steps T.

Using a larger step size (that is, fewer denoising steps) in the reverse process will have two problems. Firstly, the assumption of Gaussianity in the denoising distribution cannot be guaranteed to hold under large denoising steps and non-Gaussian data distributions. Furthermore, with each subsequent enhancement of the denoising process, the distribution for noise removal becomes increasingly intricate and exhibits multiple modes. To address this, a non-Gaussian multimodal distribution can be constructed to accurately model the denoising distribution. In light of the ability of conditional GANs to accurately simulate

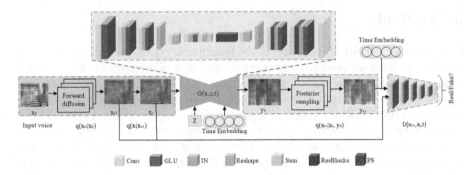

Fig. 2. The overall pipeline of DiffGAN-VC, where $G(x, z, t)$ is the generator of our model. $D(x_{t-1}, x_t, t)$ denotes the discriminator of our model, $q(x_t|x_{t-1})$ is thus the true denoising diffusion transition.

complex conditional distributions within the image domain, we adopt conditional GANs to approximate the true denoising distribution. We use CycleGAN as the base model, then modify this model to conditional GANs, and finally fuse this model with the diffusion model, forming a new denoising diffusion GANs. The overall pipeline of the DiffGAN-VC is depicted in the Fig. 2.

Within our framework, we express the training procedure as the alignment between a conditional GAN generator and its ability to execute operations $p_\theta(x_{t-1}|x_t)$ and $q(x_t|x_{t-1})$. This objective is accomplished through the utilization of an adversarial loss, aiming to decrease the divergence D_{adv} during every denoising iteration.

In order to facilitate the integration of adversarial training into our model, we introduce a discriminator that is dependent on time and denoted as $D_\theta(x_{t-1}, x_t, t) : \mathbb{R}^N \times \mathbb{R}^N \times \mathbb{R} \to [0, 1]$, where θ represents the associated parameters. This time-dependent discriminator is designed to assess the plausibility of x_{t-1} being a denoised version of x_t, taking into account their respective n-dimensional inputs. During the training process, the discriminator is optimized by following a specific objective. The aim is to bolster the discriminator's capacity to differentiate between the denoised and noisy variants of the input data, thereby ultimately enhancing its discriminatory prowess.

$$\min_{\varphi} \sum_{t \geq 1} \mathcal{L}q(x_t)\mathcal{L}q(x_{t-1}|x_t)[-\log D_\varphi(x_{t-1}, x_t, t)]$$
$$+ \mathcal{L}p\theta(x_{t-1}|x_t)[-\log(1 - D_\varphi(x_{t-1}, x_t, t))] \qquad (3)$$

Our approach involves the comparison between synthetic samples derived from the generator $p_\theta(x_{t-1}|x_t)$ and authentic samples obtained from the conditional distribution $q(x_t|x_{t-1})$. It is worth highlighting that the generation of the first expectation necessitates the sampling of data points from a distribution, namely $q(x_t|x_{t-1})$, which is not explicitly known to us. This introduces an additional challenge in effectively estimating and evaluating the expected values.

Given the discriminator, we train the generator using the following objective:

$$\max_{\theta} \sum_{t \geq 1} \mathcal{L}_{q(x_t)p_\theta(x_{t-1}|x_t)} [\log(D_\varphi(x_{t-1}, x_t, t))] \tag{4}$$

In order to achieve the denoising step, the objective of our model involves updating the generator using a non-saturating GAN objective.

The diffusion model introduces a parameterized approach through $f_\theta(x_{t-1}|x_t)$. This means that the denoising model incorporates a parameterization scheme, allowing for the estimation of x_{t-1} based on x_t. Initially, the denoising model $p_\theta(x_{t-1}|x_t)$ predicts the value of x_0. Subsequently, x_{t-1} is sampled from the posterior distribution $q(x_{t-1}|x_t, x_0)$, taking into account both x_t and the predicted x_0.

The distribution $q(x_{t-1}|x_0, x_t)$ provides an intuitive representation of the distribution of x_{t-1} during the denoising process, which occurs from x_t to x_0. Similarly, we define $p_\theta(x_{t-1}|x_t)$ in the following manner:

$$p_\theta(x_{t-1}|x_t) = G_\theta(x_t, z, t))dz \tag{5}$$

The distribution $p_\theta(x_0|x_t)$ captures the implicit nature of the GAN generator $G_\theta(x_t, z, t)$, which is conditioned on both x_t and the L-dimensional latent variable z. This generator is responsible for producing the unperturbed version x_0. However, in our specific scenario, the role of the generator is to solely predict the unperturbed x_0 and subsequently reintroduce the perturbation through the utilization of $q(x_{t-1}|x_t, x_0)$. This enables us to maintain the perturbation information while generating x_{t-1} based on the observed x_t and the predicted unperturbed x_0.

GANs are known to have mode collapse and training instability. The potential causes include the inability to generate samples directly from complicated distributions in a single step and overfitting issues when the discriminator only considers clean samples. In contrast, due to the strong conditioning on x, our model divides the generative process into multiple stages of conditional denoising steps, each of which is very simple to model. Furthermore, the diffusion process smoothes the data distribution and alleviates the overfitting problem in the discriminator. Therefore, our model will have higher pattern coverage and training stability. DDPM avoids the generator and discriminator in GAN being less stable during training and the "backward collapse" problem. Compared with GAN, diffusion model training is more stable and can generate more diverse high-quality data. However, since DPM requires hundreds of iterations to generate data, their inference speed is relatively slow.

4   Experiments

4.1   Experimental Setup

Dataset and Conversion Process: We evaluate our method objectively and subjectively on part of the Speech Conversion Challenge (VCC) 2020 data set, which is a semi-parallel speech data set containing input and output speech from two women and two men, each 70 English sentences, 20 parallel sentences, and 50 non-parallel sentences. There are a total of $4 \times 3 = 12$ distinct source-target combinations that can be derived. The number of sentences used for training, and test are 70 and 25, respectively. In our experiments, all speech signals are sampled at 16 kHz. From every sentence, we utilized the WORLD toolkit to extract various acoustic features, including 35 mel-cepstral coefficients (MCEPs), logarithmic fundamental frequencies ($\log F_0$), and aperiodicities (APs). These features provide valuable representations of the speech signal, capturing essential characteristics related to spectral information, pitch contour, and the irregular components of the signal. The extraction process involved careful analysis and computation to ensure accurate and informative representations of the speech data. We apply the DiffGAN-VC to MCEP transformation, for $\log F_0$, we use log-Gaussian normalized transformation for transformation, use Aps directly, and finally use WORLD vocoder to synthesize speech.

Table 1. Mean and standard error with different models.

Methods	intra-gender		inter-gender	
	MCD ↓	MOS ↑	MCD ↓	MOS ↑
VQVC	6.41	3.26 ± 0.36	6.71	2.89 ± 0.37
StarGAN-VC2	6.45	3.36 ± 0.42	5.85	3.35 ± 0.53
CycleGAN-VC2	6.09	3.71 ± 0.45	5.78	2.93 ± 0.47
DDPM	5.45	3.47 ± 0.43	5.76	3.47 ± 0.57
Our model	**5.99**	**3.75 ± 0.42**	**5.62**	**3.86 ± 0.51**

Experimental Details: We design the architecture based on CycleGAN-VC2, employing a 2-1-2D CNN in the generator (G) and a 2D CNN in the discriminator (D). During the experiments, the input to the network is adjusted to x_t, and temporal embedding is utilized to ensure conditioning on the time step (t). The latent variable (z) is responsible for controlling the normalization layer, where a multi-layer FC network predicts the displacement and scale parameters in the group normalization. The training procedure encompassed a total of 2×10^5 iterations. Moreover, to enhance the optimization process, a momentum term of 0.5 was incorporated, aiding in the stabilization and convergence of the training procedure.

In the conducted experiments, we categorized the 12 different combinations into two groups: inter-gender conversion and intra-gender conversion. The inter-gender conversion includes transformations from female to male and from male to female, while the intra-gender conversion includes transformations within the same gender, specifically from female to female and from male to male.

4.2   Objective Evaluation

We chose the VQVC, StarGAN, CycleGAN and DDPM based methods as the comparison for our experiments, and objectively evaluated the results to verify that the DiffGAN-VC has a certain improvement and optimization in model performance. To conduct a thorough analysis and assess the performance of the models, we employed the MCD metric, which serves as a measure of global structural dissimilarities. The MCD metric calculates the distance between the log-modulated spectra of the target and converted MCEPs. A smaller MCD value indicates a higher level of similarity between the target and transformed features, reflecting a better quality of conversion. In our evaluation, we generated a total of 300 sentences, obtained through 4×3 source-target combinations. Each combination consisted of 25 sentences. This comprehensive set of sentences allowed us to evaluate across various source-target pairs and assess the effectiveness of the conversion process. We compare DiffGAN-VC with VQVC, StarGAN-VC2, CycleGAN-VC2 and DDPM, which are listed in Table 1, respectively. In Table 1, DiffGAN-VC outperforms other models in terms of MCD. This suggests that the introduction of multimodal diffusion methods in conditional GANs models is useful for improving feature quality. These experiments confirm that the proposed method effectively reduces the distance between the transformed acoustic feature sequence and the target sequence.

A comprehensive evaluation was conducted to assess the performance of the DiffGAN-VC method in multi-domain non-parallel voice conversion. The evaluation involved multiple English-educated testers, and several listening tests were carried out. The primary focus was on evaluating naturalness and voice similarity.

To measure naturalness, a MOS test was employed. Testers assigned scores ranging from 5 (good) to 1 (bad), with higher scores indicating better naturalness. A subset of 32 sentences, with durations between 2 and 5 s, was randomly selected from the evaluation set for this test. The obtained results, including MOS scores, are presented in Table 1.

4.3   Subjective Evaluation

Additionally, we conduct Voice Similarity Score (VSS) evaluation to assess the similarity of the transformed voices to the original ones. The scoring system used in the VSS test was the same as the MOS test, with testers assigning scores from 1 to 5 based on perceived similarity. We randomly select 12 sample pairs for this test, which involved thirteen well-educated English speakers. Figure 3 illustrates the outcome of the VSS evaluation, in which the labels T and S denote the target and source speakers, respectively. Female and male speakers are represented by F and M respectively. The MOS naturalness and VSS similarity scores for the source and target speakers are presented in Table 1 and Fig. 3, respectively.

The evaluation results revealed several significant findings. Firstly, the DiffGAN-VC method demonstrated a substantial improvement over the baseline model in both inter-gender and intra-gender voice conversion tasks. More-

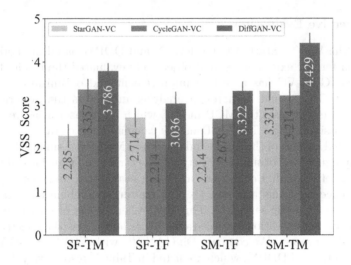

Fig. 3. VSS for speaker similarity.

over, when considering the overall performance, DiffGAN-VC exhibited significant advancements compared to the baseline CycleGAN-VC, both in terms of naturalness and speaker similarity.

5   Conclusion

The primary goal of this research is to enhance the performance of Generative Adversarial Networks (GANs) and Denoising Diffusion Probabilistic Models (DDPM) in various aspects, including speed, diversity, utilization of non-parallel training data, and the coverage and diversity of generated samples. To accomplish this, we have made modifications to the existing CycleGAN-VC2 system by incorporating the embedding of a stride denoised multimodal diffusion model. Both the generator and discriminator are conditioned on this embedding, facilitating the targeting of specific source and target speakers. DiffGAN-VC has been thoroughly evaluated using a limited training dataset. The evaluation demonstrates that DiffGAN-VC outperforms CycleGAN-VC in terms of both objective and subjective metrics. In this study, we have applied the DiffGAN-VC model to speaker identity speech conversion. It is important to note that this approach can be expanded to encompass additional tasks, such as multi-emotional voice conversion and music genre conversion. Exploring these additional applications represents a compelling direction for future research in this domain.

Acknowledgement. This paper is supported by the Key Research and Development Program of Guangdong Province under grant No. 2021B0101400003.

References

1. Chen, N., Zhang, Y., Zen, H., Weiss, R.J., Norouzi, M., Chan, W.: Wavegrad: estimating gradients for waveform generation. In: 9th International Conference on Learning Representations. OpenReview.net (2021)
2. Deng, C., Yu, C., Lu, H., Weng, C., Yu, D.: Pitchnet: unsupervised singing voice conversion with pitch adversarial network. In: 2020 IEEE International Conference on Acoustics, Speech and Signal Processing, pp. 7749–7753. IEEE (2020)
3. Ding, S., Gutierrez-Osuna, R.: Group latent embedding for vector quantized variational autoencoder in non-parallel voice conversion. In: Kubin, G., Kacic, Z. (eds.) 20th Annual Conference of the International Speech Communication Association, pp. 724–728. ISCA (2019)
4. Goodfellow, I.J., et al.: Generative adversarial networks. Commun. ACM **63**(11), 139–144 (2020)
5. Helander, E., Schwarz, J., Nurminen, J., Silén, H., Gabbouj, M.: On the impact of alignment on voice conversion performance. In: 9th Annual Conference of the International Speech Communication Association, pp. 1453–1456. ISCA (2008)
6. Ho, J., Jain, A., Abbeel, P.: Denoising diffusion probabilistic models. In: Larochelle, H., Ranzato, M., Hadsell, R., Balcan, M., Lin, H. (eds.) Advances in Neural Information Processing Systems 33: Annual Conference on Neural Information Processing Systems 2020 (2020)
7. Huang, W., et al.: Unsupervised representation disentanglement using cross domain features and adversarial learning in variational autoencoder based voice conversion. IEEE Trans. Emerg. Top. Comput. Intell. **4**(4), 468–479 (2020)
8. Jeong, M., Kim, H., Cheon, S.J., Choi, B.J., Kim, N.S.: Diff-TTS: a denoising diffusion model for text-to-speech. In: Hermansky, H., Cernocký, H., Burget, L., Lamel, L., Scharenborg, O., Motlícek, P. (eds.) 22nd Annual Conference of the International Speech Communication Association, pp. 3605–3609. ISCA (2021)
9. Kameoka, H., Kaneko, T., Tanaka, K., Hojo, N.: StarGAN-VC: non-parallel many-to-many voice conversion using star generative adversarial networks. In: 2018 IEEE Spoken Language Technology Workshop, pp. 266–273. IEEE (2018)
10. Kameoka, H., Kaneko, T., Tanaka, K., Hojo, N.: ACVAE-VC: non-parallel voice conversion with auxiliary classifier variational autoencoder. IEEE/ACM Trans. Audio Speech Lang. Process. **27**(9), 1432–1443 (2019)
11. Kaneko, T., Kameoka, H.: CycleGAN-VC: non-parallel voice conversion using cycle-consistent adversarial networks. In: 26th European Signal Processing Conference, pp. 2100–2104. IEEE (2018)
12. Kaneko, T., Kameoka, H.: CycleGAN-VC3: examining and improving CycleGAN-VCs for mel-spectrogram conversion. In: Meng, H., Xu, B., Zheng, T.F. (eds.) 21st Annual Conference of the International Speech Communication Association, pp. 2017–2021. ISCA (2020)
13. Kaneko, T., Kameoka, H., Hiramatsu, K., Kashino, K.: Sequence-to-sequence voice conversion with similarity metric learned using generative adversarial networks. In: Lacerda, F. (ed.) 18th Annual Conference of the International Speech Communication Association, pp. 1283–1287. ISCA (2017)
14. Kaneko, T., Kameoka, H., Tanaka, K., Hojo, N.: CycleGAN-VC2: improved CycleGAN-based non-parallel voice conversion. In: IEEE International Conference on Acoustics, Speech and Signal Processing, pp. 6820–6824. IEEE (2019)

15. Kaneko, T., Kameoka, H., Tanaka, K., Hojo, N.: StarGAN-VC2: rethinking conditional methods for StarGAN-based voice conversion. In: Kubin, G., Kacic, Z. (eds.) 20th Annual Conference of the International Speech Communication Association, pp. 679–683. ISCA (2019)
16. Kingma, D.P., Welling, M.: Auto-encoding variational bayes. In: Bengio, Y., LeCun, Y. (eds.) 2nd International Conference on Learning Representations (2014)
17. Kong, Z., Ping, W., Huang, J., Zhao, K., Catanzaro, B.: Diffwave: a versatile diffusion model for audio synthesis. In: 9th International Conference on Learning Representations (2021)
18. Liu, J., Li, C., Ren, Y., Chen, F., Zhao, Z.: Diffsinger: singing voice synthesis via shallow diffusion mechanism. In: Thirty-Sixth AAAI Conference on Artificial Intelligence, pp. 11020–11028. AAAI Press (2022)
19. Lu, Y., Wang, Z., Watanabe, S., Richard, A., Yu, C., Tsao, Y.: Conditional diffusion probabilistic model for speech enhancement. In: IEEE International Conference on Acoustics, Speech and Signal Processing, pp. 7402–7406. IEEE (2022)
20. Lugmayr, A., Danelljan, M., Romero, A., Yu, F., Timofte, R., Gool, L.V.: Repaint: inpainting using denoising diffusion probabilistic models. In: IEEE/CVF Conference on Computer Vision and Pattern Recognition, pp. 11451–11461. IEEE (2022)
21. Mohammadi, S.H., Kain, A.: Voice conversion using deep neural networks with speaker-independent pre-training. In: 2014 IEEE Spoken Language Technology Workshop, pp. 19–23. IEEE (2014)
22. Nakamura, K., Toda, T., Saruwatari, H., Shikano, K.: Speaking-aid systems using GMM-based voice conversion for electrolaryngeal speech. Speech Commun. **54**(1), 134–146 (2012)
23. Qian, K., Jin, Z., Hasegawa-Johnson, M., Mysore, G.J.: F0-consistent many-to-many non-parallel voice conversion via conditional autoencoder. In: 2020 IEEE International Conference on Acoustics, Speech and Signal Processing, pp. 6284–6288. IEEE (2020)
24. Qian, K., Zhang, Y., Chang, S., Yang, X., Hasegawa-Johnson, M.: Autovc: zero-shot voice style transfer with only autoencoder loss. In: Chaudhuri, K., Salakhutdinov, R. (eds.) Proceedings of the 36th International Conference on Machine Learning. Proceedings of Machine Learning Research, vol. 97, pp. 5210–5219. PMLR (2019)
25. Saharia, C., et al.: Palette: image-to-image diffusion models. In: Nandigjav, M., Mitra, N.J., Hertzmann, A. (eds.) SIGGRAPH 2022: Special Interest Group on Computer Graphics and Interactive Techniques Conference, pp. 15:1–15:10. ACM (2022)
26. Si, S., Wang, J., Zhang, X., Qu, X., Cheng, N., Xiao, J.: Boosting StarGANs for voice conversion with contrastive discriminator, pp. 355–366 (2023)
27. Sun, L., Kang, S., Li, K., Meng, H.M.: Voice conversion using deep bidirectional long short-term memory based recurrent neural networks. In: 2015 IEEE International Conference on Acoustics, Speech and Signal Processing, pp. 4869–4873. IEEE (2015)
28. Kaneko, T., Kameoka, H.: Maskcyclegan-VC: learning non-parallel voice conversion with filling in frames. In: IEEE International Conference on Acoustics, Speech and Signal Processing, pp. 5919–5923. IEEE (2021)
29. Tang, H., Zhang, X., Wang, J., Cheng, N., Xiao, J.: Emomix: emotion mixing via diffusion models for emotional speech synthesis (2023)
30. Tang, H., Zhang, X., Wang, J., Cheng, N., Xiao, J.: QI-TTS: questioning intonation control for emotional speech synthesis. In: 2023 IEEE International Conference on Acoustics, Speech and Signal Processing, pp. 1–5 (2023)

31. Tang, H., Zhang, X., Wang, J., Cheng, N., Xiao, J.: Learning speech representations with flexible hidden feature dimensions. In: 2023 IEEE International Conference on Acoustics, Speech and Signal Processing, pp. 1–5 (2023)
32. Tang, H., Zhang, X., Wang, J., Cheng, N., Xiao, J.: VQ-CL: learning disentangled speech representations with contrastive learning and vector quantization. In: 2023 IEEE International Conference on Acoustics, Speech and Signal Processing, pp. 1–5 (2023)
33. Tobing, P.L., Wu, Y., Hayashi, T., Kobayashi, K., Toda, T.: Non-parallel voice conversion with cyclic variational autoencoder. In: Kubin, G., Kacic, Z. (eds.) 20th Annual Conference of the International Speech Communication Association, pp. 674–678. ISCA (2019)
34. Toda, T., Black, A.W., Tokuda, K.: Voice conversion based on maximum-likelihood estimation of spectral parameter trajectory. IEEE Trans. Audio Speech Lang. Process. **15**(8), 2222–2235 (2007)
35. Toda, T., Nakagiri, M., Shikano, K.: Statistical voice conversion techniques for body-conducted unvoiced speech enhancement. IEEE Trans. Speech Audio Process. **20**(9), 2505–2517 (2012)
36. Wu, Z., Virtanen, T., Chng, E., Li, H.: Exemplar-based sparse representation with residual compensation for voice conversion. IEEE/ACM Trans. Audio Speech Lang. Process. **22**(10), 1506–1521 (2014)

Symbolic and Acoustic: Multi-domain Music Emotion Modeling for Instrumental Music

Kexin Zhu, Xulong Zhang, Jianzong Wang$^{(\boxtimes)}$, Ning Cheng, and Jing Xiao

Ping An Technology (Shenzhen) Co., Ltd., Shenzhen, China
jzwang@188.com

Abstract. Music Emotion Recognition involves the automatic identification of emotional elements within music tracks, and it has garnered significant attention due to its broad applicability in the field of Music Information Retrieval. It can also be used as the upstream task of many other human-related tasks such as emotional music generation and music recommendation. Due to existing psychology research, music emotion is determined by multiple factors such as the Timbre, Velocity, and Structure of the music. Incorporating multiple factors in MER helps achieve more interpretable and finer-grained methods. However, most prior works were uni-domain and showed weak consistency between arousal modeling performance and valence modeling performance. Based on this background, we designed a multi-domain emotion modeling method for instrumental music that combines symbolic analysis and acoustic analysis. At the same time, because of the rarity of music data and the difficulty of labeling, our multi-domain approach can make full use of limited data. Our approach was implemented and assessed using the publicly available piano dataset EMOPIA, resulting in a notable improvement over our baseline model with a 2.4% increase in overall accuracy, establishing its state-of-the-art performance.

Keywords: Piano emotion recognition · Music information retrieval · Multi-domain analysis

1 Introduction

The emotional aspect of music, commonly known as its affective content, holds significant importance and is often regarded as the essence of musical expression. The recognition of emotions in music, known as Music Emotion Recognition (MER), has emerged as a prominent topic and crucial objective within the field of Music Information Retrieval (MIR). This recognition process assumes paramount significance due to its widespread application in various scenarios involving emotion-driven music retrieval and recommendation. Restricted by the

---

K. Zhu and X. Zhang—These authors have equal contributions.

© The Author(s), under exclusive license to Springer Nature Switzerland AG 2023
X. Yang et al. (Eds.): ADMA 2023, LNAI 14179, pp. 168–181, 2023.
https://doi.org/10.1007/978-3-031-46674-8_12

complexity of emotion, research on MER has encountered great difficulties [9,31]. Emotion is a very complex psychological state, and different people have different emotional thresholds [32]. This makes emotional annotation more difficult and emotional data more scarce.

The recognition and understanding of the intricate interplay between various factors within music and their impact on music emotion constitute a central concern in ongoing research on MER. Investigating this matter not only facilitates the advancement of more efficient and nuanced MER techniques but also contributes to the development of comprehensive insights into the complex nature of music emotion. Existing research usually applies disentanglement or multi-domain analysis to modeling music emotion from multiple aspects. Berardinis et al. [1] applies Music Source Separation during pre-processing and analyze the emotional content in vocal, bass, drums, and other parts separately, their proposed method shows promising performance. Zhao et al. [37] provide a new perspective by modeling music emotion with both music content and music context, their proposed method applies multi-modal analysis on audio content, lyrics, track name, and artist name of the music.

To further explore the essence of music emotion, research was also carried out on instrumental music. In the field of psychology and affective computing, Laukka et al. [18] proposed a convincing music emotion perception model for instrumental music and concluded six factors that affect music emotion: Dynamics, Rhythm, Timbre, Register, Tonality, and Structure. Those factors reflect both the acoustic characteristics and the structural characteristics of the music. Laukka's model indicates the importance of incorporating both acoustic analysis and symbolic analysis for MER. Acoustic factors such as Dynamics and Timbre are highly related to the Arousal expression of the music but are not included in the symbolic representations of music. Therefore symbolic-only methods show relatively weaker performance on Arousal detection. Structural factors such as Tonality and Structure are highly related to the Valence expression. Although those factors are included in the acoustic domain, existing acoustic analysis methods can hardly learn the structural information without extra supervision. To incorporate all the important factors, both acoustic analysis and symbolic analysis are needed.

However, most existing MER methods for instrumental music are uni-domain and fail to model music emotion from multiple aspects. Existing researches mainly apply deep-learning-based methods on the acoustic domain or uses sequence-modeling methods on the symbolic domain representations of the music. In their recent publication on emotion recognition in symbolic music, Qiu et al. [30] introduced a pioneering approach utilizing the MIDIBERT model [4], a large-scale pre-trained music understanding model. At present, no existing research on Music Emotion Recognition (MER) for instrumental music integrates both acoustic and symbolic analyses. As a result, we present an innovative method in this study that encompasses music emotion modeling from both acoustic and symbolic perspectives. Given the representative nature of piano music within the instrumental domain, we implemented and conducted an evaluation of our proposed approach using the publicly available piano emotion dataset EMOPIA [16].

Our contribution can be summarized as follows:

- Inspired by existing psychology and affective computing research, we proposed a multi-domain emotion modeling method for instrumental music, which only needs audio input. Our method used a pre-trained transcription model to obtain symbolic representation, therefore can be used on each instrument that can be automatically transcribed.
- We designed a refined acoustic model with mixed acoustic features input and a transformer-based symbolic model. Both models showed promising performance.
- We implemented and evaluated our proposed method on the public piano emotion dataset EMOPIA [16]. Our method achieved state-of-the-art performance on EMOPIA with better consistency between Valence detection and Arousal detection performance.

2   Related Works

There have been many studies in the research field of MER. According to the different domains of focus, these studies include MER with acoustic-only and MER with symbolic-only studies. These works have promoted progress in MER, and there are also some points that can be improved.

2.1   MER with Acoustic-Only

In order to explore which part of the vocal or accompaniment music carries more emotional information, Xu et al. [36] used the sound source separation technology, combined with the 84-dimensional manual low-level features (such as Mel frequency cepstrum coefficient (MFCC), spectral center, spectral attenuation point, spectral flux, and other similar measures.), and then used a classifier to recognize music emotion. Coutinho et al. [6] extracted 65 Low-level Descriptors (LLDs) in a time window of 1 s and calculated their first-order difference to obtain a total of 130 low-level features, then calculated the mean and standard deviation of each LLD in one second, and finally formed a 260-dimensional feature vector, and then used Long Short-term Memory (LSTM) network to carry out regression prediction of dynamic V/A (Valence/Arousal) value. Fukayama et al. [8] proposed a method to adapt to aggregation by considering new acoustic signal input based on multi-stage regression. At the same time, a method of adjusting the aggregation weight is introduced to deal with the emotion caused by the new input that cannot be known in advance, and the deviation observed in the training data is utilized by using Gaussian process regression. Li et al. [19] introduced a novel approach to tackle dynamic emotion regression by leveraging Deep Bi-directional Long Short-term Memory (DBiLSTM) in a multi-scale regression framework. Moreover, the author also examined the influence of dissimilar sequence lengths between the training and prediction stages on the overall performance of DBiLSTM. By investigating this aspect, the study aimed to gain insights into the effects of such variations on the efficacy of the model. [23] uses the CNN network that can process local information with fewer parameters and the RNN network that can process context information, that is, the CRNN structure, which uses the least parameters than Media Eval 2015.

Other methods have achieved the best results in the dynamic regression prediction of emotion at that time. Huang et al. [14] introduced the attention mechanism into the music emotion classification task, and introduced the attention layer with short-term and short-term memory units into the deep convolution neural network for music emotion classification. Different weights are allocated on different time blocks (chunks), and the song-level emotion classification prediction is obtained through fusion. Liu et al. [22] regards music emotion recognition as a multi-label classification task, and uses convolutional neural networks and spectrum diagram to complete end-to-end classification. Chen et al. [2] considered the complementarity between CNN with different structures and between CNN and LSTM, and combined multi-channel CNN with different structures and LSTM into a unified structure (Multi-channel Convolutional LSTM, MCCLSTM) to extract advanced music descriptors. Choi et al. [3] employed a pre-trained convolutional neural network (CNN) feature, which was initially trained for music auto-tagging purposes. They then successfully transferred this CNN to various music-related classification and regression tasks, showcasing its adaptability and versatility. Similarly, Panda et al. [27] introduced a collection of innovative affective audio features to enhance emotional classification in audio music. The authors observed that conventional feature extractors primarily focus on low-level timbre-related aspects, neglecting essential elements like musical form, texture, and expressive skills. To address this limitation, the authors devised a novel set of algorithms specifically designed to capture information related to music texture and expression, effectively compensating for the significant gaps in music emotion recognition research.

2.2   MER with Symbolic-Only

Previous research employed manual extraction of statistical musical characteristics, which were subsequently inputted into machine learning classifiers to forecast the emotional aspects of notated music. Grekow et al. [10] conducted an analysis on classical music in MIDI format and extracted 63 distinct features. In a similar vein, Lin et al. [20] conducted a comparative investigation involving multiple features (audio, lyrics, and MIDI) extracted from the same music. Remarkably, they discovered that MIDI features exhibited superior performance in emotion recognition. Building upon this finding, the researchers utilized the JSymbolic library [25] to extract 112 advanced music features from MIDI files. Subsequently, Support Vector Machine (SVM) was employed to classify the data. Similarly, Panda et al. [28] employed various tools to extract features from MIDI files and utilized SVM for classification purposes.

More recent studies demonstrate a growing adoption of a symbolic music encoding technique similar to MIDI [26], which is gaining popularity among researchers. Additionally, deep learning models have emerged as the prominent approach in this field. Ferreira [7] devised a method to encode MIDI files into MIDI-like sequences, leveraging LSTM and GPT2 for sentiment classification purposes. This approach offers simplicity and efficiency. Drawing inspiration from the remarkable achievements of BERT, Chou et al. [5] introduced MidiBERTPiano, a large-scale pre-trained model utilizing CP representation. The proposed model showcases promising outcomes in various domains, including symbolic music emotion recognition. Highlighting the paramount importance of emotional expression in music's intrinsic structure, Liu et al. [30] proposed a straightforward multi-task framework for the symbolic MER task. Notably, this approach benefits from readily available labels for auxiliary tasks, eliminating the need for manual annotation of labels beyond emotion classification.

3   Methodology

The complete diagram illustrating the overall architecture of our proposed approach can be observed in Fig. 1. The structure contains two branches: the acoustic domain branch (marked in yellow) applies acoustic analysis on mixed acoustic features with a Conv-based acoustic encoder, and the symbolic domain branch (marked in blue) applies symbolic analysis on music score sequence by using a Transformer-based symbolic encoder. It is worth noting that the outputs of the two branches come from the same modality, that is, from the acoustic input, so they belong to different domains of the same modality.

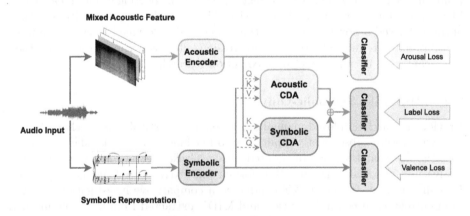

Fig. 1. The overall structure. The feature representations of the two domains are generated by the acoustic domain branch and the symbolic domain branch in the model, and the fusion process is completed in CDA.

3.1   Acoustic Domain Analysis for Arousal Modeling

For the acoustic domain analysis, we want to explicitly extract the information that relates to music emotion expressions, such as Timbre and Dynamics [18]. We use a mixed feature as input, which consists of the Mel-frequency Cepstral Coefficient (MFCC), Mel-spectrogram, Spectral Centroid (SC), and Root Mean Square Energy (RMSE) of the audio input. SC and RMSE reflect the energy distribution and changes of the audio, which is strongly correlated to music emotion expression. We use mel-spectrogram instead of STFT spectrogram because it better fits the human auditory perception process. We also calculate a 20-dimensional MFCC with librosa [24]. After these features are obtained, we resize and align them in the time dimension. The mixed feature can be obtained by splicing these features.

The processing flow of the acoustic domain branch is shown at the top of Fig. 1. We use a 2D-ConvNet module as the acoustic encoder for its great ability to encode temporal and frequency domain information simultaneously. After the feature extraction process, the extracted features are flattened and combined in the channel dimension to form the acoustic domain output. A comprehensive summary of the settings used in the experiment can be found in Table 1.

Acoustic domain analysis shows better performance on Arousal detection than symbolic domain analysis. Arousal is mainly decided by acoustic attributes such as Dynamics, Energy, and Timbre, which are not included in symbolic domain representation. Therefore we calculate an extra arousal classification loss function using Binary Cross Entropy (BCE) on the acoustic domain analysis branch during the training process.

3.2   Symbolic Domain Analysis for Valence Modeling

As mentioned above, our proposed method is designed to perform both acoustic and symbolic domain analysis with only audio input. That is to say, our symbolic part uses the automatic piano transcription module to form the symbolic domain representation instead of directly using the MIDI files in the EMOPIA dataset. This provides a common paradigm for other transcribable musical instruments. Therefore for the symbolic domain analysis branch, we use a pre-trained automatic transcription model to perform piano transcription. Specifically, we use the refined version of Onsets and Frames [11,12] proposed by Zhao et al. [38], which shows better generalizability and costs fewer computation resources. The transcribed piano score is converted into MIDI format, which includes the onset, offset, duration, and velocity of each note.

The music score is the "language" of the music and is a semantic sequence similar to natural language. Therefore the symbolic representation of the music score is similar to that of the natural language.

In this work, we use a refined MIDI-like representation for note embedding, which is shown in Fig. 2. Unlike the original MIDI-like [26] representation, we add an attribute named "harmonic" which explicitly denotes the number of sounding notes at the onset of a note. Since harmonic is an important part of musical performance, we decide to add extra information about it. Therefore, the symbolic domain representation for a single note consists of the onset time, harmonic, velocity, time shift, and offset time of the note.

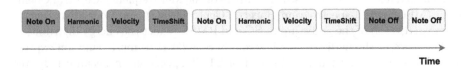

Fig. 2. The refined MIDI-like symbolic representation we used.

The structure of the symbolic domain analysis branch is shown at the bottom of Fig. 1. After the note embeddings are obtained, we input them into a Transformer encoder module [34] to extract the emotional representation of the piano score. The Transformer encoder module consists of four original Transformer encoder layers adopted in [34]. We pre-trained the encoder with the MIDI data from the MAESTRO dataset, for there are not enough samples in EMOPIA to train our Transformer encoder module.

Symbolic domain analysis mainly focuses on the high-level semantics of the note sequences, which leads to better Valence detection accuracy than acoustic analysis. As we want to make use of its advantage, we calculate an extra valence classification loss on the symbolic domain analysis branch during the training process.

3.3   Combining Symbolic and Acoustic Analysis

The final purpose of our method is to perform 4-Quadrant (4Q) classification concerning both Arousal and Valence, therefore the cross-domain feature fusion method is important. When combining extracted acoustic domain features and symbolic domain features, the Cross-domain Attention (CDA) module is used for cross-domain feature fusion. CDA has a similar mechanism to multi-head cross-modal attention [33]. In CDA module, Query and Key-Value pairs come from two different domains instead of different modalities in cross-modal attention. Each attention head can be calculated separately:

$$Attention(F_Q, F_K, F_V) = softmax(\frac{F_Q(F_K)^T}{\sqrt{d}})F_V$$

$$= softmax(\frac{F_\alpha W_Q(F_\beta W_K)^T}{\sqrt{d}})F_\beta W_V \qquad (1)$$

Let F_Q, F_K, and F_V denote the vectors for *Query*, *Key*, and *Value*, respectively. Within the attention mechanism, these input vectors are obtained by multiplying the extracted features of the α and β domains, represented as F_α and F_β, with their respective learnable weight matrices W_Q, W_K, and W_V. Here,

d represents the dimension size of the Key vector. The multi-head attention can be defined as the concatenation of each individual head:

$$MultiHead(F_\alpha, F_\beta) = Concat(head_1, ..., head_H)W_O \qquad (2)$$

$$head_i = Attention(F_\alpha W_Q^i, F_\beta W_K^i, F_\beta W_V^i) \qquad (3)$$

The learnable weight matrix W_O and the number of attention heads H play crucial roles in this multi-head attention mechanism. By leveraging multiple attention heads, this mechanism effectively highlights the significant aspects of each domain, which cannot be achieved through simple concatenation alone.

As shown in Fig. 1, in each processing procedure, our model calculates the CDA mechanism twice. We calculate an acoustic cross-domain attention mechanism and a symbolic cross-domain attention mechanism separately. This bidirectional CDA fusion strategy brings higher fusing efficiency. The output of acoustic CDA and symbolic CDA are concatenated and input into a classifier for 4Q emotion classification. During the training process, we calculate a 4Q Label loss on this classifier using Cross Entropy (CE) loss function.

4 Experiments

To assess the effectiveness of our proposed model, we conducted two primary types of experiments in this study: comparative studies and ablation studies. These experiments were designed to thoroughly evaluate and analyze the performance of our model from different perspectives.

4.1 Expriments Setup

We use the EMOPIA [16] dataset, which is an open-source dataset for piano-based emotion recognition. EMOPIA contains 1087 piano clips from 387 songs, all piano clips are annotated with their MIDI files and emotion labels. As only music metadata is available, we collect all music files by their corresponding YouTube ID with the 'youtube-dl' package. Following the configuration employed in [16], the dataset was divided into train-validation-test splits with a ratio of 7:2:1, ensuring appropriate proportions for training, validation, and testing stages. However, due to the unavailability of several music pieces on YouTube, we're only able to use approximately 90% data of the whole dataset. Similarly, we not only perform the classification of 4 quadrants but also carry out the binary classification tasks of high/low Valence and high/low Arousal. For the pre-training phase of the Automatic Piano Transcription model, we utilized the MAESTRO dataset ("MIDI and Audio Edited for Synchronous TRacks and Organization") [12], encompassing a comprehensive collection of more than 200 h of meticulously paired audio and MIDI recordings.

Table 1. Acoustic Encoder Settings.

Layer	Channel	Kernel Size	Stride	Maxpooling
conv1	64	3×3	1	2×2
conv2	128	3×3	1	2×2
conv3	256	3×3	1	2×2

During the training process, the training data is divided into mini-batches with a batch size of 64. The Adam optimizer [17] is employed, utilizing a learning rate of 0.0001. To implement all experiments, the PyTorch framework [29] is utilized.

It is important to note that MIDI files from the EMOPIA dataset were not utilized in our experiments. As our proposed model exclusively takes audio files as input, our aim is to evaluate the overall performance of the complete model, including the refined AMT module.

Table 2. Comparison with symbolic-domain methods on EMOPIA.

Method	4Q	A	V
LSTM-Attn [21]+MIDI-like [26]	.684	.882	.833
LSTM-Attn [21]+REMI [15]	.615	**.890**	.746
symbolic-LR [16]	.581	.849	.651
MIDIBERT [4]	.634	/	/
MT-MIDIBERT [4,30]	.676	/	/
proposed model	**.708**	.874	**.869**

4.2   Comparative Studies

We compared our proposed model with other existing methods on the same EMOPIA dataset. To the best of our knowledge, there is no existing multi-domain piano emotion recognition research. So we compared our model with several uni-domain symbolic-domain models proposed in [16] and [30], including two models based on BLSTM and self-attention mechanism (LSTM-Attn for short) using MIDI-like and REMI symbolic representation, a linear regression model based on hand-crafted features, and two pre-trained Bert-like models. For a fair comparison, we directly used the original results announced in their works. In [4,30], valence metrics and arousal metrics are not provided, therefore are not shown in the table.

Table 2 shows the comparison between our method and the other five symbolic-domain methods. All the methods show high and similar performance on Arousal detection, which indicates that Arousal detection is a relatively simple task. Due to the strong sequence-modeling ability of our transformer-based symbolic domain model, our method shows the highest Valence detection performance and outperforms the LSTM-Attn+MIDI-like model by 3.6%. On 4Q classification metrics, our model also achieves state-of-the-art performance and outperforms the LSTM-Attn+MIDI-like model by 2.4%.

We also compared our model with two existing acoustic-domain models, one uses linear regression on hand-crafted features and the other uses a ResNet-like network. Table 3 shows the comparison between our method and the other two acoustic-domain methods. All acoustic-domain methods show strong performance on Arousal detection as well. This is in line with common sense, because Arousal is greatly affected by energy, velocity, and dynamics, and this information is evident in acoustic information. Though our method is slightly weaker on Arousal detection, it still outperforms the Short-chunk ResNet model by 3.1% on the 4Q metrics.

Table 3. Comparison with acoustic-domain methods on EMOPIA.

Method	4Q	A	V
Audio-LR [16]	.523	**.919**	.558
Short-chunk ResNet [13,35]	.677	.887	.704
proposed model	**.708**	.874	**.869**

4.3 Ablation Studies

We designed and carried out a series of ablation studies to test the effect of our improvements. In the symbolic-only model and acoustic-only model, we use our symbolic branch and acoustic branch individually in order to test the effect of combining them. In the STFT-input model, we use an STFT spectrogram as input instead of the mixed acoustic feature. In the Single-loss model, we do not calculate the extra loss on the two branches and only calculate the Label loss.

The experimental results of the ablation studies are shown in Table 4. Compared to the two uni-domain models, our cross-domain fusion strategy costs performance loss on Arousal and Valence detection. However, our proposed model outperforms these two models by over 5% on the overall 4Q accuracy metrics. This indicates that our model is able to make better decisions by considering both symbolic and acoustic information.

The STFT-input model shows huge performance loss on Arousal metrics, which proves that using mixed acoustic features can improve Arousal detection performance. When using the STFT spectrogram as input, a deeper network is needed to extract the acoustic features. By using hand-crafted features, our method shows strong acoustic modeling ability with only three Conv layers. The single-loss model also shows over 2.5% performance loss on both 4Q and Arousal metrics, which indicates that our strategy of calculating the extra loss function works.

Table 4. Ablation studies trained and evaluated on the EMOPIA dataset.

Method	4Q	A	V
Symbolic-only	.651	.843	**.891**
Acoustic-only	.630	**.902**	.697
STFT-input	.689	.804	.871
Single-loss	.683	.845	.883
proposed model	**.708**	.874	.869

5 Conclusion

In this study, we introduce a novel multi-domain approach for piano emotion recognition, which can also be extended to other instruments with automatic transcription capabilities. Our proposed model leverages a pre-trained transcription model, enabling multi-domain analysis solely based on audio input. To the best of our knowledge, there is a lack of research specifically addressing piano emotion recognition. Our proposed model capitalizes on the complementary and redundant aspects between the acoustic and symbolic domains, leading to improved consistency in valence detection and arousal detection. Experimental results demonstrate that our proposed model surpasses the baseline approaches in terms of Valence classification and 4Q classification metrics. Moving forward, our future work will focus on designing enhanced symbolic representations for music, investigating superior cross-domain fusion strategies to enhance overall performance, and developing a universal framework for addressing the emotional aspects of transcribed musical instruments.

Acknowledgement. This paper is supported by the Key Research and Development Program of Guangdong Province under grant No. 2021B0101400003.

References

1. de Berardinis, J., Cangelosi, A., Coutinho, E.: The multiple voices of musical emotions: source separation for improving music emotion recognition models and their interpretability. In: Proceedings of the 21st International Society for Music Information Retrieval Conference, pp. 310–317 (2020)
2. Chen, N., Wang, S.: High-level music descriptor extraction algorithm based on combination of multi-channel cnns and lstm. In: Proceedings of the 18th International Society for Music Information Retrieval Conference, pp. 509–514 (2017)

3. Choi, K., Fazekas, G., Sandler, M., Cho, K.: Transfer learning for music classification and regression tasks. In: 18th International Society for Music Information Retrieval Conference, pp. 141–149. International Society for Music Information Retrieval (2017)
4. Chou, Y.H., Chen, I., Chang, C.J., Ching, J., Yang, Y.H., et al.: Midibert-piano: large-scale pre-training for symbolic music understanding. arXiv:2107.05223 (2021)
5. Chou, Y.H., Chen, I., Chang, C.J., Ching, J., Yang, Y.H., et al.: Midibert-piano: Large-scale pre-training for symbolic music understanding. arXiv:2107.05223 (2021)
6. Coutinho, E., Trigeorgis, G., Zafeiriou, S., Schuller, B.: Automatically estimating emotion in music with deep long-short term memory recurrent neural networks. In: Working Notes Proceedings of the MediaEval 2015 Workshop, vol. 1436, pp. 1–3 (2015)
7. Ferreira, L., Whitehead, J.: Learning to generate music with sentiment. In: Proceedings of the 20th International Society for Music Information Retrieval Conference, pp. 384–390 (2019)
8. Fukayama, S., Goto, M.: Music emotion recognition with adaptive aggregation of gaussian process regressors. In: 2016 IEEE International Conference on Acoustics, Speech and Signal Processing, pp. 71–75. IEEE (2016)
9. Gómez-Cañón, J.S., et al.: Music emotion recognition: toward new, robust standards in personalized and context-sensitive applications. IEEE Signal Process. Mag. **38**(6), 106–114 (2021)
10. Grekow, J., Raś, Z.W.: Detecting emotions in classical music from midi files. In: Foundations of Intelligent Systems: 18th International Symposium, pp. 261–270. Springer (2009)
11. Hawthorne, C., et al.: Onsets and frames: Dual-objective piano transcription. In: Proceedings of the 19th International Society for Music Information Retrieval Conference, pp. 50–57 (2018)
12. Hawthorne, C., et al.: Enabling factorized piano music modeling and generation with the MAESTRO dataset. In: 7th International Conference on Learning Representations (2019)
13. He, K., Zhang, X., Ren, S., Sun, J.: Deep residual learning for image recognition. In: Proceedings of the IEEE Conference on Computer Vision and Pattern Recognition, pp. 770–778 (2016)
14. Huang, Y.S., Chou, S.Y., Yang, Y.H.: Music thumbnailing via neural attention modeling of music emotion. In: 2017 Asia-Pacific Signal and Information Processing Association Annual Summit and Conference, pp. 347–350. IEEE (2017)
15. Huang, Y.S., Yang, Y.H.: Pop music transformer: beat-based modeling and generation of expressive pop piano compositions. In: Proceedings of the 28th ACM International Conference on Multimedia, pp. 1180–1188 (2020)
16. Hung, H., Ching, J., Doh, S., Kim, N., Nam, J., Yang, Y.: EMOPIA: a multi-modal pop piano dataset for emotion recognition and emotion-based music generation. In: Proceedings of the 22nd International Society for Music Information Retrieval Conference, pp. 318–325 (2021)
17. Kingma, D.P., Ba, J.: Adam: a method for stochastic optimization. In: 3rd International Conference on Learning Representations (2015)
18. Laukka, P., Eerola, T., Thingujam, N.S., Yamasaki, T., Beller, G.: Universal and culture-specific factors in the recognition and performance of musical affect expressions. Emotion **13**(3), 434 (2013)

19. Li, X., Tian, J., Xu, M., Ning, Y., Cai, L.: Dblstm-based multi-scale fusion for dynamic emotion prediction in music. In: 2016 IEEE International Conference on Multimedia and Expo, pp. 1–6. IEEE (2016)
20. Lin, Y., Chen, X., Yang, D.: Exploration of music emotion recognition based on midi. In: Proceedings of the 14th International Society for Music Information Retrieval Conference, pp. 221–226 (2013)
21. Lin, Z., et al.: A structured self-attentive sentence embedding. In: 5th International Conference on Learning Representations (2017)
22. Liu, X., Chen, Q., Wu, X., Liu, Y., Liu, Y.: Cnn based music emotion classification. arXiv:1704.05665 (2017)
23. Malik, M., Adavanne, S., Drossos, K., Virtanen, T., Ticha, D., Jarina, R.: Stacked convolutional and recurrent neural networks for music emotion recognition. arXiv:1706.02292 (2017)
24. McFee, B., et al.: librosa: audio and music signal analysis in python. In: Proceedings of the 14th Python in Science Conference, vol. 8, pp. 18–25. Citeseer (2015)
25. McKay, C., Fujinaga, I.: jsymbolic: a feature extractor for midi files. In: Proceedings of the 2006 International Computer Music Conference (2006)
26. Oore, S., Simon, I., Dieleman, S., Eck, D., Simonyan, K.: This time with feeling: learning expressive musical performance. Neural Comput. Appl. **32**(4), 955–967 (2020)
27. Panda, R., Malheiro, R., Paiva, R.P.: Musical texture and expressivity features for music emotion recognition. In: 19th International Society for Music Information Retrieval Conference, pp. 383–391 (2018)
28. Panda, R., Malheiro, R., Rocha, B., Oliveira, A., Paiva, R.P.: Multi-modal music emotion recognition: a new dataset, methodology and comparative analysis. In: International Symposium on Computer Music Multidisciplinary Research (2013)
29. Paszke, A., et al.: Pytorch: an imperative style, high-performance deep learning library. In: Advances in Neural Information Processing Systems 32 (2019)
30. Qiu, J., Chen, C., Zhang, T.: A novel multi-task learning method for symbolic music emotion recognition. arXiv:2201.05782 (2022)
31. Ru, G., Zhang, X., Wang, J., Cheng, N., Xiao, J.: Improving music genre classification from multi-modal properties of music and genre correlations perspective. In: ICASSP 2023–2023 IEEE International Conference on Acoustics, Speech and Signal Processing (ICASSP), pp. 1–5 (2023). https://doi.org/10.1109/ICASSP49357.2023.10097241
32. Tang, H., Zhang, X., Wang, J., Cheng, N., Xiao, J.: Emomix: emotion mixing via diffusion models for emotional speech synthesis. In: 24th Annual Conference of the International Speech Communication Association (2023)
33. Tsai, Y.H.H., Bai, S., Liang, P.P., Kolter, J.Z., Morency, L.P., Salakhutdinov, R.: Multimodal transformer for unaligned multimodal language sequences. In: Proceedings of the 57th Conference of the Association for Computational Linguistics, vol. 2019, p. 6558. NIH Public Access (2019)
34. Vaswani, A., et al.: Attention is all you need. In: Advances in Neural Information Processing Systems 30 (2017)
35. Won, M., Ferraro, A., Bogdanov, D., Serra, X.: Evaluation of cnn-based automatic music tagging models. arXiv:2006.00751 (2020)
36. Xu, J., Li, X., Hao, Y., Yang, G.: Source separation improves music emotion recognition. In: Proceedings of International Conference on Multimedia Retrieval, pp. 423–426 (2014)

37. Zhao, J., Ru, G., Yu, Y., Wu, Y., Li, D., Li, W.: Multimodal music emotion recognition with hierarchical cross-modal attention network. In: 2022 IEEE International Conference on Multimedia and Expo, pp. 1–6. IEEE (2022)
38. Zhao, J., Wu, Y., Wen, L., Ma, L., Ruan, L., Wang, W., Li, W.: Improving automatic piano transcription by refined feature fusion and weighted loss. In: Proceedings of the 9th Conference on Sound and Music Technology. pp. 43–53. Springer, Cham (2023). doi: https://doi.org/10.1007/978-981-19-4703-2_4

Document-Level Relation Extraction with Relational Reasoning and Heterogeneous Graph Neural Networks

Wanting Ji, Yanting Dong, and Tingwei Chen[✉]

School of Information, Liaoning University, Shenyang, China
twchen@lnu.edu.cn

Abstract. Document-level relation extraction aims to identify the relations between the entities in an unstructured text and represents them in a structured way for downstream tasks such as knowledge graphs and question answering. In recent years, graph neural network-based methods have made significant progress in relation extraction. However, these methods usually require extracting all the entities in the document first, then a classifier is used to analyze the relations between the entities regardless of whether they have any relation. This wastes a lot of time analyzing the relations of irrelevant entity pairs and reduces the classifier's attention to relevant entity pairs. To address this issue, this paper proposes a relation extraction module that integrates **R**elational **R**easoning and **H**eterogeneous **G**raph neural **N**etworks (RRHGN). The method finds a meta-path for each entity pair in a document and uses multi-hop reasoning to analyze the entities on the meta-path to determine whether there is a strong reasoning path between the entity pair. The relational reasoning module built into the method makes the classifier focus more on the relevant entity pairs in the document, thus reducing the task burden of the classifier and improving the accuracy of entity relation extraction. Experimental results on the large-scale document-level relation extraction dataset DocRED show that the proposed method achieves significant performance improvement compared with existing methods.

Keywords: Document-level relation extraction · Heterogeneous graph neural network · Multi-hop reasoning

1 Introduction

The purpose of document-level relation extraction is to extract the relations between different entities in a document and to represent them in a structured way. It plays an important role in natural language processing tasks such as information retrieval [1], question answering [2], and dialogue system [3]. Usually, document-level relation extraction involves a large number of entities, and

© The Author(s), under exclusive license to Springer Nature Switzerland AG 2023
X. Yang et al. (Eds.): ADMA 2023, LNAI 14179, pp. 182–195, 2023.
https://doi.org/10.1007/978-3-031-46674-8_13

Kungliga Hovkapellet

[1] *Kungliga Hovkapellet* (The *Royal Court Orchestra*) is a *Swedish* orchestra, originally part of the *Royal Court* in *Sweden*'s capital *Stockholm*. [2] The orchestra originally consisted of both musicians and singers. [3] It had only male members until *1727,* when *Sophia Schroder* and *Judith Fischer* were employed as vocalists; in the *1850s*, the harpist *Marie Pauline Ahman* became the first female instrumentalist.[4] From *1731* public concerts were performed at *Riddarhuset* in *Stockholm*, [5] Since 1773, when the *Royal Swedish Opera* was founded by *Gustav III* of *Sweden*, the *Kungliga Hovkapellet* has been part of the opera's company.

Subject: *Riddarhuset*
Object: *Sweden*
Relation: country
Supporting Evidence: 1, 4

Fig. 1. A example from the dataset DocRED.

these entities are sparsely distributed in multiple sentences that constitute a document. According to the statistics of human tagged corpus extracted from Wikipedia documents, more than 40.7% of entity relationship facts need to be jointly extracted from multiple sentences. Therefore, it is very necessary to study document-level relationship extraction methods [4,5]. Recently, some researches introduced graph data structure into the task of document-level relationship extraction [6–8]. The common way is to construct document-level heterogeneous graphs according to different entity types, and then encoding the graphs using attention mechanism, finally classifying the relationships among entities in the graphs using classifiers. However, such methods need to extract all the entities in a document first, and then classify the relationships among the entities. In this process, relation analysis for a large number of unrelated entity pairs not only distracts the attention of the classifier, but also reduces the efficiency of the classifier.

Figure 1 is an example in DocRED dataset. Entities (Riddarhuset, Sweden) are pairs of entities to be classified. They are located in sentences 1 and 4 respectively, and need to be obtained by relational reasoning between sentences. However, sentences 1 and 4 contain a large number of irrelevant entities (e.g., Kunglia Hovkapellet, Roval Court, 1731), and the relational classifier needs to classify them regardless of whether there is a relationship between these entities. Obviously, these irrelevant entity pairs distract the classifier. Usually, judging whether there is a relationship between two entities across sentences requires reasoning, and there is often a reasoning path for related entity pairs, such as sentences 1 and 4. We can judge that Stockholm is the capital of Sweden through the first sentence, and that Riddarhuset is an area of Stockholm through the fourth

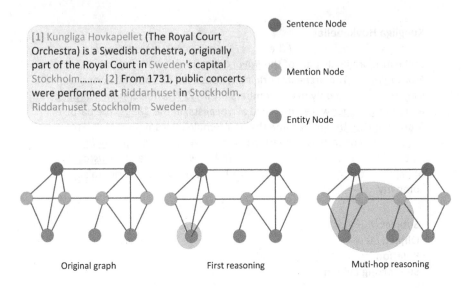

Fig. 2. The reasoning process of RRHGN.

sentence. Therefore, through reasoning between these two sentences, we can get (Riddarhuset, Sweden) that the relationship between the two entities is (country). If there is a relationship between two entities, we can find a reasoning path. However, the existing methods need to extract the relationship between entities regardless of whether there is a relationship between them, which greatly reduces the efficiency of task execution.

To solve the above problems, this paper proposes a document-level relationship extraction method based on relational reasoning and heterogeneous graph neural network (RRHGN). In this method, a relational reasoning module is built to judge whether there is a strong reasoning path between two entities, to predict the probability of relationship between these two entities. By constructing a dynamic graph structure, relational reasoning module is completed through multiple iterations on the selected meta-path. In each reasoning process, only the related nodes are reserved, and the irrelevant nodes are shielded, thus ensuring that all useful information is transmitted. Figure 2 shows the reasoning process of RRHGN, which tries to find a strong reasoning path between two entity pairs, so that the classifier pays more attention to those related entity pairs and completes the relationship extraction better. Gray denotes nodes in the reasoning process.

The main contributions of this paper are as follows:

(1) A relation reasoning module is proposed for relation extraction of graph structure to solve the problem that irrelevant entity pairs will distract the attention of entities in the process of relation reasoning using graph structure.
(2) Through multi-hop reasoning on meta-path nodes, it can be judged whether there is a strong reasoning path between entities.

(3) Experiments on DocRED, a large relational extraction dataset, show that the proposed method can accurately predict the relationships between entities.

2   Related Work

Existing document-level relation extraction methods can be roughly divided into sequence-based methods and graph-based methods.

2.1   Sequence-Based Document-Level Relationship Extraction Methods

Sequence-based document-level relationship extraction methods directly use neural networks to learn entity representations in documents, and classify all sent entity pairs. Zhou et al. [9] proposed a global context-enhanced graph convolution network model, which combined Transformer encoder with graph neural network, and considered both global and local dependencies among entities. Ye et al. [10] proposed a pre-training model based on BERT, which enhanced the reference reasoning ability of language representation by introducing reference resolution task, carried out document-level relation extraction experiments on DocRED dataset, and achieved very good results. Zeng et al. [11] proposed a model for separating intra-sentence and cross-sentence reasoning, which uses Transformer encoder to process each sentence and the whole document respectively, and uses graph convolution network to classify relations. Giorgi et al. [12] developed a sequence-to-sequence approach, seq2rel, that can learn the subtasks of DocRE (entity extraction, coreference resolution and relation extraction) end-to-end, replacing a pipeline of task-specific components. Liu et al. [13] proposed an effective structure enhanced transformer encoder model (SETE), integrating entity structural information into the transformer encoder. However, for long documents, sequence based methods are prone to losing semantic relationships and cannot effectively obtain global information.

2.2   Graph-Based Document-Level Relation Extraction Methods

Graph-based document-level relationship extraction methods often need to model documents according to the relationship between entities and sentences in documents, and use graph neural network [14] to build document graphs and learn the related information between entities. Some researches use graph convolution network (GCN) [15] to extract document-level relations, but these methods do not make full use of the global information of documents. To solve this problem, Sahu et al. [7] proposed a labeled graph convolution neural network model (GCNN), which uses cross-sentence and intra-sentence dependencies to capture local and non-local dependency information. Park et al. [16] used a graph structure and an entity attention awareness mechanism to capture the global information of documents. Hu et al. [17] proposed a multi-granularity interactive network (HAIN) to capture global information at three levels: word, sentence

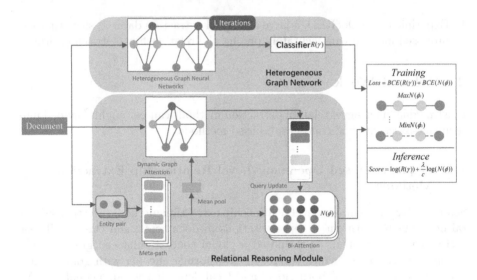

Fig. 3. Overall overview of the method.

and document-level. In addition, Zeng et al. [18] proposed a graph aggregation reasoning network (GAIN) with double graph features, considering that a single global graph can not get complete global information, which used multiple hierarchical networks to extract structured features. Nan et al. [19] used graph structure for multi-hop reasoning, which effectively solves the problems of long distance and implicit relationship. Sun et al. [20] proposed Dual-Channel and Hierarchical Graph Convolutional Networks (DHGCN), which constructed three graphs in token-level, mention-level, and entity-level to model complex interactions among different semantic representations across the document. Based on the multi-level graphs, they applied the Graph Convolutional Network (GCN) for each level to aggregate the relevant information scattered throughout the document for better inferring the implicit relations. Although these methods capture the global information well, they do not take into account that not all entity pairs need relation extraction, and some irrelevant entity pairs will distract the classifier's attention from related entity pairs. Therefore, this paper proposes a relational reasoning module based on graph structure, which tries to find a strong reasoning path through reasoning analysis on meta-path, and helps classifiers to extract relations better.

3   The Proposed Method

In this paper, a document-level relationship extraction method based on relation reasoning and heterogeneous graph neural network (RRHGN) is proposed for document-level relationship extraction. As shown in Fig. 3, the method mainly includes two parts: a heterogeneous graph network and a relational reasoning module. Heterogeneous graph network uses self-attention mechanism to encode entities in heterogeneous graph, and a multi-layer perception is used as a classifier to extract the relationship of entity pairs. Considering that most irrelevant entities will distract the attention of related entity pairs, this paper proposes a relational reasoning module, which starts from one of the entity pairs, makes reasoning analysis on the meta-path, calculates the probability of the existence of meta-path, and judges whether there is a relationship between entity pairs, so that the classifier pays more attention to related entity pairs, which increases the efficiency and accuracy of relational classification.

3.1   Heterogeneous Graph Network

Construction of Heterogeneous Graph. Referring to the construction of heterogeneous graph by Xu et al [27], this paper defines three types of nodes: Sentence node, Mention node and Entity node, and defines six types of edges: Sentence-Sentence (SS), Mention-Sentence (MS), Mention-Mention (MM), Entity-Mention (EM), Entity-Sentence (ES) and Mention-Coref- erence (CO). Therefore, a document can generate an adjacency matrix to represent the connection between nodes. The final document can be represented by a heterogeneous diagram $G = \{V, E\}$.

Encoder. Following the work of Guo et al. [21], we use graph attention network to encode each node in the heterogeneous graph to obtain an effective graph representation. Let h_n^l be the initial node, we first concatenate the outputs of all the previous l layers of $\{s_n^1, s_n^2, \cdot, \cdots, s_n^{l-1}\}$ and transform them into a fixed-dimensional vector:

$$h_n^l = W_e^l \cdot [v_n : s_n^1 : s_n^2 : \cdots : s_n^{l-1}] \tag{1}$$

where $s_n^{l-1} \in \mathbb{R}^{d_0}, W_e^l \in \mathbb{R}^{d_0 \times (l \times d_0)}$. We use self-attention mechanism[20] to extract the feature relations of C neighbor nodes $\{h_{a1}^l, h_{a2}^l, \cdots, h_{ac}^l\}$ and h_n^l connected to v_n. Here, K and V are key-value matrices determined by the types of edges of the neighbor nodes:

$$s_n^l = \text{softmax}(\frac{h_n^l K^T}{\sqrt{d_0}})V \tag{2}$$

Finally, combine the node v_n and the relation information of the document through a non-linear layer to obtain the global information of the document:

$$q_n = \text{Relu}(W_0 \cdot [v_n : s_n^1 : \cdots : s_n^l] \tag{3}$$

The heterogeneous graph is finally represented as: $G = (q_1, q_2, \cdots, q_n)$.

Classifier. Classifier is a sigmod function with a multi-layer perceptron (MLP) to compute the probability of the relation:

$$R(r) = P(r|\{e_i, e_j\} = \text{sigmod}(\text{MLP}([q_i, q_j])) \tag{4}$$

3.2   Relation Reasoning Module

The classification strategy of classifying all entity pairs by the classifier is obviously unreasonable. Therefore, the relation reasoning module is used to judge whether there is a relation between entity pairs, so that the classifier can pay more attention to the related entity pairs and improve the efficiency and accuracy of classification.

Meta-Path. When there is a relation between two entity pairs, a strong reasoning path can usually be found to prove that there is indeed a relation between the entity pairs. Conversely, when there is no relation between two entity pairs, such a strong reasoning path cannot be found.

Therefore, we need to find such a strong reasoning path to prove that there is indeed a relation between the entity pairs. Hence, this paper defines three meta-paths to infer whether there is such a strong reasoning path between the entity pairs [23].

(1) Pattern recognition: In this form of reasoning, two entities are connected by a sentence, and the relation pattern is EM-MM-ME.
(2) Logical reasoning: In this form, two entities are connected by a common entity, and the relation pattern is EM-MM-CO-MM-ME.
(3) Coreference reasoning: In this form of reasoning, two entities appear in a sentence, and the relation pattern is ES-SS-SE.

Different entity pairs have one or more meta-paths between them. Therefore, we prioritize the meta-paths according to the priority: pattern recognition > logical reasoning > coreference reasoning. Many entity pairs have multiple paths for the same meta-path type, but according to the document writing habit, the entities that appear later are often replaced by pronouns, and the entities usually appear for the first time at the beginning of the article. Therefore, we choose the first meta-path that appears in the document.

Relation Reasoning. For each entity pair $\{e_{1n}, e_{2n}\}$, a meta-path $\phi_n = \{Q^1, Q^2, \cdots, Q^t\}$ can be found. A graph neural network is used to propagate the node information of an entity to its neighboring nodes. A dynamic graph attention mechanism is employed to simulate the reasoning process. In the reasoning stage, if each node needs to propagate information to its neighboring nodes, then the more relevant two nodes are, the more information will be propagated. This paper only allows the node information that is related to the query to be propagated. An attention network between the query and the entities is used to predict a mask m_t, which obtains the starting entity in the t-th reasoning step.

$$\tilde{q}^t = \text{Meanpool}(Q^t) \tag{5}$$

$$\gamma^{(t)} = \frac{\tilde{q}^t V^{(t+1)} e_i^t}{\sqrt{d_2}} \tag{6}$$

$$m^{(t+1)} = \text{sigmod}([\gamma_1^{(t+1)}, \cdots, \gamma_N^{(t+1)}]) \tag{7}$$

$$\tilde{E}^{(t)} = [m_1^{(t+1)} e_1^t, \cdots, m_N^{(t+1)} e_N^t] \tag{8}$$

where V_t is a linear projection function that multiplies the entity node with the mask, encouraging the required initial entities, and other unnecessary entities will be penalized, so this can limit the information dissemination of irrelevant nodes.

Then the graph attention method (GAT) [24] is used to compute the attention score α between a meta-path node and its neighbors:

$$h_i^{(t+1)} = U_t e_i^{(t+1)} + b_t \tag{9}$$

$$\beta_{i,j}^{(t+1)} = \text{LeakyReLU}(W_t^T[h_i^{(t+1)}, h_j^{(t+1)}]) \tag{10}$$

$$\alpha_{i,j}^{(t+1)} = \frac{\exp(\beta_{i,j}^{(t+1)})}{\sum_k exp(\beta_{i,k}^{(t+1)})} \tag{11}$$

where $U_t \in \mathbb{R}^{d_2 \times 2d_2}$ and $W_t \in \mathbb{R}^{2d_2}$ are linear projection parameters, and α represents the proportion of neighbor information assigned to neighbor entity j in row i.

Summing each node column-wise yields a new entity containing information gathered from neighbor nodes:

$$e_i^{(t+1)} = \text{ReLU}(\sum_{j \in B_i} \alpha_{j,i}^{(t+1)} h_j^{(t+1)}) \tag{12}$$

where B_i is the set of neighbors of entity i, and finally we get the updated entity embedding$E^{(t+1)} = [e_1^{(t+1)}, \cdots, e_N^{(t+1)}]$.

Relational reasoning consists of multiple steps, and the newly visited entity in the previous step will be the starting entity for the next step. Here we use the modified Bi-Attention network [25] to update the probability of query reasoning to the next step:

$$p(Q^{(t+1)}|Q^t) = \text{Bi-Attention}(Q^{(t)}, E^{(t+1)}, Q^{(t+1)}) \tag{13}$$

The probability of the final whole meta-path is expressed as:

$$N(\phi_n) = \prod_1^C p(Q_{(c+1)}^{(t+1)}|Q_c^t) \tag{14}$$

where C is the number of probabilities reasoned on the meta-path.

3.3 Path Reasoning

Using the relational reasoning module as a relational indicator, when classifying relations, the auxiliary classifier performs relational classification:

$$S(r) = \log(R(r)) + \lambda \cdot \frac{1}{C} \log (N(\phi_n)) \tag{15}$$

where λ is a hyper-parameter that controls the importance of relational reasoning.

3.4 Loss Function

When training the proposed method, this paper uses the binary cross loss to train the triplet (subject, object, relation) in the dataset, namely $\{\{e1_n^t, e2_n^t, r_n^t\}_{n=1}^{N_t}\}_{t=1}^{T}$, to optimize the parameters of the neural network.

Loss function for the heterogeneous graph network:

$$Loss_h = -\frac{1}{\sum_{t=0}^{T} N_t} \sum_{t=1}^{T} \sum_{n=1}^{N_t} \left\{ r_n^t \log\left(R\left(r_n^t\right)\right)\right\} + \left(1 - r_n^t\right) \log\left(1 - R\left(r_n^t\right)\right) \quad (16)$$

Loss function for the relational reasoning module:

$$Loss_r = -\frac{1}{\sum_{t=0}^{T} N_t} \sum_{t=1}^{T} \sum_{n=1}^{N_t} \left\{ r_n^t \log N\left(\phi_n\right)\right\} + \left(1 - r_n^t\right) \log\left(1 - N\left(\phi_n\right)\right) \quad (17)$$

where $r_n^t \in (0, 1)$. Finally, the whole loss of RRHGN is the sum of the heterogeneous graph network loss and the relational reasoning module loss:

$$Loss = Loss_r + Loss_h \quad (18)$$

4 Experiments

4.1 Dataset

This paper uses a widely used document-level relation extraction dataset DocRED for experiments. DocRED is a large-scale human-annotated document-level relation extraction dataset built from Wikipedia and Wikidata. DocRED contains 132,375 entities and 56,354 relational facts, which are annotated on 5,053 Wikipedia documents. It is currently the largest human-annotated document-level relationship extraction Dataset [4].

4.2 Experimental Setting

All the experiments in this paper are completed on the Ubuntu20.4 platform, the CPU uses Intel(R) Xeon(R) Platinum 8358P CPU @ 2.60GHz, and the graphics card uses NVIDIA A40 GPU. The language used is Python3.9, the encoder uses GloVe embedding (100d), the model optimizer uses Adam, and the optimal parameter settings of the model are shown in Table 1.

For evaluation, on DocRED, following Yao et al. [4], we use the widely employed F1 and Ign F1 as the evaluation metrics. Ign F1 refers to excluding the relational facts shared by the training and dev/test sets. F1 is defined as:

$$F1 = \frac{2 \times P \times R}{P + R}$$

where precision(P) and recall(R) are defined as:

$$P = \frac{\text{True Positives}}{\text{True Positives} + \text{False Positives}}$$

$$R = \frac{\text{True Positives}}{\text{True Positives} + \text{False Negatives}}$$

Table 1. Model parameter values.

parameter	values
learning rate	1e-4
l	2
batch size	32
epoch	300

4.3 Baseline

This paper compares the proposed method with existing sequence-based document-level relation extraction methods (convolution neural networks (CNN) [4], bidirectional LSTM (BiLSTM) [4], Context-Aware LSTM [4], HIN-Glove [26]) and graph-based document-level relation extraction methods (GAT [24], GCNN [8], EOG [6], AGGCN [21], LSR-Glove [19], GAIN-Glove [18]) in DocRED. The performance on the DocRED dataset was compared.

4.4 Experimental Results

Table 2 presents the experimental results of different document-level relation extraction methods on the DocRED dataset.

Table 2. Comparison of Model Experiment Results.

Model Name	Dev		Test	
	Ign F1	F1	Ign F1	F1
CNN	41.58	43.45	40.33	42.26
LSTM	48.44	50.68	47.71	50.70
BiLSTM	48.87	50.94	48.78	51.06
Context-Aware	48.94	51.09	48.40	50.70
HIN-GloVe	51.06	52.95	51.15	53.30
GAT	45.17	51.44	47.36	49.51
GCNN	46.22	51.52	49.59	51.62
EOG	45.94	52.15	49.48	51.82
AGGCN	46.29	52.47	48.89	51.45
LSR-GloVe	48.82	55.17	52.15	54.18
GAIN-GloVe	53.05	55.29	52.66	55.08
RRHGN-GloVe(ours)	**54.23**	**55.80**	**53.47**	**55.54**

In the model based on Glove for word vector representation, the F1 of the method proposed in this paper is higher than that of the sequence-based and

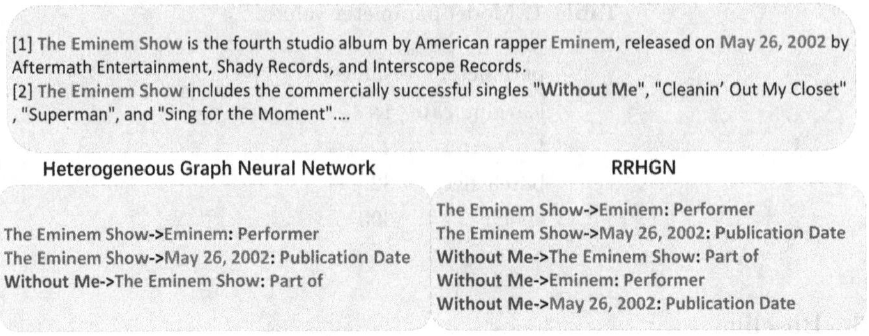

[1] The Eminem Show is the fourth studio album by American rapper Eminem, released on May 26, 2002 by Aftermath Entertainment, Shady Records, and Interscope Records.
[2] The Eminem Show includes the commercially successful singles "Without Me", "Cleanin' Out My Closet", "Superman", and "Sing for the Moment"....

Heterogeneous Graph Neural Network   **RRHGN**

The Eminem Show->Eminem: Performer
The Eminem Show->May 26, 2002: Publication Date
Without Me->The Eminem Show: Part of

The Eminem Show->Eminem: Performer
The Eminem Show->May 26, 2002: Publication Date
Without Me->The Eminem Show: Part of
Without Me->Eminem: Performer
Without Me->May 26, 2002: Publication Date

Fig. 4. The case study of our proposed RRHGN and baseline.

graph-based baseline models by 0.46-13.28 in the test set, and achieved good experimental results, reflecting the superiority of the RRHGN. The relational reasoning module in this paper judges whether there is a relationship between the entity pairs by analyzing the meta-path of the entity pair, so that the model can pay more attention to the entity pairs that have relationships, and is more conducive to relationship classification.

Table 3. Results of ablation experiments.

Model Name	F1
Heterogeneous Graph Network	53.52
Heterogeneous Graph Network + Relational Reasoning Module(ours)	55.54

4.5   Ablation Experiments

In order to verify the gain effect of the relational reasoning module on heterogeneous graph network, this paper conducted an ablation experiment on the DocRED dataset, and the experimental results are shown in Table 3.

Among them, the heterogeneous graph network means that only heterogeneous graphs are used to directly extract relationships.

Experimental results show that RRHGN proposed in this paper makes a positive contribution on the task of relation extraction. RRHGN improves the original basic model by 2.02% points. It can be seen that the RRHGN model can improve the accuracy of relation extraction by constructing meta-paths and performing reasoning analysis.

4.6   Case Study

Figure 4 shows a case study of our proposed method RRHGN and baseline. Heterogeneous graph network can identify the relationships between entity pairs within the same sentence, but their performance across sentences is not ideal. RRHGN has achieved good results in cross sentence relationship extraction through relational reasoning.

5   Conclusion

This paper proposes RRHGN to solve the problem that a large number of irrelevant entity pairs distracts the classifier from relational entity pairs. The proposed method judges whether there is a relationship between entity pairs by reasoning and analyzing the meta-paths between entity pairs, and provides a basis for the classifier. RRHGN acts as an indicator to assist classifiers when classifying relations. Experiments show that the proposed method improves the accuracy and efficiency of relation extraction. Although the model proposed in this paper has a certain degree of improvement in the task of relation extraction, how to further reduce the expenditure of computing resources, and how to update and query through the simplest and most effective method while be considered in our future work.

References

1. Kadry, A., Dietz, L.: Open relation extraction for support passage retrieval: merit and open issues. In: Proceedings of the 40th International ACM SIGIR Conference on Research and Development in Information Retrieval, pp. 1149–1152 (2017)
2. Yu, M., Yin, W., Hasan, K.S., dos Santos, C., Xiang, B., Zhou, B.: Improved neural relation detection for knowledge base question answering. In: Annual Meeting of the Association for Computational Linguistics. Association for Computational Linguistics (ACL) (2017)
3. Young, T., Cambria, E., Chaturvedi, I., Huang, M., Zhou, H., Biswas, S.: Augmenting end-to-end dialog systems with commonsense knowledge (2017). arXiv preprint arXiv:1709.05453
4. Yao, Y., et al.: Docred: a large-scale document-level relation extraction dataset. In: Proceedings of the 57th Annual Meeting of the Association for Computational Linguistics, pp. 764–777 (2019)
5. Cheng, Q., et al.: Hacred: a large-scale relation extraction dataset toward hard cases in practical applications. In: Findings of the Association for Computational Linguistics: ACL-IJCNLP 2021, pp. 2819–2831 (2021)
6. Christopoulou, F., Miwa, M., Ananiadou, S.: Connecting the dots: Document-level neural relation extraction with edge-oriented graphs. In: Proceedings of the 2019 Conference on Empirical Methods in Natural Language Processing and the 9th International Joint Conference on Natural Language Processing (EMNLP-IJCNLP), pp. 4925–4936 (2019)
7. Sahu, S.K., Christopoulou, F., Miwa, M., Ananiadou, S.: Inter-sentence relation extraction with document-level graph convolutional neural network. arXiv preprint arXiv:1906.04684 (2019)
8. Li, B., Ye, W., Sheng, Z., Xie, R., Xi, X., Zhang, S.: Graph enhanced dual attention network for document-level relation extraction. In: Proceedings of the 28th International Conference on Computational Linguistics, pp. 1551–1560 (2020)

9. Zhou, H., Xu, Y., Yao, W., Liu, Z., Lang, C., Jiang, H.: Global context-enhanced graph convolutional networks for document-level relation extraction. In: Proceedings of the 28th International Conference on Computational Linguistics, pp. 5259–5270 (2020)

10. Ye, D., et al.: Coreferential reasoning learning for language representation. In: Proceedings of the 2020 Conference on Empirical Methods in Natural Language Processing (EMNLP), pp. 7170–7186 (2020)

11. Zeng, S., Wu, Y., Chang, B.: Sire: separate intra-and inter-sentential reasoning for document-level relation extraction. In: Findings of the Association for Computational Linguistics: ACL-IJCNLP 2021, pp. 524–534 (2021)

12. Giorgi, J., Bader, G., Wang, B.: A sequence-to-sequence approach for document-level relation extraction. In: Proceedings of the 21st Workshop on Biomedical Language Processing, pp. 10–25 (2022)

13. Liu, W., Zhou, L., Zeng, D., Qu, H.: Document-level relation extraction with structure enhanced transformer encoder. In: 2022 International Joint Conference on Neural Networks (IJCNN), pp. 1–8. IEEE (2022)

14. Zhou, J., et al.: Graph neural networks: a review of methods and applications. AI Open **1**, 57–81 (2020)

15. Zhang, S., Tong, H., Xu, J., Maciejewski, R.: Graph convolutional networks: a comprehensive review. Comput. Soc. Networks **6**(1), 1–23 (2019)

16. Park, S., Yoon, D., Kim, H.: Improving graph-based document-level relation extraction model with novel graph structure. In: Proceedings of the 31st ACM International Conference on Information & Knowledge Management, pp. 4379–4383 (2022)

17. Hu, N., Zhang, T., Yang, S., Nong, W., He, X.: HAIN: hierarchical aggregation and inference network for document-level relation extraction. In: Wang, L., Feng, Y., Hong, Yu., He, R. (eds.) NLPCC 2021. LNCS (LNAI), vol. 13028, pp. 325–337. Springer, Cham (2021). https://doi.org/10.1007/978-3-030-88480-2_26

18. Zeng, S., Xu, R., Chang, B., Li, L.: Double graph based reasoning for document-level relation extraction. In: Proceedings of the 2020 Conference on Empirical Methods in Natural Language Processing (EMNLP), pp. 1630–1640 (2020)

19. Nan, G., Guo, Z., Sekulić, I., Lu, W.: Reasoning with latent structure refinement for document-level relation extraction. In: Proceedings of the 58th Annual Meeting of the Association for Computational Linguistics, pp. 1546–1557 (2020)

20. Sun, Q., et al.: Dual-channel and hierarchical graph convolutional networks for document-level relation extraction. Expert Syst. Appl. **205**, 117678 (2022)

21. Guo, Z., Zhang, Y., Lu, W.: Attention guided graph convolutional networks for relation extraction. In: Proceedings of the 57th Annual Meeting of the Association for Computational Linguistics, pp. 241–251 (2019)

22. Vaswani, A., et al.: Attention is all you need. Advances in neural information processing systems 30 (2017)

23. Sun, Y., Han, J.: Mining heterogeneous information networks: a structural analysis approach. ACM SIGKDD Explorations Newsl. **14**(2), 20–28 (2013)

24. Veličković, P., Cucurull, G., Casanova, A., Romero, A., Lio, P., Bengio, Y.: Graph attention networks. arXiv preprint arXiv:1710.10903 (2017)

25. Seo, M., Kembhavi, A., Farhadi, A., Hajishirzi, H.: Bidirectional attention flow for machine comprehension. arXiv preprint arXiv:1611.01603 (2016)

26. Tang, H., et al.: HIN: hierarchical inference network for document-level relation extraction. In: Lauw, H.W., Wong, R.C.-W., Ntoulas, A., Lim, E.-P., Ng, S.-K., Pan, S.J. (eds.) PAKDD 2020. LNCS (LNAI), vol. 12084, pp. 197–209. Springer, Cham (2020). https://doi.org/10.1007/978-3-030-47426-3_16

27. Xu, W., Chen, K., Zhao, T.: Document-level relation extraction with reconstruction. In: Proceedings of the AAAI Conference on Artificial Intelligence, vol. 35, pp. 14167–14175 (2021)

A Chinese Named Entity Recognition Method Based on Textual Information Perception Fusion

Wanting Ji, Lei Zhang, and Baoyan Song[✉]

Liaoning University, No. 66, Chongshan Middle Road, Huanggu District, Shenyang, Liaoning, China
bysong@lnu.edu.cn

Abstract. Named entity recognition (NER) aims to identify the entities with specific meanings from text, which is an important basic tool in many fields such as information extraction, question answering, and machine translation. In real-world, to perform named entity recognition tasks conveniently and quickly, it requires NER methods can quickly and accurately identify entities from the input text after being trained by a small training set for subsequent operations. The existing Chinese NER methods are usually based on BERT + LSTM + CRF models. However, when the training set is small, the recognition accuracy of these methods is relatively low. To solve this problem, this paper proposes a named entity recognition method based on Textual Information Perception Fusion (TIPF). It fully extracts the global features of the Chinese text through the Textual Information Memory Perception module, and fully fuses the global features and local features of the Chinese text using the Textual Information Adaptive Fusion module, so that the Chinese NER can be realized quickly and accurately. The proposed method is tested on several public datasets. Experimental results show that, compared with the existing methods, TIPF can achieve a higher recognition accuracy after training with a smaller training set.

Keywords: named entity recognition · deep neural network · natural language processing · data integration

1 Introduction

Named entity recognition (NER) is one of the fundamental tasks in natural language processing, which aims to identify entities with specific meanings from text [1], such as person, location, organization, etc. In the process of multi-source data integration, NER is often used in integration operations such as data cleaning and similar join to determine whether different records in one or more data sources are described as the same entity [2].

Recent research on Chinese NER mainly focuses on deep neural network-based methods [3], especially the methods based on BERT + LSTM + CRF [4] model, which contains a Bidirectional Encoder Representation from Transformers (BERT) [5] to encode the input text, a Long Short-Term Memory (LSTM) network [6] to extract features from the encoded text, and a Conditional Random Field (CRF) [7] to label entities

X. Yang et al. (Eds.): ADMA 2023, LNAI 14179, pp. 196–210, 2023.
https://doi.org/10.1007/978-3-031-46674-8_14

based on the extracted features. For example, Liu Bingyang [8] proposed a Chinese NER method incorporating global word boundary features, which suffered from the problem of word boundary errors when using the double-character combination boundary features for location recognition. Wei Zhu [9] proposed a Lex-BERT model that introduced entity type information. Although the model introduces entity information into the underlying part of BERT, which can improve recognition accuracy, the model relies heavily on high-quality vocabulary with entity type information during the recognition process, and the portability is poor.

More recently, the Lattice LSTM network [10] is proposed. It creates a thesaurus through text segmentation, and then encodes the character sequence of the text to be recognized and matches the latent words in the text according to the created lexicon. The above process is based on characters and makes full use of the word and word sequence information in the text. However, its accuracy will decrease rapidly when the number of training samples is reduced.

In the real world, some applications utilize NER methods to meet users' needs. In these applications [11], to interact with users in (near) real-time, the number of training samples of NER methods is usually reduced to reduce the training time, to ensure the high interactivity of the model. However, for the existing NER methods, reducing the training time by reducing the number of training samples will reduce the recognition accuracy of the model and even lead to under-fitting. Therefore, how to reduce the number of training samples under the premise of ensuring recognition accuracy is a problem to be solved.

To solve the above problems, this paper proposes a method based on Textual Information Perception Fusion (TIPF) for Chinese named entity recognition. It consists of two modules (a Textual Information Memory Perception module and a Textual Information Adaptive Fusion module) to extract the text-level information (global features) and the word-level information (local features). The proposed method is tested on several public datasets such as People's Daily 2014, Weibo NER, Boson, etc. Experimental results show that TIPF outperforms the state-of-the-art methods for Chinese NER, and can achieve a higher recognition accuracy than other methods after short-term training with a small number of training samples.

The main contributions of this paper are summarized as follows.

(1) A Textual Information Perception Fusion based method is proposed for Chinese named entity recognition.
(2) Based on a key-value memory network, a Textual Information Memory Perception module is proposed to extract text features to improve recognition accuracy.
(3) Based on an attention mechanism, a Textual Information Adaptive Fusion module is proposed to calculate the proportion of each information branch and improve the accuracy of information fusion.

This paper contains five chapters. Section 1 introduces the background knowledge related to named entity recognition and discusses the existing Chinese named entity recognition methods. Section 2 introduces the proposed method. Section 3 introduces the experimental environment used, and discusses and analyses the experimental results. Section 4 concludes the whole paper and gives future research directions.

2 The Proposed Method

This paper proposes a Chinese NER method based on Textual Information Memory Perception named TIPF. As shown in Fig. 1, TIPF consists of four components. The left part is the Textual Information Memory Perception module, the right part is the Lattice LSTM module, and the middle part is the Textual Information Adaptive Fusion module, the top part is the conditional random field module. Specifically, after encoding the input text sequence in character units, it is sent to the Lattice LSTM module for feature extraction, and each character generates a hidden vector that incorporates matching word information. Then, the hidden vector corresponding to each character is sent to the Textual Information Memory Perception module, stored in the key-value memory network, and the hidden vectors of all characters that are the same as the character are extracted from the key-value memory network to generate the corresponding context information. The context information of the character and the hidden vector of the character are calculated by the Textual Information Adaptive Fusion module to calculate their respective weights, and the two are fused by weighted summation. Finally, the hidden vector sequence fused with context information is sent to the conditional random field module for entity labeling. In this way, TIPF can integrate the feature information of the same words in different contexts, make full use of text-level information, and realize the memory perception of text information from a global perspective, to improve the accuracy of NER.

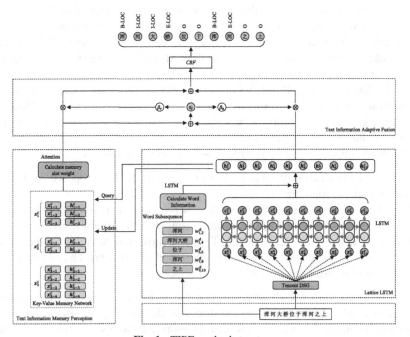

Fig. 1. TIPF method structure

2.1 Word Representation and Word Embedding

In this paper, we use an open-source Chinese word embedding data of Tencent AI Lab to encode the input text sequence [12]. To improve the labeling accuracy, character and word embedding will be fine-tuned during the training of NER.

For the text to be recognized by the named entity, let its length be m, and denote it as a sequence of characters: $C = \{c_1, c_2, \ldots, c_m\}$. For the j th character c_j in this sequence, its embedding is denoted as x_j^c:

$$x_j^c = e^c(c_j) \tag{1}$$

where e^* denotes an embedding lookup table. Using a Chinese word segmenter to divide the large raw text into words to generate the corresponding lexicon D, and form a sub-sequence of all words in the lexicon D that match the characters in the character sequence, which is represented by the form of $w_{b,j}^d$. As shown in Fig. 2, the starting values of b and j are 1, which respectively represent the start and the end subscript of the word in the character sequence. The word embedding of $w_{b,j}^d$ is:

$$x_{b,j}^w = e^w\left(w_{b,j}^d\right) \tag{2}$$

Both character sequences and word sub-sequences are used as the input of the model, and the BIOES tagging scheme [13] is used for NER tagging.

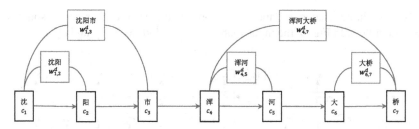

Fig. 2. Word representation

2.2 Lattice LSTM Module

This paper is based on the character-based LSTM model, and the LSTM network used to calculate the hidden vector of characters can be expressed as:

$$\begin{bmatrix} i_j^c \\ o_j^c \\ f_j^c \\ \tilde{c}_j^c \end{bmatrix} = \begin{bmatrix} \sigma \\ \sigma \\ \sigma \\ \tanh \end{bmatrix} \left(W^{c^T} \begin{bmatrix} x_j^c \\ h_{j-1}^c \end{bmatrix} + b^c \right) \tag{3}$$

$$c_j^c = f_j^c \odot c_{j-1}^c + i_j^c \odot \tilde{c}_j^c \tag{4}$$

$$h_j^c = o_j^c \odot \tanh\left(c_j^c\right) \tag{5}$$

where c_j^c serves to record recurrent information flow from the beginning of the sentence to c_j. When there is no related matching word in c_j, c_j^c can be calculated by the formula (4). h_j^c denotes the hidden vector corresponding to each character c_j. i^*, o^* and f^* denote a set of input, output and forget gates, σ represents the *sigmoid* function, W^* and b^* represent the weight and bias of the model.

When there are related matching words in c_j, the cell vector (word information) of the related matching words is calculated by the simplified LSTM model. Since the tagging process is based on characters, there is no need to calculate the hidden vector corresponding to the matching word when calculating the word information. Therefore, the output gate is removed based on the general LSTM model, and only $c_{b,j}^w$ is used to represent the recurrent state of $x_{b,j}^w$ from the beginning of the sentence, finally, a simplified LSTM model is obtained. The word embedding $x_{b,j}^w$ of the matching word is sent to the model to calculate the word information $c_{b,j}^w$.

$$\begin{bmatrix} i_{b,j}^w \\ f_{b,j}^w \\ \tilde{c}_{b,j}^w \end{bmatrix} = \begin{bmatrix} \sigma \\ \sigma \\ \tanh \end{bmatrix} \left(W^{wT} \begin{bmatrix} x_{b,j}^w \\ h_b^c \end{bmatrix} + b^w \right) \tag{6}$$

$$c_{b,j}^w = f_{b,j}^w \odot c_b^c + i_{b,j}^w \odot \tilde{c}_{b,j}^w \tag{7}$$

By combining character input and word input, each c_j^c has more sources of information flow. In Fig. 3, the input sources of c_7^c include x_7^c, $c_{6,7}^w$, and $c_{4,7}^w$.

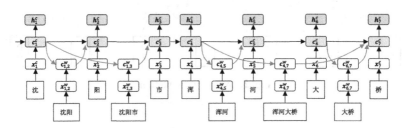

Fig. 3. Lattice LSTM module

Each $c_{b,j}^w$ is linked with $b \in \left\{ b' | w_{b',j}^d \in \mathbb{D} \right\}$ to the cell c_j^c. Since the importance of each word information $c_{b,j}^w$ to c_j^c is different, the gate control unit $i_{b,j}^c$ is used to control the proportion of each $c_{b,j}^w$, and the gate control unit i_j^c is used to control the contribution of the temporary unit state $\tilde{c}_j^c$ to c_j^c corresponding to x_j^c. Finally, normalize the $i_{b,j}^c$ and i_j^c to $\alpha_{b,j}^c$ and α_j^c. The calculation method is as follows:

$$i_{b,j}^c = \sigma\left(W^{lT} \begin{bmatrix} x_j^c \\ c_{b,j}^w \end{bmatrix} + b^l \right) \tag{8}$$

$$\alpha_{b,j}^{c} = \frac{exp\left(i_{b,j}^{c}\right)}{exp\left(i_{j}^{c}\right) + \sum_{b' \in \left\{b''|w_{b'',j}^{d} \in \mathbb{D}\right\}} exp\left(i_{b',j}^{c}\right)} \qquad (9)$$

$$\alpha_{j}^{c} = \frac{exp\left(i_{j}^{c}\right)}{exp\left(i_{j}^{c}\right) + \sum_{b' \in \left\{b''|w_{b'',j}^{d} \in \mathbb{D}\right\}} exp\left(i_{b',j}^{c}\right)} \qquad (10)$$

Therefore, when there are related matching words in c_j, c_j^c can be calculated as:

$$c_{j}^{c} = \sum_{b \in \left\{b'|w_{b',j}^{d} \in \mathbb{D}\right\}} \alpha_{b,j}^{c} \odot c_{b,j}^{w} + \alpha_{j}^{c} \odot \tilde{c}_{j}^{c} \qquad (11)$$

Finally, the LSTM hidden vector h_j^c corresponding to c_j is calculated as:

$$h_{j}^{c} = o_{j}^{c} \odot \tan h\left(c_{j}^{c}\right) \qquad (12)$$

Applying the input sequence $X = \left(x_1^c, x_2^c, \dots, x_m^c\right)$ to the bidirectional Lattice LSTM model, respectively, the bidirectional hidden vectors $\overrightarrow{h_1^c}, \overrightarrow{h_2^c}, \dots, \overrightarrow{h_m^c}$ and $\overleftarrow{h_1^c}, \overleftarrow{h_2^c}, \dots, \overleftarrow{h_m^c}$ can be obtained. For the character c_j, the hidden vector h_j^c after the fusion of related matching word information can be expressed as: $h_j^c = \left[\overrightarrow{h_j^c}, \overleftarrow{h_j^c}\right]$.

2.3 Text Information Memory Perception Module

In this paper, a key-value memory network [14] is constructed for sequence feature extraction to obtain text-level information. The key-value memory network is based on the key memory cache for addressing, and uses the value memory cache for reading, which has a good effect on storing global information.

First, the LSTM hidden vector h_j^c corresponding to each c_j is used as an extra knowledge source to help prediction [15], and the corresponding key-value memory network is generated. Key-value memory networks store text-level information based on "key-value" memory components. The character embedding x_j^c of the character c_j is used as the key, and the corresponding LSTM hidden vector h_j^c is used as the value. During entity recognition, it is stored in memory slots.

Memory slots are defined as pairs of vectors $((k_1, v_1) \dots (k_m, v_m))$, where m denotes the number of memory slots. So the same character occupies in many different memory slots because of changing representations and embeddings under different contexts. In each training epoch, each memory slot in the key-value memory network will be updated once. For example, if the value of the hidden vector corresponding to the i th character changes after calculation in a new training epoch, the value of the i th memory slot in the corresponding key-value memory network will be rewritten to the calculated value.

2.4 Attention Mechanism

For the i th character P_i in the entire training set, this paper all the contextualized representations for this character P_i in the key-value memory network through an inverted index [16] that find a subset $M_{P_i} : \left(\left(k_{s_1}, v_{s_1}\right) \ldots \left(k_{s_T}, v_{s_T}\right)\right)$ of size T, and the inverted index records the positions of the same character in the key-value memory network. T represents the number of occurrences of character P_i among the training set. In Table 1, the 16th character "河" can be found through the inverted index to find that its subscript set in the key-value memory network is [4, 10, 16,...].

Table 1. Training instance

1. 沈(0)阳(1)市(2)浑(3)河(4)大(5)桥(6)。
2. 位(7)于(8)浑(9)河(10)之(11)上(12)。
3. 沟(13)通(14)浑(15)河(16)两(17)岸(18)交(19)通(20)。
4. ...
Invert index: 河 [4,10,16...]

For the character P_i, the contribution of the memory slots in the contextualized representation M_{P_i} to the character P_i in the labelling process will be different depending on the context. Therefore, this paper needs to use the attention mechanism [17] to calculate the weight of the contribution of each memory slot in M_{P_i} to P_i.

For each character, in this paper, the key $k_j \in \left[k_{s_1}; \ldots; k_{s_J}\right]$ of the memory slot in the corresponding subset of the character is used as the attention key, the value $v_j \in \left[v_{s_1}; \ldots; v_{s_J}\right]$ of the memory slot is used as the attention value, and the character embedding x_{q_i} of the queried character serves as the attention query q_i. This paper uses Formula (13) to calculate the correlation degree $u_{ij} = o\left(q_i, k_j\right)$ of q_i and k_j:

$$o\left(q_i, k_j\right) = q_i k_j^T \tag{13}$$

Finally, according to u_{ij}, the contribution weight α_{ij} of the memory slot whose key is k_j to the character embedding x_{q_i} can be calculated:

$$\alpha_{ij} = \frac{exp\left(u_{ij}\right)}{\sum_{z=1}^{T} exp(u_{iz})} \tag{14}$$

For the character c_i, the text-level information r_i related to it can be calculated by weighted summation:

$$r_i = \sum_{j=1}^{J} \alpha_{ij} v_j \tag{15}$$

2.5 Text Information Adaptive Fusion

In order to integrate the text-level information r_i of the current character c_i with its corresponding hidden vector h_i^c more perfectly, the integration ratio of the two should be determined according to the context of each character. In this way, the more important parts of the two are better extracted, and the less important representations are discarded. In this paper, a Textual Information Adaptive Fusion module is designed, and its goal is to enable the neural network to adaptively adjust the proportion of the two fusions according to the recognition results. The basic idea is to use gates to control the information flow from text-level information r_i and hidden vector h_i^c, so that these branches carry information of different scales into the final hidden state. To achieve this goal, the gates need to integrate information from all branches, resulting in better fusion results [18].

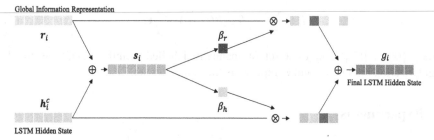

Fig. 4. Textual Information Adaptive Fusion module

The Fig. 4 shows the structure of the Textual Information Adaptive Fusion module. First, the results of the text-level information r_i and the hidden vector h_i^c are preliminary fused through element-wise summation to generate the feature descriptor s_i for precise and adaptive selection guidance:

$$s_i = h_i^c + r_i \tag{16}$$

Soft attention is used to adaptively select different dimensions of information, and β_r and β_h denote soft attention of r_i and h_i^c respectively.

$$\beta_r = \frac{exp(W_r s_i^T)}{exp(W_r s_i^T) + exp(W_h s_i^T)} \tag{17}$$

$$\beta_h = \frac{exp(W_h s_i^T)}{exp(W_r s_i^T) + exp(W_h s_i^T)} \tag{18}$$

The final hidden state g_i is obtained by multiplying r_i, h_i^c by their respective soft attention and adding them:

$$g_i = \beta_r \cdot r_i + \beta_h \cdot h_i^c \tag{19}$$

2.6 Conditional Random Field Module

The sequence $g_1, g_2, \ldots, g_m$ is sent to the standard CRF layer for decoding, and the probability of obtaining a label sequence $y = l_1, l_2, \ldots, l_m$ is:

$$P(y|s) = \frac{exp\left(\sum_i \left(W_{CRF}^{l_i} g_i + b_{CRF}^{(l_{i-1}, l_i)}\right)\right)}{\sum_{y'} exp\left(\sum_i \left(W_{CRF}^{l_{i'}} g_i + b_{CRF}^{(l_{i-1'}, l_{i'})}\right)\right)} \tag{20}$$

where y' represents an arbitrary label sequence. This paper uses the first-order Viterbi algorithm [19] to calculate the score of each label sequence, and the set of label sequences with the highest score will be used as the sequence labelling result. The loss function can be expressed as:

$$L = \sum_{i=1}^{N} log(P(y_i|s_i)) + \frac{\lambda}{2}\Theta^2 \tag{21}$$

where $\{(s_i, y_i)\}|_{i=1}^{N}$ denotes a set of manually labelled training data, λ is the L_2 regularization parameter, and Θ represents the parameter set.

3 Experiments

To demonstrate the accuracy and effectiveness of the proposed method, experiments are conducted on the People's Daily 2014 dataset, Weibo NER dataset and Boson dataset. The comparison methods include the current best-performing Lattice LSTM [10] method, the Lex-BERT [9] method, and the classic BiLSTM-CRF method [20]. In the same experimental environment, compare the performance of the TIPF method and existing methods and analyze the experimental results. The precision, recall and F1 value of are used as evaluation criteria to evaluate the performance of the method.

The specific information of the software and hardware environment of the training process is, CPU: i5-10210U; memory: 16GB; operating system: macOS 11; Python version: 3.8.12; PyTorch version: 1.10.0. The hyper-parameter settings of TIPF are shown in Table 2.

Table 2. Hyper-parameter values

Parameters	Value	Parameters	Value
char emb size	200	LSTM hidden	200
char dropout	0.1	LSTM layer	1
learning rate lr	0.015	lr decay	0.05

Table 3. Statistics of datasets

Datasets	Type	Train	Dev	Test
People's Daily 2014 (2K)	sentence	2.0k	0.3k	0.3k
	char	186.4k	20.6k	16.9k
People's Daily 2014 (22W)	sentence	22.0w	3.3w	3.3w
	char	1822.9w	268.8w	272.4w
Weibo NER	sentence	1.3k	0.3k	0.3k
	char	73.7k	14.5k	14.8k
Boson	sentence	1.8k	0.3k	0.3k
	char	412.3k	67.4k	65.2k

3.1 Experimental Datasets

This paper conducts experiments on the People's Daily 2014, Weibo NER and Boson datasets. The statistical results for each dataset are shown in Table 3. Due to the large corpus of the People's Daily, in order to simulate the environment of a small data set, this paper randomly intercepts part of the data for experiments (People's Daily 2014(2K)), and conducts experiments on the full data set (People's Daily 2014(22W)) of the People's Daily to verify the effectiveness of the Textual Information Memory Perception module in the TIPF method.

3.2 Experimental Results on Several Datasets

Table 4 shows the precision, recall and F1 value of TIPF and comparative methods for Chinese NER on the People's Daily 2014 (2K) dataset. Compared with other methods, TIPF achieves the highest performance on each item.

Table 5 shows the experimental results on the People's Daily 2014 dataset. It can be seen that after increasing the amount of training data, the recognition accuracy of the Lattice LSTM model has improved, but it is lower than the performance of the TIPF method trained with 2k pieces of data. This shows that in the case of a small dataset, TIPF can use the Textual Information Memory Perception module to fuse global and local features, thereby achieving better recognition accuracy.

Table 6 shows the recognition accuracy, recall and F1 value of TIPF and contrasting methods on the Weibo NER dataset. Compared with other methods, TIPF achieves the highest or equal performance on each item.

Table 7 shows the experimental results on the Boson dataset. Compared with other methods, TIPF achieved the highest F1 value, recognition accuracy and recall.

To sum up, comparing the experimental results of Weibo NER, People's Daily 2014 and Boson in Table 4, 5, 6 and 7, it is found that the performance improvement effect of the TIPF method on the People's Daily 2014 data set is more obvious. By analyzing the data samples in these three datasets, it is found that the People's Daily 2014 data set is based on the official news corpus of the People's Daily, and some of the sentences

Table 4. Results on the People's Daily 2014(2K) dataset

Models	Precision	Recall	F1
Lattice LSTM [10]	87.61	75.49	81.10
BiLSTM-CRF [20]	76.63	79.05	77.82
Lex-BERT [9]	89.77	76.28	82.48
TIPF	**91.97**	**80.53**	**85.87**

Table 5. Results on the People's Daily 2014 dataset

Models	Data Type	Precision	Recall	F1
Lattice LSTM [10]	People's Daily 2014 (2K)	87.61	75.49	81.10
Lattice LSTM [10]	People's Daily 2014 (22W)	90.86	78.25	84.08
TIPF	People's Daily 2014 (2K)	**91.97**	**80.53**	**85.87**

Table 6. Results on the Weibo NER dataset

Models	Precision	Recall	F1
Lattice LSTM [10]	51.69	34.52	41.39
BiLSTM-CRF [20]	49.32	35.16	41.05
Lex-BERT [9]	52.63	35.48	42.39
TIPF	**56.53**	**35.94**	**43.94**

Table 7. Results on the Boson dataset

Models	Precision	Recall	F1
Lattice LSTM [10]	72.58	65.32	68.76
BiLSTM-CRF [20]	70.00	64.24	67.00
Lex-BERT [9]	72.19	65.91	68.91
TIPF	**75.12**	**68.25**	**71.52**

to be recognized in the training set are all from the same news corpus, and the context of the sentences to be recognized is more closely related. In the Weibo NER dataset and the Boson dataset, there are fewer sentences from the same corpus, which makes the contextual connection of the sentences to be recognized relatively weak, resulting in different overall performances of TIPF on these three datasets.

The different performances of the TIPF method on the three datasets also illustrate the effectiveness of this method. In the People's Daily 2014 data set, since many sentences to be recognized are from the same news corpus and have close contextual relationships, the Textual Information Memory Perception module in the TIPF method can accurately store and memorize this context information, thereby improving the recognition efficiency of the named entity recognition. On the Weibo NER dataset and the Boson dataset, although the entity recognition accuracy of the TIPF method is generally lower than that on the People's Daily 2014 dataset, it is still higher than that of the existing methods. This shows that the TIPF method can fuse the global and local features of the data as much as possible to achieve better entity recognition results even in the face of data with weak contextual connections.

3.3 Time Analysis

The above experimental results show that the Textual Information Memory Perception module can achieve better recognition accuracy on small data sets. But for this paper, the ultimate purpose of reducing the number of training sets is to reduce the training time of the model. For this reason, we record the relevant time information in the model training process, so as to better analyze and compare the time spent by each model in obtaining a closer recognition result, and evaluate the time performance of each model.

People's Daily 2014 Dataset. As shown in Table 8, the total training time required for TIPF is much lower than that of other models to obtain the same F1 value on the People's Daily 2014 dataset. Compared with the Lattice LSTM model, the total training time required by TIPF is only 39% of the total training time of the Lattice LSTM model. Compared with the Lex-BERT model, the total training time required by TIPF is only 18% of the total training time of the Lex-BERT model. Compared with the BiLSTM-CRF model, the total training time required by TIPF is only 22% of the total training time of the BiLSTM-CRF model.

At the same time, the Lattice LSTM model achieved the highest F1 value of 84.08 after 31 rounds of training on the People's Daily 2014 (22W) dataset, while the highest F1 value of this model on the People's Daily 2014 (2K) dataset is 81.10. This shows that the recognition accuracy of the Lattice LSTM model decreases when the size of the training set decreases. However, TIPF achieved the best results after only 22 rounds of training on the People's Daily 2014 (2K) than the Lattice LSTM model trained on the full dataset (People's Daily 2014 (22W)). In addition, the total training time required by TIPF in a low-performance environment is nearly 95,000 s less than the total training time of Lattice LSTM in a high-performance environment, which is only 17% of the total training time of the Lattice LSTM model.

Weibo NER Dataset. Table 9 shows that the total training time required for TIPF is much lower than that of other models. Compared with the Lattice LSTM and BiLSTM-CRF models, the total training time required for TIPF is only 69% and 56% of the total training time of the remaining two models. In addition, compared with the Lex-BERT

Table 8. Comparison of total training time on the 2014 dataset of People's Daily

Models	Type	F1	Time(s)
Lattice LSTM [10]	People's Daily 2014 (2K)	81.10	17454.96
Time for TIPF to obtain the same F1 value			6786.81
BiLSTM-CRF [20]	People's Daily 2014 (2K)	77.82	23631.20
Time for TIPF to obtain the same F1 value			5080.26
Lex-BERT [9]	People's Daily 2014 (2K)	82.48	53238.30
Time for TIPF to obtain the same F1 value			9344.35
Lattice LSTM [20]	People's Daily 2014 (22W)	84.08	114546.83
Time for TIPF to obtain the same F1 value			18843.28

model, the total training time of the TIPF is only 83% of the total training time of the Lex-BERT model, saving nearly 1100 s of training time.

Table 9. Comparison of total training time on Weibo NER dataset

Models	F1	Time(s)
Lattice LSTM [10]	41.39	4943.24
Time for TIPF to obtain the same F1 value		3216.61
BiLSTM-CRF [20]	41.05	6037.9
Time for TIPF to obtain the same F1 value		3400.13
Lex-BERT [9]	42.39	6195.84
Time for TIPF to obtain the same F1 value		5091.17

Boson Dataset. Table 10 shows that the total training time required for TIPF is much lower than that of other models. Compared with the Lattice LSTM model, the total training time required by TIPF is only 22% of the total training time of the Lattice LSTM model. Compared with the Lex-BERT model, the total training time required by TIPF is only 34% of the total training time of the Lex-BERT model. Compared with the BiLSTM-CRF model, the total training time required by TIPF is only 21% of the total training time of the BiLSTM-CRF model.

Table 10. Comparison of total training time on the Boson dataset

Models	F1	Time(s)
Lattice LSTM [10]	68.76	43585.95
Time for TIPF to obtain the same F1 value		9452.47
BiLSTM-CRF [20]	67.00	29017.68
Time for TIPF to obtain the same F1 value		6041.27
Lex-BERT [9]	68.91	27833.21
Time for TIPF to obtain the same F1 value		9452.47

4 Conclusion

In recent years, introducing lexical information into models has been a research topic for Chinese named entity recognition tasks. In this paper, a method based on textual information memory perception is proposed for Chinese named entity recognition. The method fully integrates the global and local features of the text, and uses the LSTM hidden state corresponding to each character/word as an extra knowledge source for named entity recognition. In this way, after training with fewer training samples, the method can achieve a high recognition accuracy and meet the needs of named entity recognition for Chinese text. This paper tests the performance of the proposed method on multiple public datasets. Experimental results show that the proposed method can achieve better performance than state-of-the-art named entity recognition methods on small datasets.

References

1. Pan, L.A., Yg, A., Fw, B.: Chinese named entity recognition: the state of the art. ScienceDirect **473**(1), 37–53 (2022)
2. Jiao, K.N., Li, X., Zhu, R.C.: Overview of Chinese domain named entity recognition. Comput. Eng. Appl. **57**(16), 1–15 (2021)
3. Lauriola, I., Lavelli, A., Aiolli, F.: An introduction to deep learning in natural language processing: models, techniques, and tools. Neurocomputing **470**(1), 443–456 (2021)
4. Li, J.Z.: An improved Chinese named entity recognition method with TB-LSTM-CRF. Comput. Sci. Appl. **11**(3), 720–728 (2021)
5. Jia, C., Shi, Y., Yang, Q.: Entity enhanced BERT pretraining for Chinese NER. In: Proceedings of the 2020 Conference on Empirical Methods in Natural Language Processing, pp. 6384–6396. Association for Computational Linguistics (2020)
6. Jia, C., Zang, Y.: Multi-Cell compositional LSTM for NER domain adaptation. In: Proceedings of the 58th Annual Meeting of the Association for Computational Linguistics, pp. 5906–5917. Association for Computational Linguistics, Stroudsburg (2020)
7. Hobley, E.: Iterative named entity recognition with conditional random fields. Appl. Sci. **12**(1), 330 (2022)
8. Liu, B.Y., Wu, D.Y., Liu, X.R.: Chinese named entity recognition incorporating global word boundary features. J. Chin. Inf. Process. **31**(2), 86–91 (2017)

9. Zhu, W., Cheung, D.: Lex-BERT: enhancing BERT based NER with lexicons. In: International Conference on Learning Representations, pp. 3–7. OpenReview.net, Virtual Event (2021)
10. Yue, Z, Jie, Y.: Chinese NER using lattice LSTM. In: Proceedings of the 56th Annual Meeting of the Association for Computational Linguistics, pp. 1554–1564. Publisher, Melbourne (2018)
11. Kai, Z., Na, S.: Text mining and analysis of treatise on febrile diseases based on natural language processing. World J. Trad. Chin. Med. **6**(01), 73–79 (2020)
12. Song, Y., Shi, S., Li. J.: Directional skip-gram: explicitly distinguishing left and right context for word embeddings. In: Proceedings of the 2018 Conference of the North American Chapter of the Association for Computational Linguistics: Human Language Technologies, pp. 175–180. Association for Computational Linguistics, New Orleans (2018)
13. Ratinov, L., Roth, D.: Design challenges and misconceptions in named entity recognition. In: Proceedings of the Thirteenth Conference on Computational Natural Language Learning, pp. 147–155. Association for Computational Linguistics, Boulder (2009)
14. Luo, Y., Xiao, F., Zhao, H.: Hierarchical contextualized representation for named entity recognition. In: Proceedings of the AAAI Conference on Artificial Intelligence, vol. 34, no. 5, pp. 8441–8448 (2020)
15. Lee, H., Jin, S., Chu, H.: Learning to remember patterns: pattern matching memory networks for traffic forecasting. In: International Conference on Learning Representations, pp. 78–85. OpenReview.net, Virtual Event (2021)
16. Jian, M.A., Zhang, T., Chen, Y.: New inverted index storage scheme for Chinese search engine. J. Comput. Appl. **33**(7), 2031–2036 (2013)
17. Hu, C., Cheng, J.: Named entity recognition based on character-level language models and attention mechanism. Int. Core J. Eng. **6**(1), 196–201 (2020)
18. Li, X., Wang, W., Hu, X.: Selective kernel networks. In: CVF Conference on Computer Vision and Pattern Recognition, pp. 128–130. IEEE, Long Beach (2019)
19. Wang, D.Z., Michelakis, E., Franklin, M.J.: Probabilistic declarative information extraction. In: 2010 IEEE 26th International Conference on Data Engineering, pp. 173–176. IEEE, Long Beach (2010)
20. Yin, J.Z., Luo, S.L., Wu, Z.T.: Chinese named entity recognition with character-level BLSTM and soft attention model. J. Beijing Inst. Technol. **103**(01), 63–74 (2020)

Aspect-Based Sentiment Analysis via BERT and Multi-scale CBAM

Qingsong Wang and Nianyin Yang[✉]

School of Information, Liaoning University, Shenyang, China
2054289205@qq.com

Abstract. Aspect-based sentiment analysis (ABSA), as a fine-grained sentiment analysis task, predicts polarities for given aspects in one text. However, in the process of aspect-based sentiment analysis, there may be multiple perspectives of sentiment-prone text information, as well as implicit sentiment expressions, which can cause greatly influence on the accuracy of sentiment analysis results. To address the above problems, in this paper, we propose an aspect-based sentiment analysis model BERT-MSCBAM based on BERT and Multi-Scale Convolutional Block Attention Module (MSCBAM). Our model firstly uses BERT to encode the input text, then uses the MSCBAM module, which extracts the deep semantic and key features by its structure of two Multi-Scale Channel Attention Modules (CAMs) with different scales and one Spatial Attention Module (SAM) between them, combining with ResNet, and finally obtains the prediction results through a fully connected layer. To validate the effectiveness of our BERT-MSCBAM model, we conduct experiments on the Restaurant and Laptop datasets of SemEval2014 Task 4, and the Twitter dataset for our model in comparison with several current mainstream models. The experimental results validate the effectiveness of our model.

Keywords: Aspect-based sentiment analysis · BERT · CBAM

1 Introduction

Traditional coarse-grained sentiment analysis of product reviews identifies the sentiment expressed in the whole review. However, in some cases, one review message may cover multiple perspectives, and the sentiment from different perspectives may be inconsistent. For example, a review on one product may express the view that the quality of the product is good, but the environment of the store is average while its location is not good, on which the coarse-grained sentiment analysis cannot cover the complete sentiment tendency and draw an inaccurate conclusion. To address this situation, Aspect-based sentiment analysis (ABSA) [1] can extract the different sentiment tendencies of corresponding aspects of one review text to reach a relatively comprehensive and accurate sentiment analysis conclusion. Aspect level sentiment analysis can identify the sentiment tendency of each given aspect word in one text, and thus draw a more effective sentiment analysis conclusion to avoid information loss.

X. Yang et al. (Eds.): ADMA 2023, LNAI 14179, pp. 211–225, 2023.
https://doi.org/10.1007/978-3-031-46674-8_15

However, when performing ABSA for one certain aspect, there maybe multiple other perspectives of sentiment-prone textual information, as well as implicit sentiment expressions, which can lead to information missing during training and cause great interference in the accuracy of sentiment analysis results.

To address the above problems, in this paper, we propose a deep learning model based on BERT and the Multi-Scale Convolutional Block Attention Module (MSCBAM). The main contributions of our work are as follows:

- A new aspect-based sentiment analysis model BERT-MSCBAM is proposed. The model firstly encodes the text with BERT to obtain a matrix with contextual semantic information; then processes the matrix with the MSCBAM module we propose in this paper, which extracts the deep semantic and key features by its structure of two Multi-Scale Channel Attention Modules (CAMs) with different scales and one Spatial Attention Module (SAM) between them, combining with ResNet; finally uses a fully connected layer to obtain the prediction results.
- The performance of BERT-MSCBAM is evaluated using the Restaurant and Laptop datasets from SemEval 2014 Task 4, and the Twitter dataset, which are widely used in the field of ABSA tasks. The experimental results show that our BERT-MSCBAM model achieves better results on all three datasets.

2 Related Work

Early solutions for ABSA tasks mainly used feature engineering-based approaches, which performed sentiment analysis through supervised learning. Machine learning models such as SVM [2] and decision trees [3] achieved good classification results in ABSA tasks, which are simple in structure and effective in operation. However, they rely too much on complicated pre-processing and feature engineering.

In recent years, as deep learning continues to make breakthroughs, various studies that using deep learning models for ABSA have also achieved good results. Dong et al. [4] conveyed the sentiment of words to the corresponding aspect words by combining Recurrent Neural Network (RNN) with syntactic analysis. Tang et al. [5] spliced the aspect words with sentence on their left as well as right, respectively, and input the results into Long Short-Term Memory Network (LSTM) to capture the connection between aspect words and the context, and then spliced the two together to obtain the integrated textual information. Convolutional Neural Networks (CNNs) can also be useful in aspect-based sentiment analysis tasks. Xue et al. [6] proposed a model Gated Convolutional network with Aspect Embedding (GCAE) based on CNN and gating mechanisms, which enabled information filtering and improved computational efficiency through gated Tanh-ReLU units. Zhang et al. [7] used a multilayer CNN to continuously reinforce the contextual feature information associated with aspect words.

With the wide application of attention mechanism, many researchers have also applied attention mechanism to ABSA tasks. Wang et al. [8] proposed ATAE-LSTM (Attention-based LSTM with Aspect Embedding) model that combined attention mechanism with LSTM, which gave different weights to different words when determining the sentiment tendency of different aspect words in a sentence through the attention mechanism, thus obtaining superior results than the traditional LSTM. Zhao et al. [9]

proposed a classification model based on a bidirectional attention mechanism and Graph Convolutional Network (GCN), which improved the model performance by capturing the sentiment dependencies among multiple aspects of words in a sentence.

Pre-trained language models such as Bert [10] are also effective for aspect-based sentiment analysis tasks. Xu et al. [11] proposed an ABSA method using BERT pre-trained models, which was jointly trained to achieve reading comprehension and aspect-based sentiment analysis. Dai et al. [12] demonstrated by experiments that trees induced by RoBERTa performed better than syntactic dependency trees, thus proving the effectiveness of RoBERTa for ABSA tasks.

3 Method

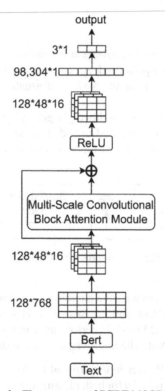

Fig. 1. The structure of BERT-MSCBAM

BERT-MSCBAM model consists of a text encoding layer, an attention focus layer and a fully connected layer, as shown in Fig. 1. The text encoding layer is responsible for converting the short text into a vector matrix by a BERT encoder; the attention focus layer extracts the deep semantic and key features from the vector matrix through the MSCAMs and the SAM of the MSCBAM, combining with ResNet; then the fully connected layer obtains the prediction results from the previous key features.

3.1 Text Encoding Layer

The dataset of ABSA task is $D = \{(s_i, a_i, y_i)\}_{i=1}^{N}$ and its tag set is $C = \{c_1, \ldots, c_k\}$. In the dataset, there is a total of N pieces of comment texts, and this ABSA task belongs to k classification problem. s_i is the i th comment text, a_i is its corresponding aspect word, and $y_i \in C$.

We first perform pre-processing by splicing each piece of comment text with each of its corresponding aspect words respectively to obtain texts with the same length d in the form of $[CLS] + s_i + [SEP] + a_i + [SEP]$, which are used as the input of the model. The specific input form is represented as $e_i = \{[CLS], w_1, w_2, \ldots, w_x, [SEP], z_1, \ldots, z_y, [SEP]\}$, e_i consists of d words.

Then each piece of text is converted into a vector matrix X^{d*h} by a BERT encoder with h hidden layers.

3.2 Attention Focus Layer

In the attention layer, we make improvement on the Convolutional Block Attention Module (CBAM) [13] to better perform deep semantic and key feature extraction of the text, enabling the model to better capture the key information in the sentence for a given aspect word.

CBAM, as shown in Fig. 2, is a lightweight general-purpose module that can be incorporated into various CNNs for end-to-end training. The module is simply and effectively designed, consisting of a CAM module and an SAM module.

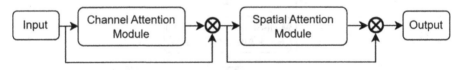

Fig. 2. The structure of CBAM

Like the original CBAM module, the attention focus layer of our model also contains the SAM, the similar CAM, and the ResNet mechanism; however, our attention focus layer improves the CBAM module performance in two aspects and we name this improved CBAM as MSCBAM. The improvement can be summarized in two aspects:

- We adjust the shared MLP layer in the original CAM module to three shared MLP layers of different sizes in parallel for better feature extraction and name the improved CAM as MSCAM;
- Compared with the original CBAM module consisting of one CAM and one SAM module to extract features serially, we add one MSCAM module after the SAM, and the scale of the Shared MLP layer of this added one is different from that of the first MSCAM.

Multi-scale Channel Attention Module. The original CAM module uses one Shared MLP layer for feature extraction, as shown in Fig. 3; while our MSCAM in this paper uses three parallel Shared MLPs instead of the original one Shared MLP layer, as shown in Fig. 4. The specific structure is as follows.

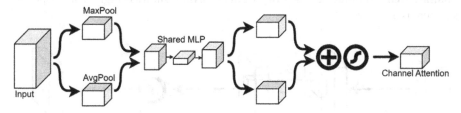

Fig. 3. The structure of CAM

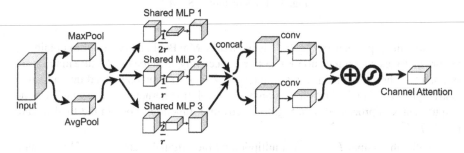

Fig. 4. The structure of MSCAM

Firstly, the input matrix $F_{c1}^{x_1*x_2*x_3}$ is passed through parallel MaxPool and AvgPool layers to get two output matrixes, in which process the size of the feature matrix is changed from $x_1*x_2*x_3$ to x_1*1*1; and then the two matrixes are passed through three Shared MLP modules of different scales in parallel, where the weights of the two are shared. In the three Shared MLP modules, the x_1 dimension is compressed into $\frac{1}{2r}$, $\frac{1}{r}$ and $\frac{2}{r}$ times from origin, respectively, and then expanded back; then three output matrixes of MaxPool and the three of AvgPool are processed by ReLU activation function and stitched together respectively, and then the two output matrixes are processed by convolution layer. Then the two obtained output matrixes are summed element by element, passed through a dropout layer, and then passed through a Sigmoid activation function to get the final output result $F_{c2}^{x_1*1*1}$ of MSCAM.

Finally, the output $F_{c2}^{x_1*1*1}$ is multiplied by the original matrix to obtain $F_{c3}^{x_1*x_2*x_3}$, transforming back to the size of $x_1*x_2*x_3$.

The equation of MSCAM is as follows:

$$
\begin{aligned}
\mathbf{M_c}(\mathbf{F}) &= \sigma\left(f^{7*7}\left[MLP_1(AvgPool(\mathbf{F})) + MLP_2(AvgPool(\mathbf{F}))+\right.\right. \\
&\quad MLP_3(AvgPool(\mathbf{F}))\right] + f^{7*7}[MLP_1(MaxPool(\mathbf{F}))+ \\
&\quad MLP_2(MaxPool(\mathbf{F})) + MLP_3(MaxPool(\mathbf{F}))]) \\
&= \sigma\left(f^{7*7}\left[MLP_{Group_r}\left(\mathbf{F_{avg}^c}\right) + MLP_{Group_r}\left(\mathbf{F_{max}^c}\right)\right]\right)
\end{aligned}
\tag{1}
$$

In the MSCAM, x_1, which is the word vector dimension of the text, is compressed and then recovered to the original size, while the dimensions x_2 and x_3, which are corresponding to the hidden layer size $x = x_2 * x_3$, is compressed to $1 * 1$. This module focuses on the meaningful information in each word vector.

Spatial Attention Module. The structure of the SAM is shown in Fig. 5, and its specific structure is as follows.

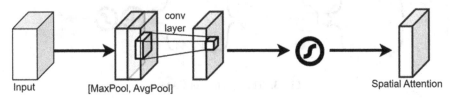

Input   [MaxPool, AvgPool]   Spatial Attention

Fig. 5. The structure of SAM

Firstly, the input matrix $F_{s1}^{x_1*x_2*x_3}$ is pooled by MaxPool and AvgPool to obtain the feature matrixes with dimension $1 * x_2 * x_3$ respectively, then the two feature matrixs are stitched together into a matrix with dimension $2 * x_2 * x_3$ and convolved into a feature matrix with size 1 in the first dimension by a convolution kernel of size $m * m$. Then the output matrix is processed by a Sigmoid activation function to obtain the final feature matrix $F_{s2}^{1*x_2*x_3}$.

Finally, the output $F_{s2}^{1*x_2*x_3}$ is multiplied by the original matrix $F_{s1}^{x_1*x_2*x_3}$ to obtain $F_{s3}^{x_1*x_2*x_3}$, transforming back to the size of $x_1 * x_2 * x_3$.

The equation of SAM is as follows:

$$\mathbf{M_s}(\mathbf{F}) = \sigma\left(f^{7*7}\left([AvgPool(\mathbf{F}); MaxPool(\mathbf{F})]\right)\right)$$
$$= \sigma\left(f^{7*7}\left[\mathbf{F_{avg}^s}; \mathbf{F_{max}^s}\right]\right) \tag{2}$$

In the SAM, the dimensions related to hidden layer keep their sizes while the dimension of the word vectors are compressed. This module focuses on the different attention weights corresponding to the different words in each piece of text.

Multi-scale Convolutional Block Attention Module. The structure of our proposed MSCBAM is shown in Fig. 6, which contains two MSCAMs of different scales and one SAM in the order of r_0 scale MSCAM, SAM, and $2 * r_0$ scale MSCAM.

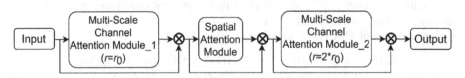

Fig. 6. The structure of MSCBAM

In the attention focus layer of our model, the text encoding matrix X^{d*h} is first transformed into $A_1^{d*h_1*h_2}$ ($h = h_1 * h_2$), and then input into the MSCBAM. Then the output of the module $A_2^{d*h_1*h_2}$ is added with the previous module input using ResNet mechanism and $A_3^{d*h_1*h_2}$ is obtained. Finally, the output of the attention focus layer $A_4^{d*h_1*h_2}$ is obtained by a ReLU activation function layer.

3.3 Output Layer

The key information obtained from the attention focus layer $A_4^{d*h_1*h_2}$ is transformed into $M^a(a = d * h_1 * h_2)$ and input into a fully connected layer to obtain the final output vector of dimension p, where p represents the number of different sentiment polarities of the ABSA task. The specific equation is as follows:

$$\hat{y} = argmax\left(M^a * W^{a*p} + b^p\right) \tag{3}$$

where W^{a*p} is a trainable parameter, b^p is the bias term; and $\hat{y}$ is the predicted sentiment polarity of the model.

ABSA is a classification task, so we choose Cross-Entropy as the loss function of our model, which is calculated as:

$$loss = -\sum_{i=1}^{c} y_i log\hat{y}_i \tag{4}$$

where y_i is the actual classification label.

4   Experiment

4.1   Experimental Datasets

To validate the effectiveness of our model, we experiment on the Laptop and Restaurant dataset from the publicly available dataset SemEval-2014 task4 [14] and the Twitter dataset [4], which contain comments and corresponding aspect words data, including Positive, Negative and Neural emotional polarities. Table 1 gives the statistics of the datasets used in our experiments (Tables 2 and 3).

Table 1. Statistical table of datasets

Dataset	Positive	Negative	Neutral
Laptop-train	994	870	464
Laptop-test	341	128	168
Restaurant-train	2164	807	637
Restaurant-test	728	196	196
Twitter-train	1561	1560	3127
Twitter-test	173	173	346

4.2   Experimental Environment

Table 2. Experimental environment

Experimental environment	Configuration
Operating System	Linux-5.15.65 + -x86_64-with-debian-bullseye-sid
Memory	13 GB
Programming Language	Python 3.7.12
GPU	Tesla P100-PCIE-16 GB
Deep Learning Framework	PyTorch 1.11.0
CUDA	11.0

Table 3. Experimental hyperparameters

Parameters	Value
BERT model	bert-base-uncased
Optimizer	adam
Word vector dimension d	128
BERT hidden layer dimension h	768
Number of Transformer layers	12
MLP descending coefficient r_0	16
SAM convolution kernel size $m * m$	3 * 3
Number of emotional polarity categories p	3
Dropout factor	0.5
Learning Rate	2e–5
Number of iterations	30
Batch Size	16

4.3 Experimental Hyperparameters

The strategy for training and optimizing our model is to use Adam optimizer to train this model. And in the optimization process, we use Cross-Entropy as the loss function.

4.4 Experimental Evaluation Indicators

We choose Acc and F1 as evaluation indicators for the experiments, and the equations for both are as follows:

$$P = \frac{TP}{TP + FP} \tag{5}$$

$$R = \frac{TP}{TP + FN} \tag{6}$$

$$Acc = \frac{TP + TN}{TP + FP + FN + TN} \tag{7}$$

$$F1 = \frac{2PR}{P + R} \tag{8}$$

4.5 Comparison Models

To verify the validity of this model, we compare our model with the following models in experiments on the Laptop14, Restaurant14 and Twitter datasets, using Acc and F1 values as evaluation metrics.

- feature-based SVM: The model uses traditional feature engineering for contextual feature extraction, and then uses support vector machine as a classifier for sentiment polarity classification.
- TD-LSTM: The model splices aspect words with their left and right sentences respectively, and then inputs the spliced texts into the LSTM network for encoding in order to capture the connection between the aspect words and their contexts, and then splices the two encoded texts to get the integrated text information.
- ATAE-LSTM: The model combines LSTM model with attention mechanism to capture the correlation between aspect words and each context word, and then makes integration to get the classification results.
- GCAE: The model is based on CNN and gating mechanism, in which gated Tanh-ReLU units are used to selectively extract sentiment information from the context based on aspect words.
- ASGCN: In the model, the syntactic dependency tree knowledge is used to construct the adjacency graph, and then the syntactic feature information of the context is extracted by the GCN network.
- AOA [15]: The model jointly extracts information on interaction features between context and aspect words through an attention-over-attention neural network structure.
- AEN-BERT [16]: The model is based on BERT, the semantic information between context and aspectual words is encoded by the attention mechanism.
- BERT-SPC [17]: The model uses BERT to conduct ABSA tasks.
- BERT4GCN [18]: The model integrates the syntactic order features of BERT PLM and the syntactic knowledge of dependency graphs.

4.6 Main Results

The main experimental results are shown in Table 4.

The models of this comparison experiment contain traditional machine learning models, LSTM models, CNN models, GCN models, attention models, and BERT models. In the above three datasets, the metrics of our model BERT-MSCBAM have all improved to a certain extent compared with the baseline models. Specifically, the accuracy of our model reaches 87.41%, 81.19%, and 75.14%, respectively, which increases by 2.95%, 1.26%, and 0.43% compared with the highest values in the baseline model, respectively;

Table 4. Model comparison results

Models	Restaurant14 dataset		Laptop14 dataset		Twitter dataset	
	Acc	F1	Acc	F1	Acc	F1
feature-based SVM	0.8106	–	0.7049	–	0.6340	0.6330
TD-LSTM	0.7563	0.6795	0.6813	0.6521	0.7080	0.6900
ATAE-LSTM	0.7720	0.7080	0.6870	0.6393	0.6864	0.6660
GCAE	0.7935	0.7052	0.7278	0.6710	0.7080	0.6766
ASGCN	0.8077	0.7202	0.7555	0.7105	0.7215	0.7040
AOA	0.7997	0.7042	0.7262	0.6752	0.7230	0.7020
AEN-BERT	0.8312	0.7376	0.7993	0.7631	0.7471	0.7313
BERT-SPC	0.8446	0.7698	0.7899	0.7503	0.7355	0.7214
BERT4GCN	0.8475	0.7711	0.7749	0.7301	0.7473	0.7376
BERT-MSCBAM	0.8741	0.8115	0.8119	0.7684	0.7514	0.7427

and the F1 values reaches 81.15%, 76.84%, and 74.27%, which increases by 4.17%, 0.53%, and 1.15% compared with the highest values in the baseline models, respectively.

Firstly, the results of our experiments prove the rationality and effectiveness of our BERT-MSCBAM model in this paper. Secondly, the results demonstrates that the SAM and CAM modules in CBAM are not only effective in the image processing field, but also can achieve good results in the natural language processing field. In our comparison experiment, the ASGCN model and the BERT-MSCBAM model both applies the combination of CNN model and attention mechanism. According to the experimental results, however, it can be revealed that the performance of our model BERT-MSCBAM significantly surpasses that of ASGCN, which shows that the combination of attention mechanism and CNN in our MSCBAM can more accurately extract the deep semantic information.

4.7 Model Analysis

Influence of Parameters. The MLP descending coefficient r_0 and the value of SAM convolution kernel size m are two important parameters of BERT-MSCBAM, so that their value selection can greatly influence the performance of the model. To explore the most suitable values of r_0 and m for BERT-MSCBAM in this paper, we conduct experiments on the Restaurant and Laptop datasets, and finally obtain the results shown in Fig. 7 and Fig. 8. It can be found from the experimental data that the best results are achieved when the value of r_0 is 16 and when the value of m is 3. Therefore, in our model, the value of r_0 is taken as 16 and the value of m is taken as 3.

Influence of Scale Sizes of MSCAM. To explore the effect of the scale sizes of MSCAM, we conduct experiments on the Restaurant and Laptop datasets. We select a total of 5 sets of combinations of different scales for the experiments, and the specific downscaling data are shown in Table 5.

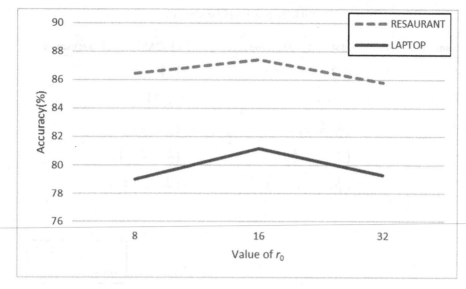

Fig. 7. Comparison experiment results with different values of r_0

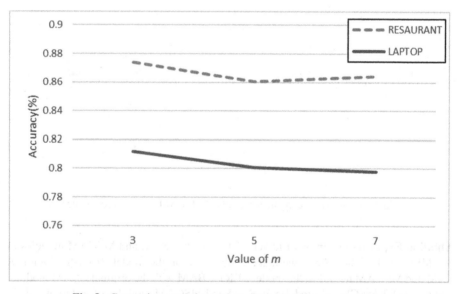

Fig. 8. Comparison experimental results with different values of m

The results of the experiments are shown in Fig. 9.

The experimental data shows that our model performs best when the number of scales is 3. Too few scales can result in insufficient information extraction of texts, while too many scales can lead to overfitting and prolong the training time. Therefore, we select 3 as the number of scales for our model MSCAM.

Table 5. Scale data for MSCAM

Number of scales	MSCAM_1 MLP downscaling	MSCAM_2 MLP downscaling
1	$\left(\frac{1}{r}\right)$	$\left(\frac{2}{r}\right)$
2	$\left(\frac{1}{r}, \frac{2}{r}\right)$	$\left(\frac{2}{r}, \frac{4}{r}\right)$
3	$\left(\frac{1}{2r}, \frac{1}{r}, \frac{2}{r}\right)$	$\left(\frac{1}{r}, \frac{2}{r}, \frac{4}{r}\right)$
4	$\left(\frac{1}{2r}, \frac{1}{r}, \frac{2}{r}, \frac{4}{r}\right)$	$\left(\frac{1}{r}, \frac{2}{r}, \frac{4}{r}, \frac{8}{r}\right)$
5	$\left(\frac{1}{4r}, \frac{1}{2r}, \frac{1}{r}, \frac{2}{r}, \frac{4}{r}\right)$	$\left(\frac{1}{2r}, \frac{1}{r}, \frac{2}{r}, \frac{4}{r}, \frac{8}{r}\right)$

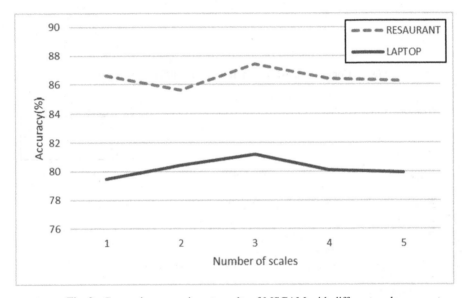

Fig. 9. Comparison experiment results of MSCAM with different scales

Ablation Experiment. In order to verify the effect of SAM and MSCAM modules in our MSCBAM of the model, we make comparison on the model BERT-SPC without SAM and MSCAM modules, the model BERT-CBAM with the original CBAM module, and the BERT-MSCBAM model with SAM and MSCAM modules of different orders. The experimental results are as shown in Table 6.

From Table 6, there are several conclusions can be drawn. Firstly, by comparing the results of BERT-CBAM model with the BERT-SPC model, it can be concluded that the introduction of CBAM module can make a small improvement in the performance of BERT model in ABSA tasks. Then, the small improvement obtained from the introduction of MSCAM module proves that our MSCAM can improve the performance of CBAM module in the ABSA tasks. And it can be demonstrated that the BERT-MSCBAM model in this paper achieves the best results compared with the BERT baseline, the baseline of the combination model of BERT and CBAM modules and all other combinations of BERT, MSCAM and SAM modules with different orders, which can indicate that our MSCBAM module can greatly improve the performance of the ABSA tasks, and that BERT-MSCBAM model can extract the deep semantic meaning of the text corresponding to the specified aspects more effectively and obtain more accurate results.

Table 6. Results of ablation experiments of BERT-MSCBAM model

Models	Restaurant14 dataset		Laptop14 dataset		Twitter dataset	
	Acc	F1	Acc	F1	Acc	F1
BERT-SPC	0.8446	0.7698	0.7899	0.7503	0.7355	0.7214
BERT-CBAM	0.8509	0.7912	0.7968	0.7461	0.7370	0.7163
BERT-MSCAM-SAM	0.8652	0.8072	0.7978	0.7531	0.7312	0.7186
BERT-MSCAM-MSCAM	0.8598	0.7947	0.7947	0.7589	0.7370	0.7268
BERT-SAM-MSCAM	0.8696	0.8118	0.7994	0.7630	0.7428	0.7367
BERT-MSCBAM	0.8741	0.8115	0.8119	0.7684	0.7514	0.7427

5   Conclusions

In this paper, we prove the effectiveness of the combination model of BERT and CBAM in ABSA tasks, propose an improvement on the CAM module in CBAM, which is named as MSCAM, and finally propose an ABSA model BERT-MSCBAM, which makes effective improvement to original CBAM module in ABSA tasks and combines the improved module MSCBAM with BERT. To verify the effectiveness of our model, we conduct extensive experiments in this paper using Restaurant14, Laptop14 and Twitter datasets, and the accuracy of the BERT-MSCBAM model on these three datasets reaches 87.41%, 81.19%, and 75.14%, respectively, with F1 values of 81.15%, 76.84%, and 74.27%, respectively, demonstrating that our model can effectively address specific ABSA tasks by comparing it with the mainstream methods of ABSA tasks in experimental results.

Although our model has achieved good results, the datasets we use in the current experiments are limited to the English language. Therefore, we will investigate the effectiveness of our model on ABSA tasks in other languages in the future works. And because the lengths of the sentences in the ABSA tasks vary greatly, key information may be lost in the pre-processing process, which may cause some impact on the ABSA results. In the future, we will further study on how to make improvement on text pre-processing methods. And in the process of experiments, we find that despite the use of dropout and other methods to prevent the overfitting of the model, a certain amount of overfitting still occurs in the training process. Therefore, we will also study on how to better avoid model overfitting in the future in order to further improve the performance of BERT-MSCBAM.

References

1. Liu, B., Zhang, L.: A survey of opinion mining and sentiment analysis. In: Mining Text Data, pp. 415–463. Springer, Heidelberg (2012). https://doi.org/10.1007/978-1-4614-3223-4_13
2. Luo, F., Li, C., Cao, Z.: Affective-feature-based sentiment analysis using SVM classifier. In: 2016 IEEE 20th International Conference on Computer Supported Cooperative Work in Design, pp. 276–281. IEEE, New York (2016)
3. Chauhan, C., Sehgal, S.: Sentiment classification for mobile reviews using KNIME. In: 2018 International Conference on Computing. Power and Communication Technologies, pp. 548–553. IEEE, New York (2018)
4. Dong, L., Wei, F., Tan, C., Tang, D., Zhou, M., Xu, K.: Adaptive recursive neural network for target-dependent twitter sentiment classification. In: Proceedings of the 52nd Annual Meeting of the Association for Computational Linguistics (Volume 2: Short Papers), pp. 49–54. Association for Computational Linguistics, Baltimore (2014)
5. Tang, D., Qin, B., Feng, X., Liu, T.: Effective LSTMs for target-dependent sentiment classification. In: Proceedings of COLING 2016, the 26th International Conference on Computational Linguistics: Technical Papers, pp. 3298–3307. The COLING 2016 Organizing Committee, Osaka (2016)
6. Xue, W., Li, T.: Aspect based sentiment analysis with gated convolutional networks. In: Proceedings of the 56th Annual Meeting of the Association for Computational Linguistics (volume 1: Long Papers), pp. 2514–2523. Association for Computational Linguistics, Melbourne (2018)
7. Zhang, C., Li, Q., Song, D.: Aspect-based sentiment classification with aspect-specific graph convolutional networks. In: Proceedings of the 2019 Conference on Empircal Methods in Natural Language Processing and the 9th International Joint Conference on Natural Language Processing, pp. 4568–4578. Association for Computational Linguistics, Hongkong (2019)
8. Wang, Y., Huang, M., Zhao, L., Zhu, X.: Attention-based LSTM for aspect-level sentiment classification. In: Proceedings of the 2016 Conference on Empirical Methods in Natural Language Processing, pp. 606–615. Association for Computational Linguistics, Stroudsburg (2016)
9. Zhao, P., Hou, L., Wu, O.: Modeling sentiment dependencies with graph convolutional networks for aspect-level sentiment classification. Knowl.-Based Syst. **193**, 105443 (2020)
10. Vaswani, A., et al.: Attention is all you need. In: Proceedings of the 31st International Conference on Neural Information Processing Systems, pp. 6000–6010. Curran Associates Inc., New York (2017)
11. Xu, H., Liu, B., Shu, L., Yu, P.: BERT post-training for review reading comprehension and aspect-based sentiment analysis. In: Proceedings of the 2019 Conference of the North American Chapter of the Association for Computational Linguistics: Human Language Technologies, pp. 2324–2335. Association for Computational Linguistics, Minneapolis (2019)
12. Dai, J., Yan, H., Sun, T., Liu, P., Qiu, X.: Does syntax matter? a strong baseline for aspect-based sentiment analysis with roberta. In: Proceedings of the 2021 Conference of the North American Chapter of the Association for Computational Linguistics: Human Language Technologies, pp. 1816–1829. Association for Computational Linguistics, Online (2021)
13. Woo, S., Park, J., Lee, J., Kweon, I.: Cbam: convolutional block attention module. In: Proceedings of the European Conference on Computer Vision, pp. 3–19. Springer, Munich (2018)
14. Pontiki, M., Galanis, D., Pavlopoulos, J., Papageorgiou, H., Androutsopoulos I., Manandhar, S.: SemEval-2014 task 4: aspect based sentiment analysis. In: Proceedings of the 8th International Workshop on Semantic Evaluation (SemEval 2014), pp. 27–35. Association for Computational Linguistics, Dublin (2014)

15. Huang, B., Ou, Y., Carley, K.M.: Aspect level sentiment classification with attention-over-attention neural networks. In: Thomson, R., Dancy, C., Hyder, A., Bisgin, H. (eds.) SBP-BRiMS 2018. LNCS, vol. 10899, pp. 197–206. Springer, Cham (2018). https://doi.org/10.1007/978-3-319-93372-6_22

16. Song, Y., Wang, J., Tao, J., Liu, Z., Rao, Y.: Attentional encoder network for targeted sentiment classification. In: 28th International Conference on Artificial Neural Networks, pp. 93–103. Springer, Cham (2019)

17. Devlin, J., Chang, M., Lee, K., Toutanova, K.: BERT: pre-training of deep bidirectional transformers for language understanding. In: Proceedings of the 2019 Conference of the North American Chapter of the Association for Computational Linguistics: Human Language Technologies, pp. 4171–4186. Association for Computational Linguistics, Minneapolis (2019)

18. Xiao, Z., Wu, J., Chen, Q., Deng, C.: BERT4GCN: using BERT intermediate layers to augment GCN for aspect-based sentiment classification. In: Proceedings of the 2021 Conference on Empirical Methods in Natural Language Processing, pp. 9193–9200. Association for Computational Linguistics, Punta Cana (2021)

A Novel Adaptive Distribution Distance-Based Feature Selection Method for Video Traffic Identification

Licheng Zhang[1,3], Shuaili Liu[1,3], Qingsheng Yang[1,3], Zhongfeng Qu[2], and Lizhi Peng[1,3(✉)]

[1] Shandong Provincial Key Laboratory of Network Based Intelligent Computing, University of Jinan, Jinan, China
plz@ujn.edu.cn
[2] School of Mathematical Sciences, University of Jinan, Jinan, China
[3] Quancheng Laboratory, Jinan, China

Abstract. The rapid proliferation of video applications in recent years has triggered an unprecedented surge in Internet video traffic, which in turn has presented substantial challenges for effective network management. However, existing methods for extracting features from video traffic primarily focus on conventional traffic attributes, resulting in suboptimal identification accuracy. Furthermore, the challenge of handling high-dimensional data is a common hurdle in video traffic identification, necessitating a robust approach to select the most pertinent features crucial for accurate identification. Despite the abundance of studies utilizing feature selection to enhance identification performance, there exists a notable lack of research that addresses the quantification of feature distributions with small or no overlap. This study proposes, firstly, the extraction of features relevant to videos, thereby assembling an expansive feature repertoire. Secondly, in the pursuit of forming an effective subset of features, the current research introduces the adaptive distribution distance-based feature selection (ADDFS) methodology. Using the Wasserstein distance metric to quantify the differences between feature distributions. To gauge the efficacy of this proposal, a dataset comprising video traffic from various platforms within a campus network environment was collected, and a series of experiments were conducted using these datasets. The experimental results indicate that the proposed method can achieve highly accurate identification performance for video traffic.

Keywords: Feature selection · Feature extraction · Video traffic identification · Network management

1 Introduction

At present, most Internet users watch videos daily, resulting in a rapid increase in video traffic. According to Ericsson's mobility report, video traffic is anticipated

---

L. Zhang and S. Liu—Contributed equally to this paper.

X. Yang et al. (Eds.): ADMA 2023, LNAI 14179, pp. 226–240, 2023.
https://doi.org/10.1007/978-3-031-46674-8_16

to comprise 80% of the complete mobile data traffic by the year 2028. Therefore, effective identification and management of video traffic, particularly game video traffic, have become an important research topic for network management.

Some researchers have explored video traffic identification over the past decade. In the early research of computer vision, video content identification often uses image shapes, textures and other features to complete [1]. However, this method is not applicable from the network traffic perspective. The rapid development of network traffic identification is helpful in solving this problem. Most existing researchers have extracted traffic features related to video transmission, such as application data unit, burst, etc., and used them thereafter to complete the prediction of video QoS and QoE, identification of video application type. Unfortunately, they did not focus on identifying video scene traffic. Additionally, cloud game, as an emerging game mode, is essentially a way of video flow transmission, which is potentially harmful to teenagers. As far as we know, there is no research reported about identifying cloud game traffic. Some researchers have also begun to focus on improving the QoE and QoS of cloud game traffic, but they have not approached it from the perspective of network traffic. Thus, to extract effective features for video identification becomes an urgent concern.

A key issue is that current research studies mainly focus on traditional features, these features can not achieve the ideal video identification effect, and further research on video scene traffic feature extraction is needed. Besides, another key problem that should be further discussed is that the quality of extracted or selected features can significantly and directly impact the performance of identification. Irrelevant or redundant features can cause unnecessary cost and time overhead, even negative impact for the model identification. Thus, a high-performance feature selection method is crucial for traffic recognition.

In response to the challenges outlined above, we present the following contributions.

- A novel method for adaptive distribution distance-based feature selection (ADDFS) is introduced.
- A new feature extraction method based on video traffic peak point is proposed, which can be used as an effective supplement of traditional packet and flow level features.
- Different kinds of video traffic data are collected, including cloud game video traffic and video scene traffic.

Roadmap: Sect. 2 introduces the related research. Sect. 3 reviews the video traffic identification method. Sect. 4 presents the experimental results. Lastly, this paper is concluded in Sect. 5.

2   Related Research

2.1   Video Traffic Identification

Three kinds of traffic identification methods have been used for video traffic identification: port-based, deep packet inspection, and machine learning algorithms.

The first two methods have become ineffective owing to the dynamic port and encryption techniques, making machine learning-based method a widely used technology. In 2012, Ameigeiras et al. [2] analyzed YouTube's video traffic generation pattern to predict the quality of video watching experience. Given that early YouTube videos were based on Flash, which is no longer used, this method is no longer effective for current video traffic. Reed et al. [3] proposed a new bit per peak feature extraction method, and used these features for classifying video stream titles.

At present, only a few researches focus on cloud gaming video traffic identification, and the existing study has primarily concentrated on the analysis and modeling of cloud gaming traffic and improving the cloud gaming experience. Suznjevic et al. [4] collected cloud gaming video samples to calculate video indicators from the time and space dimensions. Thereafter, they analyzed the relationship among game types, cloud gaming video traffic features, and video indicators. In 2015, Amiri et al. [5] proposed a paradigm for SDN controller to reduce cloud gaming delay. These studies rarely focus on identifying cloud game video traffic and video scene traffic, and this paper will focus on it.

2.2 Feature Selection

Feature selection is vital for traffic identification because all types of features are extracted from raw traffic data. Many of these features are redundant or with no contribution for identification. Therefore, researchers have attempted to develop effective methods to evaluate and select traffic features in recent years. Zhang et al. [6] and Mousselly et al. [7] used KL and JS divergence respectively to analyze the correlation and redundancy of different class labels, which can effectively deal with the fluctuation of feature samples. Nevertheless, their research did not address the issue of small overlap or no overlap between feature distributions.

Recently, certain researchers have started employing feature selection techniques for video traffic identification. Dong et al. [8] combined ReliefF and PSO to solve the excessive dimensionality problem in network traffic classification. Wu et al. [9] used a linear consistency-constrained method to select features for multimedia traffic classification and completed instance purification in the selection process. As far as we know, no study using has been conducted on distribution distance to measure the similarity between video traffic feature distributions. Therefore, this paper overcomes this drawback, by using Wasserstein distance to adaptively measure the similarity between feature distributions, and build an effective feature selection algorithm thereafter.

3   Methodology

This section describes the framework for video traffic identification, as shown in Fig. 1.

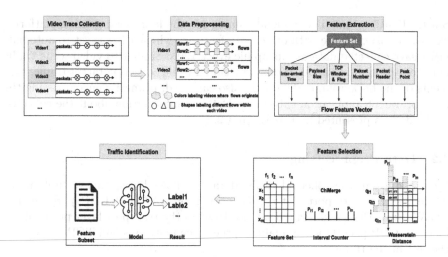

Fig. 1. The framework of the proposed video traffic identification method.

3.1 Data Collection

Only a few public video traffic data sets are available for video traffic identification research. Thus, a cloud gaming video traffic data set (CG-UJN-2022) and video scene traffic data set (VS-UJN-2022) in a controlled campus environment was collected.

Video Scene Traffic Data Collection. The collected video scene traffic data can be divided into two categories: static and action scene videos. The action scene video mainly consists of fragments from science fiction action films, such as *Pirates, Transformers, The Avengers*, etc. However, static scene videos have a simple scene, such as light music video, natural views, and class scenes. We collected both types of data from YouTube and Bilibili.

Videos from the mentioned categories will be initially downloaded to the client computer, followed by using FFmpeg to segment the original video into clips with a consistent duration of 120 s. We regard a 120 s video segment as a scene because such a segment can provide sufficient network features for coarse-grained video scene identification.

Secondly, with the Selenium library and Xpath Helper, fixed video clips are automatically uploaded to YouTube and Bilibili. With t-shark, we achieved automatic on-demand delivery of targeted videos and automatic collection of video traffic while playing the videos. During video playback on the client's computer, all other network applications are shut down to prevent the generation of extraneous traffic.

Cloud
Gaming Video Traffic Data Collection. YOWA cloud gaming, Tencent

Start, MiguPlay, and Tianyi cloud gaming are the four cloud gaming platforms we visited. To compare with other data features, wireshark was set to automatically save collected data as a.pcap file every 120 s. Similar to video scene traffic data collection, other applications were closed while collecting target traffic. Segments of the background traffic were also captured, primarily encompassing the most prevalent application categories. Detailed information about the collected traffic data is presented in Tables 1 and 2.

3.2 Data Preprocessing

First, We group the collected traffic data into flows based on five tuple information: {src IP, src port, dst IP, dst port, protocol (TCP/UDP)}. Since that YOWA, MiguPlay, and TianyiPlay use UDP as the transport layer protocol, we focus on UDP packets when analyzing the three platforms and TCP packets for the rest of the traffic.

Second, elephant flows are selected from the mice flows. Elephant flows is an important focus in this study, as video traffic is mostly elephant flows. The number of non zero payloads is used to eliminate mice flow. According to experience, those flows with under 500 packets are considered mice flows to be eliminated.

Lastly, the SNI extension field within the Client Hello packet serves the purpose of identifying whether the captured flow corresponds to the intended target flow.

Table 1. The details of video scene traffic data

Data Set	Platform	Flows	Bytes
VS-UJN-2022	YouTube	Action	4,994,062,797
		Static	3,294,957,584
	Bilibili	Action	5,956,426,782
		Static	7,672,791,660
CG-UJN-2022	START	83	3,401,167,754
	YOWA	58	2,116,354,048
	MiguPlay	61	2,548,042,936
	TianyiPlay	63	3,132,444,936

3.3 Feature Extraction

A total of 89 statistical features are extracted from preprocessed data in this study. We analyze the packet sequence features of each flow from three directions, namely upstream, downstream, and all packets. The traditional traffic features mainly include packet inter-arrival time (IAT), payload size, TCP window size, TCP flag, packet number, packet header. The detail are shown in Table 3.

Table 2. The details of background traffic data

	Flows	Bytes
Online Meeting	5	21,501,967
Chatting	56	134,386,569
PC Video	42	704,530,546
PC Live Steaming	40	397,123,864
Web Browsing	57	73,440,492
Online Shopping	42	142,277,293
File Download	40	57,409,424

Additionally, different video styles will lead to different traffic behavior patterns. Therefore, the maximum data transmission amount over a period of time will be defined as the peak point in this study.

Payload peak point (PPP). Assume there are d packets in a flow and the packets is $Pkt_1, Pkt_2, ..., Pkt_d$. Payload size of the sth packet is presented as pay_s. If $pay_s \geq pay_{s-1}$ and $pay_s \geq pay_{s+1}$ ($1 < s \leq d-1$), then payload reaches a peak in a certain period of time, which is defined as the PPP. A set of counters $c_1^l, c_2^l, ..., c_\theta^l$ was used to count the number of PPP every α s in the first β s of the lth flow, then the θ is calculated as follows:

$$\theta = \frac{\beta}{\alpha}. \tag{1}$$

Then, the count matrix CT can be obtained by traversing the entire flow sequence.

$$CT = \begin{bmatrix} c_1^1 & c_2^1 & \cdots & c_\theta^1 \\ c_1^2 & c_2^2 & \cdots & c_\theta^2 \\ \vdots & \vdots & \ddots & \vdots \\ c_1^t & c_2^t & \cdots & c_\theta^t \end{bmatrix} \tag{2}$$

Based on CT, the std and mean of the PPP of the tth flow is obtained as follows:

$$M_t = \frac{1}{\theta} \sum_{a=1}^{\theta} c_a^t, \tag{3}$$

$$Std_t = \sqrt{\frac{1}{\theta}[(c_1^t - M_t)^2 + (c_2^t - M_t)^2 + ... + (c_\theta^t - M_t)^2]}. \tag{4}$$

In a similar vein, the standard deviation and mean of PPP for all flows can be derived. Additionally, we extracted the maximum, minimum, and aggregate count of PPPs in three orientations. Nevertheless, in cases where certain scene videos are being consistently transmitted, alterations in packet payload remain insignificant. Hence, we introduce the concept of byte rate peak point (BRPP).

BRPP. Assuming that the summation of packet payloads (SPP) during period T is calculated in the following manner:

$$SPP = \sum_{b=1}^{H} pay_b, \tag{5}$$

where H is the total number of packets within T s, and pay_b is the size of the bth packet payload. Thereafter, the definition of byte rate (BR) in T seconds is as follows:

$$BR = \frac{SPP}{T}. \tag{6}$$

Similarly, If BR satisfies the criteria of being a peak point, it is labeled as BRPP. In this study, T is configured to be 1 s.

Table 3. The details of the extracted traditional features

Feature Name	Description
UIAT_*	Mean, Min, Max, Std of upstream IAT interval
DIAT_*	Mean, Min, Max, Std of downstream IAT interval
IAT_*	Mean, Min, Max, Std of all packets IAT interval
UWindow_*	Sum, Mean, Min, Max, std of upstream TCP window sizes
DWindow_*	Sum, Mean, Min, Max, std of downstream TCP window sizes
Window_*	Sum, Mean, Min, Max, std of all packets TCP window sizes
*_pnum	Number of packets for three directions
*_pnum_s	The rate of packet number for three directions
UDpnum_s	packets downstream to/packets upstream
*_cnt	TCP flag count
UDPSH,UDURG_cnt	Upstream and downstream PSH and URG count
Uhdr,Dhdr,hdr	Sum of packet header length for three directions
*_hdrR	the packet header length sum/the packet payload sum
Upay_*	Mean, Min, Max, Std of upstream payload
Dpay_*	Mean, Min, Max, Std of downstream payload
pay_*	Mean, Min, Max, Std of all packets payload

BRPP with sliding windows (BRPPSW). To catch continuous video information more accurately, we design sliding windows to extract the size of peak points as feature vectors based on BRPP. Length of the sliding window is L and offset factor is denoted by Z. In this study, L and Z are set to 3 and 0.5, respectively. For a packet sequence $(Pkt_1, Pkt_2, ..., Pkt_d)$, we calculate the sum of packet size under in time window L and use the offset factor thereafter to move the window to calculate the total packet size in turn. The sum of packet size in the zth window can be calculated as follows:

$$r_z = \sum_{pt=0}^{L} pktLen_{pt}, \tag{7}$$

where pt is the arrival time of the packet and $pktLen_{pt}$ is the packet size at ptth s. The processed sequence $R = (r_1, r_2, ..., r_n)$ is obtained, where n is the number of sliding windows. If the value in the sequence meets the definition of the preceding peak point, then the point is defined as BRPPSW. Therefore, we will obtain the sequence $R\_F = (r_1, r_2, ..., r_u)$ of BRPPSW, which is a subset of R.

We calculate the mean, std, maximum and minimum values of BRPPSW from three directions. The first, second, and third quartile of BRPPSW are also extracted as features.

3.4   Feature Selection

By the previous step, a comprehensive feature set is obtained. However, note that we do not consider whether these features are redundant or useless at the extracting process. In order to choose a feature subset that is both effective and concise, we introduce an approach called Adaptive Distribution Distance-Based Feature Selection (ADDFS).

Assuming a dataset $X = \{X_1, X_2, ..., X_n\}$, where X_i ($1 \leqslant i \leqslant$ n) represents the ith sample data, and m denotes the total number of samples. Moreover, x_{ij} denotes the value of the jth feature for the ith sample.

First, we employ Min-Max scaling to standardize all feature values across the dataset into the [0,1] interval. The formula for Min-Max scaling is as delineated below:

$$X_{ij} = \frac{X_{ij} - min(X_{.j})}{max(X_{.j}) - min(X_{.j})}, \tag{8}$$

here, $max(X_{.j})$ represents the maximum value of the jth feature, while $min(X_{.j})$ corresponds to the minimum value of the jth feature.

Second, the supervised ChiMerge algorithm [10] is used to divide each feature into multiple consecutive intervals. For each feature, we first sort all values in ascending order. Thereafter, we group the data with the same feature value into the same interval, and calculate the chi-square value of the interval. Each adjacent chi-square value is calculated and the smallest pair of intervals are merged. This step is repeated until the set maximum binning interval or chi-square stopping threshold is reached. Lastly, the chi-square binning interval of each feature is obtained. According to empirical values, the maximum binning interval and stop confidence threshold in this paper are set to 15 and 0.95, respectively. The chi-square calculation formula is as follows:

$$\chi^2 = \sum_{\gamma=1}^{G} \sum_{\psi=1}^{C} \frac{(A_{\gamma\psi} - E_{\gamma\psi})^2}{E_{\gamma\psi}}, \tag{9}$$

$$E_{\gamma\psi} = \frac{N_\gamma}{N} \times C_\psi, \tag{10}$$

where G stands for the number of intervals, and C represents the number of classes, $A_{\gamma\psi}$ represents the quantity of samples from the ψth class within the γth interval, $E_{\gamma\psi}$ is the expected frequency of $A_{\gamma\psi}$, and N, N_γ, and C_ψ denotes the overall sample count, the sample count within the γth interval, and the sample count within the ψth class, respectively.

For each feature, the number of samples of a particular feature within the chi-square binning interval in each class is counted. Take the feature F_j as an example. For class $C1$, the distribution of feature F_j within the chi-square binning intervals $(p_{11}, p_{12}, ..., p_{1k})$ can be acquired by tallying the occurrences of feature F_j across each interval, in which k is the number of chi-square binning intervals for this feature. For class $C2$, the distribution of feature F_j can be calculated as $(p_{21}, p_{22}, ..., p_{2k})$. On this basis, we can obtain the feature distribution matrix P of feature F_j on n classes. In the same manner, the feature distribution of other features on different classes can also be obtained.

$$P_{n \times k} = \begin{bmatrix} p_{11} & p_{12} & \cdots & p_{1k} \\ p_{21} & p_{22} & \cdots & p_{2k} \\ \vdots & \vdots & \ddots & \vdots \\ p_{n1} & p_{n2} & \cdots & p_{nk} \end{bmatrix} \tag{11}$$

The Wasserstein distance (EMD) is employed to quantify the distribution disparity between every pair of classes. A higher EMD value for a given feature across two classes indicates a more discerning characteristic. The computation of EMD for each class pair is conducted as follows:

$$W(P_U, P_V) = inf_{\gamma \sim \Pi(P_U, P_V)} E_{(U,V) \sim \gamma} \left[\|x - y\| \right], \tag{12}$$

where P_U and P_V are the feature distribution of a feature on two classes, $\Pi(P_U, P_V)$ denotes the set of all potential joint distributions P_U and P_V, while $W(P_U, P_V)$ signifies the mathematical lower bound of the expected value of $\gamma(x, y)$. The calculation of EMD for multi-class is detailed as follows:

$$EMD = \sum_{\kappa=1}^{C} \sum_{\lambda=\kappa+1}^{C} W(P_\kappa, P_\lambda), \tag{13}$$

Finally, calculate the EMD value for each feature. Subsequently, sort features in descending order based on their respective EMD values. The pseudo-code of ADDFS is shown in Algorithm 1.

Algorithm 1. Adaptive distribution distance-based feature selection algorithm

---

Require: Feature set F, classes C
Ensure: The selected feature subset S
1: BEGIN
2: compute F according to euqation (8);
3: for each $f \in F$:
4:      Interval=ChiMerge(f);
5:      / / The ChiMerge algorithm is used to divide each feature into multiple consecutive intervals
6:      for each $c \in C$:
7:          Pc=count(f,c,Interval);
8:          / / Calculate the number of samples of the feature in the chi-square binning interval of each class
9:      end for;
10:     for each $p \in P$:
11:         for each $p' \in P$, $p' \neq p$:
12:             compute W(p,p') according to equation; (12)
13:             EMD_f += W(p,p');
14:         end for;
15:     end for;
16: end for;
17: S= sort(EMD);
18: END;

3.5   Machine Learning Model

This study employs six machine learning models for identification. Noted that we do not focus on the actual machine learning model but on the effect of our proposed method combined with the machine learning model on video traffic identification. By comparing the identification results of different models, we can choose the model with superior performance for video traffic identification.

4   Experiment

4.1   Performance Measures

In this paper, accuracy (ACC) and F1 score can be derived as the evaluation criteria in our experiment. The accuracy (ACC) in a binary classification task can be defined as follows:

$$ACC = \frac{TP + TN}{TP + FN + TN + FP}, \tag{14}$$

Precision and recall can be defined as follows:

$$Precision = \frac{TP}{TP + FP}, \tag{15}$$

$$Recall = \frac{TP}{TP + FN}. \tag{16}$$

With precision and recall, F1 score, a widely used performance measure, can be derived as follows:

$$F1 = 2 \times \frac{Precision \times Recall}{Precision + Recall}. \tag{17}$$

4.2 Evaluation of ADDFS with Video Traffic Identification

The overall identification performance of video traffic is first evaluated by using the selected learning models and proposed feature selection algorithms. ADDFS is utilized to choose feature subsets comprising 10%, 20%, 30% ..., and 90% of the complete feature set. Thereafter, all selected learning models are used to identify both types of video traffic. The results are presented in Fig. 2.

From the perspective of the number of selected feature set, for YouTube and Bilibili, the identification effects of most of the learning models hit the optimum at 20% and 60%, respectively, of the feature set and reach a steady state thereafter. For cloud games, the recognition effect of the learning model maintains a small range of fluctuations on different feature subsets.

From a learning model perspective, Random Forest (RF), Extremely Randomized Trees (ET), and Adaptive Boosting (AdaBoost) perform well. In a stable state, RF and AdaBoost achieve accuracy levels exceeding 0.95 on YouTube. Furthermore, the accuracies of RF, ET, and AdaBoost on Bilibili and cloud gaming are above 0.92 and 0.99, respectively.

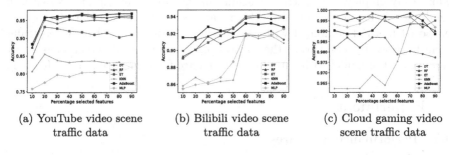

(a) YouTube video scene   (b) Bilibili video scene   (c) Cloud gaming video
      traffic data   traffic data   scene traffic data

Fig. 2. Accuracy results with varying feature number percentage selected by ADDFS

4.3 Assessment of the Efficacy of Peak Point Features

This subsection assesses the influence of various sliding window sizes and offset factors on video flow identification in cloud gaming. The Random Forest (RF) classifier is employed, and a 10-fold cross-validation approach is once again implemented.

Figure 3(a) and (b) shows the results of the comparison, in which FS is the complete feature set with peak point features, and FS-PP is the feature set without peak point features. The results of FS are observed to be better than those of FS-PP, particularly for data of video scene traffic on the YouTube platform. That is, ACC and F1 increased by over 3%. For the other two cases, the two evaluation measures also improved slightly with the joining of peak point features. Hence, the experimental outcomes unequivocally demonstrate the efficacy of the proposed peak point feature for video traffic identification.

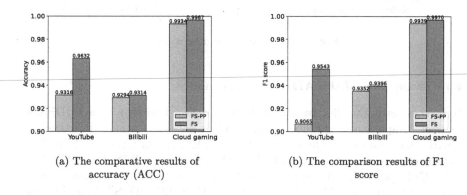

(a) The comparative results of accuracy (ACC)

(b) The comparison results of F1 score

Fig. 3. The comparison results with/without peak point features

4.4   Evaluation of the Impact of Sliding Windows

This subsection evaluates the impact of different sliding window sizes and offset factors on video flow identification on cloud gaming. RF is used as a classifier, and 10-fold cross-validation is again applied.

Figure 4(a) demonstrates the impact of different sliding window sizes on identification accuracy. Offset factor is set to 0.5. As window size grows, identification accuracy increases initially. Thereafter, it reaches the highest when window size is set to 3. Accuracy decreases thereafter as window size increases. Therefore, we obtain the empirical optimal window size of 3. Figure 4(b) shows the results with the varying offset value. Note that when offset factor is 0.5, accuracy of video traffic identification hits the highest value. When offset factor increases, recognition accuracy tends to be stable. Thus, we set window size L to 3 and the offset factor Z to 0.5 in our studies.

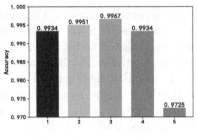 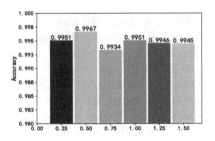

| (a) The accuracy of different sliding window sizes (s) | (b) The accuracy of different offset factor sizes (s) |

Fig. 4. The impact of sliding window

4.5 Evaluation of the ADDFS Performance

To further verify the effectiveness of the feature selection algorithm ADDFS, we conduct comparative experiments on 3 public datasets (wine, Mushroom and QSAR_biodegradat, the first one is from KEEL, the last two are from UCI) and 3 private traffic datasets (VS-UJN-2022-YouTube, VS-UJN-2022-Bilibili and CG-UJN-2022) with 5 feature selection algorithms. The five compared feature selection methods are Relief [11], Person [12], RFS [13], DDFS [14], and F-score [15]. We employ Decision Tree (DT), as the classifier and compare the ACC of the evaluated methods using 10-fold cross-validation. The classification ACC outcomes are illustrated in Fig. 5.

As shown in Fig. 5, all compared methods will receive increasing accuracy as the number of selected features increases for most data sets, and reach a relatively steady state thereafter. In cases where the count of chosen features is limited, ADDFS demonstrates superior accuracy when compared to the alternative methods. Note that it has consistently maintained efficient and stable performances for the cases of the wine, Mushroom, QSAR_biodegradat, and VS-UJN-2022-Bilibili datasets. Although there are numerous redundant and irrelevant features in the CG-UJN-2022 dataset, ADDFS can still obtain a relatively stable classification accuracy in the early stage.

5   Conclusion

A comprehensive feature set is constructed in this study for identifying video traffic. In order to obtain an efficient feature subset, a novel ADDFS method is introduced. Moreover, we collected video traffic data from different platforms in a campus network environment and used these data to conduct a set of experiments. The experimental findings demonstrate a significant enhancement in identification performance through the utilization of the proposed peak point feature. The proposed ADDFS can also be considerably applied to the task of video traffic identification.

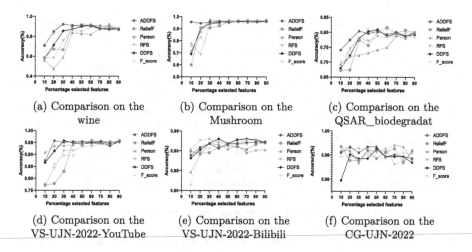

(a) Comparison on the wine

(b) Comparison on the Mushroom

(c) Comparison on the QSAR_biodegradat

(d) Comparison on the VS-UJN-2022-YouTube

(e) Comparison on the VS-UJN-2022-Bilibili

(f) Comparison on the CG-UJN-2022

Fig. 5. Results of the compared feature selection methods

Acknowledgment. This research was partially supported by the National Natural Science Foundation of China under Grant No. 61972176, Shandong Provincial Natural Science Foundation, China under Grant No. ZR2021LZH002, Jinan Scientific Research Leader Studio, China under Grant No. 202228114, Shandong Provincial key projects of basic research, China under Grant No. ZR2022ZD01, Shandong Provincial Key R&D Program, China under Grant No. 2021SFGC0401, and Science and Technology Program of University of Jinan (XKY1802).

References

1. Brezeale, D., Cook, D.J.: Automatic video classification: a survey of the literature. IEEE Trans. Syst. Man Cybern. Part C (Applications and Reviews) **38**(3), 416–430 (2008)
2. Ameigeiras, P., Ramos-Munoz, J.J., Navarro-Ortiz, J., Lopez-Soler, J.M.: Analysis and modelling of YouTube traffic. Trans. Emerg. Telecommun. Technol. **23**(4), 360–377 (2012)
3. Dubin, R., Dvir, A., Pele, O., Hadar, O.: I know what you saw last minute-encrypted http adaptive video streaming title classification. IEEE Trans. Inf. Forensics Secur. **12**(12), 3039–3049 (2017)
4. Suznjevic, M., Beyer, J., Skorin-Kapov, L., Moller, S., Sorsa, N.: Towards understanding the relationship between game type and network traffic for cloud gaming. In: 2014 IEEE International Conference on Multimedia and Expo Workshops (ICMEW), pp. 1–6. IEEE (2014)
5. Amiri, M., Al Osman, H., Shirmohammadi, S., Abdallah, M.: An SDN controller for delay and jitter reduction in cloud gaming. In: Proceedings of the 23rd ACM International Conference on Multimedia, pp. 1043–1046 (2015)
6. Zhang, Y., Li, S., Wang, T., Zhang, Z.: Divergence-based feature selection for separate classes. Neurocomputing **101**, 32–42 (2013)
7. Mousselly-Sergieh, H., Döller, M., Egyed-Zsigmond, E., Gianini, G., Kosch, H., Pinon, J.M.: Tag relatedness using Laplacian score feature selection and adapted

Jensen-Shannon divergence. In: Gurrin, C., Hopfgartner, F., Hurst, W., Johansen, H., Lee, H., (eds.) MultiMedia Modeling. MMM 2014. LNCS, vol. 8325. Springer, Cham (2014). https://doi.org/10.1007/978-3-319-04114-8_14

8. Dong, Y., Yue, Q., Feng, M.: An efficient feature selection method for network video traffic classification. In: 2017 IEEE 17th International Conference on Communication Technology (ICCT), pp. 1608–1612. IEEE (2017)

9. Wu, Z., Dong, Y., Wei, H.L., Tian, W.: Consistency measure based simultaneous feature selection and instance purification for multimedia traffic classification. Comput. Netw. **173**, 107190 (2020)

10. Kerber, R.: ChiMerge: discretization of numeric attributes. In: Proceedings of the Tenth National Conference on Artificial Intelligence, pp. 123–128 (1992)

11. Robnik-Šikonja, M., Kononenko, I.: Theoretical and empirical analysis of ReliefF and RReliefF. Mach. Learn. **53**(1), 23–69 (2003)

12. Hall, M.A.: Correlation-based feature selection for machine learning, Ph. D. thesis, The University of Waikato (1999)

13. Nie, F., Huang, H., Cai, X., Ding, C.: Efficient and robust feature selection via joint $\ell2$, 1-norms minimization. In: Advances in Neural Information Processing Systems 23 (2010)

14. Liu, S., Zhang, L., Sun, P., Bao, Y., Peng, L.: Video traffic identification with a distribution distance-based feature selection. In: 2022 IEEE International Performance, Computing, and Communications Conference (IPCCC), pp. 80–86. IEEE (2022)

15. Jaganathan, P., Rajkumar, N., Nagalakshmi, R.: A Kernel based feature selection method used in the diagnosis of Wisconsin breast cancer dataset. In: Abraham, A., Lloret Mauri, J., Buford, J.F., Suzuki, J., Thampi, S.M. (eds.) ACC 2011. CCIS, vol. 190, pp. 683–690. Springer, Heidelberg (2011). https://doi.org/10.1007/978-3-642-22709-7_66

SVIM: A Skeleton-Based View-Invariant Method for Online Gesture Recognition

Yang Zhao[1], Lanfang Dong[1(✉)], Guoxin Li[1], Yingchao Tang[1], Yuhang Zhang[1], Meng Mao[2], Guoming Li[2], and Linxiang Tan[2]

[1] University of Science and Technology of China, Hefei 230026, China
{yvngzhvo,tangyc314,yhzhang}@mail.ustc.edu.cn, lfdong@ustc.edu.cn,
guoxinli@mail.ustc.edu.cn
[2] AI Lab, China Merchants Bank, Shenzhen 518040, China
{melvinmaonn,tanlinxiang252}@cmbchina.com, lkm@cmbchina.com

Abstract. Online gesture recognition is a challenging task in practical application scenarios since the gesture is not always directly in front of the camera. In order to solve the challenges caused by multiple viewpoints of skeleton data, in this paper, we proposed a novel view-invariant method for online skeleton gesture recognition. The whole skeleton sequence data as a point set in our method and a PCA-based view-invariant data preprocessing algorithm is proposed and applied in this paper. We can transform similar skeleton data to relatively stable viewpoints by applying the PCA algorithm according to the similarity of distribution features of the point set, which can ensures the viewpoint stability of our gesture recognition model. We conduct extensive experiments on the NTU RGB+D and Northwestern-UCLA benchmark datasets which contain multiple viewpoints and the results have demonstrated the effectiveness of the method proposed in this paper.

Keywords: Online gesture recognition · View-invariant · PCA

1 Introduction

The view-invariant gesture recognition algorithm has a wide range of applications. When applying the gesture recognition model to real scenarios, the person doing the gesture action often does not happen to be standing directly in front of the camera. For example, in a robot scenario, the robot may need to respond to a user's waving gesture, and the user doing the waving gesture may not necessarily be directly in front of the robot, although he or she is within the robot's view. Another example is that when a self-driving car needs to detect the traffic police action, the location of the traffic police may not be right in front of the car's camera either.

Due to the change of viewpoint, the estimated skeleton coordinates from different viewpoints sometimes differ greatly, which seriously affects the recognition performance of the action recognition model. In addition, the movements

X. Yang et al. (Eds.): ADMA 2023, LNAI 14179, pp. 241–255, 2023.
https://doi.org/10.1007/978-3-031-46674-8_17

from different viewpoints are affected by self-occlusion, which causes the estimated skeleton to be disturbed by different degrees of noise. Therefore, gesture recognition with a constant viewpoint is a challenging problem.

The problem of view-invariance can be hardly solved by data enhancement due to the diversity of viewpoints. Some researchers try to weakens the effect of viewpoint variation on action representation by designing view-invariant features [8,15,25], but this approach can only handle small magnitude viewpoint changes. Other researcher split the skeleton into multiple parts and deal with viewpoint changes by modeling the geometric relationships and cannot really address the viewpoint change problem [23,26]. There is also literature on building new coordinate systems from the first frame or the skeleton of the previous frames [14,24], however this approach is highly sensitive to the onset motion of the gesture. Deep learning has achieved great success in many fields in recent years, and more and more researchers seek deep learning solutions. One solution is to use feature migration to seek a common feature space from data with different perspective [7], and other solution is to learn perspective-invariant representation from data [18]. However, the biggest problem with the learning-based approach is that the dataset used to train the model contains only a limited number of perspectives.

Unlike these approaches mentioned above, the basic idea of our method is that, the same action, although it may have various intra-class differences, is still composed of many similar motion states from a global perspective. If the whole motion sequence is treated as a point set, then these point sets tend to have similar shape characteristics. The difference in veiwpoint is reflected in the point set as a different in rotation direction. Therefore, if a way can be found to rotate this point set to a stable orientation, then this orientation can be considered as the standard viewpoint of this skeleton sequence. In this way the gesture recognition model can obtain an input source with a stable viewpoint. Moreover, this method only needs to be added to the preprocessing process of the data and can be applied directly to almost any existing gesture recognition method.

Based on the above ideas, we propose a view-invariant algorithm based on PCA. Principal Components Analysis(PCA) can compute a set of basis vectors from the point set that reflect the characteristics of the data distribution. In this paper, this set of basis vectors is used to transform the point set to a new basis coordinate space. Since this set of basis vectors is determined by the distribution characteristics of the data, data with similar distribution characteristics also have similar distribution characteristics in the new base coordinate space.

Our contribution is as follows:

1. For the multi-view problem of skeleton data, we propose a novel solution by applying the PCA algorithm to rotate similar skeleton sequences to relatively stable viewpoints based on the similarity of point set distribution, thus achieving view-invariant gesture recognition based on skeleton;
2. we demonstrate the effectiveness of the algorithm in several experiments on the multi-view datasets NTU RGB-D and Northwestern-UCLA, which contain multiple viewpoints for action recognition.

2   Related Work

Since the coordinates of the skeleton nodes obtained by the skeleton estimation algorithm vary greatly from viewpoint to viewpoint, and the differences in motion occlusion also cause the estimated skeleton to be disturbed by different degrees of noise. The viewpoint-invariant gesture recognition algorithm investigates the method that gestures taken from different angles from the training data can also be classified accurately.

Xia et al. [24] proposed method is to establish a spherical coordinate system in a specific direction on the skeleton. Specifically, they chose the hip center joint of the human skeleton as the midpoint, defined the horizontal reference vector as a vector projection from the left hip center joint point to the right hip center joint point onto the horizontal plane (parallel to the ground), and the zenith reference vector was defined as a vector perpendicular to the ground plane. Then they discretized the 3-dimensional space into n small intervals and discretized the joint point coordinates into these small intervals. Finally, they do probabilistic voting on these discretized coordinates to increase the stability of the features, use Linear Discriminant Analysis (LDA) to extract more discriminative features, k-Means clustering into dictionaries, and finally use Discrete Hidden Markov Model (DHMM) to do the classification. Zhang Yi et al [30] proposed to map the gesture trajectory features to be represented as global invariants based on the Centroid Distance Function (CDF). The center-of-mass distance function is the distance from each point on the trajectory point to the centroid, and the authors in the paper take the center of the hand as the centroid. Pei Xiaomin et al. [16] added the angle between the trajectory point and the centroid on top of this. Ghorbel et al. [6] proposed to independently fuse two multi-view invariant methods: the Ghorbel et al. [5] and Vemulapalli et al. [21]'s approach to perspective invariant classification. Ji et al [7] proposed using an attention mechanism to focus on the most critical joint points in the skeleton of multiple views and the relationship between them. Li et al. [14] create a new coordinate system from the first few frames of the camera view of the skeleton, and then convert the skeleton sequence to an orthographic view on this coordinate system, so that the skeleton has a stable view.

3   Method

It is known that the PCA algorithm can calculate from the data a set of basis vectors that reflect the characteristics of the data distribution, and the set of basis vectors is the eigenvectors of the covariance matrix of the data.

3.1   Calculate the Eigenmatrix

For a sequence of skeletons it can be considered as a point set $P = [p_1, p_2, \ldots, p_N] \in \mathbb{R}^{3 \times N}$. Firstly, we center the point set P, i.e., for any point $p_i \in P$

$$\hat{p}_i = p_i - \frac{1}{n}\sum_{i=1}^{n} p_i \tag{1}$$

thus forming the new point set $\bar{P} = [\hat{p}_1, \hat{p}_2, \ldots, \hat{p}_n] \in \mathbb{R}^{3 \times n}$. Then we eignde-compose the covariance matrix of the point set $\bar{P}$, i.e.

$$\bar{P}\bar{P}^T = R \Lambda R^T \tag{2}$$

Here, the matrix $R = [r_1, r_2, r_3] \in \mathbb{R}^{3 \times 3}$ is the eigenmatrix and its three eigenvectors(also called the principal axes), and the diagonal matrix $\Lambda = \text{diag}(\lambda_1, \lambda_2, \lambda_3)$ are three eigenvalues (also called the principal values) corresponding to the eigenvectors.

The point set P_{can} with rotational invariance can be obtained by aligning the principal axes with the world coordinate, that is, by computing $P_{\text{can}} = R^T P$.

Theorem 1. *The point set P_{can} is rotation invariant.*

Proof. Assume that $Q \in \text{SO}(3)$ is an arbitrary rotation matrix, then QP is the set of points after rotation of the point set P. The centerized point set $Q\bar{P}$ can be obtained from Eq. (1), then the covariance matrix of this point set can be convert to

$$\begin{aligned} Q\bar{P}(Q\bar{P})^T &= Q\bar{P}\bar{P}^T Q^T \\ &= Q(R \Lambda R^T)Q^T \\ &= (QR)\Lambda(QR)^T \end{aligned} \tag{3}$$

At this point, QR becomes the new principal axes. Therefore, after rotating the point set QP and aligning it with new main axis

$$\begin{aligned} (QP)_{\text{can}} &= (QR)^T QP \\ &= R^T Q^T QP \\ &= R^T P = P_{\text{can}} \end{aligned} \tag{4}$$

This means that rotating the matrix Q has no effect.

3.2   Ambiguity of the Feature Matrix

However, if we use the eigenmatrix as a transformation matrix directly, we will suffer from two kinds of ambiguities: sign ambiguity and order ambiguity. The sign ambiguity refers to the fact that, for a given eigenvector r_i, it can take either $+r_i$ or $-r_i$ under the condition that the eigendecompose is satisfied. By assigning the positive or negative sign to eigenvectors, then the transformed skeleton sequence will yield 8 possible perspectives. The order ambiguity refers to the problem of the order of the eigenvectors, which is not specified to be arranged in the order of the eigenvalues. In this case, 6 possible views will be generated when computing the standard view. That is, when changing the positive and negative signs of the eigenvectors or the order of the eigenvectors, it still produces

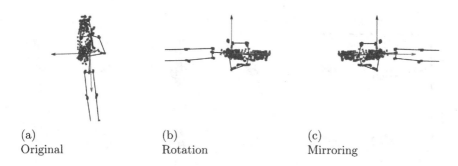

(a)   (b)   (c)
Original   Rotation   Mirroring

Fig. 1. Rotation and mirroring comparison of the eigenmatrix

a point set with many different rotations. In fact, a total of 48 views are possible after the eigenmatrix transformation.

As Li et al [12] pointed out, among the 8 sign ambiguities, some cases are in fact not rational transformations. Specifically, if the combination of some eigenvector $R = [r_1, r_2, r_3]$ and its determinant is 1, then only four of the eight ambiguities with a determinant of 1 are true rotation, and the other determinant values of -1 are a combination of rotation and mirror transformation(see Figure 1). Table 1 list all of sign ambiguities.

Table 1. The sign ambiguities of eigenmatrix

Eigenmatrix	Determinant	Rotation
$[+r_1, +r_2, +r_3]$	$+1$	Yes
$[-r_1, +r_2, +r_3]$	-1	No
$[+r_1, -r_2, +r_3]$	-1	No
$[+r_1, +r_2, -r_3]$	-1	No
$[-r_1, -r_2, +r_3]$	$+1$	Yes
$[+r_1, -r_2, -r_3]$	$+1$	Yes
$[-r_1, +r_2, -r_3]$	$+1$	Yes
$[-r_1, -r_2, -r_3]$	-1	No

If mirroring is included, then for some gestures in opposite directions, such as waving to the left and waving to the right, these actions will not be distinguishable. After removing the sign ambiguity with the mirror transformation, the PCA-based perspective algorithm produces

$$4(\text{sign ambiguity}) \times 6(\text{order ambiguity}) = 24. \tag{5}$$

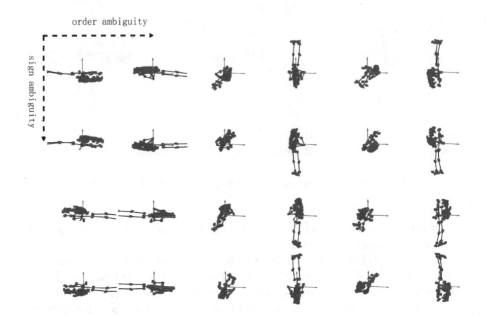

Fig. 2. The 24 ambiguities of the eigenmatrix

The ambiguities are listed in Fig. 2, which contains these 24 transformations.

3.3 Flow of Algorithm

To solve the ambiguity problem arising from PCA transformation, the solution proposed in this paper is to add as many ambiguous cases as possible to the training data during training, while only the order of the feature vectors is processed or not at all during testing. Although such an approach is similar to data augmentation of the viewpoints of the skeleton sequence, conventional multi-viewpoint data augmentation methods often enable the model to learn only a limited number of viewpoints, and in fact, it is impossible to learn all of them. In contrast, the preprocessing algorithm proposed in this paper enables the model to learn only a limited number of cases to be able to cover all viewpoint cases. The specific process is shown in Algorithms 1 and 2.

Data: Train model M, train set $D_{\text{train}} \in \mathbb{R}^{B \times 3 \times T \times N}$
Result: Models with perspective invariance M
for $X \in D_{train}$ **do**
  $P \leftarrow \mathbf{Reshape}(X, (3, -1));$ /* $P \in \mathbb{R}^{3 \times (T \times N)}$ */
  $\bar{P} \leftarrow P - \mathbf{Mean}(P);$
  $\hat{P} \leftarrow \bar{P}\bar{P}^T;$
  $L, Q \leftarrow \mathbf{Eign}(\hat{P});$ /* L is eigenvalues, Q is eigenmatrix */
  /* The function PcaAmbiguity returns the disambiguation of
     the feature matrix on demand   */
  $Rs \leftarrow \mathbf{PcaAmbiguity}(L, Q);$
  for $R \in Rs$ **do**
    $P' \leftarrow R^T P;$
    $X' \leftarrow \mathbf{Reshape}(P', (3, T, N));$
    $\mathbf{Train}(M, X');$
  end
end

Algorithm 1: Training process of PCA-based view-invariant algorithm

---

Data: Test model M, test set $D_{\text{test}} \in \mathbb{R}^{B \times 3 \times T \times N}$
for $X \in D_{test}$ **do**
  $P \leftarrow \mathbf{Reshape}(X, (-1, 3));$
  $\bar{P} \leftarrow P - \mathbf{Mean}(P);$
  $\hat{P} \leftarrow \bar{P}\bar{P}^T;$
  $L, Q \leftarrow \mathbf{Eign}(\hat{P});$ /* Optional: Sort the vectors in Q by the
  size of L */
  $P' \leftarrow Q^T P;$
  $X' \leftarrow \mathbf{Reshape}(P', (3, T, N));$
  $\mathbf{Test}(M, X');$
end

Algorithm 2: Testing procedure of PCA-based view-invariant algorithm

3.4 Experiment and Analysis

In order to verify the effectiveness of the PCA-base view-invariant algorithm proposed in this paper, this section applies the algorithm to two datasets with multiple views, NTU RGB+D and Northwestern-UCLA for experiments. The dataset Northwestern-UCLA is a relatively small action recognition dataset, while the dataset NTU RGB+D is a large action recognition dataset. This section first introduces these two multiview datasets, then presents implementation details used in this section, and finally, the experimental results and analysis are presented.

3.5 Dataset

NTU RGB+D. NTU RBG+D [17] is a large dataset designed from human action recognition, containing a total of 56880 3D skeleton sequences. It contains 60 action categories, including "drinking", "snacking", "brushing teeth", "combing hair ", "picking things up", and so on. The sample actions are performed by a total of 40 volunteers, and a maximum of 2 people in a sample is guaranteed. Each action sample was simultaneously captured by 3 different views of the Microsoft Kinect v2 camera. The different people and perspectives presented a significant challenge in discriminating between intra- and inter-class differences. NTU RGB+D is quite a challenging dataset considering the size of the dataset, the effect of similar actions and the noise in the dataset. To experiment the viewpoint invariant algorithm proposed in this paper, we use the recommended cross-view (X-View) benchmark test for this dataset: training data from camera views #2 and #3, and test data from camera view #1.

Northwestern-UCLA. The Northwestern-UCLA multi-view 3D event dataset [23] is a multi-view multimodal dataset containing RGB, depth, and human skeleton data captured by three Kinects simultaneously. The dataset contains a total of 1494 video clips, covering 10 action categories, each performed by 10 different volunteers. The full list of actions and the corresponding sample sizes are shown in Table 2. In this paper, we use the recommended evaluation of this dataset: training data from the first two cameras and test data from the latter one.

Table 2. Gesture information for Northwestern-UCLA

No.	Gesture	Amount
1	pick up with one hand	150
2	pick up with two	152
3	drop trash	141
4	walk around	173
5	sit down	148
6	stand up	149
7	donning	142
8	doffing	142
9	throw	145
10	carry	142

3.6 Implementation Details

The experimental gesture recognition model in this paper is DD-Net [27], and the Stochastic Gradient Descent(SGD) algorithm is chosen as the optimizer during training, the base learning rate is set to 0.1, each experiment is trained for 100 rounds, and the warm up strategy is used in the first 5 rounds followed by the ReduceLROnPlateau is used as the learning rate adjustment strategy to reduce the learning rate from 10^{-1} to 10^{-5}, and cross-entropy is used as the loss function. To test the effectiveness of the perspective invariant algorithm, both the training and test data are randomly rotated and scaled in this paper.

Because the dataset NTU RGB+D contains 2 skeletons per frame on some action categories, i.e., it contains actions done by two people together, and DD-Net does not consider this situation, a simple strategy is proposed here to fix this problem. Suppose the input data size is $B \times C \times T \times N \times M$, where B represents the size of a training batch, C represents the dimension of the skeleton sequence (usually 3), T represents the duration of the skeleton sequence, N represents the number of joint points of the skeleton, and M represents the number of people contained in the frame. Then, before feeding into the model, this paper first adjusts the input data to the shape of $(B \times M) \times C \times T \times N$, which is equivalent to expanding the batch size by a factor of M. Then the M in $B \times M$ is eliminated using the mean function when the data enters the final fully connected layer stage.

3.7   Experimental Results and Analysis

Different Data Preprocessing Methods. To test which of the two ambiguities, signed ambiguity or order ambiguity, is more important for the accuracy of the result, and to find a balance between improving the performance and reducing the data expansion(for a dataset with large sample size). To find a balance between improving performance and reducing data augmentation(for a dataset with a large sample size, excessive data augmentation can seriously increase training time), we first conduct several comparative experiments on Northwestern-UCLA.

In this paper, we use the DD-Net (filters=32) model to conduct comparative experiments. The way of experimentation is to design three levels of elimination schemes for sign ambiguity and order ambiguity, which also affect the expansion of the number of samples in the training set. For sign ambiguity, three expansion options are designed: randomly assigning positive and negative signs to the three eigenvectors, using all combinations of positive and negative signs for the rotation cases, and using all combinations of positive and negative signs. For order ambiguity, three expansion options are designed, namely, random order of eigenvectors, following the order of eigenvalues from smallest to largest, and all possible orders. In the test set, if the order ambiguity uses the ranking of eigenvalues, then the eigenvectors of the test set are treated similarly; otherwise, the default computed eigenvectors are used (as the default computed eigenvectors are in random order).

Specific expansion schemes and resulting expansion scale to the training dataset are listed below:

1. Direct use of feature matrix (no expansion)

2. Eigenvectors are sorted by eigenvalue only(no expansion)
3. Eigenvectors sorted by eigenvalue, positive and negative signs for all rotation cases(expanded by a factor of 4)
4. Eigenvectors sorted by eigenvalue, positive and negative signs of all eigenvectors(expanded by a factor of 8)
5. All sorting of feature vectors(expanded by a factor of 6)
6. All ordering of feature vectors, positive and negative signs of all rotation cases(expanded by a factor of 24)
7. All ordering of feature vectors, positive and negative signs of all feature vectors(expanded by a factor of 48)
8. Positive and negative signs for all rotation cases(expanded by a factor of 4)
9. Positive and negative signs of all eigenvectors(expanded by a factor of 8)
10. Control group, without any special treatment(no expansion)

Table 3. Results of different experiments on Northwestern-UCLA

No.	Sign Ambiguity			Order Ambiguity			Result/%
	Random	Rotation	All	Random	Eigenvalue	All	
1	✓			✓			91.8
2	✓				✓		91.6
3		✓			✓		94.2
4			✓		✓		92.9
5	✓					✓	92.9
6		✓				✓	92.9
7		✓				✓	94.4
8		✓		✓			94.2
9			✓	✓			93.1
10							89.7

Finally, the experimental results obtained are shown in Table 3. From the experimental results, the following conclusions can be drawn:

1. Overall, the results processed by the PCA-based view-invariant algorithm(Exp 1-9) are generally better than the results without any processing(Exp 10). Even the eigenmatrix generated directly using PCA (i.e., the order and sign of the eigenvectors are variable, Exp 1) also improves the accuracy much more than the control experiment(Exp 10). This comparison demonstrates the effectiveness of our method.
2. In terms of the importance of elimination order ambiguity(Exp 2 and 5) v.s. sign ambiguity(Exp 8 and 9), eliminating Sign ambiguity is more effective in improving the model's performance. Intuitively, order ambiguity changes the overall orientation of the point set, which has a greater impact on the overall

distribution of the point set than sign ambiguity, which is only rotation and mirroring around the axes, thus affecting the performance improvement of the classifier.

3. The results sorted by eigenvalue are essentially equivalent to those without any treatment(Exp 1 and Exp 2). This is because even for the same actions, the overall distribution of the point set is different due to intra-class differences in the actions, and the eigenmatrix sorted based on the magnitude of the eigenvalues does not allow them to have similar orientations (e.g, so the heads of the skeletons are all oriented on the z-axis)

4. The comparison between Exp 3 and Exp 4, as well as the comparison between Exp 8 and Exp 9, verified the analysis done in sect. 3.2 of this paper, where the mirror transformation of the skeleton leads to some direction-dependent actions that are indistinguishable (eg, stand up actions and sit down actions), i.e., the non-rotating eigenmatrix has a certain degree of negative impact on performance.

5. However, when eliminating the sign ambiguity along with the order ambiguity(Exp6 and Exp 7), the non-rotating eigenmatrix is much higher than the rotating eigenmatrix. Considering that Exp 6 is already a larger expansion(24 times), Exp 7 is twice as large, Exp 7 likely has a larger amount of training data resulting in a more generalized model. In fact, Exp 7 has both the most expanded data and the best results of all the experiments.

Table 4. Results of different experiments on NTU RGB-D

No.	Sign Ambiguity			Order Ambiguity			Result/%
	Random	Rotation	All	Random	Eigenvalue	All	
1	✓			✓			85.7
2	✓				✓		86.0
3		✓			✓		89.6
4			✓		✓		90.0
5	✓					✓	88.5
6		✓				✓	88.5
7			✓			✓	91.0
8		✓		✓			89.3
9			✓	✓			90.0
10							86.4

In order to test the generalizability of the above findings, the above experiments were redone on the NTU RGB-D, and the results are shown in Table 4. The main difference between this dataset and Northwestern-UCLA is in the sample data size, which is much higher than the latter. The performance impact of data expansion cannot be ignored when expanding exponentially on a dataset with an already large sample size base. In general, the larger the sample size of the dataset, the less likely the trained model is to be overfitted, resulting in better model performance. In addition, the intra-class differences of each action will be highlighted by the increased sample size.

For NTU RGB-D, the result of Exp 10 is slightly better than the result of directly using the PCA sign matrix(Exp 1) and eigenvalue ranking(Exp 2). We analyze that this is due to the intra-class variation in this dataset. The same class of actions has been transformed by the eigenmatrix due to the different distributions aggravating the differences between them, which leads to performance degradation.

The conclusion that the elimination of sign ambiguity is more important than the elimination of order ambiguity still holds for NTU RGB-D. However, the eigenmatrix of rotation in the sign ambiguity is not higher than the eigenmatrix of all symbols. We analyze that for this dataset, the data expansion have a more important impact on the performance improvement, e.g, the accuracy of experiments without data expansion is around 86% (Exp1, Exp 2, and Exp 10) the accuracy of experiment with 4 times expansion is around 89%(Exp 3 and Exp 8) accuracy of experiments with 8 times expansion is 90%(Exp 4 and Exp 9), while the accuracy of the experiment with a 48-fold expansion was 91%(Exp 7). There are some exceptions to the results for the 6-fold and 24-fold expansions, both of which have an accuracy of 88.5%, which we estimate to be due to random factors when using the model.

Combining the experimental result of both datasets, the implementation of the sign ambiguity data expansion is more helpful to improve the model performance, while the maximum expansion(48 times) gives the best results. If it is necessary to find a balance between training time, we recommend using the 4-fold expansion of Exp 3, i.e., sorting the eigenvectors by eigenvalues, with all rotated eigenvectors. In addition, although the increase in the amount of training data has an impact on the experimental results, the above comparison experiments can still fully demonstrate the effectiveness of our PCA-based view-invariant algorithm.

Table 5. Results of comparison with other methods on the dataset NTU RGB-D

Methods	X-View/%
Ind-RNN [13]	88.0
HCN [11]	91.1
ST-GCN [26]	88.3
AGC-LSTM [19]	95.0
DDGCN [9]	**97.1**
CA-GCN [29]	91.4
SGN [28]	94.5
Shift-GCN [3]	96.5
CTR-GCN [1]	96.8
DD-Net(filters=64)	89.6
DD-Net(filters=64, Exp 7)	91.2

Table 6. Results of comparison with other methods on the dataset Northwestern-UCLA

Methods	X-View/%
Actionlet ensemble [22]	76.0
Lie Group [20]	74.2
HBRNN-L [4]	78.5
Ensemble TS-LSTM [10]	89.2
AGC-LSTM [19]	93.3
Shift-GCN [3]	94.6
DC-GCN+ADG [2]	95.3
CTR-GCN [1]	**96.5**
DD-Net(filters=64)	89.7
DD-Net(filters=64, Exp 7)	94.4

Comparison with Other Methods. Tables 5 and 6 show the results of this paper's gesture recognition method before and after using PCA-based view-invariant algorithm, compared with other methods. On the large skeleton action recognition dataset, DD-Net does not stand out compared to other methods, because it is designed to be lightweight and efficient without using complex feature learning methods and deep neural networks, so the difference with the best method on the relatively small dataset Northwestern-UCLA is not as large as that on another large dataset. However, after applying the PCA-based view-invariant algorithm, the gesture recognition algorithm still has significant improvement.

4 Conclusions

In this paper, a PCA-based view-invariant algorithm is proposed. The method treats the whole skeleton sequence as a point set, obtains the feature matrix of the point set using the PCA algorithm, and then uses the feature matrix as a transformation matrix to transform the skeleton sequence to a relatively stable viewpoint. However, there are two kinds of ambiguities in the eigenmatrix generated from PCA, namely sign ambiguity and order ambiguity. In order to eliminate the effects of these two ambiguities, we propose to add all possible ambiguities to the training set to enhance the generalization ability of the model. We design several sets of experiments on two multi-view action recognition datasets to verify the effectiveness of this approach and analyze which ambiguities have a more significant impact on the performance, so as to find a balance between improving the performance and expanding the data. Finally, the PCA-based view-invariant algorithm is applied to the proposed gesture recognition model and compared with other methods. Although the gesture recognition performance of this paper still differs from the best method after applying the algorithm, there is still a significant improvement compared with that before applying the algorithm.

Acknowledgements. This work was supported by the National Key Research and Development Program of China under Grant No. 2020YFB1313602, The authors thank the reviewers for their valuable suggestions.

References

1. Chen, Y., Zhang, Z., Yuan, C., Li, B., Deng, Y., Hu, W.: Channel-wise topology refinement graph convolution for skeleton-based action recognition. In: Proceedings of the IEEE/CVF International Conference on Computer Vision, pp. 13359–13368 (2021)
2. Cheng, K., Zhang, Y., Cao, C., Shi, L., Cheng, J., Lu, H.: Decoupling GCN with DropGraph module for skeleton-based action recognition. In: Vedaldi, A., Bischof, H., Brox, T., Frahm, J.-M. (eds.) ECCV 2020. LNCS, vol. 12369, pp. 536–553. Springer, Cham (2020). https://doi.org/10.1007/978-3-030-58586-0_32
3. Cheng, K., Zhang, Y., He, X., Chen, W., Cheng, J., Lu, H.: Skeleton-based action recognition with shift graph convolutional network. In: Proceedings of the IEEE/CVF Conference on Computer Vision and Pattern Recognition, pp. 183–192 (2020)
4. Du, Y., Wang, W., Wang, L.: Hierarchical recurrent neural network for skeleton based action recognition. In: Proceedings of the IEEE Conference on Computer Vision and Pattern Recognition, pp. 1110–1118 (2015)
5. Ghorbel, E., Boutteau, R., Boonaert, J., Savatier, X., Lecoeuche, S.: Kinematic spline curves: a temporal invariant descriptor for fast action recognition. Image Vis. Comput. **77**, 60–71 (2018)
6. Ghorbel, E., et al.: A view-invariant framework for fast skeleton-based action recognition using a single RGB camera. In: Proceedings of the 14th International Joint Conference on Computer Vision, Imaging and Computer Graphics Theory and Applications, pp. 573–582 (2019)
7. Ji, Y., Xu, F., Yang, Y., Xie, N., Shen, H.T., Harada, T.: Attention transfer (ant) network for view-invariant action recognition. In: Proceedings of the 27th ACM International Conference on Multimedia, pp. 574–582 (2019)
8. Junejo, I.N., Dexter, E., Laptev, I., Pérez, P.: Cross-view action recognition from temporal self-similarities. In: Forsyth, D., Torr, P., Zisserman, A. (eds.) ECCV 2008. LNCS, vol. 5303, pp. 293–306. Springer, Heidelberg (2008). https://doi.org/10.1007/978-3-540-88688-4_22
9. Korban, M., Li, X.: DDGCN: a dynamic directed graph convolutional network for action recognition. In: Vedaldi, A., Bischof, H., Brox, T., Frahm, J.-M. (eds.) ECCV 2020. LNCS, vol. 12365, pp. 761–776. Springer, Cham (2020). https://doi.org/10.1007/978-3-030-58565-5_45
10. Lee, I., Kim, D., Kang, S., Lee, S.: Ensemble deep learning for skeleton-based action recognition using temporal sliding LSTM networks. In: Proceedings of the IEEE International Conference on Computer Vision, pp. 1012–1020 (2017)
11. Li, C., Zhong, Q., Xie, D., Pu, S.: Co-occurrence feature learning from skeleton data for action recognition and detection with hierarchical aggregation (2018)
12. Li, F., Fujiwara, K., Okura, F., Matsushita, Y.: A closer look at rotation-invariant deep point cloud analysis. In: Proceedings of the IEEE/CVF International Conference on Computer Vision, pp. 16218–16227 (2021)
13. Li, S., Li, W., Cook, C., Zhu, C., Gao, Y.: Independently recurrent neural network (indrnn): building a longer and deeper RNN. In: Proceedings of the IEEE Conference on Computer Vision and Pattern Recognition, pp. 5457–5466 (2018)

14. Li, Y., Xia, R., Liu, X.: Learning shape and motion representations for view invariant skeleton-based action recognition. Pattern Recogn. **103**, 107293 (2020)
15. Papadakis, A., Mathe, E., Spyrou, E., Mylonas, P.: A geometric approach for cross-view human action recognition using deep learning. In: 2019 11th International Symposium on Image and Signal Processing and Analysis, pp. 258–263 (2019)
16. Xiaomin, P., Fan Huijie, T.Y.: Action recognition method of spatio-temporal feature fusion deep learning network. Infrared Laser Eng. **47**(2), 55–60 (2018)
17. Shahroudy, A., Liu, J., Ng, T.T., Wang, G.: NTU RGB+D: a large scale dataset for 3D human activity analysis. In: Proceedings of the IEEE Conference on Computer Vision and Pattern Recognition, pp. 1010–1019 (2016)
18. Shao, Z., Li, Y., Zhang, H.: Learning representations from skeletal self-similarities for cross-view action recognition. IEEE Trans. Circuits Syst. Video Technol. **31**(1), 160–174 (2020)
19. Si, C., Chen, W., Wang, W., Wang, L., Tan, T.: An attention enhanced graph convolutional LSTM network for skeleton-based action recognition. In: Proceedings of the IEEE/CVF Conference on Computer Vision and Pattern Recognition, pp. 1227–1236 (2019)
20. Veeriah, V., Zhuang, N., Qi, G.J.: Differential recurrent neural networks for action recognition. In: Proceedings of the IEEE International Conference on Computer Vision, pp. 4041–4049 (2015)
21. Vemulapalli, R., Arrate, F., Chellappa, R.: Human action recognition by representing 3D skeletons as points in a lie group. In: Proceedings of the IEEE Conference on Computer Vision and Pattern Recognition, pp. 588–595 (2014)
22. Wang, J., Liu, Z., Wu, Y., Yuan, J.: Learning actionlet ensemble for 3D human action recognition. IEEE Trans. Pattern Anal. Mach. Intell. **36**(5), 914–927 (2013)
23. Wang, J., Nie, X., Xia, Y., Wu, Y., Zhu, S.C.: Cross-view action modeling, learning and recognition. In: Proceedings of the IEEE Conference on Computer Vision and Pattern Recognition, pp. 2649–2656 (2014)
24. Xia, L., Chen, C., Aggarwal, J.K.: View invariant human action recognition using histograms of 3D joints. In: 2012 IEEE Computer Society Conference on Computer Vision and Pattern Recognition Workshops, pp. 20–27 (2012)
25. Yan, P., Khan, S.M., Shah, M.: Learning 4D action feature models for arbitrary view action recognition. In: 2008 IEEE Conference on Computer Vision and Pattern Recognition, pp. 1–7 (2008)
26. Yan, S., Xiong, Y., Lin, D.: Spatial temporal graph convolutional networks for skeleton-based action recognition. In: Proceedings of the Thirty-Second AAAI Conference on Artificial Intelligence, pp. 7444–7452 (2018)
27. Yang, F., Wu, Y., Sakti, S., Nakamura, S.: Make skeleton-based action recognition model smaller, faster and better. In: Proceedings of the ACM Multimedia Asia, pp. 1–6 (2019)
28. Zhang, P., Lan, C., Zeng, W., Xing, J., Xue, J., Zheng, N.: Semantics-guided neural networks for efficient skeleton-based human action recognition. In: proceedings of the IEEE/CVF conference on computer vision and pattern recognition, pp. 1112–1121 (2020)
29. Zhang, X., Xu, C., Tao, D.: Context aware graph convolution for skeleton-based action recognition. In: 2020 IEEE/CVF Conference on Computer Vision and Pattern Recognition, pp. 14321–14330 (2020)
30. Yi, Z., Shuo, Z., Yuan, L.: View-invariant 3D hand trajectory-based recognition. J. Univ. Electr. Sci. Technol. China **43**(1), 60–65 (2014)

A Unified Information Diffusion Prediction Model Based on Multi-task Learning

Yingdan Shang[1]([✉])[ID], Bin Zhou[2][ID], Xiang Zeng[1][ID], and Kai Chen[1][ID]

[1] College of Computer Science and Technology, National University of Defense Technology, Changsha, China
altsuzy@hotmail.com
[2] Key Lab of Software Engineering for Complex Systems, School of Computer, National University of Defense Technology, Changsha, China

Abstract. The prediction of online information diffusion trends on social networks is crucial for understanding people's interests and concerns, and has many real-world applications in fields such as business, politics and social security. Existing research on information diffusion prediction has predominantly focused on either macro level prediction of the future popularity of online information or micro level user activation prediction. However, information diffusion prediction on micro or macro level only may lead to one-sided prediction results, and there is a lack of research on implementing micro and macro level diffusion prediction tasks at the same time. Since micro and macro level diffusion prediction tasks are related to each other, we propose a unified information diffusion model which can jointly predicting the micro level user activation probability and macro level information popularity based on multi-task learning framework. We utilize graph neural network to learn user representation from both the information cascades and social network structure. Comparing with micro and macro level baseline methods separately, the prediction results of our proposed model outperform all baseline methods and proves the effectiveness of the proposed method.

Keywords: Information diffusion prediction · Multi-task learning · Social network · Neural network

1 Introduction

The natural tendency of people to share information with others drives individuals to exchange and share information, spreading the online content to other users further away on the social network. When a large number of people participate in this kind of information diffusion, it will cause a huge cascade of information diffusion and lead to great social impact. The research of information diffusion prediction which aims at predicting the possible diffusion trend of

© The Author(s), under exclusive license to Springer Nature Switzerland AG 2023
X. Yang et al. (Eds.): ADMA 2023, LNAI 14179, pp. 256–267, 2023.
https://doi.org/10.1007/978-3-031-46674-8_18

online content is conducive to understanding the information diffusion mechanism and is significant in many areas such as commercial promotion, academic research and supervision of social public opinion.

Existing researches on information diffusion prediction predominantly concentrates on either micro or micro level prediction. At the micro level, information diffusion models considers more detailed user behaviors and aims at forecasting potentially infected users in the process of information diffusion. At the macro level, information diffusion models emphasis on predicting the potential final diffusion scale, namely, the popularity of information diffusion. Figure 1 illustrates a toy example of micro and macro level diffusion prediction problem.

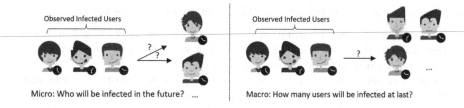

Fig. 1. Micro level diffusion prediction and macro level information diffusion prediction example

To the best of our knowledge, there has been limited research on the joint prediction of information diffusion at both the microscopic and macroscopic levels. However, constructing a unified diffusion prediction model has a natural advantage in reducing the cost of model training. Yu et al. [11] propose a networked Weibull regression model to predict the macro level information popularity by modeling micro level user behavior dynamics. Chen et al. [2] propose a novel deep learning approach based on multi-task framework to jointly predict micro and macro level information diffusion in an end-to-end manner, demonstrating the feasibility of jointly predicting information diffusion on microscopic and macroscopic level. However, this work only focuses on information cascade data and ignore the underlying social network structure which leads to poor prediction performance. Yang et al. [10] propose a multi-scale information diffusion prediction model utilizing reinforcement learning framework, the method incorporates reinforcement learning combined with recurrent neural network based microscopic prediction. However, this model predicts the macroscopic diffusion size on the basis of the results of microscopic prediction, so the quality of the prediction results of microscopic information dissemination directly affects the results of macroscopic prediction.

The microscopic social behaviors of users (such as forwarding and commenting) drives the information spread in the network, which may eventually lead to macroscopic large information cascade, in other words, micro level user behavior is the direct cause of macro level popularity. On the other hand, recent deep learning based information diffusion prediction models at micro or macro level use similar information as inputs. For example, micro level information diffusion prediction model [5,12] use cascade data and social network information as input and predict the potential users of infection, the macro level information diffusion prediction model [4] also takes information cascade data and social network information as input and predict popularity of online content.

To sum up, existing researches on jointly predicting information diffusion for both micro and macro level are quite scarce, a unified model can provide richer contextual information for information diffusion models. The academic and industrial fields have made numerous research achievements utilizing multi-task learning, such as Google's MOSE, SNR ranking model and Tencent's PLE model in the recommendation system in recent years. Therefore, we propose a unified information diffusion prediction model that employs the multi-task learning framework, the primary contributions are outlined as follows:

- We propose a unified information diffusion prediction model based on multi-task learning framework to realize the joint prediction of information diffusion trend on microscopic and macroscopic level.
- Prediction performance on multiple real-world social network datasets show that the unified information diffusion model can provide more information of the information diffusion process, the prediction accuracy on microscopic and macroscopic information diffusion prediction tasks are improved comparing with recent methods.

2   Problem Definition

Given the user set $\mathcal{U}$ and the information diffusion cascade set C, each cascade in the cascade set is defined as a list of users sorted by their infection time: $c_i = \{u_1^i, u_2^i, ..., u_l^i\}, c_i \in C$, where l represents the observed size of the cascade c_i, namely the number of infected users of the information item. Also given the underlying social network structure $\mathcal{G} = \{\mathcal{U}, \mathcal{E}\}$, the task of this paper is to jointly predict the diffusion trend of online information on both the micro and macro level. For micro level user activation prediction task, we aim to predict the next potential infection user u_{k+1}^i, for macro level popularity prediction task, we aim to forcast the incremental diffusion size of online information $p_i = |c_i| - l$.

3   Model

The general framework of the proposed model is shown in Fig. 2. The model mainly comprises of three parts: (1) The user representation layer, which uses graph neural networks to learn the user feature embedding from the dynamic heterogeneous diffusion graph. The user feature representation describes the social influence of users and the homogeneity among users. (2) The user embedding look-up layer, which query user embedding from previous graph convolution layer with time restriction. (3) The multi-task learning layer, the embeddings of early infected users are fused by multi-head self-attention neural network, then the micro and macro level information diffusion trend are predicted utilizing fused embedding simultaneously by designed multi-task loss function.

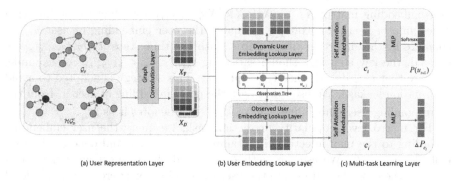

(a) User Representation Layer   (b) User Embedding Lookup Layer   (c) Multi-task Learning Layer

Fig. 2. Unified information diffusion prediction model framework

3.1 The User Representation Layer

The diffusion behavior among users reflects the rich proximity relationship between users, such as the social influence and homogeneity of users. Since in many real-world application scenarios, we only know the diffusion order of online content rather than the explicit influential relationship, we utilize the concept of k-neighbor sets in recommendation research [8] to construct the dynamic diffusion graph from the observed diffusion cascade. For an infected user in the cascade, his/her k-neighbor sets are infected users who are within k hops before him/her in the same cascade. We adopt this strategy in order to take advantage of the higher-order proximity between users.

The construction process of dynamic diffusion graph is the same with DyHGCN [12], we combine the social network structure with our constructed dynamic diffusion graph to form the dynamic heterogeneous diffusion graph and denote it as $\mathcal{HG}_D = \left[\mathcal{HG}^{t_1}, \mathcal{HG}^{t_2}, \mathcal{HG}^{t_3}, ..., \mathcal{HG}^{t_s}\right]$, where s denotes the number of time steps we use to divide our observation time window.

We employ two different graph convolution network to learn user latent representation from the social network $\mathcal{G}_F$ and the dynamic heterogeneous diffusion graph $\mathcal{HG}_D$:

$$X_F = GCN_1(A_F, H) \tag{1}$$

$$X_D = GCN_2(A_D, H), \tag{2}$$

where $H \in \mathbf{R}^{|\mathcal{U}| \times d}$ is the initial user embedding matrix, d is the dimension of user embedding. A_F represents the follow-up user relationship in social network $\mathcal{G}_F$ and A_D represents the user relationship in heterogeneous diffusion graph $\mathcal{HG}_D$. $X \in \mathbf{R}^{|\mathcal{U}| \times d}$ denotes the learned user embedding by graph convolution layer. Note that for the dynamic heterogeneous diffusion graph $\mathcal{HG}_D$, the corresponding learned user embedding matrix is denoted as $X_D = [X_D^{t_1}, X_D^{t_2}, X_D^{t_3}, ...X_D^{t_s}]$.

3.2 The User Embedding Look-Up Layer

For a cascade $c = \{(u_1, t_1), (u_2, t_2), ... (u_i, t_i)\}$, we query the user embedding used for down-stream micro level user activation prediction task from the learned user embedding matrix $X_D = [X_D^{t_1}, X_D^{t_2}, X_D^{t_3}, ... X_D^{t_s}]$ according to the infection timestamp of user. For example, to query the user embedding of user u_i, we compare his/her infection timestamp t_i with the time steps $[t_1, t_2, t_3, ..., t_s]$ we divided, the key step t_{key} should satisfying $t_{key} < t_i < t_{key+1}$. For example, suppose that $t_i \in [t_3, t_4)$, then t_{key} equals t_3. This strategy is adopted to avoid information leakage. We use t_{key} to get the user embedding of user u_i from the corresponding user embedding matrix $X_D^{t_{key}}$ and denotes it as $x_{u_i}^I$.

For macro level popularity prediction task, we query the user embedding according to the max infection timestamp in the cascade and denotes the user embedding as $x_{u_i}^A$. We use the method described above to get all user embedding for the cascade and denote the user embedding for micro and macro level prediction separately as $X_D^I = [x_{u_1}^I, x_{u_2}^I, ..., x_{u_i}^I]$ and $X_D^A = [x_{u_1}^A, x_{u_2}^A, ..., x_{u_i}^A]$.

3.3   The Multi-task Learning Layer

We use two different multi-head self-attention neural network to fuse the sequence of early infected users for micro and macro level prediction, the equations of multi-head self-attention neural network are listed below:

$$h_k = softmax \left(\frac{X_D W_k^Q (X_F W_k^K)^T}{\sqrt{d}} \right) X_D W_k^V \tag{3}$$

$$h = [h_1; h_2; ...; h_H] W^O, \tag{4}$$

where H is the head number, W^* are learning parameters.

For micro level user activation prediction task, we predict the user infection probability using fully connected neural network:

$$\hat{p}(u_{i+1}|h_{micro}) = softmax(W_c h_{micro} + b_c) \tag{5}$$

We employ the cross-entropy loss function for micro level prediction task following many previous work [12]:

$$\mathcal{L}_1 = \sum_{j=1}^{|C|} \sum_{i=1}^{l-1} -log\hat{p}(u_{i+1}^j) \tag{6}$$

For macro level popularity prediction task, we predict the popularity increment using fully connected neural network and use mean log-transformed squared error (MSLE) as the loss function:

$$\Delta P = relu(W_o h_{macro} + b_o) \tag{7}$$

$$\mathcal{L}_2 = \frac{1}{|C|} \sum_{j=1}^{|C|} (log_2 \Delta P_j - log_2 \Delta \hat{P}_j)^2 \tag{8}$$

There are large differences in the loss values of micro and macro level tasks, and their convergence speeds are different, so we use geometric combination of the final loss function to train the model:

$$\mathcal{L}_{total} = \sum_{i=1}^{n} \sqrt[n]{\mathcal{L}_i}, \tag{9}$$

n is the task number, here in our model n equals 2.

4 Experiments

4.1 Experimental Setting

Datasets. We conduct experiments on three publicly available datasets following previous work [6] to evaluate the proposed model.

- Twitter contains tweets with URLs on Twitter platform and their diffusion path during October 2010, the follow-up relationship among users is collected as the social network.
- Douban contains the sharing process of online movie or book among users on Douban social website. The behavioral relations are recorded as the social network.
- Christianity contains user interactions on topic of Christian on the Stack-Exchanges website.

To testify the validness of k-neighbor set when constructing the diffusion graph, we analyse the following statistics on datasets: assuming user u_i and u_j are infected in an information cascade with k users infected between them, how often do u_i and u_j participate in the same cascade?

Figure 3 shows the statistical results on active users who participate in at least 10 cascades in our datasets, we can see that the expectation of user co-occurrence times are quite high. The statistical results suggests the importance of applying k-neighbor set when constructing the diffusion graph in our model.

Evaluation Metrics. Similar with previous works [3,6], we use ranking metrics hits score on top k (hits@k) and mean average precision on top k (map@k) as evaluation metrics for micro level user activation prediction task. For macro level popularity prediction task, we use Mean Log-transformed Squared Error (MSLE) as evaluation metric.

Baselines. We compare our work with Deep Multi-Task Learning based Information Cascades model (DMT-LIC) [2] which has been mentioned above. Besides, we compare our model with recent macro level popularity prediction models and micro level user activation prediction models separately, the first three methods are classical macro level popularity prediction models:

- DeepCas [4] samples diffusion path using random walk, uses GRU neural network to learn user embedding and predict the popularity using fully connected neural network.

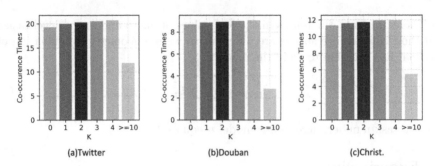

Fig. 3. Statistics of the expectation of user co-occurrence with k infected users between them on three datasets

- DeepHawkes [1] uses RNN neural network to encode the diffusion path and introduce attention mechanism as time decay function to fuse the diffusion path embeddings for popularity prediction.
- CasCN [3] uses graph convolutional network to learn user embedding and RNN neural network to capture the dynamic diffusion process.
- NDM [9] uses convolution neural network with attention mechanism to model information cascades with the social network unknown.
- SNIDSA [7] incorporate structure information from social network into RNN neural network to predict user activation.
- DyHGCN [12] combines social network with information diffusion relations to construct heterogeneous diffusion graph and learn user dynamic representation with GCN.
- MS-HGAT [6] builds diffusion hypergraph and use memory-enhanced sequential hypergraph attention network to learn user interactions from social network and diffusion graph.

Parameter Settings. We randomly divide our dataset into three parts, 80% of the data for training, 10% of the data for validation and the rest for model test. The cascade length is set to be 200 at maximum, the graph convolutional layer used to learn user relationship is a 2 layer GCN with batch normalization, the batch size is 64, the embedding dimension is set to 64, the learning rate is 0.001, and the dropout rate is 0.3. The k value for constructing diffusion graph based on k-neighbor set is 4 for all datasets.

4.2   Performance Comparison

The macro level popularity prediction results of the baseline methods and the proposed model on three datasets are shown in Table 1. It can be observed from the table that the performance of DeepCas and DeepHawkes based on diffusion path are much worse than CasCN which utilize graph neural network to model information cascade graph. The main reason is that it is inefficient to model the information diffusion graph by using sequence models such as recurrent neural networks.

Compared with the baseline method CasCN, the MSLE values of our model are reduced by about 7%, 8%, 10% on Twitter, Douban and Christianity dataset respectively. Our model extracts high-order dependencies between users from information diffusion data and social network graphs, and uses graph convolutional networks to generate dynamic user embedding representations. The high-order dependencies between users include the propagation relationship and follow-up relationship which are direct reflection of user social influence and homophily.

Our model has higher prediction accuracy compared with DMT-LIC, experimental results show that our method can effectively use both information diffusion data and social network graph to predict the popularity of information diffusion from a macro perspective.

Table 1. Model performance on macro diffusion prediction task

Datasets	Weibo Dataset	Douban	Christianity
Metric	MSLE		
T	3 h	1 month	2 months
DeepCas	2.379	2.012	1.496
DeepHawkes	2.175	1.890	1.103
CasCN	1.972	1.689	0.646
DMT-LIC	1.895	1.621	0.619
Proposed	**1.831**	**1.541**	**0.583**

For the micro level user activation prediction task, the prediction results of the proposed model and the baseline methods are shown in Table 2 and Table 3. The experimental results show that compared with the sub-optimal baseline method MS-HGAT model, the proposed model based on multi-task learning has significantly improved the metric Hits@K and Map@K on micro level prediction tasks. The NDM model uses self-attention mechanism and convolutional neural network to model the correlation between users in the same information cascade, but the model ignores the social network information and fails to fully learn the features of non-activated potential users, which reduces the prediction ability of the model. SNIDSA model integrates the social network information into the information diffusion model by using the attention mechanism, but the model essentially focuses on modeling the local mutual influence between users in the information diffusion sequence. DMT-LIC performs poorer than our work in the micro level user activation prediction task since it neglects the importance of social network.

The baseline methods DyHGCN and MS-HGAT both learn the feature representation of users using graph neural network from the social relationship graph and the propagation relationship graph constructed by the information cascade. Compared with previous models, their prediction performance has been greatly improved. However, they both model the sequential dependencies between users and ignore the possible high-order dependencies between users in the information cascade user sequence under the influence of social influence and homogeneity.

Compared with several different baseline methods on micro and macro level information diffusion prediction task, the experimental results confirm that our proposed model based on multi-task learning can simultaneously predict the macro level information popularity and micro level user activation probability, and the prediction performance of the trained model on multiple real-world datasets exceeds recent methods.

Table 2. Model performance on micro level diffusion prediction task (%) (hits@k)

Datasets	Twitter			Douban			Christianity		
Model	@10	@50	@100	@10	@50	@100	@10	@50	@100
NDM	15.21	28.23	32.30	10.00	21.13	30.14	15.41	31.36	45.86
SNIDSA	25.37	36.64	42.89	14.31	26.04	33.23	17.74	34.58	48.76
DMT-LIC	25.54	37.96	43.45	14.61	27.33	34.30	19.62	36.22	48.85
DyHGCN	27.43	45.21	54.65	17.52	30.82	38.08	27.31	48.04	49.05
MS-HGAT	28.30	45.04	55.01	19.54	**33.34**	**40.41**	26.76	47.75	49.01
Proposed	**29.44**	**47.92**	**58.20**	**19.78**	32.29	39.33	**30.53**	**45.58**	**49.17**

Table 3. Model performance on micro level diffusion prediction task (%) (map@k)

Datasets	Twitter			Douban			Christianity		
Model	@10	@50	@100	@10	@50	@100	@10	@50	@100
NDM	12.41	13.23	14.30	8.24	8.73	9.14	7.41	7.68	7.86
SNIDSA	15.34	16.64	16.89	8.42	9.01	9.06	8.69	8.94	9.72
DMT-LIC	15.39	16.70	16.94	8.86	9.41	9.52	8.75	9.08	9.96
DyHGCN	16.25	16.59	17.09	10.73	11.21	11.36	12.85	13.71	13.96
MS-HGAT	16.90	17.64	17.79	11.12	11.79	11.89	12.24	13.62	13.90
Proposed	**17.89**	**18.76**	**18.90**	**11.34**	**11.93**	**12.03**	**15.35**	**16.83**	**17.03**

Variants Comparison. To further clarify the effectiveness of the prediction model with multi-task learning, we design the multi-task training scheme and the single-task training scheme, and carry on experiments on these two training scheme. The multi-task training scheme is based on the multi-task learning framework proposed in this paper to predict the micro and macro level diffusion results at the same time. The single-task training scheme is to keep the model

Table 4. Performance: Single task learning v.s. Multi-task learning (Twitter)

Training method	MSLE	map@10	map@50	map@100	hits@10	hits@50	hits@100
micro (only)	–	16.00	16.84	16.98	27.51	45.65	54.96
macro (only)	1.862	–	–	–	–	–	–
multi-task	**1.833**	**17.89**	**18.76**	**18.90**	**29.44**	**47.92**	**58.20**

Table 5. Performance: Single task learning v.s. Multi-task learning (Douban)

Training method	MSLE	map@10	map@50	map@100	hits@10	hits@50	hits@100
micro (only)	–	10.70	11.29˙	11.38	19.69	**32.39**	38.43
macro (only)	1.618	–	–	–	–	–	–
multi-task	**1.541**	**11.34**	**11.93**	**12.03**	**19.78**	32.29	**39.33**

Table 6. Performance: Single task learning v.s. Multi-task learning (Christianity)

Training method	MSLE	map@10	map@50	map@100	hits@10	hits@50	hits@100
micro (only)	–	**16.14**	**17.56**	**17.79**	**31.70**	**46.17**	**49.93**
macro (only)	0.595	–	–	–	–	–	–
multi-task	**0.583**	15.35	16.83	17.03	30.53	45.58	49.17

unchanged, and when training the model, we only use the micro or macro level information diffusion results. Table 4, 5, 6 show the comparison of prediction performance of these two training schemes on different datasets.

As can be seen from the table, the micro and macro level prediction results of the multi-task training scheme are significantly improved compared with the single-task training scheme on the Twitter and Douban datasets. On the Christianity dataset, the micro level prediction results in the multi-task training scheme are slightly worse than that of the single-task training scheme, while the macro popularity prediction results are slightly better than that of the single-task scheme. At the same time, compared with training different model for each task separately, the multi-task model can save model calculation cost, time cost and maintenance cost as well.

5 Conclusions

Microscopic user interaction behavior will affect the macroscopic popularity of information. However, there is a lack of information diffusion models that jointly model the macro level information popularity prediction and micro level user behavior prediction. Considering the correlation between micro and macro level prediction tasks, we propose a unified information diffusion prediction model based on multi-task learning framework, which are able to simultaneously predict the popularity of online information and the user activation probability.

Compared with the classical baseline methods related to micro and macro level prediction tasks, the prediction results on multiple datasets show the effectiveness of our model. At the same time, we compare the experimental performance under the single-task training scheme and the multi-task training scheme to further verify the feasibility and effectiveness of the model. We think more research on how to jointly predicting micro and macro level information diffusion trend should be carried out in the future.

Acknowledgement. This work is supported by the National Natural Science Foundation of China No. 62172428, 61732022, 61732004.

References

1. Cao, Q., Shen, H., Cen, K., Ouyang, W., Cheng, X.: DeepHawkes: bridging the gap between prediction and understanding of information cascades. In: Proceedings of the 2017 ACM on Conference on Information and Knowledge Management, pp. 1149–1158. Association for Computing Machinery, New York (2017)
2. Chen, X., Zhang, K., Zhou, F., Trajcevski, G., Zhong, T., Zhang, F.: Information cascades modeling via deep multi-task learning. In: Proceedings of the 42nd International ACM SIGIR Conference on Research and Development in Information Retrieval, pp. 885–888. Association for Computing Machinery, New York (2019)
3. Chen, X., Zhou, F., Zhang, K., Trajcevski, G., Zhong, T., Zhang, F.: Information diffusion prediction via recurrent cascades convolution. In: IEEE 35th International Conference on Data Engineering, pp. 770–781 (2019)
4. Li, C., Ma, J., Guo, X., Mei, Q.: DeepCas: an end-to-end predictor of information cascades. In: Proceedings of the 26th International Conference on World Wide Web, pp. 577–586. International World Wide Web Conferences Steering Committee, Republic and Canton of Geneva, CHE (2017)
5. Sankar, A., Zhang, X., Krishnan, A., Han, J.: InF-VAE: a variational autoencoder framework to integrate homophily and influence in diffusion prediction. In: Proceedings of the 13th International Conference on Web Search and Data Mining, pp. 510–518. Association for Computing Machinery, New York (2020)
6. Sun, L., Rao, Y., Zhang, X., Lan, Y., Yu, S.: MS-HGAT: memory-enhanced sequential hypergraph attention network for information diffusion prediction. In: Proceedings of the AAAI Conference on Artificial Intelligence, vol. 36. pp. 4156–4164. AAAI Press (2022)
7. Wang, Z., Chen, C., LI, W.: A sequential neural information diffusion model with structure attention. In: Proceedings of the 27th ACM International Conference on Information and Knowledge Management, pp. 1795–1798. Association for Computing Machinery, New York (2018)
8. Wang, Z., Wei, W., Cong, G., Li, X.L., Mao, X.L., Qiu, M.: Global context enhanced graph neural networks for session-based recommendation. In: Proceedings of the 43rd International ACM SIGIR Conference on Research and Development in Information Retrieval, pp. 169–178. Association for Computing Machinery, New York (2020)
9. Yang, C., Sun, M., Liu, H., Han, S., Liu, Z., Luan, H.: Neural diffusion model for microscopic cascade study. IEEE Trans. Knowl. Data Eng. **33**(3), 1128–1139 (2021)

10. Yang, C., Tang, J., Sun, M., Cui, G., Liu, Z.: Multi-scale information diffusion prediction with reinforced recurrent networks. In: Proceedings of the 28th International Joint Conference on Artificial Intelligence, pp. 4033–4039. AAAI Press (2019)
11. Yu, L., Cui, P., Wang, F., Song, C., Yang, S.: From micro to macro: uncovering and predicting information cascading process with behavioral dynamics. In: Proceedings of the 2015 IEEE International Conference on Data Mining, pp. 559–568. IEEE Computer Society, USA (2015)
12. Yuan, C., Li, J., Zhou, W., Lu, Y., Zhang, X., Hu, S.: DyHGCN: a dynamic heterogeneous graph convolutional network to learn users' dynamic preferences for information diffusion prediction. In: Hutter, F., Kersting, K., Lijffijt, J., Valera, I. (eds.) ECML PKDD 2020. LNCS (LNAI), vol. 12459, pp. 347–363. Springer, Cham (2021). https://doi.org/10.1007/978-3-030-67664-3_21

Learning Knowledge Representation with Entity Concept Information

Yuanbo Xu[1], Lin Yue[2(✉)], Hangtong Xu[1], and Yongjian Yang[1]

[1] MIC Lab, College of Computer Science and Technology, Jilin University, Changchun, China
{yuanbox,yyj}@jlu.edu.cn, xuht21@mails.jlu.edu.cn
[2] The University of Newcastle, Callaghan, Australia
Lin.Yue@newcastle.edu.au

Abstract. The goal of Knowledge Representation Learning (KRL) is to learn an accurate knowledge representation that conforms to human understanding. Currently, many works have used entity multi-source information to improve the entity representation semantic precision, such as entity description, attribute, and visual information. However, few methods consider the timeliness in KRL, which significantly affects representation learning performance. In this paper, we attempt to utilize concept information with human understanding to learn an accurate, time-stable knowledge representation. Specifically, we first build a novel Knowledge Graph (KG) - *Structure Concept Graph* (SCG), which can provide entity structure and concept information jointly. Based on the SCG, we devise a novel KRL model that can embed entity concept information to ensure accuracy and timeliness of improving the KRL's effect with entity structure information. We evaluate our method on two downstream tasks: the knowledge graph completion task and the zero-shot task. Experimental results on real-world datasets show that our method outperforms other baselines by building effective entities' representations from their concept information. The source code of this paper can be obtained from https://anonymous.4open.science/r/CKRL-adma2023.

Keywords: Entity Concept · Representation learning · Knowledge Graphs

1 Introduction

Knowledge Representation Learning (KRL) methods based on entity structure information have achieved a good effect for learning an explainable embedding recently, such as TransE [1], RotatE [13], HittER [3], etc. In order to enhance the entity representation semantics accuracy, many current works use entity multi-source information for KRL. For example, DKRL [17] uses entity description information to learn entity semantic information. IKRL [18] uses visual (picture) information to capture more entity information for the entities' representations; DT-GCN [11] uses the entity multi-type attribute values to enhance the knowledge representations effect. However, the above entity semantics information has disadvantages: entity description information is embodied in a paragraph, and many words in it are mostly used to help describe

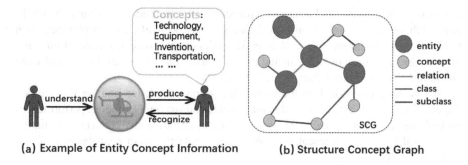

(a) Example of Entity Concept Information   (b) Structure Concept Graph

Fig. 1. Example of Entity Concept Information and Structure Concept Graph.

the current entity. But when learning the entities' embeddings, these words are also learned, which largely causes the entity representation semantic drift. In addition, entity visual information and attribute values have strong timeliness because the visual information and attribute values will change with time, which also widely affects the entity representation effectiveness. Our research found that entity concept information mainly avoids the above problems. The concept, the first step of human beings to know the world from abstract to concrete, can be formulated as an entity semantic information generated through human understanding. Once it is produced, the concept will not be changed easily (time-stable). As shown in Fig. 1(a), people can generate concepts in their minds by cognizing the airplane; on the other hand, people can also recognize other airplanes based on concept information in their minds. In this paper, we advocate using entity concept information to learn an accurate and time-stable entity representation, which provides a new idea for improving the KRL's effect.

Concept Graph (CG) is a particular type of Knowledge Graph such as AliCG [22]. These CGs make it possible to extract entity concept information. However, the current CGs only focus on constructing the relations between entities and concepts as well as concepts and concepts, while ignoring the relations between entities and entities. This negligence causes some tasks unavailable on CGs (such as knowledge graph completion and zero-shot). In this paper, we construct a novel KG - *Structure Concept Graph* (SCG), as shown in Fig. 1(b): the nodes in SCG are composed of entities and concepts. The connection between entity and entity is defined as relation, the connection between the entity and concept is defined as a class, and the connection between concept and concept is defined as a subclass to construct a Structure Concept Graph.

In order to effectively use entity concept information, we design two novel encoders which efficiently encode entity concept information to generate the entity's representation. Based on the two encoders, we propose a new joint training model *Concept Knowledge Representation Learning Model* (CKRL) for SCG, which can combine structure information and concept information jointly to improve KRL's effect. In the CKRL model, an entity's representation is responsible for the modeling of the corresponding structure information and its concept information.

For learning entity structure information, we follow a typical KRL method TransH [16] and regard the relation in each triple as a translation from head entity to tail entity.

In this way, the entities' and relations' embeddings are learned to maximize the likelihood of these translations. Meanwhile, given an entity, we will use concept information to learn its representation. We explore two encoders to encode entity concept information in entities' representations, including *Sum-based Encoder* (Sum-based) and *Convolutional Neural Network Encoder* (CNN-based).

We evaluate our model on the knowledge graph completion and zero-shot tasks. Experimental results demonstrate that our model achieves state-of-the-art performances on both tasks, which indicates that our model can encode entity concept information well into the entity's representation. We demonstrate the main contributions of this work as follows:

- We construct a novel Knowledge Graph called *Structure Concept Graph* (SCG), which not only complements the relations between entities and entities, but also provides rich entity structure information and concept information for Knowledge Representation Learning.
- We design two novel encoders *Sum-based Encoder* (sum-based) and *Convolutional Neural Network Encoder* (CNN-based), which can effectively encode entity concept information to generate the entities' representations. To the best of our knowledge, this is the first work to use entity concept information for enhancing the knowledge graph representation learning effect.
- We propose a novel KRL model *Concept Knowledge Representation Learning* (CKRL), which utilizes both entity structure information and concept information for improving KRL's effect. Moreover, both representations can complement and strengthen each other in the training process to achieve an accurate embedding.
- We evaluate the CKRL model's effectiveness on two tasks, including the knowledge graph completion and zero-shot tasks. Experimental results on real-world datasets illustrate that the CKRL model consistently outperforms other baselines on the two tasks.

We give basic definitions in the next section. Then we represent our proposed methodology and optimizations in Sect. 3. Section 4 reports details of datasets and experimental results. Related works are introduced in Sect. 5. Finally, we conclude our work in Sect. 6.

2   Problem Formulation

We will first introduce the symbols used in this paper. Given a structural triple $(h, r, t) \in T$, while $h, t \in E$ stand for entities, $r \in R$ stands for relation. $c \in C$ stands for the concept. Respectively, h and t are head entity and tail entity. T stands for the structural triple training set, E is the set of entities, R is the set of relations, and C is the set of concepts. We propose two kinds of representations for each entity to utilize structure information and concept information in CKRL.

Definition 1. Structure-Based Representations: e_s indicates the entity's structure-based representation vector, while e_{sh} and e_{st} are the structure-based representations based on the head entity and tail entity. $\mathbf{r}$ indicates the relation's representation. These representations could be learned through existing translation-based models.

Definition 2. Concept-Based Representations: e_c represents the entity's concept-based representation vector, while e_{ch} and e_{ct} are the concept-based representations based on the head entity and tail entity. These representations are obtained by encoding with two encoders.

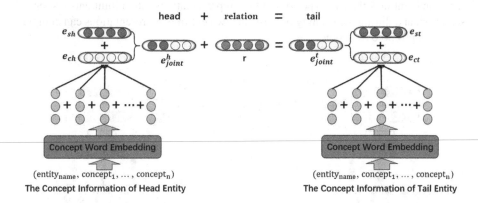

Fig. 2. The CKRL's Overall Architecture Based on Sum-based Encoder

3 Methodology

3.1 Overall Architecture

To fully use these two kinds of information, we integrate the two kinds of representations into a joint representation [4]. We assume that each dimension of structure-based and concept-based representations represents the same semantics information. Then we combine all the dimensions linearly to build a joint representation:

$$e_{joint} = \alpha * e_s + (1 - \alpha) * e_c, \tag{1}$$

where $*$ is a multiplication operator, $e_{joint} \in \mathbb{R}^{d \times 1}$ represents the entity's joint representation, $e_s \in \mathbb{R}^{d \times 1}$, $e_c \in \mathbb{R}^{d \times 1}$ where d is the dimension of representation, α is the weight of combination structure-based representation and concept-based representation, and its value domain is [0, 1]. Given a triple (h, r, t), we denote e_{joint}^h / e_{joint}^t as the joint representation of its head/tail entity and $\mathbf{r}$ as its relation's representation. The score function of our models is based on TransH:

$$S(h, r, t)_{joint} = ||(e_{joint}^h - \mathbf{w}_r^T e_{joint}^h \mathbf{w}_r) + \mathbf{r} - (e_{joint}^t - \mathbf{w}_r^T e_{joint}^t \mathbf{w}_r)||_{L1/L2}, \tag{2}$$

where $||.||_{L1/L2}$ is L1-norm or L2-norm function, $\mathbf{w}_r \in \mathbb{R}^{d \times 1}$ denotes the relation specific hyperplane for relation r.

According to the score function, the CKRL's overall architecture based on a sum-based encoder is demonstrated in Fig. 2. The CKRL's overall architecture based on a CNN-based encoder is demonstrated in Fig. 3. We gain the entities' structure-based representations and relations' representations from TransH. We get the entities' concept-based representations from the sum-based encoder and CNN-based encoder. The score function combines the two types of entities' representations into a joint entity's representation, and during the training process, the two entities' representations can complement and promote each other.

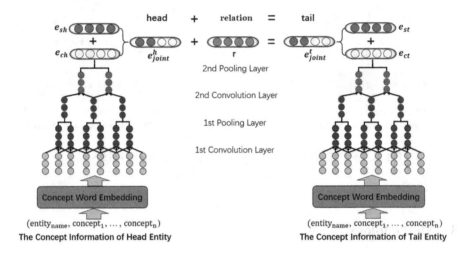

Fig. 3. The CKRL's Overall Architecture Based on CNN-based Encoder

3.2 Entity Concept Word Embedding

For each entity's concepts, we assume that they have the same number of concept words. For this, we adopt a zero-padding strategy to unify the concept word number for each entity. We randomly generate an embedding for each concept word and learn the same embedding for all zeros to prevent the zeros' negative effects in training such as the disappearance of gradients. In order to solve the problem that the different entities have the same concepts, We also learn the embeddings of their names for each entity. This can not only prevent the problem that different entities have the same concept words but also further strengthen the entity representations' semantic precision. In addition to the random initialization method, we can also use the Word2Vec [8] to get the entity concept word embeddings. Given an entity's name and an entity's n concept words: $C_i = (c_{name}, c_1, ..., c_i, ..., c_n)$, we get the following entity concept embeddings:

$$\mathbf{C}_i = (\mathbf{c}_{name}, \mathbf{c}_1, ..., \mathbf{c}_i, ..., \mathbf{c}_n). \tag{3}$$

3.3 Sum-Based Encoder

Figure 2 depicts the framework of the sum-based encoder. Then we take the entity concept word embeddings as input, and we simply sum up the concept words embeddings to generate the entity's representation:

$$e_c = c_{name} + c_1 + ... + c_i + ... + c_n, \tag{4}$$

where c_{name} is the the entity's name embedding, c_i is the i-th concept word embedding belonging to the concept set of entity e, and e_c will be used to minimize loss function.

3.4 Convolutional Neural Network Encoder

Convolutional Neural Networks (CNN) is the popular model for computer vision tasks, which also proved to be effective for natural language processing tasks (such as part-of-speech tagging, chunking, named entity recognition, and semantic role labeling) [6]. Recently CNN models are also proposed for Knowledge Representation Learning [17] and achieved good effect. The conventional Sum-based encoder has two inherent drawbacks: (1) not considering the concept word's order; (2) not fully understanding the concept information. To tackle the above issues, we exploit a CNN-based encoder to refine the representations and explore the semantic and order information.

The Framework of CNN-Based Encoder. Figure 3 depicts the overall architecture of the CNN-based encoder. The CNN-based encoder framework has five layers, including an input layer, two convolution layers, and two pooling layers. In addition, the entity's representation obtained by the CNN-based encoder will participate in minimizing the loss function.

Convolution Layer. In the convolution layer, we set $\mathbf{X}^{(l)}$ to be the input of l-th convolution layer and $\mathbf{O}^{(l)}$ to be the output of l-th convolution layer. First a size k window will slide through the input embeddings in $\mathbf{X}^{(l)}$ to gain $\mathbf{X}'^{(l)}$. For the first layer, $\mathbf{X}^{(1)}$ is a set of vectors consisting of the concept word embeddings $(c_{name}, c_1, ..., c_i..., c_n)$, and the window process is defined as follows:

$$c_i'^{(1)} = c_{i:i+k-1} = [c_i^T, c_{i+1}^T, ..., c_{i+k-1}^T]^T, \tag{5}$$

where the i-th vector of $c'^{(1)}$ is gained by aggregating k features in i-th window. Considering that the input is inconsistent in length, we use the zero-padding strategy in Entity Concept Embedding section to align the length of the sequence and endow the same embedding to all zeros. The i-th output vector of the convolution layer is defined as follows:

$$o_i^{(l)} = \sigma(\mathbf{K}^{(l)} c_i'^{(l)} + \mathbf{b}_i^{(l)}), \tag{6}$$

where $\mathbf{K}^{(l)} \in \mathbb{R}^{n_2^{(l)} \times n_1^{(l)}}$ is the convolution kernel for all input embeddings of l-th convolution layer after window process and $\mathbf{b}_i^{(l)}$ is the optional bias. $n_2^{(l)}$ is the dimension of output embeddings which could be considered as the number of feature maps.

$n_1^{(l)} = k \times n_0^{(l)}$ where $n_0^{(l)}$ is the dimension of input embeddings. σ is the activation function such as **tanh** or **ReLU**. In general, zero-filling embeddings shouldn't contribute in forward propagation nor be updated in backward propagation. However, we endow the same embedding to all zeros, and in this way, we can align the length of the input concept word embedding sequences and avoid the possible side effects of all zero-filling embeddings.

Pooling Layer. We use the pooling strategy after every convolution layer to shrink the parameter space of the CNN-based encoder and reserve all concepts' semantic information. We assume that concept information in each dimension is equally important. For the first pooling layer, in each window, we utilize the average value of every feature map to generate a new embedding. Compared with the previous model using the max-pooling strategy to compress semantics such as DKRL [17], our operation can not only reduce training parameters but also learn the entity concept information at all levels of entities in feature maps. The mean-pooling is defined to determine the average feature values in each dimension of the input embeddings within a size n window, and we default that the stride is equal to the window size:

$$\mathbf{c}_i^{(2)} = \text{mean}(\mathbf{o}_{n \cdot i}^{(1)}, ..., \mathbf{o}_{n \cdot (i+1)-1}^{(1)}). \tag{7}$$

The mean-pooling, as a non-parameter operator, can attain well-established feature dimensional reduction/complexity trade-offs. Considering that an entity contains various concepts, different semantics can be revealed from multi-angles. Thus, we choose mean-pooling at the last layer to aggregate features across all concepts. The equation can be formulated as follows:

$$\mathbf{e}_c = \sum_{(i=1,...,m)} \frac{\mathbf{o}_i^{(2)}}{m}. \tag{8}$$

We argue that the m concept embeddings contain different semantic information from different aspects, all of which contribute to the entity's representation. Due to the pooling strategies, we can not only handle the concept embeddings better and obtain fixed-length entities' representations $\mathbf{e}_c$ for each entity but also capture more semantic information.

3.5 Training

We utilize a margin-based loss function as our training objective, which is defined as follows:

$$L = \sum_{(h,r,t)\in T} \sum_{(h',r',t')\in T'} \max(\gamma + S(h,r,t)_{joint} - S(h',r',t')_{joint}, 0), \tag{9}$$

where margin γ means the artificially defined minimum distance between positive and negative examples. $S(h,r,t)_{joint}$ is the score function stated above, in which both head and tail entities have two kinds of representations. We compared the $L1$-norm with the

$L2$-norm in our experiment, and the performance of the $L1$-norm is better. T' is the negative sampling set of T, which we define as follows:

$$T' = \left\{(h',r,t)|h' \in E\right\} \cup \left\{(h,r',t)|r' \in R\right\} \cup \left\{(h,r,t')|t' \in E\right\}. \quad (10)$$

Supposing that a triple is already in T, it will not be regarded as a negative sample. Because the head entity and tail entity have two types of representations, the entities in the margin-based loss function could either be structure-based representations or concept-based representations.

3.6 Optimization and Implementation Details

The Sum-based and CNN-based encoders take concept word embeddings as input and entities' representations as output to minimize the above loss function. $\mathbf{K}^{(1)}$, $\mathbf{K}^{(2)}$ are convolution kernels, which are initialized randomly. The $\mathbf{c}$ can be initialized randomly or pre-trained through Word2Vec [8]. $\mathbf{E}$ and $\mathbf{R}$ are the sets of entities' and relations' embeddings respectively, they could either be initialized randomly or by learning from existing translation-based models such as TransE [1] and TransR [7]. We use the standard stochastic gradient descent (SGD) for the model's optimization. The chain rule is applied top-down through the CKRL model until the concept word embedding layer. For the consideration of efficiency, we use GPU to accelerate the training.

Table 1. Statistics of Datasets

SCG					
Ent	Rel	Concept	Rel tr	Train set	Test set
1242	161	6452	2014	2014	700
Zero-Shot Dataset					
IKGEs	OOKGEs	Concept	Rel tr	Train set	Test set
899	355	5483	1982	1582	656

4   Experiments

4.1   Datasets and Experiment Settings

Datasets. In our experiments, we first extract the structure triples in CASIA-KB [20] in OpenKG[1] to build a KG. Then we extract concept information related to entities of CASIA-KB in OpenCG[2] to construct a CG. Finally, we combine the KG and CG to build an SCG. The SCG details are listed in Table 1. In addition, we built a new

---

[1] http://www.openkg.cn.

[2] http://openconcepts.openkg.cn.

dataset named Zero-Shot-Dataset based on SCG to simulate a zero-shot scenario. The Zero-Shot-Dataset details are listed in Table 1. We select 899 entities as In-KG entities (IKGEs) and select 355 entities in SCG that are related to In-KG entities as Out-of-KG entities (OOKGEs). In addition, we extract structural triples containing IKGEs and OOKGEs to make up our test set. We guarantee that the OOKGEs' concept words appear in IKGEs' concept words.

Experiment Setting. In the CKRL model, we set the margin γ set among $\{1.0, 2.0, 3.0\}$. The learning rate λ set among $\{0.0005, 0.0003, 0.001\}$. The optimal configurations of CKRL are $\lambda = 0.001$, $\gamma = 1.0$. In the CNN-based encoder, we set different window sizes k among $\{1, 2, 3\}$ for different convolution layers. The dimensions of the entity's concept-based representation and the structure-based representation are set to 128. The dimension of the relation's representation is set to 128. Besides, we utilize two evaluation settings, "Raw" and "Filter", where "Filter" drop the existing triples in KGs from reconstructed triples as training triples while "Raw" does not.

Table 2. Performance on entity prediction

model	SCG				
	MR		Hits@N(%)		
	Raw	Filter	1	3	10
TransE	422	403	5.77	7.34	7.97
ComplEx	381	376	5.32	7.91	8.47
SimplE	408	395	6.75	8.95	10.26
RotatE	391	378	8.87	12.11	13.12
QuatRE	367	333	<u>9.21</u>	12.88	14.77
TransRHS	<u>265</u>	<u>252</u>	9.11	<u>12.97</u>	15.01
ParamE	321	318	8.22	12.78	14.88
DT-GCN	402	397	6.29	8.32	9.33
HittER	297	292	7.82	12.65	<u>15.23</u>
KG-BERT	281	294	7.31	12.59	12.11
StAR	277	281	8.11	12.61	13.14
CKRL_Sum(str)	384	381	8.94	11.76	12.33
CKRL_CNN(str)	332	321	8.87	11.57	13.43
Sum(con)	417	409	5.78	7.44	8.12
CNN(con)	276	272	8.32	**15.11**	15.57
CKRL_Sum(con)	361	348	6.76	9.11	10.32
CKRL_CNN(con)	357	354	6.32	9.95	11.33
CKRL_Sum(joint)	298	287	9.09	13.07	15.44
CKRL_CNN(joint)	**130**	**118**	**11.55**	14.56	**16.13**
Improv.	51.1%	52.2%	25.4%	16.5%	5.9%

Table 3. Performance on relation prediction

model	SCG				
	MR		Hits@N(%)		
	Raw	Filter	1	3	10
TransE	96	92	15.77	21.24	25.94
ComplEx	86	84	15.78	21.87	24.47
SimplE	87	82	18.63	22.55	28.26
RotatE	84	82	24.87	28.67	29.45
QuatRE	79	74	<u>25.26</u>	<u>29.98</u>	31.73
TransRHS	80	<u>71</u>	25.17	28.86	31.01
ParamE	85	81	24.45	28.78	30.83
DT-GCN	98	94	14.25	19.61	26.66
HittER	<u>78</u>	73	23.13	27.34	<u>33.63</u>
KG-BERT	81	73	23.12	28.74	31.32
StAR	79	72	24.11	28.75	31.46
CKRL_Sum(str)	79	75	21.94	26.01	29.98
CKRL_CNN(str)	77	74	21.87	28.57	30.88
Sum(con)	78	76	21.37	26.18	30.41
CNN(con)	75	71	25.96	**33.31**	34.24
CKRL_Sum(con)	77	72	21.71	30.76	34.38
CKRL_CNN(con)	73	71	22.01	30.12	35.32
CKRL_Sum(joint)	73	70	24.09	29.02	33.44
CKRL_CNN(joint)	**54**	**51**	**25.97**	32.99	**35.98**
Improv.	30.8%	28.2%	2.8%	11.1%	5.8%

4.2 Validation of Knowledge Graph Completion

Knowledge graph completion aims to learn the good entities' representations, relations' representations, and an energy function. We can predict a triple (h, r, t) with h, r or t unknown. For example, when given (h, r), we should predict the corresponding t.

Evaluation Protocol. Due to the fact that we can obtain two kinds of entities' representations, we will report 8 kinds of prediction results based on CKRL model: Sum(con) and CNN(con) only use entity concept information for training and use the entities' concept-based representations for entity and relation prediction. Under joint training based on entity structure information and concept information: CKRL_Sum(str) and CKRL_CNN(str) use the entities' structure-based representations for entity and relation prediction. CKRL_Sum(con) and CKRL_CNN(con) use the entities' concept-based representations for entity and relation prediction. CKRL_Sum(joint) and CKRL_CNN(joint) use the joint entities' representations (structure-based representation + concept-based representation) for entity and relation prediction. In our experiment, we select TransE [1], ComplEx [14], SimplE [5], RotatE [13], QuatRE [9], TransRHS [21], ParamE [2], DT-GCN [11], and HittER [3] as baselines. Besides, we employ two text-based approaches, KG-BERT [19] and StAR [15] for a plain comparison, The baselines are discussed in the Related Work section. We use two evaluation metrics: Hits@N [21] and Mean Rank (MR) [17]. We follow the evaluation settings "Raw" and "Filter".

Experimental Results. The entity and relation prediction results are shown in Table 2 and Table 3. From these results, we can observe that: (1) most CKRL models outperform all baselines on both MR, Hits@1, Hits@3 and Hit@10. It indicates that the entities' representations with entity concept information perform better in knowledge graph completion. This proves that CKRL model can effectively use entity concept information to learn an accurate entity' representation for improving KRL's effect. (2) CKRL_Sum(str) and CKRL_CNN(str) show good performance, although it is inferior to some baselines' results. After the joint training of the two kinds of information, compared with some models (such as TransE and ComplEx and SimplE) performance effects have been improved. This shows that CKRL model can not only construct an accurate concept-based representation, but also improve the effect of structure-based representations. (3) In CKRL models, the prediction results based on the CNN-based encoder are better than prediction results based on the Sum-based encoder, which proves that the CNN has a better effect in capturing entity semantics. This further proves the robustness and effectiveness of the CKRL model based on the CNN-based encoder. (4) The CKRL's results outperform baselines on MR. The MR depends on the overall quality of knowledge representations and it is sensitive to the wrong predicted results. In this paper, we use entity concept information as accurate entity semantic information to improve the entity representations' semantic accuracy. Therefore, the CKRL's results are much better than the baselines' results on MR. (5) CKRL(joint) experimental results performed well, indicating our success in combining the two representations into a joint representation, which is a good inspiration for multi-resource information learning.

4.3 Validation of Zero-Shot Task

The zero-shot task focuses on the situation when at least one of the entities in the test triples is the OOKGE. All existing models based on structure information can not handle this task because they can't learn the OOKGEs' representations directly. Many works try to use entity multi-source information to handle the zero-shot task. For example, DKRL uses entity description information to solve the zero-shot problem. However, our model can not only use concept information to deal with the zero-shot situation but also use entity concept information to learn the interpretable and understandable entities' representations for OOKGEs, which proposes a new direction for solving zero-shot tasks.

Evaluation Protocol. We choose DKRL [17], ConMask [12], and OWE [10] as our baselines for our proposed CKRL_CNN(joint) (short as CKRL). We utilize MR [17] and Hits@1,3,10 [21] for entity prediction and relation prediction. We also use the "Raw" and "Filter" settings.

Experimental Results. Table 4 and Table 5 are the zero-shot's experimental results. Based on the experimental results, our analysis is as: (1) CKRL achieves better performance than DKRL, ConMask, and OWE. This proves that using accurate entity semantic information (concept information) can learn an accurate entity's representation for OOKGEs to solve the zero-shot problem. In addition, using entities' concept-based representations to deal with the zero-shot problem can greatly reduce the risk of inaccurate expression and better capture semantic information of OOKGEs' multiple aspects. (2) The baselines use entity description information and entities' names to learn entities' representations to solve the zero-shot problem. Experimental results prove that entity concept information has superior advantages. (3) From Table 4, we can know that some CKRL's results are not very satisfactory and the reason may be that IKGEs and OOKGEs belong to two different entity spaces, separating these two entities means that it may alienate the connection between the two types of entities. Therefore, the ability to represent the OOKGEs still needs to be enhanced.

Table 4. Performance on entity prediction in Zero-Shot Task

model	Zero-Shot-Dataset				
	MR		Hits@N(%)		
	Raw	Filter	1	3	10
DKRL	376	363	2.123	5.316	7.126
ConMask	383	379	1.332	5.205	6.324
OWE	361	353	2.303	5.514	7.448
CKRL	**353**	**344**	**2.631**	**7.424**	**9.771**
Improv.	2.2%	2.5%	14.2%	34.6%	31.2%

Table 5. Performance on relation prediction in Zero-Shot Task

model	Zero-Shot-Dataset				
	MR		Hits@N(%)		
	Raw	Filter	1	3	10
DKRL	79	76	11.984	16.332	18.547
ConMask	98	93	8.854	14.413	16.318
OWE	82	75	9.281	16.669	18.554
CKRL	**74**	**71**	**12.556**	**18.244**	**20.145**
Improv.	6.3%	5.3%	4.8%	9.3%	8.6%

4.4 Efficiency Analysis

We compare training time and inferring time for knowledge completion task. Since our proposed method utilizes a new dataset extracted from CASIA-KB, we hope to explore whether our proposed dataset structure could enhance the efficiency without losing accuracy for SOTA approaches. Thus we build $CASIA_{sub}$, which contains the same entities and relations as SCG. The only difference is that SCG contains the concepts while $CASIA_{sub}$ does not. The results reported in Fig. 4 indicate that utilizing SCG consistently accelerate the existing KG approaches, especially for translation-based ones (TransRHS, RotatE), which proves the extraction ability of SCG for enhancing KG representation learning.

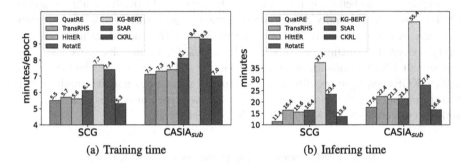

(a) Training time   (b) Inferring time

Fig. 4. Efficiency performance of the KG completion task.

4.5 Parameter Analysis

We explore our proposed CKRL's parameters, including margin γ among $\{1.0, 2.0, 3.0\}$; the learning rate λ among $\{0.0005, 0.0003, 0.001\}$. In the CNN-based encoder, we set different window sizes k among $\{1, 2, 3\}$ and the dimensions among $\{64, 128, 256\}$. The results reported in Fig. 5 indicate that the best performance of CKRL is achieved at $\{\gamma = 1.0, \lambda = 0.001, \text{dimension} = 128, k= 3\}$.

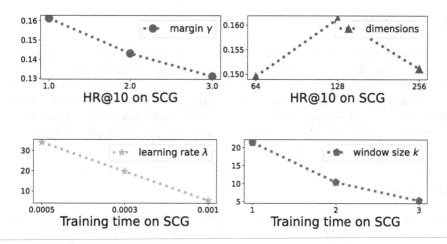

Fig. 5. Parameter analysis.

5   Related Work

5.1   Zero-Shot Learning

Only a few works have solved the zero-shot task in knowledge graph completion by using entity multi-source information. DKRL [17] uses a joint training of graph-based embeddings and text-based embeddings to handle the zero-shot problem. ConMask [12] is a text-centric approach where text-based embeddings for the head, relation, and tail are derived by an attention model over names and descriptions. OWE [10] comprehensively uses the entity's name and description information and proposes a conversion space to perform open-world link prediction.

5.2   Knowledge Representation Learning

Knowledge Representation Learning methods have achieved great success in recent years. TransE [1] regards the relation as a transformation between the head entity and the tail entity for modeling. ParamE [2] takes advantage of the nonlinearity fitting ability of neural networks and translational properties to improve the KRL. ComplEx [14] improves the knowledge embedding effect by using complex-valued embeddings. SimplE [5] uses Canonical Polyadic to improve KG embedding. RotatE [13] proposes a rotational model, taking relation as a rotation from the head entity to the tail entity in complex space. QuatRE [9] models the relation as a rotation on a 4-dimensional space (hypercomplex space) to unify ComplEx and RotatE. TransRHS [21] encodes each relation as a vector together with a relation-specific sphere in the same space. DT-GCN [11] has the advantage of fully embedding attribute values of data types information and refines data types into five primitive modalities. HittER [3] presents a deep hierarchical transformer model to learn representations of entities and relations in KGs. KG-BERT [19] and StAR [15] are two state-of-the-art text-based approaches, which utilize the abundant text information to enhance KG representation learning abilities.

6   Conclusion

In this paper, we construct a novel Knowledge Graph - SCG, and propose a novel KRL model (CKRL) that utilizes both the entity structure and concept information to learn an accurate representation. We also design two concept information encoders including a sum-based encoder and a CNN-based encoder to encode concept information. In knowledge graph completion and zero-shot tasks, CKRL's performance is comparatively better than the other baselines, which indicates its potential application ability for enhancing KG's downstream tasks.

References

1. Bordes, A., Usunier, N., García-Durán, A., Weston, J., Yakhnenko, O.: Translating embeddings for modeling multi-relational data. In: Proceedings of NIPS, pp. 2787–2795 (2013)
2. Che, F., Zhang, D., Tao, J., Niu, M., Zhao, B.: ParamE: regarding neural network parameters as relation embeddings for knowledge graph completion. In: Proceedings of AAAI, pp. 2774–2781 (2020)
3. Chen, S., Liu, X., Gao, J., Jiao, J., Zhang, R., Ji, Y.: Hitter: hierarchical transformers for knowledge graph embeddings. In: Proceedings of EMNLP, pp. 10395–10407 (2021)
4. Ding, J., Ma, S., Jia, W., Guo, M.: Jointly modeling structural and textual representation for knowledge graph completion in zero-shot scenario. In: Proceedings of Big Data, pp. 369–384 (2018)
5. Kazemi, S.M., Poole, D.: Simple embedding for link prediction in knowledge graphs. In: Proceedings of NIPS, pp. 4289–4300 (2018)
6. Kim, Y.: Convolutional neural networks for sentence classification. In: Proceedings of the EMNLP, pp. 1746–1751 (2014)
7. Lin, Y., Liu, Z., Sun, M., Liu, Y., Zhu, X.: Learning entity and relation embeddings for knowledge graph completion. In: Proceedings of AAAI Conference on Artificial Intelligence, pp. 2181–2187 (2015)
8. Mikolov, T., Chen, K., Corrado, G., Dean, J.: Efficient estimation of word representations in vector space. In: Proceedings of ICLR (2013)
9. Nguyen, D.Q., Vu, T., Nguyen, T.D., Phung, D.: QuatRE: relation-aware quaternions for knowledge graph embeddings. CoRR abs/2009.12517 (2020)
10. Shah, H., Villmow, J., Ulges, A., Schwanecke, U., Shafait, F.: An open-world extension to knowledge graph completion models. In: Proceedings of the AAAI, pp. 3044–3051 (2019)
11. Shen, Y., Li, Z., Wang, X., Li, J., Zhang, X.: Datatype-aware knowledge graph representation learning in hyperbolic space. In: Proceedings of CIKM, pp. 1630–1639 (2021)
12. Shi, B., Weninger, T.: Open-world knowledge graph completion. CoRR abs/1711.03438 (2017)
13. Sun, Z., Deng, Z., Nie, J., Tang, J.: Rotate: knowledge graph embedding by relational rotation in complex space. In: Proceedings of ICLR (2019)
14. Trouillon, T., Welbl, J., Riedel, S., Gaussier, É., Bouchard, G.: Complex embeddings for simple link prediction. In: Proceedings of ICML, pp. 2071–2080 (2016)
15. Wang, B., Shen, T., Long, G., Zhou, T., Wang, Y., Chang, Y.: Structure-augmented text representation learning for efficient knowledge graph completion. In: WWW 2021: The Web Conference 2021, Virtual Event/Ljubljana, Slovenia, 19–23 April 2021, pp. 1737–1748. ACM/IW3C2 (2021)
16. Wang, Z., Zhang, J., Feng, J., Chen, Z.: Knowledge graph embedding by translating on hyperplanes. In: Proceedings of AAAI, pp. 1112–1119 (2014)

17. Xie, R., Liu, Z., Jia, J., Luan, H., Sun, M.: Representation learning of knowledge graphs with entity descriptions. In: Proceedings of AAAI, pp. 2659–2665 (2016)
18. Xie, R., Liu, Z., Luan, H., Sun, M.: Image-embodied knowledge representation learning. In: Proceedings of IJCAI, pp. 3140–3146 (2017)
19. Yao, L., Mao, C., Luo, Y.: KG-BERT: BERT for knowledge graph completion. CoRR abs/1909.03193 (2019)
20. Zeng, Y., Zhang, T., Hao, H.: Active recommendation of tourist attractions based on visitors interests and semantic relatedness. In: Ślęzak, D., Schaefer, G., Vuong, S.T., Kim, Y.-S. (eds.) AMT 2014. LNCS, vol. 8610, pp. 263–273. Springer, Cham (2014). https://doi.org/10.1007/978-3-319-09912-5_22
21. Zhang, F., Wang, X., Li, Z., Li, J.: TransRHS: a representation learning method for knowledge graphs with relation hierarchical structure. In: Proceedings of IJCAI, pp. 2987–2993 (2020)
22. Zhang, N., et al.: AliCG: fine-grained and evolvable conceptual graph construction for semantic search at Alibaba (2021)

Domain Adaptive Pre-trained Model for Mushroom Image Classification

Yifei Shen[1]([envelope]), Zhuo Li[2], Yu Yang[3], and Jiaxing Shen[1]

[1] Department of Computing and Decision Science, Lingnan University, Tuen Mun, Hong Kong, China
sam.yifei.shen@gmail.com, jiaxingshen@ln.edu.hk
[2] School of Computer Science and Technology, Chongqing University of Posts and Telecommunications, Chongqing 400065, China
lizhuo@cqupt.edu.cn
[3] Department of Computing, The Hong Kong Polytechnic University, Hung Hom, Hong Kong, China
cs-yu.yang@polyu.edu.hk

Abstract. Mushroom is highly diverse in morphology and colors, and difficult for ordinary people to discriminate between them. The lack of high-quality labeled mushroom datasets is one of the bottlenecks restricting cutting-edge image recognition models to achieve high performance in mushroom image recognition. To address the limitation, in this paper, we will introduce a large mushroom dataset, called Mushroom-23, constructed by us. It collects over 35000 labeled mushroom images belonging to 203 popular mushroom species in Hong Kong. Along with constructing the new mushroom dataset, we also propose the domain adaptive pre-trained (DAPT) model to make the state-of-the-art Vision Transformer (ViT) adaptive to specific mushroom recognition. The DAPT is first pre-trained on ImageNet and Danish Fungi (DF20) datasets, then fine-tuned on the collected Mushroom-23 dataset to gain the capability of different categories of mushrooms. Extensive experimental results show DAPT outperforms all baseline models by a large margin in terms of Accuracy and Macro F1.

Keywords: Mushroom Recognition · Domain Adaption · Pre-trained Model · Vision Transformer

1 Introduction

Mushrooms are a highly diverse and complex group of species with rich edible, economical, medical, and nutrient values [21–23]. Some of them are poisonous that may lead to food poisoning, or even endanger people's lives. Without the expert knowledge, it is difficult for ordinary people to distinguish between the edible and poisonous mushrooms [21]. Fortunately, the fast development of deep learning-based computer vision technology makes it possible to use the AI technology to recognize different types of mushrooms. However, the awkward situation is though we have the state-of-the-art image classification model, no desirable dataset with complete species and high-quality mushroom images can be used to train the model. To the best of our knowledge, Danish Fungi 2020 (DF 20) [22,23] is one of the popular fungi datasets, but the fungi collected in datasets are from the Nordic countries and few species from East and Southeast Asia, particularly, Hong Kong, are collected. Moreover, the state-of-the-art models, such as AlexNet [19], GoogLeNet [26], ResNet [10], can achieve high classification accuracy, but they are trained on the standard publically available datasets, e.g., ImageNet [29], COCO [30], which may fail in mushroom recognition due to the lack of sufficient fine-tune on specific mushroom dataset.

To address these limitations, we first construct a large Hong Kong Mushroom-23 dataset containing 35,457 images of 203 mushroom species, which are collected and filtered by massive fungi images downloaded from the internet. In addition, we propose a domain adaptive pre-trained model (DAPT) to boost the performance of mushroom image classification. DAPT draws the successful experiences of the Vision Transformer model (ViT) [6] and follows the same design philosophy of the model architecture of ViT. DAPT divides the input mushroom images into small patches, then transforms them into serial data and input to the transformer encoder followed by a fully-connected layer to obtain the classification results. DAPT is first trained on the ImageNet dataset, then DF 20 dataset, and finally fine-tuned on the Mushroom-23 dataset.

To evaluate the DAPT's performance, we conduct extensive experiments on the Mushroom-23 dataset. The experimental results show DAPT outperforms all baseline models and achieves 82.9% and 0.788 in terms of classification accuracy and F1 score, respectively.

Simply, the contributions of this paper are summarized as follows:

- To begin with, we construct the Mushroom-23 dataset comprising a large number of different varieties of mushroom images.
- Then, we propose a transformer-based domain adaptive pre-trained (DAPT) model to perform the mushroom image classification.
- Furthermore, to make the DAPT model customized to mushroom data, transfer learning is used to first train DAPT on the ImageNet and DF 20 datasets, then fine-tuned on the Mushroom-23 dataset.

2   Related Work

Conventionally, the mushroom (fungi) image classification is based on traditional machine learning approaches such as support vector machines (SVM) and decision trees [1–3]. With the booming of deep learning technology, particularly, the great success of convolutional neural network (CNN) in image classification [10,11,19,20,26], it has also been introduced to fungi recognition [7]. Subsequently, transfer learning becomes popular in image classification tasks, which first pre-trains a model on a large dataset, then fine-tunes it on the small individualized domain. Fu et al. [25] proposed a novel recurrent attention convolutional neural network (RA-CNN) using discriminative region attention and region-based feature representation technologies. He et al. [27] proposed a two-stream model combining the vision and the language processing module for learning latent semantic representations. In addition, Yu et al. [31] employed a hierarchical framework to perform cross-layer bilinear pooling to enhance the feature representation learning. Zheng et al. [32] used a new group convolution method by first dividing channels into different groups according to their semantic meanings, then performing the bilinear pooling for each group to keep its corresponding dimension such that it can be directly integrated into any existing classification model backbone.

Recently, transformers and self-attention models have caused a huge sensation in natural language processing (NLP) [8,9,12]. Inspired by this, many attempts were made to expend the successful experiences of the transformers in NLP to the various computer vision problem, such as segmentation [14], object detection [17], and object tracking [18]. Girdhar et al. [13] first used transformers to deal with the sequential video feature extraction leveraging in the CNN backbone. Later, motivated by replacing CNN with a pure transformer structure, the vision transformer model (ViT) [6] was proposed which introduce the transformers to transform the 3D image into the serialized image patches and process it subsequently. On top of this, Transformer-based object re-identification (TransReID) was proposed which incorporated side information into the transformer along with the jigsaw patch module (JPM) to enhance the object re-identification accuracy [24]. Besides, Zheng et al. [28] proposed a pure self-attention-based encoder method named segmentation transformers (SETR) that exploits ViT as the encoder for segmentation.

3   Mushroom-23 Dataset

In this section, we will introduce the Mushroom-23 dataset constructed by us. All the images collected are searched and download from Google Image and Bing Image. As shown in Fig. 1 and Table 1, Mushroom-23 dataset contains 203 different mushroom species with a total of 35,457 images. On average, each category contains 175 images, but the distribution of the number of images per category varies widely. For example, the most representative category consists of 1,233 images but the least representative one has 30 images only. Additionally, Fig. 2 demonstrates some selected species of the mushroom collections. It gives us an intuitive impression of the plant morphological characters of different types of mushrooms, such as forms and colors, and can help us better recognize them. To ensure the data quality of the dataset, we check the clarity, lighting, and angles of the images carefully during manual screening. In addition, we also manually annotate the dataset, assigning each image a unique category, which can be used as the image label in the supervised learning task.

Table 1. The total numbers of collected mushroom species and images, and the average number of images per species in Mushroom-23 Dataset.

Number of Species	Total Number of Images	Average Number of Images per Species
203	35457	175

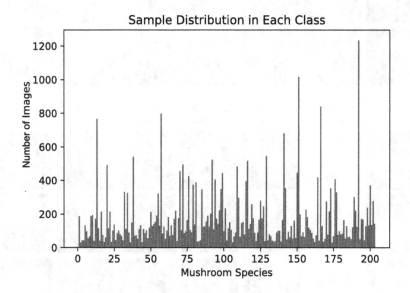

Fig. 1. The distribution of the number of collected mushroom images of different species.

Fig. 2. An illustration of some selected species from Mushroom-23 Dataset.

4   Domain Adaptive Pre-trained (DAPT) Model for Mushroom Classification

In this section, we will introduce the domain adaptive pre-trained (DAPT) model for mushroom classification in detail. First of all, we will elaborate on the model architecture and then introduce how we train the model.

4.1   Model Architecture

We propose the domain adaptive pre-trained (DAPT) model for mushroom classification, the architecture of which follows the basic framework Vision Transformer model (ViT) [6]. As illustrated in Fig. 3, the DAPT model consists of several key components, namely patch location embedding, transformer encoder, and classification result. In patch location embedding, the input image is divided into small patches of the same size, then the linear projection is performed to realign all the patches into a sequence according to their patch location. Subsequently, the serialized patch data are fed into the transformer encoder in parallel to use the attention mechanism to learn the dependency of the sequence of the patches. The transformer encoder block, which is the core of the DAPT model, consists of multi-head self-attention and feed forward network layers. Multi-head attention is composed of multiple one-head attention modules which can make the model draw attention from the different local image representations. Feed-forward network consists of two fully-connected layers followed by an activation function to learn high-level feature representations. Multi-head attention can make the model draw attention from the different local image representations. Finally, a classification result is used to map the learned feature vector to different categories of mushrooms.

4.2   Domain Adaptive Training Process

As illustrated in the top part of Fig. 3, domain adaptive training is performed to make the DAPT perform well on the mushroom image classification task. To start with, DAPT is trained on the ImageNet dataset to gain initial capabilities in recognizing a large variety of objects in our lives and have learned parameters fixed. Then, to enhance the competence of DAPT in fungi recognition, the pre-trained DAPT model is further trained on the DF20 dataset such that DAPT can discriminate between the main species of fungi. Although DF20 is a large specific fungi dataset of Demark collecting over 7,500 fungal species, including the taxonomic classification, morphology, ecology, distribution, and molecular data, most of the collections are from Northern Europe, and few species in East and Southeast Asia are included. Therefore, to address the Asian mushroom recognition issue, we further fine-tune DAPT on the Mushroom-23 dataset to make it adapted to the Asian mushroom species recognition task.

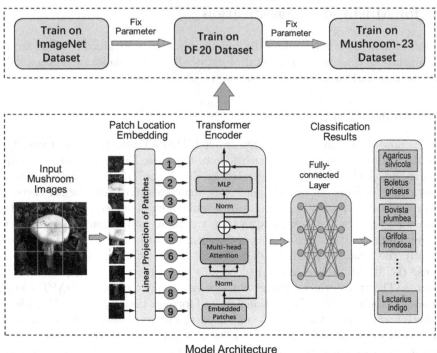

Fig. 3. An illustration of the domain adaptive pre-trained (DAPT) model for mushroom recognition.

5   Experiment

In this section, we will introduce the experimental part. First of all, we have a brief description of the dataset, then introduce the baseline models, and finally present the experimental results in detail.

5.1   Experimental Setup

In the experiment, the Mushroom-23 dataset is divided into a training set and a validation set with a ratio of 9:1, which means we use 90% of the images for training, and others for validation. Also, some image preprocessing work is conducted before the training, say, we use the OpenCV library to crop and resize each image, and perform grayscale conversion and histogram equalization. We use transformers to implement our experiment and set the batch size to 120. We train the model for 100 epochs and select the best accuracy on the validation dataset. All models are trained on 4 GTX 3090 GPUs.

5.2 Baseline Model

In the experiment, we use the following baseline models:

- **Vision Transformer** [4,6,15] is an attention-based neural network model used for image classification tasks. It uses the self-attention mechanism from the transformer to process feature vectors of image patches and generates a global feature vector of the entire image using a global pooling layer. The core idea of ViT is to treat an image as a sequence and use a transformer model to encode this sequence. ViT performs similarly to traditional convolutional neural network models on the ImageNet dataset, but is more flexible and scalable, as it can handle images of different sizes and resolutions.
- **Swin Transformer** [4,5,16] is a new transformer model designed specifically for computer vision tasks, building upon the ViT. It introduces a hierarchical structure, dividing the transformer blocks into multiple stages, each containing several transformer blocks, to reduce computation and memory consumption. Additionally, the Swin transformer introduces the position encoding and local attention mechanism to handle the relationship between local features and global features. Swin transformer achieves better performance than ViT and other models on several computer vision tasks.

5.3 Experiment Result

Table 2. Mushroom recognition performance comparison among different models.

Model	Accuracy	Macro F1
Vision Transformer	79.0%	0.732
Swin Transformer	78.7%	0.736
DAPT (ours)	82.9%	0.788

Performance Comparison Among Different Models: Table 2 presents the experimental results on Mushroom-23 dataset between DAPT and baselines in terms of Accuracy and Macro F1. The ViT model achieves an accuracy of 79.0% and a Macro F1 score of 0.732. It demonstrates that the ViT model, which has been proved to be effective in various computer vision tasks, also performs well in the mushroom classification task. The Swin transformer, another advanced transformer model, can obtain comparable results with the ViT. It achieves slightly lower accuracy, 78.7%, than the ViT model, but had a marginally higher Macro F1 score of 0.736. DAPT performs the best among all models, and achieves an accuracy of 82.9% and a Macro F1 score of 0.788. It is because DAPT introduces the domain adaptive training approach to fine-tune on mushroom images dataset to boost its recognition performance on specific mushrooms.

Table 3. Compare mushroom recognition results with different learning rates.

Model	Learning Rate on DF 20	Learning Rate on Mushroom-23	Accuracy	Macro F1
DAPT	0.0001	0.0001	81.9%	0.777
DAPT	0.0001	0.00002	82.9%	0.788

Performance Comparison with Different Learning Rates: Table 3 show-cases the results of the ViT model using different learning rates on both DF 20 and Mushroom-23 datasets. In this upper one, the DAPT model is first fine-tuned with a learning rate of 0.0001 on the DF 20 dataset and then with the same learning rate on Mushroom-23 dataset. The DAPT model achieves an accuracy of 81.9% and a Macro F1 score of 0.777. These results indicate that the two-stage fine-tuning process, using a consistent learning rate of 0.0001, enhances the model's performance in the mushroom classification task. In the lower one, the DAPT model is first fine-tuned with a learning rate of 0.0001 on the DF 20 dataset and then with a reduced learning rate of 0.00002 on Mushroom-23 dataset. The DAPT model achieves an accuracy of 82.9% and a Macro F1 score of 0.788. These results suggest that the combination of a higher learning rate during the initial fine-tuning on the DF20 dataset and a lower learning rate on our dataset leads to even better performance.

Table 4. Performance comparison between DAPT and its reduced cases.

Reduced Model	Learning Rate	Accuracy	Macro F1
Complete DAPT	0.0001 on DF 20	82.9%	0.788
	0.00002 on Mushroom-23		
Reduced DAPT (train on DF 20 only)	0.0001	76.3%	0.686
	0.00002	71.5%	0.608
Reduced DAPT (train on Mushroom-23 only)	0.0001	79.0%	0.732
	0.00002	76.3%	0.647

Performance Comparison Between DAPT and Its Reduced Cases: To show the advantage of adaptive learning, we compare the performance of the complete DAPT model and the reduced case by removing the adaptive learning module in Table 4. The complete DAPT model achieves an accuracy of 82.9% and macro F1 score of 0.788. The following two rows show the impact of different learning rates on the performance of the reduced case model without the adaptive learning module, trained on the DF 20 dataset and our customized Mushroom-23 dataset. The table shows two different learning rates tested on the DF 20 dataset: 0.0001 and 0.00002. When the model is train with a learning rate of 0.0001, its accuracy reaches 76.3%, and the macro F1 score is 0.686. However, when the learning rate is lower to 0.00002, the model's performance decreased, resulting in an accuracy of 71.5% and a macro F1 score of 0.608. This indicates that a higher learning rate of 0.0001 resulted in better performance on the DF 20 dataset. The same learning rates are also evaluate on Mushroom-23 dataset. With a learning rate of 0.0001, the model achieves an accuracy of 79.0% and a macro F1 score of 0.732. When the learning rate is lower to 0.00002, the performance decreased, resulting in an accuracy of 76.3% and a macro F1 score of 0.647. Similar to the results on the DF 20 dataset, a higher learning rate of 0.0001 also seem to produce better performance on our dataset. Overall, the DAPT model outperform the simplified case in terms of accuracy and macro F1.

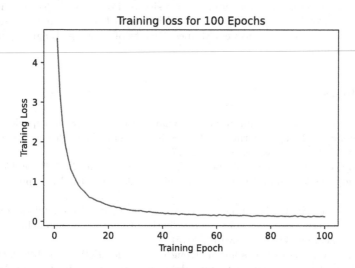

Fig. 4. The change of the training loss as the increase training iteration.

The Change of the Training Loss as the Increase Training Iteration: Figure 4 illustrates the change of the training loss as the increase of training iterations of the DAPT model. The horizontal axis represents the number of training iterations, while the vertical axis represents the value of the loss function. In the beginning, the loss function starts to decrease from a position higher than 4. As the number of training iterations increases, the value of the loss function decreases rapidly, indicating that the model is starting to learn effective features and correctly classify mushroom images. This descent phase is typically completed within the first 20 to 30 training iterations. Next, the rate of descent of the loss function gradually slows down because the model has already learned most of the effective features and requires more training to further improve accuracy. Over the next several dozen training iterations, the curve will become smoother, and the value of the loss function will gradually approach 0, indicating that the model's performance is gradually reaching optimal levels.

6   Conclusion

In this paper, we construct a large mushroom image dataset named Mushroom-23, then we propose a domain adaptive pre-trained (DAPT) model to use the pre-training and domain adaptive transfer learning methods to boost the performance of the mushroom species recognition. Next, we will incorporate DAPT into the mobile APP to realize portable mushroom recognition.

Acknowledgement. This research was partially conducted in Research Institute for Artificial Intelligence of Things (RIAIoT) at PolyU and supported by PolyU Research and Innovation Office (No. BD4A), Hong Kong, China, and the Ph.D. Scientific Research Foundation, The Chongqing University of Posts and Telecommunications (No. E012A2022026), Chongqing, China, and benefited from the financial support of the Hong Kong Institute of Business Studies, Faculty of Business, Lingnan University, Hong Kong, China.

References

1. Sabour, S., Frosst, N., Hinton, G.E.: Dynamic routing between capsules. In: 31st Conference on Neural Information Processing Systems, Long Beach, CA, USA, pp. 3859–3869. Curran Associates Inc. (2017)
2. Szegedy, C., et al.: Going deeper with convolutions. In: 2015 IEEE Conference on Computer Vision and Pattern Recognition, Boston, MA, USA, pp. 1–9. IEEE (2015)
3. Ioffe, S., Szegedy, C.: Batch normalization: Accelerating deep network training by reducing internal covariate shift. In: International Conference on Machine Learning, Lille, France, pp. 448–456. JMLR (2015)
4. Yu, J., et al.: Bag of tricks and a strong baseline for FGVC. In: 13th Conference and Labs of the Evaluation Forum, Bologna, Italy, pp. 1–16. Curran Associates Inc. (2022)
5. Liu, Z., et al.: Swin transformer: hierarchical vision transformer using shifted windows. In: Proceedings of the IEEE/CVF International Conference on Computer Vision, Montreal, QC, Canada, pp. 10012–10022 (2021)
6. Dosovitskiy, A., et al.: An image is worth 16x16 words: transformers for image recognition at scale, arXiv preprint arXiv:2010.11929 (2020)
7. Zeiler, M.D., Krishnan, D., Taylor, G.W., Fergus, R.: Deconvolutional networks. In: 2010 IEEE Computer Society Conference on Computer Vision and Pattern Recognition, San Francisco, CA, pp. 2528–2535. IEEE (2010)
8. Dai, Z., Yang, Z., Yang, Y., Carbonell, J., Le, Q.V., Salakhutdinov, R.: Transformer-XL: attentive language models beyond a fixed-length context, arXiv preprint arXiv:1901.02860 (2019)
9. Devlin, J., Chang, M.-W., Lee, K., Toutanova, K.: BERT: pre-training of deep bidirectional transformers for language understanding, arXiv preprint arXiv:1810.04805 (2018)
10. He, K., Zhang, X., Ren, S., Sun, J.: Deep residual learning for image recognition. In: Proceedings of the IEEE Conference on Computer Vision and Pattern Recognition, Las Vegas, NV, USA, pp. 770–778. IEEE (2016)

11. Xie, S., Girshick, R., Dollár, P., Tu, Z., He, K.: Aggregated residual transformations for deep neural networks. In: Proceedings of the IEEE Conference on Computer Vision and Pattern Recognition, San Francisco, CA, USA, pp. 1492–1500. IEEE (2017)
12. Vaswani, A., et al.: Attention is all you need, arXiv preprint arXiv:1706.03762 (2017)
13. Girdhar, R., Carreira, J., Doersch, C., Zisserman, A.: Video action transformer network. In: Proceedings of the IEEE/CVF Conference on Computer Vision and Pattern Recognition, pp. 244–253. CVPR (2019)
14. Xie, E., et al.: Segmenting transparent object in the wild with transformer, arXiv preprint arXiv:2101.08461 (2021)
15. Arnab, A., Dehghani, M., Heigold, G., Sun, C., Lučić, M., Schmid, C.: ViViT: a video vision transformer. In: Proceedings of the IEEE/CVF International Conference on Computer Vision, Los Alamitos, California, pp. 6836–6846. IEEE (2021)
16. Liang, J., Cao, J., Sun, G., Zhang, K., Van Gool, L., Timofte, R.: SwinIR: image restoration using swin transformer. In: Proceedings of the IEEE/CVF International Conference on Computer Vision, Los Alamitos, California, pp. 1833–1844. IEEE (2021)
17. Carion, N., Massa, F., Synnaeve, G., Usunier, N., Kirillov, A., Zagoruyko, S.: End-to-end object detection with transformers. In: Vedaldi, A., Bischof, H., Brox, T., Frahm, J.-M. (eds.) ECCV 2020. LNCS, vol. 12346, pp. 213–229. Springer, Cham (2020). https://doi.org/10.1007/978-3-030-58452-8_13
18. Sun, P., et al.: TransTrack: multiple-object tracking with transformer, arXiv preprint arXiv:2012.15460 (2020)
19. Krizhevsky, A., Sutskever, I., Hinton, G.E.: ImageNet classification with deep convolutional neural networks. Commun. ACM **60**(6), 84–90 (2017)
20. LeCun, Y., Bengio, Y., Hinton, G.: Deep learning. Nature **521**(7553), 436–444 (2015)
21. Van Horn, G., et al.: The inaturalist species classification and detection dataset. In: Proceedings of the IEEE Conference on Computer Vision and Pattern Recognition, Salt Lake City, UT, USA, pp. 8769–8778. CVPR (2018)
22. Picek, L., Šulc, M., Matas, J., Heilmann-Clausen, J.: Overview of FungiCLEF 2022: fungi recognition as an open set classification problem. In: Conference and Labs of the Evaluation Forum, Bologna, Italy, pp. 9–25. CLEF (2022)
23. Joly, A., Goëau, H., Kahl, S., Picek, L., Lorieul, T., Cole, E., Hrúz, M.: Overview of lifeCLEF 2022: an evaluation of machine-learning based species identification and species distribution prediction. In: Barrón-Cedeño, A., et al. (eds.) CLEF 2022. LNCS, vol. 13390, pp. 257–285. Springer, Cham (2022). https://doi.org/10.1007/978-3-031-13643-6_19
24. He, S., Luo, H., Wang, P., Wang, F., Li, H., Jiang, W.: TransReiD: transformer-based object re-identification, arXiv preprint arXiv:2102.04378 (2021)
25. Fu, J., Zheng, H., Mei, T.: Look closer to see better: recurrent attention convolutional neural network for fine-grained image recognition. In: Proceedings of the IEEE Conference on Computer Vision and Pattern Recognition, Honolulu, Hawaii, pp. 4438–4446. IEEE (2017)
26. Christian, S., et al.: Going deeper with convolutions. In: Proceedings of the IEEE Conference on Computer Vision and Pattern Recognition, pp. 1–9. CVPR (2015)
27. He, X., Peng, Y.: Fine-grained image classification via combining vision and language. In: Proceedings of the IEEE Conference on Computer Vision and Pattern Recognition, Honolulu, Hawaii, pp. 5994–6002. IEEE (2017)

28. Zheng, S., et al.: Rethinking semantic segmentation from a sequence-to-sequence perspective with transformers. In: Proceedings of the IEEE/CVF Conference on Computer Vision and Pattern Recognition, pp. 6881–6890. CVPR (2021)
29. Deng, J., Dong, W., Socher, R., Li, L. J., Li, K., Fei-Fei, L.: ImageNet: a large-scale hierarchical image database. In: Proceedings of 2009 IEEE Conference on Computer Vision and Pattern Recognition, pp. 248–255. CVPR (2009)
30. Lin, T.-Y., et al.: Microsoft COCO: common objects in context. In: Fleet, D., Pajdla, T., Schiele, B., Tuytelaars, T. (eds.) ECCV 2014. LNCS, vol. 8693, pp. 740–755. Springer, Cham (2014). https://doi.org/10.1007/978-3-319-10602-1_48
31. Yu, C., Zhao, X., Zheng, Q., Zhang, P., You, X.: Hierarchical bilinear pooling for fine-grained visual recognition. In: Ferrari, V., Hebert, M., Sminchisescu, C., Weiss, Y. (eds.) ECCV 2018. LNCS, vol. 11220, pp. 595–610. Springer, Cham (2018). https://doi.org/10.1007/978-3-030-01270-0_35
32. Zheng, H., Fu, J., Zha, Z.-J., Luo, J.: Learning deep bilinear transformation for fine-grained image representation, arXiv preprint arXiv:1911.03621 (2019)

Training Noise Robust Deep Neural Networks with Self-supervised Learning

Zhen Wang(ID), Shuo Jin(✉)(ID), Jiapeng Du, Linhao Li(ID), and Yongfeng Dong(ID)

School of Artificial Intelligence, Hebei University of Technology, Tianjin, China
1160772114@qq.com

Abstract. Training accurate deep neural networks (DNNs) on datasets with label noise is challenging for practical applications. The sample selection paradigm is a popular strategy that selects potentially clean data from noisy data for noise-robust training. In this study, we first analyze the sample selection models and find that the key aspects of this paradigm are the scale and degree of purity of the selected samples; however, these are restricted by the noise strength in the training set and the learning capacity of the models. Therefore, we propose a simple yet effective method called CoPL, which cross-trains two noise-robust DNNs simultaneously based on small loss criteria, along with learning accurate pseudo-labels. Benefitting from pseudo-labels, CoPL can reduce the noise strength and further promote the learning capacity and robustness of the models. Additionally, we discuss CoPL from the perspective of label smoothing to provide a theoretical guarantee of its performance. Extensive experimental results on both simulated (MNIST, CIFAR-10 and CIFAR-100) and real-world datasets (Clothing1M) demonstrate that CoPL is superior to other state-of-the-art methods and obtains a more noise-robust learning capacity.

Keywords: Pseudo label · Label noise · Noise-robust learning · Sample selection

1 Introduction

Deep neural networks (DNNs) have achieved remarkable success in areas such as computer vision and natural language processing tasks [8]. However, one limitation of these networks is the supervision of a mass of training samples with accurate labels, resulting in high labor costs and time consumption. Thus, easier and quicker annotation methods have been developed, such as crowdsourcing [23], web-crawling [12], and online queries [18]. However, such methods inevitably suffer from error-tagging issues, namely, label noise. The excellent learning capability of DNNs can result in complete learning of the resulting confused (noisy) pattern with respect to the ground-truth pattern, leading to poor generalization capability [1].

Fortunately, research into the memorization pattern of DNNs on label noise has found that DNNs first learn an easy/clean pattern and then gradually the hard/noisy pattern [1,26]. Inspired by this, a sample-selected or sample-weighted algorithm, typified by the co-teaching paradigm [3], has been proposed for label noise scenarios. In essence, the co-teaching paradigm attempts to simultaneously train two peer networks and cross-update each other on the samples selected via small loss criteria. That is, small loss samples are more likely to be clean samples; otherwise, they are noisy samples. The key to the co-teaching paradigm is deriving different learning abilities and decision boundaries from two peer networks; meanwhile, the confused pattern derived from one network can be blocked via cross-update. These features enable the co-teaching paradigm to avoid the overconfidence of the self-teaching paradigm trained on noisy data.

To further improve the robustness of DNNs, additional co-teaching-based methods [10,21,25] have been proposed. However, the co-teaching paradigm has some evident deficiencies. First, sample selection decreases the total number of training samples, inevitably degrading the performance. Learning using label noise further degrades performance. Additionally, the information contained in noisy samples can improve the robustness and generalization of DNNs. Furthermore, the selected samples inevitably include noisy samples, which causes further performance degradation. These issues will be further discussed and verified in Sect. 3.2.

To alleviate the aforementioned issues, we propose an update to the co-teaching paradigm called CoPL (**Co**-teaching with **p**seudo **l**abel). In this method two networks are trained using the supervised information provided from each other, along with the learning of pseudo-labels. This study makes two main contributions to the literature:

- We find the co-teaching paradigm is restricted by the scale and purity of the samples selected for model training, then propose the novel method CoPL to overcome these restrictions. CoPL maintains two peer networks simultaneously so that confused patterns accumulated in networks can be mutually reduced by the peer networks, and it learns pseudo-labels to reduce the noise strength.
- Comprehensive experiments on both simulated and real-world datasets verify the effectiveness of the proposed algorithm. Specifically, ablation studies indicate that the proposed method can select clean samples and provide more confidence-supervised information. Comparisons with state-of-the-art methods showed a superior classification performance against label noise.

The remainder of this paper is organized as follows. In Sect. 2, we will review related work on label noise. In Sect. 3, we introduce multiclassification in *Label Noise* and two characteristics of learning with label noise, then present our method and discuss its relationship with label smoothing. The experimental results and discussion are presented in Sect. 4. Finally, conclusions are presented in Sect. 5.

2 Related Works

In this section, let us briefly review the three main ways to solve the label noise problem: robust loss functions, loss correction, and sample selection.

Robust Loss Functions. The most direct method for dealing with label noise is to design a loss function that is not unduly affected by noisy labels. Ghosh et al. [2] provided sufficient conditions for a loss function such that the risk minimization when used in multiclass classification problems would be inherently tolerant to label noise. Inspired by these conditions, several loss functions have been designed and proven to satisfy noise-robust loss-bounded conditions. Examples include GCE [27], SL [19], and APL [11], which are derived from noise-robust losses and introduce noise-robust factors to promote CE loss. However, the conditions of noise-robust loss are very stringent, and they often encounter under-fitting issues.

Loss Correction. An alternative approach is to estimate the latent noise transition matrix to correct the training process. Many studies [5,16] have embedded the noise transition matrix into the network architecture as a noisy layer and trained the network parameters and noise transition matrix simultaneously via a stochastic gradient descent algorithm, resulting in non-convergence or trivial results. Patrini et al. [15] proposed a loss correction procedure using matrix inversion and multiplication; however, the noise transition matrix and network both must be updated, blocking the interaction between them [4,15]. To better learn the noise transition matrix, Wang et al. [20] propose Meta Loss Correction" (MLC). MLC optimizes the noise transfer matrix by meta-learning, and although it achieves good results, the meta-learning is very time-consuming.

Sample Selection. The sample selection method divides the training data into different subsets according to specific classification criteria. Different strategies are then applied to each subset to train the model. Coteaching [3] simultaneously cross-trains two networks by selecting small-loss samples and the two networks provide supervision information to each other to block the negative information. Decoupling [13] introduces the disagreement" strategy in which updates only use the instances that have different predictions from these two networks, decoupling when to update from how to update. Co-teaching+ [25] selects small-loss samples only from the prediction disagreement data to keep the two networks separate within the training epochs. DivideMix [10] trains networks on both clean and unlabeled sets, which are divided from the original training samples by dynamically fitting a Gaussian mixture model (GMM) to its per-sample loss distribution. JoCoR [21] aims to reduce the diversity of the two peer networks during training via co-regularization loss, making the predictions of each network closer to that using clean labels and peer networks. Jo-SRC [24] selects clean samples globally by adopting Jensen-Shannon divergence (JSD). It distinguishes in-distribution (ID) and out-of-distribution (OOD) noisy samples based on the prediction consistency between samples different views. PNP [17] predicts the noise type of each sample by establishing a label prediction network. The whole training set is then divided into three subsets: clean, ID, and OOD set. To address the disproportionate selection of easy and hard samples, UNICON [6] assembles the predictions of both networks to calculate the JSD. After estimating the filter rate from the JSD distribution, it takes an equal number of samples from each class as the clean sample set.

We Propose CoPL that trains two noise-robust

3   Methodology

3.1   Preliminary

For traditional multi-class classification, a pair of random variables $(x, y) \in \mathcal{X} \times \mathcal{Y}$ are drawn from distribution $p(X, Y)$, where $\mathcal{X} \in \mathbb{R}^d$ denotes the feature spaces, and $\mathcal{Y} \in [0, 1]^k$ denotes the label spaces. Let $f(\theta)$ be the multi-class classifier (parameterized by θ), which maps the feature space to label space ($\mathbb{R}^d \to [0, 1]^k$). Let $p \in \triangle^{k-1}$ denote the output of $f(\theta)$ passing through a *softmax* operator, where $\triangle^{k-1}$ is a k-dimensional simplex, e.g. $p = softmax(f(\theta))$.

While in noisy scenario, y may flip to noisy label $\tilde{y}$ with the transition probability $T_{y\tilde{y}}$, where

$$\tilde{y}_i = \begin{cases} y_i & 1 - \eta_i \\ j, & j \in [k], j \neq y_i \ \bar{\eta}_{i,j} \end{cases} \tag{1}$$

where η is the noise ratio, $\sum_{j \neq i} \bar{\eta}_{i,j} = \eta_i$, and $T \in [0, 1]^{k \times k}$ is the noise transition matrix. $(x, \tilde{y}) \in \mathcal{X} \times \mathcal{Y}$ are drawn from distribution $\tilde{p}(X, \tilde{Y})$. Then:

$$\tilde{p}(X, Y) = T^T p(X, \tilde{Y}) \tag{2}$$

The noise is termed as *Symmetric* if $\eta_i = \eta$ and $\bar{\eta}_{ij} = \frac{\eta}{k-1}$, where η is a constant; it is termed as *Asymmetric* if η_i is dependent on x.

Let L be the loss function, mapping $p \times \mathcal{Y} \to \mathbb{R}_+$. The classic loss function in multi-classification is *CrossEntropy* (CE), which can be parameterized as:

$$\mathcal{L}_{\text{CE}} = -\frac{1}{n} \sum_{i=1}^{n} \sum_{j=1}^{k} y_{ij} \log p_{ij} \tag{3}$$

We seek a optimal classifier f by minimizing the $L - risk$ of f as follows:

$$R_L(f) = \mathbb{E}_\mathcal{D} \left[L\left(f(\mathbf{x}; \theta), \tilde{y}_i\right) \right] = \mathbb{E}_{\mathbf{x}, \tilde{y}_i} \left[L\left(f(\mathbf{x}; \theta), \tilde{y}_i\right) \right] \tag{4}$$

where $\mathbb{E}$ denotes the expectation and its subscript indicates the distribution.

3.2   Learning with Label Noise

Learning Pattern on Label Noise. Figure 1(a) illustrates the standardized dynamic learning process of deep models using CE loss function under label noise. The training and test accuracy curves gradually increase as the deep model fits to the easy samples (pattern). However, instead of further increasing training accuracy, the negative gradient of the noisy labels, the test accuracy curve becomes more pronounced after the model reaches pseudo-sufficient learning, indicating that the deep model is now over-fitting the label noise in training samples. This behavior is consistent with current studies. In addition, a noisy training set can provide relatively high-confidence model parameters before over-fitting label noise.

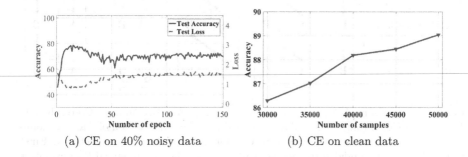

(a) CE on 40% noisy data   (b) CE on clean data

Fig. 1. (a) Test accuracy and Test loss of CE when training on noisy data; (b) Test accuracy (%) for different numbers of training samples.

Effect of Different Samples. The original Co-teaching [3] algorithm is a sample-weighting- or sample-selection-based method which screens and prunes the noisy samples to mitigate the negative effect induced by label noise. However, pruning samples means fewer samples are available to train DNNs In addition, noisy samples may still be used (more or less), resulting in under-fitting to the training samples. Figure 1(b) illustrates the test accuracy vs. numbers of clean training samples, we can find that the scale of training samples restricts the learning and classification capacities. Thus, the Co-teaching algorithm suffers from the under-fitting problem because it cannot learn samples with the ground-truth labels sufficiently, resulting in poor generalization ability in DNNs. We find that this under-fitting problem occurs across different training settings in terms of learning rate, learning rate scheduler, weight decay, and the number of training epochs.

3.3 Co-pseudo-label

We develop CoPL with two nested processes: learning with pseudo-labels and Co-teaching paradigm, as shown in Fig. 2.

Learning with Pseudo-labels. As discussed in previous section, benefiting from the warm up process, the model can provide a high confidence prediction. A straight-forward strategy is to reconstruct the training labels as the convex combination between the current training labels and the prediction. Then, the pseudo-label can be formulated using the training label s_i and the prediction with the softmax operator p_i:

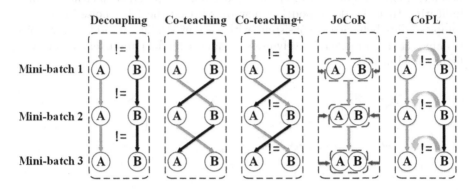

Fig. 2. Comparison of error flows among Co-teaching, Decoupling, Co-teaching+, JoCoR, and CoPL (ours). Error flows assumed from biased selection of training examples, denoted by red arrows or blue arrows for network A or B, respectively. **First Panel**: Decoupling updates the two networks with prediction-disagreed (!=) examples from a mini-batch. **Second Panel**: Co-teaching maintains two networks (A & B). In each mini-batch data, each network selects its small-loss data to teach its peer network for robust training. **Third Panel**: In Co-teaching+, each network selects its small-loss instances within prediction disagreement (!=) to teach its peer network. **Fourth Panel**: JoCoR trains two networks as a whole with a joint loss, which makes predictions of each network closer to the peer networks. **Fifth Panel**: CoPL maintains two networks (A & B). In each mini-batch data, each network selects its small-loss data to teach its peer network, and meanwhile learns the pseudo labels. (Color figure online)

$$s_i = \alpha \times s_i + (1 - \alpha) \times p_i \tag{5}$$

The trade-off term α balances the contribution from model prediction and pseudo labels. The model prediction dominates the pseudo-label learning during the second half of the training process. Notably, the training labels are initially set as the original noisy labels $s_i = \tilde{y}_i$.

Next, under empirical risk minimization, we employ the *CrossEntropy* loss between the pseudo labels and the prediction to train the networks. We reformulate the *CrossEntropy* loss as $\mathcal{L}_{CoPL}$, where:

$$\mathcal{L}_{CoPL} = -\frac{1}{n} \sum_{i=1}^{n} \sum_{j=1}^{k} s_{ij} \log p_{ij} \tag{6}$$

Learning with Sample-Selected Paradigm. The sample-selected paradigm is based on the training loss, that is to say, in each training batch, the networks are trained on the small-loss training samples provided by each other. Specifically, for two networks f and g, in each training batch, the training samples are selected based on the small $\mathcal{L}_{CoPL}$ with $R(t)$ percent, namely $\overline{\mathcal{D}}_f$ and $\overline{\mathcal{D}}_g$:

$$\overline{\mathcal{D}}_f = \arg \min_{\mathcal{D}':|\mathcal{D}'| \geq R(T)|\overline{\mathcal{D}}|} \ell\left(f, \mathcal{D}'\right) \tag{7}$$

$$\overline{\mathcal{D}}_g = \arg \min_{\mathcal{D}':|\mathcal{D}'| \geq R(T)|\overline{\mathcal{D}}|} \ell\left(g, \mathcal{D}'\right) \tag{8}$$

where $R(t)$ controls the number of samples to be selected in each batch, balancing the trade-off between the purity of clean samples and network performances. And $\overline{\mathcal{D}}$ represents the mini-batch extracted from the training set.

Subsequently, the two networks are trained on the $\overline{\mathcal{D}}_f$ and $\overline{\mathcal{D}}_g$ via stochastic gradient descent algorithm, respectively.

$$w_f = w_f - \gamma \nabla \ell\left(f, \overline{\mathcal{D}}_g\right) \tag{9}$$

$$w_g = w_g - \gamma \nabla \ell\left(g, \overline{\mathcal{D}}_f\right) \tag{10}$$

Dynamical R(t). $R(t)$ changes dynamically with the training epoch. At the beginning of the training process, CoPL selects more samples in each batch because the prediction is at low confidence, resulting in the small-loss samples not being equal to clean samples. Thus, more samples are provided to the networks to fit the pattern of the training samples. Then, as training continues and the prediction confidence increases, the number of training samples is decreased. This results in only 'clean' samples (the small-loss samples can be regarded as clean examples) being provided to networks, thereby alleviating the under-fitting issue for samples with true labels. Finally, we gradually increase $R(t)$; because the pseudo labels can provide high-confidence supervised information, they provide more high-confidence samples in each batch, resulting in the training samples used being more than that used in the previous sample-selected method. The overall procedure of the proposed method is shown in Algorithm 1.

3.4 Reviewing CoPL from the Perspective of Label Smoothing

At the beginning of training, consider the case of a single given label $\tilde{y}$, so that $q(\tilde{y}) = 1$ and $q(j) = 0$ for all $j \neq \tilde{y}$. To quantify the loss, we transfer it to the one-hot vector, however, this may result in over-fitting: if the model learns to assign full probability to the given label for each training example, it cannot be guaranteed to generalize. In addition, it encourages the gap between the largest logit and others to increase, restricting the model generalization ability. Intuitively, this occurs because the model becomes too confident in its predictions. Noisy labels can additionally aggravate the overconfidence issue.

Fortunately, during the training process with CoPL, the $q(j) = 0$ for all $j \neq \tilde{y}$ may rise, and the $q(\tilde{y})$ may decrease according to Eq. 5, hence preventing the $q(\tilde{y})$ from becoming much larger than all of the others, and the sample in-class distance decreases whereas the between-class distances increase. Therefore, the CoPL formulated in Eq. 5 can be regarded as a special case of label smoothing:

Algorithm 1: CoPL

Input: Network f and g with parameters w_f and w_g
learning rate γ
noisy training set $\tilde{D}$
pseudo label s
noisy label $\tilde{y}$
max epoch T_{max}
max iteration I_{max}

1 **for** $t = 1, 2, ..., T_{max}$ **do**
2   **Shuffle** training set $\tilde{D}$;
3   **for** $i = 1, 2, ..., I_{max}$ **do**
4   **Fetch** mini-batch $\overline{D}$ from $\tilde{D}$;
5   **Obtain** $\overline{D}_f = \arg\min_{D' : |D'| \geq R(T)|\overline{D}|} \ell(f, D')$
6   **Obtain** $\overline{D}_g = \arg\min_{D' : |D'| \geq R(T)|\overline{D}|} \ell(g, D')$
7   **Update** $s_i \leftarrow \alpha \times s_i + (1 - \alpha) \times p_i$
8   **Update** $w_f = w_f - \gamma \nabla \ell(f, \overline{D}_g)$
9   **Update** $w_g = w_g - \gamma \nabla \ell(g, \overline{D}_f)$
10   **end**
11 **end**
    Output: w_f and w_g

$$q_i = \begin{cases} 1 - \varepsilon & \text{if } i = y \\ \varepsilon/(K - 1) & \text{otherwise} \end{cases} \tag{11}$$

where ε is a small constant. CoPL transfers the hard labels to soft labels. On one hand, CoPL learns more confidence labels (pseudo labels), on the other hand, CoPL can prevent the networks from over-fitting to label noise (the characteristic derived from label smoothing).

4 Experiment and Analysis

In this section, we present the experimental settings (Sect. 4.1), the ablation study (Sect. 4.2), and experiments with other state-of-the-art methods on both simulated (Sect. 4.3) and real-world (Sect. 4.4) noise.

4.1 Experimental Setup

In this section, we describe the experimental settings implemented for the included methods in terms of datasets, baselines, noises and implementation details.

Datasets. We used two popular noise-free datasets (MNIST and CIFAR) and a large-scale noisy real-world dataset (Clothing1M), all of which have been widely used in previous label noise studies to evaluate model effectiveness under label noise.

- MNIST: The MNIST [9] dataset is annotated with 10 object categories containing 28×28 handwritten digit images. The training and test sets contain $60k$ and $10k$ images, respectively.
- CIFAR: The CIFAR [7] dataset is popular for image classification. It contains $50k$ color images for training and $10k$ images for testing with a resolution of 32×32. Two sub-datasets, CIFAR-10 and CIFAR-100 are annotated with 10 and 100 classes, respectively.
- Clothing1M: Clothing1M [22] is a large-scale real-world noisy dataset comprising images collected from online shopping websites. The labels are generated using keywords, which results in label noise. The Clothing1M dataset is annotated using 14 classes corrupted by label noise. The Clothing1M dataset comprises $1M$ noisy samples and an additional $50k$, $14k$, and $10k$ clean samples for training, validation, and testing, respectively. The noise level is approximately 40% [22].

Noise Types. For MNIST and CIFAR datasets, the clean data must be corrupted manually to simulate label noise according to the noise transition matrix. Three representative types of noise are employed in this study: (1) symmetry noise; (2) pairflip noise; (3) tridiagonal noise. The noise transition matrix of the first three types of noise are shown in Fig. 3.

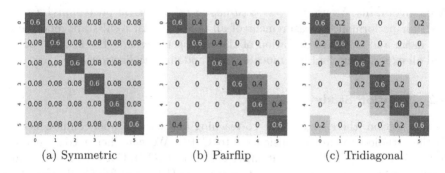

(a) Symmetric   (b) Pairflip   (c) Tridiagonal

Fig. 3. An example of noise transition matrix under 40% noise ratio (taking 6 classes as an example).

For the real-world dataset Clothing1M, the $50k$ clean training data was not used, as only noisy samples (containing noisy labels) were required during the training process. In addition, we resized the images to 256×256, cropped the middle 224×224 as input, and performed normalization for preprocessing.

State-of-the-Art Methods. We implemented the following state-of-the-art methods using the original codes releases and PyTorch [14] with default settings, and compared their performance to that of CoPL:

- Standard: Standard trains DNNs using the standard cross entropy loss.

- Co-teaching: Co-teaching [3] cross-trains two networks simultaneously via small-loss criteria.
- Co-teaching+: Co-teaching+ [25] cross-train two deep neural networks simultaneously using both disagreement-update and cross-update steps.
- JoCoR: JoCoR [21] cross-trains two networks simultaneously via joint loss, which reduces the diversity of the two networks.
- APL: APL [11] combines two loss functions that mutually enhance each other to achieve better robustness and performance.

The aforementioned methods are systematically show in Fig. 2. Although we focused on the sample selection method for label noise mitigation, to make this work more convincing, we also compared our method with other types of advanced methods. We employ methods for designing robust loss functions, i.e., APL [11]. APL combines two mutually reinforcing robust loss functions.

Implementation Details. Following recent Co-teaching studies [3,21], we employ a 9-layer CNN structure as the backbone with Adam optimizer. We set momentum $= 0.9$ with an initial learning rate of 0.001, the batch size is set to 128, and all experiments ran 200 epochs. All sample selection methods need to cross-train two networks with same structure and optimizing strategy, but different initializations. However, the two networks exhibit different characteristics owing to the highly non-convex nature of CNNs.

Measurement. We use the *test accuracy = (# of correct predictions) / (# of test)* for performance measurement of all methods. All results are reported as the mean and standard deviation of the test accuracy over the last 10 epochs, and all experiments had five repeats with random initializations.

4.2   Ablation Study

In this section, the effects of the co-teaching and pseudo-label components are further studied to provide insights into CoPL. We can then reconstruct the following models: CoPL without either component (CE), CoPL without pseudo-labels (Co-teaching), CoPL without co-teaching (PL), and full CoPL. Note that the training process remained unchanged for all options. We conducted this study using CIFAR-10 with four types of noise, and the results are listed in Table 1. These results show that CoPL performs significantly better both best and last, with negligible decreases in accuracy from best to last. This observation indicates that CoPL strongly suppresses memorization of noisy labels, and is a more efficient paradigm for clean example selection.

4.3   Results on Simulated Noise

First, we briefly summarize the compared results for the three datasets.

- All sample-selection methods performed better than the Standard method, indicating the effectiveness of the sample-selected paradigm for label noise.

Table 1. Ablation Study: Test accuracy (%) of on CIFAR-10 over the last 10 epochs

	CE	Co-teaching	PL	CoPL
Symmetric-20%	75.740.15	82.240.17	86.050.22	**87.290.14**
Symmetric-40%	58.640.48	77.160.15	81.810.18	**83.700.17**
Pairflip-20%	76.380.35	82.550.14	87.870.19	**88.040.08**
Pairflip-40%	55.030.38	75.740.20	79.540.28	**82.150.21**
Tridiagonal-20%	76.260.31	82.500.16	87.180.17	**87.580.21**
Tridiagonal-40%	58.720.61	76.280.26	81.150.19	**83.260.16**

- CoPL performs significantly better than other sample-selection methods, which demonstrates the precise and effective design of CoPL.
- The accuracy of all the methods decreased to varying degrees when the noise ratio was increased from 20% to 40%, nevertheless, CoPL achieves sustained and steady accuracy gain than others methods.

In the following sections, we detail the results and discussions for the MNIST, CIFAR-10, and CIFAR-100 datasets.

Results on MNIST. Table 2 lists the test accuracy for MNIST. For symmetric noise, as the noise strength increases the accuracy of Standard, Co-teaching and Co-teaching+ methods clearly falls, especially in the Symmetric-40% case where the Standard methods failed to work with label noise. Conversely, the accuracy of JoCoR, APL, and CoPL methods remained at very high levels, which maintain an accuracy rate of over 99%.

Table 2. Mean and standard deviations of test accuracy (%) on MNIST over the last 10 epochs

Flipping-Rate	Standard	Co-teaching	Co-teaching+	JoCoR	APL	Ours
Symmetric-20%	94.330.20	97.450.10	99.450.02	**99.600.02**	99.510.04	99.560.02
Symmetric-40%	79.880.45	94.820.20	98.810.07	**99.520.04**	99.100.05	99.450.01
Pairflip-20%	88.880.30	95.720.12	99.120.10	99.570.03	99.190.06	**99.580.02**
Pairflip-40%	64.910.31	91.670.23	95.470.24	99.360.03	80.851.30	**99.430.02**
Tridiagonal-20%	91.860.21	96.500.15	99.370.04	99.560.02	99.450.03	**99.640.01**
Tridiagonal-40%	71.250.30	92.390.17	97.450.08	99.420.03	97.200.42	**99.440.02**

Results on CIFAR-10. Table 3 shows the detailed results on the CIFAR-10 dataset. The compared methods show good performance under Symmetric 20% noise. However, when it comes to others cases, their performance drops significantly. Our CoPL still obtains the best performance, while other methods (e.g. JoCoR, APL) cannot achieve comparable performance having poor accuracy at

3% to 20%. In the Pairflip-40% case, CoPL method outperforms other methods by a large margin. Specifically, CoPL can achieve 49.2% and 8.5% improvement in accuracy over CE (baseline) and Co-teaching (second best).

Table 3. Mean and standard deviations of test accuracy (%) on CIFAR10 over the last 10 epochs

Flipping-Rate	Standard	Co-teaching	Co-teaching+	JoCoR	APL	Ours
Symmetric-20%	75.74 0.15	82.24 0.17	81.96 0.24	85.23 0.13	84.20 0.14	**87.29 0.14**
Symmetric-40%	58.64 0.48	77.16 0.15	71.49 0.48	79.77 0.13	76.19 0.13	**83.70 0.17**
Pairflip-20%	76.38 0.35	82.55 0.14	79.71 0.20	81.75 0.25	81.48 0.17	**88.04 0.08**
Pairflip-40%	55.03 0.38	75.74 0.20	58.39 0.44	68.29 0.23	52.98 0.29	**82.15 0.21**
Tridiagonal-20%	76.26 0.31	82.50 0.16	81.15 0.21	82.78 0.26	83.29 0.23	**87.58 0.21**
Tridiagonal-40%	58.72 0.61	76.28 0.26	64.79 0.23	73.53 0.21	65.80 0.50	**83.26 0.16**

Results on CIFAR-100. For the more difficult dataset CIFAR-100, CoPL can also generally perform better than other methods. All other compared methods incur a sharp decline in performance, especially under the extremely noisy Pairflip-40%, resulting in a very low confidence in the model. On the contrary, the CoPL can achieve consistent improvement with high confidence. In the Symmetric-20% case, CoPL can achieve about 12.1% accuracy gain over the second best Co-teaching+ (Table 4).

Table 4. Mean and standard deviations of test accuracy (%) on CIFAR100 over the last 10 epochs

Flipping-Rate	Standard	Co-teaching	Co-teaching+	JoCoR	APL	Ours
Symmetric-20%	42.71 0.69	50.21 0.68	52.51 0.39	51.39 0.17	45.02 0.87	**58.89 0.26**
Symmetric-40%	28.08 0.61	42.75 0.55	42.83 1.10	43.43 0.29	37.66 0.28	**52.50 0.18**
Pairflip-20%	46.22 0.35	48.78 0.35	51.36 0.37	48.68 0.19	44.14 0.39	**58.38 0.23**
Pairflip-40%	32.68 0.19	34.76 0.49	33.96 0.24	33.76 0.25	31.28 0.76	**45.61 0.19**
Tridiagonal-20%	46.40 0.34	49.99 0.46	51.93 0.37	49.08 0.25	43.55 0.37	**58.70 0.44**
Tridiagonal-40%	33.68 0.41	39.32 0.36	41.20 0.21	37.58 0.49	33.57 0.60	**52.93 0.26**

4.4   Results on Real-World Noise

We further explore the effectiveness of our method on the real-world noisy dataset Clothing1M. The test results are shown in Table 5, including results from JoCoR [21]. We report the results as best and last accuracies, which denote the best accuracy and mean accuracy of the last ten epochs of the whole training process, respectively. CoPL achieves both higher *best* and *last* accuracies, and better performance than other state-of-the-art methods, indicating the greater robustness of our proposed method to real-world noise.

Table 5. Test accuracy (%) on the Clothing1M set

Method	best	last
Standard	67.22	64.68
Decoupling	68.48	67.32
Co-teaching	69.21	68.51
Co-teaching+	59.32	58.79
JoCoR	70.30	69.79
APL	56.01	55.84
Ours	**71.80**	**71.03**

5 Conclusion

This study introduces an efficient approach for label noise called CoPL, which trains two noise-robust DNNs simultaneously via small-loss criterion and learns accurate pseudo labels in label noise scenarios. The key idea of CoPL is cross-training two networks on the 'clean' samples via small loss criteria while also improving both the scale and the degree of purity of 'clean' samples. This combination enables CoPL to overcome the restriction of the sample-selected paradigm under label noise. Abundant experiments on both simulated (MNIST, CIFAR-10 and CIFAR-100) and real-world (Clothing1M) datasets demonstrate the robustness of CoPL to label noise even under extremely noisy conditions. Ablation studies further justify the effectiveness of CoPL.

References

1. Arpit, D., et al.: A closer look at memorization in deep networks. In: International Conference on Machine Learning, pp. 233–242. PMLR (2017)
2. Ghosh, A., Kumar, H., Sastry, P.: Robust loss functions under label noise for deep neural networks. In: Proceedings of the AAAI Conference on Artificial Intelligence, vol. 31 (2017)
3. Han, B., et al.: Co-teaching: Robust training of deep neural networks with extremely noisy labels. Advances in Neural Information Processing Systems (2018)
4. Hendrycks, D., Mazeika, M., Wilson, D., Gimpel, K.: Using trusted data to train deep networks on labels corrupted by severe noise. Adv. Neural. Inf. Process. Syst. **31** (2018)

5. Jindal, I., Nokleby, M., Chen, X.: Learning deep networks from noisy labels with dropout regularization. In: 2016 IEEE 16th International Conference on Data Mining (ICDM), pp. 967–972. IEEE (2016)

6. Karim, N., Rizve, M.N., Rahnavard, N., Mian, A., Shah, M.: UNICON: combating label noise through uniform selection and contrastive learning. In: Proceedings of the IEEE/CVF Conference on Computer Vision and Pattern Recognition, pp. 9676–9686 (2022)

7. Krizhevsky, A.: Learning multiple layers of features from tiny images. Master's thesis, University of Tront (2009)

8. Krizhevsky, A., Sutskever, I., Hinton, G.E.: ImageNet classification with deep convolutional neural networks. In: Advances in Neural Information Processing Systems, vol. 25 (2012)

9. LeCun, Y.: The MNIST database of handwritten digits (1998). http://yann.lecun.com/exdb/mnist/

10. Li, J., Socher, R., Hoi, S.C.: DIVIDEMIX: learning with noisy labels as semi-supervised learning. In: International Conference on Learning Representations (2019)

11. Ma, X., Huang, H., Wang, Y., Romano, S., Erfani, S., Bailey, J.: Normalized loss functions for deep learning with noisy labels. In: International Conference on Machine Learning, pp. 6543–6553. PMLR (2020)

12. Mahajan, D., et al.: Exploring the limits of weakly supervised pretraining. In: Ferrari, V., Hebert, M., Sminchisescu, C., Weiss, Y. (eds.) ECCV 2018. LNCS, vol. 11206, pp. 185–201. Springer, Cham (2018). https://doi.org/10.1007/978-3-030-01216-8_12

13. Malach, E., Shalev-Shwartz, S.: Decoupling "when to update" from "how to update". In: Proceedings of the 31st International Conference on Neural Information Processing Systems, pp. 961–971 (2017)

14. Paszke, A., et al.: PyTorch: an imperative style, high-performance deep learning library. Adv. Neural. Inf. Process. Syst. **32**, 8026–8037 (2019)

15. Patrini, G., Rozza, A., Krishna Menon, A., Nock, R., Qu, L.: Making deep neural networks robust to label noise: a loss correction approach. In: Proceedings of the IEEE Conference on Computer Vision and Pattern Recognition, pp. 1944–1952 (2017)

16. Sukhbaatar, S., Bruna, J., Paluri, M., Bourdev, L., Fergus, R.: Training convolutional networks with noisy labels. In: 3rd International Conference on Learning Representations, ICLR 2015 (2015)

17. Sun, Z., et al.: PNP: robust learning from noisy labels by probabilistic noise prediction. In: Proceedings of the IEEE/CVF Conference on Computer Vision and Pattern Recognition, pp. 5311–5320 (2022)

18. Thomee, B., et al.: YFCC100M: the new data in multimedia research. Commun. ACM **59**(2), 64–73 (2016)

19. Wang, Y., Ma, X., Chen, Z., Luo, Y., Yi, J., Bailey, J.: Symmetric cross entropy for robust learning with noisy labels. In: Proceedings of the IEEE/CVF International Conference on Computer Vision, pp. 322–330 (2019)

20. Wang, Z., Hu, G., Hu, Q.: Training noise-robust deep neural networks via meta-learning. In: Proceedings of the IEEE/CVF Conference on Computer Vision and Pattern Recognition, pp. 4524–4533 (2020)

21. Wei, H., Feng, L., Chen, X., An, B.: Combating noisy labels by agreement: a joint training method with co-regularization. In: Proceedings of the IEEE/CVF Conference on Computer Vision and Pattern Recognition, pp. 13726–13735 (2020)

22. Xiao, T., Xia, T., Yang, Y., Huang, C., Wang, X.: Learning from massive noisy labeled data for image classification. In: Proceedings of the IEEE Conference on Computer Vision and Pattern Recognition, pp. 2691–2699 (2015)
23. Yan, Y., Rosales, R., Fung, G., Subramanian, R., Dy, J.: Learning from multiple annotators with varying expertise. Mach. Learn. **95**(3), 291–327 (2014)
24. Yao, Y., ET AL.: JO-SRC: a contrastive approach for combating noisy labels. In: Proceedings of the IEEE/CVF Conference on Computer Vision and Pattern Recognition, pp. 5192–5201 (2021)
25. Yu, X., Han, B., Yao, J., Niu, G., Tsang, I., Sugiyama, M.: How does disagreement help generalization against label corruption? In: International Conference on Machine Learning, pp. 7164–7173 (2019)
26. Zhang, C., Bengio, S., Hardt, M., Recht, B., Vinyals, O.: Understanding deep learning (still) requires rethinking generalization. Commun. ACM **64**(3), 107–115 (2021)
27. Zhang, Z., Sabuncu, M.: Generalized cross entropy loss for training deep neural networks with noisy labels. Adv. Neural Inf. Process. Syst. **31** (2018)

Path Integration Enhanced Graph Attention Network

Hui Wang[1,2], Peng Zhou[1,2], and Junbo Ma[1,2(✉)]

[1] Key Lab of Education Blockchain and Intelligent Technology, Ministry of Education, Guangxi Normal University, Guilin 541004, China
nudt_mjb@outlook.com
[2] Guangxi Key Lab of Multi-Source Information Mining and Security, Guangxi Normal University, Guilin 541004, China

Abstract. Graph attention networks are a deep learning method for processing graph data. By learning the relationships between neighbouring nodes in the graph, GATs have been widely used in many fields. However, the graph attention network has the problem of information lag in the process of information aggregation, which degrades the performance of the graph attention network. Referring to the ideas of Feynman path integral theory in physics, we proposed a new graph attention method called PaInGAT to solve the above issue by introducing a new neighbor information aggregation mechanism. Specifically, we improve the neighbour node aggregation mechanism of traditional graph attention networks by calculating the path integral from the source node to the target node to obtain the attention factor, and update the information of multi-order neighbours to the central node directly by the attention factor of the current state at each layer. Through experimental demonstration combining different downstream tasks, our method achieves excellent results on several datasets, demonstrating its effectiveness and advancement.

Keywords: Graph Attention Networks · Graph Data · Information Lag · Aggregation Mechanism

1 Introduction

In real word, data often has quite complex relationships and irregular structures, typically represented as graph structures on non-Euclidean space. The graph data structure can represent the characteristics of nodes and the relationships between nodes. It is often used in a wide range of data representations in various fields, such as index graphs between papers, wiring graphs of the circuit, membership graphs of social networks, molecular graphs of chemical substances, etc. Learning graph data representation has therefore become a popular topic of interest to researchers in recent years [23,45].

Graph neural networks (GNNs) [33] constitute an effective framework for learning graph representations and have been successfully applied to a variety

X. Yang et al. (Eds.): ADMA 2023, LNAI 14179, pp. 312–324, 2023.
https://doi.org/10.1007/978-3-031-46674-8_22

of graph-based tasks. GNNs work by iteratively updating node or subgraph representations through message passing between neighboring nodes. Each node aggregates information from its neighbors and updates its representation based on the received messages. This process is repeated multiple times, allowing nodes to propagate information throughout the graph and refine their representations. Graph Convolutional Networks (GCNs) [28] are the basis for many complex graph neural network models, including autoencoder-based models [29,42], generative models [24] and spatio-temporal networks [43]. It can be divided into two main categories [38], spectral-based and spatial-based. Spectral-based approaches [28] define graph convolution by introducing filters from the perspective of graph signal processing, where the graph convolution operation is interpreted as removing noise from the graph signal. Spatial-based approaches [3,17] represent graph convolution as the aggregation of feature information from neighbourhoods. When algorithms for graph convolution networks run at the node level, graph pooling modules can be interleaved with the graph convolution layer to coarsen the graph into high-level substructures.

One of the most popular variants of GNNs is Graph Attention Networks (GATs) [37], which addresses the shortcomings of GCNs that treat all neighbours equally. Essentially, both GCNs and GATs aggregate features from neighbouring vertices onto the central vertex, but the difference is that GCNs uses a Laplace matrix and GATs uses attention coefficients. Specifically, GATs uses a self-attention mechanism, which means that each node in the graph computes its own attention coefficients based on the similarity between its own features and the features of its neighboring nodes. These attention coefficients are then used to compute a weighted sum of the neighboring node features, which is combined with the original node features to obtain a new representation of the node [37]. Because in real-world scenarios, each of the neighbouring nodes may play a different roles in the influence on the core node, while GCNs simply ignore the correlation of spatial information between nodes and focus only on the topology of the graph when combining the features of neighbouring nodes making the model less generalisable and performance. Overall, GAT has been shown to be effective in a wide range of graph-based learning tasks, such as node classification, graph classification, and link prediction. It has also been extended to handle more complex graph structures, such as heterogeneous graphs and dynamic graphs, and has been combined with other deep learning techniques, such as convolutional neural networks, to achieve state-of-the-art performance on a variety of benchmarks.

However, GATv2 [4] argue that the attention score computed by GAT [37] is only a restricted static attention and does not compute a dynamic attention that can truly express the relevance of nodes, because the attention function computed by GAT is monotonic for any query node with respect to the key, i.e., for different query nodes, the attention score ranking of their neighbouring nodes is fixed [4]. The GATv2 method performance is improved by modifying the internal order of operations to obtain a more expressive approximation of the attention function. Although it computes more expressive attention scores, we found that GAT and GATv2 still suffer from information lag in the computation of attention scores, and that the attention scores used in the computation lagged by $K-1$ layers (K is the path length between two nodes). We will elaborate on this in Sect. 3.

To overcome these drawbacks, we propose a new path integral based graph attention network(PaInGAT). Inspired by ideas from Feynman's path integral theory in physics, we calculate a more effective attention factor by considering the influence of the path length between nodes on the weights when weighting sums in a graph attention networks aggregating information. In continuous space we calculate the transformation of the energy of the path between two points by integration, extending to the discrete space of the graph structure we use the summation operation instead of the integration operation. In addition, by increasing the nonlinear transformation of neighbourhood information, the model can aggregate the neighbourhood information more effectively and improves the expressiveness of the model. At the same time, in traditional GNNs, after propagating multiple base layers, the node information is globally over-smoothed to white noise, resulting in a severe performance degradation. This is because the message passing mechanism of GNNs is based on a plain assumption that neighbouring nodes usually have the same category information. A shallow GNN therefore allows for more cohesive information within categories. Deepening GNNs, on the other hand, means expanding the receptive field of information and inevitably absorbing much inter-category information, which leads to each node tending to be similar. Our proposed method can naturally alleviate this problem, as aggregated higher-order neighbourhood information decays with the edge length of the path, especially when the attention value between some two intermediate nodes on the path drops sharply.

In summary, our main contribution is to propose a new graph attention framework called PaInGAT. Unlike previous graph attention networks, we use node features to compute the path integral between nodes as the attention score to update node representations to obtain new graph representations. Essentially, instead of training a separate feature vector for each node, we train a new set of aggregation functions that aggregate feature information from the nodes' local neighbours of different hop numbers in a base layer. We evaluated our algorithm on four node classification benchmarks and three graph classification benchmarks that tested PaInGAT's ability to generate useful embeddings. Experimental results show that our approach is more effective than previous graph attention models, achieving more expressive graph embeddings.

2   Related Work

Graph neural networks(GNNs) [33] play a very important role for the application of non-Euclidean data in deep learning. Generating node representations that actually rely on graph structure and feature information through graph neural networks is a hot topic of interest for researchers. Various graph neural networks have been proposed in recent years, with both spectral-based approaches [5, 11,

21,31,42] represented by GCN and spatial-based approaches [1,3,8,10,18,19,26, 30,46] represented by GAT achieving outstanding graph embedding results in the field of graph representation learning. Among them, GAT uses self-attentive mechanism to achieve modelling of relations for graph data.

Attention mechanisms have been used extensively in the field of deep learning in recent years, whether for computer vision [7,36], speech processing [9,44], natural language processing [12,22] or a variety of other tasks. The operation of introducing attention mechanism in GNNs can be traced back to GAT, which introduced the self-attention mechanism into GNNs replacing the convolution operation in GCNs. Subsequently various graph attention methods have been proposed by researchers. AGNN [35] removes all intermediate fully connected layers of the GCN and calculates the attention factor by cosine similarity. The work on SuperGAT [27] summarises four attention scoring functions, respectively the original GAT function (GO), the node vector dot product function (DP), the scaled dot product function (SD) and the mixed function (MX), and adds a self-supervised task of link prediction to the model to better learn graph embeddings. There are also researchers who have introduced graph attention mechanisms into the field of heterogeneous graphs, such as RGAT [6]. GATv2 [4] computes improved dynamics of attention by modifying the order of operations of the linear mapping and nonlinear transformations of GAT.

3 Preliminaries

In this setion, we first introduce previous work on GAT and GAT2, then explain the existence of information lag in the attentional scores in GATs by analysing the process of aggregation of neighbourhood information.

In the following, we define the problem of interest and the corresponding notations that will be used in this paper. For convenience, we introduce the model on an undirected graph. Like GAT, our method can also be used for digraphs.

$\mathcal{G} = (\mathcal{V}, \mathcal{E})$ is a undirected graph, where node $\mathcal{V} = \{1, 2, ..., n\}$, edge $\mathcal{E} \subseteq \mathcal{V} \times \mathcal{V}$. In special, for undirected graphs, we consider each edge as two directed edges with opposite directions. Define the central node i and its neighbour node j, and their corresponding feature vectors are denoted by $\vec{h_i}$ and $\vec{h_j}$ ($\vec{h_i}, \vec{h_j} \subseteq R^F$) respectively, and (j, i) denotes the edge from node j to node i.

GAT. It adopts a graph attention layer to update the node representation of a graph G by successive applications of the layer. A set $h_i \in R^F (i \in V)$ and a corresponding set of edges $\varepsilon(\varepsilon \subseteq \mathcal{E})$ are taken as input to the layer, and the updated node embedding representation $h' \in RF$ is output after one or more layers of superimposed base layers.

In the graph attention layer, each node gives its own query representation for its neighbours, (its own node intermediate representation as query, and its neighbour node intermediate representation as key), i.e. for each vertex i, the similarity coefficients between his neighbours $(j \in N_j \cup i)$ and itself is computed by learning the parameters W and the mapping $a(\cdot)$ as follows:

$$e_{ij} = \text{LeakyReLU} \left(\vec{a}^T \cdot [Wh_i \| Wh_j] \right), j \in Ni_i \quad (1)$$

where $\cdot^T$ represents matrix transpose and $[\cdot \text{——} \cdot]$ denotes the concatenation operation. Then normalize them across all choices of j by using the softmax function:

$$\alpha_{ij} = softmax_j(e_{ij}) = \frac{\exp\left(\text{LeakyReLU}\left(\vec{a}^T \Delta \left[W\vec{h}_i \| W\vec{h}_j\right]\right)\right)}{\sum_{k \in N_i} \exp\left(\text{LeakyReLU}\left(\vec{a}^T \Delta \left[W\vec{h}_i \| W\vec{h}_k\right]\right)\right)} \quad (2)$$

Finally, the standardized attention coefficient is combined with the input features of the linear mapping, and then the nonlinear mapping is used to obtain the final graph embedding representation.

$$\vec{h}'_i = \sigma \left(\sum_{j \in \mathcal{N}_j} \alpha_{ij} \Delta W \vec{h}_j \right) \quad (3)$$

GATv2. The type of attention computed by GAT is restricted because the attentional scores is unconditional on the query node and the attentional function is monotonic with respect to the neighbourhood (key), thus limiting the expressiveness of the model. To address this problem, Shaked Brody et al. propose an improvement to the calculation of attention scores: changing the order of operation of the non-linear transformation and mapping $\vec{a}$ in Eq. 1.

$$e_{ij} = \vec{a}^T LeakyReLU \left(W \Delta [h_i \| h_j] \right), j \in \mathcal{N}_i \quad (4)$$

They believed that using the learning layers W and $\vec{a}$ consecutively could collapse into a single linear layer, affecting the calculation of attention scores. Like the attention mechanism in Transformer, both GAT and GATv2 also consider the multi-heads attention. The output results of multiple attention heads are obtained through the concate operation or their average value is taken to obtain the output of the GAT layer.

$$\vec{h}'_i = \mathop{\|}_{k=1}^{K} \sigma \left(\sum_{j \in N_j} \alpha_{ij} W^k \vec{h}_j \right) \quad (5)$$

Information Lag in the Aggregation Process. Taking the two-order neighbourhood information aggregation of node i in Fig. 1b as an example, according to the information aggregation formula of GAT and GATv2 we can obtain the representation of nodes i, j after the first base layer aggregation of its own and neighbourhood information as:

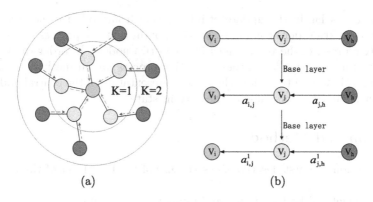

(a)   (b)

Fig. 1. Aggregation process of GATs

$$x_i^1 = \sigma(a_{i,i}Wx_i + \sum_{j\in N_i} a_{i,j}Wx_j) \tag{6}$$

$$x_j^1 = \sigma(a_{j,j}Wx_j + \sum_{h\in N_j} a_{j,h}Wx_h) \tag{7}$$

where $\sigma(\cdot)$ indicates a layer of nonlinear transformation. After the second base layer, the representation of node i is:

$$
\begin{aligned}
x_i^2 =& a_{i,i}^1 W^1 x_i^1 + \sum_{j\in N_i} a_{i,j}^1 W^1 x_j^1 \\
=& a_{i,i}^1 W^1 (a_{i,i}Wx_i + \sum_{j\in N_i} a_{i,j}Wx_j) + \sum_{j\in N_i} a_{i,j}^1 Wa^1(a_{j,j}Wx_j + \sum_{h\in N_j} a_{j,h}Wx_h) \\
=& a_{i,i}^1 W^1 a_{i,i}Wx_i + a_{i,i}^1 W^1 \sum_{j\in N_i} a_{i,j}Wx_j + \sum_{j\in N_i} a_{i,j}^1 W^1 a_{j,j}Wx_j \\
&+ \sum_{j\in N_i} a_{i,j}^1 W^1 \sum_{h\in N_j} a_{j,h}Wx_h \\
=& a_{i,i}^1 W^1 a_{i,i}Wx_i + (a_{i,i}^1 a_{i,j} + a_{i,j}^1 a_{j,j}) \sum_{j\in N_i} W^1 Wx_j \\
&+ \sum_{j\in N_i} \sum_{h\in N_j} a_{i,j}^1 a_{j,h} W^1 Wx_h
\end{aligned}
\tag{8}
$$

For ease of representation, we omit the operation of the non-linear transformation in Equation(8). We can intuitively know that our conclusions are unaffected due to the monotonically increasing nature of LeakyReLU.

It can be seen from the above calculation process that: The graph attention network uses the attention factors $a_{i,j}^1$ of the current state and $a_{j,h}$ of the previous state to calculate the representation of the node when aggregating information from the second-order neighbour node h to the central node i. This implies that there is a lag in the update of information from node i to node h. When expanding to kth order neighbours, we can obtain that the process lags by k-1st order.(K represents the path length between two nodes). In order to avoid the problem of information lag in the calculation process, in our method, we use the path integral method to update the node representation of the central node with the inter-node attention value of the current state.

4   Proposed Method

In this section, we first describe the structure of the base layer of the PaInGAT network.

Our model will be based on two illuminating assumptions:

1. Information exchange will exist between the nodes on each path in the graph.
2. Information transfer is not only a function of paths, but can be further understood as a function of path length.

We compose paths between nodes by edges on the graph. Generally, for two nodes in a graph, there are multiple paths between them. By the different paths they can be classified as neighbours of different orders. It is even possible to change into more jumping neighbours by using oneself as a folding point. Then we define the length of the path between central node i and neighbouring node j as K, i.e. set the perceptual field size of the model as K, and aggregate the 1st to Kth order neighbourhood information of the central node i through the attention mechanism.

4.1   Attention Module

In the previous graph attention methods, in order to obtain sufficient expressiveness, the input raw features need to be transformed in a learnable way to get intermediate representations. Finally the intermediate representation will be aggregated into the representation of the central node by using the attention factor as a coefficient. However, this transformation uses only a linear transformation of a fully connected layer and not a non-linear transformation, so it yields a limited expressive power. For this reason, we set up a separate feature transformation module $W(\cdot)$, which further improves the expressiveness of the features by boosting the linear transformation of the input features through two fully connected layers and adding an activation function between the two fully connected layers to do non-linear transformation (we use the LeakyReLU function with negative input slope $\alpha = 0.2$ in our experiments).

$$g_i = W(h_i) \in \mathbb{R}^{N \times F'} \tag{9}$$

We then use the transformed input features to calculate the attentional fraction between adjacent nodes. The attention score is:

$$e_{ij} = \vec{a}^T LeakyReLU \left(\left[g_i' \parallel g_j' \right] \right), \quad j \in N_i \tag{10}$$

It represents the importance, i.e. the degree of influence, of node i on adjacent node j. Where $\vec{a} \in \mathbb{R}^{2F'}$ denotes a fully connected layer. Then the attention score is normalised by the softmax function as follows.

$$a_{ij} = \frac{exp\,(e_{ij})}{\sum_{k \in N_k \cup i} e_{ik}} \tag{11}$$

4.2 Information Aggregation Module

Finally, we weight the aggregated neighbourhood information by the computed attention score. Feynman path integral theory tells us that the probability amplitude of a particle moving from A to B in continuous space is the integral of all possible paths, which is defined as summation in discrete space. Inspired by Feynman's path integral theory and PanConv [32], we extend the rule of motion of particles in space to the propagation of information in graph networks.

Similarly, then the propagation coefficient of the message from node i to its kth order neighbour node j is the sum of the coefficients on each path between them. Intuitively, we represent the attentional score $p(i, j; k)$ of a single path passing through multiple nodes as the product of the attentional scores of each edge on this path. For example, assuming that a particular path of length 2 from i to j goes through node h, then $p\,(i, j; 2) = a_{ih}a_{hj}$ for that path. By this method we can work out the attention fraction of any path between two nodes. Finally the information passed on all paths (including the node's own information) is summed to obtain the final node representation. By correcting the path length K, we can control the perceptual field of the model.

$$\vec{h}'_i = \sum_{k=1}^{K} \sum_{j \in N_{k(i)} \cup i} \sum p(i, j; k) g'_i \tag{12}$$

where $\sum p(i, j; k)$ denotes the sum of the attention scores of all paths of length k between nodes i, j. In graph data, a node can often be considered as a neighbour of a different order to the central node by following different paths, or even as a neighbour of a higher order by repeatedly passing an intermediate node in the path. Also inspired by the theoretical minimum action principle in physics, we determine the order between nodes in terms of their shortest path lengths in the real calculation process.

5 Experiments

In this section, we show test results of PaInGAT on graph datasets for node classification and graph classification tasks, and demonstrate the advanced performance of our approach by rigorously comparing it with previous graph attentive methods and other no-attention methods. All experiments were done with code written in pyTorch Geometric. Table 1 shows our dataset specifications for the node classification task, and Table 2 shows our dataset specifications for the graph classification task.

Table 1. Specification of the dataset used in the node classification task

	Cora	Citeseer	Pubmed	ogbn-arxiv
# Nodes	2708	3327	19717	169343
# Edges	5429	4732	44338	2315598
# Features	1433	3703	500	128
# Classes	7	6	3	40

Table 2. Specification of the dataset used in the grpah classification task

	PROTEINS	MUTAG	PTC
#Graphs	1113	188	344
#Nodes	∼39.1	∼17.9	∼25.6
#Edges	∼39.1	∼39.6	∼2.0
#Features	3	7	19
#Classes	2	2	2

5.1 Node Classification

We chose four commonly used graph datasets, Cora, Citeseer, Pumbed [34], and ogbn-arxiv. We tested the node classification task for graph data on the these datasets, comparing PaInGAT with GCN [28], GraphSAGE [19], Diff-Net [39], TAGCN [14], Graph-Bert [40], AGNN [30], SuperGAT [27], GAT [37], and GATv2 [4], which were previously graph neural network models. To be fair, we used the same datasets divisions as in the GAT experiments. Also, to prevent interference from other conditions, the same content was used for all parts of our code except for the model.

Table 3 shows our final experimental results. It can be seen that our model achieves the best experimental results compared with other models. In addition, we tested the effect of different attention heads for model performance on the ogbn arxiv dataset. The experimental results are shown in Table 4. It can be seen that the number of attention heads has an effect on the performance of the model.

5.2 Graph Classification

To further evaluate the effectiveness of our model, we implemented experiments on several real-world graph classification problems. PROTEINS [13] dataset is a collection of protein molecules that are classified as enzymes or non-enzymes. MUTAG [25] and PTC [20] datasets are composed by small molecule compounds. In the former dataset, the task is to identify mutagenic molecular compounds for potentially commercial drugs, while in the latter the goal is to identify chemical compounds based on their carcinogenicity in rodents. Three different sizes of graph classification datasets were chosen to validate the performance of

Table 3. Results of node classification tests on the Cora, Citeseer, Pumbed

Method	Cora	Citeseer	Pubmed
GCN	81.5% ± 0.7%	70.3% ± 1.0%	79.0% ± 0.7%
GraphSAGE	82.1% ± 0.6%	71.9% ± 0.9%	78.0% ± 0.7%
Diff-Net	85.1% ± 0.4%	72.7% ± 0.6%	78.3% ± 0.6%
TAGCN	83.3% ± 0.7%	71.4% ± 0.5%	81.1% ± 0.5%
GRAPH-BERT	84.1% ± 0.9%	71.0% ± 0.5%	79.5% ± 0.4%
AGNN	81.0% ± 0.3%	69.8% ± 0.4%	78.0% ± 0.5%
SuperGAT	84.3% ± 0.6%	75.6% ± 0.8%	81.7% ± 0.5%
GAT	83.0% ± 0.4%	72.5% ± 0.5%	79.0% ± 0.4%
GATv2	83.5% ± 0.3%	72.6% ± 0.3%	79.3% ± 0.6%
PaInGAT(ours)	**85.4% ± 0.7%**	**74.1% ± 0.3%**	**81.8% ± 0.6%**

Table 4. The effect of different heads on classification accuracy

Method	head number	ogbn-arxiv
GAT	1	70.71 ± 0.19
	8	70.73 ± 0.18
GATv2	1	71.01 ± 0.14
	8	70.91 ± 0.20
PaInGAT(ours)	1	72.13 ± 0.41
	8	**72.33 ± 0.15**

our model. We chosed no-attention method SPI-GCN [2], GCN [28], DGCNN [41], GIN [15], PANConv [32] and attentional methods hGANet [16], GAT [37], GATv2 [4] to compare with our model, the experiments results show that our method outperforms other graph attention methods under the same experimental conditions (Table 5).

Table 5. The Result of graph classification on Dataset

Method	PROTEINS	MUTAG	PTC
SPI-GCN	74.05%	85.30%	57.72%
GCN	76.23%	84.33%	63.45%
DGCNN	76.26%	85.83%	58.59%
GIN	72.32%	89.47%	64.86%
PANConv	74.60%	88.54%	65.92%
hGANet	77.64%	88.96%	64.76%
GAT	74.54%	89.63%	66.40%
GATv2	76.83%	89.94%	67.22%
PaInGAT(ours)	**78.76%**	**91.26%**	**68.58%**

6   Conclusion

In this paper, we analyse previous work on GATs and find that they fail to compute better attention scores and do not efficiently aggregate information about the higher-order neighbours of nodes. In order to solve these problems, we add a non-linearly transformed node embedding module to the process of computing attention scores, and use the attention product on the path to compute attention scores among higher-order neighbours, allowing PaInGAT to implement the approximator attention function.

It has been demonstrated experimentally that our model achieves good performance on various datasets. However, we have to admit that PaInGAT has a higher computational complexity compared to other models such as GAT, which will be the next direction of our research.

Acknowledgment. This work is supported in part by the Project of Guangxi Science and Technology with Grant Number GuiKeAB23026040, the Research Fund of Guangxi Key Lab of Multi-source Information Mining & Security with Grant Number MIMS20-04 and the Research Fund of Guangxi Key Lab of Multi-source Information Mining & Security with Grant Number 20-A-01-02.

References

1. Abu-El-Haija, S., et al.: Mixhop: higher-order graph convolutional architectures via sparsified neighborhood mixing. In: International Conference on Machine Learning, pp. 21–29. PMLR (2019)
2. Atamna, A., Sokolovska, N., Crivello, J.C.: SPI-GCN: a simple permutation-invariant graph convolutional network (2019)
3. Atwood, J., Towsley, D.: Diffusion-convolutional neural networks. In: Advances in Neural Information Processing Systems, vol. 29 (2016)
4. Brody, S., Alon, U., Yahav, E.: How attentive are graph attention networks? arXiv preprint arXiv:2105.14491 (2021)
5. Bruna, J., Zaremba, W., Szlam, A., LeCun, Y.: Spectral networks and locally connected networks on graphs. arXiv preprint arXiv:1312.6203 (2013)
6. Busbridge, D., Sherburn, D., Cavallo, P., Hammerla, N.Y.: Relational graph attention networks. arXiv preprint arXiv:1904.05811 (2019)
7. Carion, N., Massa, F., Synnaeve, G., Usunier, N., Kirillov, A., Zagoruyko, S.: End-to-end object detection with transformers. In: Vedaldi, A., Bischof, H., Brox, T., Frahm, J.-M. (eds.) ECCV 2020. LNCS, vol. 12346, pp. 213–229. Springer, Cham (2020). https://doi.org/10.1007/978-3-030-58452-8_13
8. Chen, J., Ma, T., Xiao, C.: FastGCN: fast learning with graph convolutional networks via importance sampling. arXiv preprint arXiv:1801.10247 (2018)
9. Chorowski, J., Bahdanau, D., Cho, K., Bengio, Y.: End-to-end continuous speech recognition using attention-based recurrent NN: First results. arXiv preprint arXiv:1412.1602 (2014)
10. Dai, H., Kozareva, Z., Dai, B., Smola, A., Song, L.: Learning steady-states of iterative algorithms over graphs. In: International Conference on Machine Learning, pp. 1106–1114. PMLR (2018)

11. Defferrard, M., Bresson, X., Vandergheynst, P.: Convolutional neural networks on graphs with fast localized spectral filtering. In: Advances in Neural Information Processing Systems, vol. 29 (2016)
12. Devlin, J., Chang, M.W., Lee, K., Toutanova, K.: Bert: pre-training of deep bidirectional transformers for language understanding. arXiv preprint arXiv:1810.04805 (2018)
13. Dobson, P.D., Doig, A.J.: Distinguishing enzyme structures from non-enzymes without alignments. J. Mol. Biol. **330**(4), 771–783 (2003)
14. Du, J., Zhang, S., Wu, G., Moura, J.M., Kar, S.: Topology adaptive graph convolutional networks. arXiv preprint arXiv:1710.10370 (2017)
15. Fey, M., Lenssen, J.E.: Fast graph representation learning with pytorch geometric. arXiv preprint arXiv:1903.02428 (2019)
16. Gao, H., Ji, S.: Graph representation learning via hard and channel-wise attention networks. In: Proceedings of the 25th ACM SIGKDD International Conference on Knowledge Discovery & Data Mining, pp. 741–749 (2019)
17. Gasteiger, J., Weißenberger, S., Günnemann, S.: Diffusion improves graph learning. In: Advances in Neural Information Processing Systems, vol. 32 (2019)
18. Gilmer, J., Schoenholz, S.S., Riley, P.F., Vinyals, O., Dahl, G.E.: Neural message passing for quantum chemistry. In: International Conference on Machine Learning, pp. 1263–1272. PMLR (2017)
19. Hamilton, W., Ying, Z., Leskovec, J.: Inductive representation learning on large graphs. In: Advances in Neural Information Processing Systems, vol. 30 (2017)
20. Helma, C., King, R.D., Kramer, S., Srinivasan, A.: The predictive toxicology challenge 2000–2001. Bioinformatics **17**(1), 107–108 (2001)
21. Henaff, M., Bruna, J., LeCun, Y.: Deep convolutional networks on graph-structured data. arXiv preprint arXiv:1506.05163 (2015)
22. Hu, D.: An introductory survey on attention mechanisms in NLP problems. In: Bi, Y., Bhatia, R., Kapoor, S. (eds.) IntelliSys 2019. AISC, vol. 1038, pp. 432–448. Springer, Cham (2020). https://doi.org/10.1007/978-3-030-29513-4_31
23. Hu, R., et al.: Graph self-representation method for unsupervised feature selection. Neurocomputing **220**, 130–137 (2017)
24. Hu, Z., Dong, Y., Wang, K., Chang, K.W., Sun, Y.: GPT-GNN: generative pre-training of graph neural networks. In: Proceedings of the 26th ACM SIGKDD International Conference on Knowledge Discovery & Data Mining, pp. 1857–1867 (2020)
25. Kazius, J., McGuire, R., Bursi, R.: Derivation and validation of toxicophores for mutagenicity prediction. J. Med. Chem. **48**(1), 312–320 (2005)
26. Kearnes, S., McCloskey, K., Berndl, M., Pande, V., Riley, P.: Molecular graph convolutions: moving beyond fingerprints. J. Comput. Aided Mol. Des. **30**(8), 595–608 (2016)
27. Kim, D., Oh, A.: How to find your friendly neighborhood: graph attention design with self-supervision. arXiv preprint arXiv:2204.04879 (2022)
28. Kipf, T.N., Welling, M.: Semi-supervised classification with graph convolutional networks. arXiv preprint arXiv:1609.02907 (2016)
29. Kipf, T.N., Welling, M.: Variational graph auto-encoders. arXiv preprint arXiv:1611.07308 (2016)
30. Li, R., Wang, S., Zhu, F., Huang, J.: Adaptive graph convolutional neural networks. In: Proceedings of the AAAI Conference on Artificial Intelligence, vol. 32 (2018)
31. Li, Y., Yu, R., Shahabi, C., Liu, Y.: Diffusion convolutional recurrent neural network: data-driven traffic forecasting. arXiv preprint arXiv:1707.01926 (2017)

32. Ma, Z., Xuan, J., Wang, Y.G., Li, M., Liò, P.: Path integral based convolution and pooling for graph neural networks. Adv. Neural. Inf. Process. Syst. **33**, 16421–16433 (2020)
33. Scarselli, F., Gori, M., Tsoi, A.C., Hagenbuchner, M., Monfardini, G.: The graph neural network model. IEEE Trans. Neural Networks **20**(1), 61–80 (2008)
34. Sen, P., Namata, G., Bilgic, M., Getoor, L., Galligher, B., Eliassi-Rad, T.: Collective classification in network data. AI Mag. **29**(3), 93–93 (2008)
35. Thekumparampil, K.K., Wang, C., Oh, S., Li, L.J.: Attention-based graph neural network for semi-supervised learning. arXiv preprint arXiv:1803.03735 (2018)
36. Vaswani, A., et al.: Attention is all you need. In: Advances in Neural Information Processing Systems, vol. 30 (2017)
37. Veličković, P., Cucurull, G., Casanova, A., Romero, A., Lio, P., Bengio, Y.: Graph attention networks. arXiv preprint arXiv:1710.10903 (2017)
38. Wu, Z., Pan, S., Chen, F., Long, G., Zhang, C., Philip, S.Y.: A comprehensive survey on graph neural networks. IEEE Trans. Neural Networks Learn. Syst. **32**(1), 4–24 (2020)
39. Zhang, J.: Get rid of suspended animation problem: deep diffusive neural network on graph semi-supervised classification. arXiv preprint arXiv:2001.07922 (2020)
40. Zhang, J., Zhang, H., Xia, C., Sun, L.: Graph-Bert: only attention is needed for learning graph representations. arXiv preprint arXiv:2001.05140 (2020)
41. Zhang, M., Cui, Z., Neumann, M., Chen, Y.: An end-to-end deep learning architecture for graph classification. In: Proceedings of the AAAI Conference on Artificial Intelligence, vol. vol. 32 (2018)
42. Zhang, X., Liu, H., Li, Q., Wu, X.M.: Attributed graph clustering via adaptive graph convolution. arXiv preprint arXiv:1906.01210 (2019)
43. Zheng, W., Zhu, X., Zhu, Y., Hu, R., Lei, C.: Dynamic graph learning for spectral feature selection. Multimedia Tools Appl. **77**(22), 29739–29755 (2018)
44. Zhou, P., Yang, W., Chen, W., Wang, Y., Jia, J.: Modality attention for end-to-end audio-visual speech recognition. In: ICASSP 2019–2019 IEEE International Conference on Acoustics, Speech and Signal Processing (ICASSP), pp. 6565–6569. IEEE (2019)
45. Zhu, X., Zhu, Y., Zhang, S., Hu, R., He, W.: Adaptive hypergraph learning for unsupervised feature selection. In: IJCAI, pp. 3581–3587 (2017)
46. Zhuang, C., Ma, Q.: Dual graph convolutional networks for graph-based semi-supervised classification. In: Proceedings of the 2018 World Wide Web Conference, pp. 499–508 (2018)

Graph Contrastive Learning with Hybrid Noise Augmentation for Recommendation

Kuiyu Zhu[1], Tao Qin[1(✉)], Xin Wang[1], Zhouguo Chen[2], and Jianwei Ding[2]

[1] School of Computer Science and Technology, Xi'an Jiaotong University, Xi'an, China
{kyzhu,wx508810851}@stu.xjtu.edu.cn, qin.tao@mail.xjtu.edu.cn
[2] The 30th Research Institute of China Electronics Technology Group Corporation, Cheng Du, China
czgexcel@163.com, mathe007@163.com

Abstract. Recommendation System is one of the effective tools to solve the problem of information overload in the era of big data, but the data sparsity has greatly affected its performance. Recently, contrastive learning, has attracted great attention and is expected to solve this problem. However, most of the existing graph-based contrastive learning methods perturb the original graph for data enhancement, which may affect the recommended performance. Meanwhile, studies have shown that improving the uniformity of data distribution is more important than data augmentation by graph perturbation. In this paper, in order to improve the uniformity of the data distribution, we propose a **G**raph **C**ontrastive **L**earning with **H**ybrid **N**oise **A**ugmentation for **Rec**ommendation, which is abbreviated as **GCLHANRec**. Specifically, we add uniform distribution random noise to users and normal distribution random noise to items, to improve the data uniformity while increasing the user's interest diversity for different items, thereby improving the accuracy and personalization degree of the recommendation system. Additionally, we propose Balanced Bayesian Personalized Ranking (BBPR) as the loss function for recommendation tasks, which is a modification of BPR to better make the model pay more attention to the difference between positive and negative samples, thus performing better in ranking tasks. We conducted extensive experiments using three datasets collected from actual environment, including Movielens, LastFM and Douban-book. The results show that our method outperforms several existing methods.

Keywords: Recommendation System · Graph Contrastive Learning · Data Distribution · Noise Augmentation

1 Introduction

In the network today, people can obtain anything they want from the internet, which has brought great convenience to our daily lives. However, their quality of experiences is inevitably affected by information overload [1,2]. Such as online shopping, people can not find what they want to purchase when face by thousands of similar available options. To solve this problem, recommendation systems have quickly developed and gradually become one of the most effective ways to prevent users from drowning in the amount of information [3–5]. Exploring recommendation systems for specific field and alleviating information overload has becoming hot research topic [6–8]. Many researchers employ collaborative filtering-based [9,10], content-based [11] and deep learning-based methods [12] to design different kinds of recommendation methods. But those methods usually have a basic assumption: they have enough labeled data, thus their performances are poor or can not be deployed without enough labeled data. However, on one hand, obtain enough labeled data is time-consuming and labeled data may be unavailable in many fields. On the other hand, data collected from actual environment is high sparsity, such as the long-tail distribution of items. Those characteristics have seriously affect the development of recommendation systems [13,14].

To utilize the unlabeled data, Self-Supervised Learning (SSL) has been widely used in computer vision, NLP and other fields in recently, contrastive Learning [15–17] is among the most popular ones. Its basic characteristic is the model does not require large amounts of labeled data, because it can define pseudo-labels from the data itself to supervise the model training [18]. For example, when applying contrastive learning to the image classification, the original image is usually regarded as the anchor, its augmented images are considered as positive samples, and the rest images are treated as negative samples. Then an encoder is trained to distinguish between positive and negative samples, so that the vector distance of positive samples and anchor should be as close as possible, and the vector distance of negative samples and anchor should be as far as possible [15,19]. Since contrastive learning does not rely on lager amount of labeled data, it has a natural advantage in solving the data sparsity problem.

Recently, some works on the recommendation system has been carried out around contrastive learning, which is called Self-Supervised Recommendation (SSR) [20]. For instance, DCL [21], SEPT [22], SGL [23], which was proposed to apply SSL to user-item bipartite graphs, they employ node dropout, edge dropout, and random walk to generate different representations of a node from three aspects. Then apply those data augmentation methods to contrastive learning to generate augmented versions (also known as positive samples) of the anchor, to better address the data sparsity problem.

NCL [24] is proposed to utilize neighborhood-enhanced contrastive learning process to improve the previous problem of not mining the nearest neighbor relationship between users (or items) due to random sampling. XSimGCL [25] is an extremely simple graph contrastive learning method that discards the graph augmentation strategies adopted by many state-of-the-art recommendation models. It uses a simple yet effective noise embedding augmentation as the contrastive view to learn more uniform data representations.

Despite researchers have achieved remarkable success of the above methods, SSR still suffer from two some problems: First, some works [22,23,26,27] use a strategy of randomly dropping information (e.g., node dropout, edge dropout, etc.), which may cause some important training samples to be lost while increasing the sparsity of the interactions, and in addition, this causes the integrity of the subgraph to be compromised [24]. Second, GNN-based contrastive learning method used in previous work take graph data augmentation as an indispensable option to construct different contrastive views, however, studies have shown that improving the uniformity of data distribution in the embedding space is more important than data augmentation by graph perturbation [25]. Therefore, to address the above issues and improve the uniformity of the distribution, we propose a graph contrastive learning with hybrid noise augmentation for recommendation in this paper.

Specifically, to increase the diversity of user interest in different items and improve the performance of the recommendation system, we add uniformly distributed random noise to the users. At the same time, to increase the user's preference to certain items and improve the personalization of the recommendation system, we add normally distributed random noise to the items. By adding different kinds of random noise to each user and item, we ultimately increase the uniformity of the data distribution. According to previous research [25], adding noise vectors with a very small constant can be numerically understood as points on a hypersphere with a radius of that constant, which usually does not cause deviation from the original embedding space. Therefore, we add very small constants that follow uniform and normal distributions respectively as hybrid random noise for users and items. In addition, to better make the model pay more attention to the difference between positive and negative samples, we propose a modified version of the BPR loss function called BBPR as the loss function for our recommendation task. Extensive experiments on multiple public real-world datasets show that our method improves the recommendation performance.

The remainder of this paper is organized as follows. Section 2 reviews the related work on graph-based recommendation and contrastive learning. Section 3 introduce the problem definition of recommendation system. In Sect. 4, we present our proposed model and its components. In Sect. 5, we report and analyze the experimental results. Finally, Sect. 6 provides a conclusion and future work.

2   Related Work

The related works related to our work are classified into two categories: graph-based recommendation and contrastive learning, the previous works related are reviewed as follows:

2.1   Graph-Based Recommendation

Graph-based recommendation use GCN to model higher-level collaboration signals by treating user project interaction behavior as a bipartite graph. For instance, PinSage [28] generates node representations by sampling the local neighbors of a node and aggregating node features. It is thus more suitable for large-scale graph representation learning and enhances the generalization ability. DeepICF [29] is a method capable of learning nonlinear and higher-order relationships between items from data to obtain more complex interactions to aid in modeling user decisions. NGCF [9] combines embedding learning and collaborative filtering. NGCF [9] introduces collaborative signals into the user item graph structure to exploit the higher order connections in the user item integration graph for recommendation accuracy. Further study on NGCF [9], LightGCN [30] is an optimization and improvement of the neural graph collaborative filtering algorithm. LightGCN [30] apply graph convolutional neural networks to recommendation system and the representation of the nodes is learned by smoothing the features on the graph. In addition, researchers can improve performance by removing nonlinear transformations from GCN based recommendations and replacing them with linear embedded propagation. Therefore, they designed LR-GCCF [31], which introduces residual preference learning at each layer to facilitate modeling of deeper layers, and alleviates the problem of excessive smoothing of sparse interactive data in the process of graph convolution. MixGCF [32] focuses on the synthesis of negative sampling, designing a general negative sampling method for graph neural network-based recommendation system to synthesize hard negative samples.

2.2   Contrastive Learning

Previous studies [20] shows that, according to the source of self supervision signals, the current contrastive learning work for recommendation systems mainly focuses on two aspects: structure-level contrast and model-level contrast. For instance, in order to solve the limitations of the traditional graph based recommendation model, SGL [23] introduces the idea of comparative learning based on LightGCN [30], and enhances the representation learning of nodes with the help of proxy tasks. Specifically, SGL [23] makes data enhancement on the user-items bipartite graph, and designs node dropout, edge dropout, and random walk to generate different views of the same node in three ways to improve the ability of automatically mining hard negatives. NCL [24], a contrastive learning method that adds neighborhood enhancement to graph collaborative filtering. NCL [24] obtains the structural information of high-order neighbors and the semantic information of similar neighbors on the graph by introducing structural neighbors and semantic neighbors. SimGCL [33], a model-level contrastive learning method. Their research shows that the comparison loss function plays a key role in the recommendation based on comparative learning, and the previous data enhancement is only a secondary role. Therefore, SimGCL [33] discards the enhancement of the graph, optimizes the contrastive loss function, reduces the deviation in the recommendation system scenario, and improves the uniformity of the representation distribution.

3   Problem Definition

For graph-based recommendation systems, most works are based on bipartite graphs. Generally, user-item interactions are represented in the form of a bipartite graph $\mathcal{G} = (\mathcal{V}, \mathcal{E})$. Among them, $\mathcal{V} = (\mathcal{U} \cup \mathcal{I})$ represents the collection of users and the collection of items, $\mathcal{E} = \{y_{ui} \in R^{\mathcal{U} \times \mathcal{I}} | (u, i) \in \mathcal{O}\}$ represents the collection of all observed user-item interactions, which can also be understood as the edge between the user and the item or the user may be interested in the item, where u represents the user, i represents the item, and $\mathcal{O}$ represents the observed interaction. The goal of a recommendation system is to select the Top-k items that are most likely to be interacted with by a given user u from a list of items that the user has never interacted with before, in order to improve user experience and increase user satisfaction. Graph neural networks have great advantages in mining user-item interaction relationships, and thus, GNNs have become the mainstream models in recommendation systems.

GNN-Based for Recommendation. Recently, the research community of recommendation systems has mostly adopted graph-based recommendation. For example, the classic LightGCN [30], which is an optimization based on the NGCF [9], is a lightweight graph neural network model for recommendation systems. Due to its excellent performance, it has become an indispensable baseline and a basic component of various methods (such as SGL [23], NCL [24], etc.). Our method is also built upon LightGCN [30], therefore, it is necessary to give a brief review of LightGCN [30].

$$e_u^{(l)} = f_{Agg}(e_u^{(l-1)}, \{e_i^{(l-1)} | i \in \mathcal{N}_u\}) \tag{1}$$

In Eq. 1, $e_u^{(l-1)}$ and $e_i^{(l-1)}$ represent the user u and item i embeddings of the l-1 layer respectively. $\mathcal{N}_u$ represents the set of items that user u has interacted with before. f_{Agg} is an aggregation function that aggregates the l-1 layer neighbors $\mathcal{N}_u$ of user u and the l-1 layer $e_u^{(l-1)}$.

$$e_u = f_{Com}(e_u^{(0)}, e_u^{(1)}, ..., e_u^{(l)}) \tag{2}$$

After obtaining the l-1 layer embeddings of user u through the f_{Agg}, the embeddings from 0-layer to l-layer are combined using a function f_{Com} to form the final embedding of user e_u.

$$\mathcal{L}_{BPR} = - \sum_{(u,i,j) \in \mathcal{O}} log\sigma(\hat{y}_{ui} - \hat{y}_{uj}) \tag{3}$$

During the training phase, to optimize the model, it applies Bayesian Personalised Ranking [34] $\mathcal{L}_{BPR}$ to maximise the score difference between positive and negative samples, so that the observed items score higher than the unobserved ones. $\hat{y}_{ui} = e_u^T e_i$ and $\hat{y}_{uj} = e_u^T e_j$ are predicted values for y_{ui} and y_{uj}, respectively.

4   Proposed Model

4.1   GCLHANRec Framework

In this section, we will briefly introduce the overall framework of our proposed GCLHANRec as shown in Fig. 1. In this framework, it mainly consists of two parts: recommendation task and contrastive task.

Recommendation Task. We first take the initial embedding represented by the user interaction bipartite graph as input, then apply the lightweight graph convolutional network for node representation learning. At the same time, in order to improve the uniformity of data distribution and improve the recommendation performance, we introduce different random noises for users and items respectively. Specifically, we introduce uniform distribution random noise for users and normal distribution random noise for items. In addition, we propose Balanced Bayesian Personalized Ranking (BBPR) as the loss function for recommendation tasks, which is a modification of BPR to better make the model focus more on the absolute difference between samples when predicting, improve the performance in ranking process, and thus enhance the recommendation performance.

Contrastive Task. Given a pair of samples, the augmented hybrid noise sample is regarded as the anchor sample $\tilde{e}$, while the other augmented sample can be either a positive sample or a negative sample $\tilde{e}'$ (Here, positive samples refer to augmented samples from the same node as the anchor samples, while negative samples refer to augmented samples from different nodes than the anchor samples, see Sect. 4.3). Then an encoder is trained based on LightGCN using InfoNCE loss function to distinguish between positive and negative samples.

4.2   Hybrid Noise-Based Augmentation

Inspired by the work of XSimGCL [25], improving the uniformity of data distribution in the embedding space is crucial for graph contrastive learning. Therefore, in our work, we propose a hybrid random noise augmentation method for users and items to further improve the uniformity of data distribution. Specifically, for users, we use uniform distribution random noise augmentation to increase the user's interest diversity for different items, improving coverage and quality; for items, we use normal distribution random noise augmentation to increase the user's preference and repulsion degree for different items.

After hybrid noise augmentation, we update the user embedding e_u to $\tilde{e}_u$, and similarly, we update the item embedding e_i to $\tilde{e}_i$, as shown in formula (4), where φ_1 is random noise generated using a uniform distribution and φ_2 is random noise generated using a normal distribution. In recommendation systems, to better improve the performance in ranking process, we propose a modified version of the BPR loss function called Balanced Bayesian Personalized Ranking (BBPR)

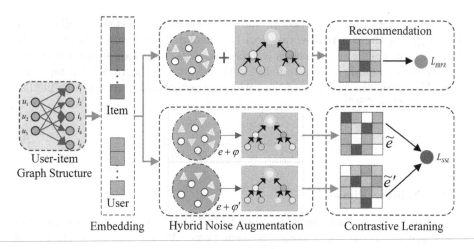

Fig. 1. The architectures of GCLHANRec.

as the loss function for our recommendation task. Specifically, BBPR first calculates the score between the user vector and item vector using dot product, then transforms these scores into probability values through the sigmoid function. Finally, it uses these probability values to compute the cross-entropy loss, thus considering the imbalance between positive and negative samples and maximizing the difference between them. Which is formulated as (6), where $\tilde{y}_{ui} = \tilde{e}_u^T \tilde{e}_i$, $\tilde{y}_{uj} = \tilde{e}_u^T \tilde{e}_j$.

$$\tilde{e}_u = e_u + \varphi_1 \qquad \tilde{e}_u' = e_u + \varphi_1' \tag{4}$$

$$\tilde{e}_i = e_i + \varphi_2 \qquad \tilde{e}_i' = e_i + \varphi_2' \tag{5}$$

$$\mathcal{L}_{BBPR} = - \sum_{(u,i,j) \in \mathcal{O}} log(\sigma(\tilde{y}_{ui})(1 - \sigma(\tilde{y}_{uj}))) \tag{6}$$

4.3 Contrastive Learning

Next, after hybrid noise augmentation, we treat the augmentations from the same node as positive pairs such as $\tilde{e}_u$ and $\tilde{e}_u'$. All augmentations from different nodes are treated as negative pairs such as $\tilde{e}_u$ and $\tilde{e}_v'$. Following most graph contrastive learning methods [23–25], we also apply InfoNCE [35] to compute the loss $\mathcal{L}_{ssl}^{user}$ of the contrastive task for users. The goal of InfoNCE is to make the positive samples $\tilde{e}_u'$ and the anchor sample $\tilde{e}_u$ closer in vector space, on the contrary, make the negative samples $\tilde{e}_v'$ and the anchor sample $\tilde{e}_u$ vector distances relatively far, and finally learn an encoder for contrastive task in the recommendation system. The goal is to make the hybrid noise-augmented positive samples and the anchor sample closer in vector space, on the contrary, make the negative samples and the anchor sample vector distances relatively far, and finally learn an encoder for contrastive task in the recommendation system. Similarly, the loss $\mathcal{L}_{ssl}^{item}$ of the contrastive task can be calculated for items. Where τ is the temperature coefficient, which is a hyperparameter that can adjust the degree of attention to hard samples to help the model learn better from hard negatives [36]. Therefore, the final loss of the entire contrastive task can be represented as $\mathcal{L}_{ssl} = \mathcal{L}_{ssl}^{user} + \mathcal{L}_{ssl}^{item}$.

$$\mathcal{L}_{ssl}^{user} = \sum_{u \in \mathcal{U}} -log \frac{exp(s(\tilde{e}_u, \tilde{e}'_u)/\tau)}{\sum_{v \in \mathcal{U}} exp(s(\tilde{e}_u, \tilde{e}'_v)/\tau)} \tag{7}$$

4.4   Training Phase

To optimize our model, we apply a multi-task learning strategy that includes contrastive learning and recommendation tasks. Where, λ is the coefficient of the contrastive task. The overall loss is defined as:

$$\mathcal{L} = \mathcal{L}_{BBPR} + \lambda\mathcal{L}_{ssl} \tag{8}$$

5   Performance Evaluation

5.1   Experimental Setup

Dataset. We evaluate the effectiveness of our proposed model on three publicy available datasets: MovieLens, Douban-book and LastFM, which is widely used in the field of recommendation systems. The detailed statistics for the three datasets are reported in Table 1. Where *density* is used to measure the sparsity of user-item interactions, and the larger the value, the sparser the interaction. It is defined as:

$$density = \frac{\#actions}{\#users \times \#items} \tag{9}$$

In Eq. 8, where $\#actions$ is the number of actions of users and items, and $\#users$ represents the number of users, $\#items$ represents the number of items.

- **MovieLens**[1] mainly includes the user's occupation, age and other information, the movie's name, genre and release year information, as well as the user's rating of the movie. It recorded 1,000,209 ratings on 3,900 movies by 6,040 users who joined MovieLens in 2000.
- **Douban-book**[2] is a collection of book ratings and reviews from douban.com. The rating range is 1–5. We discard ratings less than 4 and reset the rest to 1. Finally, we obtain 23,299 users, 41,287 items, and these user-item generate 598,420 interactions in total.

---

[1] https://grouplens.org/datasets/movielens/1m/.
[2] http://challenges.2014.eswc-conferences.org/index.php/RecSys.

- **LastFM**[3] is a music recommendation dataset provided by Last. For each user in the dataset, it contains a list of their most popular artists and the number of plays. In addition, it includes user application tags that can be used to build content vectors.

Table 1. The statistics of the datasets.

Dataset	#users	#items	#actions	density
MovieLens	6,040	3,706	1,000,029	0.04468
Douban-book	23,299	41,287	598,420	0.00622
LastFM	1,872	17,632	92,834	0.00281

Baselines. In order to demonstrate the effectiveness of our method, we compare it with three existing types of baseline methods, including classical matrix factorization models (MF), graph neural network models (LightGCN), and contrastive learning models (SGL, NCL, XSimGCL). The following is a brief overview of each baseline method.

- **MF** [37] applies matrix factorization technology to the recommendation system. At the same time, in order to improve the recommendation performance, MF introduces additional information such as implicit feedback and confidence, which is significantly better than the classic nearest neighbor method.
- **LightGCN** [30] is an optimization and improvement of the Neural Graph Collaborative Filtering (NGCF) algorithm. The representation of the nodes is learned by smoothing the features on the graph. Applying graph convolutional neural networks to recommendation system.
- **SGL** [23] is proposed to apply SSL to user-item bipartite graphs, was designed with node dropout, edge dropout, and random walk to generate different views of a node in three ways to improve the ability to automatically mine hard negatives.
- **NCL** [24] is proposed to utilize neighborhood-enhanced contrastive learning process to improve the previous problem of not mining the nearest neighbor relationship between users (or items) due to random sampling.
- **XSimGCL** [25] is an extremely simple graph contrastive learning method that discards the graph augmentation strategies adopted by many state-of-the-art recommendation models. It uses a simple yet effective noise embedding augmentation as the contrastive view to learn more uniform data representations.

---

[3] http://www.cp.jku.at/datasets/LFM-1b/.

Evaluation Metrics. To evaluate the performance of different methods, we use two evaluation metrics that are widely used in recommendation system tasks: Recall@K and Normalized Discounted Cumulative Gain@K (respectively named **Recall@K** and **NDCG@K**), which are widely used in the Top-N recommendation.

Implementation Details. All experiments are based on the following environments and configurations: Ubuntu 18.04, Intel(R) Xeon(R) Silver 4316 CPU @ 2.30 GHz, GeForce RTX 3090. In our experiment, the value of k is 20. We follow the setting of majority of recent methods [20,24,30], randomly selecting 80% of the interactive data of each user as the training set and the remaining 20% as the test set. For the training set, we randomly take 10% of the data as the validation set to tune hyper-parameters. And we optimize each methods according to the validation sets. In addition, we also follow the setting of LightGCN [30], the calculation of Recall@20 and NDCG@20 adopt all-ranking protocol, that is, all items that have not interacted with the user in each user 's interaction are candidates. Furthermore, the reported results are the average values obtained from 5 runs to ensure fairness. For baseline methods, we tune the hyper-parameters based on original papers. In our method, we use Adam [38] to optimize our method and set the learning rate to 0.001 and the batch size to 2048. In addition, we set the coefficient λ of the contrastive task to 0.2. we fine-tune the size of the hybrid noise α and temperature coefficient τ in the ranges of SPSVERBc10, 0.01, 0.05, 0.1, 0.2, 0.25, 0.5SPSVERBc2 and SPSVERBc10.05, 0.1, 0.2, 0.5, 1.0 }, respectively.

5.2   Performance Evaluation

To validate the effectiveness of our method, in this section, we compare the performance of our proposed GCLHANRec with other baselines from multiple perspectives, including different numbers of layers and different Top-k values.

Comparison with Different Layers. Table 2 shows the performance comparison of our method and other baselines on different layers (1 to 3). Bold font indicates the best results.

We can get the following conclusions based on the analysis results:

(1) In most cases, our method outperforms existing baseline methods, especially compared to the classic graph contrastive learning method SGL. On one hand, the results verify the efficiency of improving the uniformity of data distribution. On the other hand, the results also verify the efficiency of our methods. For example, on MovieLens, XSimGCL improves by 4.239% and 4.010% on the Recall and NDCG respectively compared to SGL, while our method improves by 5.318% and 4.578% respectively compared to SGL.

(2) We can also find that we can improve the recommendation performance on most methods increasing the number of layers of the model. The improvement

Table 2. The performance of different methods using different layers on three datasets.

Layer	Method	MovieLens		LastFM		Douban-book	
		Recall@20	NDCG@20	Recall@20	NDCG@20	Recall@20	NDCG@20
1-Layer	LightGCN	0.2561	0.3017	0.2629	0.2681	0.1342	0.1134
	SGL	0.2661	0.3141	0.2667	0.2765	0.1678	0.1503
	XSimGCL	0.2719	0.3192	0.2737	0.2844	**0.1703**	0.1518
	GCLHANRec	**0.2793**	**0.3288**	**0.2795**	**0.2891**	0.1697	**0.1521**
2-Layer	LightGCN	0.2620	0.3068	0.2662	0.2758	0.1452	0.1248
	SGL	0.2691	0.3188	0.2788	0.2892	0.1716	0.1518
	XSimGCL	0.2799	0.3286	0.2792	**0.2907**	0.1777	0.1606
	GCLHANRec	**0.2808**	**0.3304**	**0.2798**	0.2899	**0.1787**	**0.1613**
3-Layer	LightGCN	0.2625	0.3066	0.2676	0.2785	0.1470	0.1263
	SGL	0.2689	0.3167	0.2776	0.2896	0.1723	0.1536
	XSimGCL	0.2803	0.3294	0.2795	0.2902	0.1797	0.1617
	GCLHANRec	**0.2832**	**0.3312**	**0.2810**	**0.2907**	**0.1798**	**0.1622**

is obvious on dense datasets. For example, in terms of Recall, our method on MovieLens and LastFM datasets, the performance can improve by 0.854% and 1.396% respectively based on 2 layers and 1 layer model.

(3) Graph contrastive learning methods can help to solve the problems caused by data sparsity in recommendation system. Specifically, for the Douban-book with higher sparsity, the three graph contrastive learning methods SGL, XSimGCL and GCLHANRec achieve significant performance improvements compared to the traditional graph neural network method LightGCN (17.210%, 22.245% and 22.313% on Recall, 21.615%, 28.029% and 28.424% on NDCG). And on the dense dataset MovieLens, the three methods can also improve (2.438%, 6.781% and 7.886% on Recall, 3.294%, 7.436% and 8.023% on NDCG).

Comparison with Baselines. The performance of different methods on three datasets are shown in Table 3. We generated the Top-20 recommended items. Bolded font indicates the best results. From the experimental results, we found that our method outperforms all baselines on MovieLens and Douban-book, and significantly outperforms the classic graph neural network method LightGCN (improving Recall and NDCG by 7.886% and 8.023% respectively on MovieLens) and graph contrastive learning methods SGL (improving Recall and NDCG by 5.318% and 4.578% respectively on MovieLens) and NCL (improving Recall and NDCG by 3.736% and 2.666% respectively). However, we also observe that our method has only slightly higher Recall than the current state-of-the-art XSimGCL on Douban-book, which may be related to the sparsity of this dataset.

Comparison with Different Top-K. In Fig. 2, we evaluate Top-k recommendation with values of 1, 5, 10, 15, and 20 on the used datasets. Figure 2 (a) and

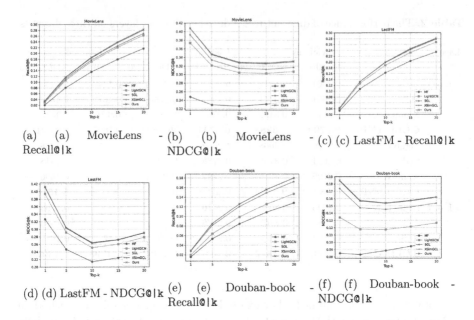

(a) (a) MovieLens - (b) (b) MovieLens - (c) (c) LastFM - Recall@|k
Recall@|k NDCG@|k

(d) (d) LastFM - NDCG@|k (e) (e) Douban-book - (f) (f) Douban-book -
Recall@|k NDCG@|k

Fig. 2. Evaluating Top-k recommendation with k values of 1, 5, 10, 15, and 20 on MovieLens, LastFM and Douban-book.

Table 3. The performance of different methods on three datasets.

Method	MovieLens		LastFM		Douban-book	
	Recall@20	NDCG@20	Recall@20	NDCG@20	Recall@20	NDCG@20
MF	0.2165	0.2378	0.2342	0.2401	0.1280	0.1003
LightGCN	0.2625	0.3066	0.2676	0.2785	0.1470	0.1263
SGL	0.2689	0.3167	0.2776	0.2896	0.1723	0.1536
NCL	0.2730	0.3226	0.2765	0.2876	0.1625	0.1414
XSimGCL	0.2803	0.3294	0.2795	0.2902	0.1797	0.1617
GCLHANRec	**0.2832**	**0.3312**	**0.2810**	**0.2907**	**0.1798**	**0.1622**

(b) show the Top-k performance comparison of various methods on MovieLens. Our method outperforms all existing baselines, and the Recall values of various methods increase as k increases, which is consistent with the basic fact that as k increases, the hit rate of items that users like is higher, thus recall is larger. In contrast, the NDCG values are decreasing, which may be due to the fact that as k increases, the model is more likely to recommend items that are irrelevant to users, thus NDCG may decrease. Figure 2 (c), (d) and Fig. 2 (e), (f) show the comparison of Top-k performance of various methods on LastFM and douban-book respectively. Our method has a larger performance improvement compared with other methods on Douban-book, and the improvement increases with the increase of k.

5.3 Hyper-parameters Analysis

The temperature parameter τ is an important hyper-parameter in our experiments. Related studies [36] show that it cannot be too large or too small. Too large τ will cause the data distribution to be more uneven, and too small τ will cause the model to focus on the nearby negative examples and ignore the potential positive examples. Therefore, we fine-tune τ in the range of SPSVERBc10.05, 0.1, 0.2, 0.5, 1.0SPSVERBc2 to determine its best value. Figure 3 shows the trend of RecallSPSVERBc720 on three datasets with different temperature parameters τ. According to the results, we finally choose $\tau = 0.2$ for MovieLens, $\tau = 0.2$ for LastFM, and $\tau = 0.2$ for Douban-book. In addition, we fine-tune the size of the hybrid noise α in the range of SPSVERBc10, 0.01, 0.05, 0.1, 0.2, 0.25, 0.5SPSVERBc2, choose $\alpha = 0.2$ for MovieLens, $\alpha = 0.25$ for LastFM, and $\alpha = 0.25$ for Douban-book. In addition, due to space limitations, we primarily focus on exploring and presenting the main hyperparameters, while the remaining hyperparameters will be discussed in detail in future studies.

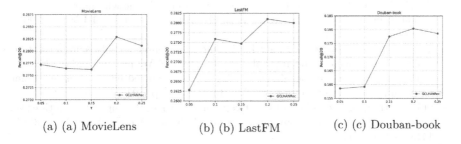

(a) (a) MovieLens   (b) (b) LastFM   (c) (c) Douban-book

Fig. 3. Evaluating Top-20 recommendation with different τ on MovieLens, LastFM and Douban-book.

6   Conclusion and Future Work

In this work, a method based on hybrid noise augmentation and graph contrastive learning is proposed to solve the data sparsity problem in recommendation system. In our method, we add different random noises to users and items respectively. Specifically, we add uniform distribution random noise to users and normal distribution random noise to items, to improve the data uniformity while increasing the user's interest diversity for different items, thereby improving the accuracy and personalization degree of the recommendation system. In addition, we propose BBPR as the loss function for recommendation tasks, which is a modification of BPR to better make the model pay more attention to the difference between positive and negative samples, thus performing better in ranking tasks. In the future, we will explore more advanced methods to improve the data uniformity in recommendation systems, to further enhance the recommendation performance.

Acknowledgments. The work reported herein was supported by National Key R&D Program (2020YFB1406900), National Natural Science Foundation of China (62172324, 62102310), Key R&D in Shaanxi Province (2023-YBGY-269, 2022QCY-LL-33HZ).

References

1. Li, Y., Zhou, T., Yang, K., et al.: Personalized recommender systems based on social relationships and historical behaviors. Appl. Math. Comput. **437**, 127549 (2023)
2. Liu, T., Wu, Q., Chang, L., et al.: A review of deep learning-based recommender system in e-learning environments. Artif. Intell. Rev. **55**(8), 5953–5980 (2022)
3. Roy, D., Dutta, M., et al.: Optimal hierarchical attention network-based sentiment analysis for movie recommendation. Soc. Netw. Anal. Min. **12**(1), 138 (2022)
4. Zhang, H., Wang, H., Wang, G., et al.: A hyperbolic-to-hyperbolic user representation with multi-aspect for social recommendation. In: ACM International Conference on Information & Knowledge Management, pp. 4667–4671 (2022)
5. Qiu, Z., Hu, Y., Wu, X.: Graph neural news recommendation with user existing and potential interest modeling. ACM Trans. Knowl. Discov. Data **16**(5), 96:1-96:17 (2022)
6. Tao, Y., Li, Y., Zhang, S., et al.: Revisiting graph based social recommendation: a distillation enhanced social graph network. In: WWW, pp. 2830–2838 (2022)
7. Wang, J., Chen, Y., Wang, Z., et al.: Popularity-enhanced news recommendation with multi-view interest representation. In: CIKM, pp. 1949–1958 (2021)
8. Shi, C., Han, X., Song, L., et al.: Deep collaborative filtering with multi-aspect information in heterogeneous networks. IEEE Trans. Knowl. Data Eng. **33**(4), 1413–1425 (2021)
9. Wang, X., He, X., Wang, M., et al.: Neural graph collaborative filtering. In: SIGIR, pp. 165–174 (2019)
10. Dong, X., Yu, L., Wu, Z., et al.: A hybrid collaborative filtering model with deep structure for recommender systems. In: AAAI, pp. 1309–1315 (2017)
11. Pérez-Almaguer, Y., Yera, R., Alzahrani, A.A., et al.: Content-based group recommender systems: a general taxonomy and further improvements. Expert Syst. Appl. 115444 (2021)
12. Peng, Y.: A survey on modern recommendation system based on big data. CoRR, abs/2206.02631 (2022)
13. Joshi, A., Wong, C., de Oliveira, D.M., et al.: Imbalanced data sparsity as a source of unfair bias in collaborative filtering. In: RecSys, pp. 531–533 (2022)
14. Elahi, E., Halim, Z.: Graph attention-based collaborative filtering for user-specific recommender system using knowledge graph and deep neural networks. Knowl. Inf. Syst. **64**(9), 2457–2480 (2022)
15. Chen, T., Kornblith, S., Norouzi, M., Hinton, G.E.: A simple framework for contrastive learning of visual representations. In: PMLR, pp. 1597–1607 (2020)
16. Zeng, J., Xie, P.: Contrastive self-supervised learning for graph classification. In: AAAI, pp. 10824–10832 (2021)
17. You, Y., Chen, T., Sui, Y., et al.: Graph contrastive learning with augmentations. In: NeurIPS (2020)
18. Jaiswal, A., Babu, A.B., Zadeh, M.Z., et al.: A survey on contrastive self-supervised learning. CoRR, abs/2011.00362 (2020)

19. Hassani, K., Khas Ahmadi, A.H.: Contrastive multi-view representation learning on graphs. In: ICML, pp. 4116–4126 (2020)
20. Yu, J., Yin, H., Xia, X., et al.: Self-supervised learning for recommender systems: a survey. CoRR, abs/2203.15876 (2022)
21. Liu, Z., Ma, Y., Ouyang, Y., et al.: Contrastive learning for recommender system. CoRR, abs/2101.01317 (2021)
22. Yu, J., Yin, H., Gao, M., et al.: Socially-aware self-supervised tri-training for recommendation. In: KDD, pp. 2084–2092 (2021)
23. Wu, J., Wang, X., Feng, F., et al.: Self-supervised graph learning for recommendation. In: SIGIR, pp. 726–735 (2021)
24. Lin, Z., Tian, C., Hou, Y., et al.: Improving graph collaborative filtering with neighborhood-enriched contrastive learning. In: WWW, pp. 2320–2329 (2022)
25. Yu, J., Xia, X., Chen, T., et al.: XSimGCL: towards extremely simple graph contrastive learning for recommendation. CoRR, abs/2209.02544 (2022)
26. Zhang, J., Gao, M., Yu, J., et al.: Double-scale self-supervised hypergraph learning for group recommendation. In: CIKM, pp. 2557–2567 (2021)
27. Zhou, X., Sun, A., Liu, Y., et al.: SelfCF: a simple framework for self-supervised collaborative filtering. CoRR, abs/2107.03019 (2021)
28. Ying, R., He, R., Chen, K., et al.: Graph convolutional neural networks for web-scale recommender systems. In: KDD, pp. 974–983 (2018)
29. Xue, F., He, X., Wang, X., et al.: Deep item-based collaborative filtering for top-n recommendation. ACM Trans. Inf. Syst. **37**(3), 33:1-33:25 (2019)
30. He, X., Deng, K., Wang, X., et al.: LightGCN: simplifying and powering graph convolution network for recommendation. In: SIGIR, pp. 639–648 (2020)
31. Chen, L., Wu, L., Hong, R., et al.: Revisiting graph based collaborative filtering: a linear residual graph convolutional network approach. In: IAAI, pp. 27–34 (2020)
32. Huang, T., Dong, Y., Ding, M., et al.: MixgCF: an improved training method for graph neural network-based recommender systems. In: KDD, pp. 665–674 (2021)
33. Yu, J., Yin, H., Xia, X., et al.: Graph augmentation-free contrastive learning for recommendation. CoRR, abs/2112.08679 (2021)
34. Rendle, S., Freudenthaler, C., Gantner, Z., et al.: BPR: Bayesian personalized ranking from implicit feedback. In: UAI, pp. 452–461 (2009)
35. Oord, A.V.D., Li, Y., Vinyals, O.: Representation learning with contrastive predictive coding. arXiv preprint arXiv:1807.03748 (2018)
36. Wang, F., Liu, H.: Understanding the behaviour of contrastive loss. In: CVPR, pp. 2495–2504 (2021)
37. Koren, Y., Bell, R.M., Volinsky, C.: Matrix factorization techniques for recommender systems. Computer 30–37 (2009)
38. Kingma, D.P., Ba, J.: Adam: a method for stochastic optimization. In: ICLR (2015)

User-Oriented Interest Representation on Knowledge Graph for Long-Tail Recommendation

Zhipeng Zhang[1], Yuhang Zhang[1], Anqi Wang[1], Pinglei Zhou[1], Yao Zhang[2(✉)], and Yonggong Ren[1(✉)]

[1] School of Computer Science and Artificial Intelligence, Liaoning Normal University, Dalian, China
ryg@lnnu.edu.cn
[2] Mechanical Engineering and Automation, Dalian Polytechnic University, Dalian, China
zhangyao@dlpu.edu.cn

Abstract. Graph neural networks have demonstrated impressive performance in the field of recommender systems. However, existing graph neural network recommendation approaches are proficient in capturing users' mainstream interests and recommending popular items but fall short in effectively identifying users' niche interests, thus failing to meet users' personalized needs. To this end, this paper proposes the User-oriented Interest Representation on Knowledge Graph (UIR-KG) approach, which leverages the rich semantic information on the knowledge graph to learn users' long-tail interest representation. UIR-KG maximizes the recommendation of long-tail items while satisfying users' mainstream interests as much as possible. Firstly, a popular constraint-based long-tail neighbor selector is proposed, which obtains the target user's long-tail neighbors by conducting high-order random walks on the collaborative knowledge graph and constraining item popularity. Secondly, a knowledge-enhanced hybrid attention aggregator is proposed, which conducts high-order aggregation of users' long-tail interest representations on the collaborative knowledge graph by comprehensively combining relationship-aware attention and self-attention mechanisms. Finally, UIR-KG predicts the ratings of uninteracted items and provides Top-N recommendation results for the target user. Experimental results on real datasets show that compared with existing relevant approaches, UIR-KG can effectively improve recommendation diversity while maintaining recommendation accuracy, especially in long-tail recommendations. Code is available at https://github.com/ZZP-RS/UIR-KG

Keywords: Long-tail recommendation · Interest representation · Knowledge graph · Graph neural network

© The Author(s), under exclusive license to Springer Nature Switzerland AG 2023
X. Yang et al. (Eds.): ADMA 2023, LNAI 14179, pp. 340–355, 2023.
https://doi.org/10.1007/978-3-031-46674-8_24

1   Introduction

With the deepening integration of the Internet into people's lives, recommender systems have been widely applied in various applications to provide personalized recommendations and effectively address the problem of information overload. Recommender systems predict user preferences based on historical interaction information and present a recommendation list to the user. However, current recommender systems are facing the problem of the long tail, where a small number of popular items are recommended frequently by the system, while a large number of niche items are rarely recommended, resulting in an "80/20" effect. However, a niche market requires diversified recommendations, which bring many benefits to users and providers. Several studies have shown that recommendation approaches that consider both the diversity and novelty of recommended content can effectively improve user satisfaction and provide long-term benefits for recommendation platforms [1]. However, current recommendation approaches tend to recommend popular items to users because these items are more likely to be interacted with by users than niche items. The long-tail recommendation has important practical significance, from the perspective of providers, the market competition for popular items is fierce and the profit is limited, while the marginal profit of niche items is relatively high. From the perspective of users, it is not possible to click on the same popular item repeatedly [2,3], and recommending niche items can bring surprises to users, increase user loyalty and satisfaction [4]. However, most current recommendation approaches either focus on improving recommendation accuracy but cannot provide diversified recommendations, or sacrifice accuracy for diversity, which makes it a challenge for researchers to effectively improve diversity while ensuring recommendation accuracy and providing personalized long-tail recommendations to users.

Based on the aforementioned issues, we propose the UIR-KG that leverages the rich semantic information in the knowledge graph (KG) and the powerful representation ability of graph convolutional network (GCN). Its advantage lies in effectively capturing the long-tail interest representation of users and recommending long-tail items that match their interests. First, UIR-KG extracts the user's collaborative paths via random walks and calculates the average popularity of the item nodes in the collaborative paths to obtain the x paths with the lowest average popularity, replacing the original first-order neighbors with the items from these paths. Then, we propose a hybrid attention mechanism that combines relationship-aware attention and self-attention to aggregate the user's long-tail interest representation. Finally, we predict the scores of uninteracted items and generate the final recommendation results. The contributions of this paper can be summarized as follows: (1) UIR-KG effectively captures the long-tail interest representation of users, allowing for the recommendation of long-tail items while still satisfying their mainstream interests. (2) We propose a popular constraint-based long-tail neighbor selector that utilizes semantic information in the KG to select potential long-tail neighbors of the target user and overcomes the problem of decreasing long-tail feature information in the high-order information aggregation process. (3) We propose a knowledge-enhanced hybrid attention aggregator that can effectively learn the user's long-tail interest representation and eliminate the noise impact in the reconstructed neighbors.

2 Related Work

As an important side information, KG contains rich semantic information about entities, and it has shown good performance in improving recommendation accuracy. Wang et al. [5] propose a method that combines collaborative signals and knowledge associations by explicitly encoding collaborative signals through collaborative propagation. Wang et al. [6] create a collaborative knowledge graph to consider user-item interactions and the edge information from the KG, enriching the representations of users and items. However, existing KG-based recommendation methods have focused on improving recommendation accuracy while neglecting recommendation diversity. It is unreasonable to solely rely on accuracy to measure the quality of a recommendation method [7]. Generating diverse long-tail recommendations is also an important metric for evaluating recommendation method performance.

To address the long-tail recommendation problem, existing methods often assign different weights to nodes based on their popularity or introduce external information to enhance the representation of long-tail items. Zhang et al. [8] propose a dual transfer learning framework that trains model parameters using popular items and learns the features of popular items by connecting them with features from tail items, thereby mitigating the influence of item popularity on long-tail recommendations. Yin et al. [9] propose a novel suite of graph-based algorithms. They represent user-item information using an undirected weighted graph and apply the hitting time algorithm for long-tail item recommendation. Park et al. [1] address the long-tail problem through clustering algorithms, where items are divided into the head and tail parts, and clustering is performed only on the tail part. However, these methods are unable to effectively model users' long-tail interests from their historical preference information, often sacrificing accuracy to improve recommendation diversity. Capturing users' long-tail interests from their historical preferences and improving recommendation diversity while maintaining accuracy remains a challenge in long-tail recommendations.

3 Proposed Approach

3.1 Problem Formulation

Here, we explain the symbols used in this paper in Table 1. Then, we continue to introduce key concepts and problem setting of the proposed UIR-KG.

User-Item Bipartite Graph: In this paper, we define these interaction data as a user-item bipartite graph $\mathcal{G}_{\text{UI}} = \{(u, y_{ui}, i) | u \in U, i \in I, y_{ui} \in Y\}$, where U is the set of users, I is the set of items, and the element value $y_{ui} = 1$ indicates that user u has interacted with item i in the past, otherwise $y_{ui} = 0$.

Table 1. Summary of Notations

Symbols	Description
U	Set of users:$\{u_1, u_2, ..., u_{|U|}\}$
I	Set of items:$\{i_1, i_2, ..., i_{|I|}\}$
$\mathcal{G}_{\text{UI}}$	User-item bipartite graph
$\mathcal{G}_{\text{KG}}$	Knowledge graph
$\mathcal{G}_{\text{CKG}}$	Collaborative knowledge graph
ϵ	Set of entities: $\{e_1, e_2, ..., e_{|\epsilon|}\}$
$\mathcal{R}$	Set of relations: $\{r_1, r_2, ..., r_{|\mathcal{R}|}\}$
$\mathcal{P}$	Set of collaborative paths
p_i	Popularity of item
ε	Set of seeds
N_u	Set of u's neighbor after restructuring
K	Sampling depth
E_u	Set of user representation
E_i	Set of item representation
Φ	The parameters of embedding layer
ω	The parameters of attentive-aware layer

Knowledge Graph: To enhance item information, we introduce a knowledge graph to provide item attribute information. The knowledge graph, as auxiliary data, is composed of entities and their relationships in the real world and can serve as item attribute information [10]. We define the knowledge graph as $\mathcal{G}_{\text{KG}} = \{(h, r, t) | h, t \in \epsilon, r \in \mathcal{R}\}$, where each triple (h, r, t) represents a connection between the head entity h and the tail entity t under the relationship r, and ϵ and $\mathcal{R}$ are the entity set and the relationship set in the knowledge graph, respectively.

Collaborative Knowledge Graph: The origin of the collaborative knowledge graph (CKG) is to merge $\mathcal{G}_{\text{UI}}$ and $\mathcal{G}_{\text{KG}}$ into a single relationship graph, which we define as $\mathcal{G}_{\text{CKG}} = \{(h, r, t) | h, t \in \epsilon', r \in \mathcal{R}'\}$, where $\epsilon' = \epsilon \cup U$, and $\mathcal{R}' = \mathcal{R} \cup \{\text{Interact}\}$.

Task Description: This study aims to obtain user embeddings that aggregate long-tail collaborative signals by exploring user behavior at the sampling level. We now customize the recommendation task in this paper:

- **Input:** User-item bipartite graph $\mathcal{G}_{\text{UI}}$, knowledge graph $\mathcal{G}_{\text{KG}}$, and a target user.
- **Output:** The top N items that are most relevant to the target user and contain as many long-tail items as possible.

Figure 1 shows the basic flowchart of our approach. Our approach consists of three main components: Popular constraint-based long-tail neighbor selector; Knowledge-enhanced hybrid attention aggregator; Model prediction and optimization.

3.2 Popular Constraint-Based Long-Tail Neighbor Selector

In recommender systems, we often model user interests based on their historical interaction with items. By utilizing the high-order connectivity of the graph, we can effectively explore users' long-tail interests. Based on this, we designed a user popular constraint-based long-tail neighbor selector, which consists of three parts: high-order random walk on CKG, popular constraint-based collaborative path extraction, and long-tail neighbor selection and reconstruction.

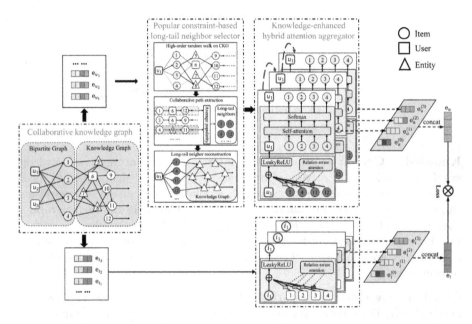

Fig. 1. The flowchart of the proposed UIR-KG approach.

High-Order Random Walk on CKG. Given a user's set of historically interacted items $\varepsilon_u^0 = \{i | i \in y_{ui} = 1\}$, we use it as the seed for propagation. We randomly sample m seed neighbors and continue to treat them as seeds, propagating further down the graph. This process generates a high-order collaborative path that starts from the target node and has a length of K. To fully exploit our sampler's capabilities, we attempt to perform collaborative signal sampling on the CKG.

Figure 2 illustrates the mechanism of the popular constraint-based long-tail neighbor selector, as shown in the figure. Firstly, for the target user u_1, its initial seed set is $\varepsilon_{u_1}^0 = \{i|(u_1, i) \in \mathcal{G}_{\mathrm{UI}}, i \in \{i|y_{u_1 i} = 1\}\}$. Based on this, its k-th order neighbors are denoted as $\varepsilon_{u_1}^k = \{t|(h, r, t) \in \mathcal{G}_{\mathrm{CKG}}, h \in \varepsilon_{u_1}^{k-1}\}$. Taking Fig. 2 as an example, for the target user u_1, its initial seed set is $\varepsilon_{u_1}^0 = \{i_1, i_2, i_3, i_4\}$. When sampling paths, we first randomly select an item from $\varepsilon_{u_1}^0$ as the starting point of the path, and then randomly select a user who has interacted with this item as the second-hop node.

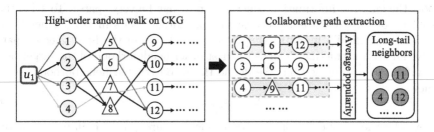

Fig. 2. An illustration of popular constraint-based long-tail neighbor selection for u_1.

Popular Constraint-Based Collaborative Path Extraction. The high-order collaborative paths obtained through the above approach already contain rich collaborative signals. Then, we need to filter these paths to ensure that the high-order collaborative paths finally retained contain long-tail collaborative signals. The characteristic of long-tail items is that they have less interaction information with users. Based on this feature, we calculate the average item popularity of each collaborative path, and the high-order collaborative paths with the lowest average popularity are the ones we need. Here, we denote the set of paths as $\mathcal{P}$, and the calculation formula for average popularity is defined as follows:

$$\bar{P} = \frac{\sum_{i=1}^{N} a_i p_i}{N'}, \tag{1}$$

in the equation, N is the total number of nodes in the collaborative path, N' is the number of item nodes in the path, a_i is a parameter indicating whether the ith node should be accumulated, with a value of 0 or 1 determined by whether the node is an item node, and p_i is the popularity of node i. For example, consider the path $u_1 \rightarrow i_2 \rightarrow e_5 \rightarrow i_{10}$, with an average item popularity $\bar{P} = \frac{p_{i_2} + p_{i_{10}}}{2}$, and the path $u_1 \rightarrow i_4 \rightarrow e_8 \rightarrow i_{11}$, with an average item popularity $\bar{P} = \frac{p_{i_4} + p_{i_{11}}}{2}$. Clearly, the path with smaller $\bar{P}$ is more in line with our selection because, under other conditions being the same, the probability that a path with lower average item popularity contains long-tail information is higher than a path with higher average item popularity.

Long-Tail Neighbor Selection and Reconstruction. If we use these collaborative paths directly for information propagation, the target user node can aggregate the long-tail information we want. However, our goal is to explore the long-tail interests of the target user, and it is not enough for the target user node to only obtain a small amount of long-tail information. Therefore, there is a problem with the above process: we do not know which node in the path is the long-tail item node. If we directly use GCN for information propagation, the long-tail item information at the end of the path will be constantly lost during the propagation process. When the information is transmitted to the target user node, the long-tail information contained in it may be very sparse. To solve this problem, our approach is to reconstruct the target user's first-order neighborhood ε_u^0, specifically by extracting all item nodes from the final selected path above and reconstructing them as the first-order neighborhood of the target user for information propagation. We define the reconstructed neighborhood set as follows:

$$N_u = \{i|i \in \mathcal{P}\}, \tag{2}$$

where $\mathcal{P}$ is the set of collaborative paths selected through the high-order random walk on CKG, replacing the user's original first-order neighbors with the reconstructed neighbors allows for effective information propagation and ensures that the user can aggregate a certain amount of long-tail information. This enables the system to make effective long-tail recommendations.

3.3 Knowledge-Enhanced Hybrid Attention Aggregator

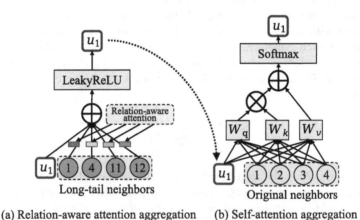

(a) Relation-aware attention aggregation   (b) Self-attention aggregation

Fig. 3. An illustration of knowledge-enhanced hybrid attention aggregation for u_1.

An important approach for maintaining the structure of a graph is to construct entity embeddings and relation embeddings. We start by randomly initializing node embeddings. Then, we employ a learnable projection layer to enrich their representations:

$$e_i = W_e c_i + b, \tag{3}$$

c_i is a vector randomly generated for entities or relations in the CKG with dimension d_o, while $W_e \in R^{d_o \times d_e}$ and $b \in R^{d_e}$ are trainable parameters. Then, we designed a knowledge-enhanced hybrid attention aggregator for high-order propagation and aggregation. Figure 3 illustrates the mechanism of the knowledge-enhanced hybrid attention aggregator, which includes two parts: relation-aware attention aggregation with reconstructed neighbors, which calculates the attention weights between the head and tail nodes based on their relationship, and self-attention aggregation with original neighbors, which is used to aggregate the original neighbor information of the user.

Relation-Aware Attention Aggregation with Reconstructed Neighbors. We define the relation-aware attention as follows, where $N_u^c = \{(u, r, t)|(u, r, t) \in \mathcal{G}_{CKG}, u \in U\}$ represents the set of triples in the CKG with the target user u as the head entity:

$$e_{N_u} = \sum_{(h,r,t) \in N_u^c, e_t \in \varepsilon_u} g(u, r, t) e_t, \tag{4}$$

where $g(u, r, t)$ represents the attention weight of tail entity t under the relation r for the target user u. We use a relation-aware attention mechanism to calculate this importance:

$$g'(u, r, t) = (W_r e_t)^\top \tanh(W_r e_u + e_r), \tag{5}$$

$$g(u, r, t) = \mathrm{Softmax}(g'(u, r, t)), \tag{6}$$

where $W_r \in R^{d_r \times d_e}$ is used to map entity vectors from a d_e-dimensional vector space to a d_r-dimensional relationship vector space. The same operation is performed on item nodes to obtain e_{N_i}. Based on this, the user's embedding already contains a large amount of long-tail information, which can be used for long-tail recommendations. However, there are some issues with this approach. Firstly, the reconstructed first-order neighbors will inevitably bring some noise (as verified in the ablation experiments section). Secondly, blindly pursuing long-tail recommendations is not advisable. Therefore, inspired by the self-attention mechanism, we deployed an attention mechanism on the user's original neighbors to aggregate their mainstream interests while still recommending long-tail items to the user.

Self-attention Aggregation with Original Neighbors. We use $\varepsilon_u^0 = \{i | i \in y_{ui} = 1\}$ to represent the original neighbors set of user u. For each input node embedding e, we transform it into three vectors using three transformation matrices:

$$q = W_q e, \ k = W_k e, \ v = W_v e, \tag{7}$$

the self-attention score between any two nodes is obtained by the dot product of the q vector and the k vector. The self-attention score between user u and item i is given by:

$$\alpha_{u,i} = \text{Softmax}(q_u k_i). \tag{8}$$

Finally, the output vector is obtained by weighted summation of the attention scores and v vector:

$$e_u = \sum_{j \in \{\varepsilon_u^0 \cup e_u\}} \alpha_{u,j} v_j. \tag{9}$$

Note that we only perform the above operations for users.

After completing the information propagation, we use bi-interaction aggregation [6] to aggregate the user's own features and the features of their neighbors, as follows:

$$e_h^{l+1} = f(e_h^{(l)}, e_{N_h}^{(l)}), \tag{10}$$

$$\begin{aligned} f(e_h^{(l)}, e_{N_h}^{(l)}) =&\text{LeakyReLU}(W_1(e_h^{(l)} + e_{N_h}^{(l)})) \\ &+ \text{LeakyReLU}(W_2(e_h^{(l)} \odot e_{N_h}^{(l)})), \end{aligned} \tag{11}$$

where $W_1, W_2 \in R^{d_l \times d_e}$ are trainable weight matrices, $\odot$ denotes element-wise multiplication, l represents the l-th layer of propagation, and h is a placeholder for u, i. After completing L layers of propagation, we can obtain a collection of user-item representations of different orders:

$$E_u = \{e_u^{(0)}, e_u^{(1)}, \ldots \ldots e_u^{(L)}\}, \tag{12}$$

$$E_i = \{e_i^{(0)}, e_i^{(1)}, \ldots \ldots e_i^{(L)}\}. \tag{13}$$

3.4   Model Prediction and Optimization

Model Prediction. Each layer's representation can be interpreted as the hierarchical latent influence, emphasizing different high-order connectivity and preference similarity. Given the composite representations of the user E_u and item E_i, we connect the representations of each layer and finally obtain a single vector:

$$e_u^* = e_u^{(0)} || e_u^{(1)} || \ldots || e_u^{(L)}, \tag{14}$$

$$e_i^* = e_i^{(0)} || e_i^{(1)} || \ldots || e_i^{(L)}. \tag{15}$$

Then, we predict the user preference score by taking the inner product of the user and item representations:

$$\hat{y}_{ui} = e_u^{*\top} e_i^*. \tag{16}$$

Model Optimization. In this paper, we employed two types of optimizers, namely collaborative-aware learner and knowledge-aware learner.

The goal of collaborative-aware learner is to learn multi-faceted preferences from CKG. Here, we also use the BPR loss to understand the preferences of each user. Specifically, it assumes that observed interactions should be assigned higher predicted values than unobserved interactions, indicating stronger user preferences.

$$\mathcal{L}_{\mathrm{CF}} = \sum_{(u,i,j)\in O} -\ln\sigma(e_u^{*\top}e_i^* - e_u^{*\top}e_j^*), \tag{17}$$

here, $O = \{(u,i,j)|(u,i) \in O^+, (u,j) \in O^-\}$, where O^+ represents the set of observed interactions between users and items in the training set, i.e., positive samples, and O^- represents the set of unobserved user-item interactions in the training set, i.e., negative samples generated by the model through random sampling. The sigmoid function is denoted by σ.

The knowledge-aware learner models entity representations based on triplets in CKG. For a given triplet, its energy score is defined as follows:

$$g(h,r,t) = ||W_r e_h + e_r - W_r e_t||_2^2, \tag{18}$$

here, e_h, e_r, and e_t represent the embedding vectors of the head entity, relation entity, and tail entity, respectively. $W_r \in R^{d_r \times d_e}$ is a transformation matrix based on the relation r, used to transform e_h and e_t from a d_e-dimensional vector space to a d_r-dimensional vector space. Its meaning is to use the head entity vector and relation vector to approach the tail entity vector. Therefore, the lower the score, the more reliable the triplet is, and vice versa. Based on this, BPR loss is also utilized through negative sampling to encourage real triplets.

$$\mathcal{L}_{\mathrm{KG}} = \sum_{(h,r,t,t')\in T} -\ln\sigma(g(h,r,t') - g(h,r,t)), \tag{19}$$

where $T = \{(h,r,t,t')|(h,r,t) \in \mathcal{G}_{\mathrm{KG}}, (h,r,t') \notin \mathcal{G}_{\mathrm{KG}}\}$ is the set of negative samples used in training TransR, which is constructed by randomly replacing one of the entities in a real sample. It should be noted that the negative triples used for training must not be in the positive triple set.

Finally, we have the objective function to learn Eqs. (17) and (19) jointly:

$$\mathcal{L} = \mathcal{L}_{\mathrm{CF}} + \mathcal{L}_{\mathrm{KG}} + \lambda||\theta||_2^2, \tag{20}$$

where $\theta = \{E, W_r, W_1^{(l)}, W_2^{(l)}, \forall l \in \{1,...,L\}\}$ is the model parameter set, and E is the embedding table for all entities and relations; L_2 regularization parameterized by λ on θ is conducted to prevent overfitting.

4   Experiments and Evaluations

In this section, we first introduce an experimental setup, which includes an introduction to the datasets and evaluation metrics. Next, we will present our experimental results on real-world datasets.

4.1 Experimental Setup

Datasets and Parameters Settings. To evaluate the performance of UIR-KG, we conducted experiments on the Last-FM and Amazon-book datasets. The Last-FM dataset is a music dataset collected from the Last.fm online music system. We followed the processing steps in [6] for the dataset. Additionally, we derived knowledge graphs from Free-base, which have been widely used in previous research [5,6]. The overall statistics of the datasets are summarized in Table 2. Additionally, we selected 1500 high-order collaborative paths by random sampling and set the number of paths with the lowest average item popularity to 3.

Evaluation Metrics. We consider all items i that a user u has not interacted with as negative samples. Then, each approach outputs the preference scores of a user on all items, excluding those that have been rated in the training set. To evaluate the accuracy and diversity of our proposed approaches, we use precision and recall as metrics for recommendation accuracy and aggregate diversity (AD) [7] as a metric for recommendation diversity. Additionally, since short-head and long-tail are relative concepts, to effectively evaluate the performance of our model in long-tail recommendations, we use average recommendation popularity (ARP) [11] of the Top-N recommended items as a metric.

Table 2. Datasets

Datasets		Last-FM	Amazon-book
User-item interation	Users	23,566	70,679
	Items	48,123	24,915
	Interation	3,034,796	847,733
Knowledge graph	Entities	58,266	88,572
	Relations	9	39
	Triplets	464,567	2,557,746

The precision metric indicates the percentage of recommended items that are relevant to the target user among all recommended items. In addition, the recall metric represents the proportion of relevant recommended items for the target user among all relevant items. It is worth noting that the higher the values of both metrics, the better the recommendation accuracy. Assuming $Top_u(N)$ represents the set of N recommended items for user u, and $Rec_u(N)$ represents the Top-N items with the highest ratings that the user interacted with in the test set, the precision and recall metrics can be expressed as follows:

$$Precision = \frac{1}{|U|} \sum_{u \in U} \frac{|Top_u(N) \cap Rec_u(N)|}{|Top_u(N)|}, \tag{21}$$

$$\text{Recall} = \frac{1}{|U|} \sum_{u \in U} \frac{|Top_u(N) \cap Rec_u(N)|}{|Rec_u(N)|}. \tag{22}$$

The AD metric calculates the total number of distinct items recommended to all users. The ARP metric represents the average popularity of items recommended to users. Obviously, the higher the recommendation diversity, the larger the value of AD, and vice versa, while the lower the ARP value, the lower the average popularity of the recommended items. Assuming there are two users, u_x and u_y, and the sets of N recommended items for them are represented by $Top_{u_x}(N)$ and $Top_{u_y}(N)$ respectively, the AD metric can be represented as follows:

$$AD = |\bigcup_{u \in U} \{i \in Top_u(N)\}|. \tag{23}$$

The ARP metric can be represented as follows:

$$ARP = \frac{\sum_{i \in Top_u(N)} p_i}{N}. \tag{24}$$

4.2 Performance Comparisons

To validate the effectiveness of our proposed approach, we selected the following approaches for comparison:

- **BPRMF** [12]: A basic matrix factorization approach is used to rank items of interest for each user according to their preferences.
- **CKE** [13]: A typical regularization-based approach that enhances matrix factorization using semantic embeddings derived from TransR.
- **CFKG** [14]: A model applies TransE to a unified graph that contains users, items, entities, and relations, transforming the recommendation task into the likelihood prediction of the $(u, interactive, i)$ triple.
- **KGAT** [6]: A novel propagation-based model that utilizes an attention mechanism to distinguish the importance of neighbors in the CKG.

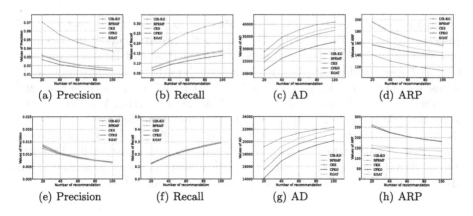

Fig. 4. Comparison results on (a-d) Last-FM dataset and (e-h) Amazon-book dataset.

Figure 4 shows the performance comparison of UIR-KG and other baseline approaches on the Last-FM and Amazon-book datasets. It is evident from the figure that UIR-KG achieves comparable performance with other comparative approaches in terms of accuracy metrics such as precision and recall on the Amazon-book dataset. However, on the Last-FM dataset, UIR-KG outperforms the closest competitor, KAGT, and obtains the highest values. Thus, it can be concluded that UIR-KG performs better than other related approaches in terms of recommendation accuracy, particularly on the Last-FM dataset. Furthermore, in terms of diversity metrics such as AD and ARP, UIR-KG outperforms other related approaches for both the Last-FM and Amazon-book datasets. These results indicate that our proposed UIR-KG approach can greatly enhance recommendation diversity while maintaining at least the same level of recommendation accuracy. Notably, UIR-KG shows significant improvement in ARP, which can reflect the ability to recommend long-tail items, indicating that it can recommend more long-tail items for target users. In summary, based on the experimental results, we can conclude that UIR-KG is a promising approach for recommendation systems, as it can significantly improve recommendation diversity while maintaining or even improving recommendation accuracy.

4.3   Ablation Study

To further validate the effectiveness of two core components in UIR-KG (i.e., popular constraint-based long-tail neighbor selector and knowledge-enhanced hybrid attention aggregator), we retain only one component individually in UIR-KG to generate two simplified versions. Then, we compare UIR-KG with two variations as well as the base version.

- **Base**: A basic GCN approach without using selector and aggregator.
- **Base(Sel)**: A simplified version that removes the popular constraint-based long-tail neighbor selector from UIR-KG.
- **Base(Agg)**: A simplified version that removes the knowledge-enhanced hybrid attention aggregator from UIR-KG.

Figure 5 displays the results of our ablation experiments on Last-FM and Amazon datasets. UIR-KG outperformed Base(Agg) in accuracy metrics (i.e., precision and recall) on both datasets, surpassed Base and Base(Sel) on Last-FM, and was roughly on par with Base and Base(Sel) on Amazon-book. In terms of diversity metrics (i.e., AD and ARP), UIR-KG outperformed Base and Base(Sel) on both datasets but performed similarly to Base(Agg) on AD and lower than Base(Agg) on ARP. This is because our sampling strategy captures the long-tail interests of target users but may also introduce noisy information. Meanwhile, the self-attention module integrates the user's historical interaction information into the user embedding, thereby enhancing user representation but also weakening some of the long-tail information introduced by sampling.

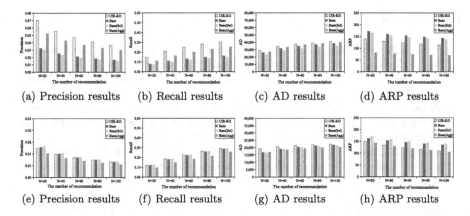

(a) Precision results   (b) Recall results   (c) AD results   (d) ARP results

(e) Precision results   (f) Recall results   (g) AD results   (h) ARP results

Fig. 5. Experimental results of ablation study on (a-d) Last-FM dataset and (e-h) Amazon-book dataset.

In conclusion, the popular constraint-based long-tail neighbor selector plays a greater role in improving recommendation diversity, while the knowledge-enhanced hybrid attention aggregator contributes more to enhancing recommendation accuracy.

5   Conclusions and Future Work

In this paper, we present a novel approach to improving long-tail recommendation diversity while maintaining recommendation accuracy by proposing a user-oriented interest representation. The approach involves selecting potential long-tail neighbors for users from the collaborative knowledge graph using a popular constraint-based long-tail neighbor selector. Then, a knowledge-enhanced hybrid attention aggregator is proposed, which combines relationship-aware attention and self-attention mechanisms to obtain user representations for long-tail interests. The experiments validate the proposed approach's effectiveness in long-tail recommendations. While the approach has the potential to address the cold-start problem in recommender systems, the challenge of recommending new users or items with little or no interaction records remains. Future work will focus on improving the sampling module to provide more neighbor information for cold-start users or items and finding ways to address the recommendation difficulty for new users or items in the system.

Acknowledgements. This work was supported in part by the National Science Foundation of China (No.61976109); Liaoning Revitalization Talents Program (No. XLYC2006005); The Scientific Research Project of Liaoning Province (No. LJKZ0963); Liaoning Province Ministry of Education (No. LJKQZ20222431); China Scholarship Council Foundation (No. 202108210173).

References

1. Park, Y.J., Tuzhilin, A.: The long tail of recommender systems and how to leverage it. In: Conference on Recommender Systems, RecSys 2008, New York, pp. 11–18. Association for Computing Machinery (2008). https://doi.org/10.1145/1454008.1454012

2. Zhang, Z., Kudo, Y., Murai, T., Ren, Y.: Improved covering-based collaborative filtering for new users' personalized recommendations. Knowl. Inf. Syst. **62**, 3133–3154 (2020). https://doi.org/10.1007/s10115-020-01455-2

3. Zhang, Z., Zhang, Y., Ren, Y.: Employing neighborhood reduction for alleviating sparsity and cold start problems in user-based collaborative filtering. Inf. Retrieval J. **23**, 449–472 (2020). https://doi.org/10.1007/s10791-020-09378-w

4. Zhang, Z., Dong, M., Ota, K., Kudo, Y.: Alleviating new user cold-start in user-based collaborative filtering via bipartite network. IEEE Trans. Comput. Soc. Syst. **7**(3), 672–685 (2020). https://doi.org/10.1109/TCSS.2020.2971942

5. Wang, Z., Lin, G., Tan, H., Chen, Q., Liu, X.: CKAN: collaborative knowledge-aware attentive network for recommender systems. In: Proceedings of the 43rd International ACM SIGIR Conference on Research and Development in Information Retrieval, SIGIR 2020, New York, pp. 219–228. Association for Computing Machinery (2020). https://doi.org/10.1145/3397271.3401141

6. Wang, X., He, X., Cao, Y., Liu, M., Chua, T.: KGAT: knowledge graph attention network for recommendation. In: Proceedings of the 25th ACM SIGKDD International Conference on Knowledge Discovery and Data Mining, KDD 2019, New York, pp. 950–958. Association for Computing Machinery (2019). https://doi.org/10.1145/3292500.3330989

7. Zhang, Z., Dong, M., Ota, K., Zhang, Y., Ren, Y.: LBCF: a link-based collaborative filtering for over-fitting problem in recommender system. IEEE Trans. Comput. Soc. Syst. **8**(6), 1450–1464 (2021). https://doi.org/10.1109/TCSS.2021.3081424

8. Zhang, Y., Cheng, D.Z., Yao, T., Yi, X., Hong, L., Chi, E.H.: A model of two tales: dual transfer learning framework for improved long-tail item recommendation. In: Proceedings of the Web Conference 2021, WWW 2021, New York, NY, USA, pp. 2220–2231. Association for Computing Machinery (2021). https://doi.org/10.1145/3442381.3450086

9. Yin, H., Cui, B., Li, J., Yao, J., Chen, C.: Challenging the long tail recommendation. Proc. VLDB Endowment **5**(9), 896–907 (2012). https://doi.org/10.14778/2311906.2311916

10. Zhang, Z., Dong, M., Ota, K., Zhang, Y., Kudo, Y.: Context-enhanced probabilistic diffusion for urban point-of-interest recommendation. IEEE Trans. Serv. Comput. **15**(6), 3156–3169 (2022). https://doi.org/10.1109/TSC.2021.3085675

11. Wan, Q., He, X., Wang, X., Wu, J., Guo, W., Tang, R.: Cross pairwise ranking for unbiased item recommendation. In: Proceedings of The Web Conference 2022, WWW 2022, New York, pp. 2370–2378. Association for Computing Machinery (2022). https://doi.org/10.1145/3485447.3512010

12. Rendle, S., Freudenthaler, C., Gantner, Z., Schmidt-Thieme, L.: BPR: Bayesian personalized ranking from implicit feedback. In: Proceedings of the Twenty-Fifth Conference on Uncertainty in Artificial Intelligence, UAI 2009, Arlington, Virginia, USA, pp. 452–461. AUAI Press (2009). https://doi.org/10.5555/1795114.1795167

13. Zhang, F., Yuan, N.J., Lian, D., Xie, X., Ma, W.Y.: Collaborative knowledge base embedding for recommender systems. In: Proceedings of the 22nd ACM SIGKDD International Conference on Knowledge Discovery and Data Mining, KDD 2016,

New York, pp. 353–362. Association for Computing Machinery (2016) . https://doi.org/10.1145/2939672.2939673

14. Ai, Q., Azizi, V., Chen, X., Zhang, Y.: Learning heterogeneous knowledge base embeddings for explainable recommendation. Algorithms **11**(9), 137 (2018). https://doi.org/10.3390/a11090137

Multi-Self-Supervised Light Graph Convolution Network for Social Recommendation

Yunsheng Zhou[1], Yinying Zhou[1], and Dunhui Yu[1,2](✉)

[1] College of Computer and Information Engineering, Hubei University, Wuhan 430062, China
202031112012017@stu.hubu.edu.cn, yumhy@163.com
[2] Education Informationization Engineering and Technology Center of Hubei Province,
Wuhan 430000, China

Abstract. Self-supervised learning is a novel paradigm between unsupervised and supervised learning, which can use the ground-truth samples automatically generated from the raw data for self-identification based contrastive learning and effectively alleviate the problem of data sparsity. Its ingenious combination with graph convolution network is a major breakthrough in the current recommendation field. However, mostly existing methods fail to fully consider the interference factors in social relations, fail to effectively improve the deviation of predicting scores only relying on the user and item representations, and fail to take into account the time cost of model training while improving the recommendation performance. To address these issues, this paper proposes a social recommendation framework based on multi-self-supervised light graph convolution network, named as MSR. Technically, MSR first utilizes the user-item sampling interaction graph to enhance the learning of user/item interaction representations, and then further enhances the learning of user social representations by fusing social relations through two-stage encoding. Finally, the interaction representations, social representations, and item bias are combined to predict scores. Our experiments on two real-world datasets LastFM and Douban-Book demonstrate that, compared with the current mainstream models, MSR improves the Precision@10 at least 2.35% and 10.41%, respectively, and reduces the time consumption at least 35.35% and 76.48%, respectively.

Keywords: self-supervised learning · social recommendation · graph convolution network

1 Introduction

With the rapid development of information technology and mobile Internet, people are facing the problem of information overload while enjoying the convenience and fast services brought by big data [1]. As an effective way to alleviate the problem of information overload, recommender systems have been widely used in e-commerce [2], social media [3] and other fields [4–6]. However, traditional recommender systems only rely on the user-item interactions and are susceptible to data sparsity and cold-start problems [7]. With the widespread application of social platforms, it is an effective way to alleviate the problem of data sparsity and cold start by deeply exploring existing social

X. Yang et al. (Eds.): ADMA 2023, LNAI 14179, pp. 356–370, 2023.
https://doi.org/10.1007/978-3-031-46674-8_25

relations and combining social relations with interactions. In real life, users usually prefer to get information from friends [8]. Therefore, social recommendation, which aims to incorporate social relations into recommender systems, has received more and more attention in recent years [9–11].

The core idea of social recommendation is to make full use of user-item interactions and social relations to mine the representations of users and items. However, some works [12, 13] fail to effectively deal with the importance of interactions and social relations when integrating social relations. Some works [14, 15] analyze the correlation between users from multiple dimensions to improve the recommendation performance, but also lead to a great increase in time complexity. Therefore, this paper proposes a multi-self-supervised social recommendation algorithm, which uses multi-feature self-supervised contrastive learning to alleviate data sparsity, uses two-stage encoding to integrate social relations, and introduces item bias to improve the deviation of predicting scores. In the end, the purpose of reducing the training time of the model is achieved while ensuring the recommendation performance.

2 Related Work

Classical Social Recommendation. Classical social recommendation methods mostly use matrix factorization (MF) to model user-item interactions to learn the representations of users and items, and then combined with social relations to enhance user representations. For example, CUNE-BPR [16] analyzes the similarity between users through network embedding, and then uses semantic social feedback to constrain MF; SocialIT [17] models users from the perspective of truster and trustee to obtain implicit similarity in trust. IF-BPR [18] adaptively merges different numbers of similar users as implicit friends of each user to reduce the adverse consequences of unreliable social relations. However, MF pays more attention to the direct interactions between objects, resulting in recommendation performance susceptible to data sparsity. In addition, recommendation can be divided into ratings prediction [19, 20] and items ranking [21, 22] according to evaluation metrics. In between, ratings prediction mostly relies on bias or friend representations to assist in predicting ratings, while items ranking only relies on user and item representations to predict scores. Therefore, most existing items ranking methods have a large deviation in predicting scores.

Graph Convolution Network in Social Recommendation. Recently, graph convolution network (GCN) [23] has achieved great success in the field of graph representation learning and its applications [24, 25]. Since user-item interactions can be naturally represented as a user-item bipartite graph structure, some research works [26–29] have propose a series of graph neural recommendation models for learning more accurate user/item representations. For example, GC-MC [30] applies graph neural network to recommendation task for the first time, and learns node representations based on user/item first-order neighbors. NGCF [31] combines graph neural network and collaborative filtering algorithm to encode high-order graph structure information into node representations through neighbor aggregation, which can alleviate the problem of data sparsity. LightGCN [32] abandons the complex nonlinear activation and transformation functions on the basis of NGCF, which significantly improvs the performance and becomes the most popular graph convolution network model at present.

In social recommendation, interactions and social relations are usually regarded as equally important, and then both are modeled simultaneously by GCN [33–35], and finally enhanced user representations are obtained through operations such as splicing and weight transformation. However, there are often more uncertainties in social relations than reliable user-item interactions. In addition, GCN does not perform well when labeled data are severely scarce, and too many propagation layers can lead to over-smoothing.

Self-Supervised Learning in Social Recommendation. Self-Supervised Learning (SSL) [36] is a deep learning method that uses the structure and feature information of the data samples to mine the information of the raw data itself. SSL can create supplementary views through uniform node/edge dropout, random feature/attribute shuffling/masking, and subgraph sampling using random walk for the purpose of reinforcing representation learning. Previous studies [37–41] have shown that SSL can make better use of the inherent information of the data than the supervised learning trained with sparse label information, and has achieved excellent results on a variety of tasks. This property of SSL makes it another way to alleviate the problem of data sparsity in recommender systems and successfully combined with GCN [42–45]. For example, SGL [43] uses three methods of node dropout, edge dropout and random walk to generate different views for the same node, so as to achieve the effect of enhancing the raw data; MHCN [46] uses hypergraphs to model complex correlations among users to enhance user representations; SEPT [47] utilizes two complementary views to improve the self-supervision signals, and combines the masked unlabeled synthetic views for contrastive learning of user representations. However, these methods mostly focus on the enhancement of user representations, ignoring the attention to items, and the construction and modeling of a large number of supplementary views seriously affect the speed of model training.

After researching the above mainstream recommendation methods, we find that they generally have four problems: 1) Failure to make reasonable use of social relationships, which leads to the introduction of too many interference factors; 2) Most of the focus is on users and not enough attention to items; 3) In the items ranking, it only relies on the user and item representations to predict scores, and there is a large deviation; 4) The time complexity of the model is too large, which affects the speed of model training.

3   MSR: Proposed Framework

3.1   Problem Formulation

Based on the user-item interactions and the user-user social relations, this paper recommends the most interesting items for the target user. Therefore, the following formal definitions are given for the problem in this paper. Suppose there are m users $U = \{u_1, u_2, \ldots, u_m\}$ and n items $I = \{i_1, i_2, \ldots, i_n\}$ in social recommender systems. The interactions of m users on n items form an interaction matrix $\mathbf{R} \in \mathbb{R}^{m \times n}$. If user u has consumed/clicked item $i, r_{u,i} = 1$, otherwise $r_{u,i} = 0$. The social relations between

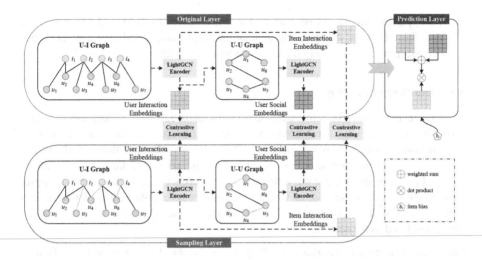

Fig. 1. Overview of the proposed multi-self-supervised social recommendation framework.

users form a social matrix $\mathbf{S} \in \mathbb{R}^{m \times m}$. If user u trusts/follows user v, $s_{u,v} = 1$, otherwise $s_{u,v} = 0$.

For the two behaviors of users, the user-item interactions are represented as an undirected bipartite graph $G_R = (U \cup V, \mathbf{R})$, and the user-user social relations are represented as a directed bipartite graph $G_S = (U, \mathbf{S})$. In order to perform multiple contrastive learning, this paper follows existing works [42,43] to perturb the raw graph with edge dropout at a certain probability ρ to construct two sampling graphs $\tilde{G}_R = (U \cup V, \tilde{\mathbf{R}} \subseteq \mathbf{R})$ and $\tilde{G}_S = (U, \tilde{\mathbf{S}} \subseteq \mathbf{S})$. The sampling graphs are regenerated at each iteration to fully learn the representations of users/items.

3.2 Model Framework

In order to obtain reliable representations of users/items and improve the recommendation performance, this paper proposes a multi-self-supervised social recommendation framework, called MSR. The overview of MSR is illustrated in Fig. 1.

The whole framework consists of three parts: original layer, sampling layer and prediction layer. Original layer and sampling layer have the same structure, and interactions and social relations are modeled respectively through two-stage encoding, and then prediction layer combines interaction representations, social representations and item bias to predict scores for items ranking. The major contributions are given as follows:

- We propose a multi-self-supervised social recommendation framework, which integrates social relations through two-stage encoding to reduce the influence of interference factors, and then alleviates data sparsity and strengthens representations through multi-feature self-supervised contrastive learning.
- Item bias is introduced to reflect the popularity of items, so as to improve the deviation that only depends on user/item representations when predicting scores, thereby improving the recommendation performance and reducing the model training time.

- The experimental results on two real-world datasets LastFM and Douban-Book show that MSR has better recommendation performance, and less time cost than the existing mainstream social recommendation algorithms.

3.3 Two-Stage Encoding

In social recommender systems, user-item interactions can truly reflect users' preferences, but user-user social relations contain many uncertain factors. One user trusting/following another does not mean similar interests between them, perhaps just because they are from the same school. Therefore, in order to make more rational use of social relations and reduce the influence of interference factors, this paper proposes a two-stage encoding method, in which only user-item reliable historical interactions are considered in the first stage, and the influence of social relations on user interest are only considered in the second stage.

User-Item Interaction Graph Encoding. Let d be the dimension of latent factors, $\mathbf{E}_U^{(0)} \in \mathbb{R}^{m \times d}$ and $\mathbf{E}_I^{(0)} \in \mathbb{R}^{n \times d}$ respectively denote the initial embedding matrices of users and items, L denote the number of propagation layers. The first stage takes user and item initial embeddings as input, and performs convolution operation on user-item interaction graphs G_R and $\tilde{G}_R$, respectively. Taking the user-item original interaction graph as an example, the output of each propagation layer obtained through the Light-GCN is defined as:

$$\mathbf{e}_u^{(l+1)} = \sum_{i \in N_u^R} \frac{1}{\left|\sqrt{N_u^R}\right| \left|\sqrt{N_i^R}\right|} \mathbf{e}_i^{(l)}, \mathbf{e}_i^{(l+1)} = \sum_{u \in N_i^R} \frac{1}{\left|\sqrt{N_i^R}\right| \left|\sqrt{N_u^R}\right|} \mathbf{e}_u^{(l)} \quad (1)$$

where $\mathbf{e}_u^{(l)}$ and $\mathbf{e}_i^{(l)}$ respectively denote the embedding of user u and item i after $l = \{1, 2, \ldots, L\}$ layers propagation, N_u^R denotes the set of items interacted by user u, N_i^R denotes the set of users interacting with item i, In order to speed up model training and improve performance, L2 normalization is performed on the output by each propagation layer. This process can be formulated as:

$$\mathbf{z}_u^{(l)} = \frac{\mathbf{e}_u^{(l)}}{\left\|\mathbf{e}_u^{(l)}\right\|_2}, \mathbf{z}_i^{(l)} = \frac{\mathbf{e}_i^{(l)}}{\left\|\mathbf{e}_i^{(l)}\right\|_2} \quad (2)$$

Then, we combine the initial embedding and the normalized outputs of all propagation layers to form the user/item interaction embedding:

$$\mathbf{e}_u^R = \mathbf{e}_u^{(0)} + \sum_{l=1}^{L} \mathbf{z}_u^{(l)}, \mathbf{e}_i^R = \mathbf{e}_i^{(0)} + \sum_{l=1}^{L} \mathbf{z}_i^{(l)} \quad (3)$$

Do the same for the user-item sampling interaction graph $\tilde{G}_R$ to get the corresponding user and item interaction embedding $\tilde{\mathbf{e}}_u^R$ and $\tilde{\mathbf{e}}_i^R$, respectively.

User-User Social Graph Encoding. The second stage takes the output of the first stage as input and performs convolution operation on user-user social graphs G_S and $\tilde{G}_S$, respectively. The difference is that we only combine the normalized output of each propagation layer to form the user social embedding:

$$e_u^S = \sum_{l=1}^{L} z_u^{(l)} \tag{4}$$

Do the same for the user-user sampling social graph $\tilde{G}_S$ to get the corresponding user social embedding $\tilde{e}_u^S$. Finally, the user interaction embedding and social embedding are linearly fused with a certain weight α to obtain the final user embedding:

$$e_u^C = (1 - \alpha)e_u^R + \alpha e_u^S \tag{5}$$

where $\alpha \in (0, 1)$ denotes the importance of social relations in constituting the final user embedding, and its value is determined in subsequent experiment.

3.4 Constructing Self-Supervision Signals

Some existing works [46,47]have shown that SSL can alleviate the problem of data sparsity and improve recommendation quality by learning with the automatically generated supervisory signals fromtherawdata. However, these methods only focus on users and does not comprehensively consider the impact of items. Therefore, this paper constructs three self-supervision signals $\langle E_u^R, \tilde{E}_u^R \rangle$, $\langle E_i^R, \tilde{E}_i^R \rangle$, and $\langle E_u^S, \tilde{E}_u^S \rangle$, and conducts contrastive learning from the perspectives of user, item, and social, respectively.

Taking the user interaction embeddings $\langle E_u^R, \tilde{E}_u^R \rangle$ as an example. We predict the semantically positive examples of a given user u in the user-item sampling interaction graph $\tilde{G}_R$ using the user interaction embeddings from G_R. Let y_{u+}^R denote the predicted probability of each user being the semantically positive example of user u in the corresponding view. It can be calculated as follows:

$$y_{u+}^R = \text{Softmax}\left(\phi\left(\tilde{E}_U^R, e_u^R\right)\right) \tag{6}$$

where $\phi(\cdot)$ is a discriminator function, which can be any similarity calculation formula (this paper selects the cosine operation). Finally, according to the probabilities, we select K user nodes with the highest reliability as the positive samples of user u.

$$P_{u+}^R = \left\{u_k \mid k \in \text{TopK}\left(y_{u+}^R\right)\right\} \tag{7}$$

Do the same for the remaining two self-supervision signals $\langle E_i^R, \tilde{E}_i^R \rangle$, and $\langle E_u^S, \tilde{E}_u^S \rangle$.

3.5 Multi-feature Contrastive Learning

The core idea of contrastive learning is to maximize the consistency between it and its positive samples and minimize the consistency between it and its negative samples for a given node. Taking the self-supervision signal $\langle E_u^R, \tilde{E}_u^R \rangle$ as an example, we follow the previous studies [37,43,47] to adopt InfoNCE [48] as the loss function:

$$L_{SSL}^U = -\frac{1}{|U|} \sum_{u \in U} \log \frac{\sum_{p \in P_{u+}^U} \psi\left(e_u^R, \tilde{e}_p^R\right)}{\sum_{p \in P_{u+}^U} \psi\left(e_u^R, \tilde{e}_p^R\right) + \sum_{v \in U / P_{u+}^U} \psi\left(e_u^R, \tilde{e}_p^R\right)} \tag{8}$$

where $\psi\left(\mathbf{e}_u^R, \tilde{\mathbf{e}}_p^R\right) = \exp\left(\phi\left(\mathbf{e}_u^R, \tilde{\mathbf{e}}_p^R\right)/\tau\right)$, τ is a coefficient used to amplify the discrimination effect, and we follow the previous work [47] to set it to 0.1. Finally, the contrastive learning loss functions of the three supervisory signals are combined according to the weights to obtain the overall loss function as:

$$L_{SSL} = \beta^U L_{SSL}^U + \beta^I L_{SSL}^I + \beta^S L_{SSL}^S \tag{9}$$

where β^U, β^I, and β^s denote the comparative learning weights of users, items, and social, respectively. For the simplicity of the model, we set $\beta^U = \beta^I$, which means that users and items are equally important.

Table 1. Experimental datasets statistics.

Dataset	#User	#Item	#Interaction	#Relation	Density
LastFM	1892	17632	92834	25434	0.278%
Douban-Book	13024	22347	792062	169150	0.272%

3.6 Model Prediction and Optimization

The previous works [33–35,43,46,47] only rely on the representations of users and items when predicting scores, resulting in large deviations. Therefore, this paper introduces item bias $\mathbf{b} \in \mathbb{R}^n$ to reflect the popularity of items. It can be understood that if an item is more popular, the user may rate the item higher. The final score prediction formula is defined as:

$$\hat{r}_{ui} = \left(\mathbf{e}_u^C\right)^T \mathbf{e}_i^R + b_i \tag{10}$$

Correspondingly, we add item bias regularization to the BPR loss function [49] to obtain the updated recommendation task loss function as:

$$L_R = \sum_{i \in N_u^R, j \notin N_u^R} -\log\sigma\left(\hat{r}_{ui} - \hat{r}_{uj}\right) + \lambda\left(\left\|\mathbf{E}_U^{(0)}\right\|_2^2 + \left\|\mathbf{E}_I^{(0)}\right\|_2^2 + \|\mathbf{b}\|_2^2\right) \tag{11}$$

Finally, we combine multi-feature contrastive learning as an auxiliary task on the recommendation task to improve the recommendation performance. The overall optimization objective of joint learning is defined as:

$$L_C = L_R + L_{SSL} \tag{12}$$

It should be emphasized that the model only optimizes Eq. (11) in the early stage, and optimizes Eq. (12) in the later stage, and the boundary is $\frac{1}{3}$ of the number of iterations.

4 Experiments

4.1 Experimental Settings

Datasets. Two real-world datasets:LastFM[1] and Douban-Book[2] are used in our experiments to evaluate MSR and baselines. The statistics of the experimental datasets are

---

[1] http://files.grouplens.org/datasets/hetrec2011/.
[2] https://github.com/librahu/HIN-Datasets-for-Recommendation-and-NetworkEmbedding.

shown in Table 1. We follow the previous researches [46,47] to remove records with rating less than 4 in Douban-Book, which consists of explicit ratings on a scale of 1–5, and then set the remaining ratings to 1. For the reliability of experimental results, we conduct 5-fold cross-validation in all experiments and present the average results.

Baselines. To verify the effectiveness of MSR, we compare it with the following five recent methods:

- **DiffNet++** [33]: is a GCN-based social recommendation method that can attentively aggregate user representations from three aspects: users themselves, influence diffusion and interest diffusion.
- **LightGCN** [32]: is the most popular graph convolution model at present that is a simplified model obtained by abandoning the nonlinear activation and transformation functions on the basis of NGCF.
- **MHCN** [46]: is a self-supervised social recommendation method based on hypergraph convolution network that utilizes hypergraphs to analyze complex correlations among users from three motifs of social, joint and purchase.
- **SEPT** [47]: is a multi-view-based self-supervised recommendation method that refines supervisory signals with supplementary views to improve recommendation performance.
- **SocialLGN** [34]: is a LightGCN-based social recommendation method that integrates social relations in the process of user interest propagation.

Metrics. For the top N items ranking evaluation, we select three widely used metrics *Precision*, *Recall* and *NDCG*, and the values are presented in percentage. Specifically, *Precision* and *Recall* reflect the hit rate of recommended items, and *NDCG* puts more emphasis on top-ranked items.

Settings. For a fair comparison, we use Adam to optimize these models with an initial learning rate of 0.001 and training batch size of 2000, and set the embedding dimension to 50, the regularization parameter to 0.001. The remaining parameters are set according to the best parameter values reported in the original papers and the results after multiple tunings. In Sect. 4.5, we analyze the impact of MSR's hyperparameters on recommendation performance, and use the best parameters in Sects. 4.2, 4.3 and 4.4. In addition, the test platform (Cloud server from Matpool[3]) is Intel(R) Xeon(R) Platinum 8260L CPU @ 2.30 GHz with 40 GB Memory.

4.2 Performance Comparison with Baselines

In order to verify the robustness of MSR and baselines, we conduct experiments on the complete test set and cold-start test set which only consists of users with less than 5 interaction records in the train set. The performance of each model on the two datasets is shown in Table 2 and Table 3. By analyzing the experimental results, we can draw the following conclusions:

---

[3] https://matpool.com/host-market/cpu.

- LightGCN, which only relies on user-item interactions, does not perform as well as MHCN, SEPT, and SocialLGN that integrate social relations in all metrics, indicating that using the user preference information implied in social relations can effectively improve the performance of recommendation model.
- DiffNet++ and SocialLGN, which model user-item interactions and social relations at the same time, do not perform as well as MHCN and SEPT that analyze social relations at deeper levels, indicating that rational use of social relationships is of great significance to improve the precision of recommendation model.

Table 2. Performance of MSR and baselines on the complete test set.

Method	LastFM			Douban-Book		
	P@ 10	R@ 10	N@ 10	P@ 10	R@ 10	N@ 10
DiffNet++	17.674	17.918	21.269	6.591	8.221	96.753
LightGCN	18.495	18.725	22.497	7.011	8.690	10.218
MHCN	19.716	19.960	24.369	8.254	10.670	12.730
SEPT	19.958	20.188	24.594	8.288	10.663	12.535
SocialLGN	18.631	18.862	22.643	7.211	8.720	10.375
MSR	**20.427**	**20.690**	**25.255**	**9.150**	**12.192**	**14.160**
Improv.	2.35%	2.49%	2.69%	10.40%	14.34%	12.96%

- Compared with baselines, the recommendation performance of MSR is significantly improved, mainly because MSR performs self-supervised comparative learning from three dimensions of user, item, and society, and improves the accuracy of score prediction by using user social representations and item bias.

Table 3. Performance of MSR and baselines on the cold-start test set.

Method	LastFM			Douban-Book		
	P@ 10	R@ 10	N@ 10	P@ 10	R@ 10	N@ 10
DiffNet++	0.787	8.934	5.424	1.164	8.759	5.297
LightGCN	0.952	9.524	5.575	1.382	9.029	5.403
MHCN	1.852	18.524	12.239	1.655	10.778	6.826
SEPT	1.602	16.024	9.236	1.696	10.963	7.247
SocialLGN	1.134	11.256	6.352	1.412	9.186	5.134
MSR	**1.852**	**18.524**	**12.255**	**2.038**	**13.540**	**8.882**
Improv.	15.61%	15.60%	32.69%	20.17 %	23.51%	22.56%

4.3   Effectiveness Analysis of Each Concept

In order to analyze the effectiveness of each concept in MSR, the basic model that only includes user and item supervisory signals is called MSR-UI, the model after introducing item bias is called MSR-B, and the model after integrating social relations is called MSR-S. During the experiments, we choose SEPT with the best performance in all the baselines as a reference, and compare the performance on P@10, number of iterations and time consumption respectively. The results are shown in Fig. 2. According to the results, we can draw a conclusion that each concept proposed in this paper has a positive contribution to the final recommendation effect of MSR, and with the introduction of item bias, the training time of MSR is greatly reduced.

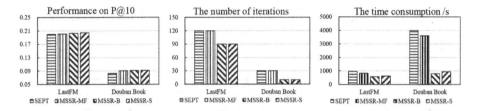

Fig. 2. Effectiveness analysis from different dimensions.

4.4   Analysis of Time Complexity

We summarize the time complexity of each model in Table 4. In order to make readers a more intuitive understanding of the training speed of each model, this paper also lists the approximate time (seconds) required for each model to perform 5-fold cross-validation on two datasets. According to the results, we can draw the following conclusions:

- LightGCN has the smallest time complexity, but its slower convergence speed leads to longer training time.
- The sharp increase in the training time of MHCN on Douban-Book is because MHCN analyzes the correlation between users from 10 dimensions and performs a large number of matrix operations. It is worth noting that the user scale of the Douban-Book is about 7 times that of LastFM.
- The time complexity of SEPT is close to MSR proposed in this paper. Since MSR converges faster after introducing item bias, the training time is greatly reduced.

where $\mathbf{A}_R$ and $\mathbf{A}_S$ respectively denote the symmetrically normalized matrix corresponding to the user-item interaction graph and the user-user social graph, and $|\mathbf{A}^+|$ is the number of non-zero elements in the corresponding matrix. In MHCN, the subscripts S, J and P represent three relation hypergraphs: social, joint and purchase; in SEPT, the subscripts F, J and UL represent three relation views: friend, sharing and unlabeled sampling.

Table 4. Time complexity comparison.

Model	Time Complexity	LastFM	Douban-Book
DiffNet++	$O\left(L\left(\left|\mathbf{A}_R^+\right|+\left|\mathbf{A}_S^+\right|\right)d\right)$	2678	19246
LightGCN	$O\left(L\left(\left|\mathbf{A}_R^+\right|\right)d\right)$	1816	13029
MHCN	$O\left(L\left(\left|\mathbf{A}_R^+\right|+\left|\mathbf{A}_P^+\right|+\left|\mathbf{A}_J^+\right|+\left|\mathbf{A}_S^+\right|\right)d+md^2\right)$	1093	16513
SEPT	$O\left(L\left(\left|\mathbf{A}_R^+\right|+\left|\mathbf{A}_F^+\right|+\left|\mathbf{A}_S^+\right|+\left|\mathbf{A}_{UL}^+\right|\right)d\right)$	959	3988
SocialLGN	$O\left(L\left(\left|\mathbf{A}_R^+\right|+\left|\mathbf{A}_S^+\right|+md\right)d\right)$	2245	16505
MSR	$O\left(L\left(\left|\mathbf{A}_R^+\right|+\left|\mathbf{A}_S^+\right|+\left|\tilde{\mathbf{A}}_R^+\right|+\left|\tilde{\mathbf{A}}_S^+\right|\right)d\right)$	**620**	**938**
Improv.		35.35%	76.48%

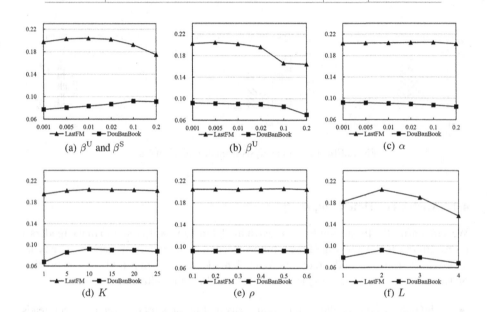

(a) β^U and β^S (b) β^U (c) α

(d) K (e) ρ (f) L

Fig. 3. Influence of different hyperparameters.

4.5   Sensitivity Analyses of Hyperparameters

There are six important hyperparameters used in MSR: β^U and β^I - the interaction signal weight, β^S - the social signal weight, α - the social fusion weight, K - the number of positive examples, ρ - the edge dropout rate, L - the number of propagation layers. During the experiments in this stage, unless otherwise specified, except for the parameters to be verified, the experiments in the later stage are all carried out on the best values of the previous stage, and the default values $\left(\beta^S=\alpha=0.001, K=10, \rho=0.3, L=2\right)$ are used for the rest of the parameters, and Take P@10 as the main judgment metric. The experimental results are shown in Fig. 3. It can be seen that contrastive learning and social relations as auxiliary means usually lead to better performance with small weights. MSR is not sensitive to the value of, and even taking a large value (e.g., 0.6) can create useful self-supervision signals, indicating that self-supervision has strong applicability. The number of propagation layers directly affects the performance of MSR. When L is small, the encoder cannot well aggregate high-order neighbor information, and conversely, it will cause the generalization of node features.

5 Conclusion

In order to solve the problems of ignoring interference factors when integrating social relations, deviations in predicting scores, and long training time in existing methods, this paper proposes a social recommendation framework based on multi-self-supervised light graph convolution network, named MSR. MSR enhances the representations of users and items through multi-feature self-supervised contrastive learning, integrates social relations through two-stage convolution operation, and finally introduces item bias for predicting scores. Extensive experiments demonstrate the effectiveness of MSR, which not only has better recommendation performance, but also has less time consumption.

References

1. Das, J., Banerjee, M., Mali, K., Majumder, S.: Scalable recommendations using clustering based collaborative filtering. In: 2019 International Conference on Information Technology (ICIT), pp. 279–284. IEEE (2019)
2. Pan, H., Zhang, Z.: Research on context-awareness mobile tourism e-commerce personalized recommendation model. J. Signal Process. Syst. **93**, 147–154 (2021)
3. Pongpaichet, S., Unprasert, T., Tuarob, S., Sajjacholapunt, P.: SGD-rec: a matrix decomposition based model for personalized movie recommendation. In: 2020 17th International Conference on Electrical Engineering/Electronics, Computer, Telecommunications and Information Technology (ECTI-CON), pp. 588–591. IEEE (2020)
4. Choudhury, S.S., Mohanty, S.N., Jagadev, A.K.: Multimodal trust based recommender system with machine learning approaches for movie recommendation. Int. J. Inf. Technol. **13**, 475–482 (2021)
5. Feng, S., Meng, J., Zhang, J.: News recommendation systems in the era of information overload. J. Web Eng. **20**, 459–470 (2021)
6. Zhang, C., Wu, X., Yan, W., Wang, L., Zhang, L.: Attribute-aware graph recurrent networks for scholarly friend recommendation based on internet of scholars in scholarly big data. IEEE Trans. Industr. Inf. **16**(4), 2707–2715 (2019)
7. Feng, J., Xia, Z., Feng, X., Peng, J.: RBPR: a hybrid model for the new user cold start problem in recommender systems. Knowl.-Based Syst. **214**, 106732 (2021)
8. Chen, T., Zhu, Q., Zhou, M., Wang, S.: Trust-based recommendation algorithm in social network. J. Software **28**(3), 721–731 (2018)
9. Li, W., et al.: Social recommendation using Euclidean embedding. In: 2017 International Joint Conference on Neural Networks (IJCNN), pp. 589–595. IEEE (2017)
10. Liu, G., Meng, K., Ding, J., Nees, J.P., Guo, H., Zhang, X.: An entity-association-based matrix factorization recommendation algorithm. Comput. Mater. Continua **58**(1), 101–120 (2019)
11. Zhang, T.W., Li, W.P., Wang, L., Yang, J.: Social recommendation algorithm based on stochastic gradient matrix decomposition in social network. J. Ambient Intell. Hum. Comput. **11**, 601–608 (2020)

12. Wu, L., Sun, P., Fu, Y., Hong, R., Wang, X., Wang, M.: A neural influence diffusion model for social recommendation. In: Proceedings of the 42nd International ACM SIGIR Conference on Research and Development in Information Retrieval, pp. 235–244 (2019)

13. Li, Y., Mu, K.: Heterogeneous information diffusion model for social recommendation. In: 2020 IEEE 32nd International Conference on Tools with Artificial Intelligence (ICTAI), pp. 184–191. IEEE (2020)

14. Yu, J., Yin, H., Li, J., Gao, M., Huang, Z., Cui, L.: Enhancing social recommendation with adversarial graph convolutional networks. IEEE Trans. Knowl. Data Eng. **34**(8), 3727–3739 (2020)

15. Fu, B., Zhang, W., Hu, G., Dai, X., Huang, S., Chen, J.: Dual side deep context-aware modulation for social recommendation. In: Proceedings of the Web Conference 2021, pp. 2524–2534 (2021)

16. Zhang, C., Yu, L., Wang, Y., Shah, C., Zhang, X.: Collaborative user network embedding for social recommender systems. In: Proceedings of the 2017 SIAM International Conference on Data Mining, pp. 381–389. SIAM (2017)

17. Pan, Y., He, F., Yu, H.: Social recommendation algorithm using implicit similarity in trust. Chin. J. Comput. **41**(1), 65–81 (2018)

18. Yu, J., Gao, M., Li, J., Yin, H., Liu, H.: Adaptive implicit friends identification over heterogeneous network for social recommendation. In: Proceedings of the 27th ACM International Conference on Information and Knowledge Management, pp. 357–366 (2018)

19. Ying, W., Yu, Q., Wang, Z.: Social recommendation combining implicit information and rating bias. In: 2021 IEEE 24th International Conference on Computer Supported Cooperative Work in Design (CSCWD), pp. 1087–1092. IEEE (2021)

20. Xu, C., Han, K., Gui, F., Xu, J.: Similarmf: a social recommender system using an embedding method. In: 2019 IEEE 21st International Conference on High Performance Computing and Communications; IEEE 17th International Conference on Smart City; IEEE 5th International Conference on Data Science and Systems (HPCC/SmartCity/DSS), pp. 1328–1334. IEEE (2019)

21. Wang, X., Lu, W., Ester, M., Wang, C., Chen, C.: Social recommendation with strong and weak ties. In: Proceedings of the 25th ACM International on Conference on Information and Knowledge Management, pp. 5–14 (2016)

22. Wang, M., Zheng, X., Yang, Y., Zhang, K.: Collaborative filtering with social exposure: a modular approach to social recommendation. In: Proceedings of the AAAI Conference on Artificial Intelligence, vol. 32 (2018)

23. Kipf, T.N., Welling, M.: Semi-supervised classification with graph convolutional networks. arXiv preprint arXiv:1609.02907 (2016)

24. Hamilton, W., Ying, Z., Leskovec, J.: Inductive representation learning on large graphs. In: Advances in Neural Information Processing Systems, vol. 30 (2017)

25. Li, R., Wang, S., Zhu, F., Huang, J.: Adaptive graph convolutional neural networks. In: Proceedings of the AAAI Conference on Artificial Intelligence, vol. 32 (2018)

26. Ying, R., He, R., Chen, K., Eksombatchai, P., Hamilton, W.L., Leskovec, J.: Graph convolutional neural networks for web-scale recommender systems. In: Proceedings of the 24th ACM SIGKDD International Conference on Knowledge Discovery & Data Mining, pp. 974–983 (2018)

27. Wu, S., Tang, Y., Zhu, Y., Wang, L., Xie, X., Tan, T.: Session-based recommendation with graph neural networks. In: Proceedings of the AAAI Conference on Artificial Intelligence, vol. 33, pp. 346–353 (2019)

28. Jin, B., Gao, C., He, X., Jin, D., Li, Y.: Multi-behavior recommendation with graph convolutional networks. In: Proceedings of the 43rd International ACM SIGIR Conference on Research and Development in Information Retrieval, pp. 659–668 (2020)

29. Wu, L., Yang, Y., Chen, L., Lian, D., Hong, R., Wang, M.: Learning to transfer graph embeddings for inductive graph based recommendation. In: Proceedings of the 43rd International ACM SIGIR Conference on Research and Development in Information Retrieval, pp. 1211–1220 (2020)
30. Berg, R.v.d., Kipf, T.N., Welling, M.: Graph convolutional matrix completion. arXiv preprint arXiv:1706.02263 (2017)
31. Wang, X., He, X., Wang, M., Feng, F., Chua, T.S.: Neural graph collaborative filtering. In: Proceedings of the 42nd International ACM SIGIR Conference on Research and Development in Information Retrieval, pp. 165–174 (2019)
32. He, X., Deng, K., Wang, X., Li, Y., Zhang, Y., Wang, M.: LightGCN: simplifying and powering graph convolution network for recommendation. In: Proceedings of the 43rd International ACM SIGIR Conference on Research and Development in Information Retrieval, pp. 639–648 (2020)
33. Wu, L., Li, J., Sun, P., Hong, R., Ge, Y., Wang, M.: Diffnet++: a neural influence and interest diffusion network for social recommendation. IEEE Trans. Knowl. Data Eng. 34(10), 4753–4766 (2020)
34. Liao, J., et al.: Sociallgn: light graph convolution network for social recommendation. Inf. Sci. 589, 595–607 (2022)
35. Song, C., Wang, B., Jiang, Q., Zhang, Y., He, R., Hou, Y.: Social recommendation with implicit social influence. In: proceedings of the 44th International ACM SIGIR Conference on Research and Development in Information Retrieval, pp. 1788–1792 (2021)
36. Liu, X., et al.: Self-supervised learning: generative or contrastive. IEEE Trans. Knowl. Data Eng. 35(1), 857–876 (2021)
37. Chen, T., Kornblith, S., Norouzi, M., Hinton, G.: A simple framework for contrastive learning of visual representations. In: International Conference on Machine Learning, pp. 1597–1607. PMLR (2020)
38. Han, T., Xie, W., Zisserman, A.: Self-supervised co-training for video representation learning. Adv. Neural. Inf. Process. Syst. 33, 5679–5690 (2020)
39. Lan, Z., Chen, M., Goodman, S., Gimpel, K., Sharma, P., Soricut, R.: Albert: a lite Bert for self-supervised learning of language representations. arXiv preprint arXiv:1909.11942 (2019)
40. Oord, A.V.D., Li, Y., Vinyals, O.: Representation learning with contrastive predictive coding. arXiv preprint arXiv:1807.03748 (2018)
41. Bachman, P., Hjelm, R.D., Buchwalter, W.: Learning representations by maximizing mutual information across views. In: Advances in Neural Information Processing Systems, vol. 32 (2019)
42. Jin, W., et al.: Self-supervised learning on graphs: deep insights and new direction. arXiv preprint arXiv:2006.10141 (2020)
43. Wu, J., et al.: Self-supervised graph learning for recommendation. In: Proceedings of the 44th International ACM SIGIR Conference on Research and Development in Information Retrieval, pp. 726–735 (2021)
44. You, Y., Chen, T., Sui, Y., Chen, T., Wang, Z., Shen, Y.: Graph contrastive learning with augmentations. Adv. Neural. Inf. Process. Syst. 33, 5812–5823 (2020)
45. Lee, D., Kang, S., Ju, H., Park, C., Yu, H.: Bootstrapping user and item representations for one-class collaborative filtering. In: Proceedings of the 44th International ACM SIGIR Conference on Research and Development in Information Retrieval, pp. 317–326 (2021)
46. Yu, J., Yin, H., Li, J., Wang, Q., Hung, N.Q.V., Zhang, X.: Self-supervised multi-channel hypergraph convolutional network for social recommendation. In: Proceedings of the Web Conference 2021, pp. 413–424 (2021)

47. Yu, J., Yin, H., Gao, M., Xia, X., Zhang, X., Viet Hung, N.Q.: Socially-aware self-supervised tri-training for recommendation. In: Proceedings of the 27th ACM SIGKDD Conference on Knowledge Discovery & Data Mining, pp. 2084–2092 (2021)
48. Hjelm, R.D., et al.: Learning deep representations by mutual information estimation and maximization. arXiv preprint arXiv:1808.06670 (2018)
49. Rendle, S., Freudenthaler, C., Gantner, Z., Schmidt-Thieme, L.: BPR: Bayesian personalized ranking from implicit feedback. arXiv preprint arXiv:1205.2618 (2012)

A Poisoning Attack Based on Variant Generative Adversarial Networks in Recommender Systems

Hongyun Cai[1,2], Shiyun Wang[1,2(✉)], Yu Zhang[1,2], Meiling Zhang[1,2], and Ao Zhao[1,2]

[1] School of Cyber Security and Computer, Hebei University, Baoding 071000, Hebei, China
[2] Key Laboratory on High Trusted Information System in Hebei Province, Hebei University, Baoding 071000, Hebei, China
wang_shiyun@163.com

Abstract. The emergence of poisoning attacks brings significant security risks to recommender systems. Injecting a well-designed set of fake user profiles into these systems can severely impact the quality of recommendations. However, existing attack models against recommender systems struggle to balance the imperceptibility and harmfulness of the generated fake user profiles. In this paper, we propose a novel poisoning attack model based on variant GAN. Firstly, we construct a candidate item set and divide each user rating behavior sequence into multiple shorter sequences. By identifying high-impact short sequences and calculating the proportion of high-impact sequences in each user rating sequence, we can determine the template profiles. Secondly, we design two different generators to simulate the fake user profiles, and employ a discriminator to enhance the generation of higher-quality fake user profiles. Finally, experimental results on three real datasets demonstrate that the proposed method outperforms state-of-the-art attacks in terms of attack performance, and the generated fake profiles are more difficult to detect when compared to the baselines.

Keywords: Recommender System · Poisoning Attack · Generative Adversarial Network

1 Introduction

Due to the rapid development of computer technology, massive information is flooded in networks. It becomes difficult for users to obtain valuable information. In order to alleviate the problem of information overload, recommender systems (RSs) are widely employed in various web services such as online shopping, video-on-demand. The openness of RSs brings convenience to users, but also gave rise to malicious motivations for attackers, i.e., shilling attacks. Shilling attack is also called data poisoning [1] or profile injection attack [2]. Various studies have

© The Author(s), under exclusive license to Springer Nature Switzerland AG 2023
X. Yang et al. (Eds.): ADMA 2023, LNAI 14179, pp. 371–386, 2023.
https://doi.org/10.1007/978-3-031-46674-8_26

shown that Collaborative Filtering, as the most commonly used recommendation algorithm, is vulnerable to data poisoning attacks. By injecting a number of well-crafted fake rating profiles into RS, attackers can promote or demote the recommendation ranking of target items [3]. Based on this situation, we focus on the study of poisoning attack based on Generative Adversarial Networks (GANs) [4] against recommender systems, hoping to provide insights and strategies for defending against malicious attackers from the opposite direction.

The previous data poisoning attacks focused on building hand-crafted fake profiles following different strategies such as random, popular, love-hate and bandwagon attacks [5], which limited the attack effect. In 2016, Li et al. [1] proposed the first ML-optimized attack on factorization-based RSs. Since then, novel data poisoning attacks or adversarial attacks appear successively, which can achieve better attack effect than the previous attacks [6]. For these state-of-the-art attacks, some focused on optimizing the defined attack objects and described the process of a poisoning attack as a bi-level optimization problem [7–9], the others focused on optimizing the profiles of fake users [10–13] by GAN to learn data distributions. The poisoning attacks using GAN in the RSs mainly improve the training process, because attackers can select favorable template users to generate attack profiles, which can help to improve the attack effect. However, few of these attacks paid attention to learning the complex data distribution from sparse and diverse user ratings, so that the rating features of generated fake user profiles are different from that of real user profiles, resulting in the generated attack profiles being detected.

To address the aforementioned limitations, we propose a novel poisoning attack framework named PAGUP. In PAGUP, we first group each user's rating item sequence into two different impact sequences, based on which, we select template user profiles. Then, a variant GAN is presented to generate fake user profiles. For illustrative purposes, we named the variant GAN as RDGAN. RDGAN consists of two generator neural networks and a discriminator neural network. The two generators are trained based on different purposes, i.e., attack efficiency and undetectability, respectively. This can help to achieve a better attack effect and make the generated fake profiles closer to the real profiles, indicating that the generated fake profiles are more difficult to be detected. The main contributions of our work are as follows.

(1) To select template user profiles, we construct the candidate item set and group each user's rating sequence into multiple short sequences. By calculating the overlapping degree between candidate item set and user short sequences, high impact sequences can be spotted, based on the proportion of high impact sequences in user rating sequence, the user profile can be determined whether it is a template.

(2) To generate more high-quality fake user profiles, we propose a variant GAN named RDGAN to learn the rating distribution of real users in RSs. By designing two different generators, we improve the convergence phenomenon of ratings on filler items given by attack users, so the rating distribution of

generated fake user profiles are more consistent with that of genuine user profiles.

(3) We evaluate our method on three real-world datasets. The experimental results show that the proposed poisoning attack model is superior to the state-of-the-art attacks in attack performance, and it is more difficult to be detected than the base-line attack models.

2   Related Work

In this section, we review some poisoning attacks and related technologies in RSs that are directly related to our work.

Poisoning Attacks in RSs. Data poisoning attacks have been studied in RSs for many years. Early data poisoning attacks, e.g., random attack, average attack and popular attack, mainly focused on hand-crafted fake user profiles, which have poor attack performance and are easily detected [14]. With the development of deep learning techniques, data poisoning attacks against specific types of RSs have been designed in recent years, e.g., graph-based RSs [15], and matrix-factorization-based RSs [1]. This kind of attack can have more serious attack efficiency on the specific recommendation algorithm, but it requires prior knowledge of RSs. In recent years, many researchers have applied GANs to generate diverse fake user profiles in RSs [10–12]. Lin et al. [10] proposed a novel Augmented Shilling Attack (AUSH), AUSH uses GANs to generate ratings for selected items instead of random generate ratings for selected items. Wu et al. [11] proposed a novel shilling attack called GOAT based on graph convolution and GAN (GOAT). They utilize CNN to consider correlations between items to smoothen ratings and generate fake user profiles. In some cases, these smooth ratings can affect recommendations. Zhang et al. [12] proposed the RecUP attack based on Generative Adversarial Network to generate ratings for all items in fake user profiles.

Generative Adversarial Networks (GANs). GAN is a popular deep learning model. The model consists of two basic neural networks, i.e., a generator network (G) and a discriminator network (D), where G is used to generate contents, and D is used to discriminate the generated contents. Inspired by zero-sum games, GAN treats the generation problem as an adversarial game between the discriminator and the generator [16]. The generator generates synthetic data from a given noise (usually a uniform or normal distribution), and the discriminator distinguishes the generator output from the real data. The former tries to generate data that is closer to the real data, and correspondingly, the latter tries to perfectly distinguish real data from generated data. The two networks progress in the confrontation and continue to fight after the progress. The data obtained by the generative adversarial network is becoming more and more perfect, so that the desired data (pictures, videos, etc.) can be generated.

3   The Framework of PAGUP

To improve the attack effectiveness and enhance the difficulty of detection, we propose a novel data poisoning attack against RSs, i.e., PAGUP. We first select some genuine rating profiles in a particular way as the templates, based on which, the presented variant GAN is trained. The framework of PAGUP is shown in Fig. 1, which mainly consists of two parts, i.e., template profiles selection and fake attack profiles generation based on RDGAN.

For easy discussions, let $R = [r_{ij}]_{|U| \times |I|}$ denote the user-item rating matrix. U and I denote the sets of users and items, r_{uj} is the rating provided by user u for item j.

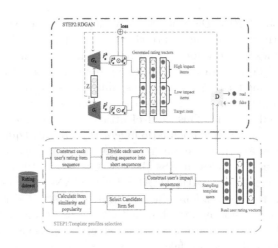

Fig. 1. The framework of PAGUP.

3.1   Threat Model

Threat model includes attack goal, attack prior knowledge, and the ability of manipulating the training data.

Attack goal: From the attack purpose, the poisoning attacks in RSs can be categorized into push attacks and nuke attacks. Push(nuke) attacks aim to increase(decrease) the probability that the target items will be recommended. Similar to the recent works [11], we only focus on push attacks for simplicity in this work. It is noted that nuke attacks can also be realized in the similar way.

Attack Prior Knowledge: Rating matrices based on RSs are pretty much universal in practice including the neighborhood-based RSs, the matrix-factorization-based RSs and the deep-learning based RSs. In view of the fact that full knowledge of the target RSs is hard to acquire, we only assume that the attacker knows the rating matrix of all users and items, but do not know the specific recommendation algorithms and parameters of the target recommender system.

Attack capability: Given the rating matrix $R = [r_{ij}]_{n \times m}$, $\alpha \times m \in N$ fake profiles are injected into the training matrix, where α refers to the attack ratio or attack size. Moreover, to simulate the real rating profiles, each fake user profile can rate at most σ filler items.

3.2 Template Profiles Selection

It is important to select some proper rating profiles as template profiles from real user rating profiles. Inspired by the idea of sequence recommendations [17], we group each user's ratings sequences according to their interaction habits (generally, users' behaviors are similar and related within a short time), so as to improve the invisibility of fake user profiles. As filler items and their ratings can reflect user's preferences, the first step in template profiles selection is to analyze the correlation between the target item and items rated by real users.

Definition 1 (User Rating Item Sequence, URIS). The rating item sequence of user $u \in U$ refers to an ordered series of items rated by user u in the ascending order of the corresponding rating time, which is denoted by

$$URIS_u = \{i_1^u, i_2^u, \ldots, i_s^u\} \tag{1}$$

where i_x^u represents the xth item rated by user u. User Rating Item Sequence (URIS) contains users' rating behaviors.

For $\forall u \in U$ and $\forall i \in URIS_u$, the pearson correlation coefficient between i and target item j is denoted as w_{ij} and calculated by

$$w_{ij} = \frac{\sum_{u \in U_{ij}} (r_{ui} - \overline{r_i})}{\sqrt{\sum_{u \in U_{ij}} (r_{ui} - \overline{r_i})^2} \sqrt{\sum_{u \in U_{ij}} (r_{uj} - \overline{r_j})^2}} \tag{2}$$

where $\overline{r_i}$ and $\overline{r_j}$ represent the average ratings of item i and item j. Let $U_{ij} = U_i \cap U_j$ represent the set of users who have rated both item i and item j.

Inspired by [18], we also need to calculate the popularity of each item rated by any user, which is denoted as τ_i and calculated by

$$\tau_i = \frac{|U_i|}{\sum_{\hat{i} \in I_u} |U_{\hat{i}}|} \tag{3}$$

Definition 2 (Candidate Item Set, CIS). The candidate item set refers to the set of items having a high similarity with the target item or having a high popularity, which is denoted by

$$CIS = \{i \in I : w_{ij} > \gamma \ or \ \tau_i > \rho\} \tag{4}$$

where γ and ρ represent the threshold for similarity and popularity, respectively. In this paper, we set γ to 0.6. The settings of ρ will be explained in the experiment.

376   H. Cai et al.

Definition 3 (User Short Sequence Set, USSS). For $\forall u \in U$ and $\forall i \in URIS_u$, we divide $URIS_u$ into multiple short sequences according to a time interval Δt, the set of these short sequences is referred as $USSS_u = \{S_1^u, S_2^u, \ldots\}$. Let i_x^u, i_{x+1}^u be any two adjacent items in $URIS_u$, and the corresponding rating time is t_x^u, t_{x+1}^u, respectively. If $t_{x+1}^u - t_x^u \leq \Delta t$, items i_x^u, i_{x+1}^u will be classified into the same short sequence; otherwise, i_{x+1}^u will belong to the next short sequence.

Next, for all short sequences in $USSS_u$, we divide them into high impact sequences and low impact sequences based on the extent of overlapping of each sequence with the candidate item set CIS. Let $OR_{u,i}$ be the extent of overlapping of the ith short sequence in $USSS_u$ with CIS, which is calculated by

$$OR_{u,i} = \frac{|CIS \cap S_i^u|}{|S_i^u|} \tag{5}$$

The short sequence S_i^u will be referred as a high impact sequence if $OR_{u,i} > \lambda$; otherwise, it means a low impact sequence. The threshold λ is the partition baseline, which is decided by extensive experiments. For $\forall u \in U$, we will regard its rating profile as a template profile if the proportion of high impact sequences in $USSS_u$ exceeds the predefined threshold (We set it to 0.25 in the experiment based on the work [12]).

3.3 Fake Attack Profiles Generation Based on RDGAN

The distribution characteristics of generated user profiles are not consistent with the real user characteristics, resulting in fake user profiles easily to be detected. Therefore, we present a variant GAN named RDGAN to generate fake user profiles. The details of RDGAN are shown in Figure 1. In the following, we will present the design details of the generators and discriminator. Z corresponding to the Gaussian distribution is the noise injected into the generator; G_h and G_l are two neural networks as generators, and D is the discriminator.

Generator. We inject Z into the two generators as input. Through two neural networks G_h and G_l, the noise Z is converted into $\hat{r}_u^h$, $\hat{r}_u^l$, an n-dimensional vector that represents a possible rating vector of the user u. Taking the training process of G_h as an example, we explain the output settings of generators. To simulate the diversity of real user profiles, We mimic the masking setting in CFGAN [19] to generate fake user profiles. The purpose of masking is to remove ratings on those items which were not rated by the real users. The high impact vectors are calculated by $\hat{r}_u^h \odot e_u^h$, where e_u^h is an n-dimensional vector and $\odot$ stands for element-wise multiplication. If u has rated the item i in user u's high impact sequence, $e_{u,i}^h = 1$; otherwise, $e_{u,i}^h = 0$. The process of G_l is the same as G_h. The generated rating vectors refer to the union of high impact vectors and low impact vectors, where rating items in high impact vectors are called high impact items and rating items in low impact vectors are called low impact items. The two generators share the same network structure, but they generate data for different purposes.

The training process of G_h can be understood as filling ratings for selected items. Our purpose is to generate profiles with high ratings that is beneficial to the realization of our attack target. Therefore, we applied the loss function in formula 6 to optimize parameters in G_h.

$$J^{Gh} = \sum_u \log\left(1 - D\left(\hat{r}_u^h \odot e_u^h\right)\right) + \mathcal{L}_{\text{Shill}} \tag{6}$$

where $\mathcal{L}_{\text{Shill}} = \sum_J (Q - \hat{x}_{uj})^2$. Q is a higher rating score in the rating matrix, e.g.,4 in the five-grade rating dataset. $\hat{x}_{uj}$ denotes the rating given to item j by the generated user u. J denotes the set of items in $\hat{r}_u^h \odot e_u^h$ whose user rating is not zero.

The training process of G_l can be understood as filling ratings for filled items. Therefore, our goal is to generate profiles that are as similar as possible to real rating profiles. Based on the ability of GAN to learn data distribution, the loss function is Adversarial loss, which is denoted by

$$J^{Gl} = \sum_u \log\left(1 - D\left(\hat{r}_u^l \odot e_u^l\right)\right) \tag{7}$$

The parameter settings are similar to those in Eq. 6.

Discriminator. The discriminator is used to distinguish fake user profiles from real user profiles and to help generators generate more realistic fake user profiles. In this paper, the input values of the discriminator are the generated profiles and the template user profiles. The neural network converts the input value into an n-dimensional vector, where n is the number of generated fake user profiles, and then we will calculate the loss value. The loss value of the discriminator is the sum of the loss that fake user profiles are classified as fake and real user profiles are classified as true. The objective function of Discriminator is as follows:

$$\begin{aligned} J^D &= -\mathbf{E}_{\mathbf{x} \sim P_{\text{data}}}[\log D(\mathbf{x})] - \mathbf{E}_{\hat{\mathbf{x}} \sim P_\phi}[\log(1 - D(\hat{\mathbf{x}}))] \\ &= -\sum_u \left(\log D\left(r_u\right) + \log\left(1 - D\left(\hat{r}_u^h \odot e_u^h\right) \cup \left(\hat{r}_u^l \odot e_u^l\right)\right)\right) \end{aligned} \tag{8}$$

After the training of two generators and the discriminator, we randomly select α users from template users, and generate $\alpha \times m$-dimensional fake profiles through the two generators G_h and G_l.

4 Experiments and Analysis

4.1 Experimental Setup

Dataset. We use three real-world datasets widely used in previous works, including ML-100K[1](MovieLens-100K), ML-1M (see Footnote 1) (MovieLens-1M), and FilmTrust[2] to evaluate our attack. The rating scores of FilmTrust are on the scale of 0.5–4 with a step 0.5, and we scale up these ratings into 1–5 for convenience. Since our dataset involves short sequence dividing, we randomly populate the FilmTrust dataset with interaction times. The descriptions for these datasets are listed in Table 1.

---

[1] https://grouplens.org/datasets/movielens/.
[2] https://guoguibing.github.io/librec/datasets.html.

Attack Baseline. We compare PAGUP with the classic shilling attacks and the advanced attacks, including Random attack [20], Average attack [20], Bandwagon attack [20], AUSH [10], RecUP [11], GOAT [12]. The details of these attacks are mentioned in Related Work.

Table 1. Descriptions of three experimental datasets

Dataset	User	Item	Ratings	Density
ML-100K	943	1682	100000	93.7%
ML-1M	6040	3706	1000209	95.53%
FilmTrust	1508	2071	35497	98.86%

Detection Method. Poisoning attack detection is usually considered as a binary classification problem. From the perspective of machine learning, poisoning attack detection techniques can be divided into supervised-learning attack detection, semi-supervised learning attack detection and unsupervised-learning attack detection. Since it is difficult to obtain labels in real scenarios, supervised attack detection and semi-supervised attack detection have great limitations. Here, we use two state-of-the-art attack detection methods based on unsupervised-learning: CBS [21] and UD-HMM [22], which can achieve excellent detection performance under various types of existing attacks.

Evaluation Metrics. We evaluate performances of our method from two aspects: attack efficiency and undetectability. For the evaluation of attack efficiency, we use the metric of $HR@N$, which refers to the proportion of the target item being recommended in top-N recommender lists of real users, and calculate by

$$HR@N = \frac{1}{|U|} \sum_{u \in U} \delta(u) \tag{9}$$

where $\delta(u)$ is an indicator function, it equals 1 if the target item occurs in the top-N recommendation list of user u; otherwise, it equals 0. To evaluate the undetectability of the proposed poisoning attack model, we use the metrics of Precision, Recall and F1-measure, which are described as

$$Precison = \frac{TP}{TP + FP} \tag{10}$$

$$Recall = \frac{TP}{TP + FN} \tag{11}$$

$$F1\_measure = \frac{2 \times Recall \times Precision}{Recall + Precision} \tag{12}$$

where TP is the number of fake profiles being identified correctly, FP is the number of real profiles misclassified as fake ones, FN is the number of fake profiles misclassified as real ones.

Parameter Settings. In the experiments, we use Pytorch1.11.0 as the deep learning framework and define a neural network framework for RDGAN. Adam is used for optimization. Two generators and one discriminator have three hidden layers, the output layer size is the number of all items for the generator.

The default parameters during the experiment are set as follows: The learning rate u_G, u_D of generators and discriminator is set to 0.0002; $step_G = 5$, $step_D = 2$; The epoch of adversarial training is 10. The internal time Δt is set to 30 mins following [17]. The attack size is set to 1–5%. The N in $HR@N$ is set to 10. We did not set an exact value for the profile size, because we use the real user profiles masking to generate a fake user profile. This process ensures the diversity of the generated fake user profiles.

The parameter λ is used to realize the division of high and low impact sequences. The attack size is set to 5% and the detection method is UD-HMM. We set λ to different values and observe the attack efficiency of PAGUP and detection performance of UD-HMM against PAGUP. Take the ML-100K dataset as example, the experiment results are shown in Fig. 2. As observed from Fig. 2, when λ is set to 0, all sequences are regarded as high impact sequences. In this case, only fake user profiles with high ratings will be generated, which are easy to be detected by UD-HMM. When λ is set above 0.7, there will be few high impact sequences that meet the conditions, that is, most sequences are considered as low similarity sequences, and the attack efficiency begins to gradually decrease. Therefore, we set $\lambda = 0.5$ on ML-100K datasets. For other datasets, the experimental results are similar.

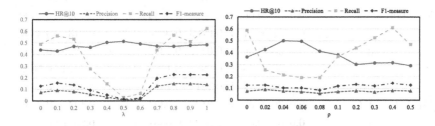

Fig. 2. Attack efficiency and detection performance under different λ and ρ on ML-100K dataset

The parameter ρ is used to determine how many popular items are selected as selected items. The reason for selecting popular items is that there are fewer items related to the target item when the target item has fewer interactions with users. If the target item has few similar items, it will lead to insufficient high impact profiles. The training of RDGAN will be degraded to only utilize the G_l generator for training. We know that popular items can also help target items to be recommended [18]. We consider both similar items and popular items to realize the generation of user behavior sequences. When the proportion of popular items selected is high, the training process of RDGAN will be degraded to only use G_h to generate false user profiles. Determining how many popular items to be chosen is a problem that needs to consider during the experiment. We injected fake user profiles with 5% attack size into the ML-100k dataset to observe the changes of attack efficiency and undetectability when different proportions of popular items were selected. UD-HMM is used to detect fake profiles. The experimental results are shown as Fig. 2. Based on the experimental results, ρ was set to 0.06 on ML-100K dataset, For other datasets, the experimental results are similar.

4.2   Performance Evaluation

Attack Performance. Our goal is to recommend the target item to as many users as possible and maximize $HR@10$. We first test the attack efficiency of PAGUP and compare it with other attack methods on three datasets. We randomly selected ten groups of generated fake user profiles to test their attack efficiency, and then calculated the average value of these data as our experimental data. The experimental results are shown in Table 2. Compared with other attacks, PAGUP has the best attack efficiency on ML-100K, ML-1M and FilmTrust. For three traditional shilling attacks, i.e., random attack, average

Table 2. HR@10 of seven attack methods under different attack sizes on three datasets

Dataset	Attack Size	Attack Method						
		Ran	Ave	Band	AUSH	RecUP	GOAT	PAGUP
ML-100K	1%	0.0659	0.0628	0.0687	0.0734	0.1956	0.1067	0.2480
	2%	0.0721	0.0750	0.0763	0.0844	0.2227	0.2134	0.3457
	3%	0.0735	0.0652	0.0842	0.1375	0.2448	0.3029	0.4793
	4%	0.0986	0.0802	0.0878	0.1760	0.2834	0.3424	0.5089
	5%	0.1229	0.1124	0.1212	0.1851	0.3327	0.3668	0.5378
ML-1M	1%	0.0176	0.0429	0.0126	0.0324	0.1048	0.1445	0.2631
	2%	0.0562	0.0854	0.0620	0.0851	0.1513	0.2587	0.3733
	3%	0.1082	0.1261	0.1030	0.1194	0.2838	0.2772	0.4310
	4%	0.1908	0.1562	0.1237	0.1556	0.2969	0.3911	0.4901
	5%	0.1754	0.1784	0.1811	0.1785	0.3652	0.3900	0.5014
FilmTrust	1%	0.0121	0.0121	0.0165	0.0723	0.1418	0.0866	0.2900
	2%	0.0134	0.0508	0.0508	0.0943	0.1916	0.1294	0.3987
	3%	0.0698	0.0565	0.0639	0.1559	0.2217	0.1942	0.4659
	4%	0.0732	0.0632	0.0689	0.1700	0.2257	0.2046	0.4953
	5%	0.0719	0.0722	0.0675	0.1905	0.2662	0.2255	0.5297

attack and bandwagon attack, these attacks do not work well on three datasets especially under a smaller attack size. As the filled items and their ratings in the fake profiles are randomly generated, which requires enough attack profiles to be able to manipulate the recommendation result. For AUSH, although it is a GAN-based attack, its performance was a little better than early shilling attacks. This is because, it directly fills filled items with real user ratings so that the attack performance may be weakened. The $HR@10$ of Goat and RecUP is higher than that of AUSH, but their attack efficiency is obviously inferior to that of PAGUP. Compared with the most advanced methods, $HR@10$ of PAGUP on the three datasets is always the highest. For example, it is improved by 17.1%, 13.06% and 26.36% when the attack size is 5%. Therefore, we can conclude that the attack performance of PAGUP outperforms the other six attacks.

Performance Against Attack Detection. Limited by the length of the paper, we only took the experimental results on the ML-100K dataset as an example, we applied two advanced unsupervised detection methods [21,22] to test the undetectability of different attack models. We randomly selected three groups of generated fake user profiles to inject real user profiles for detection. For each dataset, we conducted five repeated experiments. Data in Figs. 3, 4 is the average value of experimental results. The smaller the value of detection performance is, the better attack models are.

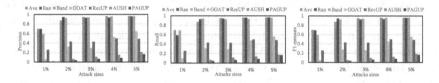

Fig. 3. Detection performance of CBS on ML-100K dataset

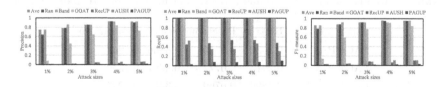

Fig. 4. Detection performance of UD-HMM on ML-100K dataset

We can observe that these two detection algorithms do not perform well in detecting fake user profiles generated based on PAGUP, indicating the advantages of PAGUP in generating fake profiles. The detection results of traditional attacks (e.g. Ave, Ran, Band) are higher than that of other comparison attack methods. This means that the fake profiles generated by traditional shilling attacks can be detected easily. In comparison, fake users generated by new poisoning attack models cannot be detected effectively. Note that some of the results shown in Figs. 3, 4 are zero, which means that the injected fake profiles are not identified by the corresponding detection method. Although under the CBS detection algorithm, when the attack size is 1%, detection metric against AUSH is better than that of PAGUP. However, with the increase of attack sizes, the malicious target of the attacker is gradually exposed, and the detection metric against PAGUP gradually becomes optimal. At the same time, it can be seen that the detection efficiency against AUSH is lower than that of GOAT and RecUP when using the CBS detection method. The detection method UD-HMM is based on user behavior characteristics, so the detection efficiency against new poisoning attacks is low. We also find that the detection efficiency under GOAT is higher than that of other attacks based on GAN. The reason may be that GOAT is a multi-target attack method, which is easier to be detected than other single-target attack methods.

Impact of Components in PAGUP. In this section, we evaluated how much each component of PAGUP contributed to the efficiencies of the attack. We removed or changed some components of PAGUP and investigated performance changes. $PAGUP_{Gh}$ means that the G_l component in RDGAN is removed, $PAGUP_{Gl}$ indicates that the G_h component in RDGAN is not considered and template profiles selection process is omitted, and $PAGUP_{ran}$ means that we randomly select template users to generate false user profiles. Take the ML-100K dataset and UD-HMM as example, experimental results of these variant methods and PAGUP are shown in Table 3. The $HR@10$ of PAGUP is higher than three variant methods, indicating that the template profiles selection process and two generators in PAGUP can help to improve the attack efficiency. Moreover, the detection results of UD-HMM is the lowest when detecting PAGUP. These results can show that the generated profiles by PAGUP are more difficult to be detected than those attack profiles generated by other three variants. Therefore, we can conclude that PAGUP is superior to three variant methods.

Table 3. Comparison of PAGUP and three variant methods on ML-100K dataset

Attack method	HR@10	Precision	Recall	F1-measure
$PAGUP_{Gl}$	0.3630	0.0750	0.5866	0.1264
$PAGUP_{Gh}$	0.2913	0.0777	0.4681	0.1269
$PAGUP_{ran}$	0.4806	0.0769	0.3404	0.1247
PAGUP	0.5378	0.0160	0.1064	0.0277

Performance of Convergence. In order to verify whether the training of GANs can converge when two generators are used for training, we experimentally show the changes of losses in the generators and discriminator during training. The change of losses in the generators and discriminator is shown in Fig. 5.

The losses of both the generators and discriminator gradually decrease as the number of iterations increases. For the discriminator, maximizing $\max V(D, G)$ is equivalent to $\min[-V(D, G)]$. Therefore, the training of the discriminator is convergence. The losses of the generators are gradually stable at 0.25 and 0.15, which indicates that the learning ability of the generator is improving.

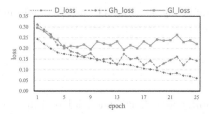

Fig. 5. Changes of losses in PAGUP on ML-100K dataset

The Rating Distribution of Generated Users Profiles and Genuine Data. PAGUP tries to make a trade-off between mimicking the genuine data and achieving effective attacks. To demonstrate the rating distribution characteristics of the fake user profiles generated in current advanced attack methods, we conduct comparison experiment between genuine data and generated data by different attack models. The results are shown in Fig. 6. As seen from Fig. 6, we found that the rating distribution of fake user profiles generated by AUSH is very similar to that of the real user. The reason is that AUSH only uses GANs to generate ratings for selected items. The ratings for other items are filled with ratings from real users. However, when the ratings of real users are filled directly, the filled items will affect the possibility of the target item being recommended, resulting in low attack efficiency. At the same time, we can see that the ratings generated by GAN in AUSH are between 3 and 4 on ML-100K, which is very different from that of generated profiles on ML-1M. The reason may be that AUSH samples based on users' ratings and the ratings number on ML-1M dataset is more, which causes the ratings on ML-1M scattered. RecUP and GOAT use GAN to generate all ratings of fake user profiles. Although their attack efficiency is good, the rating distribution of fake user profiles is concentrated, which may increase the possibility of fake profiles being detected. The rating distribution of fake user profiles in PAGUP is more similar with genuine data than that of RecUP and GOAT, which effectively improve the invisibility of fake user profiles. In addition, PAGUP do not choose ratings for filled items from real users, it uses the variant GAN to generate all ratings instead of choosing ratings for filled items from real users, which effectively reduces the impact of filled items on recommended results and improves the efficiency of the attack.

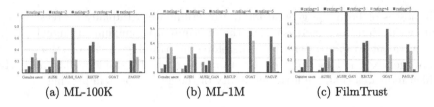

(a) ML-100K   (b) ML-1M   (c) FilmTrust

Fig. 6. The distribution of ratings generated by different poisoning attacks and Genuine users' rating distribution

5   Conclusion

The research of poisoning attack is beneficial to improve the security of recommender systems. In this paper, we propose a new framework for poisoning attacks based on variant GAN, which can create subtle fake user profiles and evade detection algorithms. Our core idea is to simulate the characteristics of real users based on the minimax game of GAN. We use two neural networks with different loss functions to achieve the attacker's goal. Instead of selecting all real rating profiles as templates to optimize fake user profiles, the proposed method can further enhance the attack effect by selecting template profiles beneficial to the target based on the internal characteristics of users. We conduct a large number of experimental studies to verify the effectiveness and superiority of PAGUP in attack performance and evading attack detection. In the future, we will continue to optimize the attack model and investigate how to improve the robustness of the recommender systems.

Acknowledgements. This research is funded by Science and Technology Project of Hebei Education Department (ZD2022105), Hebei Natural Science Foundation (F2020201023), and the high-level personnel starting project of Hebei University (521100221089).

References

1. Li, B., Wang, Y., Singh, A., Vorobeychik, Y.: Data poisoning attacks on factorization-based collaborative filtering. In: Proceedings of the 30th International Conference on Neural Information Processing Systems, Barcelona, Spain, December 5–10, 2016, pp. 1893–1901. Curran Associates Inc., Red Hook, NY, USA (2016)
2. Wang, S., Zhang, P., Wang, H., Yu, H., Zhang, F.: Detecting shilling groups in online recommender systems based on graph convolutional network. Inf. Process. Manage. **59**(5), 103031 (2022)
3. Wilson, D.C., Seminario, C.E.: When power users attack: assessing impacts in collaborative recommender systems. In: Seventh ACM Conference on Recommender Systems, pp. 427–430. ACM, New York, NY, USA (2013)
4. Aggarwal, A., Mittal, M., Battineni, G.: Generative adversarial network: an overview of theory and applications. Int. J. Inf. Manage. Data Insights **1**(1), 100004 (2021)

5. Wang, Z., Gao, M., Li, J., Zhang, J., Zhong, J.: Gray-box shilling attack: an adversarial learning approach. ACM Trans. Intell. Syst. Technol. **13**, 82:1-82:21 (2022)
6. Deldjoo, Y., Noia, T.D., Merra, F.A.: A survey on adversarial recommender systems: from attack/defense strategies to generative adversarial networks. ACM Comput. Surv. **54**(2), 35:1-35:38 (2022)
7. Fan, W., et al.: Attacking black-box recommendations via copying cross-domain user profiles. In: 37th IEEE International Conference on Data Engineering, pp. 1583–1594. IEEE (2021)
8. Song, J., Li, Z., Hu, Z., Wu, Y., Li, Z., Li, J., Gao, J.: PoisonREC: an adaptive data poisoning framework for attacking black-box recommender systems. In: 36th IEEE International Conference on Data Engineering, pp. 157–168. IEEE (2020)
9. Tang, J., Wen, H., Wang, K.: Revisiting adversarially learned injection attacks against recommender systems. In: Proceedings of the 14th ACM Conference on Recommender Systems, pp. 318–327. Association for Computing Machinery, New York, NY, USA (2020)
10. Lin, C., Chen, S., Li, H., Xiao, Y., Li, L., Yang, Q.: Attacking recommender systems with augmented user profiles. In: The 29th ACM International Conference on Information and Knowledge Management, pp. 855–864. Association for Computing Machinery, New York, NY, USA (2020)
11. Zhang, X., Chen, J., Zhang, R., Wang, C., Liu, L.: Attacking recommender systems with plausible profile. IEEE Trans. Inf. Forensics Secur. **16**, 4788–4800 (2021)
12. Wu, F., Gao, M., Yu, J., Wang, Z., Liu, K., Wang, X.: Ready for emerging threats to recommender systems? A graph convolution-based generative shilling attack. Inf. Sci. **578**, 683–701 (2021)
13. Wu, C., Lian, D., Ge, Y., Zhu, Z., Chen, E.: Triple adversarial learning for influence based poisoning attack in recommender systems. In: The 27th ACM SIGKDD Conference on Knowledge Discovery and Data Mining, pp. 1830–1840. Association for Computing Machinery, New York, NY, USA (2021)
14. Lin, C., Chen, S., Zeng, M., Zhang, S., Gao, M., Li, H.: Shilling black-box recommender systems by learning to generate fake user profiles. CoRR abs/2206.11433 (2022)
15. Fang, M., Yang, G., Gong, N.Z., Liu, J.: Poisoning attacks to graph-based recommender systems. In: Proceedings of the 34th Annual Computer Security Applications Conference, pp. 381–392. Association for Computing Machinery, New York, NY, USA (2018)
16. Goodfellow, I., et al.: Generative adversarial networks. Commun. ACM **63**(11), 139–144 (2020)
17. Feng, Y., et al.: Deep session interest network for click-through rate prediction. In: Proceedings of the Twenty-Eighth International Joint Conference on Artificial Intelligence, pp. 2301–2307. AAAI Press, Palo Alto, California, USA (2019)
18. Seminario, C.E., Wilson, D.C.: Attacking item-based recommender systems with power items. In: Eighth ACM Conference on Recommender Systems, pp. 57–64. ACM, New York, NY, USA (2014)
19. Chae, D., Kang, J., Kim, S., Lee, J.: CFGAN: A generic collaborative filtering framework based on generative adversarial networks. In: Proceedings of the 27th ACM International Conference on Information and Knowledge Management, pp. 137–146. Association for Computing Machinery, New York, NY, USA (2018)
20. Fang, M., Gong, N.Z., Liu, J.: Influence function based data poisoning attacks to top-n recommender systems. In: WWW '20: The Web Conference 2020, pp. 3019–3025. Association for Computing Machinery, New York, NY, USA (2020)

21. Zhang, Y., Tan, Y., Zhang, M., Liu, Y., Tat-Seng, C., Ma, S.: Catch the black sheep: Unified framework for shilling attack detection based on fraudulent action propagation. In: Proceedings of the Twenty-Fourth International Joint Conference on Artificial Intelligence, pp. 2408–2414. AAAI Press (2015)
22. Zhang, F., Zhang, Z., Zhang, P., Wang, S.: UD-HMM: an unsupervised method for shilling attack detection based on hidden Markov model and hierarchical clustering. Knowl. Based Syst. **148**, 146–166 (2018)

Label Correlation Guided Feature Selection for Multi-label Learning

Kai Zhang[1,2], Wei Liang[1,2], Peng Cao[1,2,3(✉)], Jinzhu Yang[1,2,3(✉)], Weiping Li[4], and Osmar R. Zaiane[5]

[1] Computer Science and Engineering, Northeastern University, Shenyang, China
[2] Key Laboratory of Intelligent Computing in Medical Image of Ministry of Education, Northeastern University, Shenyang, China
[3] National Frontiers Science Center for Industrial Intelligence and Systems Optimization, Shenyang 110819, China
caopeng@mail.neu.edu.cn, yangjinzhu@cse.neu.edu.cn
[4] School of Software and Microelectronics, Peking University, Beijing, China
[5] Alberta Machine Intelligence Institute, University of Alberta, Edmonton, AB, Canada

Abstract. Multi-label learning has received much attention due to its wide range of application domains. Multi-label data often has high-dimensional features, which brings more challenges to classification algorithms. Feature selection based on sparse learning is one of the most important approaches, which can select discriminative features to alleviate the curse of dimensionality. However, most of these methods do not consider feature redundancy and label correlation simultaneously. In this work, we propose a multi-label feature selection method that takes feature redundancy and label correlation into account. Specifically, the proposed method first divides features into groups according to feature correlation and then encourages competition within each group but relaxes competition between groups to eliminate redundant and irrelevant features. Moreover, we propose an approach to capture label correlation and exploit it to improve the feature selection. Finally, an iterative optimization algorithm is designed to obtain the feature weights for multi-label feature selection. Extensive experiments on various multi-label datasets demonstrate the superiority of the proposed method compared with some state-of-the-art multi-label feature selection methods.

Keywords: Feature selection · Multi-label learning · Label correlation · Feature redundancy · Sparse learning

1 Introduction

Conventional supervised learning mainly focuses on addressing the problem where an instance is associated with a single predefined label. Actually, one instance might be assigned multiple labels simultaneously in a wide range of

X. Yang et al. (Eds.): ADMA 2023, LNAI 14179, pp. 387–402, 2023.
https://doi.org/10.1007/978-3-031-46674-8_27

applications, such as image annotation [9], text categorization [14] and biological data analysis [2]. Traditional single-label learning methods can not effectively solve the problem of multiple labels, and thus multi-label learning framework came into being. Multi-label learning algorithms aim to learn a mapping from the feature space $X \subseteq \mathbb{R}^d$ to the label space $Y \subseteq \{0,1\}^m$, where d denotes the dimension of features and m denotes the number of predefined labels. However, multi-label learning data often has high-dimensional features, which brings major challenges, resulting in a significant increase in computational cost and memory storage, machine learning models being prone to overfitting and limiting the usage of these algorithms in real-world applications.

In order to deal with these challenges, a large number of multi-label dimensionality reduction methods have been proposed. These methods can be divided into feature extraction and feature selection. Specifically, feature extraction maps the original feature space to a lower dimensional space, while feature selection directly selects a small feature subset from all features. The difference between them is that the former creates new features that cannot be explained while the latter can preserve the meaning of the original features. Hence, feature selection in multi-label learning has attracted more attention. In recent years, multi-label feature selection methods based on sparse regression have been proven to be able to obtain the discriminant feature subset [7]. These methods generally employ a least squares regression model with a sparse regularization term and some other constraint conditions, and then the regression coefficients are used to measure the importance of each feature. However, most of these methods mainly focus on finding relevant features while ignoring redundant and irrelevant features, resulting in redundant and irrelevant features still existing in the final selected feature subset, which reduces the generalization performance of the model. Thus, considering the redundancy of features is necessary to ensure the diversity of features to improve multi-label classification performance. Additionally, these methods assume that labels are linearly related, which is not consistent with the complex correlations among labels in real scenarios. Label correlation is beneficial to the learning of correlated labels and provides useful information, especially when there are insufficient training examples for some labels. Therefore, how to learn and exploit the label correlation for improving feature selection is important and challenging in multi-label learning.

To address the above issues, we propose a Label Correlation guided Feature Selection for multi-label data (named LCFS), which attempts to leverage label correlation to alleviate the problems of high-dimensional features and select discriminative features across multiple labels. Specifically, we first divide features into groups based on the feature correlation, and then force competition within each group, but relax competition between groups to eliminate redundant and irrelevant features. In addition, we employ a simple RNN (SRN) to capture the non-linear label correlation better and then exploit the learned label correlation to select discriminative features across multiple labels. The contributions of this paper are summarized as follows:

- Developing an algorithm to learn non-linear label correlation and exploiting it to improve feature selection in a principled way.
- Presenting a new multi-label feature selection method that effectively integrates label correlation learning with feature redundancy analysis.
- Designing an effective iterative optimization algorithm to estimate the feature weights for multi-label feature selection.
- Conducting experiments on multiple benchmark datasets to demonstrate the effectiveness of the proposed LCFS.

2   Related Work

Our work is most related to multi-label feature selection based on sparse learning and label correlation technologies. Therefore, in this section, we briefly review some related work on these two aspects.

Over the past decades, plenty of feature selection methods have been proposed. Recently, multi-label feature selection based on sparse learning has received increasing attention due to its excellent performance and interpretability [10]. $\ell_{2,1}$-norm regularization can effectively guarantee the sparsity of feature coefficients, thus it has been widely used in feature selection. For example, Cai et al. [1] propose a feature selection method that combines feature manifold learning and sparse regularization, employing $\ell_{2,1}$-norm to capture the sparse terms of feature selection. Zhang et al. [20] introduce a multi-label feature selection framework, which maps the original feature space into a low-dimensional embedding to construct local and global label correlation, and incorporates $\ell_{2,1}$-norm regularization to search for discriminant features. However, these methods based on sparse regression model only lay emphasis on the discriminative features while neglecting the correlations among different features. This may trigger the phenomenon that some strongly correlated features tend to be selected together, leading to information redundancy and degradation in performance. To alleviate this phenomenon, we consider feature redundancy by clustering strongly related features into the same group and then imposing sparse within the group to reduce redundant information.

In multi-label learning, labels are commonly related and interdependent. A common approach is calculating the co-occurrence frequency or cosine similarity of labels in output space. For example, Chou et al. [5] aim to capture the relations between labels by calculating a co-occurrence statistical matrix. Zan et al. [22] utilize cosine similarity to calculate label correlation and then exploit it to guide feature selection. However, most of these methods obtain the second-order label correlation by calculation, which fail to fully mine the label information, resulting in limited performance. In order to better model label correlation, some graph-based methods are proposed to model high-order label correlation. For example, Huang et al. [8] learn an embedding matrix from the predefined label correlation graph via graph embedding to model label correlation. Chen et al. [3] introduce a label-aware graph representation learning to learn more reliable and discriminative graph feature representation. A multi-label classification algorithm, named ML-GCN [4], is proposed to leverage the advantages

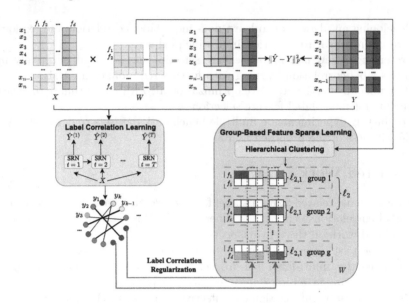

Fig. 1. The architecture of our proposed LCFS.

of graph convolutional networks to capture correlations among labels. However, most graph-based methods have a common drawback: graph structure for modeling the label correlation is pre-defined via mining their co-occurrence patterns, which hinders the capability of methods in modeling more accurate correlations via an end-to-end way under supervision. To address the above problem, we propose to consider the high-order label correlation as a sequence learning way by taking advantage of the internal memory characteristic in SRN.

3   Method

In this section, we present our proposed LCFS algorithm in detail. We first introduce an overview and preliminary of our proposed method. Then we discuss how to model the discriminant and sparsity of features. Next, we introduce how to learn and incorporate the label correlations into our multi-label feature selection model. Finally, we provide an iterative update algorithm to solve the optimization problem of the proposed LCFS.

3.1   Overview

The architecture of our proposed method called LCFS is shown in Fig. 1. It mainly consists of two components: (*i*) Group-Based Feature Sparse Learning for selecting the relevant features while reducing feature redundancy. (*ii*) Label Correlation Learning for capturing the label correlation to improve the performance of multi-label feature selection.

Table 1. Description of Notations

Notations	Meanings
$\mathcal{X}$	Input space with d dimensions $\mathcal{X} \in \mathbb{R}^d$
$\mathcal{Y}$	Set of m predefined labels $\mathcal{Y} = \{y_1, y_2, ..., y_m\}$
X	Feature matrix of all training instances $X = \{x_i \mid 1 \leqslant i \leqslant n\} \in \mathbb{R}^{n \times d}$
Y	Label matrix of all training instances $Y = \{Y_i \mid 1 \leqslant i \leqslant n\} \in \mathbb{R}^{n \times m}$
$\mathcal{D}$	Multi-label training set $\mathcal{D} = \{(x_i, Y_i) \mid 1 \leqslant i \leqslant n\}$
W	Weight matrix of global features $W \in \mathbb{R}^{d \times m}$
W^g	Weight matrix of the g-th group features $W^g \in \mathbb{R}^{d \times m}$
$W_{i\cdot}, W_{\cdot j}$	The i-th row of W, the j-th column of W
$Tr(\cdot)$	The trace of a matrix
$\|\cdot\|_F, \|\cdot\|_2, \|\cdot\|_{2,1}$	The Frobenius norm, ℓ_2-norm and $\ell_{2,1}$-norm of a matrix

3.2 Preliminary

The correlations between different features are simply ignored, resulting in some strongly correlated features tending to be selected or deselected together. Related features may have similar properties, revealing redundant or overlapping information. Therefore, we propose the LCFS algorithm to eliminate redundant information by enforcing the sparsity of strongly correlated features. The proposed LCFS is a multi-label feature selection method based on the sparse regression model. The common objective function of sparse regression is defined as follows:

$$\min_{W} \ \|XW - Y\|_F^2 + R(W) \tag{1}$$

$R(W)$ is a regularization term to impose sparsity on W. W is the weight matrix of features. Each row of W reflects the importance of the corresponding feature in the dataset, which can be used for feature selection [1]. In addition, the important notations used in this paper are summarized in Table 1.

3.3 Group-Based Feature Sparse Learning

In order to reduce feature redundancy, the proposed LCFS first divides features into G groups and then forces sparse within each group to eliminate highly correlated features. In this paper, the popular FINCH Clustering method [16] is adopted to divide features into groups according to their cosine similarity due to the effectiveness and parameter-free property of FINCH.

After the features are divided into groups, our objective is to achieve feature sparsity within each group. This can be formulated as solving an exclusive group lasso problem [13], which encourages competition within each group but relaxes competition between groups. To be specific, W^g is the weight of the feature of the g-th group and $W^g = diag(I_g) \times W$, where $diag(I_g)$ is a diagonal matrix and $I_g \in \{0,1\}^d$ is the group index indicator such that $I_{g,i} = 1$ if i-th feature belongs to group g, otherwise $I_{g,i} = 0$. Mathematically, an $\ell_{2,1}$-norm penalty is first imposed on each W^g for joint feature sparsity across m different class labels. Then, an ℓ_2-norm penalty is imposed on the inter-group level for non-sparsity. Thus, the exclusive group lasso regularization is formulated as $\sum\limits_{g=1}^{G} \|W^g\|_{2,1}^2$.

3.4   Label Correlation Learning

In multi-label learning, different labels are typically not independent but inherently correlated. Therefore, it is important to learn and exploit label correlation to improve multi-label feature selection. Recurrent Neural Network (RNN) is a class of neural network model commonly used to solve sequence prediction problems. In order to capture the correlation among labels, we use the SRN [11] model, which is a basic variation of RNN. It can iteratively learn the label correlation through its memory structure. Formally, label correlation learning is formulated as a sequence prediction problem as follows:

$$\hat{Y}^{(1)} = \sigma(XU) \tag{2}$$

$$\hat{Y}^{(t)} = \sigma\left(XU + \hat{Y}^{(t-1)}V\right) \tag{3}$$

where $\hat{Y}^{(t)}$ denotes the output, $U \in \mathbb{R}^{d\times m}$ is used to transform the feature vector into the output space, d is the dimension of data, m is the number of labels and $V \in \mathbb{R}^{m\times m}$ is used to transform the output of the previous iteration into the same output space as the output of U. We set the iteration number of the SRN layer as T. In the first iteration, the sequence learning produces a prediction based on Eq. (2) without considering other labels. From the second iteration, the sequence learning begins to exploit the output of the previous iteration to make better predictions. Particularly, the memory term $\hat{Y}^{(t-1)}V$ in Eq. (3) serves as the model for label correlation by taking in the previous output and transforming it to the same output space as the output of U. Therefore, the final prediction $\hat{Y}^{(T)}$ is obtained through T iterations, and the label correlation is captured by V.

Then, we exploit the learned label correlation by assuming that strongly correlated labels share more features than weakly correlated labels. That is, if label y_i and label y_j have a strong correlation, the corresponding weight vectors $W_{\cdot i}$ and $W_{\cdot j}$ will be similar, and vice versa. Thus, label correlation guided regularization is formulated as:

$$\sum_{i,j=1}^{m} V_{ij}\|W_{\cdot i} - W_{\cdot j}\| = Tr(WFW^T) \tag{4}$$

where $F = V^* - V$ indicates the $m \times m$ label Laplacian matrix of V, V^* is the diagonal matrix and $V_{i,i}^* = \sum_{j=1}^{m} V_{i,j}$ and $V_{i,j}$ is the label correlation between label y_i and label y_j.

In summary, by combining label correlation regularization with exclusive group lasso regularization, our objective function can be written as:

$$\min_{W} \|XW - Y\|_F^2 + \alpha \sum_{g=1}^{G} \|W^g\|_{2,1}^2 + \beta \mathrm{Tr}(WFW^T) \tag{5}$$

where α and β are trade-off parameters to control the contribution of sparsity and label correlation to the multi-label feature selection, respectively.

3.5 Optimization Solution

Due to the $\ell_{2,1}$-norm regularization term on W^g, the objective function of LCFS in Eq. (5) is not smooth. As a result, it is difficult to directly obtain the closed solution of the optimization problem. Therefore, we propose to solve the optimization problem by an iterative optimization algorithm as described below.

The derivative of $\|XW - Y\|_F^2$ w.r.t. W is:

$$\frac{\partial \|XW - Y\|}{\partial W} = 2X^T(XW - Y) \tag{6}$$

$\alpha \sum_{g=1}^{G} \|W^g\|_{2,1}^2$ is an exclusive group lasso penalty term, which can be reformulated as:

$$\alpha \sum_{g=1}^{G} \|W^g\|_{2,1}^2 = \alpha \sum_{g=1}^{G} (\sum_{k=1}^{d} \|W_{k\cdot}^g\|_2) \|W^g\|_{2,1} = \alpha \sum_{g=1}^{G} \sum_{k=1}^{d} \|W_{k\cdot}^g\|_2^2 \frac{\|W^g\|_{2,1}}{\|W_{k\cdot}^g\|_2}$$

$$= \alpha Tr(\sum_{g=1}^{G} \sum_{k=1}^{d} (W_{k\cdot}^g)^T \frac{\|W^g\|_{2,1}}{\|W_{k\cdot}^g\|_2} W_{k\cdot}^g) = \alpha Tr(W^T Q W) \tag{7}$$

where $Q \in \mathbb{R}^{d \times d}$ is a diagonal matrix and its diagonal element $Q_{i,i}$ is:

$$Q_{i,i} = \sum_{g=1}^{G} \frac{I_{g,i} \|W^g\|_{2,1}}{\|W_i^g\|_2 + \varepsilon} \tag{8}$$

where ε is a very small constant. As mentioned in Sect. 3.1, $I_g \in \{0,1\}^d$ is the group index indicator such that $I_{g,i} = 1$ if i-th feature belongs to group g, otherwise $I_{g,i} = 0$. Thus, the derivative of $\alpha \sum_{g=1}^{G} \|W^g\|_{2,1}^2$ w.r.t. W is:

$$\frac{\partial \alpha \sum_{g=1}^{G} \|W^g\|_{2,1}^2}{\partial W} = \frac{\partial \alpha Tr(W^T Q W)}{\partial W} = 2\alpha Q W \tag{9}$$

The derivative of $\beta \mathrm{Tr}(WLW^T)$ w.r.t. W is:

$$\frac{\partial \beta \mathrm{Tr}(WLW^T)}{\partial W} = 2\beta W F \tag{10}$$

According to Eqs. 6, 9 and 10, the solution of W can be obtained by setting the derivative of our objective function w.r.t. W to zero. We have:

$$2X^T(XW - Y) + 2\alpha QW + 2\beta LW = 0$$
$$\Rightarrow (X^T X + \alpha Q)W + W\beta F = X^T Y \tag{11}$$

Since Q is related to W, it is difficult to solve for W directly. To address this problem, an iterative method is presented to solve it. To be specific, we obtain Q with W and obtain the optimal solution until the iterative optimization converges. Equation (11) is a *Sylvester* equation with the form of $AW + WB = C$, where $A = X^T X + \alpha Q$, $B = \beta F$ and $C = X^T Y$. For solving the mathematical issue, several existing methods [15, 18] can be employed to obtain the numerical solution w.r.t. W. In this paper, we use the *Lyapunov* function to obtain the solution of W. Generally, the optimization process of LCFS can be summarized as Algorithm 1. The procedure repeats until convergence of the algorithm. Finally, the importance of each feature is evaluated based on the value of $\|W_{i\cdot}\|_2$ $(1 \leqslant i \leqslant d)$.

Algorithm 1. The optimization of LCFS

Input: Multi-label training set $\mathcal{D} = \{(x_i, Y_i) | 1 \leqslant i \leqslant n\}$, trade-off parameters α, β
Output: Weight matrix W
1: Perform clustering to group features into G groups
2: Obtain label correlation matrix V by label correlation learning
3: Construct label Laplacian matrix F according to V
4: Initialize W, Q
5: **repeat**
6:     Compute Q according to Eq. (8)
7:     Update W by solving Eq. (11)
8: **until** convergence
9: Obtain W for feature selection

---

4   Experiment

4.1   Experimental Setup

Datasets. To thoroughly verify the effectiveness of the proposed algorithm, eight multi-label datasets from different domains are employed in experiments. These datasets are available online at Mulan [17] and are commonly used for multi-label learning. Among these datasets, the number of instances is from 593 to 13770, the number of labels is from 5 to 164 and the number of features is from 72 to 1449. Therefore, the selected datasets have a certain representativeness. The detailed properties of datasets are presented in Table 2.

Table 2. Description of the multi-label classification datasets.

Datasets	Instances	Features	Labels	Domain
Emotions	593	72	6	Music
Birds	645	260	19	Audio
Medical	978	1449	45	Text
Yeast	2417	103	14	Biology
Image	2000	294	5	Image
Enron	1702	1001	53	Text
Corel16k1	13770	500	153	Image
Corel16k2	13760	500	164	Image

Evaluation Metrics. Six widely used multi-label evaluation metrics are employed in the experiment, including *Macro-F1*, *Micro-F1*, *Hamming loss*, *Ranking loss*, *One-error* and *Average precision* (*AP*). Detailed evaluation metric definitions can be found in [12,19]. These metrics can evaluate the performance of multi-label algorithms from various aspects. For *Macro-F1*, *Micro-F1* and *Average precision*, larger values indicate better performance, while for the other metrics, smaller values indicate better performance. In addition, ten-fold cross-validation is used in our experiments.

Compared Algorithms. Four multi-label feature selection methods are selected to compare. All of them are based on the sparse learning, including GLFS [22], MDFS [20], MSSL [1] and MIFS [10]. The above methods are used to select features and then ML-KNN ($K = 10$) [21] is used as the classifier to evaluate the performance of the selected features.

Hyper-parameters. For a fair comparison, we tune the regularization parameters for all methods by a grid-search strategy from $\{10^{-3}, 10^{-2},..., 10^{3}\}$. Moreover, we set the number of selected features as $\{10, 20, 30,...,100\}$ for all datasets except Emotions dataset since it doesn't have enough features. For Emotions dataset, the number of selected features is $\{5, 10, 15,..., 50\}$. When tuning parameters, we select different numbers of features as above and the average *AP* is recorded to determine the optimal parameters.

4.2 Experimental Result

To evaluate the overall performance of our proposed method LCFS, we compare it with four state-of-the-art multi-label feature selection methods on eight public datasets in terms of six widely used metrics. Furthermore, we choose the top-k features ($k \in \{10, 20, ..., 100\}$) as the optimal feature subsets and record the average result with 10 groups of feature subsets. The detailed experimental results are shown in Tables 3 and 4, where the best result among all the methods on

Table 3. Comparison results of multi-label feature selection methods on eight datasets in terms of *Macro F*1, *Micro F*1 and *Average precision* (Higher is better).

Method	*Macro F*1 ↑								Rank
	Emotions	Birds	Medical	Yeast	Image	Enron	Corel16k1	Corel16k2	
LCFS	**0.6178**	0.1322	**0.3868**	**0.4390**	**0.5912**	**0.6624**	**0.0092**	**0.0081**	1.13
GLFS	0.5671	0.1284	0.3615	0.3920	0.5858	0.6372	0.0083	0.0068	2.75
MDFS	0.5915	0.1270	0.3608	0.4227	0.5843	0.6518	0.0033	0.0059	3.38
MIFS	0.5918	**0.1392**	0.3515	0.4253	0.5306	0.6139	0.0080	0.0068	2.88
MSSL	0.5526	0.1265	0.3330	0.3911	0.5329	0.5970	0.0058	0.0052	4.75

Method	*Micro F*1 ↑								Rank
	Emotions	Birds	Medical	Yeast	Image	Enron	Corel16k1	Corel16k2	
LCFS	**0.6198**	**0.3051**	**0.7047**	**0.6379**	**0.5871**	**0.6375**	**0.0287**	**0.0345**	1.00
GLFS	0.5843	0.2941	0.6567	0.6139	0.5794	0.6103	0.0253	0.0287	2.75
MDFS	0.6025	0.2754	0.6951	0.6357	0.5784	0.6249	0.0217	0.0203	2.88
MIFS	0.5979	0.2849	0.6330	0.6263	0.5238	0.5939	0.0238	0.0231	3.50
MSSL	0.5683	0.2579	0.6281	0.6068	0.5276	0.5795	0.0193	0.0151	4.88

Method	*Average precision* ↑								Rank
	Emotions	Birds	Medical	Yeast	Image	Enron	Corel16k1	Corel16k2	
LCFS	**0.6132**	**0.3017**	**0.2894**	**0.6092**	**0.6377**	**0.2442**	**0.0333**	**0.0337**	1.00
GLFS	0.6007	0.2470	0.2549	0.5874	0.6291	0.2239	0.0307	0.0295	2.75
MDFS	0.6001	0.2589	0.2087	0.6065	0.6297	0.2400	0.0295	0.0266	3.13
MIFS	0.5979	0.2628	0.2440	0.6003	0.5848	0.2047	0.0305	0.0284	3.38
MSSL	0.5830	0.2412	0.2145	0.5821	0.5867	0.1700	0.0287	0.0252	4.75

each dataset is highlighted in boldface. Furthermore, the last column of Tables 3 and 4, "Rank", represents the average ranking of the method on all datasets. Based on these experimental results, we have the following observations: (1) Our proposed method LCFS outperforms the state-of-the-art methods on most datasets. Especially in the datasets (Enron, Medical) with higher feature dimensions, LCFS consistently performs better than other compared methods on all the evaluation metrics. (2) On all the eight datasets, LCFS achieves the best performance on all the datasets with respect to *Micro F*1, *Average precision* and *Ranking loss*, seven datasets on *Macro F*1 and *One-error*, and six datasets on *Hamming loss*. (3) The proposed method LCFS performs the best on all the six evaluation metrics in terms of average ranking. The main reason why LCFS is superior to other compared methods is that we not only reduce feature redundancy, but also fully exploit label correlation to improve feature selection, thus capturing more useful information.

To further analyze the performance among all the methods, we use the Friedman test to conduct the statistical significance test [6]. As shown in Table 5, it can be observed that the null hypothesis, which follows the principle that all the methods have equal performance, is clearly rejected on each metric at a

Table 4. Comparison results of multi-label feature selection methods on eight datasets in terms of *Hamming loss, One-error* and *Ranking loss* (Lower is better).

Method	Hamming loss ↓								Rank
	Emotions	Birds	Medical	Yeast	Image	Enron	Corel16k1	Corel16k2	
LCFS	**0.2224**	**0.0521**	**0.0279**	**0.2039**	**0.1801**	**0.0495**	0.0188	0.0178	**1.38**
GLFS	0.2341	0.0615	0.0312	0.2124	0.1828	0.0524	0.0190	0.0187	3.88
MDFS	0.2291	0.0526	0.0389	0.2043	0.1823	0.0524	0.0190	0.0182	3.25
MIFS	0.2299	0.0582	0.0297	0.2075	0.1971	0.0512	**0.0187**	**0.0175**	2.50
MSSL	0.2446	0.0525	0.0318	0.2165	0.1980	0.0526	0.0188	0.0178	3.88
Method	One-error ↓								Rank
	Emotions	Birds	Medical	Yeast	Image	Enron	Corel16k1	Corel16k2	
LCFS	**0.3712**	0.8462	**0.5370**	**0.3609**	**0.3851**	**0.7337**	**0.9551**	**0.9487**	**1.13**
GLFS	0.4054	0.8615	0.5778	0.3849	0.4010	0.7412	0.9593	0.9581	3.63
MDFS	0.3875	0.8592	0.6250	0.3701	0.3950	0.7396	0.9644	0.9685	2.88
MIFS	0.3949	**0.8438**	0.5778	0.3719	0.4431	0.7806	0.9611	0.9619	3.13
MSSL	0.3964	0.8633	0.5752	0.4010	0.4429	0.8324	0.9666	0.9705	4.25
Method	Ranking loss ↓								Rank
	Emotions	Birds	Medical	Yeast	Image	Enron	Corel16k1	Corel16k2	
LCFS	**0.1533**	**0.0344**	**0.0174**	**0.0693**	**0.1406**	**0.0163**	**0.0065**	**0.0061**	**1.00**
GLFS	0.1547	0.0369	0.0185	0.0727	0.1438	0.0166	0.0066	0.0062	2.88
MDFS	0.1563	0.0357	0.0192	0.0697	0.1430	0.0166	0.0066	0.0062	2.88
MIFS	0.1570	0.0376	0.0188	0.0705	0.1577	0.0170	0.0066	**0.0061**	3.50
MSSL	0.1628	0.0362	0.0201	0.0731	0.1571	0.0171	0.0066	0.0062	4.38

significance level of 0.05. Accordingly, we further use the Nemenyi test [6] to analyze the performance differences between the proposed method LCFS and all compared methods, where $CD = 2.1564$ ($k = 5$, $N = 8$). LCFS is considered to have significantly different performance from one comparison method if their average ranks differ by at least one CD. Figure 2 shows the CD diagrams on different evaluation metrics, where average ranking is drawn along the axis and methods connected by a line indicate no significant difference in performance. From Fig. 2, it can be observed that LCFS significantly performs better compared with MSSL. Compared with MIFS, LCFS significantly performs better on *Macro F1, Average precision* and *Ranking loss*. Compared with MDFS, LCFS significantly performs better on *Macro F1*, which also outperforms GLFS on *Hamming loss* and *One-error*. Therefore, these results further confirm that our proposed LCFS achieves highly competitive performance against the other four compared feature selection algorithms.

4.3 Influence of Selected Features

To explore the influence of selected features, we report the result of selecting different number of features on the Enron dataset. The performance of each

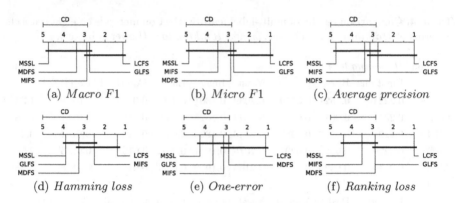

Fig. 2. Comparison between LCFS and the other four algorithms under the Nemenyi test (CD = 2.1565 at 0.05 significance level).

Table 5. Summary of the Friedman statistics.

Evaluation metric	F_F	Critical value (0.05)
Macro F1	9.9057	2.714
Micro F1	25.4638	
Average precision	18.7471	
Hamming loss	4.1721	
One-error	8.6643	
Ranking loss	4.5167	

metric is calculated by varying the number of selected features. As shown in Fig. 3, it can be found that the performance of all methods first improves, and then keeps stable or even degrades, which demonstrates that it is meaningful to conduct feature selection for multi-label learning. In addition, when the number of selected features is less than 50, the proposed method LCFS is obviously superior to other methods, which demonstrates that LCFS can obtain good classification performance even if a few features are selected.

4.4 Parameter Sensitivity Study

In this section, we investigate the sensitivity of the proposed method with respect to its two hyper-parameters, α and β. α measures the contribution of eliminating redundant features, while β controls the contribution of label correlation. To study how do α and β affect the test performance, we conduct experiment on the Enron dataset. Specifically, we turn the parameters α and β from $\{10^{-3}, 10^{-2},...,10^3\}$, and vary one parameter while the other is fixed as 0.1. The experimental result is shown in Fig. 4, in which Fig. 4(a) and (b) depict the influence of α and β, respectively. It can be observed that the performance doesn't change greatly when vary α or β, which demonstrates that LCFS is not very sensitive to these model parameters. Therefore, it is safe to tune them in a wide range, which is suitable for real-world scenarios.

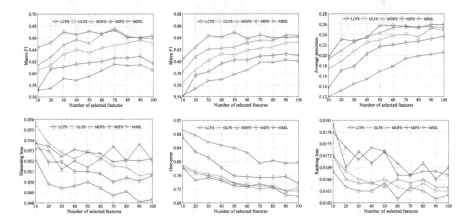

Fig. 3. Influence of selected feature number on Enron dataset.

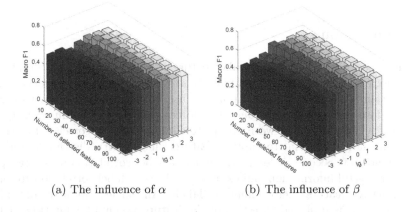

(a) The influence of α (b) The influence of β

Fig. 4. Macro F1 of LCFS on Enron dataset with respect to different α, β.

4.5 Ablation Study

To valid the effect of eliminating redundant features and exploiting label correlation, we consider two model variants of (1) LCFS-NoFR, which doesn't consider feature redundancy (i.e., α is set to zero); (2) LCFS-NoLC, which doesn't take label correlation into account (i.e., β is set to zero). In Fig. 5, we show the experimental results of LCFS and its two variants on the Enron, Image and Yeast

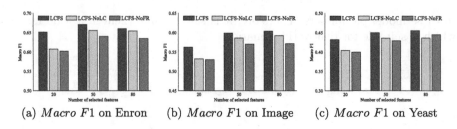

(a) *Macro F*1 on Enron   (b) *Macro F*1 on Image   (c) *Macro F*1 on Yeast

Fig. 5. Ablation study of LCFS on three datasets in terms of *Macro F*1.

datasets. It can be observed that LCFS performs better than the two variants on three datasets regardless of the number of selected features, which demonstrates that the proposed method integrates these components in a principled manner to exploit the strengths of each part. LCFS outperforms LCFS-NoFR on all data sets, showing the importance of eliminating redundant features in feature selection. LCFS-NoLC is inferior to LCFS indicating that label correlation can effectively improve feature selection. Additionally, we can see that the LCFS improved more than the other two variants when the number of select features is 20. This observation shows that eliminating redundant features and exploiting label correlation can play a more important role when a few number of features are selected.

5   Conclusion

In this paper, we present a multi-label feature selection method LCFS that considers both feature redundancy and label correlation. The proposed method has two appealing properties. First, it makes use of feature correlation to divide features into different groups, and then forces competition within each group but relaxes competition between groups to eliminate highly correlated features. Therefore, it is beneficial to capture feature subsets with discrimination and low redundancy. Second, it learns the inherent label correlation which can provide more useful information and exploits the learned label correlation to select discriminative features across multiple labels. An efficient iterative optimization algorithm is designed to estimate the feature weights for multi-label feature selection. Experiments on eight benchmark multi-label datasets in terms of six evaluation metrics demonstrate that the proposed method can significantly improve the performance with feature selection.

Acknowledgment. This research was supported by the National Key Research and Development Program of China (No. 2020YFC0833302), the National Natural Science Foundation of China (No. 62076059), the Science Project of Liaoning Province (2021-MS-105) and the 111 Project (B16009).

References

1. Cai, Z., Zhu, W.: Multi-label feature selection via feature manifold learning and sparsity regularization. Int. J. Mach. Learn. Cybern. **9**, 1321–1334 (2018)
2. Chen, L., et al.: Predicting gene phenotype by multi-label multi-class model based on essential functional features. Mol. Genet. Genom. **296**(4), 905–918 (2021)
3. Chen, Y., Zou, C., Chen, J.: Label-aware graph representation learning for multi-label image classification. Neurocomputing **492**, 50–61 (2022)
4. Chen, Z.M., Wei, X.S., Wang, P., Guo, Y.: Multi-label image recognition with graph convolutional networks. In: Proceedings of the IEEE/CVF Conference on Computer Vision and Pattern Recognition, pp. 5177–5186 (2019)
5. Chou, H.C., Lee, C.C., Busso, C.: Exploiting co-occurrence frequency of emotions in perceptual evaluations to train a speech emotion classifier. In: Proceedings of Interspeech, vol. 2022, pp. 161–165 (2022)
6. Demšar, J.: Statistical comparisons of classifiers over multiple data sets. J. Mach. Learn. Res. **7**, 1–30 (2006)
7. Hu, L., Li, Y., Gao, W., Zhang, P., Hu, J.: Multi-label feature selection with shared common mode. Pattern Recogn. **104**, 107344 (2020)
8. Huang, J., Xu, Q., Qu, X., Lin, Y., Zheng, X.: Improving multi-label learning by correlation embedding. Appl. Sci. **11**(24), 12145 (2021)
9. Jia, X., Sun, F., Li, H., Cao, Y., Zhang, X.: Image multi-label annotation based on supervised nonnegative matrix factorization with new matching measurement. Neurocomputing **219**, 518–525 (2017)
10. Jian, L., Li, J., Shu, K., Liu, H.: Multi-label informed feature selection. IJCAI **16**, 1627-33 (2016)
11. Jordan, M.I.: Serial order: a parallel distributed processing approach. In: Advances in Psychology, vol. 121, pp. 471–495. Elsevier (1997)
12. Kashef, S., Nezamabadi-pour, H., Nikpour, B.: Multilabel feature selection: a comprehensive review and guiding experiments. Wiley Interdiscip. Rev. Data Mining Knowl. Disc. **8**, e1240 (2018)
13. Kong, D., Liu, J., Liu, B., Bao, X.: Uncorrelated group lasso. In: Proceedings of the AAAI Conference on Artificial Intelligence, vol. 30 (2016)
14. Lee, J., Yu, I.: Park: memetic feature selection for multilabel text categorization using label frequency difference. Inf. Sci. **485**, 263–280 (2019)
15. Liu, Y., Wen, K., Gao, Q., Gao, X., Nie, F.: SVM based multi-label learning with missing labels for image annotation. Pattern Recogn. **78**, 307–317 (2018)
16. Sarfraz, S., Sharma, V., Stiefelhagen, R.: Efficient parameter-free clustering using first neighbor relations. In: Proceedings of the IEEE/CVF Conference on Computer Vision and Pattern Recognition, pp. 8934–8943 (2019)
17. Tsoumakas, G., Spyromitros-Xioufis, E., Vilcek, J., Vlahavas, I.: Mulan: a java library for multi-label learning. J. Mach. Learn. Res. **12**, 2411–2414 (2011)
18. Wu, B., Liu, Z., Wang, S., Hu, B.G., Ji, Q.: Multi-label learning with missing labels. In: 2014 22nd International Conference on Pattern Recognition, pp. 1964–1968. IEEE (2014)

19. Wu, X.Z., Zhou, Z.H.: A unified view of multi-label performance measures. In: International Conference on Machine Learning, pp. 3780–3788. PMLR (2017)
20. Zhang, J., Luo, Z., Li, C., Zhou, C., Li, S.: Manifold regularized discriminative feature selection for multi-label learning. Pattern Recogn. **95**, 136–150 (2019)
21. Zhang, M.L., Zhou, Z.H.: ML-KNN: a lazy learning approach to multi-label learning. Pattern Recogn. **40**(7), 2038–2048 (2007)
22. Zhang, Z., Liu, L., Li, J., Wu, X.: Integrating global and local feature selection for multi-label learning. ACM Trans. Knowl. Discov. Data **17**(1), 1–37 (2023)

Iterative Encode-and-Decode Graph Neural Network

Linxuan Song, Siwei Wang, Sihang Zhou, and En Zhu$^{(\boxtimes)}$

National University of Defense Technology, Changsha, China
enzhu@nudt.edu.cn

Abstract. Graph neural networks (GNNs) have been extensively explored due to semi-supervised learning on graphs, which uses few labels to complete tasks without employing costly labeling information. Related methods are dedicated to mitigating the over-smoothing phenomenon to generate reliable node representations. However, existing methods lack correct guidance for neighbors and links from graph characteristics to node representations, resulting in incorrect neighbor information aggregation and poor representation discriminability. In this paper, we introduce a novel encoding and decoding framework that correctly leverages structure guided by labels and uses features for self-supervision of representations to alleviates the over-smoothing phenomenon, dubbed as Iterative Encode-and-Decode Graph Neural Network (IEDGNN). First, we offer a central component reconstruction module to correct the category centers of node representations, lowering the likelihood of aggregating neighbor information across categories. Then, we propose a feature self-reconstruction module that enables node representations to contain valid original attributes, making representations more informative in downstream classification tasks. We also theoretically analyze the impact of different encoder-decoder combinations on representation generation in our design. Extensive experiments demonstrate that our IEDGNN outperforms the state-of-the-art models on eight graph benchmark datasets with three label ratios.

Keywords: Graph attributes mining · High-order neighborhood exploration · Semi-supervised node classification

1 Introduction

Semi-supervised learning is gaining increasing prominence in research owing to the remarkable growth of data volume and the high cost associated with label acquisition. The methodology efficiently extracts meaningful information from a limited set of labels in order to apply the acquired knowledge to downstream tasks. In the context of graph-based semi-supervised learning, effective techniques are developed to leverage the information in graphs, with the goal of generating high-quality node embeddings. However, due to limited availability of high-quality ground-truth labels, graph-based models tend to oversmooth when

© The Author(s), under exclusive license to Springer Nature Switzerland AG 2023
X. Yang et al. (Eds.): ADMA 2023, LNAI 14179, pp. 403–417, 2023.
https://doi.org/10.1007/978-3-031-46674-8_28

Table 1. We generate node representations using edge connections of the same category (Category Link) and the original adjacency matrix (Original Link) for experiments and evaluate the accuracy in a node classification task. Edge Number indicates the number of edge connections of the current type.

Dataset	Edge Number	Accuracy (%)	Edge Type
Cora_ML	2285	73.12	Category Link
	4570	91.80	Category Link
	7981	85.83	Original Link
CiteSeer	1532	90.87	Category Link
	3064	95.94	Category Link
	3668	76.42	Original Link

exploring graph data. To mitigate this issue, popular graph convolutional neural networks (GCNs) adopt the Laplace smoothing method, which allows relevant information to be extracted from input. Nonetheless, applying GCNs with multiple layers may compromise performance by combining information from distinct node categories and interpreting the combined information as part of the current node's representation. This phenomenon, known as oversmoothing, can significantly hinder the ability of the model to differentiate between nodes.

The oversmoothing phenomenon [13, 26] is a critical challenge that confronts researchers who work on semi-supervised graph neural networks. Two main perspectives have been proposed to address this challenge, revolving around structural enhancements and feature engineering. Structural methods such as GAE [10], APPNP [11], and ADAGCN [17] strive to improve node embeddings by increasing the relevance of the correlation between graph nodes while feature engineering methods like GraphCompletion and AttributeMask eliminate the dependence on irrelevant features. Although the aforementioned methods can alleviate some oversmoothing problems, none of them can fully exploit the essential information in the graph to improve the quality of node embeddings.

From the perspective of utilizing structural information, most current models can only exploit the first- or second-order neighbor information. Recent studies [5] have shown that multi-hop messaging is more expressive than one-hop messaging when incorporating higher-order neighbor information, which can alleviate the oversmoothing problem. To incorporate multi-hop neighbor information and avoid oversmoothing, ADAGCN eliminates the use of multi-layer convolution and adopts a simple MLP for feature extraction. However, the accuracy of edge connectivity when using higher-order neighbor information cannot be guaranteed. Figure 1(a) demonstrates that the higher-order adjacency matrix tends to be fully connected, potentially connecting nodes of different classes. Correct edge connections, as shown in Table 1, can significantly enhance downstream task accuracy and provide additional evidence that an increased amount of noise in the original adjacency matrix can negatively impact the quality of node embeddings, thus exacerbating the oversmoothing problem, especially when considering

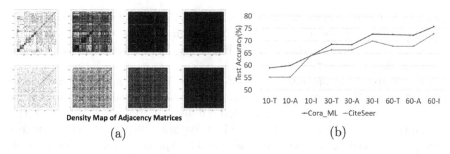

Density Map of Adjacency Matrices

(a)   (b)

Fig. 1. Figure (a) represents the adjacency matrix density plots from first order to fourth order; the first row is the Cora_ML dataset, and the second row is CiteSeer. Figure (b) displays the performance of ADAGCN ('-T' and '-A') and IEDGNN ('-I') models. The number prefix shows the training samples used. For ADAGCN, '-T' uses only labeled samples, and '-A' includes all samples but restricts training to the number prefix. Accuracy rates help determine if ADAGCN efficiently utilizes unlabeled sample information. Furthermore, the results demonstrate that IEDGNN significantly outperforms ADAGCN in terms of effectiveness.

higher-order neighbor relationships. The ADAGCN model's loss function solely employs labeled data to generate node representations, disregarding the potential value of unlabeled data. However, our proposed IEDGNN approach leverages both labeled and unlabeled data to generate more effective node embeddings. As illustrated in Fig. 1(b), ADAGCN using cross-entropy loss produce node representations influenced by labeled samples only, neglecting the potential of unlabeled data which remains untapped. Our IEDGNN, in contrast, maximizes feature utilization and outperforms ADAGCN by using all available information.

To address the limitations of existing models, we propose the Iterative Encode-and-Decode Graph Neural Network (IEDGNN), which features an encoder-decoder structure and optimizes node representation from both structural and feature perspectives. The IEDGNN comprises two main components: a central component reconstruction module and a feature self-reconstruction module. The former module accurately extracts neighbor information while eliminating inter-cluster noise. We also generate a cluster center representation matrix with limited labels, which we align with a unit matrix for proofreading neighbours. Moreover, to extract informative attributes and enhance the discriminability of node representations in classification tasks, we introduce a self-supervision module that associates node representations with graph features. We also align the correlation matrix of the decoder's masked output with the original feature correlation matrix for deeper characterisation. Experimental evaluation demonstrates that IEDGNN surpasses 14 state-of-the-art models on 8 datasets with varying labeling rates, underscoring the effectiveness of our proposed approach.

The following are the paper's contributions,

- We propose IEDGNN, an iteratively trained coding-decoding framework that generates high-quality node representations in semi-supervised tasks. Our model reduces information noise from different classes of edge connections in higher-order neighborhoods and utilizes graph features to self-supervise the generation of node representations, improving their correctness and uniqueness in downstream tasks.

- We introduce two optimization modules that improve node representations significantly. The central component reconstruction module enhances accuracy of cluster centers with limited labels and improves representation quality across different structures. The feature self-reconstruction module employs a masking mechanism to align node representations with their original attributes, increasing their informativeness. Our innovative approach enhances node representations by correcting multi-hop neighbor information and utilizing self-supervision of features.
- We evaluate our proposed model on eight datasets with varying labeling ratios and demonstrate that IEDGNN outperforms all compared models. Additionally, we conduct ablation experiments with 16 pairs of encoder-decoder combinations and determine that MLP and GATv2 deliver the optimal combination for iterative training.

2 Related Work

2.1 Graph Neural Networks

GNNs have recently attracted significant interest due to their capability to process complex data structures, resulting in the development of various notable models for graph-based semi-supervised tasks. GCN-Cheby [4] employs CNNs and spectral graph theory to create fast, localized convolutional filters on graphs. GATs [19] utilize masked self-attentional layers in graph convolutions. JK-Net [25] extracts knowledge via jump connections during final aggregation. GPRGNNs [3] adaptively train GPR weights for enhanced feature extraction, regardless of node label homophily or heterophily. MixupForGraph [21] leverages node neighbor representations for graph convolutions. SGC [22] streamlines models by reducing nonlinearities and collapsing weight matrices. GWNN [24] enhances Graph Fourier-based CNN techniques using graph wavelet transform. Furthermore, PPNP and APPNP [11] apply PageRank to obtain deeper adjacency information, challenging to extract with conventional GNNs.

2.2 Oversmoothing Phenomenon

Currently, the oversmoothing problem in graph neural networks is the most researched topic, initially proposed in [12]. As multiple neighbor feature aggregations that utilize GCN layers are performed, the current node's representation increasingly resembles those of its adjacent nodes, leading to collapse. This phenomenon makes it challenging to distinguish nodes and affects downstream activities. While the graph powering-based approach [7] implicitly leverages more spatial information than standard spectral graph theory, it does not

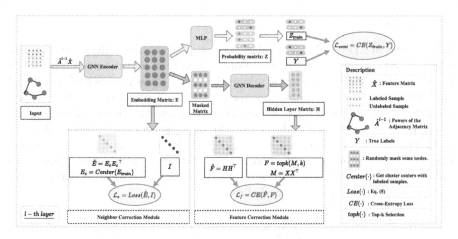

Fig. 2. Depiction of IEDGNN. This illustration presents an example of the l-layer for clarity. The model employs sequential iteration for training, with shared parameters across layers. Furthermore, it incorporates two optimization modules to enhance results. The inputs consist of a normalized attribute matrix and various order adjacency matrices, while the output yields a predicted cluster probability matrix.

focus on model depth. To address this issue, FAGCN [2] aggregates high- and low-frequency information, generating frequency innovations by aggregating the first-order neighbors of different frequencies. This approach enhances the central node's informativeness and distinguishability from other nodes.

3   Method

This section introduces the model's symbolic meaning and goal definition, our motivation, and IEDGNN's modules and details.

3.1   Preliminaries

Definition. Given an undirected graph $\mathcal{G} = (V, \mathcal{E}, X)$, with adjacency matrix $\mathbf{A} \in \mathbb{R}^{n \times n}$, where V is the set of nodes which has n samples. $\mathcal{E}$ is the set of edges. $\mathbf{X} \in \mathbb{R}^{n \times d}$ is the feature matrix. We use $\hat{\mathbf{A}} = \tilde{\mathbf{D}}^{-\frac{1}{2}} \tilde{\mathbf{A}} \tilde{\mathbf{D}}^{-\frac{1}{2}} \in \mathbb{R}^{n \times n}$ as a frequency filter where $\tilde{\mathbf{A}} = \mathbf{A} + \mathbf{I}$. $\tilde{\mathbf{D}}$ is the degree matrix of $\tilde{\mathbf{A}}$ and $\mathbf{I}$ denotes the identity matrix. We first normalize $\mathbf{X}$ by calculating $\hat{\mathbf{X}} = Normalize(\mathbf{X})$ and take $\hat{\mathbf{X}}$ matrix as input.

Task. IEDGNN is a semi-supervised graph neural network that utilizes an encoder-decoder architecture. The model constructs node representations by taking in $\hat{\mathbf{X}}$ and $\hat{\mathbf{A}}$ as inputs and employing a central component reconstruction module and a feature self-reconstruction module. Through multiple layer iterations, the model trains and refines its weights, eventually producing a final output $\mathbf{Z}$ that is an average of the output from each layer.

3.2   Motivation

The analysis of ADAGCN reveals that current models cannot effectively suppress edge connection noise from the multi-order structure, and the node representations lack direct correlation to the original features. These factors lead to reduced node representation discrimination and downstream performance. Specifically, the fully connected multi-hop adjacency matrix has the potential to decrease the rate of edge-connected nodes within the same category, leading to noise generation from neighboring nodes, causing oversmoothing, and ultimately resulting in reduced expressiveness of the node representations. Furthermore, ADAGCN's first layer only takes $\mathbf{X}$ input to utilize features of labeled samples according to the loss function, which is shown below.

$$\mathcal{L}_{semi} = -\frac{1}{N} \sum \mathbf{w}_i [y_i \ln \hat{y}_i + (1 - y_i) \ln(1 - \hat{y}_i)], \qquad (1)$$

where N represents the number of labeled samples, y_i and $\hat{y}_i$ represent the true label and the predicted label, respectively. The adjacency matrix indirectly links node representations to sample features, which may result in node representations lacking important information for a large number of sample features.

3.3   The Design of IEDGNN

In this paper, we address the challenges of optimizing node representations using a multi-layer iterative encoder-decoder architecture called IEDGNN, which includes a central component reconstruction module and a feature self-reconstruction module. The model's parameters are shared between layers. In this subsection, we introduce the two modules and show the overall structure of IEDGNN. Figure 2 illustrates our model.

Central Component Reconstruction Module. To effectively utilize multi-order neighborhoods and avoid edge connection noise between different clusters during aggregation of node representations, we provide a module capable of providing right neighborhood guidance.

To construct each layer, the feature matrix $\hat{\mathbf{X}}$ is the sole input in the first layer, and the adjacency matrix $\hat{\mathbf{A}}$ product with the feature matrix is used as the input of the second layer. The number of hops in the adjacency matrix is incrementally increased in the succeeding phase to effectively use high-order neighbor information, as shown below.

$$\mathbf{A}^{(l)} = \begin{cases} \hat{\mathbf{X}}, & l = 1 \\ \hat{\mathbf{A}} \mathbf{A}^{(l-1)}, & l > 1 \end{cases}, \quad \mathbf{E}^{(l)} = g(\mathbf{A}^{(l)}; w), \qquad (2)$$

where l denotes the model's layer number and $l = 1, ..., L$. $\mathbf{A}^{(l)} \in \mathbb{R}^{n \times d}$ and $\mathbf{E}^{(l)} \in \mathbb{R}^{n \times h'}$ denote the input and representation matrix of the l-th layer, respectively. $g(\cdot; w)$ represents a GNN encoder.

Cluster centers are computed by taking the average representation of nodes with labels. This approach is used to eliminate intra-cluster distance and inter-cluster representation noise, allowing the model to train node representations with information on the correct cluster relationships, as shown in the following equations.

$$\mathbf{e}'^{(l)}_i = \frac{\sum_{j \in i}^b \mathbf{e}_j^{(l)}}{b}, \tag{3}$$

$$\hat{\mathbf{E}}^{(l)}_{ij} = \mathbf{e}'^{(l)}_i \mathbf{e}'^{(l)}_j{}^\top, \tag{4}$$

where i denotes the current cluster and b denotes the number of labeled samples. $\mathbf{e}'^{(l)}_i$ represents the representation class center of the i-th cluster and $\hat{\mathbf{E}}^{(l)} \in \mathbb{R}^{C \times C}$ is cluster center correlation matrix. The corresponding loss function of the module is shown below.

$$\mathcal{L}_c = \frac{\sum_i^C (1 - \frac{\hat{\mathbf{E}}^{(l)}_{ii}}{C})^2}{C} + \lambda \frac{\sum_i^C \sum_{j \neq i}^C (\frac{\hat{\mathbf{E}}^{(l)}_{ij}}{C})^2}{C(C-1)}, \tag{5}$$

where C represents the quantity of clusters and λ is a hyperparameter designed to balance intra-class distance and inter-class noise. We address the problem of unreliable neighbor information by adjusting the category center in the semi-supervised task, which enables the model to ensure the reliability of neighbor information in the aggregation process.

Feature Self-reconstruction Module. Even though the central component reconstruction module has proven to be highly effective at maximizing the use of neighborhood information, the use of features is still lacking. In order to add original characteristics to representations, we build a module capable of self-supervision to make node representations more discriminative and improve classification performance.

Specifically, we use the regularized feature matrix $\hat{\mathbf{X}}$ and calculate the cosine distance between node features to generate the feature similarity matrix $\mathbf{M}$. Since the underlying graph structure is sparse, we define a threshold to filter out values that do not conform to the criteria; typically, is $1e - 6$. Then, we conduct topk selection of $\mathbf{M}$. We select k nodes closest to the current node and record them as 1 in the feature correlation matrix $\mathbf{F}$, while the remaining nodes are recorded as 0, formulated as:

$$\mathbf{M}_{ij} = \frac{\hat{\mathbf{x}}_i}{\|\hat{\mathbf{x}}_i\|_2} \cdot \frac{\hat{\mathbf{x}}_j}{\|\hat{\mathbf{x}}_j\|_2}, \tag{6}$$

$$\mathbf{F} = topk(\mathbf{M}, k), \tag{7}$$

where $\cdot$ denotes the dot product and $\|$ is the L_2 parametrization.

It is evident from [27] that the usage of models for representation generation is frequently accompanied by high dimensionality. However, it may lead to learned

410   L. Song et al.

node representations containing more redundant and useless information [20]. The mask mechanism has been validated as an effective solution in [6], and also widely used in CV and NLP. Inspired by this, we apply the mask mechanism to the module to achieve self-supervision.

Formally, we set different mask nodes for each layer and use a uniformly distributed random sampling strategy to select the masked nodes $\widetilde{V}^{(l)} \subset V$. The node embedding $\widetilde{\mathbf{e}}_i^{(l)}$ for $v_i \in V$ in the masked embedding matrix $\widetilde{\mathbf{E}}^{(l)}$ can be defined as:

$$\widetilde{\mathbf{e}}_i^{(l)} = \begin{cases} 0 & v_i \in \widetilde{V}^{(l)} \\ \mathbf{e}_i & v_i \notin \widetilde{V}^{(l)} \end{cases} \tag{8}$$

To minimize the extraneous memory consumption resulting from matrix alignment, a decoder is utilized to lessen the dimension of the embeddings. The formula is defined as follows:

$$\mathbf{h}_i^{(l)} = h(\widetilde{\mathbf{E}}^{(l)}; \theta), \tag{9}$$

where $h(\cdot; \theta)$ is a GNN decoder and $h_i^{(l)}$ denotes the output of the l-th layer decoder. To fully leverage the original features of the data, we obtain the embedding correlation matrix $\hat{\mathbf{F}}^{(l)}$ using the following formula:

$$\hat{\mathbf{F}}_{ij}^{(l)} = \sigma(\mathbf{h}_i^{(l)}\mathbf{h}_j^{(l)\top}), \tag{10}$$

where σ denotes the relu activation function. To correlate features with node representations, we calculate the Kullback-Leibler scatter between $\mathbf{F}$ and $\hat{\mathbf{F}}^{(l)}$. Due to the constant nature of $\mathbf{F}$, its entropy value remains unchanged. Thus, it can be represented as a cross-entropy loss function in the following manner:

$$\mathcal{L}_f = CE(\hat{\mathbf{F}}^{(l)}, \mathbf{F}), \tag{11}$$

where i denotes the i-th node. $\mathcal{L}_f^{(l)}$ calculates the l-th layer module loss function.

Overview of IEDGNN. We define the formal architecture of IEDGNN using the following formulas.

$$\begin{cases} \mathbf{E}^{(l)} = g(\mathbf{A}^{(l)}; w), \\ \mathbf{Z}^{(l)} = Softmax(MLP(\mathbf{E}^{(l)})), \\ \mathbf{Z} = \frac{\mathbf{Z}^{(1)}+\cdots+\mathbf{Z}^{(l)}+\cdots+\mathbf{Z}^{(L)}}{L}, \end{cases} \tag{12}$$

where $\mathbf{Z}$ is the output of the model. Note that we initialize the weight matrix of the $(l+1)$-th layer with the optimal learning weight matrix of the l-th layer. The final equation for the loss function is as follows:

$$\mathcal{L} = \mathcal{L}_{semi} + \mathcal{L}_c + \beta * \mathcal{L}_f. \tag{13}$$

β is a hyperparameter to coordinate the objectives.

3.4   Theoretical Analysis of Encoders and Decoders

This section presents an analysis of encoders and decoders using GNN structures such as MLP, GCN, GAT, and GATv2. In IEDGNN, the input comprises a feature matrix and an adjacency matrix that together convey information about the nodes' neighbors through edge connections. A common issue while generating node representations in encoders is over-smoothing, where similarities between nodes become too high. To address this, MLP is chosen for encoding, as it focuses solely on the nodes' own attributes and does not perform repeated aggregation that could lead to collapse. For decoders, the performance of the existing node representations and their neighbors is a key consideration. In this regard, GAT and GATv2 are preferred over GCN, which relies on the original graph topology. The reason being that neighbors in this context comprise not only edge-connected nodes, but also those with similar representations generated by the encoder. Overall, our analysis shows that the choice of the encoder and decoder can have a significant impact on the performance of GNN models in various applications.

4   Experiments

4.1   Experimental Setup

Datasets. In order to demonstrate the validity of our proposed IEDGNN in a more convincing manner, we choose eight datasets, including DBLP[1], ACM[2], AMAP [18], AMAC [18], Cora_ML [1], CiteSeer [16], PubMed [14], and MS_Academic [15]. In addition, Table 2 summarizes the data statistics.

Table 2. Dateset Statistics

Dataset	Nodes	Edges	Features	Classes
DBLP	4507	7056	334	4
ACM	3025	26 256	1870	3
AMAP	7650	287 326	745	8
AMAC	13 752	491 722	767	10
Cora_ML	2810	7981	2879	7
CiteSeer	2110	3668	3703	6
PubMed	19 717	44 324	500	3
MS_Academic	18 333	81 894	6805	15

---

[1] https://dblp.uni-trier.de.
[2] https://dl.acm.org/.

Baselines. To better compare with other models, we select fourteen representative semi-supervised methods to verify the effectiveness of IEDGNN, including GCN (with early stopping), V.GCN [9], GCN-Cheby [4], GAT [8], FAGCN [2], SGC [22], MixupForGraph [21], JK-Net [25], PPNP, APPNP [11], GWNN [24], NodeFormer [23], GPRGNN [3] and ADAGCN [17].

Training Settings. The hidden layer units of the encoder and decoder are selected from [32, 64, 128, 256, 512, 5000], and the scope of the multi-headed attention mechanism is [1, 2, 4]. Topk is set to 10, λ is set to 1, and weight decay is chosen from [1e$-$3, 1e$-$4, 1e$-$5, 1e$-$6]. The embedding matrix's mask ratio is chosen from [0.1, 0.3, 0.5, 0.7, 0.9], and β is chosen from 0.1 to 1 with an interval of 0.1. Model training has an early stop mechanism, patience steps of 200, and a maximum of 2000 rounds. The validation set has 500 samples, and the test set has the remaining samples. We choose node classification precision as the evaluation metric.

4.2   Performance Comparison

We train each cluster using 20, 16, and 8 samples, and evaluate the model's performance on 8 datasets and 14 baselines to validate IEDGNN. Table 3 displays the outcome. The results indicate that IEDGNN performs better than other models in all three labeling rates and demonstrates more stable experimental outcomes. This finding suggests that IEDGNN can effectively leverage graph information from both structure and features to optimize node representations for downstream tasks.

4.3   Ablation Experiments

In this section, we perform experiments on different combinations of encoder-decoder and do effect validation on different parts of the loss function.

Analysis of Encoder and Decoder. We have utilized four structures, namely, MLP, GCN, GAT, and GATv2, as encoder and decoder in IEDGNN, and the results of our experiments are presented in Table 4. The model's optimal effect is obtained using MLP as the encoder and GATv2 as the decoder. Moreover, both the encoder and decoder using MLP result in a better effect than using GCN as both encoder and decoder. Thus, we can draw the following three conclusions: 1) An aggregation model such as GCN, as an encoder, results in node representations of poor quality when the input data contains information after aggregating neighbors. 2) The decoder needs to be chosen taking into account the node's representations as well as the representations of other nodes that are similar to it. Dynamic GATv2 can effectively serve this goal.

Table 3. This is the result of using 20, 16, and 8 training samples per cluster for all datasets. We evaluate the node classification tasks using the average accuracy (%) and standard deviation as a criterion. Experiments use 10 random seeds. "OOM" denotes out of memory. Red results are optimal, and blue results are suboptimal.

Label	Model	DBLP	ACM	AMAP	AMAC	Cora_ML	CiteSeer	PubMed	MS_Academic
20	GCN	78.79 0.54	91.85 0.40	81.15 1.05	58.96 0.96	82.13 0.72	74.42 0.73	76.89 0.54	92.01 0.08
	V.GCN	78.32 0.78	91.06 0.44	81.41 1.47	58.71 0.76	82.61 0.76	74.07 0.69	76.70 0.63	91.83 0.16
	GCN-Cheby	80.33 0.97	91.01 0.56	91.56 0.75	77.31 1.18	82.97 0.67	72.21 0.54	75.01 0.74	OOM
	GAT	79.97 0.58	90.16 0.52	92.38 0.14	79.04 0.58	83.53 0.16	72.17 0.73	77.85 0.26	89.47 0.20
	FAGCN	79.98 0.89	90.38 0.86	90.06 1.04	80.41 0.62	83.72 0.64	71.97 1.05	77.00 0.84	90.53 0.20
	SGC	79.27 1.07	91.75 0.48	90.17 0.75	77.48 0.79	83.65 1.16	73.16 1.07	79.30 0.60	89.79 0.83
	MixupForGraph	74.29 0.71	89.33 0.59	87.28 1.08	59.65 0.71	78.76 0.74	68.68 0.91	72.70 0.40	85.59 1.26
	JK-Net	79.62 0.58	90.56 0.56	91.68 0.82	75.95 1.07	82.93 0.38	72.78 0.44	76.54 0.57	89.08 0.05
	PPNP	80.40 0.42	91.64 0.44	86.27 0.74	66.15 0.74	85.29 0.30	75.42 0.27	OOM	OOM
	APPNP	79.62 0.74		86.02 0.76	64.99 1.25	85.09 0.25	75.51 0.42	78.63 0.65	92.12 0.09
	GWNN	79.27 0.54	90.91 0.49	90.75 1.26	75.84 0.87	83.84 0.55	74.06 0.80	78.69 0.94	90.27 0.53
	NodeFormer	77.36 0.58	89.43 0.38	87.41 0.89	72.12 0.60	81.32 0.93	71.56 0.48	75.33 0.69	88.34 0.41
	GPRGNN	80.92 0.20	91.35 0.35	93.08 0.26	80.71 0.33	85.07 0.24	74.09 0.41	77.38 0.48	90.47 0.18
	ADAGCN	80.45 0.83	92.04 0.62	85.13 0.62	67.58 0.99	85.67 0.59	76.24 0.43	79.38 0.63	93.11 0.23
	IEDGNN(Ours)	82.70 0.34	93.47 0.34	94.85 0.28	81.65 0.47	87.32 0.51	78.39 0.47	80.68 0.50	93.73 0.37
16	GCN	77.86 0.78	90.81 0.99	79.79 0.71	56.80 1.57	81.93 1.39	72.09 0.80	75.65 1.22	90.80 0.78
	V.GCN	77.50 1.15	90.60 0.30	79.54 1.01	56.49 0.74	81.91 1.16	72.54 1.07	75.76 1.15	90.77 0.86
	GCN-Cheby	77.96 0.82	90.90 1.01	90.22 1.16	75.43 0.61	80.80 0.47	71.39 1.17	75.55 0.22	OOM
	GAT	78.43 0.23	90.01 0.95	92.36 0.50	78.98 0.41	81.31 0.29	71.94 1.12	78.05 0.46	88.23 0.48
	FAGCN	79.74 0.83	90.26 0.68	89.74 1.26	80.65 0.95	83.28 0.90	69.84 1.43	76.31 1.51	90.22 0.83
	SGC	79.21 1.37	91.11 0.89	89.87 1.28	76.66 0.96	82.90 0.57	72.23 0.84	77.90 0.45	89.37 1.04
	MixupForGraph	73.22 0.66	87.85 1.09	86.51 1.13	58.15 1.27	77.50 1.10	68.69 1.09	72.60 0.85	84.03 1.52
	JK-Net	78.91 0.51	90.39 0.40	91.50 0.41	74.11 0.96	80.74 1.02	73.07 0.54	76.11 0.67	89.04 0.21
	PPNP	80.09 1.07	91.29 0.46	84.37 1.22	65.66 1.02	82.58 1.03	75.41 0.63	OOM	OOM
	APPNP	79.18 0.54		85.60 0.50	65.22 1.18	82.90 1.18	75.37 0.93	76.98 1.61	92.08 0.77
	GWNN	78.08 1.18	91.01 0.40	89.99 1.37	75.09 1.13	81.85 0.97	72.13 0.82	76.76 1.36	90.20 0.84
	NodeFormer	75.38 0.67	88.82 0.44	86.54 0.53	71.43 0.54	80.65 0.67	70.10 0.43	74.53 0.67	86.65 0.36
	GPRGNN	80.88 0.58	92.19 0.22	92.70 0.11	80.13 0.72	84.14 0.31	73.68 0.27	77.62 0.24	90.30 0.22
	ADAGCN	79.38 0.83	92.09 0.64	85.39 0.98	61.83 1.50	84.69 0.41	76.17 0.73	78.04 0.47	92.66 0.17
	IEDGNN(Ours)	82.42 0.56	93.14 0.38	94.65 0.63	81.30 0.74	87.01 0.58	77.84 0.73	80.12 0.68	93.20 0.27
8	GCN	76.39 1.69	89.38 1.53	79.65 0.97	54.32 0.98	79.96 0.86	69.34 0.76	71.83 1.46	89.30 1.12
	V.GCN	75.95 1.08	90.12 1.24	78.72 1.24	54.38 0.76	79.75 0.62	69.36 0.65	71.57 1.46	90.04 0.83
	GCN-Cheby	76.04 1.10	89.62 0.63	89.88 0.79	74.16 0.82	75.99 0.65	68.31 1.14	68.10 0.30	OOM
	GAT	77.33 0.31	89.66 0.78	90.77 0.75	77.21 0.56	78.67 0.45	70.86 0.97	68.94 0.28	87.86 0.66
	FAGCN	77.59 1.25	89.75 1.45	87.56 0.63	79.41 1.13	81.37 1.09	67.46 1.51	74.30 1.56	89.11 1.18
	SGC	77.30 1.79	90.46 0.52	88.33 1.69	73.21 1.68	79.18 1.02	66.15 1.80	72.82 1.67	88.36 0.61
	MixupForGraph	69.54 0.96	85.24 0.69	85.20 1.19	57.56 1.43	75.30 1.18	65.96 1.66	67.47 1.62	84.33 1.26
	JK-Net	78.08 1.08	89.94 0.56	89.42 0.86	72.45 1.22	77.72 0.71	70.65 0.23	67.92 0.31	88.32 0.25
	PPNP	79.91 1.32	90.90 0.40	83.26 1.59	64.89 0.95	81.01 1.47	72.38 1.05	OOM	OOM
	APPNP	79.38 1.42	90.49 1.19	83.67 1.18	65.19 1.04	80.94 1.16	72.93 1.50	76.30 1.63	OOM
	GWNN	75.53 1.07	90.45 0.59	88.92 1.22	74.32 0.78	80.83 1.08	70.63 0.93	74.50 0.82	89.84 0.32
	NodeFormer	76.12 0.42	88.31 0.87	86.24 0.65	70.20 0.87	79.54 0.74	70.45 0.26	72.76 0.87	86.75 0.79
	GPRGNN	80.83 0.65	90.68 0.19	91.32 0.37	79.49 0.77	83.18 0.63	73.60 0.30	69.58 0.19	88.90 0.22
	ADAGCN	77.61 1.78	92.44 1.58	84.73 0.50	59.48 1.32	82.08 0.99	72.95 1.38	74.17 1.11	91.81 0.49
	IEDGNN(Ours)	81.68 0.79	92.67 0.58	92.89 0.64	80.74 0.80	86.01 0.69	75.88 0.70	79.22 0.56	92.57 0.42

Analysis of Loss Function. This section analyzes the effectiveness of our proposed modules by omitting the objective functions in Eq. (13). The results are presented in Fig. 3, from which we can make the following conclusions. Firstly, due to the complete utilization of structural and feature information, both of our proposed modules allow for a significant improvement in the model's performance. Secondly, using just one module is inferior to using both modules simultaneously, indicating that optimal node representation should be considered from both structural and feature perspectives.

4.4 Parameter Sensitivity Analysis

This section presents a comparison of various mask ratios, and the results are exhibited in Fig. 4. The optimal mask ratio is dependent on the dataset in question. For Cora_ML, a higher mask ratio of 0.7 yields better results, possibly due

Table 4. We experiment on the combination of different encoders and decoders (encoder & decoder) in IEDGNN, we choose MLP, GCN, GAT, and GATv2 as encoder and decoder, respectively. The models are trained using 10 random seeds. The red ones are the optimal results, and the blue ones are suboptimal.

Combination	DBLP	ACM	AMAP	AMAC	Cora_ML	CiteSeer	PubMed	MS_Academic
MLP & MLP	80.09₀.34	91.23₀.76	92.08₁.22	80.66₀.93	85.68₀.74	77.81₀.78	80.14₀.45	93.34₀.57
MLP & GCN	79.67₀.58	88.73₀.33	91.68₀.71	76.79₀.43	84.34₁.16	77.43₀.89	80.05₀.94	92.18₀.85
MLP & GAT	81.77₀.43	93.01₀.52	93.49₀.34	80.93₀.39	86.43₀.47	77.48₀.16	80.34₀.86	93.75₀.30
MLP & GATv2	82.70₀.34	93.47₀.34	94.85₀.28	82.15₀.47	87.32₀.51	78.39₀.47	80.68₀.50	94.73₀.37
GCN & MLP	78.19₀.47	92.42₀.52	87.57₁.12	77.58₀.76	84.80₀.68	77.32₀.43	78.89₀.71	91.06₀.67
GCN & GCN	78.72₀.93	92.09₀.39	86.32₀.72	76.33₀.68	85.16₀.72	76.94₀.54	78.77₀.86	91.71₀.93
GCN & GAT	74.24₀.68	92.22₀.38	88.37₀.47	76.62₀.89	84.93₀.96	77.05₀.51	78.96₀.77	92.33₀.34
GCN & GATv2	74.32₀.83	92.16₀.32	88.93₀.41	77.53₀.85	84.80₀.99	77.21₀.98	79.06₀.63	91.61₀.54
GAT & MLP	77.32₀.35	91.69₀.59	91.44₀.52	76.32₁.04	84.51₀.65	76.55₀.78	78.63₀.59	90.09₀.60
GAT & GCN	77.42₀.54	91.43₁.31	87.23₀.89	75.62₀.84	83.53₀.82	76.28₀.67	78.50₀.70	89.32₀.84
GAT & GAT	79.32₀.87	91.63₁.11	92.95₀.93	76.04₀.51	84.56₀.45	77.27₀.85	78.79₀.62	90.32₀.68
GAT & GATv2	78.35₀.78	91.76₀.72	89.83₀.60	76.82₀.37	84.89₀.49	77.42₀.73	78.67₀.61	89.87₀.54
GATv2 & MLP	80.02₀.51	91.83₀.72	92.03₀.56	77.32₀.67	84.34₀.38	76.94₁.03	78.34₀.78	90.49₀.45
GATv2 & GCN	79.33₀.62	91.56₁.05	86.46₁.26	75.38₀.73	83.91₁.04	76.61₀.67	78.29₀.64	90.37₀.79
GATv2 & GAT	79.81₀.33	91.50₁.18	83.07₁.16	76.81₀.53	84.37₀.89	76.74₀.71	78.46₀.70	90.77₀.49
GATv2 & GATv2	79.67₀.94	91.76₀.65	92.78₀.35	75.88₀.49	84.69₀.43	76.72₀.96	78.64₀.78	89.16₀.64

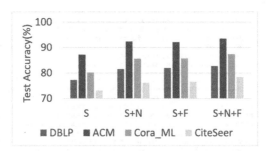

Fig. 3. This plot illustrates the validation of the objective function in IEDGNN. The legend indicates that "S" represents the retention of only $\mathcal{L}_{semi}$, "S+N" represents the retention of $\mathcal{L}_{semi}$ and $\mathcal{L}_c$, "S+F" represents the retention of $\mathcal{L}_{semi}$ and $\mathcal{L}_f$, and "S+N+F" represents the complete function in Eq. (13).

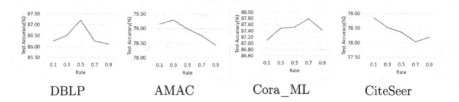

Fig. 4. The performance of four datasets with varying mask ratios is presented in the graphs. The x-axis of each graph represents the mask rate, while the y-axis represents the node classification accuracy.

to the redundancy of the data. In such cases, the missing node features can be substituted by their neighboring nodes. Conversely, with less redundant data such as CiteSeer, high mask ratios may not be sufficient for complete node feature recovery. It is evident that the optimal mask ratio is reliant on the amount of redundant information present in the dataset.

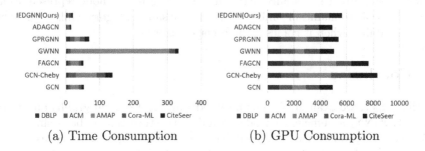

(a) Time Consumption   (b) GPU Consumption

Fig. 5. The figure on the left shows time consumption, with the horizontal axis representing the time (ms) to train an epoch for each model. The figure on right shows GPU consumption, with the horizontal axis representing the maximum GPU occupation (MB) during model training. The vertical coordinates of both plots represent the different models being compared.

4.5 Complexity Analysis

This section compares our IEDGNN model with five other models regarding their time and space consumption under the same conditions. The results are presented in Fig. 5, from which the following conclusions can be drawn: 1) IEDGNN falls under the category of relatively less time-consuming models. This could be attributed to the model's simple structure and our optimized node representation method, which does not result in an overly complex model structure. 2) In terms of GPU consumption, IEDGNN does not use memory-intensive complex convolutions, and its encoder and decoder selection is relatively flexible.

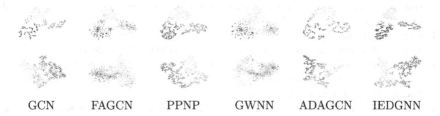

GCN   FAGCN   PPNP   GWNN   ADAGCN   IEDGNN

Fig. 6. t-SNE visualization plots for different models. The first row is the ACM dataset, and the second row is the DBLP dataset.

4.6 Visualization Analysis

Figure 6 demonstrates the use of the *t*-SNE algorithm for visualizing different models. The results attest to the superior classification and category center separation ability of our IEDGNN model. By way of illustration, when dealing with the ACM dataset in the first row, IEDGNN can effectively separate the cluster centers, extending the distance between them. As a result, node representations are better able to maintain similarity within the same category whilst enhancing discriminability when it comes to different categories.

5 Conclusion

In this paper, we examine semi-supervised learning tasks on graphs and introduce IEDGNN, an efficient encoder-decoder structure that is both simple and effective. We propose two key modules, the central component reconstruction module, and the feature self-reconstruction module, to optimize node representations, enhance the representations' discriminability, and mitigate noise from neighboring information. By using the original graph attributes to self-supervise the representations, we produce more precise downstream task representations. Using benchmark tests, we demonstrate IEDGNN surpasses other related models. In conclusion, the two effective modules in our proposed IEDGNN framework could serve as a reference for future GNN models, particularly in semi-supervised tasks. We plan to expand experimental data to improve IEDGNN's scalability in future research.

Acknowledgements. This work is supported by National Key R&D Program of China under Grant No. 2022ZD0209103.

References

1. Abu-El-Haija, S., Kapoor, A., Perozzi, B., Lee, J.: N-GCN: multi-scale graph convolution for semi-supervised node classification. In: Uncertainty in Artificial Intelligence, pp. 841–851. PMLR (2020)
2. Bo, D., Wang, X., Shi, C., Shen, H.: Beyond low-frequency information in graph convolutional networks. In: Proceedings of the AAAI Conference on Artificial Intelligence, vol. 35, pp. 3950–3957 (2021)
3. Chien, E., Peng, J., Li, P., Milenkovic, O.: Adaptive universal generalized PageRank graph neural network. arXiv preprint arXiv:2006.07988 (2020)
4. Defferrard, M., Bresson, X., Vandergheynst, P.: Convolutional neural networks on graphs with fast localized spectral filtering. In: Advances in Neural Information Processing Systems, vol. 29 (2016)
5. Feng, J., Chen, Y., Li, F., Sarkar, A., Zhang, M.: How powerful are K-hop message passing graph neural networks. arXiv preprint arXiv:2205.13328 (2022)
6. Hou, Z., Liu, X., Dong, Y., Wang, C., Tang, J., et al.: GraphMAE: self-supervised masked graph autoencoders. arXiv preprint arXiv:2205.10803 (2022)

7. Jin, M., Chang, H., Zhu, W., Sojoudi, S.: Power up! Robust graph convolutional network against evasion attacks based on graph powering. arXiv e-prints, arXiv-1905 (2019)
8. Kim, D., Oh, A.: How to find your friendly neighborhood: graph attention design with self-supervision. arXiv preprint arXiv:2204.04879 (2022)
9. Kipf, T.N., Welling, M.: Semi-supervised classification with graph convolutional networks. arXiv preprint arXiv:1609.02907 (2016)
10. Kipf, T.N., Welling, M.: Variational graph auto-encoders. arXiv preprint arXiv:1611.07308 (2016)
11. Klicpera, J., Bojchevski, A., Günnemann, S.: Predict then propagate: graph neural networks meet personalized PageRank. arXiv preprint arXiv:1810.05997 (2018)
12. Li, Q., Han, Z., Wu, X.M.: Deeper insights into graph convolutional networks for semi-supervised learning. In: Thirty-Second AAAI Conference on Artificial Intelligence (2018)
13. Liu, Y., et al.: Deep graph clustering via dual correlation reduction. In: Proceedings of the AAAI Conference on Artificial Intelligence, vol. 36, pp. 7603–7611 (2022)
14. Namata, G., London, B., Getoor, L., Huang, B., Edu, U.: Query-driven active surveying for collective classification. In: 10th International Workshop on Mining and Learning with Graphs, vol. 8, p. 1 (2012)
15. Sen, P., Namata, G., Bilgic, M., Getoor, L., Galligher, B., Eliassi-Rad, T.: Collective classification in network data. AI Mag. **29**(3), 93 (2008)
16. Shchur, O., Mumme, M., Bojchevski, A., Günnemann, S.: Pitfalls of graph neural network evaluation. arXiv preprint arXiv:1811.05868 (2018)
17. Sun, K., Zhu, Z., Lin, Z.: AdaGCN: AdaBoosting graph convolutional networks into deep models. In: International Conference on Learning Representations (2021)
18. Tu, W., Zhou, S., Liu, Y., Liu, X.: Siamese attribute-missing graph auto-encoder. arXiv preprint arXiv:2112.04842 (2021)
19. Veličković, P., Cucurull, G., Casanova, A., Romero, A., Lio, P., Bengio, Y.: Graph attention networks. arXiv preprint arXiv:1710.10903 (2017)
20. Vincent, P., Larochelle, H., Bengio, Y., Manzagol, P.A.: Extracting and composing robust features with denoising autoencoders. In: Proceedings of the 25th International Conference on Machine Learning, pp. 1096–1103 (2008)
21. Wang, Y., Wang, W., Liang, Y., Cai, Y., Hooi, B.: MixUp for node and graph classification. In: Proceedings of the Web Conference 2021, pp. 3663–3674 (2021)
22. Wu, F., Souza, A., Zhang, T., Fifty, C., Yu, T., Weinberger, K.: Simplifying graph convolutional networks. In: International Conference on Machine Learning, pp. 6861–6871. PMLR (2019)
23. Wu, Q., Zhao, W., Li, Z., Wipf, D., Yan, J.: NodeFormer: a scalable graph structure learning transformer for node classification. In: Advances in Neural Information Processing Systems (2022)
24. Xu, B., Shen, H., Cao, Q., Qiu, Y., Cheng, X.: Graph wavelet neural network. arXiv preprint arXiv:1904.07785 (2019)
25. Xu, K., Li, C., Tian, Y., Sonobe, T., Kawarabayashi, K.I., Jegelka, S.: Representation learning on graphs with jumping knowledge networks. In: International Conference on Machine Learning, pp. 5453–5462. PMLR (2018)
26. Yang, X., Hu, X., Zhou, S., Liu, X., Zhu, E.: Interpolation-based contrastive learning for few-label semi-supervised learning. IEEE Trans. Neural Netw. Learn. Syst. (2022)
27. Zhang, S., Liu, Y., Sun, Y., Shah, N.: Graph-less neural networks: teaching old MLPs new tricks via distillation. arXiv preprint arXiv:2110.08727 (2021)

Community Detection in Temporal Biological Metabolic Networks Based on Semi-NMF Method with Node Similarity Fusion

Xuanming Zhang[1,2,3,4], Jianxing Yu[1,2,3,4(✉)], Miaopei Lin[1,2,3,4],
Shiqi Wang[1,2,3,4], Wei Liu[1,2,3,4], and Jian Yin[1,2,3,4]

[1] School of Artificial Intelligence, Sun Yat-sen University, Zhuhai, China
{zhangxm236,linmp3,wangshq25}@mail2.sysu.edu.cn,
{liuw259,issjyin}@mail.sysu.edu.cn
[2] Guangdong Key Laboratory of Big Data Analysis and Processing,
Guangzhou, China
[3] Pazhou Lab, Guangzhou 510330, China
[4] Key Laboratory of Sustainable Tourism Smart Assessment Technology, Ministry of
Culture and Tourism, Zhuhai, China
yujx26@mail.sysu.edu.cn

Abstract. This paper aims to study the topological structure of biological metabolic networks, such as detecting their inherent communities to learn the useful patterns and characteristics. By partitioning a metabolic network into multiple communities and subgraphs, we can reveal their local structure and organization. That helps researchers better understand the structure and function of metabolic networks, identify metabolic pathways and functional modules, and discover mechanisms and patterns of metabolic regulation. However, community detection remains a challenging problem due to the dynamic nature of metabolic network structure and data sparsity. To address this problem, we propose a new temporal community detection method for biological metabolic networks based on a semi-NMF framework with node similarity fusion, called SNF-MNet. This method grasps the dynamic nature of metabolic networks by introducing temporal smoothness to estimate network evolution. Furthermore, SNF-MNet adopts a node similarity fusion method to comprehensively consider the relations between nodes in the network, effectively solving the limitations and data sparsity issues of existing methods. SNF-MNet can effectively reduce the impact of noise and improve discriminability in small communities, providing more accurate and reliable results. Experiments on multiple synthetic and real-world datasets demonstrate the effectiveness of this model in community detection and temporal evolution tracking, with an increase of NMI 1%–3.2%.

Keywords: Temporal networks · Biological metabolic networks ·
Node similarity fusion · Community detection · Semi-NMF

© The Author(s), under exclusive license to Springer Nature Switzerland AG 2023
X. Yang et al. (Eds.): ADMA 2023, LNAI 14179, pp. 418–432, 2023.
https://doi.org/10.1007/978-3-031-46674-8_29

1 Introduction

Metabolic networks, composed of numerous metabolic reactions and compounds, play a crucial role in cellular function. The structural and organizational properties of metabolic networks have a significant impact on cellular processes, such as cell growth and development. Analyzing the organization and function of metabolic networks is essential for understanding the underlying biological mechanisms of cellular processes. It has been widely used to detect the community, which partitions a network into densely connected subnetworks. That provides valuable insights into the modular structure and functional organization of complex biological systems. However, community detection in biological metabolic networks is still a challenging task due to its dynamic nature and sparsity of data. The networks are updating instead of statically fixed. The metabolic processes and interactions between metabolites and enzymes are constantly changing over time. In addition, the connectivity between nodes and edges in metabolic networks is highly sparse. Most metabolites interacting with only a few others rather than all metabolites. Previous studies primarily focus on analyzing metabolic pathways but neglect to analyze the community structure of metabolic networks. By identifying communities in metabolic networks, researchers can better understand the underlying biological processes, discover novel metabolic pathways and modules, and investigate the regulatory mechanisms and patterns. Despite the growing importance of community detection in metabolic networks, existing methods still face many shortcomings. For example, they are mainly used to deal with static networks, and most rely on sufficient data, which is difficult to handle the dynamic sparse metabolic networks.

To address these issues, we propose a new model, SNF-MNet, for community detection in biological metabolic networks. SNF-MNet is based on the semi-nonnegative matrix factorization framework with an evolutionary clustering technique. The Semi-NMF method is effective in reducing the impact of noise, improving the discriminative power of small communities, and providing more accurate and reliable community detection results in the context of biological metabolic networks. Considering the dynamics of metabolic networks, we introduce time smoothness to better capture the process of network evolution. In addition, the sparsity of complex networks is often overlooked in the case of metabolic networks. By incorporating node similarity in the community detection process, we can improve the overall consideration of the relationships between nodes in the network. That can effectively overcome the limitations of existing methods and data sparsity issues. In summary, the main contributions of our work are as follows:

1. We propose a novel semi-NMF method based on an evolutionary clustering framework. it can effectively reduce the impact of noise and enhances the discriminability of small communities.
2. We develop a temporal smoothness technique to capture the dynamic evolution of metabolic networks. In addition, we design a node similarity fusion mechanism to effectively address the sparsity issue in metabolic networks.

That can improve the accuracy and reliability of community detection in biological metabolic networks.

3. We proposed an iterative optimization algorithm to solve the objective function. The convergence and correctness of the algorithm were proved.
4. We conduct extensive experiments to fully evaluate our proposed model. On both synthetic and real-world networks. Significant outperformance is obtained against several typical baselines in all metrics. Compared to other algorithms, it can exhibit better community detection performance in dynamic metabolic networks.

The rest of this paper is organized as follows. In Sect. 2, we review the relevant works on temporal community detection and metabolic network community detection. Section 3 presents the framework of SNF-MNet. Section 4 describes the experimental results on both synthetic and real-world datasets. Finally, conclusions and discussions are presented in Sect. 5.

2   Related Work

Our work relates to temporal community detection in metabolic networks. Next, we survey the works on these topics.

2.1   Temporal Community Detection

Temporal community detection is an important research area in network analysis, which aims to identify groups of nodes that are highly interconnected in different time points or stages. Many methods have been developed for this purpose, such as clustering-based, network-based, and model-based methods.

Clustering-based methods, such as hierarchical clustering and k-means clustering, are commonly used for temporal community detection. For example, the work by Vangimalla et al. [17] applied hierarchical clustering to identify temporal communities in a human brain functional network. Similarly, the study by Manipur et al. [15] utilized k-means clustering to detect temporal communities in a protein-protein interaction network.

Network-based methods, such as modularity optimization and random walk, are also popular for temporal community detection. For instance, the work by Traag et al. [16] proposed an algorithm based on modularity optimization to detect temporal communities in a mobile phone call network. Similarly, the study by Xin et al. [18] applied a random walk algorithm to identify temporal communities in an online social network.

In addition, some studies have developed model-based methods for temporal community detection. For example, the work by Zhao et al. [2] proposed a stochastic block model to detect temporal communities in a citation network. Similarly, the study by Appel et al. [1] developed a dynamic latent space model for temporal community detection in a protein interaction network.

These methods have been successfully applied to various fields, including social networks, biological networks, and communication networks, and have shown promising results in detecting temporal communities.

2.2   Metabolic Network Community Detection

Community detection in biological metabolic networks has garnered significant attention in recent years as it offers a powerful way to unravel the intricate relationships between metabolites in biological systems. A plethora of studies have been conducted to explore various approaches for identifying communities or clusters of metabolites in metabolic networks. Clustering-based approaches, such as hierarchical clustering and k-means clustering, have been widely used in this context. For instance, Zhang et al. [22] employed k-means clustering to identify metabolite modules in a liver cancer metabolic network. Similarly, Huang et al. [7] utilized hierarchical clustering to detect metabolic subnetworks in human brain metabolism.

Network-based methods, such as modularity optimization and random walk, have also been proposed as effective community detection approaches. Yan et al. [20] proposed a modularity-based method to identify metabolite modules in a breast cancer metabolic network. Additionally, Li et al. [9] utilized a random walk algorithm to discover metabolic communities in a human gut microbiota metabolic network.

Furthermore, some studies have explored the incorporation of additional information, such as gene expression data or pathway information, to improve community detection in metabolic networks. For example, Yuan et al. [21] integrated metabolite concentration data and gene expression data to identify metabolic modules in a liver cancer metabolic network. In a similar vein, Liu et al. [13] utilized pathway information to enhance the detection of metabolic subnetworks in human brain metabolism. These studies represent a step forward in understanding the intricate relationships among metabolites in biological systems, and further work in this direction is likely to uncover additional insights that could have significant implications for biomedical research.

3   Methodology

In this section, we provide a detailed description of our proposed methodology, which includes the modeling of biological metabolic networks, the Semi-NMF Method with Node Similarity Fusion (SNF-MNet) algorithm, and its specific implementation. Specifically, we introduce a time-smoothness constraint and incorporate node degree and shared neighbor information to fuse node similarities, thereby preserving both micro-structural and meso-structural community structures. We present the iterative formulation and solution procedure and conclude by presenting the pseudocode of the algorithm.

3.1   Metabolic Network Modeling

Biological metabolic networks are commonly modeled as reaction networks, where metabolites and enzymes are represented as nodes and metabolic reactions as edges, establishing a static network model. However, to capture the dynamic nature of metabolic processes, temporal information is introduced by modeling the metabolic reactions as a time series and simulating the network evolution accordingly. Specifically, time is discretized into multiple time steps, and the changes in metabolite concentrations and enzyme activities are considered, resulting in the establishment of a dynamic biological metabolic network model. In this model, the states of metabolite and enzyme nodes change over time, and the occurrence or absence of metabolic reactions also alters the network's topological structure over time.

For any biological metabolic network $G = (V, E)$, where $V = (v_1, v_2, \ldots, v_n)$ denotes the set of vertices and $E = (e_1, e_2, \ldots, e_m)$ represents the set of links, the network is dynamic and each moment contains a snapshot of the network. These snapshots are represented using a binary adjacency matrix X_t, where t corresponds to the corresponding network slice. The elements of the complex network matrix, $X_{t,ij}$, indicate whether node i and node j are directly connected in snapshot t, i.e., whether there exists an edge connecting them. If an edge exists, $X_{t,ij} = 1$, otherwise $X_{t,ij} = 0$.

3.2   Node Similarity Fusion

To accurately quantify the similarity between nodes in a network, we considered how to measure their relationships in network snapshot analysis. Specifically, for each node in the network snapshot, we computed its set of neighboring nodes at different time points, and calculated the similarity between the sets of neighboring nodes at different time points using various graph-based similarity metrics, such as cosine similarity. We then integrated the node similarities at different levels to obtain a comprehensive similarity measure for each node. Finally, we replaced the original adjacency matrix with this measure as the input data for our algorithm, thereby enabling a more precise depiction of the relationships and similarities between nodes.

In social networks, first-order proximity $S^{(1)}$ is an important similarity measure that quantifies the degree of similarity between nodes. Specifically, first-order proximity $S^{(1)}$ measures the most intuitive probability of closeness between two nodes i and j, namely the probability that they are directly connected. When there is a connection between nodes i and j, their similarity score $S_{ij}^{(1)}$ is positive, indicating a strong association between them. Conversely, if there is no connection between nodes i and j, their similarity score is 0, indicating no direct connection between them. In this paper, for ease of computation, we assume that the adjacency matrix X is equivalent to the first-order adjacency matrix $S^{(1)}$.

However, in practical scenarios, the edges in complex networks are often sparse, meaning that there are only a few edges between nodes. Thus, using first-order proximity to describe node similarity may be too simplistic. To address this, the second-order proximity matrix $S^{(2)}$ is introduced to provide a more accurate description of the similarity between nodes. Specifically, if two nodes do not have a direct edge between them but share many common neighbor nodes, they can still be considered similar.

To determine the second-order proximity, $N_i = (S_{i1}^{(1)}, S_{i2}^{(1)}, \ldots, S_{in}^{(1)})$ is defined as the set of first-order proximities between node i and all other nodes. Here, the cosine similarity is used to compute the second-order proximity matrix. For two nodes i and j, where $S_{ij}^{(2)} = \frac{N_i N_j}{\|N_i\|\|N_j\|}$, and $\|N\|$ denotes the norm of N. Based on the properties of cosine similarity, the values of the second-order proximity matrix are between 0 and 1. To balance the first and second-order proximity properties, a weight variable η is introduced to connect them, i.e. $S = S^{(1)} + \eta S^{(2)}$, where $\eta > 0$.

3.3   SNF-MNet Formulation

This paper deals with a data network matrix that is unconstrained and based on semi-nonnegative matrix factorization for community structure modeling. Specifically, we design a cluster centroid matrix $F = (f_1, f_2, \ldots, f_k)$, where each column represents the center node of the corresponding community and usually has positive and negative attributes. The community membership matrix G is a non-negative matrix that represents whether s_i belongs to the c_k cluster. If the element g_{ik} is equal to 1, then s_i belongs to the c_k cluster, otherwise, it is 0. Assuming K-means clustering is performed on the input matrix S, the objective function of K-means clustering can be expressed as:

$$\sum_{i=1}^{n} \sum_{k=1}^{K} g_{ik} \|s_i - f_k\|^2 = \|S - FG^T\|^2 \tag{1}$$

In order to account for the mutual influence between adjacent time steps, we introduce a temporal smoothness constraint and incorporate it into the community membership matrix. Specifically, we define a temporal cost that represents the probability of significant changes occurring in the community membership structure between t−1 and t. By adding the temporal cost of the smoothness constraint to the temporal cost of the network topology, we define a cost function that is the sum of the quality of the community discovery and the historical cost. To address this problem, we maximize the quality of the community discovery at the current time and minimize the historical cost, thereby determining the cost function. As such, the resulting cost function is formulated as follows:

$$\min_{S_t \geq 0, G_t \geq 0} \|S_t - F_t G_t^T\|_F^2 + \alpha \|S_{t-1} - F_{t-1} G_t^T\|_F^2 \tag{2}$$

The weight parameter α for the time smoothness constraint is a crucial parameter, which typically ranges from 0 to 1, balancing the changes in community membership structure and the cost of time.

3.4   Optimization

Using the gradient descent method, we solve the objective function (2) with the exclusion of the first time slice to obtain an iterative algorithm. The algorithm employs relevant update rules and convergence of the objective function value as the termination condition for iterations, leading to the detection of community membership structures. The specific update rules and their proof are presented below.

Due to the nature of non-negative matrix factorization, its solution space is non-convex and discrete, and there is no global optimal solution. However, under the condition of fixing a parameter, local optimal solutions can be found. When G_t is fixed, updating F_t to minimize Eq. (2) can be simplified to the following form:

$$\min_{S_t \geq 0, G_t \geq 0} \|S_t - F_t G_t^T\|_F^2 \tag{3}$$

To meet the non-negativity constraint, we introduce a Lagrange multiplier matrix λ and incorporate it into the objective function to obtain the optimal solution. When the Lagrange multiplier is set to zero, the objective function can be simplified to a form without constraints. In this process, the matrix λ constrains the non-negativity of S_t, allowing us to obtain stable and feasible results during the solution process. The resulting objective function is as follows:

$$L(F_t) = tr(-S_t^T F_t G_t^T - G_t F_t^T S_t + G_t F_t^T F_t G_t^T) \tag{4}$$

Let $\frac{\partial L(F_t)}{\partial F_t} = 0$,

$$-2S_t^T G_t + 2F_t G_t^T G_t = 0 \tag{5}$$

Introducing the KKT stationary point condition, we can obtain:

$$-2S_t^T G_t + 2F_t G_t^T G_t = \lambda_{ij} F_{t,ij} = 0 \tag{6}$$

Based on our derivation, we can perform an equivalent transformation on Eq. (6).

$$(-2S_t^T G_t + 2F_t G_t^T G_t)(F_{t,ij})^2 = 0 \tag{7}$$

Therefore, the updated formula for the variable F_t can be expressed as:

$$F_{t,ij} \leftarrow F_{t,ij} \left(\frac{(S_t^T G_t)_{ij}}{(F_t G_t^T G_t)_{ij}} \right) \tag{8}$$

Similarly, when F_t is fixed and G_t needs to be updated, the non-negative Lagrangian multiplier matrix λ is introduced to constrain the non-negative property of G_t. Assuming the partial derivative of G_t is 0, we can use a series of mathematical operations, such as introducing the KKT stationary conditions, to further simplify and optimize the updating process of G_t. Through these operations, we can obtain the updating formula of G_t as follows:

$$G_{t,ij} \leftarrow G_{t,ij} \left(\frac{(\alpha F_t^T S_t + (1-\alpha)F_{t-1}^T S_{t-1})_{ij}}{(\alpha F_t^T F_t G_t + (1-\alpha)F_{t-1}^T F_{t-1} G_t)_{ij}} \right) \tag{9}$$

Based on the aforementioned updating rules, an iterative algorithm can be derived, and the optimization algorithm for the SNF-MNet model can be succinctly expressed via the pseudocode outlined in Algorithm 1.

Algorithm 1. The SNF-MNet model

Input: X_t: The adjacency matrix of the complex network at time t; α: The weight parameter of temporal smoothness; η: The tradeoff parameters for the composite proximity matrix.

Output: F_t: Cluster centroid matrix; G_t: Community membership matrix.

1: Compute the composite proximity matrix S_t based on the adjacency matrix X_t over all timesteps.;

2: **if** $t = 1$ **then**

3:     Obtain the initial F_1 and G_1 through the random non-negative matrix factorization of X_1.

4: **else**

5:     **while** the objective function has not converged. **do**

6:         $F_{t,ij} \leftarrow F_{t,ij}\left(\frac{(S_t^T G_t)_{ij}}{(F_t G_t^T G_t)_{ij}}\right)$

7:         $G_{t,ij} \leftarrow G_{t,ij}\left(\frac{(\alpha F_t^T S_t + (1-\alpha)F_{t-1}^T S_{t-1})_{ij}}{(\alpha F_t^T F_t G_t + (1-\alpha)F_{t-1}^T F_{t-1} G_t)_{ij}}\right)$

8:         Normalize such that $\sum_j F_{ij} = 1, \forall i$.

9:     **end while**

10: **end if**

11: return F_t, G_t

---

4   Experimental Results

In this section, we comprehensively evaluate our proposed SNF-MNet algorithm and compare it with other widely recognized methods, including DECS [12], ECD [19], DYNMOGA, FacetNet [14], AFEECT-spectral, and AFFECT-kmeans [3], as well as the NMF-based original community detection method [4]. These methods have high reputation and usage in the field of community detection. DECS and ECD are genome-dependent models, DYNMOGA employs a multi-objective optimization method based on local smoothing, and FacetNet and AFFECT are two classical algorithms that adopt the principle of evolutionary clustering. To evaluate the performance of the SNF-MNet algorithm, we select multiple complex network datasets for testing, including three synthetic and real-world temporal complex networks. The selection of these networks is to demonstrate the accuracy and feasibility of the algorithm and verify its performance on a wide range of datasets.

To obtain unbiased results, we used the source code provided by the authors with default parameters as reported in their original publication. In our experiments, we set the parameters α for both SNF-MNet and Semi-NMF to 0.3 and η for SNF-MNet to 5. As some of the compared algorithms may converge to local minima, we ran 20 experiments and averaged the results to improve the accuracy of our findings.

4.1   Performance Metrics

To quantitatively evaluate the performance of different community detection algorithms, we employed two widely recognized performance metrics. The first metric, accuracy (AC) [11], is a measure of the dissimilarity between the community membership matrix obtained by the algorithm and the ground truth membership matrix of the network. In essence, AC can be interpreted as an error rate, and it is defined as follows:

$$AC = \|GG^T - G'G'^T\|_F \tag{10}$$

In this context, we use the Frobenius norm to measure the difference between the community membership matrix obtained by the dynamic community detection algorithm, denoted as G, and the ground truth community membership matrix of the network denoted as G'. The AC value is defined as the Frobenius norm of the difference between G and G'. In this process, a smaller AC value indicates a more accurate result obtained by the model.

Normalized Mutual Information (NMI) [8] is used as one of the accuracy metrics to evaluate the community detection models. This metric is only applicable to static networks with ground-truth communities, therefore, we compute NMI at each time step. The similarity between the obtained community partition and the ground-truth community structure can be measured by the following formula:

$$NMI(G, G') = \frac{\sum_{G,G'} p(G, G') log \frac{G,G'}{p(G)p(G')}}{max(H(G), H(G'))} \tag{11}$$

In the equation, $H(G)$ and $H(G')$ represent the entropy of G and G' respectively, with values ranging from 0 to 1. The closer the NMI value is to 1, the higher the degree of matching between the model and the ground truth.

4.2   Synthetic Dataset: Dynamic-GN Metabolic Networks

The concept of synthesizing GN networks was originally proposed by Girvan et al. [5]. In their work, GN networks were utilized for community detection in static complex networks, where each synthesized network comprised 128 nodes and 4 communities. Expanding upon this notion, Lin et al. [10] introduced the notion of dynamic GN networks, which integrated the temporal evolution of communities within the network. Their methodology involved the random redistribution of a certain proportion of vertices from each community to other communities at specific time intervals, resulting in a total of 10 time steps to ensure consistent and stable dynamic changes. By manipulating the parameter settings, it becomes possible to generate dynamic biological metabolic networks. The outcomes of these networks serve as a valuable resource for assessing and comparing various community detection algorithms. The utilization of dynamic GN networks in real-world scenarios has the potential to yield more precise and dependable results, thereby enhancing the overall effectiveness of the analysis.

In the Dynamic-GN metabolic network, we set the data parameters as follows: the transmission rate nc is [0.1, 0.3], the average degree d is fixed at 20, and network noise z is set to 3. Each method was run 20 times, and the average values were obtained as results. The NMI results, shown in Fig. 1, demonstrate that the network becomes increasingly complex over time, leading to poor performance for all methods. However, SNF-MNet, Semi-NMF, and DECS outperform other methods in terms of NMI performance. Nevertheless, due to the influence of noise, the NMI of DECS and ECD fluctuates within some time steps, as shown in Fig. 1(a) at the 8th and 9th time steps, and in Fig. 1(b) at the 6th and 9th time steps, while methods based on NMF are less affected by it. It is worth noting that SNF-MNet performs better than all other methods at almost all time steps, and our approach is more stable compared to DECS and ECD. The introduced time smoothing technique helps eliminate the influence of individual time steps and to some extent makes the adjacent time steps correlated, thus making our results more robust. Overall, compared to Semi-NMF, SNF-MNet shows NMI improvements ranging from 1% to 3.2%.

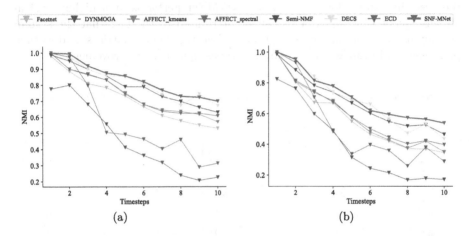

Fig. 1. The average NMI result of the Dynamic-GN Metabolic Networks is represented by (a, b), with the following parameter values: (a) nc = 10%; and (b) nc = 30%.

4.3 Synthetic Dataset: Dynamic-LFR Metabolic Networks

In order to create synthetic datasets with greater representativeness of real-world scenarios, we employed the improved LFR model developed by Greene et al. [6] This model can simulate the continuous changes of nodes and edges in the network to reflect the dynamic nature of real biological metabolic networks. By setting appropriate parameters in the LFR model, dynamic networks similar to real biological metabolic networks can be generated. Additionally, the LFR model can also be used to generate metabolic networks of different sizes and complexities. To evaluate the performance of our method, we constructed a synthetic network containing 1000 nodes and 36 communities, evolving over 10-time steps, with different synthetic networks created by varying the model parameters. Specifically, we set the following base parameters: μ is a mixing parameter that controls the edge density between communities, set at 0.1 and 0.7; p is the probability of transition between nodes within communities, set at 0.5; the average degree of nodes is set to 10; and timesteps T is set to 10.

By utilizing the LFR model to generate a temporal network, we obtained exemplary results as presented in Fig. 2. Experimental findings demonstrate that SNF-MNet performs better in network generation than other benchmark methods. Specifically, as observed in Fig. 2(a), while other methods yield the highest normalized mutual information (NMI) value of 0.965 on this dataset, SNF-MNet consistently attains NMI values within the range of 0.966 to 0.970. As for the initial few time steps shown in Fig. 2(b), SNF-MNet exhibits comparatively weaker performance, but its performance gradually improves with the increase in time steps. Between the 5th and 10th time steps, SNF-MNet's NMI results exceed the maximum values of other methods by 1% to 2.8%. With the exception of the first time step's lower accuracy, SNF-MNet performs at a higher level in all subsequent time steps. This can be attributed to the fact that the system initially utilized the random initialization factorization of matrices, and as time progresses and evolves, the system undergoes significant improvements.

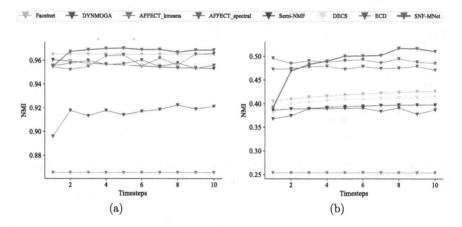

Fig. 2. The average NMI results on the dynamic LFR dataset are presented, where the dataset parameters are set as follows: (a) mixing parameter μ of 0.1; (b) mixing parameter μ of 0.7.

4.4 Real-World Datasets with Metabolic-Similar Structure: KIT Networks

Based on the email communication records of the Information Technology Department at Karlsruhe Institute of Technology (KIT), we have compiled the KIT-email1 dataset[1], which includes 1,097 email accounts and 27,887 messages

---

[1] http://i11www.iti.uni-karlsruhe.de/en/projects/spp1307/emaildata.

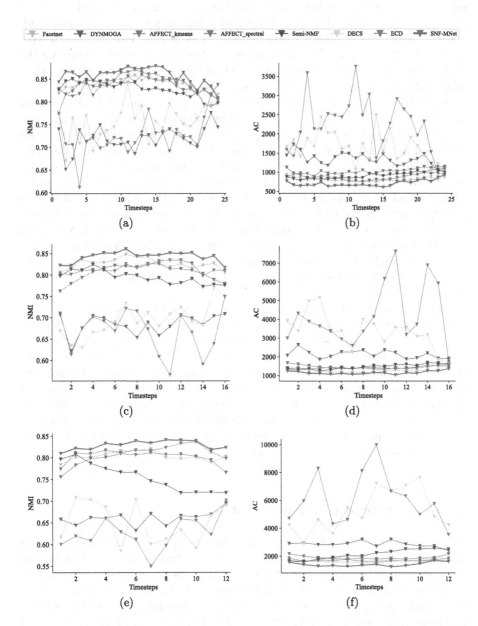

Fig. 3. The average results of NMI and AC on the KIT Dataset. Each row of the image represents a separate dataset, with snapshots taken every 2, 3, and 4 months from top to bottom.

sent. In this network, vertices represent email senders or receivers, and edges represent the connection between two vertices when an email is sent. We used this dataset to construct time networks with intervals of 2, 3, and 4 months, and calculated the number of communities in each time interval, which were 23, 25, and 25, respectively. This dataset has been widely used to evaluate the performance of dynamic community detection algorithms and to improve their adaptability to complex networks in the real world.

Currently, there is a lack of complete study and compilation of biological metabolic networks. Furthermore, existing metabolic network datasets may suffer from incomplete and inaccurate data due to limitations and errors in data collection methods. Therefore, in the absence of a complete and accurate biological metabolic network, a metabolic network can be defined as a network that describes the concentration and transformation relationships of metabolites within a biological organism. The KIT dataset, on the other hand, describes a contact network formed by KIT Information Department emails. Although they have different physical characteristics, both systems are complex systems composed of nodes and edges in terms of network structure. Therefore, we can analogize the KIT-mail dataset as experimental data for metabolic networks to study issues such as similarity, dynamic changes, and community structure in metabolic networks.

With reference to Fig. 3, the results of the three KIT datasets are presented. Specifically, we consider snapshots of the KIT datasets at 2, 3, and 4 months. The KIT datasets demonstrate that the basic evolutionary clustering framework is better suited, outperforming genome fusion methods by comparison. Additionally, SNF-MNet emerges as the most effective method across the majority of time steps for almost all datasets. For instance, in terms of the normalized mutual information (NMI) index, SNF-MNet outperforms other methods by an average of 1.7% in the best case, and likewise for the accuracy (AC) index, as well as exhibiting superior enhancement effects compared to Semi-NMI. These findings corroborate the efficacy of the composite proximity matrix that we have introduced for real-world datasets.

5   Conclusion and Future Work

As complex networks continue to evolve and develop, community detection has increasingly become a research hotspot and mainstream direction. Conducting community detection in metabolic networks helps us understand the functionality of metabolic pathways and the interaction modes between metabolites, which has significant application value in studying the physiological and biochemical processes of metabolites, exploring the pathogenesis of metabolic diseases, and discovering drug targets.

To address the instability of community structures in dynamic networks, this paper proposes a new community detection model for metabolic networks called SNF-MNet. The model introduces the concept of temporal smoothness to improve the stability and accuracy of community structures. At the same time, SNF-MNet comprehensively considers the topological structure and similarity relationship between nodes in metabolic networks. It adopts the idea of evolutionary clustering and semi-nonnegative matrix factorization to integrate node similarity into network analysis, effectively solving the problem of insufficient community division in other methods. By processing the micro-structure of metabolic networks, the model can also effectively solve the problem of data sparsity and make the connections between nodes within communities more compact. Moreover, by using the semi-nonnegative matrix factorization (Semi-NMF) method, SNF-MNet can effectively improve the distinguishability between communities, reduce the impact of noise, and improve the accuracy and reliability of community detection.

To verify the performance and accuracy of SNF-MNet, this paper conducts comparative experiments with various widely recognized temporal network community detection algorithms. The experimental results show that SNF-MNet exhibits superior community detection performance in dynamic metabolic networks compared to other algorithms. In the future, we will strive to address the high-time complexity problem of community detection in large-scale complex networks and plan to expand our model to adapt to more application scenarios.

Acknowledgements. This work is supported by the National Natural Science Foundation of China (62276279), Key-Area Research and Development Program of Guangdong Province (2020B0101100001), the Tencent WeChat Rhino-Bird Focused Research Program (WXG-FR-2023-06), and Zhuhai Industry-University-Research Cooperation Project (2220004002549).

References

1. Appel, A.P., Cunha, R.L.F., Aggarwal, C.C., Terakado, M.M.: Temporally evolving community detection and prediction in content-centric networks. In: Berlingerio, M., Bonchi, F., Gärtner, T., Hurley, N., Ifrim, G. (eds.) ECML PKDD 2018. LNCS (LNAI), vol. 11052, pp. 3–18. Springer, Cham (2019). https://doi.org/10.1007/978-3-030-10928-8_1
2. Bickel, P.J., Chen, A., Zhao, Y., Levina, E., Zhu, J.: Correction to the proof of consistency of community detection. Ann. Stat., 462–466 (2015)
3. Blackman, L., Venn, C.: Affect. Body Soc. **16**(1), 7–28 (2010)
4. Ding, C.H., Li, T., Jordan, M.I.: Convex and semi-nonnegative matrix factorizations. IEEE Trans. Pattern Anal. Mach. Intell. **32**(1), 45–55 (2008)
5. Girvan, M., Newman, M.E.: Community structure in social and biological networks. Proc. Natl. Acad. Sci. **99**(12), 7821–7826 (2002)
6. Greene, D., Doyle, D., Cunningham, P.: Tracking the evolution of communities in dynamic social networks. In: 2010 International Conference on Advances in Social Networks Analysis and Mining, pp. 176–183. IEEE (2010)
7. Huang, S., Chaudhary, K., Garmire, L.X.: More is better: recent progress in multi-omics data integration methods. Front. Genet. **8**, 84 (2017)
8. Jing, L., Ng, M.K., Huang, J.Z.: An entropy weighting K-means algorithm for subspace clustering of high-dimensional sparse data. IEEE Trans. Knowl. Data Eng. **19**(8), 1026–1041 (2007)

9. Li, J., et al.: An integrated catalog of reference genes in the human gut microbiome. Nat. Biotechnol. **32**(8), 834–841 (2014)
10. Lin, Y.R., Chi, Y., Zhu, S., Sundaram, H., Tseng, B.L.: FacetNet: a framework for analyzing communities and their evolutions in dynamic networks. In: Proceedings of the 17th International Conference on World Wide Web, pp. 685–694 (2008)
11. Lin, Y.R., Chi, Y., Zhu, S., Sundaram, H., Tseng, B.L.: Analyzing communities and their evolutions in dynamic social networks. ACM Trans. Knowl. Discovery from Data (TKDD) **3**(2), 1–31 (2009)
12. Liu, F., Wu, J., Xue, S., Zhou, C., Yang, J., Sheng, Q.: Detecting the evolving community structure in dynamic social networks. World Wide Web **23**, 715–733 (2020)
13. Liu, Z.P.: Identifying network-based biomarkers of complex diseases from high-throughput data. Biomark. Med. **10**(6), 633–650 (2016)
14. Ma, X., Dong, D.: Evolutionary nonnegative matrix factorization algorithms for community detection in dynamic networks. IEEE Trans. Knowl. Data Eng. **29**(5), 1045–1058 (2017)
15. Manipur, I., Giordano, M., Piccirillo, M., Parashuraman, S., Maddalena, L.: Community detection in protein-protein interaction networks and applications. IEEE/ACM Trans. Comput. Biol. Bioinf. (2021)
16. Traag, V.A., Van Dooren, P., De Leenheer, P.: Dynamical models explaining social balance and evolution of cooperation. PLoS ONE **8**(4), e60063 (2013)
17. Vangimalla, R.R., Sreevalsan-Nair, J.: Comparing community detection methods in brain functional connectivity networks. In: Balusamy, S., Dudin, A.N., Graña, M., Mohideen, A.K., Sreelaja, N.K., Malar, B. (eds.) ICC3 2019. CCIS, vol. 1213, pp. 3–17. Springer, Singapore (2020). https://doi.org/10.1007/978-981-15-9700-8_1
18. Xin, Y., Xie, Z.Q., Yang, J.: An adaptive random walk sampling method on dynamic community detection. Expert Syst. Appl. **58**, 10–19 (2016)
19. Xu, X., Gu, R., Dai, F., Qi, L., Wan, S.: Multi-objective computation offloading for internet of vehicles in cloud-edge computing. Wireless Netw. **26**, 1611–1629 (2020)
20. Yan, K.K., Lou, S., Gerstein, M.: MrTADFinder: a network modularity based approach to identify topologically associating domains in multiple resolutions. PLoS Comput. Biol. **13**(7), e1005647 (2017)
21. Yuan, Q., et al.: Integration of transcriptomics, proteomics, and metabolomics data to reveal HER2-associated metabolic heterogeneity in gastric cancer with response to immunotherapy and neoadjuvant chemotherapy. Front. Immunol. **13**, 951137 (2022)
22. Zhang, A., Sun, H., Yan, G., Han, Y., Ye, Y., Wang, X.: Urinary metabolic profiling identifies a key role for glycocholic acid in human liver cancer by ultra-performance liquid-chromatography coupled with high-definition mass spectrometry. Clin. Chim. Acta **418**, 86–90 (2013)

UKGAT: Uncertain Knowledge Graph Embedding Enriched KGAT for Recommendation

Quanyang Leng(iD), Wanting Jiang(iD), and Nan Guo$^{(\boxtimes)}$(iD)

School of Computer Science and Engineering, Northeastern University, Shenyang
110000, Liaoning, China
guonan@mail.neu.edu.cn

Abstract. Heterogeneous Information Networks (HINs) with complex
structures and rich semantics, which show great flexibility in modeling
the heterogeneous data of HINs, are widely used in recommendation systems. Effective embedding of HINs not only represents the complex and
heterogeneous auxiliary data in the graph, facilitating efficient extraction
and utilization of information by the recommender system, but also has
the advantage of retaining rich semantics, which provides more accurate,
varied, and interpretable recommendation results. An uncertain knowledge graph embedding method based on the expanded meta-path semantics of KGAT was proposed, which not only inherits the higher-order connectivity among nodes in KGAT in an end-to-end approach with recursive delivery of embedded information (either users, items, or attributes)
from neighboring nodes but also enables the relationships in HINs to
be uncertain by expanding the meta-path semantics of HINs to better
capture the underlying semantic relationships among entities and then
incorporate the structured knowledge they contain into machine learning. Finally, the effectiveness of the UKGAT was demonstrated through
experiments on two real-world datasets, which outperformed state-of-the-
art models in capturing semantic relationships among entities, it could
be better utilized in many knowledge-driven applications.

Keywords: Uncertain Knowledge Graph · Heterogeneous Information
Networks · Meta Path · Semantic Augmentation · High-order
Connectivity

1 Introduction

Knowledge Graphs (KGs), a structured representation of human knowledge, have
aroused much attention in academia and industry. Real-world datas often exist
in the form of graph structures, such as search engines, recommendation systems, the World Wide Web, etc. Graph Neural Networks (GNN), as a model for
learning graph structure and understanding path semantics, have shown excellent performance in network analysis and recommendation systems, and thus

X. Yang et al. (Eds.): ADMA 2023, LNAI 14179, pp. 433–447, 2023.
https://doi.org/10.1007/978-3-031-46674-8_30

have received a lot of attention from scholars. For example, Deep Neural Networks [18] are used to learn the features and structural representation of nodes in a graph.

Recently, attention mechanisms based on deep learning have attracted widespread research interest, which can not only handle variable volume data but also allow models to focus on the most important parts of data. Attention mechanisms have shown outstanding capabilities for Data Mining and various real-world applications, such as text analysis, knowledge graphs, and other fields. The growth of the Internet has been accompanied by an exponential increase in the volume of data that needs to be stored. In traditional recommendation algorithms, although collaborative filtering algorithms [14] can predict users' preferences from their historical behaviors, collaborative filtering-based recommendation methods suffer from data sparsity and cold-start problems. To integrate user ID and item ID, supervised learning (SL) [20] can convert them into a generic feature vector and compute a prediction score, but it does not take into account the interaction between users and items. So it is not possible to extract collaborative signals related to attributes from a large number of users with a large number of behaviors.

To address the limitations of SL-based recommendations, Knowledge Graph is introduced into recommendations. Knowledge graphs provide structured representations of real-world entities and relationships, and they can be classified into deterministic and uncertain knowledge graphs according to relations between entities in KGs. Deterministic KGs, i.e., all entities, attributes, and relationships in a knowledge graph are deterministic without any uncertainty or incompleteness. However, an uncertain knowledge graph refers to the existence of uncertainty or incompleteness of entities, attributes, or relationships, including but not limited to the problems of missing entity attributes, unclear entity relationships, entity ambiguity, and incomplete knowledge. In deterministic KGs, many wonderful Translational models have been generated, such as TransE, TransH, TransR, and TransD; and bilinear models also have a place, such as DistMult and ComplEx, they have made promising progress in many tasks, such as link prediction, relation extraction, relation learning, etc. Chen et al. [4] proposed a model UKGE for embedding uncertain knowledge graphs, which effectively exploits the uncertainty of relationships between entities, preserving both structure and uncertainty information of relational facts in the embedding space. The approach, which differs from the previously used binary classification techniques to characterize the relational facts, learns embeddings based on the confidence scores of uncertain relationship facts so that it improves the accuracy of UKGE by probabilistically assigning confidence scores during training.

Considering the paucity of research on augmenting deterministic knowledge graphs with uncertain information, this paper firstly proposes a method for embedding uncertain knowledge based on deterministic knowledge graphs, which not only expands the entities in heterogeneous information networks but also unearths more relationships between entities, which are with probabilities. The model is known as the Uncertain Knowledge Graph Attention Network model (UKGAT).

The main contributions of this work are summarized as follows:

- Proposed a semantic extension theory for meta-paths in deterministic KGs and provided theoretical support for knowledge-driven applications based on KGs, including recommender systems.
- Proposed a model UKGAT for embedding uncertain knowledge based on deterministic KGs, which provides more diverse, accurate, and interpretable recommendation results for recommendation models.
- The effectiveness of the embedding model was demonstrated by conducting extensive experiments on two real datasets, Amazon-book and Yelp2018.

The rest of the paper is organized as follows. Firstly, we review the related work in Sect. 2 and explain the model framework in Sect. 3. Then the experimental details and result are presented in Sect. 4. Finally, the paper is concluded in Sect. 5.

2 Related Work

A knowledge graph is a knowledge base that uses a triple structure to describe specific knowledge and semantic information, as well as directed graphs for knowledge representation and storage. To solve the problems of low computational efficiency and poor portability caused by using triple structure, an important research method called knowledge graph embedding (KGE) is proposed, which has a wide range of applications in structure complementation [12], relationship extraction [15], question-answer systems [2], social network recommendation [13] of knowledge graphs, etc.

Usually, whether a knowledge graph has factual certainty is used as a classification criterion to distinguish between deterministic knowledge graphs and uncertain knowledge graphs. The NELL knowledge base uses fact-based triples to learn inference rules, which contains a real-time knowledge base of 120 million different confidence-weighted beliefs. The ConceptNet [10] is composed of relational knowledge in the form of a triad, which implements the interconnection of various types of information in natural language. According to current advances in uncertain knowledge graph embedding techniques, the PASSLEAF framework [5] cleverly used scoring functions with different types of confidence to predict the confidence scores between entities and each relationship then used sample pools to reduce the number of spurious samples Li et al. [8] applied uncertain knowledge graph ProBase to the relationship extraction (RE) task and proposed a multi-view framework MIUK based on uncertain knowledge graphs. Brockmann et al. [3] used uncertain knowledge graphs to represent the relationships between link networks in the supply chain link prediction problem. Meanwhile, the attention mechanism is introduced in the graph convolutional neural network to implement the weight distribution among different entities. DisenKGAT [16] model uses disentangled representation Learning to effectively reduce the independence restriction between sub-representations in different semantic spaces in a complex knowledge graph complementation task. AttnPath [11] applies LSTM and graph attention networks to path inference of knowledge graphs, using deep reinforcement learning techniques instead of traditional deep learning, and training reinforcement learning intelligence by trial and error so that the model's effectiveness does not depend on pre-training. The GGAE [9] model uses the attention mechanism of entity and relationship graphs to learn features for entity and relationship embedding.

MTRec [1] is a multi-task learning framework using a self-attentive mechanism to learn the semantic information related to meta-paths in heterogeneous information networks for modeling the interaction information between users, items, and meta-paths in recommendation systems. IntentGC [19] introduced social relationships among users as auxiliary information that can fully utilize display preferences and heterogeneous relationships. ACKRec [7] uses meta-paths in heterogeneous information networks to capture semantic information related to student preferences, solving the problem of information sparsity that exists in recommender systems. MEIRec [6] adequately captures information about the user's intention when interacting with items in recommendations by performing object embedding on meta-path-guided heterogeneous graph neural networks.

In the traditional recommendation system, there may be issues of data sparsity and cold start, which can be solved by making full use of the auxiliary information in the knowledge graph to expand the data. As shown in Fig. 1 below, "Entourage" and "Chaplin" are recommended to Jane. And this knowledge graph contains various entities such as recommended users, actors, directors, movie titles, and movie genres. However, friends, couples, family, enjoyed, acted, directed, act, direct, and genre belong to the interaction relationships. The potential interaction relationships between the entities can improve the accuracy of the recommendation and also make the recommendation result interpretable, e.g., the reason for recommending Chaplin to Jane is because Kent has seen the movie Chef with the same actor.

The ability of the embedding approach of uncertain KGs for deterministic KGs to expand the information and semantics of meta-paths in heterogeneous information networks is mainly explored. The relevant work focuses on the following:

Uncertain Knowledge Graphs and Uncertain Datasets. To achieve semantic expansion of all entities of meta-paths, knowledge graphs and datasets with uncertainty and covering a wide range of knowledge are preferred, such as yelp_dataset, Amazon_review_data etc., because there is a large amount of overlap between the knowledge of ConceptNet and the keywords extracted from Amazon_review_data or yelp_dataset, these two datasets are selected as the benchmarks.

Construction and Optimization of Amazon Dictionary. This part completes the construction of the Amazon lexicon, which requires extracting the review data in the Amazon dataset to determine the keywords, embedding the

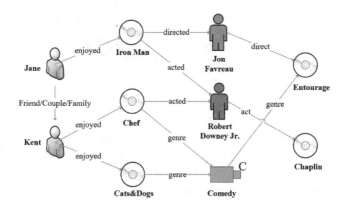

Fig. 1. An example of a knowledge graph-based recommendation.

uncertain knowledge graph ConceptNet to find its related words for the keywords, and forming triple between words. When using Amazon data for model training, a two-array AC automaton is used for speedup.

Uncertain Knowledge Graph Attention Network. Based on the Amazon Dictionary construction of the heterogeneous information network method proposed in the appeal, the heterogeneous information network based on the Amazon dataset is semantically expanded using the uncertain information present in the uncertain knowledge graph ConceptNet to obtain richer semantic information.

3 Methodology

This paper proposes an uncertain knowledge embedding model (UKGAT) based on deterministic KGs. The model framework, shown in Fig. 2, consists of four main parts:

1) Semantic Expansions for Mate-Paths in HINs: In this step this paper firstly propose a meta-path semantic augmentation theory, which leverages ConceptNet to expand the semantics of meta-paths in HINs to reveal potential relationships and connections between entities.

2) Confidence Score Modeling of Embedded Relationship Facts: Considering the uncertainty of the KG embedding model, this paper explicitly models the confidence score of each triplet to reasonably compare the predicted and true values.

3) Attentional Embedding Propagation Layers: In this part, higher-order connections between nodes are simulated in an end-to-end manner with deterministic and uncertain relations in the connection space. And then representation of each node is updated by cascadingly propogating the attention weights of each node.

4) Prediction Layer: The representation of a user node and an item from all propogations layers is aggregated and the predicted match score is outputed.

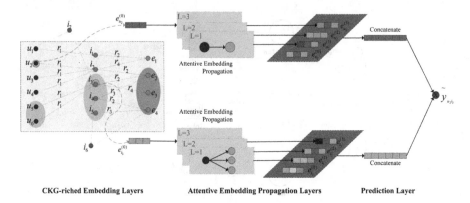

Fig. 2. Structure diagram of the UKGAT model, which consists of the expanded CKG model, attention embedding propagation layer, and prediction layer, respectively. u_i is the i-th target user, r_i is the corresponding relation, i_i is an item and e_i is a kind of attribute.

3.1 Semantic Expansions for Mate-Paths in HINs

Semantic expansion of meta-paths in HINs requires enriching the properties of one or more entities and items in a meta-path to make it a heterogeneous information network containing more information. As an example, there is a simple meta-path $Kent \xrightarrow{watch} Cats\&Dogs$, indicates that the user Kent used to watch the movie Cats&Dogs, and the semantics of this meta-path can be expanded like this: $Kent \xrightarrow{watch} Cats\&Dogs \xleftarrow{genre} Comedy$, this meta-path indicates that Kent watched a movie "Cats&Dogs" in the genre of comedy. To make the expanded items or entities have a strong relevance to the user, this paper uses reviews of the product being interacted with in the Amazon_datasets or Yelp_datasets, which contain user's opinions about products. Extracting keywords from the user's reviews can often correlate with more information.

This subsection focuses on completing the modeling of HINs based on deterministic knowledge graphs and introducing the ConceptNet uncertain knowledge graph for embedding. Because there are multiple uncertain relationships of entities in ConceptNet, it is used to semantically expand the meta-paths in the heterogeneous information network formed by these two datasets. The effect of the semantically expanded heterogeneous information network is shown in Fig. 3. ConceptNet can be accessed at https://conceptnet.io/.

3.2 Confidence Score Modeling of Embedded Relationship Facts

UKGE was used as the embedding model for uncertain knowledge graphs and was improved for experiments. In contrast to traditional deterministic KGs, uncertain KGs establish a confidence score for each triple and compare the prediction with the true score to provide a more comprehensive understanding.

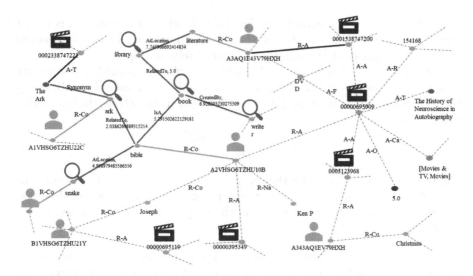

Fig. 3. The effect of the semantic expansion of heterogeneous information networks.

Reasonable Value. Given a triple l, the reasonable value $g(l) \in \mathbb{R}$ of this triple indicates how likely the fact of the relationship represented by this triple holds. A larger plausible value indicates a higher confidence score for this triple. The reasonableness of a triple is calculated by using the following formula:

$$g(l) = e_r \cdot (e_h \bullet e_t), \quad g(l) \in \mathbb{R} \tag{1}$$

where $e_h, e_t, e_r \in \mathbb{R}^k$, $\bullet$ denotes the element-wise product and $\cdot$ denotes the inner product. And e_h, e_t, e_r are used to represent the embedding vectors for head h, tail t and relation r respectively. The function is cited from DistMult, which captures the relationship between e_h and e_t conditional on the relationship r.

From Reasonable Value to Confidence Scores. To convert reasonable value to confidence scores, this paper employs a logistic mapping function $\phi(x)$. The transformation from plausible scores $g(l)$ to confidence scores $f(l)$ is achieved as follows:

$$f(l) = \phi(g(l)) \in [0, 1] \tag{2}$$

where $\phi(x)$ denotes the Logistic function, which is a common S-curve:

$$\phi(x) = \frac{1}{1 + e^{-(wx+b)}} \in [0, 1] \tag{3}$$

where w is a weight and b is a bias.

Reasoning About Unobservable Relational Facts Based on Confidence Scores. In deterministic KGs, unobserved relational facts are often assigned 0 for confidence score. However, this is unreasonable. Even if a direct relationship does not exist between two entities, it may be inferred that an indirect relationship between two entities might exist through the transfer of relationships.

For a given triple l that can be observed, their observed confidence scores are used for assignment; and for a unseen triple l, this paper uses the embedding-based confidence score function $f(l)$ for l, this paper makes $\mathcal{I}(l)$ as the interaction $\mathcal{I}$ of a triple l:

$$\mathcal{I}(l) = \begin{cases} s_l, & l \in \mathcal{L}^+ \\ f(l), & l \in \mathcal{L}^- \end{cases} \tag{4}$$

where $\mathcal{I}(l) \in [0, 1]$, $\mathcal{L}^+$ denotes the set of observed triples, while $\mathcal{L}^-$ denotes the set of unobserved triples, s_l denotes the confidence score of the observed triples, and $f(l)$ denotes the confidence score of the unobserved triples obtained by the mapping function.

Embedding Uncertain Knowledge Graph. There are deterministic triples and uncertain triples in the network after expanding the heterogeneous information network, for the deterministic triples, this paper can reflect the embedding ability of the model by calculating the variance of the true confidence score s_l of the triples and the predicted confidence score $f(l)$.

$$\zeta^+ = \sum_{l \in \mathcal{L}^+} |f(l) - s_l|^2 \tag{5}$$

And for the unobserved triple l, minimizing the confidence score $f(l)$ obtained for each triad according to the logical inference PSL, specifically using the variance of the distance as a loss function:

$$\zeta^- = \sum_{l \in \mathcal{L}^-} \sum_{\gamma \in \Gamma_l} |\Psi_\gamma(f(l))|^2 \tag{6}$$

where $\Psi_\gamma(x)$ denotes the weighted distance in UKGE, and γ denotes a given PSL rule. Γ_l is the set of ground rules with l as the rule head. By associating Eq. 5 and Eq. 6, the joint objective function $\mathcal{H}(l)$ can be obtained:

$$\mathcal{H}(l) = \sum_{l \in \mathcal{L}^+} |f(l) - s_l|^2 + \sum_{l \in \mathcal{L}^-} \sum_{\gamma \in \Gamma_l} |\Psi_\gamma(f(l))|^2 \tag{7}$$

During training this paper samples unseen triples by corrupting the head h and the tail t for observed relations facts to generate $\mathcal{L}^-$.

3.3   Attentional Embedding Propagation Layers

Firstly the graph convolutional neural network is used to recursively propagate the embedding information of the nodes, and then the graph attention mechanism is utilized to discover the weights of neighbor nodes of each node to reveal the importance of each neighbor node. In this regard, there are three main parts in this layer, which are information propagation, meta-path attention mechanism, and information aggregation.

Information Dissemination Layer. Considering an entity h and using $S_h = \{(h,r,t)|(h,r,t) \in \mathcal{G})\}$ to denote the set of triples, where $\mathcal{G} = \{(h,r,t)|h,t \in \mathcal{E}', r \in \mathcal{R}'\}$, where $\mathcal{E}'$ is the set of entities and $\mathcal{R}'$ is the set of relations. Also h is the head entity, termed self-network. To characterize the first-order connectivity structure of entity h, the linear structure of the self-network of entity h is calculated using the following equation:

$$e_{S_h} = \sum_{(h,r,t) \in S_h} \pi(h,r,t)e_t \tag{8}$$

where $\pi(h,r,t)$ denotes the decay factor of each propagation on the edge of (h,r,t) and represents the amount of information propagated from t to h through r each time.

Attentional Mechanisms. To calculate the decay factor $\pi(h,r,t)$, it is proposed to use the tanh nonlinear activation function, which is calculated as follows:

$$\pi(h,r,t) = (W_r e_t)^{\mathrm{T}} \cdot tanh(W_r e_h + e_r) \tag{9}$$

where $W_r \in \mathbb{R}^{k \times d}$ is the transformation matrix of relation r, which projects the entities from the d-dimensional entity space into the k-dimensional relationship space. From this formula, the attention score depends on the distance between e_h and e_t in the relation r-space. Thus, closer entities can propagate more information, and only the inner product is used on these representations for simplicity.

Information Aggregation. The entity representation e_h and its self-network representation e_{S_h} are used as new representation of the entity h, more formally, $e_h^{(1)} = f(e_h, e_{S_h})$. This paper uses these following three kinds of aggregators to implement $f(\cdot)$:

- GCN Aggregator applies a non-linear operation after addition, as follows:

$$f_{GCN} = LeakyReLU(W(e_h + e_{S_h})) \tag{10}$$

where $W \in \mathbb{R}^{d' \times d}$, and d' is the transformation size. LeakyReLU is a kind of activation function.
- GraphSage Aggregator concatenates e_h and e_{S_h}, followed by LeakyReLU.

$$f_{GraphSage} = LeakyReLU(W(e_h \| e_{S_h})) \tag{11}$$

where $\|$ is the concatenation operation.
- Bi-Interaction Aggregator fully considers interactions feature between e_h and e_{S_h}, as follows:

$$f_{Bi-Interaction} = LeakyReLU(W_1(e_h + e_{S_h})) \\ + LeakyReLU(W_2(e_h \bullet e_{S_h})) \tag{12}$$

where $W_1, W_2 \in \mathbb{R}^{d' \times d}$ are trainable weight matrices.

Higher-Order Connectivity. Higher-order connectivity information is explored by stacking more propagation layers to collect information propagated from higher-hop neighbors. In the l-th step, the representation of entities is recursively formulated as:

$$
\begin{aligned}
e_h^{(l)} &= f(e_h^{(l-1)}, e_{S_h}^{(l-1)}) \\
e_{S_h}^{(l-1)} &= \sum_{(h,r,t)\in S_h} \pi(h,r,t) e_t^{(l-1)}
\end{aligned}
\tag{13}
$$

$e_t^{(l-1)}$ is the representation of entity t generated from the previous information propagation steps, which stores the information from its $(l-1)$-hop neighborhoods. $e_h^{(0)}$ is set as e_h at the initial information propagation iteration, which further contributes to the representation of entity h at layer l.

3.4 Model Prediction

After executing L layers, multiple representations of user node u are obtained, i.e., $\{e_u^{(1)}, \cdots, e_u^{(L)}\}$, for item node i, $\{e_i^{(1)}, \cdots, e_i^{(L)}\}$, this paper adopts the layer-aggregation mechanism [17] to concatenate the representations of each step into a vector, finally, the inner product of user and item representations is performed to predict their matching scores as follows:

$$
\widetilde{y}(u,i) = e_u^{*\,\mathrm{T}} e_i^{*}
\tag{14}
$$

where $e_u^* = e_u^{(0)} || \cdots || e_u^{(L)}$, $\quad e_i^* = e_i^{(0)} || \cdots || e_i^{(L)}$.

4 Experiments

4.1 Datasets

To evaluate the effectiveness of the UKGAT model, two benchmark datasets, Amazon-Book and Yelp2018, were used for the experiments.

Amazon-Book is a widely used data for recommendation systems and knowledge graphs. To ensure the quality of the experiments, a 10-core setting in Amazon-Book was chosen, which means that all user-item interactions have at least 10 items.

Yelp2018 is from the 2018 Yelp Challenge, in which restaurants and bars are the items in dataset. The experiment uses a 10-core setup to ensure that the user-item interactions are also at least 10.

For each dataset, the dataset is randomly divided into 70%, 20%, and 10%, which are set as the training set, validation set, and test set, respectively. For each observable user-item interaction, it is treated as a positive instance. The unseen instances were then used to predict the confidence scores of the interactions of unseen entities using PSL logical inference.

Table 1. Overall performance comparison.

Datasets	Amazon-Book		Yelp2018	
Model	recall@20	ndcg@20	recall@20	ndcg@20
FM	0.1287	0.0866	0.0612	0.0753
NFM	0.1311	0.0899	0.0662	0.0810
CKG	0.1322	0.0881	0.0665	0.0796
CFKG	0.1147	0.0776	0.0513	0.0639
MCRec	0.1121	0.0756	-	-
KGAT	**0.1485**	**0.1003**	**0.0711**	**0.0867**
UKGAT	**0.1532**	**0.1102**	**0.0789**	**0.0934**
%Improv	3.16%	9.87%	10.97%	7.73%

4.2 Experimental Setup

Evaluation Metrics. For each user in the test set, this work does not treat all items that do not interact with the user as negative items, as KGAT does, reasons out the confidence scores of all unobserved entity-item interactions by PSL logical inference instead. A user preference score for all items is then output for each method. To evaluate the validity of top-K recommendations and preference ranking, two widely used evaluation protocols, recall@K, and ndcg@K are used. Similarly, K is set to 20 and the average metric of all users in the test set is reported.

Baseline Model. To demonstrate the effectiveness of the improved model, the proposed UKGAT model will be used to compare with the baseline models, which includes supervised learning-based FM and NFM models, regularization-based CFKG and CKE models, and path-based MCRec. The models are described as follows:

- **FM**: This is a factorization model proposed to solve the problem of feature combination with sparse data. It has good generality for both continuous and discrete features
- **NFM**: An improved model based on the FM model, which models the relationship between features of a higher order through a deep network.
- **CKE**: A representative regularization-based approach that exploits the semantic embedding of TransR to enhance matrix decomposition.
- **CFKG**: An TransE-based approach that can represent the intrinsic connection of users, items, entities, and relationships in the representable graph and treat the recommendation task as the plausibility prediction of a triple (h, r, t) in the graph.
- **MCRec**: A Top-N recommendation model based on contextual and neural co-attentive models of meta-paths, which extracts qualified meta-paths for representing the connections between users and items.

4.3 Performance Comparison

Overall Performance Comparison. The proposed UKGAT is implemented by Pytorch, and the embedding size of all models is fixed to 64. This work choose Adam optimizer to optimize all models and its batch_size is set to 1024. The learning rate is set to 0.0001, For the KGAT and UKGAT, their dropout rate is the same as 0.1. The structures of the hidden layers are set to 512,256,128,64, respectively. Then, the layer L of KGAT and UKGAT is searched in $\{1, 2, 3, 4\}$ and the dimensions of the fully connected layers for feature cross-pooling are 64,32, and 16, respectively. Finally, remaining parameters of the model are kept at their default values. The overall performance comparison results are shown in Table 1 and the following observations were reached:

- The UKGAT consistently outperforms the baseline model on all data-sets. In particular, the recall@20 of the UKGAT improved by 3.16% and 10.97% compared to the original KGAT on the Amazon-Book and Yelp2018 datasets, respectively. By semantically expanding the deterministic knowledge graph, the originally unobservable samples are not treated as negative samples and have a certain confidence score as positive samples, which effectively assists in the capture of collaborative signals between entities.
- In most cases, supervised learning-based methods (i.e., FM and NFM) obtain better performance than CFKG and CKE-based models. This might indicate the inability of regularization-based methods to use the knowledge of items fully. Additionally, FM and NFM are embeddings that can be used to connect entities to enrich the representation of items on deterministic and uncertain edges in heterogeneous graphs. The intersection feature of both is a representation of second-order connectivity between users and entities. CKF and CFKG only use embeddings of their aligned entities, and although they model connectivity at the granularity of triples, their higher-order connectivity remains constant.

Effect of Model Depth. To explore whether the layer L of the model has the same effect on the model UKGAT as KGAT, the layer L of UKGAT is changed to investigate whether the expansion of the HINs affects the performance of multiple embedded propagation layers in UKGAT. Four different layers L were purposely set for the experiments, and the results are reported in Table 2 below.

- Among the four layers of KGAT, for the Amazon-Book dataset, both KGAT and UKGAT show better performance with increasing layer number. However, in the Yelp2018 dataset, the performance of KGAT-2 is higher than that of KGAT-3, but this phenomenon does not occur in UKGAT. The UKGAT with multiple embedding propagation layers exhibits better performances than the KGAT. Hence, it indicates that the semantic expansion of HINs is more beneficial to the propagations of information from multiple embedding layers.
- Analyzing the results of UKGAT-3 and UKGAT-4, it can be observed that the performance of UKGAT-4 is improved but only by a little, thus it shows

that the third order relationship between entities is sufficient to capture the signal of collaboration.

- Jointly analyzing Tables 1 and 2, we observe that the performance of UKGAT-1 is always better than the other baseline models in most cases. This validates the effectiveness of attention propagation and empirically shows that it can better model first-order relationships.

Table 2. Effect of the number of embedded layers L on UKGAT.

Datasets	Amazon-Book		Yelp2018	
Model	recall@20	ndcg@20	recall@20	ndcg@20
KGAT-1	0.1288	0.0866	0.0697	0.0851
KGAT-2	0.1316	0.0942	0.0734	0.0884
KGAT-3	**0.1485**	**0.1003**	**0.0711**	**0.0867**
KGAT-4	0.1517	0.1106	0.0775	0.0923
UKGAT-1	0.1326	0.0936	0.0692	0.0844
UKGAT-2	0.1454	0.1002	0.0752	0.0896
UKGAT-3	**0.1532**	**0.1102**	**0.0789**	**0.0934**
UKGAT-4	0.1547	0.1126	0.0791	0.0949

Influence of Model Clustering Approach. To explore the performance of aggregators on UKGAT, a variant of UKGAT-1 with different settings and a variant of KGAT-1 with different settings were used. More specifically, the aggregators used are GCN, GraphSage, and Bi-Interaction, respectively.

From Table 3, it can be seen that the performance of the GCN aggregator is better than that of the GraphSage aggregator in both the KGAT and the UKGAT. One possible reason for this is that GraphSage abandons the interaction between the entity representation e_h and its self-network representation e_{S_h}, and thus the feature interaction when performing information aggregation and propagation importance may be less effective. Bi-Interaction combines additional feature interactions and therefore improves the capability of the representation, again illustrating the rationality and effectiveness of the bidirectional interaction aggregator.

Among the three aggregators mentioned above, the performance of UKGAT is not significantly improved. This indicates that the expansion of uncertain knowledge to a deterministic knowledge graph does not affect the aggregation capability of the aggregator.

5 Summary

This work investigates the semantic expansion of meta-paths for deterministic KGs and uses it for recommendation in graph attention networks. A novel

Table 3. Impact of Aggregators.

Datasets	Amazon-Book		Yelp2018	
Aggregators	recall@20	ndcg@20	recall@20	ndcg@20
KGAT$_{GCN}$	0.1378	0.0931	0.0687	0.0831
KGAT$_{GraphSage}$	0.1366	0.0927	0.0634	0.0825
KGAT$_{Bi-Interaction}$	**0.1391**	**0.0933**	**0.0693**	**0.0844**
UKGAT$_{GCN}$	0.1381	0.0936	0.0685	0.0832
UKGAT$_{GraphSage}$	0.1366	0.0925	0.0633	0.0826
UKGAT$_{Bi-Interation}$	**0.1392**	**0.0933**	**0.0695**	**0.0844**

framework, UKGAT, is mainly designed, which combines not only the embedding capability of the UKGE for uncertain information in uncertain KGs but also the KGAT modeling higher-order connectivity in CKGs presented in an end-to-end manner. In the core attention embedding propagation layer, it can adaptively propagate the embedding information from the original neighbor nodes, the embedding information goes from the neighbor nodes after expanding the semantics of meta-paths to make the embedding information of a node richer. The two real-world datasets of extensive experiments show the plausibility and feasibility of UKGAT. This work demonstrates that applying uncertain knowledge graphs to graph neural networks enriches the semantic information of meta-paths in graphs, which also provides unlimited possibilities for interpretative recommendations.

References

1. Bi, Q., Li, J., Shang, L., Jiang, X., Liu, Q., Yang, H.: MTRec: multi-task learning over BERT for news recommendation. In: Findings of the Association for Computational Linguistics, ACL 2022, pp. 2663–2669 (2022)
2. Bordes, A., Chopra, S., Weston, J.: Question answering with subgraph embeddings. arXiv preprint arXiv:1406.3676 (2014)
3. Brockmann, N., Elson Kosasih, E., Brintrup, A.: Supply chain link prediction on uncertain knowledge graph. ACM SIGKDD Explor. Newsl. **24**(2), 124–130 (2022)
4. Chen, X., Chen, M., Shi, W., Sun, Y., Zaniolo, C.: Embedding uncertain knowledge graphs. In: Proceedings of the AAAI Conference on Artificial Intelligence, vol. 33, pp. 3363–3370 (2019)
5. Chen, Z.M., Yeh, M.Y., Kuo, T.W.: PASSLEAF: a pool-based semi-supervised learning framework for uncertain knowledge graph embedding. In: Proceedings of the AAAI Conference on Artificial Intelligence, vol. 35, pp. 4019–4026 (2021)
6. Fan, S., et al.: Metapath-guided heterogeneous graph neural network for intent recommendation. In: Proceedings of the 25th ACM SIGKDD International Conference on Knowledge Discovery & Data Mining, pp. 2478–2486 (2019)
7. Gong, J., et al.: Attentional graph convolutional networks for knowledge concept recommendation in MOOCs in a heterogeneous view. In: Proceedings of the 43rd International ACM SIGIR Conference on Research and Development in Information Retrieval, pp. 79–88 (2020)

8. Li, B., Ye, W., Huang, C., Zhang, S.: Multi-view inference for relation extraction with uncertain knowledge. In: Proceedings of the AAAI Conference on Artificial Intelligence, vol. 35, pp. 13234–13242 (2021)
9. Li, Q., Wang, D., Feng, S., Niu, C., Zhang, Y.: Global graph attention embedding network for relation prediction in knowledge graphs. IEEE Trans. Neural Netw. Learn. Syst. **33**(11), 6712–6725 (2021)
10. Speer, R., Chin, J., Havasi, C.C.: 5.5: an open multilingual graph of general knowledge. In: Proceedings of the Thirty-First AAAI Conference on Artificial Intelligence, pp. 4444–4451, December 2016
11. Wang, H., Li, S., Pan, R., Mao, M.: Incorporating graph attention mechanism into knowledge graph reasoning based on deep reinforcement learning. In: Proceedings of the 2019 Conference on Empirical Methods in Natural Language Processing and the 9th International Joint Conference on Natural Language Processing (EMNLP-IJCNLP), pp. 2623–2631 (2019)
12. Wang, Q., Mao, Z., Wang, B., Guo, L.: Knowledge graph embedding: a survey of approaches and applications. IEEE Trans. Knowl. Data Eng. **29**(12), 2724–2743 (2017)
13. Wang, X., He, X., Cao, Y., Liu, M., Chua, T.S.: KGAT: knowledge graph attention network for recommendation. In: Proceedings of the 25th ACM SIGKDD International Conference on Knowledge Discovery & Data Mining, pp. 950–958 (2019)
14. Wang, X., He, X., Wang, M., Feng, F., Chua, T.S.: Neural graph collaborative filtering. In: Proceedings of the 42nd International ACM SIGIR Conference on Research and Development in Information Retrieval, pp. 165–174 (2019)
15. Wang, Z., Zhang, J., Feng, J., Chen, Z.: Knowledge graph embedding by translating on hyperplanes. In: Proceedings of the AAAI Conference on Artificial Intelligence, vol. 28 (2014)
16. Wu, J., et al.: DisenKGAT: knowledge graph embedding with disentangled graph attention network. In: Proceedings of the 30th ACM International Conference on Information & Knowledge Management, pp. 2140–2149 (2021)
17. Xu, K., Li, C., Tian, Y., Sonobe, T., Kawarabayashi, K.i., Jegelka, S.: Representation learning on graphs with jumping knowledge networks. In: International Conference on Machine Learning, pp. 5453–5462. PMLR (2018)
18. Zhang, C., Song, D., Huang, C., Swami, A., Chawla, N.V.: Heterogeneous graph neural network. In: Proceedings of the 25th ACM SIGKDD International Conference on Knowledge Discovery & Data Mining, pp. 793–803 (2019)
19. Zhao, J., et al.: IntentGC: a scalable graph convolution framework fusing heterogeneous information for recommendation. In: Proceedings of the 25th ACM SIGKDD International Conference on Knowledge Discovery & Data Mining, pp. 2347–2357 (2019)
20. Zhou, G., et al.: Deep interest network for click-through rate prediction. In: Proceedings of the 24th ACM SIGKDD International Conference on Knowledge Discovery & Data Mining, pp. 1059–1068 (2018)

Knowledge Graph Link Prediction Model Based on Attention Graph Convolutional Network

Junkang Shi[1,2], Ming Li[3], and Jing Zhao[1,2(✉)]

[1] Key Laboratory of Computing Power Network and Information Security, Ministry of Education, Shandong Computer Science Center (National Supercomputer Center in Jinan), Qilu University of Technology (Shandong Academy of Sciences), Jinan, China
[2] Shandong Provincial Key Laboratory of Computer Networks, Shandong Fundamental Research Center for Computer Science, Jinan, China
zjstudent@126.com
[3] School of intelligence and Information Engineering, Shandong University of Traditional Chinese Medicine, Jinan, China

Abstract. The main purpose of knowledge graph embedding (KGE) is to complete the missing part of triplets. By learning from existing models, we found that adding graph convolutional networks (GCN) to KGE models achieved good performance. These models mainly use the local structural characteristics of data. However, their main goal is to learn an aggregator rather than a feature vector for each node. To address the problems with existing models, we proposed a knowledge graph link prediction model based on attentional relational graph convolutional networks (ARGCN). First, we capture the relationship attribute features between nodes by adding an attention mechanism and extract important node attribute features for visualization. Then, through a relational graph convolutional neural network, we capture the associated attributes between nodes and extract the feature information that can best represent nodes. Finally, we use DistMult as the scoring function. Our experiments show that our proposed is effective on standard FB15k-237 and WN18RR datasets. Compared with other models, it has relative improvements in Hits@1, Hits@3 and Hits@10.

Keywords: Knowledge graph embedding · Graph convolutional networks · Attention

1 Introduction

Knowledge graph link prediction refers to predicting missing links based on known triplets in a knowledge graph. Currently, knowledge graph link prediction has been successfully applied to many downstream tasks and applications such as intelligent question-answering systems [10], information retrieval [4], and visual relationship detection [5].

In existing knowledge graph link prediction models, the main research direction is to learn the distributed representation of entities and relationships in knowledge graphs. For example, existing translation models (TransE [1], TransH [8], etc.), tensor decomposition models (DistMult [9], Complex [7], etc.) and neural network models (ConvE [3], ConvR [2], etc.). However, the existing model structures are relatively simple and have poor expressive power. They can't fully transform the feature representation of nodes. Moreover, entity attributes and associations are complex and often have different features for different entities.

In this paper, we propose a knowledge graph link prediction model based on attentional relational graph convolutional networks (ARGCN). The introduction of attention mechanism allows the model to focus on learning the required information. The introduction of relational graph convolutional networks can explore the internal relationships between different nodes after adding attention mechanisms, thereby extracting more critical node information to update node feature representations. At the same time, in order to reduce the error of node information, we use average feature information for fusion when fusing neighbor node information. Finally, we use DistMult as the decoder and use negative sampling to train the model. For each observable example, we sample ω negative examples. Sampling is performed by randomly shuffling the subject or object of each positive instance while optimizing the cross-entropy loss, with positive triplets scoring higher than negative triplets.

Our contributions are mainly as follows:

- In response to the problems existing in existing models, we propose the ARGCN model, which can focus more on effective information in complex neighborhood information of nodes by introducing attention mechanism.
- Using an end-to-end model structure, we use attentional relational graph convolutional networks (AGCN) with attention added as the encoder and decoder DistMult. The experimental results show the effectiveness of the model structure.

2   Model Description

In this section, we will describe the ARGCN model in detail. The overall architecture of the model is shown in the figure, which mainly includes the encoder AGCN and the decoder DistMult. The encoder focuses on updating node feature representations by introducing an attention mechanism of the relationship graph convolutional network. At the same time, in order to avoid the loss of node features, we added a single self-connection of a special relationship type to each node in the data. The decoder uses DistMult decomposition as a scoring function. When used alone, it performs well in standard link prediction benchmarks.

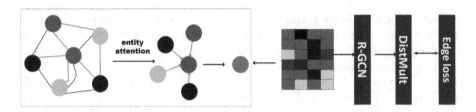

Fig. 1. The ARGCN model description is: ARGCN consists of two parts. First, the feature representation of nodes is updated through attention mechanisms, and the updated nodes are passed into a relation graph convolutional network. Then, the model is scored using DistMult.

2.1 AGCN

The translation mentions a graph structure used $\mathcal{G} = (\mathcal{V}, \mathcal{E})$, where $\mathcal{V}$ is the node set and $\mathcal{E}$ is the triple of nodes-edges-nodes (s, r, o). The node features are represented by f_a and f_b, respectively. After feature weight calculation, e_{ab} represents the attention weight of node b to node a, that is, the degree of association between nodes. At this time, attention is assigned to all nodes but does not highlight the structural information of nodes with interactive relationships. ∂ is the attention mechanism parameter (weight), with a dimension of $\mathbb{R}^{n \times n}$, and $\mathbf{W}$ is the weight parameter with a dimension of $\mathbb{R}^{b \times b}$.

$$e_{ab} = \partial(\mathbf{W}f_a, \mathbf{W}f_b) \tag{1}$$

The ∂_{ab} is obtained by normalizing all adjacent nodes b of node a through the introduced Softmax function, indicating the proportion of adjacent node b to all adjacent nodes c of node a, and the attention mechanism focuses on nodes that have interactive relationships through the adjacency matrix $\mathbf{A}$. The formula is as follows

$$\partial_{ab} = \text{Soft max}_b(e_{ab}) = \frac{\exp(e_{ab})}{\sum_{i=1}^{c} \exp(e_{ac})} \tag{2}$$

$$F_i = \partial_{ab} f_i \tag{3}$$

Formula (2) takes the dot product of the attention mechanism ∂_{ab} and F_i to obtain $F_i \in \mathbb{R}^{n \times d}$, which can enhance or weaken the expression of feature information, thereby adaptively optimizing the influence of relevant features and improving the model's ability to extract important feature factors of nodes. We pass the updated nodes obtained after the attention layer into the relation graph convolutional network. In the encoder, we map each entity $v_i \in \mathcal{V}$ to a real-valued vector $e_i \in \mathbb{R}^d$, relying on vertex representations to reconstruct the edges of the graph. The key to our work is the reliance on an encoder. Although most previous methods directly optimize a single real-valued vector e_i for every $v_i \in \mathcal{V}$ in training, we compute representations through a relation graph convolutional network encoder with $e_i = h_i^{(L)}$. Our full link prediction model is schematically depicted in Fig. 1.

2.2 DistMult

We use DistMult as the decoder of the model. Existing experiments have shown that when used alone, DistMult decomposition performs well as a scoring function in standard link prediction benchmarks.

In DistMult, each relationship r is associated with a diagonal matrix $R_r \in \mathbb{R}^{d \times d}$ and a triple (s, r, o), and scored as

$$f(s, r, o) = e_s^T R_r e_o \qquad (4)$$

For each observed example, we sample ω negative examples, and optimize the cross-entropy loss so that the model scores higher on observed triplets than on negative triplets:

$$\mathcal{L} = -\frac{1}{(1 + \omega)|\hat{\mathcal{E}}|} \sum_{(s, r, o, y) \in \mathcal{T}} y \log l\big(f(s, r, o)\big) + (1 - y) \log\big(1 - l\big(f(s, r, o)\big)\big) \qquad (5)$$

where $\mathcal{T}$ is the total set of actual and corrupted triples, $y = 1$ for positive triples and $y = 0$ for negative triples.

3 Experiments

3.1 Benchmark Datasets

This experiment used four benchmark datasets to evaluate the performance of link prediction. These datasets include FB15k, WN18, FB15k-237 and WN18RR. Among them, FB15k is a commonly used dataset for link prediction. This dataset contains entities that are mentioned more than 100 times in FreeBase. WN18 is extracted from WordNet3. To construct WN18, the authors used WordNet3 as a starting point and then iteratively filtered out entities and relationships that were mentioned too few times. FB15k-237 is a subset of the FB15k dataset, inspired by observations of test leakage suffered by FB15k, including test data seen by the model during training. WN18RR is a subset of WN18 and was also observed after test leakage in WN18.

3.2 Evaluation Protocol

We use four standard metrics to evaluate the performance of link prediction on knowledge graphs, which is a task of inferring missing facts from existing entities in knowledge graphs. These four metrics are Mean Reciprocal Rank (MRR) and Hits@N for N = 1, 3 and 10. MRR is the average of the reciprocal scores of the correctly predicted samples in all test samples, and Hits@N is the proportion of the correctly predicted samples whose scores are not lower than the N-th score in all test samples. For each test set (h, r, t), we replace h and t with all entities in the data set to calculate the scores, and use the scoring function to score them. Then we set up a filter to remove the valid triplets that already exist in the training, validation and test sets from the ranking. For the link prediction task on knowledge graphs, we strictly and fairly handle triplets with the same score. Finally, we sort the positive and negative triplets by score in descending order.

3.3   Experimental

We follow a two-step training process, where we first train AGCN to update node feature representations and then train DistMult as the decoder model to perform relation prediction tasks. We divide the dataset into three groups, including training set, validation set and test set, and train the model by Adaptive Moment Estimation (Adam) algorithm (Kingma and Ba 2014). After comparing different experimental parameter settings, we found that the following experimental parameter settings are more effective: for the data set FB15k-237, set the dropout value to 0.2, set the number of cores to 100, set the learning rate to 0.001, and set the embedding size to 300, the best experimental results were obtained.

Table 1 shows the Hits@10, Hits@3, Hits@1 and MRR test results of our model and other knowledge graph link prediction models on standard datasets FB15k-237, WN18, FB15k-237 and WN18RR. The highest score is shown in bold and the second highest score is shown with an underline. We only provided test results for some of the datasets. It can be clearly seen from the results that, on the dataset FB15k-237, the model proposed in this paper outperforms other models on four indicators; on WN18RR, our model surpasses other models on three indicators. The improvement of MRR value proves that our model can correctly represent the triple relationships. Compared with KBGAT with the same model architecture, on the dataset FB15k-237, our model improves MRR by 3.5%, and Hits@N values are also significantly higher than other models. Compared with the convolutional network model ConvE, on the dataset FB15k-237, our model improves MRR by 6.9%, and Hits@1 and Hits@3 values are also significantly higher than other models. These results show that our model can better achieve link prediction of knowledge graphs.

Table 1. Link prediction for FB15k-237 and WN18RR

Model	FB15K-237				WN18RR			
	Hits			MRR	Hits			MRR
	10	3	1		10	3	1	
TransE [1]	0.465			0.279	0.501			0.243
DisMult [9]	0.42	0.28	0.16	0.25	0.51	0.46	0.41	0.44
ConvE [3]	0.49	0.35	0.24	0.32	0.48	0.43	0.39	0.46
KBGAT [6]	0.331	_		0.35	**0.554**			0.412
ARGCN	**0.559**	**0.398**	**0.28**	**0.363**	0.538	**0.506**	**0.455**	**0.492**

4   Conclusion and Future Work

We introduce a knowledge graph link prediction model based on an attentional relation graph convolutional network (ARGCN) encoder-decoder architecture.

By adding attention mechanisms on top of relation graph convolutional neural networks, we can more effectively update the feature representations of central entities and more efficiently utilize the connectivity structure of knowledge graphs. Our experiments also demonstrate that ARGCN has achieved higher efficiency performance. In the future, we hope to incorporate the idea of neighbor selection into our training framework and consider the importance of neighbors when aggregating vector representations of neighbors. We also want to extend our model to have a larger knowledge graph.

Acknowledgements. This work is supported in part by The Key R&D Program of Shandong Province (2021SFGC0101), The 20 Planned Projects in Jinan (202228120)

References

1. Bordes, A., Usunier, N., Garcia-Duran, A., Weston, J., Yakhnenko, O.: Translating embeddings for modeling multi-relational data. In: Advances in Neural Information Processing Systems, vol. 26 (2013)
2. Çavuşoğlu, I., Pielka, M., Sifa, R.: Adapting established text representations for predicting review sentiment in Turkish. In: 2020 IEEE 7th International Conference on Data Science and Advanced Analytics (DSAA), pp. 755–756. IEEE (2020)
3. Dettmers, T., Minervini, P., Stenetorp, P., Riedel, S.: Convolutional 2D knowledge graph embeddings. In: Proceedings of the AAAI Conference on Artificial Intelligence, vol. 32 (2018)
4. Gaur, M., Gunaratna, K., Srinivasan, V., Jin, H.: ISEEQ: information seeking question generation using dynamic meta-information retrieval and knowledge graphs. In: Proceedings of the AAAI Conference on Artificial Intelligence, vol. 36, pp. 10672–10680 (2022)
5. Hu, Y., Chen, S., Chen, X., Zhang, Y., Gu, X.: Neural message passing for visual relationship detection. arXiv preprint arXiv:2208.04165 (2022)
6. Nathani, D., Chauhan, J., Sharma, C., Kaul, M.: Learning attention-based embeddings for relation prediction in knowledge graphs. arXiv preprint arXiv:1906.01195 (2019)
7. Trouillon, T., Welbl, J., Riedel, S., Gaussier, É., Bouchard, G.: Complex embeddings for simple link prediction. In: International Conference on Machine Learning, pp. 2071–2080. PMLR (2016)
8. Wang, Z., Zhang, J., Feng, J., Chen, Z.: Knowledge graph embedding by translating on hyperplanes. In: Proceedings of the AAAI Conference on Artificial Intelligence, vol. 28 (2014)
9. Yang, B., Yih, W., He, X., Gao, J., Deng, L.: Embedding entities and relations for learning and inference in knowledge bases. arXiv preprint arXiv:1412.6575 (2014)
10. Yin, Y., Zhang, L., Wang, Y., Wang, M., Zhang, Q., Li, G.: Question answering system based on knowledge graph in traditional Chinese medicine diagnosis and treatment of viral Hepatitis B. Biomed. Res. Int. **2022**, 7139904 (2022)

Knowledge Graph Embedding with Relation Rotation and Entity Adjustment by Quaternions

Wen Sun[1], Qingqiang Wu[1,2,3(✉)], Xiaoli Wang[1(✉)], Junfeng Yao[2], and Zhifeng Bao[4]

[1] School of Informatics, Xiamen University, Xiamen 361005, China
{wuqq,xlwang}@xmu.edu.cn
[2] School of Film, Xiamen University, Xiamen 361005, China
yao0010@xmu.edu.cn
[3] Key Laboratory of Digital Protection and Intelligent Processing of Intangible Cultural Heritage of Fujian and Taiwan Ministry of Culture and Tourism, Xiamen University, Xiamen, China
[4] RMIT University, Melbourne, VIC 3000, Australia
zhifeng.bao@rmit.edu.au

Abstract. Recent knowledge graph embedding models have shown promising results on link prediction, by employing operations on quaternion space to capture correlations between entities. However, they only used three quaternion embeddings for rotation calculation that fails to capture the interaction between entities and relations. The single relation quaternion to rotate the head entity also makes the connection between the head and tail entities weak. To address the problem, this paper proposes a novel knowledge graph embedding model denoted as QuatPE, to utilize paired relations to simultaneously rotate the head and tail entities in a triple. We employ the adjustment vectors to adjust the position of the same entity in a quaternion space when facing different triples. With paired relations, QuatPE can strengthen the connection between the head and tail entities which enhances the representation capabilities. By using the adjustment vectors, QuatPE also helps to better handle complex relation patterns, such as 1-to-N, N-to-1, and N-to-N. Experimental results show that QuatPE can achieve significant performance on well-known datasets for link prediction. Researchers can reproduce our results by following the source codes at https://github.com/galaxysunwen/QuatPE-master.

Keywords: knowledge graph embedding · Quaternion · link prediction

1 Introduction

By storing structured information about the facts of the real world in the form of $(head, relation, tail)$, knowledge graphs (KGs) are playing important roles

X. Yang et al. (Eds.): ADMA 2023, LNAI 14179, pp. 454–469, 2023.
https://doi.org/10.1007/978-3-031-46674-8_32

in information retrieval, recommendation systems, and question answering [1]. Despite billions of triples have been stored in KGs, the incompleteness of KGs is still an open issue, which limits the effectiveness of the downstream tasks. To address this issue, much effort has been taken in predicting missing links using knowledge graph embedding (KGE) [2–4]. These methods project the relations and entities into a continuous low-dimensional semantic space, for discovering novel facts. Among them, TransE [2] is an effective method that regards the relation as a translation between the head entity and the tail entity in the triple.

By using the real-valued vector space, learning the KGE in a complex space has been widely used and reported to be a highly effective inductive bias [6].

A recent method denoted by QuatE [6] explores the hypercomplex space for learning the KGE and shows promising performance. However, QuatE directly uses rigorous operations on quaternion space for capturing the interaction between entities and the relation in a triple without considering the connections between entities and relations. For example, in a triple of $(Tom, attacked\_by, flu)$, QuatE does not take much account of the connection between the information of the head entity, such as age and gender, and the information of the tail entity, such as the mode of infection and the susceptible population. Moreover, QuatE only uses three quaternion embedding vectors for rigorous rotation calculation to obtain the triple score, which makes it relatively weak in capturing the feature interaction between entities and relations, which makes it cannot handle complex relation patterns (1-to-N, N-to-1, and N-to-N).

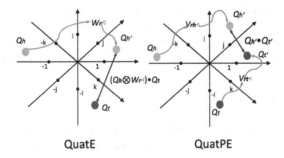

Fig. 1. Simple illustration of QuatPE with the part of paired relations versus QuatE, where Q_h and Q_t represent entities in the triplet. $V_{rh}^{\triangleleft}$, $V_{rt}^{\triangleleft}$, and $W_r^{\triangleleft}$ are all unit quaternion, which represents the relation in the triplet.

To address this issue, this paper proposes an effective model, denoted by QuatPE, with paired relations to enforce the relation-aware correlations between entities. In Fig. 1, QuatPE can simultaneously rotate the quaternion embedding of the head and tail entities to strengthen the connection between them, which can further enhance the representation capabilities. Moreover, QuatPE uses the adjustment vectors to adjust the same entity's quaternion space position which involves different triplets, as shown in Fig. 2. It can better handle complex relation patterns (1-to-N, N-to-1, N-to-N). Our main contributions are as follows:

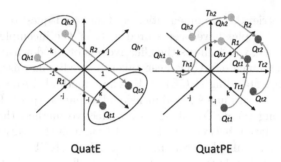

QuatE   QuatPE

Fig. 2. An illustration of QuatPE with the adjustment vectors versus QuatE. Q_{h1} and Q_{t1} represent entities in the triples. R_1, R_2, T_{h1}, T_{h2}, T_{t1}, and T_{t2} are unit quaternions, where R_1 and R_2 are relations in the triple, T_{h1}, T_{h2}, T_{t1}, and T_{t2} are adjustment vectors to the entities.

- We propose an effective model, denoted by QuatPE, to embed paired relations for simultaneously rotating the head and tail entities. It can enhance the connection between the head and tail entities to obtain better representation capabilities.
- QuatPE uses the adjustment vectors of the head and tail entities to adjust the position of the same entity in different triples, for better handling complex relation patterns.
- To verify the effectiveness of QuatPE in downstream tasks, we also develop a multi-task learning model which uses knowledge graph embeddings for recommendation systems.
- We conduct extensive experiments and the results show that QuatPE can significantly enhance the quality of knowledge graph embeddings.

2 Related Work

The KGE methods for the link prediction task can be divided into three groups: the translational distance-based models, the bilinear models, and the neural network-based models, as shown in Table 1.

2.1 Translational Distance-Based Models

Translational distance-based models treat the relation as a translation between the head and the tail entities in a triple, which suppose they satisfy $h + r \approx t$ when (h, r, t) holds, where h, r, and $t \in \mathbb{R}^k$. For example, TransE [2] defines the corresponding score function as $f_r(h, t) = -||h + r - t||_{l1/l2}$. It opens a line of translational distance-based methods.

TransE is widely used for its simplicity. However, it still has drawbacks in modeling 1-to-N and N-to-1 relation patterns. For example, TransE might learn a similar vector for *Chicago* and *Boston* which are all cities in *the_United_States*. To address this issue, entities are allowed to have different representations when

Table 1. Existing KGE models.

Category	Model	Entity embedding	Relation embedding	Scoring Function $f_r(h,t)$
Translational Distance	TransE [2]	$\mathbf{h},\mathbf{t}\in\mathbb{R}^d$	$\mathbf{r}\in\mathbb{R}^d$	$-\|\mathbf{h}+\mathbf{r}-\mathbf{t}\|_{1/2}$
	TransR [7]	$\mathbf{h},\mathbf{t}\in\mathbb{R}^d$	$\mathbf{r}\in\mathbb{R}^k,\mathbf{M}_r\in\mathbb{R}^{k\times d}$	$-\|\mathbf{M}_r\mathbf{h}+\mathbf{r}-\mathbf{M}_r\mathbf{t}\|_2^2$
	TransH [9]	$\mathbf{h},\mathbf{t}\in\mathbb{R}^d$	$\mathbf{r},\mathbf{w}_r\in\mathbb{R}^d$	$-\left\|(\mathbf{h}-\mathbf{w}_r^\top\mathbf{h}\mathbf{w}_r)+\mathbf{r}-(\mathbf{t}-\mathbf{w}_r^\top\mathbf{t}\mathbf{w}_r)\right\|_2^2$
	TransD [11]	$\mathbf{h},\mathbf{t},\mathbf{w}_h,\mathbf{w}_t\in\mathbb{R}^d$	$\mathbf{r},\mathbf{w}_r\in\mathbb{R}^d$	$-\left\|(\mathbf{w}_r\mathbf{w}_h^\top+\mathbf{I})\mathbf{h}+\mathbf{r}-(\mathbf{w}_r\mathbf{w}_t^\top+\mathbf{I})\mathbf{t})\right\|_2^2$
Polar coordinate	HAKE [10]	$\mathbf{h}_m,\mathbf{t}_m\in\mathbb{R}^k$ $\mathbf{h}_p,\mathbf{t}_p\in[0,2\pi)^k$	$\mathbf{r}_m\in\mathbb{R}_+^k$ $\mathbf{r}_p\in[0,2\pi)^k$	$-\|\mathbf{h}_m\circ\mathbf{r}_m-\mathbf{t}_m\|_2-\lambda\|\sin((\mathbf{h}_p+\mathbf{r}_p-\mathbf{t}_p)/2)\|_1$
Complex vector	ComplEx [4]	$\mathbf{h},\mathbf{t}\in\mathbb{C}^d$	$\mathbf{r}\in\mathbb{C}^d$	$\mathrm{Re}\left(<\mathbf{r},\mathbf{h},\bar{\mathbf{t}}>\right)=\mathrm{Re}\left(\sum_{k=1}^K r_k h_k \bar{t}_k\right)$
	RotatE [3]	$\mathbf{h},\mathbf{t}\in\mathbb{C}^d$	$\mathbf{r}\in\mathbb{C}^d$	$\|\mathbf{h}\circ\mathbf{r}-\mathbf{t}\|$
	QuatE [6]	$\mathbf{h},\mathbf{t}\in\mathbb{H}^d$	$\mathbf{r}\in\mathbb{H}^d$	$\mathbf{h}\otimes\frac{\mathbf{r}}{|\mathbf{r}|}\cdot\mathbf{t}$
	QuatRE [12]	$\mathbf{h},\mathbf{t}\in\mathbb{H}^d$	$\mathbf{r},\mathbf{r}_1,\mathbf{r}_2\in\mathbb{H}^d$	$((\mathbf{h}\otimes\frac{\mathbf{r}_1}{|\mathbf{r}_1|})\otimes\frac{\mathbf{r}}{|\mathbf{r}|})\cdot(\mathbf{t}\otimes\frac{\mathbf{r}_2}{|\mathbf{r}_2|})$
	QuatDE [13]	$\mathbf{h},\mathbf{p_h},\mathbf{p_t}\in\mathbb{H}^d$	$\mathbf{r},\mathbf{v}_r\in\mathbb{H}^d$	$(\mathbf{h}\otimes\frac{\mathbf{p_h}}{|\mathbf{p_h}|}\otimes\frac{\mathbf{v}_r}{|\mathbf{v}_r|})\otimes\frac{\mathbf{r}}{|\mathbf{r}|}\cdot(\mathbf{t}\otimes\frac{\mathbf{p_t}}{|\mathbf{p_t}|}\otimes\frac{\mathbf{v}_r}{|\mathbf{v}_r|})$
	DualE [1]	$\mathbf{h},\mathbf{t}\in\mathbb{H}^d$	$\mathbf{r}\in\mathbb{H}^d$	$<(\mathbf{h}\otimes\frac{\mathbf{r}}{|\mathbf{r}|},\mathbf{t}>$
	HopfE [14]	$\mathbf{h},\mathbf{t}\in(\mathbb{R}^{3^*})^d$	$\mathbf{r}\in\mathbb{H}^d$	$\|M(\mathbf{W}_r\otimes\mathbf{h}\otimes\overline{W_r},\mathbf{A}^h)-M(\mathbf{t},\mathbf{A}^t)\|$
Hyperbolic Space	AttH [15]	$\mathbf{h},\mathbf{t}\in\mathbb{B}_c^d,b_h,b_t\in\mathbb{R}$	$\mathbf{r}\in\mathbb{B}_c^d$	$-d_c^{c_r}\left(\mathrm{Att}\left(\mathbf{q}_{\mathrm{Rot}}^H,\mathbf{q}_{\mathrm{Ref}}^H;\mathbf{a}_r\right)\oplus^{c_r}\mathbf{r}_r^H,\mathbf{e}_t^H\right)^2+b_h+b_t$
Semantic Matching	DistMult [16]	$\mathbf{h},\mathbf{t}\in\mathbb{R}^d$	$\mathbf{r}\in\mathbb{R}^d$	$\mathbf{h}^\top\mathrm{diag}(\mathbf{M}_r)\mathbf{t}$
	SimplE [17]	$\mathbf{h},\mathbf{t}\in\mathbb{R}^d$	$\mathbf{r},\mathbf{r}'\in\mathbb{R}^d$	$\frac{1}{2}(\mathbf{h}\circ\mathbf{r}\mathbf{t}+\mathbf{t}\circ\mathbf{r}'\mathbf{t})$
	RESCAL [18]	$\mathbf{h},\mathbf{t}\in\mathbb{R}^d$	$\mathbf{M}_r\in\mathbb{R}^{d\times d}$	$\mathbf{h}^\top\mathbf{M}_r\mathbf{t}$
	TuckER [19]	$\mathbf{h},\mathbf{t}\in\mathbb{R}_e^d$	$\mathbf{r}\in\mathbb{R}_r^d$	$W\times_1\mathbf{h}\times_2\mathbf{r}\times_3\mathbf{t}$
Neural Networks	ConvE [5]	$\mathbf{M}_h\in\mathbb{R}^{d_w\times d_h},\mathbf{t}\in\mathbb{R}^d$	$\mathbf{M}_r\in\mathbb{R}^{d_w\times d_h}$	$\sigma\left(\mathrm{vec}\left(\sigma\left([\mathbf{M}_h;\mathbf{M}_r]*\omega\right)\right)\mathbf{W}\right)\mathbf{t}$
	ConvKB [8]	$\mathbf{h},\mathbf{t}\in\mathbb{R}^d$	$\mathbf{r}\in\mathbb{R}^d$	$\mathrm{concat}\left(\sigma\left([h,r,t]*\omega\right)\right)\cdot\mathbf{w}$
	InteractE [20]	$\mathbf{e}_s,\mathbf{e}_o\in\mathbb{R}^d$	$\mathbf{e}_r\in\mathbb{R}^d$	$g(\mathrm{vec}(f(\phi(\mathcal{P}_k)\otimes\mathbf{w}))W)\mathbf{e}_o$

they are involved in different relations. Therefore, TransH [9] regards relations in triples as translations on relation-specific hyperplanes and projects entities onto them. The entity may have different distances to different relations. The score function is $f_r(h,t) = -\|h - w_r^T h w_r + r - (t - w_r^T t w_r)\|_{l1/l2}$. TransR [7] models entities and relations in entity space and relation space, and judges the distances between the head and tail entities by projecting them from entity space to relation space. TransD [11] further simplifies the model, using two embeddings to represent entities and relations. The first embedding represents the entity or relation, and the other embedding will be used to construct the projection matrix. TransAP [21] introduces position-aware entity embedding and attention mechanism to capture the different semantics of triples for solving the problem that translation-based scoring functions cannot handle the structure of hierarchy and circle.

More recently, RotatE [3] assumes that each relation r is a rotation from the head entity h to the tail entity t., which can encode symmetry, antisymmetry, inverse, and composition relation patterns. HAKE [10] projects entities into a polar coordinate system, which can well solve the problem of entity hierarchy modeling in knowledge graphs.

2.2 Bilinear Models

Bilinear models use similarity-based scoring functions. They match latent semantics of entities and relations embedding vectors by product-based score functions. For example, RESCAL [18] uses a vector to represent each entity and a full rank matrix to represent each relation. Its score function is defined as $f_r(h,t) = h^T M_r t$. Because it is prone to overfitting due to the full rank matrix,

DistMult [16] uses a diagonal matrix to replace a full rank matrix to simplify RESCAL. Based on the issues of modeling antisymmetric relations, ComplEx [4] introduces complex-valued embedding. Some tensor decomposition models have also achieved good results, such as SimplE [17] and TuckER [19], which use CP decomposition and Tucker decomposition, respectively.

More recently, several models project entities or relations into other spaces for better performance. ATTH [15] and ROTH [15] use hyperbolic embedding that can capture both hierarchical and logical patterns. QuatE [6] uses the hyper-complex vector space to enrich the key relation encoding abilities. It uses the quaternion space with the Quaternion Multiplication to learn entity and relation embeddings. QuatDE [13], QuatRE [12], DualE [1], and HopfE [14] all focus on improving the QuatE model. Rotate3D [22] projects entities to the 3D space and represents relation as a rotation from head entity to tail entity in a triple to model the non-commutative composition pattern.

2.3   Neural Network-Based Models

Recently, deep learning models show promising performance. ConvE [5] and ConvKB [8] splice and reshape the entity and relation embeddings using the convolutional neural network. InteractE [20] improves the convolution step based on ConvE, by adding Feature Permutation, Checkered Reshaping, and Circular Convolution, which can better capture the interaction of entities and relations. CompGCN [23] uses a Graph Convolutional framework to embed nodes and relations. Although these methods have moderate performance on link prediction tasks, they often have poor interpretability.

3   Our Model

3.1   Preliminaries

Quaternion is one of the most representative hypercomplex systems, which is an expansion of the traditional complex system [13]. Because the Hamilton product is the interaction of all the real and imaginary components of the two quaternions, it is non-commutative and expressive. In addition, the Hamilton product can also be regarded as scaling spatial rotation, so it can be used as rotating parameterization well. Therefore, quaternion and its Hamilton product can provide promising performance for knowledge graph embedding.

The quaternion Q can be defined as $Q = a + b\mathbf{i} + c\mathbf{j} + d\mathbf{k}$, where a, b, c, $d \in \mathbb{R}$, and i, j, k are imaginary units following $i^2 = j^2 = k^2 = ijk = -1$. Some basic definitions and common operations of quaternion algebra are as follows:

Quaternion Normalization: The normalized quaternion vector $Q^{\triangleleft}$ of Q is computed as:

$$Q^{\triangleleft} = \frac{a + b\mathbf{i} + c\mathbf{j} + d\mathbf{k}}{\sqrt{a^2 + b^2 + c^2 + d^2}} \qquad (1)$$

Inner Product: The quaternion inner product between Q_α and Q_β can be described as:

$$Q_\alpha \cdot Q_\beta = <a_\alpha, a_\beta> + <b_\alpha, b_\beta> + <c_\alpha, c_\beta> + <d_\alpha, d_\beta> \tag{2}$$

Hamilton Product (Quaternion Multiplication): The Hamilton Product between Q_α and Q_β can be described as:

$$Q_\alpha \otimes Q_\beta = (a_\alpha a_\beta - b_\alpha b_\beta - c_\alpha c_\beta - d_\alpha d_\beta) + (a_\alpha b_\beta + b_\alpha a_\beta + c_\alpha d_\beta - d_\alpha c_\beta)\mathbf{i}$$
$$+ (a_\alpha c_\beta - b_\alpha d_\beta + c_\alpha a_\beta + d_1 b_\beta)\mathbf{j} + (a_\alpha d_\beta + b_\alpha c_\beta - c_\alpha b_\beta + d_\alpha a_\beta)\mathbf{k} \tag{3}$$

3.2   The Proposed QuatPE

A knowledge graph (denoted as $\mathcal{G}$) consists of a large number of triplets of facts, and its format is $(head, relation, tail)$. The KGE models calculate the score of each triplet so that valid triplets get higher scores than invalid triplets.

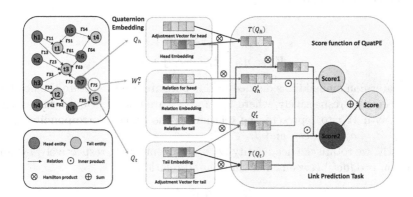

Fig. 3. The framework of QuatPE.

By utilizing quaternion space to represent the embedding of entities and relations, we propose a novel model called QuatPE. It uses paired relations to simultaneously rotate the embedding vectors of the head and tail entities, which can enhance the connection between them. Furthermore, it uses the adjustment vectors to adjust the spatial position to make the same entity have different representations when facing different triples, which can better handle various relation patterns. Figure 3 illustrates the structure of our model in detail.

We assume that triples follow the semantic matching principle, i.e., the latent semantics of entities and relations can be matched by appropriate rotations. Therefore, we utilize the Hamiltonian product to implement rotation and matching operations. Specifically, given a triple (h, r, t), the quaternion embeddings of the head entity Q_h, the tail entity Q_t and relation $W_r^\triangleleft$ are defined as:

$$Q_h = a_h + b_h \mathbf{i} + c_h \mathbf{j} + d_h \mathbf{k}$$
$$Q_t = a_t + b_t \mathbf{i} + c_t \mathbf{j} + d_t \mathbf{k}$$
$$W_r^\triangleleft = \frac{a_r + b_r \mathbf{i} + c_r \mathbf{j} + d_r \mathbf{k}}{\sqrt{a_r^2 + b_r^2 + c_r^2 + d_r^2}} \tag{4}$$

The quaternion embeddings defined in this part are used to represent the latent semantics of the corresponding entities and relations. Where Q_h, Q_t and $W_r^\triangleleft \in \mathbb{H}^k$, and $W_r^\triangleleft$ is a unit quaternion. The coefficient a_h, b_h, c_h, d_h, a_t, b_t, c_t, d_t, a_r, b_r, c_r, and $d_r \in \mathbb{R}^k$.

To enhance the relation-aware correlations between entities in a KG, we define the paired relations. That is, for a relation, in addition to the embedding used to represent the latent semantic information, there are also corresponding paired relations. When performing a rotation operation, both the head and tail entities in the same triple can be rotated at the same time, and different rotations can be performed for the head and tail entities. This solves the problem of weak connection between head and tail entities caused by only using the relation semantics embedding to rotate the head entity, the paired relations are represented by V_{rh} and V_{rt}:

$$V_{rh}^\triangleleft = a_{rh} + b_{rh} \mathbf{i} + c_{rh} \mathbf{j} + d_{rh} \mathbf{k}$$
$$V_{rt}^\triangleleft = a_{rt} + b_{rt} \mathbf{i} + c_{rt} \mathbf{j} + d_{rt} \mathbf{k} \tag{5}$$

The quaternion embeddings defined in this part are used to rotate the head and the tail entities respectively, where $V_{rh}^\triangleleft$ and $V_{rt}^\triangleleft$ are both unit quaternion, which is used with $W_r^\triangleleft$ to represent the relation in the triple. Correspondingly, a_{rh}, b_{rh}, c_{rh}, d_{rh}, a_{rt}, b_{rt}, c_{rt}, and $d_{rt} \in \mathbb{R}^k$.

Firstly, we rotate the head entity Q_h and tail entity Q_t, respectively, by doing Hamilton product between them and paired relations:

$$Q_h' = Q_h \otimes V_{rh}^\triangleleft$$
$$Q_t' = Q_t \otimes V_{rt}^\triangleleft \tag{6}$$

where symbol $\otimes$ defines the Hamilton product operation.

Then, we capture the interaction between head and tail entities after rotation by doing the inner product between them and get the first part of QuatPE:

$$s_{r,p}(h,t) = (Q_h \otimes V_{rh}^\triangleleft) \cdot (Q_t \otimes V_{rt}^\triangleleft) \tag{7}$$

where symbol $\cdot$ defines the inner product. The inner product can match the latent semantics. Through the inner product, we can get a score value of the first part, which is mainly used to enhance the connection between the head entity and the tail entity.

Moreover, we use adjustment vectors to adjust the spatial position of the head entity and tail entity of each triple when an entity faces different triples. Thus, entities can be dynamically mapped against triples to enhance interactions between triples. So that it can better handle complex relations. That is, for an entity, in addition to the embedding used to represent the latent semantic information, there is also a corresponding adjustment vector. When faced with different triples, the adjustment vector can adjust the same entity's spatial position. The specific operation is as follows:

$$T(Q_h) = Q_h \otimes T_h^\lhd$$
$$T(Q_t) = Q_t \otimes T_t^\lhd \qquad (8)$$

where $T_h^\lhd = a_{th} + b_{th}\mathbf{i} + c_{th}\mathbf{j} + d_{th}\mathbf{k}$ and $T_t^\lhd = a_{tt} + b_{tt}\mathbf{i} + c_{tt}\mathbf{j} + d_{tt}\mathbf{k}$ are the normalized adjustment vectors corresponding to head and tail, respectively, and $a_{th}, b_{th}, c_{th}, d_{th}, a_{tt}, b_{tt}, c_{tt}, d_{tt} \in \mathbb{R}^k$. The quaternion embeddings defined in this part are used to adjust the head and tail entities respectively.

Thus, we rotate the adjusted head entity by doing Hamilton product between it and $W_r^\lhd$, and then computer the quaternion inner product with $T(Q_t)$ and get the second part of QuatPE:

$$s_{r,a}(h,t) = ((Q_h \otimes T_h^\lhd) \otimes W_r^\lhd) \cdot (Q_t \otimes T_t^\lhd) \qquad (9)$$

Formally, we can define the score function of QuatPE as follows:

$$f_r(h,t) = s_{r,p}(h,t) + \lambda s_{r,a}(h,t) \qquad (10)$$

where $\lambda \in \mathbb{R}$ is a parameter that is used to balance the role of paired relations and adjustment vectors.

3.3 Learning of QuatPE

We utilize the Adagrad optimizer to minimize the following logistic loss [4]:

$$\mathcal{L} = \sum_{(h,r,t) \in \mathcal{G} \cup \mathcal{G}'} log(1 + exp(-y_{(h,r,t)} \cdot f_r(h,t))) + \lambda_1 ||\theta||_2^2$$

$$\text{in which, } y_{(h,r,t)} = \begin{cases} 1 & for(h,r,t) \in \mathcal{G} \\ -1 & for(h,r,t) \in \mathcal{G}' \end{cases} \qquad (11)$$

We use the l_2 norm with the regularization rate λ_1 and θ are model parameters. $\mathcal{G}$ is a collection of valid triples, respectively. $\mathcal{G}'$ is sampled from $\mathcal{G}$ by corrupting valid triples using negative sampling strategies of Bernoulli sampling [9].

For parameter initialization, to speed up model convergence and efficiency, our initialization is consistent with QuatE utilizing a specialized initialization scheme.

4 Experiments

4.1 Experimental Setup

Datasets Description: We evaluate our proposed model on two widely used benchmarks including WN18RR [5] and FB15k-237 [24]. Table 2 describes the details of the dataset. Furthermore, WN18RR and FB15k-237 are subsets of WN18 [2] and FB15k [2], respectively. Because WN18 and FB15k have the leakage problem of the test set which is pointed out by [5] and [24], therefore, we choose WN18RR and FB15k-237 as the benchmark datasets.

Table 2. The statistics of datasets.

Dataset	$|\mathcal{E}|$	$|\mathcal{R}|$	Train	Valid	Test
WN18RR	40,943	11	86,835	3,034	3,134
FB15K-237	14,541	237	272,115	17,535	20,466

Baselines: We select some representative models as baselines for our study. They can be divided into two categories: the most well-known classic KGE methods and the recent KGE models. TransE [2], HAKE [10], DistMult [16], ComplEx [4], ConvE [5], NKGE [25], RotatE [3], and QutatE [6] are classic models in the first group. InteractE [20], Rotae3D [22], QuatDE [13], ATTH [15], ROTH [15], CompGCN [23], GFA-NN [26], DualE [1], and HopfE [14] are recent strong baselines in the second group. These models include translation distance-based models, bilinear models, graph convolution network-based models, and models using hypercomplex space and hyperbolic space, etc.

Evaluation Metrics: Following existing work [2], we use the link prediction task for evaluation. In particular, we employ three popular evaluation metrics:

- Mean Rank (MR): It is calculated as $\frac{1}{|\mathcal{S}|}\sum_{i=1}^{|\mathcal{S}|} \text{rank}_i$, where a smaller value of MR tends to infer better results;
- Mean reciprocal ranking (MRR): It is computed by the average reciprocal ranks $1/|\mathcal{S}|\sum_{i=1}^{|\mathcal{S}|}\frac{1}{\text{rank}_i}$, where rank_i is a set of ranking results;
- Hit@n (n = 1,3,10): Hit ratio with cut-off values $n = 1, 3, 10$. It is the percentage of appearance in top-n: $1/|\mathcal{S}|\sum_{i=1}^{|\mathcal{S}|} \mathbb{I}(\text{rank}_i < k)$, where $\mathbb{I}(\cdot)$ is the indicator function.

Implementation: We implement our model using PyTorch. The embedding size k is in $\{50, 100, 150, 200, 250, 300\}$. The number of negative triples for each triple is $\{1, 5, 10\}$. The parameters λ is searched in $\{0, 0.05, 0.1, 0.2, 0.3, 0.4, 0.5\}$. The learning rate is in $\{0.01, 0.05, 0.1\}$ and the regularization rate λ_1 is in $\{0, 0.01, 0.05, 0.1, 0.2\}$. Table 3 shows the default settings of our parameters.

Table 3. Hyper-parameters specifications. n denotes the number of negative triples for each triple.

Dataset	k	λ	λ_1	n	v	epochs
WN18RR	50	0.4	0.1	0.1	10	9000
FB15K-237	300	0.05	0.3	0.1	10	4800

4.2   Results and Analysis

We report the empirical results on the two benchmarking datasets in Tables 4. Our QuatPE model shows the best performance compared to strong baselines. From the results, we have the following observations:

(1) Compared to all the baselines, QuatPE and the models using only paired relations or adjustment vectors can perform competitively well across all metrics on two datasets. This indicates that employing adjustment vectors or paired relations can enhance the representation abilities.
(2) Compared with strong baselines, the models based on convolutional neural networks are relatively weak. These methods do not consider the complex relational patterns existing in the KG, which leads to limited performance.

Table 4. Link prediction results on WN18RR and FB15K-237. In the Cat column (i.e., Category), T denotes translational distance models, S denotes semantic matching models, N denotes neural network-based models, and H denotes models that use hyperbolic Space. The best results are in bold and the second-best results are underlined. The results of TransE are taken from [8]. Results of DistMult and ComplEx are taken from [5]. Other results are taken from the best results reported in each paper. QuatPE[1]: with paired relations and adjustment vectors; QuatPE[2]: only use the paired relations; QuatPE[3]: only use the adjustment vectors.

Cat	Model	WN18RR					FB15K-237				
		MR	MRR	Hits@10	Hits@3	Hits@1	MR	MRR	Hits@10	Hits@3	Hits@1
T	TransE	3384	0.226	0.501	-	-	357	0.294	0.465	-	-
	HAKE	-	**0.497**	0.582	<u>0.516</u>	**0.452**	-	0.346	0.542	0.381	0.250
S	DistMult	5110	0.430	0.490	0.440	0.390	254	0.241	0.419	0.263	0.155
	ComplEx	5261	0.440	0.510	0.460	0.410	339	0.247	0.428	0.275	0.158
	RotatE	3340	0.476	0.571	0.492	0.428	177	0.338	0.533	0.375	0.241
	QuatE	2314	0.488	0.582	0.508	0.438	**87**	0.348	0.550	0.382	0.248
	Rotate3D	3328	0.489	0.579	0.505	0.442	165	0.347	0.543	0.385	0.250
	QuatDE	1977	0.489	<u>0.586</u>	0.509	0.438	90	0.365	0.563	<u>0.400</u>	0.268
	DualE	2270	0.492	0.584	0.513	0.444	91	0.365	0.559	<u>0.400</u>	0.268
	HopfE	2885	0.472	<u>0.586</u>	0.500	0.413	212	0.343	0.534	0.379	0.247
N	ConvE	4187	0.430	0.520	0.440	0.400	244	0.325	0.501	0.356	0.237
	NKGE	4170	0.450	0.526	0.465	0.421	237	0.33	0.510	0.365	0.241
	CompGCN	-	0.479	0.546	0.494	0.443	-	0.355	0.535	0.390	0.264
	InteractE	5202	0.463	0.528	-	0.430	172	0.354	0.535	-	0.263
	GFA-NN	3390	0.486	0.575	-	-	186	0.338	0.522	-	-
H	ATTH	-	0.486	0.573	0.449	0.443	-	0.348	0.540	0.384	0.252
	ROTH	-	<u>0.496</u>	<u>0.586</u>	0.514	<u>0.449</u>	-	0.344	0.535	0.380	0.246
	QuatPE[1]	**1875**	0.494	**0.592**	**0.521**	0.445	**87**	**0.369**	**0.567**	**0.404**	**0.272**
	QuatPE[2]	2098	0.492	0.584	0.515	0.442	<u>88</u>	<u>0.368</u>	<u>0.566</u>	**0.404**	**0.272**
	QuatPE[3]	<u>1951</u>	0.487	0.585	0.511	0.435	89	0.365	0.563	<u>0.400</u>	<u>0.269</u>

Relation Analysis: To verify that our model can deal with the complex relation patterns effectively in the KGs, we analyze the experimental results of all relations on WN18RR. The score of MRR for each relation on WN18RR is shown in Table 5. From the results, our model has obtained better results, which further confirms that our model has superior performance in different relation patterns.

Table 5. The MRR score for the models of QuatE and QuatPE tested on each relation of WN18RR.

Relation Names	Relation Category	QuatE	QuatPE
verb_group	1-to-1	0.924	**0.944**
similar_to	1-to-1	**1.000**	**1.000**
member_meronym	1-to-N	0.232	**0.240**
has_part	1-to-N	0.233	**0.236**
member_of_domain_usage	1-to-N	**0.441**	0.430
member_of_domain_region	1-to-N	0.193	**0.338**
hypernym	N-to-1	0.173	**0.177**
instance_hypernym	N-to-1	0.364	**0.376**
synset_domain_topic_of	N-to-1	0.468	**0.481**
derivationally_related_form	N-to-N	0.953	**0.957**
also_see	N-to-N	0.629	**0.639**

Furthermore, we analyze the experimental results of QuatE and QuatPE for each relation category on WN18RR and FB15K-237. The scores of MRR and Hits@10 for link prediction concerning each relation category on the datasets of WN18RR and FB15K-237 are shown in Fig. 4. Our model typically outperforms QuatE in these relations categories. In particular, our proposed model is superior to the complex relation patterns (1-to-N, N-to-1, and N-to-N). This indicates that our model can indeed handle different relation patterns better than QuatE including complex relations.

Correlation Analysis: To demonstrate that our proposed model can enhance the correlation between entities, following existing work [12], we use t-SNE [27] to visualize entities embeddings on the dataset of WN18RR learned from QuatE mode and QuatPE model respectively. We select all head and tail entities in all triples containing the relations of "*instance_hypernym*" and "*synset_domain_topic_of*". Then we use a vector to concatenate the four components of a quaternion embedding that have been projected into the same semantic space. Figure 5 shows that for entities related to the same relation, the distribution of entities in QuatPE is more concentrated than that in QuatE, and the distribution of head and tail entities is more intensive, which means that under the same relation association, the relation-aware correlations between entities are stronger, that is, QuatPE enhances the connections between the head and tail entities.

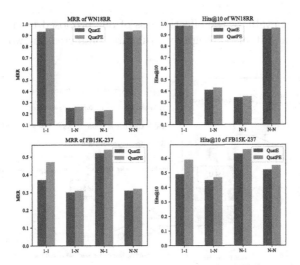

Fig. 4. Comparison of the experimental results of QuatE and QuatPE on the test set for each relation category.

Ablation Study: We conduct ablation studies on the effectiveness of paired relations and adjustment vectors on QutaPE. Table 4 shows the results on two datasets, where QuatPE[2] only uses paired relations and QuatPE[3] only uses the adjustment vectors. The results show that the models using adjustment vectors or paired relations outperform other models without using them. The models use both adjustment vectors and paired relations further boost the performance.

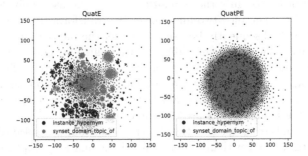

Fig. 5. Visual comparison of entity embeddings on WN18RR.

4.3   Application on Recommendation Systems

To substantiate the effectiveness of our model in downstream tasks, we have designed a multi-task learning approach based on KGs to enhance recommendation systems. Figure 6 illustrates the overall architecture of our proposed model. The model is jointly trained for both tasks to accomplish the recommendation task. Finally, we evaluate the effectiveness of our proposed model in improving recommendation systems through experiments conducted on two datasets.

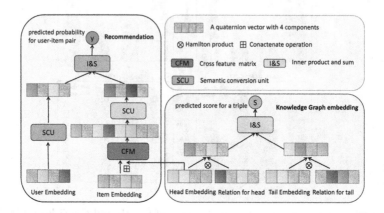

Fig. 6. Schematic diagram of a multi-task learning model based on KGs to enhance recommendation systems.

In Fig. 6, the KGE component of the model trains the KG of items using the QuatPE model only with paired relations. Subsequently, the item embedding obtained from the KGE model is concatenated with the embedding in the user-item interaction graph after feature interaction through the feature cross matrix, which is structured as a multilayer perceptron for fusing the two types of information. Since the item embedding partly contains the information of the KG at this point, in order to unify the semantic space of the user embedding and the item embedding, we employ two linear layers to transform their semantic spaces so that they can subsequently be used to model the interaction between them. Finally, the user and item embeddings with unified semantic spaces undergo inner product and summation operations to obtain the probability of the existence of a relation between them. We conduct experiments on the Last.FM and Book-Crossing datasets [28], and evaluate the results using the Precision@K and Recall@K metrics. The results are presented in Fig. 7.

Here, our aim is to investigate the effectiveness of our proposed KGE model in downstream tasks, rather than to establish a benchmark by implementing state-of-the-art (SOTA) models on recommendation systems. To this end, we have selected several typical recommendation algorithms, namely Wide&Deep [29], RippleNet [25], and LibFM [30], as baselines. In our model, the use of QuatE as the KGE component of the joint training model is indicated by w/ QuatE, while w/ QuatPE denotes the use of QuatPE as the KGE component. The absence of a KGE model is indicated by w/o KGE.

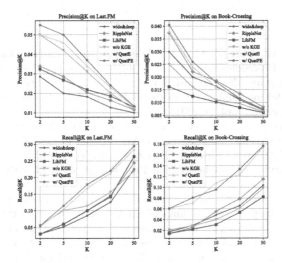

Fig. 7. Comparison of experimental results of models.

The experimental results show that our method effectively utilizes information from the KG for recommendation systems. Compared to baselines without the KGE component, the joint training approach produces more competitive results, indicating that the KGE can enhance the recommendation system. Moreover, our proposed KGE model outperforms QuatE, indicating its effectiveness in learning the structural information of the KG.

5 Conclusion

In this paper, we propose a KGE model named QuatPE, which represents entities as Quaternion embedding vectors and relations as rotations in the quaternion space. QuatPE uses paired relations to simultaneously rotate the quaternion vectors of the head and tail entities, to strengthen the connection between them for enhancing the representation capability. In addition, QuatPE further uses adjustment vectors to better handle complex relations. Experimental results on two well-known benchmarking datasets show that QuatPE can achieve better performance on link prediction tasks.

Acknowledgements. This research was supported by the School of Film Xiamen University - Dongbo Future Artificial Intelligence Research Institute Co., Ltd. Joint Laboratory for created the Metaverse (School Agreement No. 20223160C0026), School of Film Xiamen University - Xiaozhi Deep Art Artificial Intelligence Research Institute Co., Ltd. Computational Art Joint Laboratory (School Agreement No. 20213160C0032), and School of Informatics Xiamen University - Xiamen Yinjiang Smart City Joint Research Center (School Agreement No. 20213160C0029). Xiaoli Wang was supported by the Natural Science Foundation of Fujian Province of China (No. 2021J01003).

References

1. Cao, Z., Xu, Q., Yang, Z., Cao, X., Huang, Q.: Dual quaternion knowledge graph embeddings. Proceedings of the AAAI, vol. 35, no. 8, pp. 6894–6902 (2021)
2. Bordes, A., Usunier, N., Garcia-Duran, A., Weston, J., Yakhnenko, O.: Translating embeddings for modeling multi-relational data. In: Advances in Neural Information Processing Systems, vol. 26 (2013)
3. Sun, Z., Deng, Z.-H., Nie, J.-Y., Tang, J.: RotatE: knowledge graph embedding by relational rotation in complex space. arXiv preprint arXiv:1902.10197 (2019)
4. Trouillon, T., Welbl, J., Riedel, S., Gaussier, É., Bouchard, G.: Complex embeddings for simple link prediction. In: Proceedings of ICML 2016, pp. 2071–2080 (2016)
5. Dettmers, T., Minervini, P., Stenetorp, P., Riedel, S.: Convolutional 2D knowledge graph embeddings. In: Proceedings of AAAI, vol. 32, no. 1 (2018)
6. Zhang, S., Tay, Y., Yao, L., Liu, Q.: Quaternion knowledge graph embeddings. In: Advances in Neural Information Processing Systems, vol. 32 (2019)
7. Lin, Y., Liu, Z., Sun, M., Liu, Y., Zhu, X.: Learning entity and relation embeddings for knowledge graph completion. In: Proceedings of AAAI (2015)
8. Nguyen, D.Q., Nguyen, T.D., Nguyen, D.Q., Phung, D.: A novel embedding model for knowledge base completion based on convolutional neural network. In: Proceedings of NAACL-HLT (2018)
9. Wang, Z., Zhang, J., Feng, J., Chen, Z.: Knowledge graph embedding by translating on hyperplanes. In: Proceedings of AAAI, vol. 28, no. 1 (2014)
10. Zhang, Z., Cai, J., Zhang, Y., Wang, J.: Learning hierarchy-aware knowledge graph embeddings for link prediction. Proc. AAAI **34**(03), 3065–3072 (2020)
11. Ji, G., He, S., Xu, L., Liu, K., Zhao, J.: Knowledge graph embedding via dynamic mapping matrix. In: Proceedings of ACL-IJCNLP, pp. 687–696 (2015)
12. Nguyen, D.Q., Vu, T., Nguyen, T.D., Phung, D.: QuatRE: relation-aware quaternions for knowledge graph embeddings. In: Proceedings of WWW (2022)
13. Gao, H., Yang, K., Yang, Y., Zakari, R.Y., Owusu, J.W., Qin, K.: QuatDE: dynamic quaternion embedding for knowledge graph completion. arXiv preprint arXiv:2105.09002 (2021)
14. Bastos, A., Singh, K., Nadgeri, A., Shekarpour, S., Mulang, I.O., Hoffart, J.: HopfE: knowledge graph representation learning using inverse hopf fibrations. In: Proceedings of CIKM, pp. 89–99 (2021)
15. Chami, I., Wolf, A., Juan, D.-C., Sala, F., Ravi, S., Ré, C.: Low-dimensional hyperbolic knowledge graph embeddings. In: Proceedings of ACL (2020)
16. Yang, B., Yih, W.-T., He, X., Gao, J., Deng, L.: Embedding entities and relations for learning and inference in knowledge bases. In: Proceedings of ICLR (2015)
17. Kazemi, S.M., Poole, D.: Simple embedding for link prediction in knowledge graphs. In: Advances in Neural Information Processing Systems, vol. 31 (2018)
18. Nickel, M., Tresp, V., Kriegel, H.-P.: A three-way model for collective learning on multi-relational data. In: Proceedings of ICML, pp. 809–816 (2011)
19. Balažević, I., Allen, C., Hospedales, T.: "Tucker": tensor factorization for knowledge graph completion. In: Proceedings of EMNLP-IJCNLP, pp. 5185–5194 (2019)
20. Vashishth, S., Sanyal, S., Nitin, V., Agrawal, N., Talukdar, P.: InteractE: improving convolution-based knowledge graph embeddings by increasing feature interactions. In: Proceedings of the AAAI, vol. 34, no. 03, pp. 3009–3016 (2020)
21. Zhang, S., Sun, Z., Zhang, W.: Improve the translational distance models for knowledge graph embedding. J. Intell. Inf. Syst. **55**(3), 445–467 (2020)

22. Gao, C., Sun, C., Shan, L., Lin, L., Wang, M.: Rotate3D: representing relations as rotations in three-dimensional space for knowledge graph embedding. In: Proceedings of CIKM, pp. 385–394 (2020)
23. Vashishth, S., Sanyal, S., Nitin, V., Talukdar, P.: Composition-based multi-relational graph convolutional networks. In: Proceedings of ICLR (2019)
24. Toutanova, K., Chen, D.: Observed versus latent features for knowledge base and text inference. In: Proceedings of CVSC, pp. 57–66 (2015)
25. Wang, K., Liu, Y., Xu, X., Lin, D.: Knowledge graph embedding with entity neighbors and deep memory network. arXiv preprint arXiv:1808.03752 (2018)
26. Sadeghi, A., Collarana, D., Graux, D., Lehmann, J.: Embedding knowledge graphs attentive to positional and centrality qualities. In: Oliver, N., Pérez-Cruz, F., Kramer, S., Read, J., Lozano, J.A. (eds.) ECML PKDD 2021. LNCS (LNAI), vol. 12976, pp. 548–564. Springer, Cham (2021). https://doi.org/10.1007/978-3-030-86520-7_34
27. Van der Maaten, L., Hinton, G.: Visualizing data using t-SNE. J. Mach. Learn. Res. 9(11), 2579–2605 (2008)
28. Wang, H., Zhang, F., Zhao, M., Li, W., Xie, X., Guo, M.: Multi-task feature learning for knowledge graph enhanced recommendation. In: Proceedings of WWW, pp. 2000–2010 (2019)
29. Cheng, H.-T., et al.: Wide & deep learning for recommender systems. In: Proceedings of DLRS, pp. 7–10 (2016)
30. Rendle, S.: Factorization machines with libFM. ACM Trans. Intell. Syst. Technol. (TIST) 3(3), 1–22 (2012)

Towards Time-Variant-Aware Link Prediction in Dynamic Graph Through Self-supervised Learning

Guangqi Wen[1,2], Peng Cao[1,2,3(✉)], Zhiyong Jin[1], Ruoxian Song[1,2], Xiaoli Liu[4], Jinzhu Yang[1,2,3(✉)], and Osmar R. Zaiane[5]

[1] Computer Science and Engineering, Northeastern University, Shenyang, China
{caopeng,yangjinzhu}@cse.neu.edu.cn
[2] Key Laboratory of Intelligent Computing in Medical Image of Ministry of Education, Northeastern University, Shenyang, China
[3] National Frontiers Science Center for Industrial Intelligence and Systems Optimization, Shenyang 110819, China
[4] DAMO Academy, Alibaba Group, Hangzhou, China
[5] Alberta Machine Intelligence Institute, University of Alberta, Edmonton, AB, Canada

Abstract. Dynamic graph link prediction is a challenging problem because the graph topology and node attributes vary at different times. A purely supervised learning scheme for the dynamic graph data usually leads to poor generalization due to insufficient supervision. As a promising solution, self-supervised learning can be introduced to dynamic graph analysis tasks. However, the self-supervised learning paradigm for dynamic graph learning has not been sufficiently investigated due to the complicated properties of the evolving graphs. We assume that the dynamic graphs consist of three independent types of key factors, i.e., graph time-variant information, time-invariant information and noise. Based on this assumption, we propose the Self-supervised Decoupling for Dynamic Graph (SDDG) framework for explicitly decoupling the representation which characterizes these three factors, thus enhancing the interpretability of the learned representation and link prediction performance. More specifically, we design an encoder-decoder architecture to sufficiently exploit the dynamic graph itself with multiple regularizations, so that the time-variant embedding of dynamic graph data can be effectively decoupled from the perspectives of node and structure for time-variant-aware link prediction. Experiments on five benchmark link prediction tasks demonstrate the significant improvement of our SDDG over the state-of-the-art methods. For example, SDDG achieves 98.2% and 97.3% Top-1 AUC on Reddit-B and Reddit-T, outperforming the DDGCL model, by 1.9% and 1.8%, respectively.

Keywords: Dynamic graph learning · Self-supervised learning · Time-variant embedding · Link prediction

ⓒ The Author(s), under exclusive license to Springer Nature Switzerland AG 2023
X. Yang et al. (Eds.): ADMA 2023, LNAI 14179, pp. 470–485, 2023.
https://doi.org/10.1007/978-3-031-46674-8_33

1 Introduction

The purpose of the dynamic graph link prediction is to predict future links of nodes in a given network based on historical information [5,22,25]. This fundamental task has attracted increasing attention and has a wide range of applications such as social network analysis, computer vision and recommender systems, etc. [7]. For example, in social media [16,22], new interactions happen between users over time. In financial networks [1], transactions are streaming and supply chain relations are continuously evolving. Learning the temporal representations in dynamic graphs is a crucial problem and capturing the node embeddings to reflect this variation is desired. However, a purely supervised learning scheme for the dynamic graph data usually leads to limited generalization due to insufficient supervision for modeling the co-occurrence of spatio-temporal dependencies [12,13].

Recently, self-supervised learning has emerged as a principled way of obtaining useful representation without the need for human annotations [4,12,20]. In self-supervised learning, representations are learned via a pretext task derived from unlabeled data. The learned representations obtained from this pretext task are used in a downstream task. However, it is non-trivial to transfer the pretext tasks designed for the specific computer vision and natural language processing tasks [6,9] to the dynamic graph data analysis, due to the complex graph structure and temporal correlation within the dynamically evolving graph. Unfortunately, few work has studied self-supervised learning on dynamic graphs. More recently, Tian et al. [20] first extended self-supervised learning to dynamic graphs and designed a customized contrast learning for temporal graphs by contrasting two nearby snapshots of the same node identity, and design a better negative sampling process to improve the generalizability of contrastive learning. However, they ignore capturing time-variant-related embeddings and fail to exploit the graph structure, preventing it from achieving better performance.

Hence, we propose a fundamental assumption that the dynamic graphs potentially consist of three key factors: graph time-variant information, graph time-invariant information as well as the inherent noise, and the time-variant information is the key factor for dynamic link prediction. As illustrated in Fig. 1, we aim to disentangle the time-variant information for link prediction, which differs from the conventional methods. For dynamic graphs, the inherent node properties are regarded as time-invariant factors, whereas the variation in each snapshot is considered as time-variant ones. For example, dynamic graph link prediction in social networks is to predict the association between users at a certain moment with historical information. The dynamic graph is composed of several factors (e.g., social relationship trend (time-variant), personal information (time-invariant) and noise, etc. [2,18]). The trending information on topics that User A and User B follow on social networks benefits predicting whether an interaction occurs between A and B in the future moment. Specifically, if User A and User B have shown interest in talk shows over a period of time, they are likely to connect at a certain time in the future.

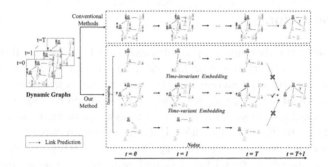

Fig. 1. Illustration of the three key factors in the dynamic graph. We assume that three independent factors, i.e. time-variant information varying over time, time-invariant information inherently remaining statistic, and noise connections, are entangled and should be disentangled in the dynamic graph. We assume that the time-variant information is essential in the link prediction.

Taking all of the above into consideration, we pose the problem of link prediction in the dynamic graph as learning of the decoupled representations for dynamic graph data. Thus, the major challenge lies in: *How to decouple the time-variant information from the time-invariant information and the potential noise in the pretext task?*

The dynamic graph-structured data is often more complex than other domains (e.g., image and text), and the complicated correlation and structure provide rich information that enables us to design pretext tasks from the perspectives of temporal and structural information. To this end, we propose the Self-supervised Decoupling for Dynamic Graph (SDDG), for decoupling the critical representation characterizing the various inherent factors, thus enhancing the interpretability of the learned representations and link prediction performance. More specifically, we decouple the entangled representations into graph time-variant embedding, graph time-invariant embedding and graph noise by the node-level encoder, edge-level encoder and noise encoder, from both aspects of node and structure. We further reconstruct them into the original graph structure and noise by cross decoder and noise decoder to separate the time-variant embedding. To further encourage the representation decoupling, the overall procedure involves the contrast loss forcing the time-variant of neighboring snapshot dissimilar and time-invariant embeddings similar as well as the discrepancy loss constraining the reconstructed graphs of re-combining the learned time-variant embeddings from the same snapshot with the learned time-invariant embeddings decoupled from nearby snapshot consistent in the graph spaces. Finally, the decoupled time-variant embedding is utilized for link prediction. Experiments on five real-world datasets for link prediction demonstrate the significant improvement of our framework over existing alternatives. For example, SDDG achieves 98.2% and 97.3% Top-1 AUC on Reddit-B and Reddit-T, outperforming the debiased dynamic graph contrastive learning (DDGCL) model [20], by 1.9% and 1.8%, respectively.

Our contributions are as follows:

1. We proposed a Self-supervised Decoupling Dynamic Graph network to disentangle the time-variant embedding, time-invariant embedding and graph noise. To the best of our knowledge, this is the first work to model the three kinds of information jointly in a self-supervised learning scheme for link prediction in dynamic graphs. Moreover, the proposed method is easily expanded to other dynamic graph analysis tasks, e.g. dynamic node classification and dynamic graph classification.
2. Unlike other self-supervised dynamic graph learning methods that only focus on node embedding learning, we propose a series of encoders and multiple regularizations to capture the comprehensive representation of the dynamic graphs from both the aspects of the node and structure.
3. Our SDDG demonstrates that the representation disentanglement learning provides a good solution to self-supervised dynamic graph pre-training. The models pre-trained with our SDDG significantly outperform those trained from scratch and several state-of-the-art methods by a large margin, which validates the effectiveness and generality of our method and suggests self-supervised learning with well-designed pretext tasks is a promising paradigm for dynamic graph data.

2 Related Work

2.1 Dynamic Graph Embedding Learning

The dynamic graph link prediction problem is a very important research topic in dynamic graph representation learning. Many static graph learning methods like GCN [24], GAT [21] and GraphSage [8] only take spatial information into consideration, which fails to integrate complex temporal information into prediction methods. TGCN [27] internalizes a GNN into the GRU cell by replacing linear transformations in GRU with graph convolution operators to capture the spatio-temporal correlation representations. EvolveGCN [15] utilizes an RNN to dynamically update weights of internal GNNs, which allows the GNN model to change during the test time. DySAT [17] captures node embedding through joint self-attention along the two dimensions of the structural neighborhood and temporal dynamics. However, a purely supervised learning scheme for the dynamic graph data usually leads to poor generalization due to insufficient supervision.

2.2 Self-supervised Dynamic Graph Embedding Learning

Self-supervised dynamic graph learning focuses on establishing dynamic graph pre-training tasks to capture features with generalized representations, which are then used for link prediction in downstream tasks. For example, DDGCL [20] is the first work to extend self-supervised learning to dynamic graphs. They designed a customized contrast learning for temporal graphs by contrasting two

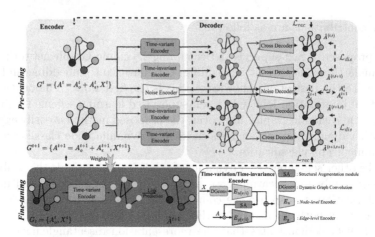

Fig. 2. Illustration of the proposed SDDG. In the pre-training stage, our model includes an encoder-decoder architecture to exploit the dynamic graph itself with self-supervised learning paradigm. We aim to decouple the entangled representations into time-variant embedding, time-invariant embedding and noise by node-level/edge-level encoder and noise encoder. Then, we design a cross decoder for reconstructing the graph structure and a noise decoder for modeling the noise. Moreover, the contrast loss and the discrepancy loss are introduced to better encourage the discrimination of time-variant and time-invariant. In the fine-tuning stage, only the decoupled time-variant embedding is utilized for the dynamic link prediction.

nearby snapshots of the same node identity and propose a better negative sampling process to improve the generalizability of contrastive approaches. However, it ignores the modeling of time-variant embedding in pre-train tasks. Hence, they only yield a minimal performance gain for link prediction (Fig. 2).

3 Methodology

3.1 Notation and Problem Statement

Consider a dynamic graph $\mathcal{G} = \{G^1, G^2, \cdots, G^T\}$ where $G^t = (\mathcal{V}^t, \mathcal{E}^t)$ denotes the graph at time step t. Let $\mathcal{V}^t = \{v_1^t, v_2^t, \cdots, v_N^t\}$ represent a set of nodes and $\mathcal{E}^t = \{e_{ij}^t | v_i^t, v_j^t \in \mathcal{V}^t\}$ represent a set of edges where nodes $v_i, v_j \in \mathcal{V}^t$ are connected at a time step t. Both $\mathcal{V}^t$ and $\mathcal{E}^t$ change across time and the cardinality in each time step may differ. Let $\mathcal{N}^t$ denote the number of nodes in G^t. Likewise, $\mathcal{A}_o = \{A_o^1, A_o^2, \cdots, A_o^T\}$ denotes the original adjacency matrices where $A_o^t \in \mathbb{R}^{\mathcal{N}^t \times \mathcal{N}^t}$. $\mathcal{X} = \{X^1, X^2, \cdots, X^T\}$ represents a sequence of node features. The task of dynamic graph future link prediction is to predict the connections between nodes at time step $t+1$ using the node embeddings trained on graph snapshots up to time step t.

3.2   Method

In order to extract the time-variant embedding of dynamic graphs, we propose the Self-supervised Decoupling Dynamic Graph network (SDDG) to decouple three key factors including time-variant embedding, time-invariant embedding and noise. Unlike the existing self-supervised learning methods, SDDG designs a representation disentanglement method for fixing the critical factors without any auxiliary supervisory signals. This design not only makes different factors to be distinguished in the representation space, but also encourages the learned representation to incorporate factor-level information for disentanglement.

Encoder. To fully exploit the dynamic graph features, the encoder stage includes a noise encoder, a time-variant encoder and a time-invariant encoder.

1) **Noise Encoder**
Real-world data is usually noisy, which leads to the degeneration of model performance. We assume that structural noise exits in the dynamic graph and affects the disentanglement of the representation related to time-variant. To decouple the noise in the dynamic graph, we first generate a noise matrix A_ϵ by randomly dropping and adding edges in the original adjacency matrix A_o for each graph. Specifically, the process is formulated as follows:

$$A^t = D^{\text{drop}} \odot A_o^t + D^{\text{add}}, A_\epsilon^t = A^t - A_o^t, \tag{1}$$

where D^{drop} as well as D^{add} denote the indicator matrices of the connections dropping or adding ratios, and each element in D^{drop} and D^{add} are independently sampled from the Bernoulli distribution.

Given the mixed matrix A^t, the noise embedding is obtained by the noise encoder as follows:

$$F_\epsilon^t = E_\epsilon(X^t, A^t; W_\epsilon) = \mathbf{relu}(X^t A^t W_\epsilon), \tag{2}$$

where F_ϵ^t is the noise embeddings at t, and W_ϵ is the linear transformation layer in the noise encoder.

2) **Node-Level Time-Variant/Time-Invariant Encoder**
To capture the dynamic node embeddings, we first propose the dynamic graph convolution layer $\mathbf{DGconv}(\cdot)$:

$$F_c^t = \mathbf{DGconv}(G^t, F_c^{t-1}; W_{c1}, W_{c2}) = \mathbf{relu}(\mathbf{concat}(A^t X^t W_{c1}, F_c^{t-1}) W_{c2}), \tag{3}$$

where W_{c1} and W_{c2} are the trainable weight matrices.

Through $\mathbf{DGconv}(\cdot)$, we obtain the node embeddings F_c^t at t. Then, we propose two independent node-level encoders to capture the time-variant and time-invariant embeddings as follows:

$$F_{n\{v/i\}}^t = E_{n\{v/i\}}(F_c^t; W_{n\{v/i\}}) = \mathbf{relu}(F_c^t W_{n\{v/i\}}), \tag{4}$$

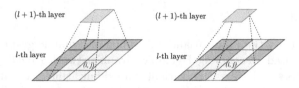

Fig. 3. Illustration of the comparison on spatial locality between images (left) and graphs (right). The difference is that the locality in images denotes the elements that are close together (for the (i, j)-th element, the locality is the 3×3 neighbourhood), while the locality in the graph refers to the local connectivity structure associated with each edge (for the (i, j)-th edge, the locality is the edge set of $e_{i\cdot}$ and $e_{\cdot j}$ associated the i-th node and j-th node).

where $E_{n\{v\}}/E_{n\{i\}}$ are the node-level time-variant/time-invariant encoders, $F^t_{n\{v\}}/F^t_{n\{i\}}$ represent the node-level time-variant and time-invariant embeddings at t, and $W_{n\{v\}}/W_{n\{i\}}$ denote the linear transformation.

3) Edge-Level Time-Variant/Time-Invariant Encoder

Existing dynamic graph learning methods mainly focus on the node embedding learning and can not sufficiently capture the graph topological information, which is crucial for link prediction. To address this issue, we propose edge-level encoders to capture the time-variant and time-invariant structural embeddings by learning the graph topology. Firstly, to introduce more structural information, we design a Semantic Augmentation module, named SA, which enables to generate an enhanced representation of the graph structure. The enhanced graph structure reflects the two aspects of node similarity from the explicit interaction relation and the implicit semantic relation. Specifically, we combine the original graph structure and the semantic augmented graph structure: $\hat{A}^t_{v/i} = A^t \odot S^t_{v/i}$, where S^t_v/S^t_i denote the semantic augmented graph structure of time-variant/time-invariant embeddings, and $\hat{A}^t_v/\hat{A}^t_i$ denote the time-variant/time-invariant enhanced graph structures at t.

To capture the essential topological representation, we leverage an edge convolution layer to design the local connectivity in the graphs by aggregating the features of the connection associated with the nodes at the two ends of the edge. Our edge convolution layer involves edge aggregation (E_{g_e}) with multiple cross-shaped filters for the spatial locality in the graph and node aggregation (E_{g_n}) for aggregating the associated edge embeddings. In contrast to the spatial locality in CNN, the locality in our method refers to the local connectivity structure associated with the edge, as illustrated in Fig. 3. To facilitate computation on large-scale graphs, we apply graph partitioning to divide the graph into multiple subgraphs with $\mathcal{M}$ nodes for each subgraph [3]. The cross-shaped filters in edge aggregation involve a combination of $1 \times \mathcal{M}$ and $\mathcal{M} \times 1$ basis filters that are less computationally expensive filters with horizontal and vertical orientations, which are defined as:

$$F^t_{g_e\{v/i\}} = E_{g_e\{v/i\}}(\hat{A}^t_{v/i}; w^r_{v/i}, w^h_{v/i}) = \sum_{u=0}^{\mathcal{M}}\sum_{k=0}^{\mathcal{M}} \hat{A}^{[(u,\cdot),t]}_{v/i} w^r_{v/i} + \hat{A}^{[(\cdot,k),t]}_{v/i} w^h_{v/i},$$

(5)

where $w^r_v \in \mathbb{R}^{1\times\mathcal{M}}$ and $w^h_v \in \mathbb{R}^{\mathcal{M}\times 1}$ denote the learned vectors of the horizontal and vertical convolution kernel for time-variant embedding learning. $w^r_i \in \mathbb{R}^{1\times\mathcal{M}}$ and $w^h_i \in \mathbb{R}^{\mathcal{M}\times 1}$ are the learned vectors of the horiozntal and vertical convolution kernels for time-invariant embedding learning. $F^t_{g_e\{v\}}/F^t_{g_e\{i\}}$ are the time-variant/time-invariant edge embeddings. $E_{g_e\{v\}}/E_{g_e\{i\}}$ denote the time-variant/time-invariant edge aggregation encoders, respectively.

With the edge embedding learned by E_{g_e}, we further learn the node embedding by aggregating the associated edges with the nodes with a learnable layer. More specifically, the E_{g_n} takes the enhanced edge embedding by E_{g_e} as the inputs, and maps them to generate a node embedding from a node-wise view by a 1D convolutional filter, which is defined as,

$$F^t_{g\{v/i\}} = E_{g_n\{v/i\}}(F^t_{g_e\{v/i\}}; w^m_{v/i}) = \sum_{u=0}^{\mathcal{M}} F^{[(u,\cdot),t]}_{g_e\{v/i\}} w^m_{v/i},$$

(6)

where $F^t_{g\{v\}}/F^t_{g\{i\}}$ are the time-variant/time-invariant edge-level embeddings, $E_{g_n\{v\}}/E_{g_n\{i\}}$ denote the time-variant/time-invariant node aggregation encoders, and $w^m_v \in \mathbb{R}^{1\times\mathcal{M}}/w^m_i \in \mathbb{R}^{1\times\mathcal{M}}$ are the learned vector of the filter for time-variant/time-invariant embedding learning, respectively.

Finally, we obtain the time-variant/time-invariant embeddings $F^t_{v/i}$ from both the node-level and the edge-level encoders, e.g. $F^t_{v/i} = F^t_{n\{v/i\}} + F^t_{g\{v/i\}}$.

Decoder. To reconstruct the original graph structure and noise, we propose cross decoder and noise decoder, which are shown as follows:

1) Cross Decoder

The purpose of the cross decoder is to reconstruct the graph structure at t with the combination of the time-variant and time-invariant embeddings. We assume that the decoupled representation satisfies the following conditions: 1) It is expected that an original graph should be reconstructed from its disengaged time-variant and time-invariant embedding; 2) The combination of the time-variant information at the same snapshot and the time-invariant at the nearby snapshot should be similar. Therefore, the cross decoder receives two inputs, including the combination of the time-variant embedding and the time-invariant embedding at t, or the time-variant embedding at t and the time-invariant embedding at $t + 1$, which is defined as:

$$\hat{F}^{\{t,t\}} = \mathbf{concat}(F^t_v, F^t_i)W_d, \hat{F}^{\{t,t+1\}} = \mathbf{concat}(F^t_v, F^{t+1}_i)W_d,$$

(7)

where $\hat{F}^{\{t,t\}}$ and $\hat{F}^{\{t,t+1\}}$ are the combination of the same time-variant embedding at t as well as the different time-invariant embedding at t and $t + 1$, and W_d is the linear transformation for cross decoder.

478   G. Wen et al.

Then, two nearby outputs $\tilde{A}^{\{t,t\}}$ and $\tilde{A}^{\{t,t+1\}}$ of the decoders are obtained by a non-linear transformation via a learnable matrix W_p.

2) Noise Decoder

In order to decouple the graph noise, we reconstruct the noise information by the noise embedding F_ϵ^t, i.e., the noise matrix A_ϵ^t, which means that we wish our model to infer A_ϵ^t from the A^t. The noise decoder is defined as:

$$\tilde{A}_\epsilon^t = \text{softmax}(\text{concat}(F_{\epsilon,u}^t, F_{\epsilon,k}^t)W_\epsilon), \forall u, \forall k \in \mathcal{V}^{(t)}, \tag{8}$$

where $\tilde{A}_\epsilon^t$ is the predicted noise matrix at t, and W_ϵ is the linear transformation for noise decoder.

Loss. To decouple the time-variant, time-invariant and noise in dynamic graphs. At first, we argue that the time-variant information at the neighboring snapshot is dissimilar, whereas the time-invariant information is similar. To constrain it, we design a time-variant/time-invariant contrast loss as follows:

$$\mathcal{L}_{ct} = \sum_t^T \left[\left\| F_i^t - F_i^{t-1} \right\|_2^2 - \left\| F_v^t - F_v^{t-1} \right\|_2^2 + \alpha \right], \tag{9}$$

where α is the value of margins in two terms.

Then, to enforce the similarity of original graph and reconstructed graphs with the decoupled time-variant/time-invariant embedding during graph reconstruction, we design a reconstruction loss defined as

$$\mathcal{L}_{rec} = -\frac{1}{\mathcal{N}^t}(A_o^t \log(\tilde{A}^{\{t,t\}} + (1 - A_o^t) \log(1 - \tilde{A}^{\{t,t\}}))), \tag{10}$$

Besides, we assume that the reconstruction from the combination of time-variant embeddings in the same snapshot and time-invariant embeddings in the nearby snapshot should be consistent with the combination of time-variant embedding and time-invariant embedding from the same snapshot in the graph space. Based on the motivation, we propose a discrepancy loss to regularize the disentanglement process as follow:

$$\mathcal{L}_{dis} = \frac{1}{\mathcal{N}^t(\mathcal{N}^t - 1)} \sum_u \sum_{k \neq u} \mathbf{k}\left(\tilde{A}_u^{\{t,t\}}, \tilde{A}_k^{\{t,t\}}\right) - 2\frac{1}{\mathcal{N}^t \cdot \mathcal{N}^t} \sum_u \sum_k \mathbf{k}\left(\tilde{A}_u^{\{t,t\}}, \tilde{A}_k^{\{t,t+1\}}\right)$$
$$+ \frac{1}{\mathcal{N}^t(\mathcal{N}^t - 1)} \sum_u \sum_{k \neq u} \mathbf{k}\left(\tilde{A}_u^{\{t,t+1\}}, \tilde{A}_k^{\{t,t+1\}}\right), \tag{11}$$

where $\mathbf{k}(\cdot)$ is the Gaussian kernel function.

Furthermore, to decouple the noise in graphs, we design a noise reconstruction loss $\mathcal{L}_\epsilon = \|\tilde{A}_\epsilon^t - A_\epsilon^t\|_F^2$ for identifying the given noise information. Finally, the overall loss is defined as:

$$\mathcal{L}_{overal} = \mathcal{L}_{ct} + \lambda_1 \mathcal{L}_{rec} + \lambda_2 \mathcal{L}_{dis} + \lambda_3 \mathcal{L}_\epsilon, \tag{12}$$

where $\lambda_1 = \lambda_3 = 0.01$ and $\lambda_2 = 1$ denote the contribution weights to balance the losses in pre-training.

After pre-training, we leverage only both the node-level and edge-level time-variant encoder to extract the time-variant information for link prediction.

Table 1. Comparison with the state-of-the-art methods on five datasets. The best results are in bold and the second best results are in bolditalic. The average link prediction performances of MRR and AUC are reported.

Model	Reddit-T		Reddit-B		UCI-M		Bitcoin-O		Bitcoin-A	
	MRR	AUC	MRR	AUC	MRR	AUC	MRR	AUC	MRR	AUC
GAT [21]	0.271	0.655	0.198	0.614	0.088	0.656	0.076	0.547	0.145	0.573
GraphSAGE [8]	0.331	0.681	0.243	0.633	0.103	0.688	0.088	0.566	0.151	0.603
GCRN [19]	0.338	0.729	0.217	0.679	0.089	0.712	0.173	0.641	0.210	0.648
EvolveGCN [15]	0.351	0.697	0.431	0.879	0.146	0.769	0.142	0.651	0.234	0.793
TGCN [27]	0.491	0.653	0.551	0.821	0.130	0.740	0.183	0.630	0.169	0.750
DGCRN [26]	0.661	0.958	0.573	0.951	0.239	0.865	0.194	0.779	0.351	0.859
CAW-N [23]	0.468	0.941	0.460	0.931	0.284	**0.927**	0.249	0.801	0.292	0.750
DySAT [17]	0.602	0.928	0.602	0.907	*0.293*	0.839	*0.262*	*0.827*	0.268	0.873
DDGCL [20]	0.643	*0.963*	*0.641*	0.955	0.271	0.858	0.188	0.713	*0.354*	*0.882*
SDDG -w/o Pre-training	*0.663*	0.958	0.612	*0.966*	0.239	0.868	0.241	0.809	0.310	0.833
SDDG	**0.692**	**0.982**	**0.678**	**0.973**	**0.295**	*0.903*	**0.288**	**0.854**	**0.374**	**0.895**

Table 2. Effectiveness of the proposed components in SDDG. The best results are in bold.

Model	Losses			Components		Reddit-T		Reddit-B		UCI-M		Bitcoin-O		Bitcoin-A	
	$\mathcal{L}_{ct}$	$\mathcal{L}_{dis}$	$\mathcal{L}_{rec}$	SA	ND	MRR	AUC	MRR	AUC	MRR	AUC	MRR	AUC	MRR	AUC
SDDG -w/o Pre-training	-	-	-	-	-	0.663	0.958	0.612	0.966	0.239	0.868	0.241	0.809	0.310	0.833
SDDG	✓					0.665	0.960	0.616	0.967	0.244	0.861	0.255	0.818	0.334	0.847
	✓	✓				0.674	0.969	0.638	0.969	0.261	0.877	0.271	0.822	0.350	0.848
	✓	✓	✓			0.682	0.974	0.654	0.970	0.279	0.894	0.276	0.831	0.351	0.851
	✓	✓	✓	✓		0.690	0.980	0.667	0.972	0.280	0.895	0.287	0.854	0.360	0.871
	✓	✓	✓	✓	✓	**0.692**	**0.982**	**0.678**	**0.973**	**0.295**	**0.903**	**0.288**	**0.854**	**0.374**	**0.895**

4 Experiments

To analyze dynamic graph embedding quality from different perspectives, we raise five questions to guide our experiments:

Q1: How does SDDG perform compared to state-of-the-art methods on link prediction?

Q2: Are the proposed components effective in SDDG?

Q3: Is the time-variant information essential in link prediction?

Q4: Can SDDG learn the potential and transferable node embedding, which is beneficial for different datasets via pre-training?

Q5: How does the pre-training benefit the link prediction?

4.1   Datasets and Evaluation Metrics

We perform experiments using five different datasets. Reddit-T and Reddit-B are networks of hyperlinks in titles and bodies of Reddit posts, respectively [10]. UCI-M dataset consists of private messages sent on an online social network system among students [14]. Bitcoin-O and Bitcoin-A contain who-trusts-whom networks of people who trade on the OTC and Alpha platforms [11]. In our comparison, we choose mean reciprocal rank (MRR) and AUC as evaluation metrics. For each node v with positive (true) edge (v, u) at $t + 1$, we randomly sample 100 negative (nonexistent) edges emitting from v and identify the rank of edge (v, u)'s prediction score among all other negative edges. We randomly split all connections into 80% (train set):10% (validation set):10% (test set). We run each experiment with 3 random seeds following the experimental setting in [26].

4.2   Compared Methods

To comprehensively evaluate the performance of dynamic graph link prediction, we compare our SDDG with state-of-the-art models. Specifically, the comparable methods can be grouped into three categories: static graph embedding learning (i.e. Graph Attention Networks (GAT) [21] and GraphSAGE [8]), supervised dynamic graph embedding learning (i.e. Graph Convolutional Recurrent Networks (GCRN) [19], Evolving Graph Convolutional Networks (EvolveGCN) [15], Temporal Graph Convolutional network (TGCN) [27], DGCRN [26], CAW-N [23] and DySAT [17]) and self-supervised dynamic graph embedding leaning (i.e. Debiased Dynamic Graph Contrastive Learning (DDGCL) [20]).

4.3   Link Prediction Comparison (Answer for Q1)

The results in Table 1 show that our method consistently outperforms the existing static and supervised/self-supervised dynamic graph embedding learning algorithms by a clear margin across all the datasets with both the MRR and AUC metrics, achieving a new state-of-the-art on all five datasets. The results demonstrate that the proposed self-supervised learning paradigm can provide a promising solution to address the challenge of insufficient supervision in the dynamic graph learning. We highlight the following observations:

1) Our experiments demonstrate the effectiveness of the proposed self-supervised learning framework. We can easily observe that a model pre-trained with SDDG can significantly outperform from-scratch models for all five benchmarks with an average improvement of 5.2% and 3.4% on MRR and AUC, respectively.
2) Our SDDG and DDGCL are both self-supervised dynamic graph embedding learning methods. It can be seen that SDDG outperforms DDGCL on all the datasets. Comparatively, our encoder exploits the greater capacity for learning the representation from both the node and edge levels, and develops more well-designed pretext tasks, which enables our model to learn more

informative representations from the data itself to achieve better representation quality and performance. The result confirms the necessity of developing more well-designed self-supervised learning for link prediction in the dynamic learning.

4.4 Ablation Study (Answer for Q2)

We show an ablation analysis of different losses and the key components in SDDG in Table 2. From the first and subsequent three rows, we can clearly see that disentangling time-variant information gives better results on all datasets. The fourth row shows that with the help of multiple losses, it can boost the decoupling of time-variant information and capture the effective representations in pre-training. Furthermore, we are also interested in the effectiveness of the Structural Augmentation (SA) and Noise Decoupling (ND). It can be observed that SA yields a significant improvement to our model, especially improvements of 2.1% MRR and 1.4% AUC are obtained by adding SA module on the Reddit-T dataset, respectively. The results demonstrate that graph topology learning is necessary for dynamic graph learning. We also find that ND can further improve the model performance from the last row, verifying our hypothesis that the structural noise in dynamic graph leads to the degeneration of model performance.

4.5 Discussion

Node-Level Encoder and Edge-Level Encoder. To justify the effectiveness of the node-level and edge-level encoders, a careful ablation study is conducted on the Reddit-T and Reddit-B datasets (the first part of Table 3). It indicates that the node-level and edge-level are both effective and necessary for graph embedding learning.

The Effectiveness of Time-Variant (Answer for Q3). To answer the Q3, we evaluate the prediction performance with time-variant and time-invariant embeddings on Reddit-T and Reddit-B datasets, respectively. As shown in Table 3, compared with the "SDDG (time-invariant embedding)", our SDDG with time-variant embedding achieves an improvement of 2.7%/2.5% and 6.7%/0.9% in terms of MRR/AUC on Reddit-T and Reddit-B datasets, indicating that the time-variant information in dynamic graphs is indeed the essential factor for link prediction.

Transfer Learning (Answer for Q4). To further investigate the generalization ability of SDDG in representation learning, we compare the pre-trained models from pre-trained individually on the same and different datasets with the model from scratch. We find that the pre-trained models can achieve better link prediction performance on the Reddit-T and Reddit-B datasets, which is shown in the second part of Table 3. The results again demonstrate that self-supervised

Table 3. Comprehensive ablation experiments to verify the effectiveness of the proposed components.

Model	Reddit-T		Reddit-B	
	MRR	AUC	MRR	AUC
SDDG -w/o E_n	0.658	0.947	0.651	0.970
SDDG -w/o E_g	0.671	0.962	0.672	0.971
SDDG -w/o Pre-training	0.663	0.958	0.612	0.966
SDDG - Pre-training on Reddit-T	-	-	0.642	0.971
SDDG - Pre-training on Reddit-B	0.671	0.975	-	-
DDGCL (linear probing)	0.453	0.621	0.439	0.618
DDGCL (fine-tuning)	0.663	0.958	0.612	0.966
SDDG (time-invariant embedding)	0.665	0.957	0.611	0.962
SDDG (linear probing)	0.504	0.633	0.459	0.626
SDDG (fine-tuning)	**0.692**	**0.982**	**0.678**	**0.973**

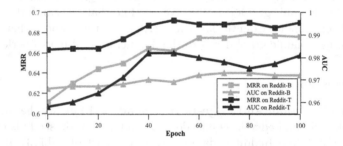

Fig. 4. The effect of pre-training epochs in the link prediction.

learning enables the model to learn more informative knowledge through well-designed pretext tasks from data itself to achieve better performance. We also find that model pre-trained from the Reddit-T/Reddit-B for link prediction on Reddit-B/Reddit-T achieves better performance than the model from scratch, demonstrating that the SDDG can learn more transferable representations to achieve better generalization and robustness via pre-training.

Partial Fine-Tuning. We evaluate the pre-training performance of SDDG and DDGCL with two strategies including fine-tuning and linear probing. The results in Table 3 demonstrate that our SDDG surpasses DDGCL on Reddit-T and Reddit-B datasets in both fine-tuning and linear probing, which indicates that SDDG is capable of capturing the spatio-temporal characteristics of the data itself in the pre-training and yield better generalized representations regardless of linear probing and fine-tuning. The results also demonstrate that the self-supervised learning can mitigate the insufficient learning problem of supervised learning.

Pre-training Epochs (Answer for Q5). We also evaluate the influence of the pre-training epochs on the link prediction performance and show the convergence curve of our model pre-training in Fig. 4. It can be found that the performance improves with the pre-training epochs increasing, which demonstrates that more sufficient pre-training help capture the dynamic node embedding. Nevertheless, as the pre-training epochs further increases, the link prediction performance no longer fluctuates significantly, indicating that the benefits of pre-training may have plateaued and excessive pre-training epochs are not beneficial for dynamic node embedding learning.

5   Conclusion

The traditional link prediction models for graphs have been mostly designed for static graphs. However, the dynamic graphs introduce great challenges for learning and prediction since nodes, attributes and edges vary over time. In this study, we introduce a Self-supervised Decoupling Dynamic Graph network, named SDDG for dynamic link prediction. Specifically, SDDG learns dynamic node representations and decouples the time-variant embedding through the dual decoupler in pre-training. We design a contrast loss, discrepancy loss and reconstruction loss to constrain the decoupling of time-variant embedding, time-invariant and noise. Then, the time-variant embedding is used for the dynamic link prediction. Our experimental results on five real-world datasets indicate significant performance gains for SDDG over several state-of-the-art static and dynamic graph embedding learning methods, and suggest self-supervised learning is a promising and trending learning paradigm for dynamic graph data.

Acknowledgements. This research was supported by the National Key Research and Development Program of China (No. 2020YFC0833302), the National Natural Science Foundation of China under Grant (No. 62076059), the Science Project of Liaoning province under Grant (2021-MS-105) and the 111 Project (B16009).

References

1. Castilho, D., Gama, J., Mundim, L.R., de Carvalho, A.C.P.L.F.: Improving portfolio optimization using weighted link prediction in dynamic stock networks. In: Rodrigues, J.M.F., et al. (eds.) ICCS 2019. LNCS, vol. 11538, pp. 340–353. Springer, Cham (2019). https://doi.org/10.1007/978-3-030-22744-9_27
2. Cheng, J., Liu, Y., Zhang, H., Wu, X., Chen, F.: A new recommendation algorithm based on user's dynamic information in complex social network. Math. Prob. Eng. **2015**, 1–6 (2015)
3. Chiang, W.L., Liu, X., Si, S., Li, Y., Bengio, S., Hsieh, C.J.: Cluster-GCN: an efficient algorithm for training deep and large graph convolutional networks. In: Proceedings of the 25th ACM SIGKDD International Conference on Knowledge Discovery & Data Mining, pp. 257–266 (2019)
4. Eldele, E., et al.: Time-series representation learning via temporal and contextual contrasting. arXiv preprint arXiv:2106.14112 (2021)

5. Farnoodian, N., Nijssen, S., Aversano, G.: Link prediction on CV graphs: a temporal graph neural network approach (2022)
6. Gomez, L., Patel, Y., Rusinol, M., Karatzas, D., Jawahar, C.: Self-supervised learning of visual features through embedding images into text topic spaces. In: Proceedings of the IEEE Conference on Computer Vision and Pattern Recognition, pp. 4230–4239 (2017)
7. Gu, S., Wang, X., Shi, C., Xiao, D.: Self-supervised graph neural networks for multi-behavior recommendation. In: International Joint Conference on Artificial Intelligence (IJCAI) (2022)
8. Hamilton, W., Ying, Z., Leskovec, J.: Inductive representation learning on large graphs. In: Advances in Neural Information Processing Systems, vol. 30 (2017)
9. Jing, L., Tian, Y.: Self-supervised visual feature learning with deep neural networks: a survey. IEEE Trans. Pattern Anal. Mach. Intell. **43**(11), 4037–4058 (2020)
10. Kumar, S., Hamilton, W.L., Leskovec, J., Jurafsky, D.: Community interaction and conflict on the web. In: Proceedings of the 2018 World Wide Web Conference, pp. 933–943 (2018)
11. Kumar, S., Spezzano, F., Subrahmanian, V., Faloutsos, C.: Edge weight prediction in weighted signed networks. In: 2016 IEEE 16th International Conference on Data Mining (ICDM), pp. 221–230. IEEE (2016)
12. Liu, X., et al.: Self-supervised learning: generative or contrastive. IEEE Trans. Knowl. Data Eng. **35**, 857–876 (2023)
13. Liu, Y., et al.: Graph self-supervised learning: a survey. IEEE Trans. Knowl. Data Eng. **35**, 5879–5900 (2022)
14. Panzarasa, P., Opsahl, T., Carley, K.M.: Patterns and dynamics of users' behavior and interaction: network analysis of an online community. J. Am. Soc. Inform. Sci. Technol. **60**(5), 911–932 (2009)
15. Pareja, A., et al.: EvolveGCN: evolving graph convolutional networks for dynamic graphs. In: Proceedings of the AAAI Conference on Artificial Intelligence, vol. 34, pp. 5363–5370 (2020)
16. Perozzi, B., Al-Rfou, R., Skiena, S.: DeepWalk: online learning of social representations. In: Proceedings of the 20th ACM SIGKDD International Conference on Knowledge Discovery and Data Mining, pp. 701–710 (2014)
17. Sankar, A., Wu, Y., Gou, L., Zhang, W., Yang, H.: DySAT: deep neural representation learning on dynamic graphs via self-attention networks. In: Proceedings of the 13th International Conference on Web Search and Data Mining, pp. 519–527 (2020)
18. Santoro, N., Quattrociocchi, W., Flocchini, P., Casteigts, A., Amblard, F.: Time-varying graphs and social network analysis: temporal indicators and metrics. arXiv preprint arXiv:1102.0629 (2011)
19. Seo, Y., Defferrard, M., Vandergheynst, P., Bresson, X.: Structured sequence modeling with graph convolutional recurrent networks. In: Cheng, L., Leung, A.C.S., Ozawa, S. (eds.) ICONIP 2018. LNCS, vol. 11301, pp. 362–373. Springer, Cham (2018). https://doi.org/10.1007/978-3-030-04167-0_33
20. Tian, S., Wu, R., Shi, L., Zhu, L., Xiong, T.: Self-supervised representation learning on dynamic graphs. In: Proceedings of the 30th ACM International Conference on Information & Knowledge Management, pp. 1814–1823 (2021)
21. Velickovic, P., Cucurull, G., Casanova, A., Romero, A., Lio, P., Bengio, Y.: Graph attention networks. Stat **1050**, 20 (2017)
22. Wang, T., He, X.S., Zhou, M.Y., Fu, Z.Q.: Link prediction in evolving networks based on popularity of nodes. Sci. Rep. **7**(1), 1–10 (2017)

23. Wang, Y., Chang, Y.Y., Liu, Y., Leskovec, J., Li, P.: Inductive representation learning in temporal networks via causal anonymous walks. arXiv preprint arXiv:2101.05974 (2021)
24. Welling, M., Kipf, T.N.: Semi-supervised classification with graph convolutional networks. In: International Conference on Learning Representations (2017)
25. Wu, X., Cheng, Q.: Stabilizing and enhancing link prediction through deepened graph auto-encoders. In: Proceedings of the IJCAI, vol. 2022, pp. 3587–3593. NIH Public Access (2022)
26. You, J., Du, T., Leskovec, J.: ROLAND: graph learning framework for dynamic graphs. In: Proceedings of the 28th ACM SIGKDD Conference on Knowledge Discovery and Data Mining, pp. 2358–2366 (2022)
27. Zhao, L., et al.: T-GCN: a temporal graph convolutional network for traffic prediction. IEEE Trans. Intell. Transp. Syst. 21(9), 3848–3858 (2019)

Adaptive Heterogeneous Graph Contrastive Clustering with Multi-similarity

Chao Liu, Bing Kong$^{(\boxtimes)}$, Yiwei Yu, Lihua Zhou, and Hongmei Chen

School of Information Science and Engineering, Yunnan University, Kunming, China
chaoliu@mail.ynu.edu.cn, kongbing@ynu.edu.cn

Abstract. With the proliferation of interactive systems, heterogeneous graph clustering has become an important research topic in the field of unsupervised learning. However, the existing methods generally have one or more of the following problems: 1) they fail to fully mine the similarity between nodes in heterogeneous graphs; 2) they cannot effectively deal with heterogeneous graphs without node attribute information; 3) the predicted labels generated during model iterations are not used as guidance information for subsequent iterations. To address the above problems, we propose an Adaptive Heterogeneous graph Contrastive clustering with Multi-Similarity (AHCMS) model. The model adaptively learns a high-level representation containing specific semantic information through a feature extraction module and an attention mechanism. Secondly, the feature enhancement module is used to extract the consistency information between different meta-paths from two aspects of attribute information and topology structure, so as to encourage the adjacent nodes of different meta-paths to be as similar as possible and reduce the dependence on the attribute information. Moreover, the topological similarity contained in the semantic information is fully explored in the high-order proximity module, making the high-level representation more discriminative. In addition, AHCMS also introduces a self-supervised clustering mechanism to guide the high-level representation to become clustering task-oriented representations. Extensive experimental results on four heterogeneous datasets show that the model's clustering performance consistently outperforms most baseline methods.

Keywords: Contrastive learning · Attention mechanism · Heterogeneous graph clustering

1 Introduction

Graph clustering aims to analyze the association between nodes by integrating the topology and node attribute information of the graph. The goal is to

Supported by the National Natural Science Foundation of China (62062066, 61762090, 61966036 and 62276227), Yunnan Fundamental Research Projects (202201AS070015), Yunnan Key Laboratory of Intelligent Systems and Computing (202205AG070003), Program for Young and Middle-aged Academic and Technical Reserve Leaders of Yunnan Province (202205AC160033).

X. Yang et al. (Eds.): ADMA 2023, LNAI 14179, pp. 486–501, 2023.
https://doi.org/10.1007/978-3-031-46674-8_34

divide the nodes into several unrelated clusters so that the nodes within a cluster have high similarity and those between clusters have low similarity. Due to its advantage of not requiring manual annotation, graph clustering has been successfully applied to downstream tasks such as recommendation systems. However, most existing graph clustering research has focused on homogeneous information networks (HON) [5], which contain only one type of node and one type of relationship in the graph. Nevertheless, real-world graph data is often complex. Therefore, heterogeneous information networks (HIN) or heterogeneous graphs (HG) [23], which consist of multiple types of nodes and/or linking relationships, have become a powerful tool for modeling complex interactions. Compared to HON, HIN contain complex topologies and rich semantic information [30] and can provide more comprehensive guidance for clustering tasks.

Thanks to the great success of graph clustering methods in HON, HDGI [20] learns feature representations by applying DGI [24] to HIN, which uses an attention mechanism on meta-paths to deal with graph heterogeneity and maximizes mutual information to handle the representation learning problem and then uses the K-means [8] algorithm for the learned the low-dimensional representations for clustering tasks. To fully exploit the topological and node attribute information in the HIN, O2MAC [6] learns node representations from the most informative meta-paths through a multilayer graph convolutional neural network (GCN) [13] encoder and reconstructs all views through multiple graph decoders. SHGNN [29] first uses the GCN module to obtain information about the neighbor nodes of the nodes in each meta-path, and then uses the tree-based attention module to aggregate the information on different meta-paths and finally uses a meta-path aggregator to fuse the information aggregated from different meta-paths.

Although heterogeneous graph clustering has been improved from different aspects, existing methods still have three main problems that need to be addressed: 1) Most methods ignore the correlation of the topological structure of nodes in different meta-paths. Although nodes contain different neighborhood information in different meta-paths, they can also be used to portray the similarity between nodes and propagate useful information for learning of low-dimensional representations. 2) The existing methods cannot handle heterogeneous graphs without node attribute information well. Because GCN-based models can be interpreted as aggregating the attribute information of neighboring nodes guided by the graph structure [11], they are difficult to perform the clustering task better when the attribute information of nodes is missing. 3) Many methods learn low-dimensional representations that lack relevant guidance for clustering tasks, which may result in the learned node embeddings not being fit for the subsequent clustering task [4].

To address the limitations of research related to heterogeneous graph clustering, we propose an Adaptive Heterogeneous graph Contrastive clustering with Multi-Similarity (AHCMS) model. The model fully exploits the deep nonlinear structure of the high-level representation in each meta-path with the interaction of GCN encoder and semantic-level attention mechanism, and then uses contrast learning to maximize the consistency information between the low-dimensional representations of different meta-path attribute graphs and the high-level representation to obtain high-quality the high-level representation. And the influence of information redundancy and noise between different features is reduced by the feature decoupling strategy, so that the model can better handle heterogeneous graphs without node attribute information. In addition, a self-supervised clustering mechanism is introduced to make the high-level representation satisfy the required clustering structure. Our major contributions can be summarized as follows: 1) We propose an Adaptive Heterogeneous graph Contrastive clustering with Multi-Similarity (AHCMS) model. The model considers the similarity between nodes at the node, meta-path, and attribute levels, uses this to generate a consistent representation containing specific semantic information, and finally reverse-supervises the model using high-confidence clustering labels. 2) We force the high-level representation to capture the topological relationships of nodes in different meta-paths using a collaborative contrastive strategy and mitigate the feature dependency of the model using a feature decoupling strategy so that the model can focus on the structural information of the HIN in a heterogeneous graph without node attributes and alleviate the interference caused by invalid attributes. 3) We use attribute similarity to capture the topological structure implied in the semantic information, thereby incorporating the structural information contained in the node attributes into the high-level representation. 4) Extensive experiments on four benchmark datasets show that AHCMS achieves excellent performance in clustering tasks compared to the baseline approach.

2   Related Work

2.1   Contrastive Learning

Since graph data is rich in information, which allows researchers to construct comparison tasks from various aspects, such as underlying graph structure [21] and graph heterogeneity [10], more and more research Personnel began to focus on graph-contrastive learning. DGI [24] is an early contrastive learning method, which performs contrastive learning by maximizing the mutual information between the global representation and the low-dimensional representation. DMGI [19] migrates DGI to HIN to integrate node embeddings from different meta-paths by introducing a consistent regularization framework that minimizes the discrepancy between node embeddings of different relation types and a general discriminator. FastGCL [27] constructs weighted aggregated and non-aggregated neighborhood information as positive samples and negative samples respectively, and then identifies potential semantic information of nodes without interfering with graph topology and node attributes.

However, if graph contrastive learning lacks a well-designed negative sample extraction strategy, the contrastive effect of graphs may appear random. So MoCL [22] uses domain knowledge for molecular graph enhancement, thus introducing transformations without changing molecular properties too much. IDCRN [16] adds two different Gaussian noises to the attribute matrix to generate two forms of the same matrix, and constructs two enhanced graph topologies through graph diffusion and similarity-based KNN graphs [3], Siamese neural networks are then used to improve the discriminative ability of node embeddings in terms of samples and features. AFGRL [14] discovers nodes that can serve as positive samples by exploring nodes that share local structural information and global semantics with the graph, thereby generating alternative views of the graph. However, heterogeneous graphs often have complex nonlinear structures and rich semantic information, so it is still an urgent problem to adopt an effective contrastive learning strategy to generate low-dimensional representations that are beneficial to subsequent the clustering task.

2.2   Heterogeneous Graph Clustering

Because HIN can make a more complete and natural abstraction of the real world, more and more researchers have begun to use complementary information in different views to complete clustering tasks. HAN [25] adopts a dual-attention mechanism to aggregate the attribute information of neighbor nodes for the target node. SESIM [17] builds supervision information by predicting the number of hops between nodes on each meta-path, and uses it as an auxiliary task to improve the performance of the main task. MVGC [28] uses Euler transforms to efficiently construct a new view descriptor for node attribute information, while imposing block-diagonal representation constraints on the self-expressive coefficient matrix to explore cluster structures.

In order to reduce the influence of noise contained in node information in HIN, MAGCN [2] designed a dual-path encoder for mapping graph embedding and learning the consistency information of views. MvAGC [15] first uses graph filters to denoise node features, then designs a new strategy to select a few anchors to reduce computational complexity, and finally proposes a new regularization strategy to mine high neighborhood information. However, they all rely on auxiliary information brought by node attributes, and it is difficult to directly apply them to heterogeneous graphs without attribute information.

3   Preliminary

Definition 1. *Heterogeneous Information Network. Heterogeneous Information Network (HIN) is defined as a network $G=\{V, E, \boldsymbol{F}, Y, U\}$. where $\boldsymbol{F} \in \mathbb{R}^{n \times d}$ is the feature matrix of the target node attribute information, n is the number of target nodes, and d is the feature dimension of the node. The mapping function of node type is $\mathcal{Y} : V \rightarrow Y$, which means that each node of the target node set V has a specific node object type in the set Y. The mapping function of edge type is $\mathcal{U} : E \rightarrow U$, which means that each edge of the edge set E has a specific edge object type in the set U. If $|Y| + |U| > 2$, the attribute graph is called a heterogeneous information network.*

Definition 2. *Meta-path. A meta-path $\boldsymbol{\Phi}$ is an ordered sequence of node types and edge types defined in HIN $G=\{V, E, \boldsymbol{F}, Y, U\}$. $\boldsymbol{\Phi} = Y_1 \xrightarrow{U_1} Y_2 \xrightarrow{U_2} \cdots \xrightarrow{U_l} Y_{l+1}$ (abbreviated as $\boldsymbol{\Phi} = Y_1 Y_2 \cdots Y_{l+1}$), where l is the length of the $\boldsymbol{\Phi}$.*

Definition 3. *Meta-path topology graph. The meta-path topology graph A^p is the meta-path topology structure extracted from the meta-path Φ_p.*

Definition 4. *Meta-path attribute graph. The meta-path attribute graph $G^p = (A^p, F)$ is composed of the meta-path topology graph A^p and the public feature matrix F.*

4   The Proposed Model

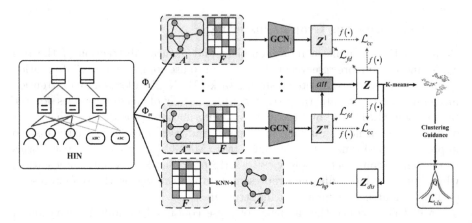

Fig. 1. The overall architecture of our proposed AHCMS.

The AHCMS model first constructs attribute graphs and meta-path attribute graphs based on the similarity between node features and each meta-path, and then aggregates the neighbor nodes of the target node in each meta-path attribute graph using the GCN module, so as to achieve feature extraction of node attribute information and fuse node representations based on the importance of specific semantic information generated by different meta-paths. It maximizes the consistency of information between different meta-path attribute graphs using a collaborative contrastive strategy to learn information-rich and consistent the high-level representation. In addition, a self-supervised clustering mechanism is introduced to supervise nodes with high confidence in the high-level representation, making the potential feature space more suitable for clustering tasks. For clarity, the overall architecture of AHCMS is illustrated in Fig. 1.

4.1   Feature Extraction Module

We use multiple GCN encoders to capture low-dimensional representations of each meta-path attribute graph:

$$Z^p = \sigma\left(\left(\hat{D}^p\right)^{-\frac{1}{2}} \hat{A}^p \left(\hat{D}^p\right)^{-\frac{1}{2}} F W^p\right) \tag{1}$$

where $\hat{A}^p = A^p + I$, $I \in \mathbb{R}^{n \times n}$ is the identity matrix, which adds self-connections to each node; $\hat{D}^p$ is the degree matrix of nodes in $\hat{A}^p$; W^p is the weight matrix of G^p; σ is a non-linear activation function; and Z^p is the low-dimensional representation learned by G^p.

Then, we use semantic-level attention mechanism to adaptively learn the weight coefficients of different meta-path attribute graphs:

$$(\alpha_1, \alpha_2 \cdots, \alpha_m) = att\left(Z^1, Z^2 \cdots, Z^m\right) \tag{2}$$

where att is the semantic-level attention mechanism, and α_p is the weight coefficient of Z^p.

Specifically, we use the similarity of the transformed embedding to the attention vector q to measure the importance of a particular semantic embedding:

$$e^p = \frac{1}{n} \sum_{i=0}^{n-1} q^T \cdot \mathrm{Tanh}\left(W z_i^p + b\right) \tag{3}$$

where $\mathrm{Tanh}(\cdot)$ is the hyperbolic tangent activation function, q^T is the transpose of q; W is the parameter matrix, b is the bias vector, and z_i^p is the low-dimensional representation of node v_i in G^p.

In order to maintain the relative relationship between different meta paths, we use the $\mathrm{softmax}(e^p)$ function to normalize the weights α_p.

Finally, the high-level representation $Z = \sum_{p=1}^{m} \alpha_p \cdot Z^p$ is obtained by weighted summation of the low-dimensional representations.

4.2 Feature Enhancement Module

Collaborative Contrastive Strategy. We let the different embeddings map to the space where the comparative loss is calculated:

$$\begin{aligned} H^p &= f\left(Z^p\right) = W_{(2)} \delta\left(W_{(1)} Z^p + b_{(1)}\right) + b_{(2)} \\ H &= f\left(Z\right) = W_{(2)} \delta\left(W_{(1)} Z + b_{(1)}\right) + b_{(2)} \end{aligned} \tag{4}$$

where $f(\cdot)$ is a small neural network projection head; H^p and H are the projection representations of Z^p and Z respectively; δ is a nonlinear activation function; $\{W_{(1)}, W_{(2)}\}$ is the parameter matrix, $\{b_{(1)}, b_{(2)}\}$ is the bias vector.

Then we adopt a positive selection strategy to select positive and negative samples. The specific strategies are as follows:

First, construct the direct neighbor node set N_i of node v_i in all meta-paths, and connect nodes v_i according to the number of meta-paths between node v_i and node v_j in descending order for nodes:

$$N_i = \left\{ j | j \in V \& \sum_{p=1}^{m} \mathbb{1}\left(j \in \mathcal{N}_i^p\right) \neq 0 \right\} \tag{5}$$

where $1\left(\cdot\right)$ is the indicator equation, and the result is 1 if the condition inside $(\cdot)$ is true, otherwise the result is 0.

Second, we select positive samples by setting the threshold T_{pos}. If $|N_i| > T_{pos}$, the first T_{pos} neighbor nodes from N_i are selected as the positive sample set $\mathbb{P}_i$ of node v_i, otherwise, all neighbor nodes are kept as $\mathbb{P}_i$; the unselected neighbor nodes are automatically added to the negative sample set $\mathbb{N}_i$ of node v_i .

$$\mathcal{L}_{cc} = -\frac{1}{|V|} \sum_{i \in V} \sum_{p=1}^{m} \log \frac{\sum_{j \in \mathbb{P}_i} \exp\left(sim\left(h_i, h_j^p\right)/\tau\right)}{\sum_{k \in (\mathbb{P}_i \cup \mathbb{N}_i)} \exp\left(sim\left(h_i, h_k^p\right)/\tau\right)} \tag{6}$$

where $sim\left(h_i, h_j^p\right) = \frac{h_i^T h_j^p}{\|h_i\|\|h_j^p\|}$ is the cosine distance between h_i and h_j^p, with smaller values indicating closer distance and higher similarity; $\|\cdot\|$ is a 2-norm, and τ is a temperature parameter.

Feature Decoupling Strategy. First, we construct the feature similarity matrix $T^p \in \mathbb{R}^{d' \times d'}$ of the meta-path attribute graph G^p:

$$t_{ij}^p = \frac{\left(s_i^p\right)^T s_j}{\|s_i^p\|\|s_j\|}, \forall i, j \in [1, d'] \tag{7}$$

where $s_i^p = \left(z_i^p\right)^T$ is the i-th dimension feature representation of all nodes in G^p, $s_j = \left(z_j\right)^T$ is the j-th dimension feature representation of all nodes in Z; d' is the embedding feature dimension of the node.

Then, in order to make the feature representations of the same dimension in Z^p and Z as similar as possible, and the feature representations of different dimensions as different as possible, we force the feature similarity matrix T^p to approximate the identity matrix $\tilde{I} \in \mathbb{R}^{d' \times d'}$:

$$\mathcal{L}_{fd} = \sum_{p=1}^{m} \frac{1}{d'^2} \sum \left(T^p - \tilde{I}\right)^2 \tag{8}$$

4.3 High-Order Proximity Module

We construct an attribute topology graph A_f based on the similarity between node attributes:

$$A_f = \text{KNN}\left(\frac{f_i \cdot f_j}{\|f_i\|\|f_j\|}\right) \tag{9}$$

where $\cdot$ is the inner product between the two vectors, $\text{KNN}(\cdot)$ denotes the K-nearest neighbors algorithm.

Then, we calculate the high-order similarity between any two nodes based on the high-order probability transition matrix:

$$\tilde{A}^r = \underbrace{\bar{A} \cdots \bar{A}}_{r} \tag{10}$$

where $\underbrace{\bar{A} \cdots \bar{A}}_{r}$ is the matrix product of r $\bar{A}$, $\bar{A}$ is the row normalized adjacency matrix of A, and $\tilde{a}_{ij}^r$ is the probability that node v_j in A^r is a neighbor of node v_i of order r, indicating their similarity within order r, and the larger the value, the higher the similarity.

We use cosine distance to explore the correlation between any two nodes in high-level representations from a topological perspective.

$$z_{dis}^{ij} = sim\left(z_i, z_j\right) \tag{11}$$

where z_{dis}^{ij} is the topological distance between node v_i and node v_j in the high-level representation; z_i is the high-level representation vector of node v_i.

We select r-order similar nodes for node v_i as positive samples, and the rest of the nodes as negative samples. Then the high-order proximity loss $\mathcal{L}_{hp}$ can be expressed as:

$$\mathcal{L}_{hp} = -\frac{1}{n}\sum_{i=1}^{n}\log\frac{\sum \tilde{a}_i^r \exp\left(z_{dis}^i/\tau'\right)}{10^{-8} + \sum \exp\left(z_{dis}^i/\tau'\right)} \tag{12}$$

where $\tilde{a}_i^r$ is the r-order similarity between node v_i and other nodes, z_{dis}^i is the topological distance between node v_i and other nodes in the high-level representation; τ' is the temperature parameter.

4.4 Self-supervised Clustering Mechanism

We use high-confidence nodes as soft labels to supervise the clustering process and help the high-level representations get closer to the center of the clusters.

$$\mathcal{L}_{clu} = \mathrm{KL}\left(P\,\|Q\right) = \sum_i\sum_j p_{ij}\log\frac{p_{ij}}{q_{ij}} \tag{13}$$

where $\mathrm{KL}\left(\cdot\|\cdot\right)$ is the KL divergence between the two distributions; Q is the soft class assignment distribution of nodes, P is the target distribution. In our study, Q is calculated by the Student's t-distribution:

$$q_{ij} = \frac{\left(1 + \|z_i - \mu_j\|^2/\xi\right)^{-\frac{\xi+1}{2}}}{\sum_{j'}\left(1 + \|z_i - \mu_{j'}\|^2/\xi\right)^{-\frac{\xi+1}{2}}} \tag{14}$$

where q_{ij} is the similarity between the high-level representation z_i of node v_i and the cluster center μ_j of the j-th category, and ξ is the degrees of freedom of the Student's t-distribution.

Target distribution P obtains a denser distribution by raising x to the power of two, with more emphasis on high-confidence nodes.

$$p_{ij} = \frac{q_{ij}^2/\bar{q}_j}{\sum_{j'} q_{ij'}^2/\bar{q}_{j'}} \tag{15}$$

where $\bar{q}_j = \sum_i q_{ij}$ is the soft cluster probability, which is used to normalize the loss contribution of each cluster center to prevent classes with heavy weights from distorting the hidden feature space.

4.5 Objective Function

The overall objective function of AHCMS includes four parts: collaborative contrastive loss, feature decoupling loss, high-order proximity loss, and self-supervised clustering loss.

$$\mathcal{L} = \gamma\mathcal{L}_{cc} + \beta\mathcal{L}_{fd} + \lambda\mathcal{L}_{hp} + \mathcal{L}_{clu} \tag{16}$$

where γ, β, λ are adjustable trade-off parameters.

5 Experiments

5.1 Datasets

We employ the following four benchmark datasets: ACM [1], DBLP [1], Freebase [26] and AMiner [26], summarized in Table 2.

Table 2. The statistics of the datasets.

Dataset	Node	Relation	Meta-path	Class
ACM	paper(P):4019	P-A:13407	PAP	3
		P-S:4019	PSP	
	author(A):7167			
	subject(S):60			
DBLP	author(A):4057	P-A:19645	APA	4
		P-C:14328	APCPA	
		P-T:85810	APTPA	
	paper(P):14328			
	conference(C):20			
	term(T):7723			
Freebase	movie(M):3492	M-A:65341	MAM	3
		M-D:3762	MDM	
		M-W:6414	MWM	
	actor(A):33401			
	direct(D):2502			
	writer(W):4459			
AMiner	paper(P):6564	P-A:18007	PAP	4
		P-R:58831	PRP	
	author(A):13329			
	reference(R):35890			

Table 1. The clustering results for module ablation study on four benchmark datasets.

Method	Metric	BL	BL-HP	BL-FD	BL-HP-FD
ACM	ACC	86.87 ± 0.41	86.97 ± 0.58	89.16 ± 0.55	89.34 ± 0.76
	NMI	63.75 ± 1.30	64.51 ± 0.84	66.34 ± 1.34	66.50 ± 1.87
	ARI	63.74 ± 0.96	64.00 ± 1.32	69.27 ± 1.35	69.75 ± 1.84
	F1	87.62 ± 0.40	87.76 ± 0.49	89.55 ± 0.54	89.67 ± 0.76
DBLP	ACC	92.87 ± 0.07	92.98 ± 0.08	92.97 ± 0.17	93.28 ± 0.04
	NMI	77.32 ± 0.18	77.60 ± 0.21	77.42 ± 0.44	78.40 ± 0.15
	ARI	82.78 ± 0.18	83.06 ± 0.17	82.98 ± 0.38	83.71 ± 0.11
	F1	92.38 ± 0.08	92.49 ± 0.08	92.52 ± 0.19	92.85 ± 0.05
Freebase	ACC	54.26 ± 1.12	54.39 ± 0.83	60.13 ± 0.99	60.41 ± 1.27
	NMI	16.62 ± 1.10	16.72 ± 1.32	17.36 ± 1.09	18.26 ± 1.28
	ARI	18.08 ± 1.81	19.33 ± 1.55	19.53 ± 1.34	20.07 ± 1.28
	F1	50.86 ± 1.33	51.58 ± 1.55	51.32 ± 1.53	51.99 ± 2.72
AMiner	ACC	62.81 ± 2.96	64.12 ± 1.65	72.29 ± 2.68	72.78 ± 1.65
	NMI	39.72 ± 2.56	39.94 ± 3.08	40.24 ± 4.95	43.86 ± 3.20
	ARI	39.40 ± 3.28	40.85 ± 5.89	50.36 ± 6.66	57.85 ± 4.65
	F1	49.66 ± 3.92	52.72 ± 3.25	58.30 ± 4.60	59.47 ± 1.89

Table 3. The clustering results of meta-path ablation study on the DBLP datasets.

Method	ACC	NMI	ARI	F1
DBLP-1	62.27 ± 1.50	29.24 ± 1.01	27.60 ± 1.35	61.62 ± 1.63
DBLP-2	68.06 ± 2.72	36.95 ± 2.12	35.20 ± 3.59	66.25 ± 3.66
DBLP-3	91.19 ± 0.28	73.62 ± 0.61	78.78 ± 0.73	90.67 ± 0.29
DBLP-12	71.86 ± 1.55	39.89 ± 1.69	41.63 ± 2.27	70.06 ± 2.06
DBLP-13	92.67 ± 0.09	76.86 ± 0.22	82.38 ± 0.19	92.16 ± 0.10
DBLP-23	93.08 ± 0.07	77.98 ± 0.23	83.28 ± 0.18	92.63 ± 0.08
DBLP-123	93.28 ± 0.04	78.40 ± 0.15	83.71 ± 0.11	92.85 ± 0.05

5.2   Baseline Methods

We compared AHCMS with seven baseline methods to demonstrate its superiority, including three HON methods (DGI [24], MVGRL [9], GCC [7]) and four HIN methods (DMGI [19], HDMI [12], HeCo [26], SNMH [1]).

5.3   Evaluation Metrics

Four widely used evaluation metrics, namely, ACCuracy (ACC) [7], Normalized Mutual Information (NMI) [26], Adjusted Rand Index (ARI) [26] and macro F1-score(F1) [7] are employed to evaluate the clustering performance.

5.4   Implementation Details

All of our experiments were implemented and completed based on the deep learning framework PyTorch 1.9.0, and run on Ubuntu 18.04.6 with an i9-11900K CPU and NVIDIA GeForce RTX 3090 GPU. For all baseline methods, we followed the settings in the original paper to handle the various parameters that emerged from the experiments, and tuned the parameters accordingly depending on the data set in order to achieve the best results for that paper. In order to prevent extreme situations in the code running, we run the experiments of each paper 10 times in the same environment and obtain the average value and corresponding standard deviation; for the three HON methods (DGI[5], MVGRL[28], GCC[29]), we will test each meta-path attribute graph of the dataset, and select the best result from it as the final clustering result for this dataset. For AHCMS, the learning rate of ACM and AMiner is set to 0.005, the learning rate of DBLP is set to 0.001, and the learning rate of Freebase is set to 0.008, and the early stopping strategy is adopted to avoid overfitting; The embedded feature dimension d' of ACM and AMiner is set to 128, d' of DBLP and Freebase is set to 64; the trade-off parameter $\{\gamma, \lambda, \beta\}$ of ACM is set to $\{1, 1, 1\}$; the trade-off parameter $\{\gamma, \lambda, \beta\}$ of DBLP is set to $\{1, 0.1, 1\}$; the trade-off parameter $\{\gamma, \lambda, \beta\}$ of Freebase and AMiner is set to $\{0.1, 1, 10\}$.

5.5   Experiment Results

Table 4. The clustering results on four benchmark datasets.

Dataset	Metric	DGI	MVGRL	GCC	DMGI	SNMH	HDMI	HeCo	AHCMS
ACM	ACC	88.79 ± 0.30	86.71 ± 0.84	86.16 ± 0.00	80.65 ± 2.31	70.47 ± 2.35	86.50 ± 2.04	85.73 ± 2.97	89.34 ± 0.76
	NMI	66.32 ± 0.49	60.48 ± 1.59	61.21 ± 0.00	54.01 ± 4.32	43.10 ± 2.24	63.96 ± 2.09	59.73 ± 4.20	66.50 ± 1.87
	ARI	68.18 ± 0.74	63.02 ± 2.07	62.01 ± 0.00	50.90 ± 4.26	38.38 ± 3.75	62.83 ± 4.67	61.69 ± 6.29	69.75 ± 1.84
	F1	89.25 ± 0.27	87.09 ± 0.76	86.84 ± 0.00	81.84 ± 2.27	70.32 ± 0.93	87.34 ± 1.74	86.06 ± 2.79	89.67 ± 0.76
DBLP	ACC	89.56 ± 0.19	76.74 ± 4.01	90.43 ± 0.00	89.00 ± 2.04	81.95 ± 8.15	88.91 ± 0.63	90.11 ± 0.31	93.28 ± 0.04
	NMI	70.57 ± 0.32	57.42 ± 4.09	72.63 ± 0.00	69.92 ± 1.64	63.54 ± 8.44	69.48 ± 0.75	71.68 ± 0.59	78.40 ± 0.15
	ARI	76.11 ± 0.42	56.28 ± 4.66	77.44 ± 0.00	75.99 ± 1.69	65.23 ± 9.57	74.89 ± 1.19	77.18 ± 0.64	83.71 ± 0.11
	F1	88.55 ± 0.27	74.75 ± 4.34	89.53 ± 0.00	88.93 ± 0.93	79.97 ± 9.95	87.96 ± 0.66	89.22 ± 0.36	92.85 ± 0.05
Freebase	ACC	61.53 ± 1.11	54.30 ± 0.46	64.06 ± 0.95	52.67 ± 1.12	55.38 ± 1.88	56.11 ± 1.00	57.86 ± 1.74	60.41 ± 1.27
	NMI	19.00 ± 0.63	16.69 ± 0.07	19.92 ± 1.47	14.23 ± 1.14	19.35 ± 1.04	13.92 ± 1.44	17.65 ± 1.19	18.26 ± 1.28
	ARI	19.07 ± 0.65	17.26 ± 0.08	20.05 ± 1.61	12.87 ± 1.10	13.89 ± 1.51	15.32 ± 1.20	19.70 ± 1.37	20.07 ± 1.28
	F1	49.07 ± 0.58	52.95 ± 0.25	47.20 ± 0.54	50.55 ± 1.09	41.05 ± 1.45	50.70 ± 1.43	51.14 ± 1.06	51.99 ± 2.72
AMiner	ACC	46.24 ± 0.53	30.89 ± 1.03	44.93 ± 0.35	32.66 ± 1.09	43.74 ± 1.81	47.01 ± 3.56	67.44 ± 5.43	72.78 ± 1.65
	NMI	2.07 ± 0.14	3.35 ± 0.32	2.10 ± 0.13	5.21 ± 0.56	3.45 ± 1.21	16.79 ± 2.36	40.86 ± 5.17	43.86 ± 3.20
	ARI	8.25 ± 0.65	0.34 ± 0.37	8.90 ± 0.26	1.11 ± 1.29	6.03 ± 1.66	14.09 ± 2.43	42.73 ± 9.41	57.85 ± 4.65
	F1	29.17 ± 1.55	26.61 ± 0.89	31.39 ± 0.52	26.87 ± 0.37	30.65 ± 1.85	42.67 ± 3.92	59.03 ± 4.04	59.47 ± 1.89

The experiment results on the four benchmark datasets are summarized in Table 4, we can conclude that 1) On the four datasets, AHCMS demonstrated good performance compared to seven baseline methods across four evaluation metrics. Taking the AMiner dataset as an example, AHCMS showed improvements of 2.85%, 5.77%, 6.27%, and 3.32% in ACC, NMI, ARI, and F1 respectively compared to the best baseline method HeCo. 2) Compared to the three HON methods, AHCMS achieved better clustering results. This is because HON methods only used one meta-path of the dataset to perform the clustering task, ignoring the importance of other meta-paths for the clustering task. In contrast, AHCMS utilized a collaborative contrastive strategy to comprehensively consider the influence of multiple meta-paths and adopted a semantic-level attention mechanism to evaluate the importance of each meta-path, thereby improving clustering performance. 3) Although the four HIN methods demonstrated good clustering performance by integrating multiple meta-paths, they did not consider the association between nodes from the attribute level. At the same time, they overly emphasized the similarity between different learned representations, resulting in too much redundant information contained in the low-dimensional representation, which is not conducive to subsequent clustering tasks. In contrast, AHCMS utilized high-order proximity modules to fully capture the topological similarity between node attributes and combined feature decoupling strategy to enable the model to learn more discriminative representations, thereby improving clustering performance. 4) On the AMiner dataset, there was a significant gap between all baseline methods and AHCMS. This is because the dataset lacks attribute information for target nodes, making it difficult to complete the clustering task solely based on topological information. In contrast, AHCMS utilized a collaborative contrasting strategy to capture the topological correlation of nodes in HIN, allowing more neighborhood information to be included in high-level representations. Additionally, through the feature decoupling strategy, the model reduced the correlation between node attribute information, forcing nodes to focus on their own information and thereby alleviating the noise problem caused by the lack of node attribute information. 5) AHCMS did not achieve the best clustering performance on the Freebase dataset. This is because the meta-path MAM of this dataset contains the main topological structure information of target nodes, and the structural correlation is large. When multiple meta-paths are used for clustering tasks, it is equivalent to adding a lot of noise to the structural information, resulting in unsatisfactory clustering results. Despite this, AHCMS utilized a semantic-level attention mechanism to encourage the model to give greater weight to MAM and provided more comprehensive guidance for the clustering task through a self-supervised clustering mechanism. As a result, the high-level representation learned by the model has stronger discriminative power, enabling it to achieve the best clustering performance among HIN methods.

5.6   Ablation Study

Module Ablation Study. We analyzed the importance of the high-order proximity module (HP) and feature decoupling strategy (FD) in AHCMS by independently removing them. First, we used the framework without these two modules as the baseline (BL), the other methods represent some combination of them. The clustering results of the module ablation experiments are shown in Table 1, we can see that 1) BL-HP-FD achieves optimal results on four datasets because it both reduces the interference of redundant information on the clustering task and captures the high-order similarity of node attributes. 2) On the AMiner dataset, BL-FD showed 9.48%, 0.52%, 10.96%, and 8.64% improvement over BL on ACC, NMI, ARI, and F1, respectively. These results demonstrate that the model can enhance the discriminative power of the potential space by feature decoupling strategy, and thus better handle the dataset without node attribute information.

Meta-Path Ablation Study. To evaluate the contribution of different meta-paths in the AHCMS training process, we measure the impact of different meta-path combinations on the clustering effect of AHCMS through meta-path ablation study using the DBLP dataset as an example. The specific descriptions are as follows: DBLP-1, DBLP-2, and DBLP-3 indicate that the meta-paths used by AHCMS are APTPA, APA, and APCPA, the other methods represent some combination of them. The clustering results of the meta-path ablation experiments are shown in Table 3, it can be seen that: 1) In the method using only a single meta-path, the clustering effect of DBLP-3 is far better than that of DBLP-1 and DBLP-2, this indicates that AHCMS has different contributions when extracting the topology of HIN using different meta-paths. 2) In the method of using multiple meta-paths simultaneously, DBLP-123 achieved the best results in all four evaluation metrics, indicating that different meta-paths can provide complementary information in clustering tasks, thereby improving clustering performance. 3) The clustering performance of DBLP-3 is better than that of DBLP-12, indicating that clustering performance is not only influenced by the number of meta-paths but also depends on the importance of meta-paths in HIN.

5.7 Featureless Analysis

To evaluate the effectiveness of AHCMS in heterogeneous graphs without node attribute information, we take the ACM dataset as an example and replace the feature matrix of paper nodes with a one-hot feature vector encoding for each paper node. The clustering results are shown in Fig. 2, it can be seen that: 1) When node attribute information is missing, AHCMS achieves the best clustering performance compared to six baseline methods in four evaluation metrics. Compared with the optimal baseline method HeCo, it improves by 6.24%, 9.72%, 11.3%, and 2.87% in ACC, NMI, ARI, and F1, respectively. It is shown that AHCMS can capture the topological information of nodes in different meta-paths through the collaborative contrastive strategy, and use the feature decoupling strategy to encourage nodes to focus on their own attribute information, thereby better dealing with datasets without node attribute information. 2) In particular, DGI obtained a clustering effect second only to AHCMS when node attribute information was considered. When the attribute information of the nodes is missing, the clustering effect is reduced by 35.97%, 59.95%, 57.05%, and 44.88% on ACC, NMI, ARI, and F1, respectively, compared to that when the attribute information is considered. This is because in the process of maximizing mutual information, DGI needs to use node attribute information as positive samples to construct negative samples. When facing datasets with highly correlated attribute information between nodes, missing attribute information makes it difficult for DGI to learn meaningful low-dimensional representations, resulting in a sharp decline in clustering performance.

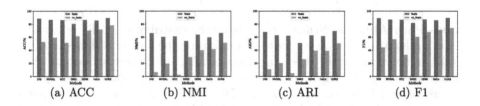

(a) ACC   (b) NMI   (c) ARI   (d) F1

Fig. 2. The clustering results of seven methods on the ACM dataset.

5.8 Visualization

To more intuitively verify the effectiveness of AHCMS, we use the t-distributed stochastic neighbor embedding (t-SNE) algorithm [18] on the DBLP dataset to visualize the low-dimensional representations learned by four methods in a two-dimensional space. The visualization results are shown in Fig. 3, where different colors represent different categories. As can be seen in Fig. 3, AHCMS has the best visualization result, which can separate different types of nodes more correctly and the boundaries are relatively obvious, so it has a clearer clustering structure and is more friendly for subsequent clustering tasks.

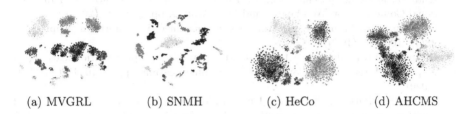

(a) MVGRL   (b) SNMH   (c) HeCo   (d) AHCMS

Fig. 3. The t-SNE visualization on the DBLP dataset.

6   Conclusion

We propose an adaptive heterogeneous graph contrastive clustering with multi-similarity (AHCMS) model, which learns the node representations of each meta-path attribute graph in the feature extraction module. Then, it adaptively integrates specific semantic information using a semantic-level attention mechanism and explores consistency information between different meta-paths through an enhancement module to obtain richer high-level representations. Secondly, using high-order proximity modules to improve the structural reliability of the original graph. Additionally, We make the high-level representation contain clustering-related discriminative information by driving it continuously closer to the cluster center. Finally, experimental results on four benchmark datasets verify the effectiveness of AHCMS.

References

1. Chen, Z., Luo, L., Li, X., Jiang, B., Guo, Q., Wang, C.: Siamese network based multiscale self-supervised heterogeneous graph representation learning. IEEE Access **10**, 98490–98500 (2022)
2. Cheng, J., Wang, Q., Tao, Z., Xie, D., Gao, Q.: Multi-view attribute graph convolution networks for clustering. In: Proceedings of The Twenty-ninth International Conference on International Joint Conferences on Artificial Intelligence, pp. 2973–2979 (2021)
3. Du, G., Zhou, L., Li, Z., Wang, L., Lü, K.: Neighbor-aware deep multi-view clustering via graph convolutional network. Inform. Fusion **93**, 330–343 (2023)
4. Du, G., Zhou, L., Yang, Y., Lü, K., Wang, L.: Deep multiple auto-encoder-based multi-view clustering. Data Sci. Eng. **6**(3), 323–338 (2021)

5. Ezugwu, A.E., et al.: A comprehensive survey of clustering algorithms: State-of-the-art machine learning applications, taxonomy, challenges, and future research prospects. Eng. Appl. Artif. Intell. **110**, 104743 (2022)
6. Fan, S., Wang, X., Shi, C., Lu, E., Lin, K., Wang, B.: One2multi graph autoencoder for multi-view graph clustering. In: Proceedings of the Web Conference 2020, pp. 3070–3076 (2020)
7. Fettal, C., Labiod, L., Nadif, M.: Efficient graph convolution for joint node representation learning and clustering. In: Proceedings of the Fifteenth ACM International Conference on Web Search and Data Mining, pp. 289–297 (2022)
8. Hartigan, J.A., Wong, M.A., et al.: A k-means clustering algorithm. Appl. Stat. **28**(1), 100–108 (1979)
9. Hassani, K., Khasahmadi, A.H.: Contrastive multi-view representation learning on graphs. In: International Conference on Machine Learning, pp. 4116–4126. PMLR (2020)
10. Hwang, D., Park, J., Kwon, S., Kim, K., Ha, J.W., Kim, H.J.: Self-supervised auxiliary learning with meta-paths for heterogeneous graphs. Adv. Neural. Inf. Process. Syst. **33**, 10294–10305 (2020)
11. Jin, D., Huo, C., Liang, C., Yang, L.: Heterogeneous graph neural network via attribute completion. In: Proceedings of the Web Conference 2021, pp. 391–400 (2021)
12. Jing, B., Park, C., Tong, H.: Hdmi: High-order deep multiplex infomax. In: Proceedings of the Web Conference 2021, pp. 2414–2424 (2021)
13. Kipf, T.N., Welling, M.: Semi-supervised classification with graph convolutional networks. arXiv preprint arXiv:1609.02907 (2016)
14. Lee, N., Lee, J., Park, C.: Augmentation-free self-supervised learning on graphs. In: Proceedings of the AAAI Conference on Artificial Intelligence, vol. 36, pp. 7372–7380 (2022)
15. Lin, Z., Kang, Z.: Graph filter-based multi-view attributed graph clustering. In: IJCAI, pp. 2723–2729 (2021)
16. Liu, Y., Zhou, S., Liu, X., Tu, W., Yang, X.: Improved dual correlation reduction network. arXiv preprint arXiv:2202.12533 (2022)
17. Ma, S., Liu, J.w., Zuo, X.: Self-supervised learning for heterogeneous graph via structure information based on metapath. arXiv preprint arXiv:2209.04218 (2022)
18. Van der Maaten, L., Hinton, G.: Visualizing data using t-sne. J. Mach. Learn. Res. **9**, 2579–2605 (2008)
19. Park, C., Han, J., Yu, H.: Deep multiplex graph infomax: attentive multiplex network embedding using global information. Knowl.-Based Syst. **197**, 105861 (2020)
20. Ren, Y., Liu, B., Huang, C., Dai, P., Bo, L., Zhang, J.: Heterogeneous deep graph infomax. arXiv preprint arXiv:1911.08538 (2019)
21. Sun, K., Lin, Z., Zhu, Z.: Multi-stage self-supervised learning for graph convolutional networks on graphs with few labeled nodes. In: Proceedings of the AAAI Conference on Artificial Intelligence, vol. 34, pp. 5892–5899 (2020)
22. Sun, M., Xing, J., Wang, H., Chen, B., Zhou, J.: Mocl: contrastive learning on molecular graphs with multi-level domain knowledge. arXiv preprint arXiv:2106.04509 (2021)
23. Sun, Y., Han, J., Yan, X., Yu, P.S., Wu, T.: Heterogeneous information networks: the past, the present, and the future. Proc. VLDB Endow. **15**(12), 3807–3811 (2022)
24. Veličković, P., Fedus, W., Hamilton, W.L., Liò, P., Bengio, Y., Hjelm, R.D.: Deep graph infomax. arXiv preprint arXiv:1809.10341 (2018)

25. Wang, X., et al.: Heterogeneous graph attention network. In: The World Wide Web Conference, pp. 2022–2032 (2019)
26. Wang, X., Liu, N., Han, H., Shi, C.: Self-supervised heterogeneous graph neural network with co-contrastive learning. In: Proceedings of the 27th ACM SIGKDD Conference on Knowledge Discovery & Data Mining, pp. 1726–1736 (2021)
27. Wang, Y., Sun, W., Xu, K., Zhu, Z., Chen, L., Zheng, Z.: Fastgcl: Fast self-supervised learning on graphs via contrastive neighborhood aggregation. arXiv preprint arXiv:2205.00905 (2022)
28. Xia, W., Wang, S., Yang, M., Gao, Q., Han, J., Gao, X.: Multi-view graph embedding clustering network: Joint self-supervision and block diagonal representation. Neural Netw. **145**, 1–9 (2022)
29. Xu, W., Xia, Y., Liu, W., Bian, J., Yin, J., Liu, T.Y.: Shgnn: structure-aware heterogeneous graph neural network. arXiv preprint arXiv:2112.06244 (2021)
30. Zhou, L., Wang, J., Wang, L., Chen, H., Kong, B.: Heterogeneous information network representation learning: a survey. Chin. J. Comput. **45**(1), 160–189 (2022)

Multi-teacher Local Semantic Distillation from Graph Neural Networks

Haibo Liu, Di Zhang$^{(\boxtimes)}$, Liang Wang, and Xin Song

School of Cyber Security and Computer, Hebei University, Baoding 071000, China
dd2696857228@gmail.com

Abstract. Knowledge Distillation (KD) as a model compression technique has been widely used in Graph Neural Networks (GNNs). Recently research has demonstrated that GNNs with shallow or deep hidden layers have different representational abilities on nodes with low or high degrees. The use of multi-teacher models can provide diverse knowledge sources for the student model. However, a significant amount of information is missing when the knowledge is transferred through the deep hidden layers in the multi-teacher models. It is difficult for multi-teacher models to transfer the information of each hidden layer to the student model while preserving the local semantics of their respective models. To solve this problem, we propose a novel multi-teacher knowledge distillation method, Multi-teacher Local Semantic distillation from Graph Neural Networks (MLS-GNN). MLS-GNN is able to agglomerate the representation capabilities of shallow and deep GNN teacher models on nodes with different degrees. Not only that, the local semantics of each hidden layer in multi-teacher models is preserved through nodes and edges. Finally, the adaptive weights are assigned to different teacher models to stabilize the knowledge acquisition process and summarize the local semantics. We conducted experiments on five public datasets for node classification task, and the results show that MLS-GNN has excellent performance and better than other baseline methods.

Keywords: Knowledge Distillation · Multi-Teacher Model · Local Semantics · GNN · Adaptive Weights

1 Introduction

Recently, deep neural networks have achieved greatly success in many fields [1, 2]. However, during the training process of deep learning models, the large computing resource consumption and high storage requirements make it difficult for the models to be deployed in specific applications. Knowledge Distillation (KD) is a viable model compression technique that can effectively address these challenges [3]. After pre-training a cumbersome and massive teacher model, a lightweight student model extracts knowledge from it and achieves good performance.

The original KD method is primarily utilized in traditional Convolutional Neural Networks(CNNs) [4, 5], with early research focusing mainly on distilling knowledge

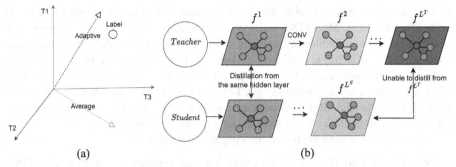

(a)   (b)

Fig. 1. (a) Comparison of average weight direction and adaptive weight direction in three-dimensional space. If most teacher predictions are biased, averaging the weights of individual teachers may deviate from the correct direction, and the weights we designed can be adjusted adaptively through knowledge learning. (b): Teacher model and student model have different number of hidden layers. Consequently, one-to-one matching of hidden layers between teacher and student models is not feasible, which restricts the ability to distill the complete local semantics of the teacher model.

from the last layer of the teacher model [6], which has a risk of overfitting when distilling knowledge [7]. Hence, researchers have shifted their focus to extracting knowledge from the hidden layers of the teacher model [8, 9]. Recently, KD has been wildly applied to Graph Neural Network (GNN) [10–12], and can be classified into two categories based on the number of teacher models: (1) Single GNN teacher model: A well-trained GNN is used as the teacher model, and the student model can either be a GNN, Multi-Layer Perception (MLP) or a combination of MLP and Label Propagation(LP) [11, 13–15]. (2) Multiple GNN teacher models: Different GNNs have their unique advantages. Multiple teacher models ensemble the prediction outputs through either fixed weights or weights designed based on the predicted logits output, and transfers the integrated information to the student model [12, 16–18].

However, on the one hand, it should be noted that GNNs differ from CNNs in terms of their structure. While CNNs are designed to process grid data, GNNs are mainly intended for non-grid data [14]. As a result, the method of KD in CNNs cannot be directly applied to GNNs. On the other hand, it has been proved that GNNs with different number of hidden layers have different classification abilities on nodes with different degrees [17]. The shallow model is applicable to nodes with low degrees and the deep model is applicable to nodes with high degrees. Moreover, multi-teacher model with shallow or deep layers has better prediction accuracy than single teacher model. Besides, as illustrated in Fig. 1(a) and (b), there are some shortcomings in multi-teacher model, such as Fig. 1(a) assigning weights using label-free strategies that may not leads student to learn in a relatively correct direction [8, 17], and Fig. 1(b) losing knowledge during transfer through deep hidden layers.

In this paper, we propose Multi-teacher Local Semantic distillation from Graph Neural Networks (MLS-GNN). In our work, the information in each hidden layer of the multi-teacher model is considered as local semantics. The preservation of local semantics in the multi-teacher model is achieved through the consolidation of information from all hidden layers, and a refined student model is trained by combining multiple teachers with different layers. MLS-GNN relies on the structure of the graph to compute node similarity in different hidden layers and preserve the similarity to the edges. The entire local semantics of a single teacher is obtained by aggregating edge information from various hidden layers across the entire graph. Adaptive weights are calculated using label combination strategies and assigned to each teacher, then the local semantics and prediction results of multiple teachers are integrated to obtain the final fused outputs. Finally, the student model learns knowledge from the fused outputs.

In summary, the contributions of this paper are as follows:

(1) We propose a novel approach for extracting local semantics of multi-teacher GNN models. By preserving all hidden layer information in the local structure, we enable the student model to flexibly learn local semantics by measuring the difference of local knowledge distributions between the hidden layers of teachers and the hidden layers of student.

(2) We design a new mechanism to calculate the adaptive weights. It can distinguish high-quality and low-quality teachers based on the degree of matching between teachers and students, and the resulting adaptive weights enable the teacher model to guide the learning of the student model in a relatively correct direction.

(3) To verify the effectiveness of MLS-GNN, we conduct experiments on five public datasets via the node classification task. The results show that our method achieves excellent performance compared to all other methods. Moreover, on some datasets, the performance of the student model even surpasses that of the teacher model.

2   Related Work

Graph Neural Network (GNN). As a deep learning method based on graph structure, GNN has become the subject of extensive contemporary research. Graph Convolutional Network (GCN) [19], as one of the representative methods of GNN, is applied to the downstream tasks of GNN based on the information of nodes and edges, and the applicable scenarios are more general and effective. Graph Attention Network (GAT) [20] is a classical GNN model for node classification. Each node in the graph can assign different weights to neighboring nodes by their similarities. GraphSAGE [21] enhances GCN by sampling neighborhood nodes and improves classification accuracy. Appnp [22] introduces the PageRank algorithm into GNN and achieves linear computational complexity by approximating topic-aware PageRank through power iteration. In this paper, we focus on how to transfer the local semantics of the shallow and deep multi-teacher GNN models to the student model in the process of knowledge distillation, rather than designing a new GNN.

Knowledge Distillation (KD). KD [3, 9, 23, 24] is a widely used model compression method that can extract a more refined and lightweight student model with fewer parameters from a more complex teacher model with more parameters. This technique has gained widespread adoption in several industries due to its simplicity and effectiveness. TinyGNN [13] samples neighborhood nodes among peers, allowing shallow student models to mimic the performance of deep teacher models. LSP [14] first proposed to handle non-grid data by preserving the local structure of the model. CPF [11] takes any fully trained GNN as the teacher model, considering the combination of Multi-Layer Perception (MLP) and Label Propagation (LP) as the student model. GLNN [15] distills the perfect GNN teacher model into MLP. However, although TinyGNN and LSP both learn the local structure of the teacher model, their student models do not necessarily have fewer layers. Both CPF and GLNN use MLP as their student model, which cannot make full use of the graph structure. All the above models use a single Neural Network as their teacher model, which has limited access to information for the student model and is not capable of effectively dealing with nodes with high and low degrees in the graph.

Knowledge Amalgamation (KA). Multi-teacher knowledge distillation is a branch of knowledge distillation that aims to diversify the learning content of student models [8, 18, 25–28]. Jing et al. [16] designs a topological attribution map and learns topological semantics from multiple teachers, but the weight of each teacher in this method is equal. MSKD [17] designs different weights through the attention mechanism and distills different teacher models at different epochs. However, if the epoch settings are not reasonable, the model is easy to overfitting. Although MSKD extracts the local semantics of the hidden layer, it does not fully extract the local semantics of each layer. Our work is also a kind of KA. It not only preserves all the hidden layers information of the graph to the edge, but also designs adaptive weights to ensemble multiple teachers to further improve the distillation performance.

3 Method

In this section, we introduce a novel KD model, Multi-teacher Local Semantic distillation from Graph Neural Networks (MLS-GNN), and the training method.

3.1 Problem Formalization

Throughout this paper, we consider an undirected graph $G = (V, E)$, where $V = \{v_1, v_2, \ldots, v_n\}$ denotes the node set, $E \subseteq V \times V$ denotes the edge set, V_L is the labeled set, and V_U is the unlabeled set. The graph consists of N nodes, and $X = \{x_1, x_2 \ldots, x_n\} \in R^{N \times d}$ denotes the feature matrix, where x_i denotes the feature vector of node v_i. In this paper, we aim to transfer the local semantics of all hidden layers and the knowledge of the output layer from multiple GNNs teachers into a compact student model.

3.2 Multi-teacher Distillation Framework

The workflow of MLS-GNN is illustrated in Fig. 2. Initially, the adaptive weights of the teacher model are determined by utilizing the soft labels predicted by both the teacher and student models [8]. Subsequently, we introduce our proposed method for transferring the local semantics of the model. Finally, we describe the transfer of knowledge in the output layer and the calculation of the overall loss function.

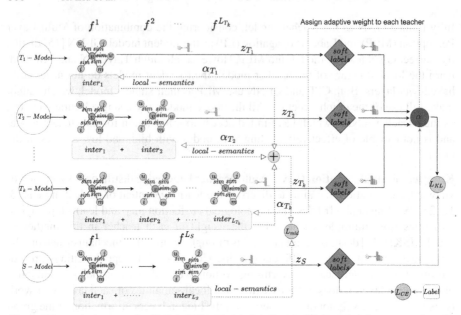

Fig. 2. The main framework of MLS-GNN. T_1-Model, T_2-Model, T_k-Model correspond to different teacher models, S-Model is the student model, f^{LT_k}, f^{LS} ($f^{LS} \leq f^{LT_k}$) correspond to different number of hidden layers (f^{LS} is the total number of hidden layers for the student model and f^{LT_k} is the total number of hidden layers for the k_{th} teacher model), and the weights are assigned adaptively by the soft labels of teachers and student.

Simultaneously, it is also critical to select the teacher set $T = \{T_1, T_2, \ldots, T_K\}$ (K is the number of teacher models). If the graph contains numerous nodes with low degrees, over-smoothing [12] can occur with an increase in the depth of the hidden layer of the teacher model. In such a scenario, if there are many teacher models with a large number of hidden layers the final prediction outputs of the integrated teacher models may be relatively poor. Therefore, the number of teachers and the number of hidden layers for each teacher need to be carefully selected according to different data characteristics.

3.3 Multi-teacher Adaptive Weight Design

In order to enable the student model to learn the correct knowledge from each teacher model, we devise adaptive weights for each teacher. Let z be the predicted output of the teacher model, and let $f = [f^1, f^2, \ldots f^L]$ represent the hidden layer, where L is the number of hidden layers. The weight for each teacher is determined based on the predictions of both the teacher and student models:

$$\alpha_{T_k} = \frac{Mean(log\_softmax(\frac{z_S * (z_{T_k})^T}{d_{z_S}}))}{\sum_{k=1}^{K} Mean(log\_softmax(\frac{z_S * (z_{T_k})^T}{d_{z_S}}))}, \tag{1}$$

where z_S is the prediction output of the student in the last layer, z_{T_k} is the prediction output of the k_{th} teacher last layer, d_{z_S} is the dimension of z_S, $Mean(\cdot)$ is an averaging function that returns the average of all elements, α_{T_k} is the weight of each teacher. Similarities are calculated by prediction results of teacher and student model, then $log\_softmax$ and average processing are performed on the similarity results to obtain the final adaptive weight α_{T_k}. Enlightened by the attention mechanism [29], the output similarities of teachers and student are divided by d_{z_S}, which can prevent the variance of the similarities from becoming large. The adaptive weights are designed so that multiple teachers have their own discriminability in the student feature spaces, and represent the measure of the difference between the student and multiple teachers.

3.4 Hidden Layer Local Semantic Transferring

In terms of local structural information, the hidden layer of the student model should show a similar distribution to the teacher model as much as possible [4, 8, 14, 17]. To maintain the stability of the student model in the process of knowledge transfer in the hidden layer, we design the measurement L_{mid} that measures the distance between the hidden layer representations of the student model and multiple teacher models:

$$L_{mid} = KL\left(\sum_{l=1}^{L_S} inter\left(f_S^l\right), \sum_{k=1}^{K} \alpha_{T_k} \sum_{l=1}^{L_{T_k}} inter\left(f_{T_k}^l\right)\right)$$

$$= \sum_{l=1}^{L_S} inter(f_S^l) log \frac{\sum_{l=1}^{L_S} inter(f_S^l)}{\sum_{k=1}^{K} \alpha_{T_k} \sum_{l=1}^{L_{T_k}} inter\left(f_{T_k}^l\right)}, \tag{2}$$

$$inter\left(f_S^l\right) = \sum_{(i,j)} inter\left(f_S^{ij}\right)$$

$$= softmax\left(\sum_{i=1}^{n} \frac{1}{|N_i|} e^{\mu * \sum_{j \in N_i} \left|f_S^i - f_S^j\right|}\right), \tag{3}$$

$$inter\left(f_{T_k}^l\right) = \sum_{(i,j)} inter\left(f_{T_k}^{ij}\right)$$

$$= softmax\left(\sum_{i=1}^{n} \frac{1}{|N_i|} e^{\mu * \sum_{j \in N_i} \left|f_{T_k}^i - f_{T_k}^j\right|}\right). \tag{4}$$

As shown in Fig. 2, the $inter(\cdot)$ function is used to measure the structural information of the hidden layers in both the *student and teacher* models. The number of hidden layers in the student and teacher models may differ, f_S^l is the l_{th} hidden layer in the student model, and $f_{T_k}^l$ is the l_{th} hidden layer in the k_{th} teacher model. In any hidden layer of a teacher, the Euclidean Distance between node v and its adjacent node j is used to represent its similarity. The similarities of all nodes and their neighbors in different hidden layers is assigned to the edge. Therefore, even the same edge retains different similarities in different hidden layers. Next, the similarities are aggregated across the teacher model, and their adaptive weights α_{T_k} are used to fuse the knowledge of each teacher, yielding the local structure information of all teachers. Finally, we measure the variability between the teacher and student models using the Kullback-Leibler (KL) divergence.

3.5 Output Layer Knowledge Transferring

The combination of Cross-Entropy (CE) and KL divergence are employed to calculate the loss of the student in the distillation process. The distribution distance between the student and the integrated teachers is calculated by KL divergence which can make the student to learn from the teachers, and the student is also guided by the ground truth through CE loss between student prediction label and real label.

The distillation loss function of the student in the output layer is:

$$L_{KD} = L_{KL}\left(\sigma\left(\frac{z^S}{\tau}\right), \sum_{k=1}^{K} \alpha_{T_k} * \sigma\left(\frac{z^{T_k}}{\tau}\right)\right) + \lambda L_{CE}\left(\sigma\left(z^S\right), y\right), \quad (5)$$

where, σ represents the *SoftMax* function. The temperature τ represents a hyperparameter, which is used to enlarge and shrink the information of negative labels. y represents the real one-hot label, and λ is the balance hyperparameter.

The overall loss function is as follows:

$$L = L_{KD} + \varphi L_{mid}, \quad (6)$$

where the coefficient $\varphi > 0$ is used as a balancing factor.

4   Experiment

In this section, we first introduce the experiment datasets, then provide the baselines we compare, and present our parameter settings and experiment results. Furthermore, we analyze the parametric efficiency of the model and baselines. Finally, we performed ablation experiments and visualization analysis based on the above experiments.

4.1 Datasets

We conducted experiments for node classification [30] on 5 public datasets, including 3 citation network: Cora [31], CiteSeer [31], PubMed [32], and two co-authored graphs: Amazon-computer [33], Amazon-photo [33]. We partition the data sets as same as CPF [11], and 20 nodes are selected for each class as the training set, 30 nodes are selected as the validation set, and the rest of the nodes are used as the test set. Cora, CiteSeer and PubMed are citation network, in which node acts as publication and edge represents citation relationship. There are no isolate nodes in the graphs. Amazon-computer and Amazon-photo are extracted from Amazon purchase graph, in which node acts as commodity and edge represents purchase together.

The details of the datasets are as follows (Table 1):

Table 1. Statistic of the five datasets

	Node	Edge	Features	Class
Cora	2485	5069	1433	7
CiteSeer	2110	3668	3703	6
PubMed	19,717	44,324	500	3
A-Photo	7487	119,043	745	8
A-Computer	13,381	245,778	767	10

4.2 Baselines and Evaluation Metrics

We compare MLS-GNN with four KD methods in nodes classification task:

- CPF [11]: CPF uses an arbitrary GNN as the single teacher model, and the student model is a combination of feature propagation and label propagation.
- GLNN [15]: GLNN is a lightweight model with single teacher that distills knowledge from GNNs to MLPs.
- Jing et. al. [16]: A topological attribution map (TAM) scheme is proposed to learn the topological semantics of multi-teachers.
- MSKD [17]: MSKD is a multi-teacher-based model which is distilled by different training stages. The teacher models extract knowledge of the penultimate layer to provide for the student model.

We also compare MLS-GNN with two none-KD models in node classification task:

- GraphSAGE [21]: GraphSAGE improves GCN by sampling neighborhood nodes, so that GCN can be extended to large graphs.
- Appnp [22]: Appnp improved GCN by introducing the PageRank algorithm.
- According to CPF, we employ the Accuracy (ACC) as evaluation metric.

4.3 Experimental Settings

To ensure fair comparison between MLS-GNN[1] and the baseline models, we use grid-search or follow the original papers to get the optimal hyperparameters. For consistency, in the baseline of the multi-teacher model, the student model uses consistent parameters. We employ GCN and GAT as the teacher and student model and the number of teachers is set 2. MLS-GCN and MLS-GAT represent MLS-GNN models with GCN and GAT as teachers and students, respectively. Different teachers have different numbers of hidden layers on the same dataset. It is worth noting that the total parameters of the teacher models are always larger than those of the student model. In teacher GCN, the hidden layer dimensions of the teacher models are 256 dimensions. In student GCN, the hidden layer dimensions of the student model are 128 dimensions. In teacher GAT, each hidden layer has 6 attention heads and 64 hidden features, and the output layer has 1 attention heads and C (the class of the dataset) hidden features. In student GAT, each hidden layer

---

[1] Our code is publicly available at https://github.com/Oddttfz/MLS_GNN.

has 6 attention heads and 32 hidden features, and the output layer has 1 attention heads and C hidden features. The number of hidden layers of the teacher model is {1, 2, 3}, the number of hidden layers of the student model is 1. We have used Adam optimization and trained each model for 200 epochs on all datasets. All results are reported in means and standard deviations over 3 runs with different random seeds. In different datasets, the hyperparameters of the models are shown in Table 2.

Table 2. Model hyperparameter design in each dataset

	Cora	CiteSeer	PubMed	A-Photo	A-Computer
Learning Rate	0.003	0.003	0.01	0.005	0.005
Weight Decay	0.01	0.037	0.03	0.01	0.01
Temperature	5	3	3	2	2

4.4 Experimental Results

The experimental results presented in Table 3 demonstrate that MLS-GNN outperforms all the other methods in terms of accuracy. Especially on the PubMed dataset, MLS-GCN surpasses the best baseline by nearly 1.4%. On the Amazon-Computer dataset, MLS-GCN also achieves a notable improvement of approximately 0.6%, while demonstrating various degrees of improvement on the other datasets as well, the performance of MLS-GAT on most datasets is approximately similar to that of MLS-GCN.

Compared with non-KD methods, GraphSAGE and APPNP, the application of GNN based KD exhibit superior performance (e.g., CPF, MSKD). This demonstrates the effectiveness of KD in improving the performance of GNN models for node classification tasks. The use of soft labels as prediction outputs of the teacher during training helps the student to better understand complex concepts and mimic the teacher with higher accuracy. As illustrated in Fig. 3, the choice of temperature during the training phase is also an important factor that affects the model performance. A lower temperature setting is generally preferred for models with fewer parameters, as it helps the student to focus on the most informative parts of the knowledge from the teacher.

Compared with CPF and GLNN which employ GNN as the single teacher and MLP as the main structure, multi-teacher distillation has shown promising results in improving performance by leveraging diverse sources of knowledge from multiple teachers (e.g., MLS-GCN). However, CPF and GLNN still have their advantages and outperform some multi-teacher models (e.g., Jing et al, MSKD) on several datasets. On one hand, Jing et al.[16] did not learn from ground truth, leading to evident performance drawbacks when compared with the CPF lightweight model, especially in cases where the teacher model prediction results exhibit obvious fluctuations. On the other hand, MSKD may suffers from the overthinking problem [34] due to having too many teachers with deep hidden layers.

Compared with the multi-teacher model Jing et al. [16] and MSKD, MLS-GNN (MLS-GCN and MLS-GAT) integrates knowledge from both teachers and ground truth,

Table 3. Accuracy (%) results of baseline models and MLS-GNN in five datasets

	Cora	CiteSeer	PubMed	A-Photo	A-Computer
Optimal Teacher	85.2 ± 0.6	76.5 ± 0.4	81.0 ± 0.4	93.4 ± 0.3	**86.9 ± 0.6**
GraphSAGE	82.9 ± 0.3	74.3 ± 0.3	81.0 ± 0.6	91.5 ± 0.2	84.6 ± 0.5
APPNP	85.3 ± 0.2	75.4 ± 0.1	81.4 ± 0.5	93.2 ± 0.1	85.2 ± 0.8
CPF	85.3 ± 0.3	76.5 ± 0.1	81.1 ± 0.4	93.4 ± 0.1	84.6 ± 0.3
GLNN	84.5 ± 0.3	69.6 ± 0.2	79.9 ± 0.2	93.4 ± 0.2	85.2 ± 0.4
Jing et al.	83.6 ± 0.5	75.2 ± 0.4	79.2 ± 0.6	91.1 ± 0.6	83.2 ± 0.4
MSKD	83.5 ± 0.3	74.5 ± 0.3	80.1 ± 0.3	92.6 ± 0.2	85.9 ± 0.5
MLS-GCN	**85.5 ± 0.5**	**76.7 ± 0.3**	**82.8 ± 0.3**	**93.6 ± 0.3**	86.5 ± 0.5
MLS-GAT	85.4 ± 0.2	76.5 ± 0.2	81.9 ± 0.3	93.4 ±0.2	86.1 ± 0.3

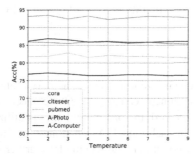

Fig. 3. The impact of different temperatures on the acc of MLS-GCN, presented in percentage (%).

which reduces the probability of teacher-learning-error being propagated to the student. Moreover, MLS-GNN strikes a balance between the number of model parameters and performance, leading to a clear advantage. In contrast, Jing et al.[16] disregards the importance of different types of knowledge, and MSKD fails to establish a reasonable teacher ratio, as an excessive number of teachers does not necessarily improve knowledge understanding. Additionally, MSKD does not fully assimilate the local semantics of each hidden layer in the teacher model, while MLS-GNN incorporates these local semantics to yield superior prediction outcomes.

It is noteworthy that, on the Cora, CiteSeer, PubMed and Amazon-Photo datasets, MLS-GCN (which can be regarded as the student model only) exhibits better performance than even the Optimal Teacher (the best teacher in MLS-GNN), providing strong evidence of the effectiveness of our proposed approach. Our student with simple structure not only assimilates the teacher's knowledge but also mitigates the overthinking issues that arise in large models with deep hidden layers. Additionally, L_{mid} serves as a regularization that enhances the learning ability of the student model.

Table 4. Parameters of MLS-GNN and its teachers and each baseline in different datasets

	Cora	CiteSeer	PubMed	A-Photo	A-Computer
Teacher$_1$-GCN	0.37M	0.95M	0.13M	0.19M	0.20M
Teacher$_2$-GCN	0.44M	1.02M	0.19M	0.26M	0.27M
Teacher$_1$-GAT	0.55M	1.43M	0.19M	0.29M	0.30M
Teacher$_2$-GAT	0.70M	1.57M	0.34M	0.44M	0.45M
GraphSAGE	0.37M	0.95M	0.13M	0.19M	0.20M
APPNP	0.37M	0.95M	0.13M	0.19M	0.20M
CPF	0.05M	0.25M	0.04M	0.06M	0.07M
GLNN	0.18M	0.47M	0.06M	0.10M	0.10M
Jing et al.	0.28M	0.71M	0.1M	0.15M	0.15M
MSKD	0.18M	0.47M	0.06M	0.10M	0.10M
MLS-GCN	0.18M	0.47M	0.06M	0.10M	0.10M
MLS-GAT	0.28M	0.71M	0.1M	0.15M	0.15M

4.5 Parameter Efficiency

The proposed KD approach aims to mitigate the memory footprint of the model, which is crucial in resource-constrained scenarios to enable successful deployment of real-world applications. In this regard, the parameter count is utilized as a metric to assess the memory consumption of each model. As illustrated in Table 4, the non-KD model exhibits a larger parameter count than the KD model. Furthermore, the KD model that employs MLP (e.g., CPF, GLNN) as the main structure have the smallest parameter count. By reducing the dimension of the hidden layers and the number of network layers, MLS-GCN and MLS-GAT reduces the model size of the teacher GNN to less than 25% and 35% on average while maintaining high performance, respectively. It is important to note that each approach has its strengths and weaknesses, and the choice of method should be based on the specific task and dataset.

4.6 Ablation Study

To demonstrate the effectiveness of our different components, we conduct ablation studies on the CiteSeer and PubMed datasets. We specifically investigate the impact of adaptive weights, local semantic learning, and multi-teacher vs single teacher approaches.

1. Fixed weights: To examine the impact of fixed weights, we employed a fixed weight of $1/N_T$ for each teacher instead of the adaptive weights (N_T denotes the number of teachers). The results presented in Table 5 demonstrate that assigning equal weights to CiteSeer and PubMed leads to a notable decline in accuracy by 0.3% and 0.9% for MLS-GCN and 0.5% and 1.5% for MLS-GAT, respectively. These findings suggest that it is crucial to ensemble multiple teacher models using adaptive weights. Our approach on adaptive weights enables us to differentiate the quality of the teachers.

Table 5. Ablation study (%) on CiteSeer and PubMed

GNN-Type	Model Components	CiteSeer	PubMed
MLS-GCN	Fixed weights	76.4 ± 0.2	81.9 ± 0.7
	MLS-GNN-no $-L_{mid}$	76.3 ± 0.2	82.2 ± 0.4
	MLS-GNN -one-Teacher	75.3 ± 0.8	82.0 ± 0.5
	MLS-GNN	76.7 ± 0.3	82.8 ± 0.3
MLS-GAT	Fixed weights	76.0 ± 0.5	80.4 ± 0.3
	MLS-GNN-no $-L_{mid}$	76.1 ± 0.3	80.7 ± 0.4
	MLS-GNN -one-Teacher	75.6 ± 0.6	79.7 ± 0.7
	MLS-GNN	76.5 ± 0.2	81.9 ± 0.2

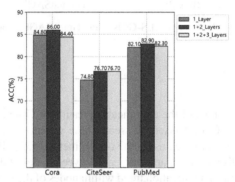

Fig. 4. MLS-GCN Acc (%) results learned from different numbers of teacher models. "1_Layer" represents a teacher with one hidden layer, while "1 + 2_Layers" refers to a model with both a single hidden layer and a two-hidden-layer network. Similarly, "1 + 2 + 3_Layers" denotes a teacher with one hidden layer and a teacher with two hidden layers and a teacher with three hidden layers.

2. MLS-GNN $-no-L_{mid}$: This ablation study aims to evaluate the impact of local semantic learning in MLS-GNN. We remove the L_{mid} measurement from the total components to obtain the variant of MLS-GNN that lacks local semantic learning. As shown in Table 5, the removal of L_{mid} leads to a substantial decrease in the accuracy of the student model, underscoring the effectiveness of the knowledge transfer through local semantic learning in hidden layers. This finding demonstrates that hidden layers contain crucial information that needs to be transferred to the student model, and our proposed method successfully leverages this information for knowledge distillation.
3. MLS-GNN -one-Teacher: In this setting, we train the MLS-GNN model using a single teacher. As presented in Table 5, the performance of the student model is clearly lower when learning from a single teacher, indicating the importance of learning from multiple teacher models and incorporating diverse knowledge. In MLS-GNN, teacher with a single hidden layer is suitable for processing nodes with low degrees, while teacher with two hidden layers is appropriate for handling nodes with high

degrees. Compared to a single teacher model, MLS-GNN has better capability in handling nodes with varying degrees. The results are further supported by Fig. 4 on the Cora and PubMed datasets, which highlights the crucial role of determining the optimal number of teacher models for effective knowledge transfer. It is important to note that having an excessive number of teachers does not necessarily improve the training outcome of the student.

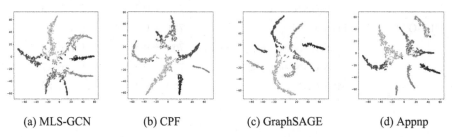

(a) MLS-GCN   (b) CPF   (c) GraphSAGE   (d) Appnp

Fig. 5. The t-SNE visualization of MLS-GCN, CPF, GraphSAGE, Appnp on Cora dataset.

4.7 Visualization

We generated visualizations of the classification results of GraphSAGE, APPNP, CPF and MLS-GCN on the Cora dataset. To create these visualizations, we embedded the nodes into a 2-dimensional space using the t-SNE method [35]. The results are shown in Fig. 5, which reveals a clear separation trend between nodes of different categories in MLS-GCN, along with an aggregation trend within nodes of the same category. Overall, the node visualization of MLS-GCN showcases its superior performance compared to the other models.

5 Conclusion

Compared with the existing single-teacher model, we can learn more comprehensive knowledge from multiple teacher models. Compared with the existing multi-teacher model, we distill the local semantics of all hidden layers of the teacher model, and give the teacher model adaptive weights for integration, making the teacher model more powerful and further reducing the difference of local knowledge distributions between the student model and the teacher model. Results show that MLS-GNN has excellent node classification ability. However, distillation from the local structure may not be possible if the local structure of the student model differs dramatically from that of the teacher model, so our method has some limitations. For this issue, we hope to obtain expansion in our future research career.

Acknowledgements. This research is supported by the Research Found for Talented Scholars of Hebei University under the grant No. 521100221088.

References

1. He, K., Zhang, X., Ren, S., Sun, J.: Deep residual learning for image recognition. In: Proceedings of the IEEE Conference on Computer Vision and Pattern Recognition, pp. 770–778 (2016).
2. Silver, D., et al.: Mastering the game of go without human knowledge. Nature **550**, 354–359 (2017)
3. Hinton, G., Vinyals, O., Dean, J.: Distilling the knowledge in a neural network. arXiv preprint arXiv:1503.02531 (2015)
4. Romero, A., Ballas, N., Kahou, S.E., Chassang, A., Gatta, C., Bengio, Y.: Fitnets: Hints for thin deep nets. arXiv preprint arXiv:1412.6550 (2014)
5. Chen, P., Liu, S., Zhao, H., Jia, J.: Distilling knowledge via knowledge review. In: Proceedings of the IEEE/CVF Conference on Computer Vision and Pattern Recognition, pp. 5008–5017 (2021)
6. Feng, T., Wang, M.: Response-based Distillation for Incremental Object Detection. arXiv preprint arXiv:2110.13471 (2021)
7. Sun, S., Cheng, Y., Gan, Z., Liu, J.: Patient knowledge distillation for bert model compression. arXiv preprint arXiv:1908.09355 (2019)
8. Zhang, H., Chen, D., Wang, C.: Confidence-aware multi-teacher knowledge distillation. In: ICASSP 2022–2022 IEEE International Conference on Acoustics, Speech and Signal Processing (ICASSP), pp. 4498–4502. IEEE (2022)
9. Haidar, M.A., Anchuri, N., Rezagholizadeh, M., Ghaddar, A., Langlais, P., Poupart, P.: Rail-kd: random intermediate layer mapping for knowledge distillation. arXiv preprint arXiv:2109.10164 (2021)
10. Yao, H., et al.: Graph few-shot learning via knowledge transfer. In: Proceedings of the AAAI Conference on Artificial Intelligence, pp. 6656–6663 (2020)
11. Yang, C., Liu, J., Shi, C.: Extract the knowledge of graph neural networks and go beyond it: an effective knowledge distillation framework. In: Proceedings of the web conference 2021, pp. 1227–1237 (2021)
12. Zhang, W., et al.: ROD: reception-aware online distillation for sparse graphs. In: Proceedings of the 27th ACM SIGKDD Conference on Knowledge Discovery & Data Mining, pp. 2232–2242 (2021)
13. Yan, B., Wang, C., Guo, G., Lou, Y.: Tinygnn: learning efficient graph neural networks. In: Proceedings of the 26th ACM SIGKDD International Conference on Knowledge Discovery & Data Mining, pp. 1848–1856 (2020)
14. Yang, Y., Qiu, J., Song, M., Tao, D., Wang, X.: Distilling knowledge from graph convolutional networks. In: Proceedings of the IEEE/CVF Conference on Computer Vision and Pattern Recognition, pp. 7074–7083 (2020)
15. Zhang, S., Liu, Y., Sun, Y., Shah, N.: Graph-less neural networks: Teaching old mlps new tricks via distillation. arXiv preprint arXiv:2110.08727 (2021)
16. Jing, Y., Yang, Y., Wang, X., Song, M., Tao, D.: Amalgamating knowledge from heterogeneous graph neural networks. In: Proceedings of the IEEE/CVF conference on computer vision and pattern recognition, pp. 15709–15718 (2021)
17. Zhang, C., Liu, J., Dang, K., Zhang, W.: Multi-scale distillation from multiple graph neural networks. In: Proceedings of the AAAI Conference on Artificial Intelligence, pp. 4337–4344 (2022)
18. Zhu, J., et al.: Ensembled CTR prediction via knowledge distillation. In: Proceedings of the 29th ACM International Conference on Information & Knowledge Management, pp. 2941–2958 (2020)

19. Kipf, T.N., Welling, M.: Semi-supervised classification with graph convolutional networks. arXiv preprint arXiv:1609.02907 (2016)
20. Veličković, P., Cucurull, G., Casanova, A., Romero, A., Lio, P., Bengio, Y.: Graph attention networks. arXiv preprint arXiv:1710.10903 (2017)
21. Hamilton, W., Ying, Z., Leskovec, J.: Inductive representation learning on large graphs. Adv. Neural Inf. Process. Syst. **30** (2017)
22. Gasteiger, J., Bojchevski, A., Günnemann, S.: Predict then propagate: Graph neural networks meet personalized pagerank. arXiv preprint arXiv:1810.05997 (2018)
23. Lin, S., et al.: Knowledge distillation via the target-aware transformer. In: Proceedings of the IEEE/CVF Conference on Computer Vision and Pattern Recognition, pp. 10915–10924 (2022)
24. Ba, J., Caruana, R.: Do deep nets really need to be deep? Adv. Neural Inf. Process. Syst. **27** (2014)
25. Shen, C., Wang, X., Song, J., Sun, L., Song, M.: Amalgamating knowledge towards comprehensive classification. In: Proceedings of the AAAI Conference on Artificial Intelligence, pp. 3068–3075 (2019)
26. Asif, U., Tang, J., Harrer, S.: Ensemble knowledge distillation for learning improved and efficient networks. arXiv preprint arXiv:1909.08097 (2019)
27. You, S., Xu, C., Xu, C., Tao, D.: Learning from multiple teacher networks. In: Proceedings of the 23rd ACM SIGKDD International Conference on Knowledge Discovery and Data Mining, pp. 1285–1294 (2017)
28. Luo, S., Wang, X., Fang, G., Hu, Y., Tao, D., Song, M.: Knowledge amalgamation from heterogeneous networks by common feature learning. arXiv preprint arXiv:1906.10546 (2019)
29. Vaswani, A., et al.: Attention is all you need. Adv. Neural Inf. Process. Syst. **30** (2017)
30. Wu, Z., Pan, S., Chen, F., Long, G., Zhang, C., Philip, S.Y.: A comprehensive survey on graph neural networks. IEEE Trans. Neural Netw. Learn. Syst. **32**, 4–24 (2020)
31. Sen, P., Namata, G., Bilgic, M., Getoor, L., Galligher, B., Eliassi-Rad, T.: Collective classification in network data. AI Mag. **29**, 93 (2008)
32. Namata, G., London, B., Getoor, L., Huang, B., Edu, U.: Query-driven active surveying for collective classification. In: 10th International Workshop on Mining and Learning with Graphs, p. 1 (2012)
33. Shchur, O., Mumme, M., Bojchevski, A., Günnemann, S.: Pitfalls of graph neural network evaluation. arXiv preprint arXiv:1811.05868 (2018)
34. Shi, H., et al.: Revisiting Over-smoothing in BERT from the Perspective of Graph. arXiv preprint arXiv:2202.08625 (2022)
35. Van der Maaten, L., Hinton, G.: Visualizing data using t-SNE. J. Mach. Learn. Res. **9**, 2579–2605 (2008)

AutoAM: An End-To-End Neural Model for Automatic and Universal Argument Mining

Lang Cao[✉]

University of Illinois Urbana Champaign, Champaign, USA
langcao2@illinois.edu

Abstract. Argument mining is to analyze argument structure and extract important argument information from unstructured text. An argument mining system can help people automatically gain causal and logical information behind the text. As argumentative corpus gradually increases, like more people begin to argue and debate on social media, argument mining from them is becoming increasingly critical. However, argument mining is still a big challenge in natural language tasks due to its difficulty, and relative techniques are not mature. For example, research on non-tree argument mining needs to be done more. Most works just focus on extracting tree structure argument information. Moreover, current methods cannot accurately describe and capture argument relations and do not predict their types. In this paper, we propose a novel neural model called AutoAM to solve these problems. We first introduce the argument component attention mechanism in our model. It can capture the relevant information between argument components, so our model can better perform argument mining. Our model is a universal end-to-end framework, which can analyze argument structure without constraints like tree structure and complete three subtasks of argument mining in one model. The experiment results show that our model outperforms the existing works on several metrics in two public datasets.

Keywords: Argument Mining · Information Extraction · Natural Language Processing

1 Introduction

Argument mining (AM) is a technique for analyzing argument structure and extracting important argument information from unstructured text, which has gained popularity in recent years [12]. An argument mining system can help people automatically gain causal and logical information behind the text. The argument mining techniques benefit plenty of many fields, like legal [31], public opinions [19], finance, etc. Argument mining is beneficial to human society, but there is still much room for development. Argument mining consists of several tasks and has a variety of different paradigms [12]. In this paper, we focus on

X. Yang et al. (Eds.): ADMA 2023, LNAI 14179, pp. 517–531, 2023.
https://doi.org/10.1007/978-3-031-46674-8_36

the most common argument structure of monologue. It is an argumentative text from one side, not an argument from two sides. The microscopic structure of argumentation is the primary emphasis of the monologue argument structure, which primarily draws out the internal relations of reasoning.

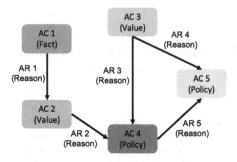

Fig. 1. An example of argument mining result after extraction in the CDCP dataset [19]. It forms an argument graph. In this graph, every node AC represents an argument component. Fact, Value, and Policy are three types of ACs. Every edge AR denotes argument relation, and Reason is one type of AR.

In this setting, an argumentative paragraph can be viewed as an argument graph. An argument graph can efficiently describe and reflect logical information and reasoning paths behind the text. An example of AM result after extraction is shown in Fig. 1. The two important elements in an argument graph are the argument component (AC) and the argument relation (AR). ACs are nodes in this graph, and ARs are edges. The goal of an AM system is to construct this argument graph from unstructured text automatically. The process of the AM system definition we use is as following steps:

1. Argument Component Identification (ACI): Given an argumentative paragraph, AM systems will detect ACs from it and separate this text.
2. Argument Component Type Classification (ACTC): AM systems will determine the types of these ACs.
3. Argument Relation Identification (ARI): AM systems will identify the existence of a relationship between any ACs.
4. Argument Relation Type Classification (ARTC): AM systems will determine the type of ARs, which are the existing relations between ACs.

Subtask 1) is a token classification task, which is also a named entity recognition task. This task has a large amount of research work on it. Most of the previous argument mining works [3,10,25] assume that the subtask 1) argument component identification has been completed, which is the argument component has been identified and can be obtained from the argumentative text. Therefore, the emphasis of argument mining research is placed on other subtasks. Following previous works, we also make such assumptions in this paper. On this basis, we design an end-to-end model to complete ACTC, ARI, and ARTC subtasks simultaneously.

ARI and ARTC are the hardest parts of the whole argument mining process. An AR is represented by two ACs. It is difficult to represent AR precisely and capture this relation. Most ACs pairs do not have a relationship at all, which leads to a serious sample imbalance problem. Among the whole process, ARI and ARTC are parts of ACI and ACTC, so the performance of these tasks will be influenced. Due to these reasons, many previous works give up and ignore the classification of ARs. Besides, much research imposes some argument constraints to do argument mining. In most normal cases, they assume argument information is a tree structure, and they can use the characteristic of the tree to extract information. Tree structure argument information is common in an argumentative essay. However, argument information with no constraints is more normal in the real world, like a huge amount of corpus on social media. This information is just like the general argument graphs mentioned before and needs to be extracted in good quality.

In this paper, we solve the above problems with a novel model called **AutoAM** (the abbreviation of **Auto**mactic and Universal **A**rgument **M**ining Model). This is an efficient and accurate argument mining model to complete the entire argument mining process. This model does not rely on domain-specific corpus and does not need to formulate special syntactic constraints, etc., to construct argument graphs from argumentative text. To improve the performance of non-tree structured argument mining, we first introduce the argument component attention mechanism (**ArguAtten**) in this model, which can better capture the relevant information of argument components in an argumentative paragraph. It benefits the overall performance of argument mining. We use a distance matrix to add the key distance feature to represent ARs. A stratified learning rate is also a critical strategy in the model to balance multi-task learning. To the best of our knowledge, we are the first to propose an end-to-end universal AM model without structure constraints to complete argument mining. Meanwhile, we combine our novelty and some successful experience to achieve the state of the art in two public datasets.

In summary, our contributions are as follows:

- We propose a novel model **AutoAM** for argument mining which can efficiently solve argument mining in all kinds of the argumentative corpus.
- We introduce **ArguAtten** (argument component attention mechanism) to better capture the relation between argument components and improve overall argument mining performance.
- We conduct extensive experiments on two public datasets and demonstrate that our method substantially outperforms the existing works. The experiment results show that the model proposed in this paper achieves the best results to date in several metrics. Especially, there is a great improvement over the previous studies in the tasks of ARI (argument relation identification) and ARTC (argument relation type classification).

2   Related Work

Since argument mining was first proposed [16], much research has been conducted on it. At first, people used rule-based or some traditional machine learning methods. With the help of deep learning, people begin to get good performance on several tasks and start to focus on non-tree structured argument mining. We discuss related work following the development of AM.

2.1   Early Argument Mining

The assumption that the argument structure could be seen as a tree or forest structure was made in the majority of earlier work, which made it simpler to tackle the problem because various tree-based methods with structural restrictions could be used. In the early stage of the development of argument mining, people usually use rule-based structural constraints and traditional machine learning methods to conduct argumentative mining. In 2007, Moens et al. [16] conducted the first argument mining research on legal texts in the legal field, while Kwon et al. [11] also conducted relevant research on commentary texts in another field. However, the former only identified the content of the argument and did not classify the argument components. Although the latter one further completed the classification of argument components, it still did not extract the relationship between argument components, and could not explore the argument structure in the text. It only completed part of the process of argument mining.

2.2   Tree Structured Argument Mining with Machine Learning

According to the argumentation paradigm theory of Van Eemeren et al. [6], Palau and Moens [15] modeled the argument information in legal texts as a tree structure and used the hand-made Context-Free Grammar (CFG) to parse and identify the argument structure of the tree structure. This method is less general and requires different context-free grammars to be formulated for different structural constraints of argument. By the Stab and Gurevych [27] [28] tree structure of persuasive Essay (Persuasive Essay, PE) dataset has been in argument mining has been applied in many studies and practices. In this dataset, Persing and Ng [23] and Stab and Gurevych [28] used the Integer Linear Programming (ILP) framework to jointly predict the types of argument components and argument relations. Several structural constraints are defined to ensure a tree structure. The arg-micro text (MT) dataset created by Peldszus [21] is another tree-structured dataset. In studies using this dataset, decoding techniques based on tree structure are frequently used, such as Minimum Spanning tree (MST) [22], and ILP [1].

2.3   Neural Network Model in Argument Mining

With the popularity of deep learning, neural network models have been applied to various natural language processing tasks. For deep learning methods based on neural networks, Eger et al. [7] studied argument mining as a sequence labeling problem that relies on parsing multiple neural networks. Potash et al. [25] used sequence-to-sequence pointer network [30] in the field of argument mining and identified the different types of argument components and the presence of argument relations using the output of the encoder and decoder, respectively. Kuribayashi et al. [10] developed a span representation-based argumentation structure parsing model that employed ELMo [24] to derive representations for ACs.

2.4 Recent Non-tree Structured Argument Mining

Recently, more works have focused on the argument mining of non-tree structures. The US Consumer Debt Collection Practices (CDCP) dataset [18,19] greatly promotes the development of non-tree structured argument mining. The argument structures contained in this dataset are non-tree structures. On this dataset, Niculae et al. [18] carry out a structured learning method based on a factor graph. This method can also handle the tree structure of datasets. It can also be used in the PE dataset, but the factor diagram needs a specific design according to the different types of the argument structure. Galassi et al. [8] used the residual network on the CDCP dataset. Mor IO et al. [17] developed an argument mining model, which uses a task-specific parameterized module to encode argument. In this model, there is also a bi-affine attention module [5] to capture the argument. Recently, Jianzhu Bao et al. [2] tried to solve both tree structure argument and non-tree structure argument by introducing the transformation-based dependency analysis method [4,9]. This work gained relatively good performance on the CDCP dataset but did not complete the ARTC task in one model and did not show the experiment results of ARTC.

However, these methods either do not cover the argument mining process with a good performance or impose a variety of argument constraints. There is no end-to-end model for automatic and universal argument mining before. Thus, we solve all the problems above in this paper.

3   Methodology

As shown in Fig. 2, we propose a new model called AutoAM. This model adopts the joint learning approach. It uses one model to simultaneously learn the ACTC, ARI, and ARTC three subtasks in argument mining. For the argument component extraction, the main task is to classify the argument component type, and the argument component identification task has been completed by default on both the PE and the CDCP datasets. For argument relation extraction, the model regards ARI and ARTC as one task. The model classifies the relationship between the argument components by a classifier and then gives different prediction results for two tasks by post-processing prediction labels.

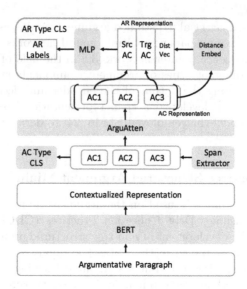

Fig. 2. The framework of our proposed model called AutoAM.

3.1 Task Formulation

The input data contains two parts: a) A set of n argumentative text $T = \{T_1, T_2, ..., T_n\}$, b) for the ith argumentative text, there are m argument component spans $S = \{S_1, S_2, ..., S_m\}$, where every span marks the start and end scope of each AC $S_i = (start_i, end_i)$. Our aim is to train an argument mining model and use it to get output data: a) types of m ACs provided in the input data $ACs = \{AC_1, AC_2, ..., AC_m\}$, b) k existing ARs $ARs = \{AR_1, AR_2, ..., AR_k\}$ and their types, where $AR_i = (AC_a \rightarrow AC_b)$.

3.2 Argument Component Extraction

By default, the argument component identification task has been completed. The input of the whole model is an argumentative text and a list of positional spans corresponding to each argument component $S_i = (start_i, end_i)$.

We input argumentative text T into pre-trained language models (PLMs) to get contextualized representations $H \in \mathbb{R}^{m \times d_b}$, where d_b is the dimension of the last hidden state from PLMs. Therefore, we represent argumentative text as $H = (h_1, h_2, ..., h_m)$, where h_i denotes the ith token contextualized representation.

We separate argument components from the paragraph using argument component spans S. In the PE dataset, the argument components do not appear continuously. We use mean pooling to get the representation of each argument component. Specifically, the i argument component can be represented as:

$$AC_i = \frac{1}{end_i - start_i + 1} \sum_{j=start_i}^{end_i} h_i, \tag{1}$$

where $AC_i \in \mathbb{R}^{d_b}$. Therefore, all argument components in the argumentative text can be represented as $ACs = (AC_1, AC_2, ..., AC_n)$. For each argument component, we input it into AC Type Classifier MLP_a in order. This classifier contains a multi-layer perceptron. A Softmax layer is after it. The probability of every type of argument component can be get by:

$$p(y_i|AC_i) = Softmax(MLP_a(AC_i)), \tag{2}$$

where y_i represent the predicted labels of the ith argument component. We get the final predicted label of its argument component as:

$$\hat{y}_i = Argmax(p(y_i|AC_i)). \tag{3}$$

3.3   Argument Relation Extraction

This model views ARI and ARTC as having the same task and distinguish them by post-processing predictions. We classify every argument component pair $(AC_i \rightarrow AC_j)$. Argument component pairs are different of $(AC_i \rightarrow AC_j)$ and $(AC_j \rightarrow AC_i)$. We add a label, 'none' here. 'none' represents that there is no relation of $AC_i \rightarrow AC_j$.

In the argument relation extraction part, we use the enumeration method. We utilize output results from the ACTC step. We combine two argument components and input them into AR Type Classifier to get the predicted output.

First, the model uses ArguAtten (Argument Component Attention mechanism) to enhance the semantic representation of argument components. The self-attention mechanism is first proposed in the Transformer [29]. The core of this mechanism is the ability to capture how each element in a sequence relates to the other elements, i.e., how much attention each of the other elements pays to that element. When the self-attention mechanism is applied to natural language processing tasks, it can often capture the interrelationship of all lexical elements in a sentence and better strengthen the contextual semantic representation. In the task of argument mining, all argument components in an argumentative text also meet this characteristic. The basic task of argument mining is to construct an argument graph containing nodes and edges, where nodes are argument components and edges are argument relations. Before the argument relation extraction task, the self-attention mechanism of argument components can be used to capture the mutual attention of argument components. It means that it can better consider and capture the argument information of the full text. This mechanism is conducive to argument relations extraction and the construction of an argument graph. We define ArguAtten as:

$$ArguAtten(Q, K, V) = Softmax(\frac{QK^T}{\sqrt{d_k}}) \times V, \tag{4}$$

where Q, K, V are got by multiplying ACs with W_Q, W_K, W_V. They are three parameter matrices $W_Q, W_K, W_V \in \mathbb{R}^{d_b \times d_k}$, and d_k is the dimension of attention layer. Besides, we also use ResNet and layer normalization (LN) after the attention layer to avoid gradient explosion:

$$ResNetOut = LN(ACs + ArgutAtten(ACs)). \tag{5}$$

Through the self-attention of argument components, we obtain a better contextualized representation of argument components and then begin to construct argument pairs to perform argument relation extraction.

We consider that the relative distance between two argument components has a decisive influence on the type of argument relations between them. By observing the dataset, we can find that there is usually no argument relation between the two argument components, which are relatively far apart. It can significantly help the model to classify the argument relation types. Therefore, we incorporate this feature into the representation of argument relations. At first, the distance vector is introduced, and the specific definition is shown as:

$$V_{dist} = (i - j) \times W_{dist}, \tag{6}$$

where $(i-j)$ represents a relative distance, it can be positive or negative. $W_{dist} \in \mathbb{R}^{1 \times d_{dist}}$ is a distance transformation matrix, and it can transform a distance scalar to a distance vector. d_{dist} is the length of the distance vector.

For each argument relation, it comes from the source argument component (Src AC), the target argument component (Trg AC), and the distance vector (Dist Vec). We concatenate them to get the representation of an argument relation as:

$$AR_{i,j} = [AC_i, AC_j, V_{dist}], \tag{7}$$

where $AR_{i,j} \in \mathbb{R}^{d_b \times 2 + d_{dist}}$, d_{dist} is the length of distance vector.

Therefore, argument relations in an argumentative text can be represented as $ARs = (AC_{1,2}, AC_{1,3}, ..., AC_{n,n-1},)$, contains $n \times (n-1)$ potential argument relations in total. We do not consider self-relation like $AR = (AC_i \rightarrow AC_i)$.

For each potential argument relation, we separately and sequentially input them into the AR Type Classifier MLP_b. The classifier uses a Multi-Layer Perceptron (MLP) containing a hidden layer of 512 dimensions. The output of the last layer of the Multi-layer Perceptron is followed by a Softmax layer to obtain the probability of an argument relationship in each possible type label, as shown in:

$$p(y_{i,j}|AR_{i,j}) = Softmax(MLP_b(AR_{i,j})), \tag{8}$$

where $y_{i,j}$ denotes the predicted label of the argument relation from the ith argument component to the jth argument component. The final predicted labels are:

$$\hat{y}_{i,j} = Argmax(p(y_{i,j}|AR_{i,j})). \tag{9}$$

To get the predicted labels of ARI and ARTC, we post-processed the prediction of the model. The existence of an argument relation in the ARI task is defined as:

$$\hat{y}_{ARI} = \begin{cases} 0 & \text{if } \hat{y}_{AR} = 0 \\ 1 & \text{if } \hat{y}_{AR} \neq 0 \end{cases} \tag{10}$$

where $\hat{y}_{AR}$ is the predicted label from the model output.

When we gain the type of an existing argument relation in the ARTC task, we assign the probability of 'none' to zero and select the other label with the higher probability. They are represented as:

$$\hat{y}_{ARTC} = Argmax(p(y_{AR}|AR_{i,j})), \quad y^{none} = 0, \tag{11}$$

where y^{none} is the model output of the label 'none'.

3.4 Loss Function Design

This model jointly learns the argument component extraction and the argument relation extraction. By combining these two tasks, the training objective and loss function of the final model is obtained as:

$$L(\theta) = \sum_i log(p(y_i|AC_i)) + \sum_{i,j} p(y_{i,j}|AR_{i,j}) + \frac{\lambda}{2}||\theta||^2, \tag{12}$$

where θ represents all the parameters in the model, and λ represents the coefficient of L2 regularization. According to the loss function, the parameters in the model are updated repeatedly until the model achieves better performance results to complete the model training.

4 Experiments

4.1 Datasets

We evaluate our proposed model on two public datasets: Persuasive Essays (PE) [28] and Consumer Debt Collection Practices (CDCP) [18].

The PE dataset only has tree structure argument information. It has three types of ACs: *Major-Claim*, *Claim*, and *Premise*, and two types of AR: *support* and *attack*.

The CDCP dataset has general structure argument information, not limited to a tree structure. It is different from the PE dataset and is more difficult. The argument information in this dataset is more similar to the real world. There are five types of ACs (propositions): *Reference*, *Fact*, *Testimony*, *Value*, and *Policy*. Between these ACs, there are two types of ARs: *reason* and *evidence*.

We both use the original train-test split of two datasets to conduct experiments.

4.2 Setups

In the model training, roberta-base [13] was used to fine-tune, and AdamW optimizer [14] was used to optimize the parameters of the model during the training. We apply a stratified learning rate to obtain a better semantic representation of BERT context and downstream task effect. The stratified learning rate is important in this task because this multi-task learning is complex and have three subtasks. The ARI and ARTC need a relatively bigger learning rate to learn the data better. The initial learning rate of the BERT layer is set as 2e-5. The learning rate of the AC extraction module and the AR extraction module is set as 2e-4 and 2e-3, respectively. After BERT output, the Dropout Rate [26] is set to 0.2. The maximum sequence length of a single piece of data is 512. We cut off ACs and ARs in the over-length text. The batch size in each training step is set to 16 in the CDCP dataset and 2 in the PE dataset. The reason is that there are more ACs in one argumentative text from the PE dataset than in the CDCP dataset.

In training, we set an early stop strategy with 5 epochs. We set the minimum training epochs as 15 to wait for the model to become stable. We use $MacroF1_{ARI}$ as monitoring indicators in our early stop strategy. That is because AR extraction is our main improvement direction. Furthermore, the ARI is between the ACTC and the ARTC, so we can better balance the three tasks' performance in the multi-task learning scenario.

The code implementation of our model is mainly written using PyTorch [20] library, and the pre-trained model is loaded using Transformers [32] library. In addition, model training and testing were conducted on one NVIDIA GeForce RTX 3090.

4.3   Compared Methods

We compare our model with several baselines to evaluate the performance:

- **Joint-ILP** [28] uses Integer Linear Programming (ILP) to extract ACs and ARs. We compare our model with it in the PE dataset.
- **St-SVM-full** [18] uses full factor graph and structured SVM to do argument mining. We compare our model with it in both the PE and the CDCP datasets.
- **Joint-PN** [25] employs a Pointer Network with an attention mechanism to extract argument information. We compare our model with it in the PE dataset.
- **Span-LSTM** [10] use LSTM-based span representation with ELMo to perform argument mining. We compare our model with it in the PE dataset.
- **Deep-Res-LG** [8] uses Residual Neural Network on AM tasks. We compare our model with it in the CDCP dataset.
- **TSP-PLBA** [17] introduces task-specific parameterization and bi-affine attention to AM tasks. We compare our model with it in the CDCP dataset.
- **BERT-Trans** [2] use transformation-based dependency analysis method to solve AM problems. We compare our model with it in both the PE and the CDCP datasets. It is also the state of the art on two datasets.

4.4   Performance Comparison

The evaluation results are summarized in Table 1 and Table 2. In both tables, '-' indicates that the original paper does not measure the performance of this

Table 1. The results of comparison experiments on the CDCP dataset. All numbers in the table are f1 scores (%). The best scores are in bold. '-' represents that the original paper does not report.

Methods	ACTC						ARI			ARTC			AVG
	Macro	Value	Policy	Testi.	Fact	Refer.	Macro	Rel.	Non-rel.	Macro	Reason	Evidence	
St-SVM-strict	73.2	76.4	76.8	71.5	41.3	100.0	–	26.7	–	–	–	–	–
Deep-Res-LG	65.3	72.2	74.4	72.9	40.3	66.7	–	29.3	–	15.1	30.2	0.0	–
TSP-PLBA	78.9	–	–	–	–	–	–	34.0	–	–	–	18.7	–
BERT-Trans	82.5	83.2	86.3	84.9	58.3	100.0	67.8	37.3	98.3	–	–	–	–
AutoAM (Ours)	**84.6**	**85.0**	**86.8**	**86.1**	**65.9**	**100.0**	**68.4**	**38.5**	**98.4**	**71.3**	**98.1**	**44.4**	**74.8**

Table 2. The results of comparison experiments on the PE dataset. All numbers in the table are f1 scores (%). The best results are in bold. The second best results are in italics. '-' represents that the original paper does not report.

Methods	ACTC				ARI			ARTC			AVG
	Macro	MC	Claim	Premise	Macro	Rel.	Non-rel.	Macro	Support	Attack	
Joint-ILP	82.6	89.1	68.2	90.3	75.1	58.5	91.8	68.0	94.7	41.3	75.2
St-SVM-strict	77.6	78.2	64.5	90.2	–	60.1	–	–	–	–	–
Joint-PN	84.9	89.4	73.2	92.1	76.7	60.8	92.5	–	–	–	–
Span-LSTM	85.7	91.6	73.3	92.1	80.7	68.8	93.7	*79.0*	*96.8*	**61.1**	81.8
BERT-Trans	*88.4*	**93.2**	*78.8*	*93.1*	**82.5**	**70.6**	*94.3*	**81.0**	–	–	**83.4**
AutoAM (Ours)	**88.7**	*91.9*	**80.3**	**93.9**	*81.6*	*65.8*	**98.5**	75.4	**97.6**	*53.2*	*81.9*

metric for its model. The best results are in bold, and the second-best results are in italics.

On the CDCP dataset, we can see our model achieves the best performance on all metrics in ACTC, ARI, and ARTC tasks. We are the first to complete all the tasks and get ideal results on the CDCP dataset. Our model outperforms the state of the art with an improvement of 2.1 in ACTC and 0.6 in ARI. The method BERT-Trans does not perform ARTC with other tasks at the same time, and it does not report results of ARTC, maybe due to unsatisfactory performance. In particular, compared with the previous work, we have greatly improved the task performance of ARTC and achieved ideal results.

On the PE dataset, our model also gets ideal performance. However, we get the second-best scores in several metrics. The first reason is that the PE dataset is tree-structured, so many previous work impose some structure constraints. Their models incorporate more information, and our model assumes they are general argument graphs in contrast. Another reason is that the models BERT-Trans, Span-LSTM, and Joint-PN combine extra features to represent ACs, like paragraph types, BoW, position embedding, etc. This information will change in the different corpus, and we want to build an end-to-end universal model. For example, there is no paragraph type information in the CDCP dataset. Therefore, we do not use them in our model. Even if our model does not take these factors into account, we achieve similar results to the state of the art.

4.5 Ablation Study

Table 3. The results of ablation experiments on the CDCP dataset. All numbers in the table are f1 scores (%). The best results are in bold.

Methods	ACTC		ARI		ARTC		AVG	
	Macro	▽	Macro	▽	Macro	▽	Macro	▽
AutoAM (Ours)	**84.6**	–	68.4	–	**71.3**	–	**74.8**	–
w/o stratified learning rate	76.9	−7.7	66.2	−2.2	49.3	−22.0	64.2	−10.6
w/o ArguAtten	82.9	−1.7	**69.8**	+1.4	57.7	−13.6	70.1	−4.7
w/o Distance Matrix	82.9	−1.7	62.9	−5.5	59.3	−12.0	68.3	−6.5

The ablation study results are summarized in Table 3. We conduct an ablation study on the CDCP dataset to see the impact of key modules in our model. It can be observed that the stratified learning rate is the most critical in this model. It verifies the viewpoint that multi-task learning is complex in this model and ARs extraction module needs a bigger learning rate to perform well. We can see ArguAtten improve the ACTC and ARTC performance by 1.7 and 13.6. However, the ARI matric decreases a little bit. Even though the numbers are small, we think that the reason is the interrelationship between ACs has little impact on the prediction of ARs' existence. ArguAtten mainly plays an effect in predicting the type of ARs. From this table, we can also find that the distance matrix brings the important distance feature to AR representation with an overall improvement of 6.5.

5 Conclusion and Future Work

In this paper, we propose a novel method for argument mining and first introduce the argument component attention mechanism. This is the first end-to-end argument mining model that can extract argument information without any structured constraints and get argument relations of good quality. In the model, ArguAtten can better capture the correlation information of argument components in an argumentative paragraph so as to better explore the argumentative relationship. Our experiment results show that our method achieves the state of the art. In the future, we will continue to explore designing a better model to describe and capture elements and relationships in argument graphs.

References

1. Afantenos, S., Peldszus, A., Stede, M.: Comparing decoding mechanisms for parsing argumentative structures. Argum. Comput. **9**, 1–16 (2018). https://doi.org/10.3233/AAC-180033

2. Bao, J., Fan, C., Wu, J., Dang, Y., Du, J., Xu, R.: A neural transition-based model for argumentation mining. In: Proceedings of the 59th Annual Meeting of the Association for Computational Linguistics and the 11th International Joint Conference on Natural Language Processing (Volume 1: Long Papers), pp. 6354–6364. Association for Computational Linguistics, Online (Aug 2021). https://doi.org/10.18653/v1/2021.acl-long.497, https://aclanthology.org/2021.acl-long.497

3. Chakrabarty, T., Hidey, C., Muresan, S., McKeown, K., Hwang, A.: AMPERSAND: argument mining for persuasive online discussions. In: Proceedings of the 2019 Conference on Empirical Methods in Natural Language Processing and the 9th International Joint Conference on Natural Language Processing (EMNLP-IJCNLP), pp. 2933–2943. Association for Computational Linguistics, Hong Kong, China (Nov 2019). https://doi.org/10.18653/v1/D19-1291, https://aclanthology.org/D19-1291

4. Chen, D., Manning, C.: A fast and accurate dependency parser using neural networks. In: Proceedings of the 2014 Conference on Empirical Methods in Natural Language Processing (EMNLP), pp. 740–750. Association for Computational Linguistics, Doha, Qatar (Oct 2014). https://doi.org/10.3115/v1/D14-1082,https://aclanthology.org/D14-1082

5. Dozat, T., Manning, C.D.: Simpler but more accurate semantic dependency parsing. In: Proceedings of the 56th Annual Meeting of the Association for Computational Linguistics (Volume 2: Short Papers), pp. 484–490. Association for Computational Linguistics, Melbourne, Australia (Jul 2018). https://doi.org/10.18653/v1/P18-2077,https://aclanthology.org/P18-2077

6. Eemeren, F.H.v., Grootendorst, R.: A Systematic Theory of Argumentation: The pragma-dialectical approach. Cambridge University Press (2003). https://doi.org/10.1017/CBO9780511616389

7. Eger, S., Daxenberger, J., Gurevych, I.: Neural end-to-end learning for computational argumentation mining. In: Proceedings of the 55th Annual Meeting of the Association for Computational Linguistics (Volume 1: Long Papers), pp. 11–22. Association for Computational Linguistics, Vancouver, Canada (Jul 2017). https://doi.org/10.18653/v1/P17-1002,https://aclanthology.org/P17-1002

8. Galassi, A., Lippi, M., Torroni, P.: Argumentative link prediction using residual networks and multi-objective learning. In: Proceedings of the 5th Workshop on Argument Mining, pp. 1–10. Association for Computational Linguistics, Brussels, Belgium (Nov 2018). https://doi.org/10.18653/v1/W18-5201,https://aclanthology.org/W18-5201

9. Gómez-Rodríguez, C., Shi, T., Lee, L.: Global transition-based non-projective dependency parsing. In: Proceedings of the 56th Annual Meeting of the Association for Computational Linguistics (Volume 1: Long Papers), pp. 2664–2675. Association for Computational Linguistics, Melbourne, Australia (Jul 2018). https://doi.org/10.18653/v1/P18-1248,https://aclanthology.org/P18-1248

10. Kuribayashi, T., Ouchi, H., Inoue, N., Reisert, P., Miyoshi, T., Suzuki, J., Inui, K.: An empirical study of span representations in argumentation structure parsing. In: Proceedings of the 57th Annual Meeting of the Association for Computational Linguistics, pp. 4691–4698. Association for Computational Linguistics, Florence, Italy (Jul 2019). https://doi.org/10.18653/v1/P19-1464,https://aclanthology.org/P19-1464

11. Kwon, N., Zhou, L., Hovy, E., Shulman, S.W.: Identifying and classifying subjective claims, pp. 76–81. dg.o '07, Digital Government Society of North America (2007)

12. Lawrence, J., Reed, C.: Argument mining: a survey. Comput. Linguist. **45**(4), 765–818 (2020). https://doi.org/10.1162/coli_a_00364,https://direct.mit.edu/coli/article/45/4/765-818/93362

13. Liu, Y., et al.: Roberta: a robustly optimized BERT pretraining approach. CoRR abs/ arXiv: 1907.11692 (2019)

14. Loshchilov, I., Hutter, F.: Fixing weight decay regularization in adam. CoRR abs/ arXiv: 1711.05101 (2017)

15. Mochales, R., Moens, M.F.: Argumentation mining: the detection, classification and structure of arguments in text, pp. 98–107 (Jan 2009). https://doi.org/10.1145/1568234.1568246

16. Moens, M.F., Boiy, E., Mochales, R., Reed, C.: Automatic detection of arguments in legal texts, pp. 225–230 (Jan 2007). https://doi.org/10.1145/1276318.1276362

17. Morio, G., Ozaki, H., Morishita, T., Koreeda, Y., Yanai, K.: Towards better non-tree argument mining: proposition-level biaffine parsing with task-specific parameterization. In: Proceedings of the 58th Annual Meeting of the Association for Computational Linguistics, pp. 3259–3266. Association for Computational Linguistics, Online (Jul 2020). https://doi.org/10.18653/v1/2020.acl-main.298,https://aclanthology.org/2020.acl-main.298

18. Niculae, V., Park, J., Cardie, C.: Argument mining with structured SVMs and RNNs. In: Proceedings of the 55th Annual Meeting of the Association for Computational Linguistics (Volume 1: Long Papers), pp. 985–995. Association for Computational Linguistics, Vancouver, Canada (Jul 2017). https://doi.org/10.18653/v1/P17-1091,https://aclanthology.org/P17-1091

19. Park, J., Cardie, C.: A Corpus of eRulemaking user comments for measuring evaluability of arguments. In: chair), N.C.C., Choukri, K., et al. (eds.) Proceedings of the Eleventh International Conference on Language Resources and Evaluation (LREC 2018). European Language Resources Association (ELRA), Miyazaki, Japan, 7–12 May (2018)

20. Paszke, A., et al.: Pytorch: an imperative style, high-performance deep learning library (2019). https://doi.org/10.48550/ARXIV.1912.01703,https://arxiv.org/abs/1912.01703

21. Peldszus, A.: Towards segment-based recognition of argumentation structure in short texts. In: Proceedings of the First Workshop on Argumentation Mining, pp. 88–97. Association for Computational Linguistics, Baltimore, Maryland (Jun 2014). https://doi.org/10.3115/v1/W14-2112,https://aclanthology.org/W14-2112

22. Peldszus, A., Stede, M.: Joint prediction in MST-style discourse parsing for argumentation mining. In: Proceedings of the 2015 Conference on Empirical Methods in Natural Language Processing, pp. 938–948. Association for Computational Linguistics, Lisbon, Portugal (Sep 2015). https://doi.org/10.18653/v1/D15-1110,https://aclanthology.org/D15-1110

23. Persing, I., Ng, V.: End-to-end argumentation mining in student essays. In: Proceedings of the 2016 Conference of the North American Chapter of the Association for Computational Linguistics: Human Language Technologies, pp. 1384–1394. Association for Computational Linguistics, San Diego, California (Jun 2016). https://doi.org/10.18653/v1/N16-1164,https://aclanthology.org/N16-1164

24. Peters, M.E., et al.: Deep contextualized word representations. In: Proceedings of the 2018 Conference of the North American Chapter of the Association for Computational Linguistics: Human Language Technologies, Volume 1 (Long Papers), pp. 2227–2237. Association for Computational Linguistics, New Orleans, Louisiana (Jun 2018). https://doi.org/10.18653/v1/N18-1202,https://aclanthology.org/N18-1202

25. Potash, P., Romanov, A., Rumshisky, A.: Here's my point: joint pointer architecture for argument mining. In: Proceedings of the 2017 Conference on Empirical Methods in Natural Language Processing, pp. 1364–1373. Association for Computational Linguistics, Copenhagen, Denmark (Sep 2017). https://doi.org/10.18653/v1/D17-1143,https://aclanthology.org/D17-1143

26. Srivastava, N., Hinton, G., Krizhevsky, A., Sutskever, I., Salakhutdinov, R.: Dropout: a simple way to prevent neural networks from overfitting. J. Mach. Learn. Res. 15(56), 1929–1958 (2014). http://jmlr.org/papers/v15/srivastava14a.html

27. Stab, C., Gurevych, I.: Annotating argument components and relations in persuasive essays. In: Proceedings of COLING 2014, the 25th International Conference on Computational Linguistics: Technical Papers, pp. 1501–1510. Dublin City University and Association for Computational Linguistics, Dublin, Ireland (Aug 2014). https://aclanthology.org/C14-1142

28. Stab, C., Gurevych, I.: Parsing argumentation structures in persuasive essays. Comput. Ling. 43(3), 619–659 (2017). https://doi.org/10.1162/COLI_a_00295,https://aclanthology.org/J17-3005

29. Vaswani, A., et al.: Attention is all you need. In: Guyon, I., Luxburg, U.V., Bengio, S., Wallach, H., Fergus, R., Vishwanathan, S., Garnett, R. (eds.) Advances in Neural Information Processing Systems, vol. 30. Curran Associates, Inc. (2017). https://proceedings.neurips.cc/paper/2017/file/3f5ee243547dee91fbd053c1c4a845aa-Paper.pdf

30. Vinyals, O., Fortunato, M., Jaitly, N.: Pointer networks. In: Cortes, C., Lawrence, N., Lee, D., Sugiyama, M., Garnett, R. (eds.) Advances in Neural Information Processing Systems, vol. 28. Curran Associates, Inc. (2015). https://proceedings.neurips.cc/paper/2015/file/29921001f2f04bd3baee84a12e98098f-Paper.pdf

31. Walker, V.R., Foerster, D., Ponce, J.M., Rosen, M.: Evidence types, credibility factors, and patterns or soft rules for weighing conflicting evidence: argument mining in the context of legal rules governing evidence assessment. In: Proceedings of the 5th Workshop on Argument Mining, pp. 68–78. Association for Computational Linguistics, Brussels, Belgium (Nov 2018). https://doi.org/10.18653/v1/W18-5209,https://aclanthology.org/W18-5209

32. Wolf, T., et al.: Huggingface's transformers: State-of-the-art natural language processing (2019). https://doi.org/10.48550/ARXIV.1910.03771,https://arxiv.org/abs/1910.03771

Rethinking the Evaluation of Deep Neural Network Robustness

Mingyuan Fan[1], Fuyi Wang[2(✉)], and Bosheng Yan[2]

[1] School of Data Science and Engineering, East China Normal University, Shanghai, China
[2] School of Information Technology, Deakin University, Geelong, Australia
wong_fuyi@outlook.com, yanbo@deakin.edu.au

Abstract. Evaluating the robustness of deep neural networks (DNNs) is crucial for ensuring the reliability and security of machine learning systems. Prior approaches quantify the probability of a DNN being compromised under a specified constraint. Despite their utility, these techniques suffer from low efficiency and effectiveness in evaluating the robustness of DNNs. The paper presents a promising evaluation approach, named typeII-EvaA, for accurately and efficiently assessing the robustness of DNNs against adversarial attacks. The typeII-EvaA overcomes the limitations of existing evaluation methods by devising several new assessment methods, called typeII-AssMs, which use attack success rate (ASR) constraints to minimize perturbation magnitudes. Additionally, we introduce a more effective human imperceptibility metric, CIEDE2000, which aligns with the human vision system to probe almost all human-imperceptible areas for obtaining the most threatening adversarial examples. Extensive experimental results corroborate that typeII-EvaA has practical implications for improving the security of DNN-based systems. And typeII-AssMs can achieve 100% ASR against various defense mechanisms. Our intention is for the typeII-EvaA to serve as a benchmark for future efforts toward developing robust DNNs that can withstand adversarial examples.

Keywords: Deep neural networks · Robustness evaluation · Adversarial attacks · Human imperceptibility metric

1 Introduction

Deep neural networks (DNNs) have garnered significant attention in recent years due to their superior performance and ability to address complex tasks across various domains. However, their vulnerability to attacks from adversaries has limited their deployment in security-critical applications [21]. Specifically, adversarial attacks exploit the susceptibility of DNNs by introducing **human-imperceptible** adversarial perturbations to natural examples, resulting in misclassifications from state-of-the-art (SOTA) DNNs [14,15,21]. As a result, there

---

M. Fan and F. Wang—The authors contribute equally to this work.

© The Author(s), under exclusive license to Springer Nature Switzerland AG 2023
X. Yang et al. (Eds.): ADMA 2023, LNAI 14179, pp. 532–547, 2023.
https://doi.org/10.1007/978-3-031-46674-8_37

are significant incentives for researchers to explore the robustness of DNNs against adversarial attacks [1,7,9,18].

The task of exploring the robustness of DNNs against adversarial attacks primarily involves developing effective defense mechanisms and evaluation approaches that are analogous to the training and validation methods used for DNNs [2,12,14,17,21]. On the one hand, an ideal evaluation approach should be capable of accurately and efficiently assessing the ground-truth robustness of DNNs against adversarial attacks [2,14]. Evaluation methods that lack rigor are inadequate for evaluating the effectiveness of defenses and can yield misleading results, hampering progress in this field. On the other hand, evaluation methods that are overly resource-intensive are impractical for real-world use due to their high computational overhead.

This paper rethinks the effectiveness and efficiency of the existing evaluation approaches, named typeI-EvaAs. TypeI-EvaAs generally involve the following two steps: 1) generating threatening adversarial perturbations as much as possible via maximizing an attack effectiveness metric, under a certain distance constraint (known as the human perceptibility metric) of magnitude ϵ, where ϵ has to enable the resulting perturbations to be human-imperceptible; 2) adopting the perturbations to produce adversarial examples and estimating the corresponding probability of the DNNs being tricked, i.e., attack success rate (ASR). In brief, typeI-EvaAs report the probability of the DNNs being fooled under a given constraint, and we call these kinds of assessment methods typeI-AssMs. However, typeI-EvaAs are of low efficiency and poor effectiveness.

Low Efficiency. Pre-setting an appropriate constraint magnitude ϵ is necessary for typeI-EvaAs: a larger constraint magnitude allows a more broad search space to be navigated that can in general raise attack effectiveness, but also easily results in visually noticeable adversarial perturbations, i.e., viotibility; vice versa. To determine the proper magnitude, empirical observations are typically used. Evaluators observe and compare the crafted adversarial examples under various constraint magnitudes and then manually select an optimal magnitude. It is vastly cumbersome and computationally intensive, as crafting a single adversarial example may require backpropagation up to hundreds or thousands of times. Furthermore, many evaluators may not have the necessary expertise to efficiently tune ϵ, which can lead to additional overhead, particularly for large-scale datasets with modern ultra-huge DNNs.

Poor Effectiveness. TypeI-EvaAs commonly use norm-based constraints, specifically ∞-norm constraint, to resultant adversarial perturbations human-imperceptible. However, many more threatening and imperceptible adversarial perturbations are beyond the ∞-norm constraint. Specifically, Fig. 1(b,c) shows adversarial examples by common attacks with ∞-norm constraint of 16 and 32. As can be seen, some adversarial ones are considerably different from the original ones in vision and also cannot completely mislead models. In contrast, TypeII-EvaA crafts adversarial ones (Fig. 1(d)) which perturbation beyond 16 and 32 but are also significantly effective to models and quite similar to original ones. In fact, ∞-norm distance treats perturbations of different images, or even different

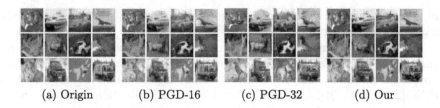

(a) Origin   (b) PGD-16   (c) PGD-32   (d) Our

Fig. 1. We craft adversarial examples for images from the leftmost column. (b) and (c) are produced by PGD with perturbation budgets of 16 and 32. (d) is crafted by TypeII-EvaA (inner-joint-optim version).

pixels in an image, as equally important, and this is barely established for the human perceptual system. Therefore, typeI-EvaAs tend to overlook the area with low human-imperceptible distance and high ∞-norm distance from the original images, resulting in the failure to detect many threatening adversarial examples and leading to a false sense of robustness.

To address the two above-mentioned limitations, we revisit typeI-EvaAs which report ASR with a specified perturbation constraint (called typeI-AssMs) for the robustness of DNNs. And a natural idea is that can we assess the robustness of DNNs by estimating how much perturbations need to be imposed to reach a given ASR? To achieve a specific ASR, we identify the most vulnerable sample combination with minimal perturbations to induce full misclassification. However, finding the most vulnerable combination is challenging as enumerating all combinations equals an NP-hard problem. We simplify the problem into finding minimal threatening adversarial perturbations for each sample independently with linear time complexity and the overhead is considerably lower than typeI-AssMs that craft adversarial perturbations for each constraint magnitude over all samples. For a specific ASR, the attackers are at least required to add adversarial perturbations above the magnitude to reach the ASR. In a nutshell, our technical contributions are threefold.

- We develop a novel evaluation approach, dubbed typeII-EvaA, that can effectively and efficiently reap the accurate robustness estimations of DNNs. TypeII-EvaA is the *first* work to explore novel and remarkably efficient assessment methods called typeII-AssMs with a specific ASR. For the practicality of typeII-EvaA, we craft fresh attack paradigms that minimize perturbation magnitudes with ASR constraints.
- For the effectiveness of typeII-EvaA, we explore an effective human imperceptibility metric compared to *norm*-based metrics, i.e., CIEDE2000[1], that aligns well with the human vision system, such that the adversarial attacks with CIEDE2000 are allowed to probe almost all human-imperceptible areas for obtaining most threatening adversarial examples.

---

[1] CIEDE2000 is a perceptual color distance recently released by International Commission on Illumination.

- We design several proxy functions for finding adversarial examples. Additionally, we devise four search algorithms with various strategies in order to determine the perturbations of adversarial examples. We systematically evaluate these designs and show that typeII-EvaA can comprehensively evaluate the efficacy of defense mechanisms.

The rest of the paper is organized as follows. In the following section, we introduce the background, e.g., DNNs and adversarial examples. Section 3 formulates the challenge for typeII-AssMs. Section 4 develops the human imperceptibility and attack effectiveness metrics. We design search algorithms consisting of three ingredients: initialization strategy, search direction, and step size in Sect. 5 followed by experimental evaluation of typeII-EvaA in the large-scale CIFAR10 and ImageNet datasets in Sect. 6. Finally, the conclusion of this paper is made in Sect. 7.

2 Background

2.1 Deep Neural Networks

Deep neural networks (DNNs) are highly intricate mathematical models composed of numerous layers. Each layer is typically comprised of a linear function and an activation function. We denote a DNN with parameter θ by $F_\theta(\cdot) \in \mathbb{R}^m$ and $F_\theta(x)[i]$ ($i = 1, 2, \cdots, m$) denotes the prediction confidence of the DNN for classifying x into i-th category, where m is the total number of categories. In order to obtain a probability distribution over all potential categories as the final prediction result, it is frequently customary to incorporate a softmax function to standardize the level of confidence in the prediction. Softmax function outputs the probability of i-th category as follows:

$$\text{softmax}(F_\theta(x))[i] = \frac{e^{F_\theta(x)[i]}}{\sum_{i=1}^m e^{F_\theta(x)[i]}}. \tag{1}$$

Given a dataset $D = \{(x_i, y_i), i = 1, 2, \cdots, n, x_i \in \mathbb{R}^{c \times h \times w}, y_i \in \mathbb{R}^m\}$ where c, h, w denote the channel, height, width of the input image, and the performance of DNNs is quantitated by accuracy as follows:

$$acc = \frac{\sum_{i=1}^n \mathbb{I}((\arg \max_{j=1, \cdots, m} F_\theta(x_i)[j]) = y_i)}{n}, \tag{2}$$

where $\mathbb{I}(\cdot)$ is a characteristic function that outputs 1 when the condition holds and otherwise outputs 0. Then, to make the optimal performance of the DNN over D, standard practice is leveraging the mini-batch gradient descent algorithm or its variants to optimize θ associated with accuracy in an end-to-end fashion. But the gradient-based optimization algorithms cannot be directly applied to optimize θ since accuracy blocks the gradient propagation process. Thus, accuracy generally is replaced with a differentiable objective function while the most frequently-used objective function is the cross-entropy function shortened by $CE(\cdot,\cdot)$.

$$CE(x,y) = -\log(\text{softmax}(F_\theta(x))[y]). \tag{3}$$

The quality of objective functions has a huge influence on the resultant DNN and an inferior objective function probably leads to a worse θ.

2.2 Adversarial Examples

Adversarial examples are malicious inputs that are artificially synthesized with clean inputs and specific perturbations crafted by the attacker. With adversarial examples, the attacker can fool the target DNN model to output attacker-chosen (or random) predictions, as defined in Definition 1.

Definition 1 (Adversarial Attacks). *Given a DNN model $F_\theta(x)$ with parameter θ and an input x, the adversarial attack aims to find a specific perturbation δ for x that satisfies:*

$$\delta = \arg\min_\delta \mathcal{M}(x,x+\delta), \quad s.t., \arg\max_{j=1,\cdots,m} F_\theta(x+\delta)[j] \neq y, \tag{4}$$

where m is number of classes, the function $\mathcal{M}(\cdot,\cdot)$ evaluates the distance between x and $x+\delta$, which also reflects the human imperceptibility of the generated adversarial example [12].

Generally, most adversarial attacks follow the below paradigm to approximately resolve Eq. 4:

$$\delta = \arg\max_\delta L(F_\theta(x+\delta),y), \quad s.t., \mathcal{M}(x,x+\delta) \leq \epsilon, \tag{5}$$

where $L(\cdot,\cdot)$ is a proxy function of $\arg\max_{j=1,\cdots,m} F_\theta(x+\delta)[j] \neq y$ like $CE(\cdot,\cdot)$, and ϵ is the perturbation budget that constraints the distance between x and $x+\delta$. Commonly, $L(\cdot,\cdot)$ is positively correlated with the misclassification rate of the model and referred to as the attack effectiveness metric.

Definition 2 (Adversarial Perturbations and Examples). *Given an input x with the ground-truth label y and a target DNN, if perturbations δ are crafted by adversarial attacks, δ and $x+\delta$ are referred to as adversarial perturbations and adversarial examples. Furthermore, if the target DNN misidentifies $x+\delta$, the δ is threatening adversarial perturbations; otherwise, the δ is weak adversarial perturbations.*

If adversarial perturbations are described as minimal for input x, this implies that such perturbations result in the lowest possible value for $\mathcal{M}(x,x+\delta)$.

Different adversarial attacks can be reduced by solving Eq. 4. An intuitive idea of solving Eq. 4 is to impulse samples to move in the direction that makes the loss of the sample higher as possible, i.e., Eq. 5, and gradient directions can effectively match the direction. The fast gradient sign method (FGSM) harnesses the idea and approximately solves Eq. 5 by directly setting $\delta = \epsilon \nabla_x L(F_\theta(x), y)$, where $L(\cdot, \cdot)$ commonly is $CE(\cdot, \cdot)$ [21]. As suggested in its name, the main merit of FGSM is efficiently crafting adversarial examples due to backpropagation only being required to implement once. But, when a bigger tolerance for perturbations is allowed, FGSM performs poorly, as ϵ is seemingly too big and the gradient direction only works around the small neighborhood of x. Accordingly, the basic iterative method (BIM) [11] and projected gradient descent (PGD) [12] improve FGSM by using an iterative way with a small step size to solve Eq. 4. In detail, given total iterations T, BIM crafts $\delta = \delta_T$ by iteratively updating $\delta_t = Clip_\epsilon(\delta_{t-1} + \nabla_{x+\delta_{t-1}} L(F_\theta(x + \delta_{t-1}), y))$ $(t = 0, 1, 2, \cdots, T)$, where $Clip_\epsilon(\cdot)$ draws the perturbations back to the constraint domain, where the initial perturbations δ_0 are full-zero vectors. Due to the local extreme points in the vicinity of x, the PGD incorporates a randomized perturbation into the initial perturbation δ_0 to evade these local extreme points [12]. Apart from the above adversarial attack methods, another famous and effective adversarial attack is C&W attack [2]. Rather than optimizing perturbations subject to constraints, the C&W attack approach entails simultaneous optimization of both the loss function and perturbations, formulating various loss functions and choosing the optimal one experimentally to replace the traditional cross-entropy loss function [3].

3   Assessment Method

3.1   Problem Formulation

The objective of typeII-AssMs is to obtain the least adversarial perturbations on the dataset D to achieve the specified ASR p. This can be formulated as optimizing the following task to obtain adversarial perturbations $\delta_1, ..., \delta_n$:

$$\delta_1, \cdots, \delta_n = \arg\min_{\delta_1, \ldots, \delta_n} \sum_{i=1}^{n} I_i \cdot \mathcal{M}(x_i, x_i + \delta_i)$$

$$s.t., \sum I_i(\arg\max_{j=1, \cdots, m} F_\theta(x_i)[j] \neq y_i) = n \cdot p, \tag{6}$$

$$I_1 + \cdots I_n = n \cdot p \text{ and } I_i = 0 \text{ or } 1,$$

where $n \cdot p$ is assumed to be an integer and I_i is an indicator function that outputs 1 if the input condition establishes otherwise outputs 0. After obtaining the solution for Eq. 6, $d = \mathcal{M}(x_1, x_1 + \delta_1) + \cdots + \mathcal{M}(x_n, x_n + \delta_n)$ can be used to assess the robustness of DNNs against adversarial attacks.

3.2   Solution to Equation 6

Before developing the solution to Eq. 6, we consider a special case of it, where $p = 100\%$. If $p = 100\%$, there is $I_i = 1$ for $\forall i$ and we then search for the adversarial perturbations δ_i that cause the misclassification of x_i from the DNN

$F_\theta(\cdot)$ and minimize $\mathcal{M}(x_i, x_i + \delta_i)$. Since searching for adversarial perturbations δ_i for different x_i is independent, Eq. 6 can be simplified to solve the following optimization task for each x_i.

$$\delta_i = \arg\min_{\delta_i} \mathcal{M}(x_i, x_i + \delta_i)$$
$$s.t., \arg\max_{j=1,\cdots,m} F_\theta(x_i)[j] \neq y_i. \tag{7}$$

Assuming that the adversarial perturbations δ_i for $\forall i$, are obtained by solving Eq. 7 with $p = 100\%$. The objective is to find the most vulnerable combination of $n \cdot p'$ instances after resetting p to a new value p'. To achieve this, one can greedily set $n \cdot (p - p')$ elements with the maximum $\mathcal{M}(x_i, x_i + \delta_i)$ in $\{\delta_1, \cdots, \delta_n\}$ to zero, which results in a minimal $\sum_i^n \mathcal{M}(x_i, x_i + \delta_i)$ among all combinations of size $n \cdot p'$ as the generation of each δ_i is independent. The resulting perturbations $\{\delta_1, \cdots, \delta_n\}$ are exactly the solution of Eq. 6 with p'. Furthermore, shrinking δ_i results in an increase in $\mathcal{M}(x_i, x_i + \delta_i)$, implying that the model $F_\theta(\cdot)$ will correctly identify x_i.

Time Complexity Comparison. Supposing that the time complexity of generating δ_i is $O(1)$. Then, the expected time complexity of solving Eq. 6 with our method is $O(n)$, whereas the expected time complexity of directly solving Eq. 6 is $O(C_n^k) = O(n!)$.

The Relationship to typeI-AssMs. We demonstrate that the results of typeI-AssMs can be readily obtained from the results of typeII-AssMs. Specifically, for typeI-AssMs, the objective is to determine the maximum achievable ASR under a given perturbation budget ϵ. By leveraging the fact that typeII-AssMs with ASR=100% generates the minimal magnitude of threatening adversarial perturbations for each instance x_i, imposing perturbations below this magnitude ensures that x_i is classified correctly. Therefore, the maximum achievable ASR can be computed as the ratio of samples for which the minimal magnitude of the threatening adversarial perturbations is smaller than ϵ. Consequently, we conclude that leveraging typeII-AssMs for evaluations is always preferable to typeI-AssMs, as the latter can be effortlessly derived from the former, but not vice versa. Furthermore, typeII-AssMs eliminate the significant burden of tuning the hyperparameter ϵ.

3.3   Solution to Equation 7

To simplify the notation, we omit the subscript i in Eq. 7. The problem we need to solve can be stated as follows:

$$\delta = \arg\min_{\delta} \mathcal{M}(x, x + \delta)$$
$$s.t., \arg\max_{j=1,\cdots,m} F_\theta(x)[j] \neq y. \tag{8}$$

Intuitively, except directly optimizing $\mathcal{M}(x, x + \delta)$ under the optimization constraint, an alternative is to slack the constraint, putting

$\arg\max_{j=1,\cdots,m} F_\theta(x)[j] \neq y$ into objective function as a punishment term, formulated as follows:

$$\delta = \arg\min_\delta \mathcal{M}(x, x + \delta) - \alpha L(F_\theta(x + \delta), y), \tag{9}$$

where $F_\theta(x + \delta) \neq y$ is substituted by differentiable $L(F_\theta(x + \delta), y)$ for making gradient-based optimization methods applicable to this task.

Equation 9 is a more efficient and effective way of searching for δ compared to Eq. 8 for generating adversarial examples that are both effective and imperceptible to humans. The search direction[2] used in Eq. 9 is informed by both the attack effectiveness metric and human imperceptibility metric, while the search direction in Eq. 8 only considers one of the two metrics. Therefore, crafting δ via Eq. 9 appears to be a better option.

However, a significant challenge in solving Eq. 9 is determining an appropriate value of α that balances the effectiveness of the attack and the human imperceptibility metrics. We discuss the corresponding solution to this challenge in Sect. 5. In the next section, we define the metrics for measuring the effectiveness of the attack and the human imperceptibility of the perturbation.

4   Metric Design

Threatening adversarial examples possess two crucial characteristics: human imperceptibility and attack effectiveness, which dominate the quality of resultant adversarial examples.

4.1   Human Imperceptibility Metrics

The fundamental objective of human imperceptibility metrics is to approximate the ground-truth human perception distance[3] between two different images. However, most previous works have conveniently adopted norm-based distance functions as the similarity distance function as $\mathcal{M}(x, y) = ||x - y||_a$. The ∞-norm distance function is the most widely used method, which calculates the maximum absolute difference between the elements of two images $|x - y|$.

The primary flaw of norm-based distance functions is insufficiently aligned closely with the human perceptible distance function. Therefore, we introduce CIEDE2000, which has been shown to have better alignment with human perception than norm-based distance functions, to replace norm-based distance functions [19]. CIEDE2000 maps the two images from RGB space to CIELAB

---

[2] Gradient-based optimization methods are commonly used and effective for solving such tasks and we also follow it. Furthermore, the search direction of optimization methods is the gradient direction of the objective function.

[3] The similarity distance function in this paper is a loose version of the distance measure defined in mathematics, as a strict distance measure should satisfy non-negativity, symmetry, and triangle inequality but sometimes human perception distance may violate triangle inequality.

space since the human perceptible distance between two images is not uniformly affected by the RGB space distance. Specifically, it computes the distance between the two images as a weighted sum of the differences in lightness, chroma, and hue in CIELAB space. This mapping results in a distance metric that more accurately reflects the human visual system's response to differences in color and brightness.

4.2 Attack Effectiveness Metrics

Proxy functions for ASR considerably influence the crafted adversarial examples, motivating us to explore a variety of potential proxy functions to get better results. We design 7 proxy functions, expressed as follows:

$$
\begin{aligned}
f_1(x, y) &= F(x)[y], \\
f_2(x, y) &= \text{softmax}(F(x))[y], \\
f_3(x, y) &= \log(f_2(x, y)), \\
f_4(x, y) &= \frac{1}{1 - f_2(x, y)} f_2(x, y), \\
f_5(x, y) &= f_2(x, y) - \arg\max_{j \neq y}\{f_2(x, j)\}, \\
f_6(x, y) &= \max\{f_4(x, y) + C, 0\}, \quad C \geq 0, \\
f_7(x, y) &= \frac{f_2(x, y)}{\arg\max_{j \neq y, j = 1, \cdots, m} f_2(x, j)}.
\end{aligned}
\tag{10}
$$

Functions f_1 and f_2 directly penalize the prediction confidence, normalized prediction confidence, and probability of the ground-truth label for input x. Function f_3 is a negative cross-entropy loss function commonly used in many adversarial attacks, such as FGSM, BIM, and PGD. Function f_4 is an improved version of f_2, taking into account the observation that higher values of $f_2(x, y)$ indicate a higher probability that x is correctly classified by the DNN. To account for this, we scale the magnitude of $f_2(x, y)$ by a regulatory factor $\frac{1}{1 - f_2(x, y)}$, which amplifies the value of $f_2(x, y)$ when it is high. This weight tuning can be interpreted as implicitly adjusting the step size during the search process.

The proxy functions $f_1 \sim f_4$ have a limitation in that they only take into account the correct category of the input and do not consider other category information that could guide the search direction for effective adversarial examples. This can be addressed by incorporating similar information between categories. Therefore, we propose proxy functions $f_5 \sim f_7$, which consider the category most similar to the ground-truth label y that the model predicts as the target category for the adversarial attack. To further improve the performance of the proxy functions, we introduce an adaptive magnitude function in f_6 and f_7 that takes into account the model's confidence C in its misclassification of x. Specifically, f_6 disregards the attack effectiveness if the model confidently misclassifies x while f_7 always considers the attack effectiveness throughout the search process.

Notably, we derive $f_4 \sim f_7$ based on f_2 instead of f_1 or f_3 because we can easily tune the hyperparameter C and observe the prediction change trend of the model for x when the prediction is in probability form.

5 Search Algorithm Design

In the context of solving Eq. 8 and Eq. 9, the design of search algorithms is a crucial step that involves three main components: initialization strategy, search direction, and step size. The initialization strategy plays a critical role in determining the success of the search algorithm. There are two main types of initialization strategies: interior initialization and exterior initialization. For Eq. 8, we use inner-optim and outer-optim to refer to the search algorithm with interior initialization and exterior initialization, respectively. Similarly, the terms, inner-joint-optim and outer-joint-optim, are used for Eq. 9.

5.1 Inner-Optim

Initialization Strategy. In the inner-optim search algorithm, the initialization of adversarial perturbations δ needs to conform to the restrict condition $F_\theta(x + \delta) \neq y$, which implies that the model should identify $x + \delta$ as belonging to other categories. To achieve this, a simple way is to initialize δ such that $x + \delta$ becomes a sample belonging to a category different from y. Here, x' can be extracted from D_{train} and then $\delta = x' - x$.

Search Direction. We employ the gradient descent algorithm to move δ in the direction that $M(x, x + \delta)$ decreases the most, i.e., the negative gradient direction of $M(x, x + \delta)$ with respect to δ. However, simply using this algorithm can cause a violation of the optimization constraint since the similarity between x and $x + \delta$ increases with the number of iterations, leading to the increasing probability of x being correctly identified by the model. To prevent this issue, before updating δ in each iteration, the algorithm examines whether this update can result in $F_\theta(x + \delta) = y$. If $F_\theta(x + \delta) = y$, the update is abandoned, and the search process is terminated. Otherwise, the algorithm runs normally.

Adative Step Size Strategy. The appearance of $F_\theta(x + \delta) = y$ may be attributed to the large initialization step size and the smaller step size is worth exploring for searching more human-imperceptible δ. Therefore, the adaptive step size strategy is introduced into the search process and the strategy allows decreasing step size to implement more fine-grained search. In detail, if a certain update leads to $F_\theta(x + \delta) = y$, the step size will be reduced to half of the original one, and then examining the condition again. Also, the procedure is usually implemented several times. If all attempts fail, the search process is terminated.

5.2 Outer-Optim

Initialization Strategy. In the outer-optim algorithm, the initialization of δ should ensure that $F_\theta(x + \delta) = y$ for δ, and not allow $F_\theta(x + \delta) \neq y$. A simple approach to achieve this is to set δ as a full-zero vector, so that $x + \delta = x$ and $F_\theta(x + \delta) = y$.

Search Direction. With the above initialization strategy, the objective function can directly obtain the optimal value 0, but the perturbations are not threatening. Hence, outer-optim algorithm should move the perturbations towards the direction that induces $F_\theta(x + \delta) \neq y$ with minimal perturbations and this direction should be the optimal search direction. However, the gradient direction of $\mathcal{M}(x, x + \delta)$ alone is not sufficient to suggest the optimal direction because the gradient direction of $\mathcal{M}(x, x + \delta)$ not contain any information about $F_\theta(\cdot)$, considering the desired perturbations can give rise to $F_\theta(x + \delta) \neq y$. There are two alternatives to intuitively approximate the optimal direction. The first one is to jointly optimize two metrics for attack effectiveness and human imperceptibility and this is our inner-joint-optim and outer-joint-optim search algorithms; the last one is alternatively optimizing the two metrics and we discuss it in the next section. Here we more focus on leveraging the gradient direction of one of the two metrics as the search direction. The initial δ is the perturbations that enable x and $x + \delta$ to be most similar and thus we should attach more attention to the constraint, i.e., how move δ to obtain $F_\theta(x + \delta) \neq y$. If $F_\theta(x + \delta) \neq y$ is differentiable, the most effective direction is its gradient direction, but, unfortunately, it is not differentiable; thus, we use the gradient direction of the proxy function of $F_\theta(x + \delta) \neq y$. In addition, if $x + \delta$ is misclassified by the model, the search process should be ended as earlier as possible, because intuitively the move probably can increase $M(x, x + \delta)$.

Step Size Strategy. Similarly, we employ the adaptive step size strategy discussed in Sect. 5.1 to efficiently search for better adversarial perturbations.

5.3   Inner-Joint-Optim and Outer-Joint-Optim

Equation 9 generally performs better than individually optimizing one of the metrics. However, a key challenge is determining an appropriate value for the weighting parameter α. Setting a small α prioritizes human imperceptibility over attack effectiveness, potentially leading to ineffective adversarial perturbations. For instance, if $\alpha = 0$, the algorithm will exclusively focus on making $\delta = 0$. Intuitively, there are two approaches to solving the problem: 1) Setting a large α focuses solely on attack effectiveness and ignores human imperceptibility, resulting in overly perceptible perturbations; 2) An alternative approach of alternating between optimizing the two metrics based on whether the adversarial example is correctly classified has been proposed. Specifically, if the adversarial example is correctly classified, we optimize the attack effectiveness metric, otherwise, we optimize the similarity metric. This method still fails to fully explore the relationship between the two metrics.

We propose an adaptive method to find the optimal value of α that balances attack effectiveness and human imperceptibility. As α is increased, the model transitions from correctly classifying the sample to misclassifying it. This indicates that there is a tipping point to cause misclassification and the tipping point is the optimal value for α. With the optimal value for α, the produced adversarial perturbations are most human-imperceptible and also remain threatening. However, the optimal value of α is unknown in advance. Therefore, in each iteration, we increase the weight of the human-imperceptible metric if the sample with adversarial perturbations is misclassified, and we increase the weight of the attack effectiveness metric otherwise. This adaptive approach enables us to determine the optimal value of α and generate the optimal adversarial perturbations.

Initialization Strategy and Step Size. We introduce two variations of the joint optimization approach: inner-joint-optim when using the initialization strategy of inner-optim, and outer-joint-optim otherwise. We additionally incorporate a step size tuning strategy into the optimization process, which linearly decreases the step size to zero with iterations.

6 Experimental Evaluation

We implement a PyTorch-based prototype of typeII-EvaA based on CIEDE2000 to evaluate its performance on two commonly used *benchmark datasets*, CIFAR10 [10] and ImageNet [5]. We assess the effectiveness of typeII-EvaA attack methods against *SOTA defense mechanisms*, as Huang2021Exploring [8], Sridhar2021Robust [20], Pang2022Robustness [13] and Dai2021Parameterizing [4] for CIFAR10, and Standard, Wong2020Fast [22], Salman2020Do [16], and Engstrom2019Robustness [6] for ImageNet. To ensure fairness, the experimental model settings are consistent with those used in prior works and the step size is 0.005. Additionally, four advanced attacks are considered as *baselines* to estimate the effectiveness of our typeII-EvaA: FGSM [21], BIM [11], PGD [12], and C&W [2]. For all experiments, ASR, indicating the accuracy success rate, is regarded as the evaluation metric of typeII-AssMs. The typeII-AssMs' goal is to maximize ASR.

Evaluation of Attacks. We leverage four perturbation δ search algorithms of typeII-EvaA to evaluate the performance of SOTA defense mechanisms. We report the ASR in Fig. 2 over CIFAR10 and Fig. 3 over ImageNet along with various proxy functions. The results of Fig. 2 almost reaffirm the fact that, as the perturbation δ increases, the ASR of typeII-EvaA also increases. Obviously, The inner-joint-optim and outer-joint-optim outperform the inner-optim and outer-optim, respectively. This means that the strategy of slacking the constraint is more effective when against the defense mechanisms. For ImageNet dataset, we concentrate on inner-joint-optim and outer-joint-optim with the f_4 and f_7 functions. The effectiveness of inner-joint-optim with f_4 is dramatically improved. More specifically, when $\delta = 6.16$, the ASR of inner-joint-optim search algorithm with f_7 is 0.46, while when $\delta = 1.13$, the ASR of inner-joint-optim search algorithm with f_4 is up to 0.89. Additionally, the trend in ASR of outer-joint-optim search algorithm is the same as that over CIFAR10.

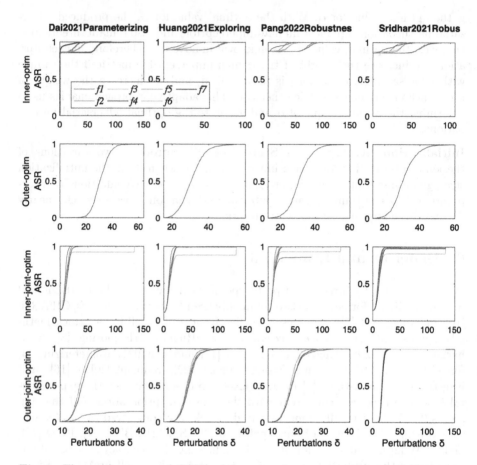

Fig. 2. The performance of SOTA defense mechanisms against typeII-EvaA over CIFAR10 dataset.

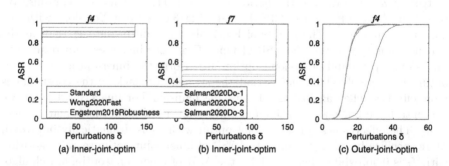

Fig. 3. The performance of SOTA defense mechanisms against typeII-EvaA over ImageNet dataset.

Table 1. Evaluation between typeII-EvaA and advanced attacks with various perturbation parameter (para.) δ and proxy functions (ASR: %).

CIFAR10	Para.	[4]	[8]	[13]	[20]	Attack	Para.	[4]	[8]	[13]	[20]
FGSM	$\delta = 1$	14.96	11.27	14.84	17.3	PGD	$\delta = 1$	14.96	11.27	14.84	17.3
	$\delta = 2$	16.18	14.06	15.74	19.08		$\delta = 2$	16.18	14.17	15.85	19.2
	$\delta = 4$	20.76	17.86	20.65	21.99		$\delta = 4$	20.98	18.97	21.54	22.66
	$\delta = 8$	29.24	26.79	29.58	30.8		$\delta = 8$	33.93	33.26	33.37	36.05
	$\delta = 12$	40.4	33.04	39.4	38.17		$\delta = 12$	50.22	53.68	48.88	50.56
	$\delta = 16$	47.1	40.07	45.76	44.53		$\delta = 16$	65.74	70.31	67.3	66.18
BIM	$\delta = 1$	14.96	11.27	14.84	17.3	C&W	$\delta = 1$	15.18	11.72	15.07	17.52
	$\delta = 2$	16.18	14.06	15.85	19.2		$\delta = 2$	16.52	14.4	16.63	19.31
	$\delta = 4$	20.98	18.97	21.54	22.66		$\delta = 4$	21.76	19.87	22.54	23.21
	$\delta = 8$	33.82	33.15	33.15	35.71		$\delta = 8$	35.94	34.6	36.5	38.5
	$\delta = 12$	50.22	52.79	48.66	50.33		$\delta = 12$	52.23	57.03	53.79	54.13
	$\delta = 16$	65.62	69.53	66.96	65.62		$\delta = 16$	68.86	75.22	71.54	70.65
Inner-joint	f_1	100	100	100	99.33	Outer-joint	f_1	100	100	100	99.89
	f_2	100	100	100	97.43		f_2	100	100	100	100
	f_3	100	100	100	99.33		f_3	100	100	100	99.89
	f_4	90.4	100	97.54	99.78		f_4	91.29	100	100	100
	f_5	100	100	100	97.77		f_5	99.89	100	100	100
	f_6	96.88	98.1	100	96.65		f_6	99.89	100	100	100
	f_7	100	100	100	98.44		f_7	98.88	100	100	100
ImageNet	Para.	Standard	[22]	[6]	[16]	Attack	Para.	Standard	[22]	[6]	[16]
FGSM	$\delta = 16$	93.15	92.64	88.51	94.46	PGD	$\delta = 16$	100	98.89	98.89	99.19
BIM	$\delta = 16$	100		98.89	98.79	99.19	C&W	$\delta = 16$	100	99.4	99.6	99.7
Inner-joint	f_4	100	100	100	100	Outer-joint	f_4	100	100	100	100
	f_7	100	99.9	99.9	100		f_7	100	100	100	100

Comparison with SOTA. We report the evaluation results compared with existing attacks against defense mechanisms and demonstrate the superiority of the typeII-EvaA in Table 1. We evaluate the quality of the adversarial examples found on the CIFAR10 and ImageNet datasets. The parameters, like proxy functions and perturbation δ are identical between the two datasets, so for brevity, we report partial results for ImageNet. For CIFAR10, all of the previous attacks fail to find adversarial examples. In contrast, our inner-joint-optim and outer-joint-optim can achieve 100% ASR when against various defense mechanisms. For ImageNet, prior work achieves approximate 99% ASR against advanced defense mechanisms when $\delta = 16$. While our inner-joint-optim and outer-joint-optim succeed with 100% success probability for each of the seven proxy functions.

7 Conclusion

The vulnerability of deep learning models to adversarial examples presents a major challenge for their practical application in security-critical domains. In order to ensure the reliability and safety of these models, it is crucial to evaluate their robustness against adversarial attacks. In this paper, we propose powerful attacks typeII-EvaA that defeat advanced defense mechanisms, demonstrating that typeII-EvaA more generally can be used to evaluate the efficacy of potential defenses. By systematically comparing many attack approaches, we settle on one that can consistently find better adversarial examples than all existing approaches with linear time complexity. We encourage those who create defenses to perform the four evaluation approaches with various proxy functions we use in this paper.

References

1. Bastani, O., Ioannou, Y., Lampropoulos, L., Vytiniotis, D., Nori, A., Criminisi, A.: Measuring neural net robustness with constraints. In: Advances in Neural Information Processing Systems 29 (2016)
2. Carlini, N., Wagner, D.: Towards evaluating the robustness of neural networks. In: 2017 IEEE Symposium on Security and Privacy (SP), pp. 39–57. IEEE (2017)
3. Chen, K., Zhu, H., Yan, L., Wang, J.: A survey on adversarial examples in deep learning. J. Big Data 2(2), 71 (2020)
4. Dai, S., Mahloujifar, S., Mittal, P.: Parameterizing activation functions for adversarial robustness. In: 2022 IEEE Security and Privacy Workshops (SPW), pp. 80–87. IEEE (2022)
5. Deng, J., Dong, W., Socher, R., Li, L.J., Li, K., Fei-Fei, L.: Imagenet: a large-scale hierarchical image database (2009). https://doi.org/10.1109/CVPR.2009.5206848
6. Engstrom, L., Ilyas, A., Santurkar, S., Tsipras, D., Tran, B., Madry, A.: Adversarial robustness as a prior for learned representations. arXiv preprint arXiv:1906.00945 (2019)
7. Gu, S., Rigazio, L.: Towards deep neural network architectures robust to adversarial examples. arXiv preprint arXiv:1412.5068 (2014)
8. Huang, H., Wang, Y., Erfani, S., Gu, Q., Bailey, J., Ma, X.: Exploring architectural ingredients of adversarially robust deep neural networks. Adv. Neural. Inf. Process. Syst. 34, 5545–5559 (2021)
9. Huang, R., Xu, B., Schuurmans, D., Szepesvári, C.: Learning with a strong adversary. arXiv preprint arXiv:1511.03034 (2015)
10. Krizhevsky, A., Nair, V., Hinton, G.: The CIFAR-10 dataset 55(5) (2014). https://www.cs.toronto.edu/kriz/cifar.html
11. Kurakin, A., Goodfellow, I.J., Bengio, S.: Adversarial examples in the physical world. In: Artificial intelligence safety and security, pp. 99–112. Chapman and Hall/CRC (2018)
12. Madry, A., Makelov, A., Schmidt, L., Tsipras, D., Vladu, A.: Towards deep learning models resistant to adversarial attacks. arXiv preprint arXiv:1706.06083 (2017)
13. Pang, T., Lin, M., Yang, X., Zhu, J., Yan, S.: Robustness and accuracy could be reconcilable by (proper) definition. In: International Conference on Machine Learning, pp. 17258–17277. PMLR (2022)
14. Papernot, N., McDaniel, P., Goodfellow, I., Jha, S., Celik, Z.B., Swami, A.: Practical black-box attacks against deep learning systems using adversarial examples. arXiv preprint arXiv:1602.02697 (2016)
15. Papernot, N., McDaniel, P., Goodfellow, I., Jha, S., Celik, Z.B., Swami, A.: Practical black-box attacks against machine learning. In: Proceedings of the 2017 ACM on Asia Conference on Computer and Communications Security, pp. 506–519 (2017)

16. Salman, H., Ilyas, A., Engstrom, L., Kapoor, A., Madry, A.: Do adversarially robust imagenet models transfer better? Adv. Neural. Inf. Process. Syst. **33**, 3533–3545 (2020)
17. Samangouei, P., Kabkab, M., Chellappa, R.: Defense-gan: protecting classifiers against adversarial attacks using generative models. In: Proceedings of the IEEE Conference on Computer Vision and Pattern Recognition, pp. 933–941 (2018)
18. Shaham, U., Yamada, Y., Negahban, S.: Understanding adversarial training: increasing local stability of neural nets through robust optimization. arXiv preprint arXiv:1511.05432 (2015)
19. Sharma, G., Wu, W., Dalal, E.N.: The CIEDE2000 color-difference formula: implementation notes, supplementary test data, and mathematical observations. Color Res. Appli. **30**(1), 21–30 (2005)
20. Sridhar, K., Sokolsky, O., Lee, I., Weimer, J.: Improving neural network robustness via persistency of excitation. In: 2022 American Control Conference (ACC), pp. 1521–1526. IEEE (2022)
21. Szegedy, C., et al.: Intriguing properties of neural networks. arXiv preprint arXiv:1312.6199 (2013)
22. Wong, E., Rice, L., Kolter, J.Z.: Fast is better than free: revisiting adversarial training. arXiv preprint arXiv:2001.03994 (2020)

A Visual Interpretation-Based Self-improved Classification System Using Virtual Adversarial Training

Shuai Jiang[1](✉), Sayaka Kamei[1], Chen Li[2], Shengzhe Hou[3],
and Yasuhiko Morimoto[1]

[1] Graduate School of Advanced Science and Engineering, Hiroshima University,
Hiroshima, Japan
jiangshuai@hiroshima-u.ac.jp
[2] Graduate School of Informatics, Nagoya University, Nagoya, Japan
[3] College of Computer Science and Engineering, Shandong University of Science and
Technology, Qingdao, China

Abstract. The successful application of large pre-trained models such as BERT in natural language processing has attracted more attention from researchers. Since the BERT typically acts as an end-to-end black box, classification systems based on it usually have difficulty in interpretation and low robustness. This paper proposes a visual interpretation-based self-improving classification model with a combination of virtual adversarial training (VAT) and BERT models to address the above problems. Specifically, a fine-tuned BERT model is used as a classifier to classify the sentiment of the text. Then, the predicted sentiment classification labels are used as part of the input of another BERT for spam classification via a semi-supervised training manner using VAT. Additionally, visualization techniques, including visualizing the importance of words and normalizing the attention head matrix, are employed to analyze the relevance of each component to classification accuracy. Moreover, brand-new features will be found in the visual analysis, and classification performance will be improved. Experimental results on Twitter's tweet dataset demonstrate the effectiveness of the proposed model on the classification task. Furthermore, the ablation study results illustrate the effect of different components of the proposed model on the classification results.

Keywords: Visual Interpretation · Self-Improved Classification · Spam Detection · Virtual Adversarial Training

1 Introduction

Deep learning is a machine learning technique that has been widely applied in various fields, such as natural language processing (NLP) [8,17], recommendation systems [16,21], and prediction tasks [15,18]. In the field of spam email classification, deep learning models such as Recurrent Neural Networks, especially the Long Short-Term Memory [31] and Gated Recurrent Unit [19], have made significant progress in classifying emails and identifying spam.

© The Author(s), under exclusive license to Springer Nature Switzerland AG 2023
X. Yang et al. (Eds.): ADMA 2023, LNAI 14179, pp. 548–562, 2023.
https://doi.org/10.1007/978-3-031-46674-8_38

Finding suitable labels for domain-specific classification models trained using deep learning is challenging for researchers. Previous research has shown that sentiment analysis can be used in combination with pre-trained models for spam detection in tweets, as spammers often use emotional expressions to increase users' trust in their messages [29]. However, determining effective tags for other types of social media content remains a challenge in this field [7]. In recent years, large pre-trained language models such as the BERT have achieved high accuracy when fine-tuning supervised tasks [5]. Additionally, some past work has partly studied the learning of linguistic features and examined the internal vector by probing the classifier [25].

This paper uses spam detection as an example scenario. First, a fine-tuned BERT model is used for the sentiment classification of tweet texts. The sentiment label is then input for another BERT model to determine whether it is a spam tweet. In the training process of this model, a semi-supervised training approach using virtual adversarial training (VAT) is introduced to improve accuracy and system robustness. Ablation experiments demonstrate its effectiveness. Secondly, relevant tools are used to interpret the internal workings of the BERT model. By comparing various models used in the experiment and visualizing word importance attribution, the contribution of each token in each layer, and the attention matrices of each layer, the reasons for the accuracy differences between different models are found, explaining the models. Furthermore, more suitable URL tags are identified through internal analysis of the model. Further training of the model results in system improvement, as demonstrated by experiments. The main contributions of this paper are as follows:

- **A BERT-based model for semi-supervised learning**: The first BERT is employed for text sentiment classification. The obtained sentiment tags are combined with the text in another BERT for spam classification via VAT.
- **Self-improved visual interpretation**: Word attention scores are analyzed with visual interpretation tools, and the parts with high feature weights are used to improve the system's classification performance.
- **Performance improvement**: Experimental results and ablation studies on the Twitter tweets dataset have demonstrated the effectiveness of the proposed model for spam classification in a semi-supervised learning task.

2   Related Work

2.1   Spam Detection

Major social media sites (e.g., Twitter, Facebook, Quora) face a massive dilemma as many fall victim to spam. This information induces users to click on malicious links or uses bots to spread false news, seriously adding to the chaos in the Internet space. In recent years, many studies have been on spam detection for tweets, and many suitable and outdated features have been summarized [12]. Many studies have shown that sentiment analysis technology can enhance the differentiation of spam tweets [2]. Therefore, many studies have used traditional machine learning methods to detect spam tweets based on sentiment features [23,24,27]. In [1], the authors use an LDA model to find the sentiment and topic of tweets, suggest features that identify spam tweets more accurately than previous methods, and predict how widely spam spreads on Twitter. In [11], the author first used the pre-trained BERT model to perform sentiment analysis on tweets, extracting various sentiment features. Then, an unsupervised GloVe model was used for Twitter bot detection, resulting in high accuracy. In addition, adversarial training has also been widely used in spam detection tasks. For example, in [6], the author utilized several adversarial strategies to enhance the spam classifier and achieved good results, laying the foundation for adversarial training in classification tasks. [9] used an attention mechanism for movie review spam detection and employed GAN models for adversarial training, achieving state-of-the-art results.

2.2 Model Interpretability

In today's era of widespread use of deep learning and neural network technology, the demand for their interpretability is also gradually increasing. Such models are usually black boxes in their organizational structure, where users input specific information into the model and can obtain specific outputs. However, the model still needs to answer how the outputs are obtained. Model interpretability aims to transform black-box models into white-box models so that users can understand why the model makes relevant predictions and identify ways to improve its validity. In addition, it eliminates ethical issues when AI models are used on a large scale in society.

Interpretability on machine learning models has long been proposed, such as SHAP [20] and LIME [26]. The SHAP (SHapley Additive exPlanations) model produces a prediction value for each prediction sample, and the SHAP value is the value assigned to each feature in that sample. Lime (Local Interpretable Model-Agnostic Explanations) is an approach that uses a trained local proxy model to explain individual samples. However, when dealing with large pre-trained deep learning models with hundreds of hidden states, the situation becomes different, and simple local explanations of the model are difficult to fit. The BERT model, for example, introduced the attention mechanism [3], which became a very successful neural network component but increased the difficulty of interpreting the model. Clark et al. [4] strongly emphasize analyzing the attention head in BERT. They studied its behavior and directly extracted sentence representations from the BERT model without fine-tuning. They discovered that the attention head exhibits recognizable patterns, such as focusing on a fixed position offset or paying attention to the entire sentence.

3   Model

To identify spam tweets, we used the public dataset "Spam Detection on Twitter" [30] posted by YASH, which contains 82,469 legitimate tweets and 97,276 spam

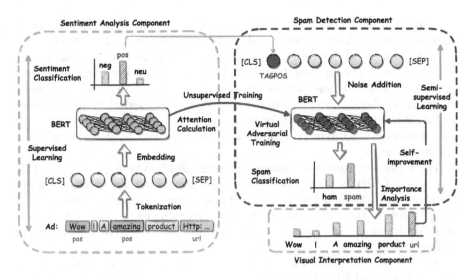

Fig. 1. Overview architecture of the proposed spam classification system. First, a fine-tuned BERT (the Sentiment Analysis Component) is used as a classifier to classify the sentiment of the text. Then, the predicted sentiment classification labels are used as part of the input of another BERT (the Spam Detection Component) for spam classification via a semi-supervised training manner using VAT. Additionally, visualization techniques, including visualizing the importance of words and normalizing the attention head matrix, are employed to analyze the relevance of each component to classification accuracy. Moreover, brand-new features are detected in the Visual Analysis Component. Finally, the classification system can be self-improved via the newly imported.

tweets. To classify the sentiment of tweets, we utilized the Sentiment140 dataset [10], which comprises 1.6 million tweets, half positive and half negative. The model architecture is shown in Fig. 1. First, the tweets will be tagged with sentiment labels after a BERT model fine-tuned by the Sentiment140 dataset, the Sentiment Analysis Component. Then, after two fully connected layers, the Spam Detection Component will output whether the tweet is spam. Semi-supervised learning and VAT are used here to improve the training accuracy. We utilize the Twitter dataset as unlabeled data for semi-supervised learning in sentiment analysis. Finally, the interpretation method will interpret the models.

3.1   Sentiment Analysis Component

Sentiment analysis of tweets helps to comprehend public opinion on topics prevalent on social media. Twitter's usage has increased as users share news and personal experiences. Hence, analyzing tweet sentiment is crucial. Despite its popularity, sentiment analysis of tweets is challenging due to the 280-character limit and irregularities in tweets (e.g., spelling variations and abbreviations).

BERT is a pre-trained deep bidirectional transformer, a powerful model for language understanding. We employed BERT for sentiment polarity classification using the Sentiment140 dataset. This dataset contains tweet content and is for a binary sentiment classification task. It includes 1.6 million tweets collected from the Twitter API and annotated as negative (0) or positive (4), making it useful for sentiment detection. Unlike previous work [11] by Heidari et al., who used the SST-2 movie review dataset for training, the Sentiment140 dataset is in the same domain as the target task, thus leading to improved accuracy in sentiment analysis of tweets.

The Sentiment Analysis Component categorizes tweets as positive, neutral, or negative. BERT is trained on 1.6 million tweets. After the tokens are input into the model, the model performs word embedding processing. Among the 12 hidden layers in the BERT model, the next layer's multi-head attention calculates the attention scores of each word in the previous layer. The BERT model's training results will be presented in the next section.

This component finally extracts sentiment features from the text of tweets through fine-tuned BERT. At the end of the model, the softmax layer outputs the sentence's sentiment polarity value score X ($0 < X < 1$). Then it divides the sentence into three categories: positive, neutral, and negative, according to this value. If $X < 0.3$, the tweet sentiment is negative. If $X > 0.7$, the tweet sentiment is positive. The rest are neutral.

3.2   Spam Detection Component

The Spam Detection Component also uses the fine-tuned BERT to determine if the tweet is spam. We use the *SpamDetectionOnTwitter* dataset in [30] to learn whether a tweet is a spam. To embed emotional features into tweets and better explain the model, we add sentiment tags into each piece of data, namely *TAGPOS, TAGNEU,* and *TAGNEG,* and add these three tags to the dictionary of the BERT model. After exporting the BERT model, it includes two fully connected layers. The final layer with softmax will output the final result indicating whether the tweet is spam.

At the same time, we also use several adversarial learning methods for training enhancement to find the best one. Adversarial training can be summarized as the following max-min formula:

$$\min_{\theta} \mathbb{E}_{(x,y)\sim\mathcal{D}}[\max_{||\delta||\leq\varepsilon} L(f_{\theta}(X + \delta), y)]. \tag{1}$$

The inner layer (in square brackets) is a maximization, where X represents the input representation of the sample, δ represents the perturbation superimposed on the input, $f_{\theta}()$ is the neural network function, y is the label of the sample, and $L(f_{\theta}(X + \delta), y)$ represents the loss obtained by superimposing a disturbance δ superimposed on the sample X. $\max(L)$ is the optimization goal, that is, to find the disturbance that maximizes the loss function. The outside minimization refers to finding the most robust parameters θ, such that the predicted distribution conforms to the distribution of the original dataset.

The Fast Gradient Method (FGM) is implemented by $L2$ normalization, which divides the value of each gradient dimension by the $L2$ norm of the gradient. Theoretically, $L2$ normalization preserves the direction of the gradient.

$$\delta = \epsilon \bullet \left(g / | \, |g| \, |_2 \right).$$ (2)

Among them, $g = \nabla_X \left(L \left(f_\theta \left(X \right), y \right) \right)$ is the gradient of the loss function L with respect to the input X. Unlike a normal FGM that only performs iteration once, PGD performs multiple iterations to find the optimal perturbation. Each iteration projects the disturbance into a specified range each time a small step is taken. The formula of the loss function in step t in PGD is shown as follows:

$$g_t = \nabla X_t \left(L \left(f_\theta \left(X_t \right), y \right) \right).$$ (3)

Although PGD is simple and effective, there is a problem that it is not computationally efficient. Without adversarial training, m iterations will only have m gradient calculations, but for PGD, each gradient descent must correspond to the K steps of gradient boosting. Therefore, PGD needs to do $m(K+1)$ gradient calculations compared with the method without adversarial training. In VAT, the loss function for adversarial training can be expressed as [22]:

$$L_{adv}(x_l, \theta) := D[q(y|x_l), p(y|x_l + r_{adv}, \theta)],$$ (4)

$$r_{adv} := \arg\max_{r; ||r|| \leq \epsilon} D[q(y|x_l), p(y|x_l + r, \theta)],$$ (5)

where $D[p, q]$ is a non-negative function that measures the divergence between two distributions p and q. The function $q(y|x_l)$ is the true distribution of the output label, which is unknown. This loss function aims to approximate the true distribution $q(y|x_l)$ by a parametric model $p(y|x_l, \theta)$ that is robust against adversarial attack to labeled input x_l. A "virtual" label generated by the $p(y|x, \theta)$ probability is used in VAT to represent the user-unknown $p(y|x, \hat{\theta})$ label, and the adversarial direction is calculated based on the virtual label. Unlabeled input x_{ul} and labeled input x_l will be unified as x_*. The formula is calculated as follows:

$$LDS(x_*, \theta) := D[p(y|x_*, \hat{\theta}), p(y|x_* + r_{qadv}, \theta)],$$ (6)

$$r_{qadv} := \arg\max_{r; ||r||_2 \leq \epsilon} D[p(y|x_*, \hat{\theta}), p(y|x_* + r)],$$ (7)

where the loss function of $LDS(x_*, \theta)$ indicates the virtual adversarial perturbation. This function can be considered a negative indicator of the local smoothness of the current model at each input data point x. A reduction in this function would result in a smoother model at each data point.

3.3 Visual Interpretation Component

In this part, we begin by creating visual representations of the significance of individual words in differentiating between spam and non-spam content. Additionally, we normalize the attention head matrix to visualize all attention matrices and identify distinctions between various models, thereby demonstrating the efficacy of our proposed model. We will delve into further details in the following part, accompanied by relevant examples.

Word Importance Attribution. Integrated Gradients [28] are used to compute attributions concerning the *BertEmbeddings* layer to obtain the importance of words. In simple terms, Integrated Gradients define the attribution of the ith feature of the input as the path integral of the straight line path from the baseline x_i' to the input x_i from [28]:

$$\text{IG}_i(x)::= (x_i - x_i') \cdot \int_{\alpha=0}^{1} \frac{\partial F(x' + \alpha(x - x'))}{\partial x_i} d\alpha, \tag{8}$$

where $\frac{\partial F(x)}{\partial x_i}$ is the gradient of F along the ith dimension at input x and baseline x'. In the NLP task described in this paper, we use the zero vector as the baseline.

Attribution in Attention Matrix. We visualize the attention probabilities of 12 attention heads in all 12 layers, totaling 144. It represents the softmax normalized dot product of key and query vectors. In [4], it is an essential indicator, indicating how related a token is to another token in the text.

4 Experiments

To demonstrate the effectiveness of the proposed model, this section first provides empirical evidence through ablation experiments, demonstrating the effectiveness of the relevant components, including the sentiment analysis component and the adversarial training component. Secondly, visualization tools are used to analyze the model's interpretability to identify the reasons for the effectiveness of the relevant components. Finally, using the above analysis, more suitable labels are identified, and the system is further trained to achieve improvements.

4.1 Dataset

Spam Dataset. We use the SpamDetectionOnTwitter dataset in [30] to learn whether a tweet is spam. This dataset contains 82,469 legitimate tweets and 97,276 spam tweets. Each tweet is tagged with user_id, tweet_id, tweet_text, time, and spam_label to show whether it is a spam tweet. Here we only select the text and spam_label for training. We select 68,919 legitimate tweets and 58,866 spam tweets as the training set, and the rest is divided into a validation set and a test set.

Sentiment Dataset. We used the Sentiment140 dataset [10] as the training dataset for the part of the tweet sentiment polarity analysis component. This dataset contains 1.6 million sentiment-labeled tweets, half positive and half negative, and each tweet is accompanied by tweet_id, time, username, and tweet_text. Similar to the last part, only the tweet_text and spam_label are selected for training at this stage. We select 1.46 million tweets as the training set, and the rest are divided into the validation and test sets.

4.2 Hyperparameter Setting

In both Sentiment Analysis Component and Spam Detection Component, we used the bert-base-multilingual-cased model, which is Google's new and recommended BERT model. We set the batch size to 16 and the dropout rate to 0.1 in both Sentiment Analysis and Spam Detection Components. The learning rates of the Adam optimizer are 2e-5 in the sentiment part and 1e-5 in the spam part. According to the size of the two datasets, we set the steps of the sentiment part as 10000, while 1000 in the spam part. All experiments used Pytorch version 1.13.1, bert4torch 0.2.4, and Captum 0.6.0.

4.3 Spam Detection

After fine-tuning, Sentiment Analysis is performed on the existing spam dataset, and the sentiment distribution of the dataset can be seen, as shown in Fig. 2. The number of spam tweets with positive sentiment is the largest, followed by neutral sentiment and the least negative sentiment, with 42547, 32291, and 7629, respectively. In the non-spam tweets, the tweets with neutral sentiment are the most, and the positive and negative sentiment is both less, among which the negative sentiment is the least, the numbers 27619, 52049, and 17606, respectively.

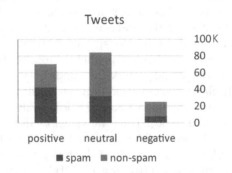

Fig. 2. Sentiment distribution in spam dataset.

To show the effectiveness of the proposed model, we performed ablation experiments on the model proposed in this paper, using the same training parameters and random seed. Since PGD is an evolutionary algorithm of FGM, here we omit the experiment of FGM and only keep PGD. The results are shown in Table 1. It can be seen that the precision of the proposed model is the highest, proving its effectiveness. This part only analyzes the model's effectiveness from the model experimentation perspective. Although the proposed model has achieved the highest accuracy and recall, as a black box model, we cannot know the reasons for the differences in the experiment results inside the different models. This issue leads us to introduce model interpretability (or XAI). The next part will study this issue in depth.

Table 1. Experiment results of all models.

	Precision	Accuracy	Recall	F1 Score
BERT	76.06	75.32	79.95	77.96
BERT+Sentiment	79.14	77.43	79.65	79.39
BERT+Sentiment+PGD	83.21	76.83	72.10	77.25
BERT+Sentiment+VAT	82.61	76.68	79.54	78.47
Proposed Model (1 dense layer)	82.58	77.81	75.41	78.83
Proposed Model	85.97	77.60	74.92	78.49

4.4 Visual Interpretation

In this section, we use the Captum [14] tool to perform visual interpretability analysis on the BERT model in the Spam Detection Component of all six models in the previous chapter. This tool is a Pytorch-based model interpretation library released by Facebook. The library provides interpretability for many new algorithms (such as ResNet, BERT, and some semantic segmentation networks), helping everyone better understand the specific features, neurons, and neural network layers that affect the model's prediction results. For text translation and other problems, it can visually mark the importance of different words and use a heat map to display the correlation between words.

Here, we first visualize the importance of words to distinguish which words play a role in judging spam or not. Then, we visualize all the attention matrices to find the differences between different models to prove the effectiveness of the proposed model. Here is an example of the tweet "*19 year old genius shares Twitter tool free. Nice guys rock!* http://ow.ly/Ul1t", a spam tweet with positive sentiment.

Word Importance Attribution. With the formula (8), we can obtain the Word Importance Attribution of the input sentence, as shown in Fig. 3. In this case, the actual label of the input sentence is "Spam." This figure's "Predicted Label" represented the model output result and predicted probability. The rightmost is a visual explanation of the contribution value of the input sentence, green represents a positive contribution to the Predicted Label, and red represents the opposite. Additionally, the deeper the color, the higher the level of contribution, and vice versa.

It can be seen from the figure that when the sentiment tag is not added, only the word *year* is a positive contribution, so the final probability is only 0.52, and the model is difficult to distinguish. After adding the sentiment tag, although the contribution of the tag is less, the contribution of many words, especially the URL part, is significantly improved, thereby increasing the final probability.

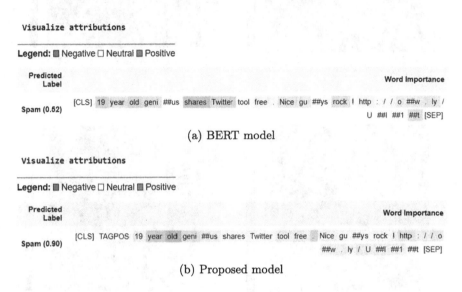

Fig. 3. Word importance attribution.

Attribution in Attention Matrix. The order of the following figures is simple BERT model and the proposed model. The x-axis and y-axis of the matrix are tokens, and each cell represents the attention score between different tokens, that is, the degree of attention obtained from the weight of attention. The brighter the cell, the higher the attention score, or level of attention, from the token on the x-axis to the token on the y-axis, and vice versa.

In most attention heads, the overall trend of words is to pay more attention to themselves or the next word. Still, in Head 2–5, 3–3, 3–4, 3–6, 3–8, 3–12, and 4–2, the [CLS] token will pay more attention to the added sentiment tag, which we can see the brightest point appears in the upper left corner, such as Head 3–6 (Fig. 4). According to [13], each sequence's initial [CLS] token is used as the sentence representation in a labeled classification task. Therefore, in the Spam Detection Component of this paper, the [CLS] token represents whether the sentence is spam. Therefore, these results show the validity of the sentiment tag. Entering layer 9, it can be seen that, as shown in Fig. 5, the attention matrix divides the token into two parts, the text, and the URL. This

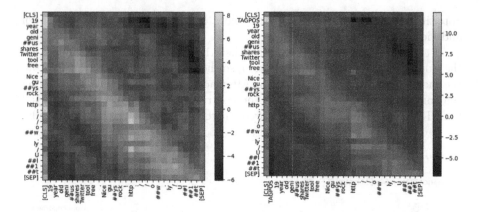

Fig. 4. Attention Head 3–6 of BERT and the proposed model.

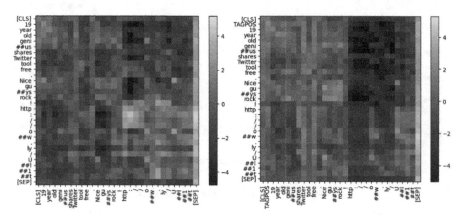

Fig. 5. Attention Head 9–12 of BERT and the proposed model.

phenomenon is more evident in Head 9–12, and the attention matrix is divided into four prominent parts, especially in the model after adding the adversarial training method.

Other examples can also demonstrate the effectiveness of the proposed model. In Head 2–12, shown in Fig. 6, we can see the part of the URL that pays more attention to *http://*, indicating that the model has detected the URL.

4.5   Model Improvement

Based on the analysis in the previous section, we found that the URL part can be detected in some attention heads. Inspired by this, we operate similarly to the sentiment tags above for the URL part, using a regular expression to extract the URL part as separate data labels. To distinguish between short and long links, we set the URL label with less than 24 characters as *TAGURLS*, indicating short

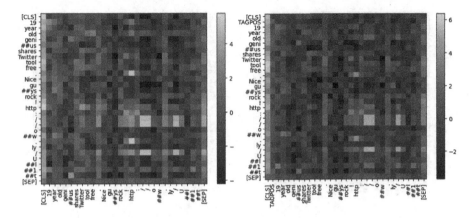

Fig. 6. Attention Head 2–12 of BERT and the proposed model.

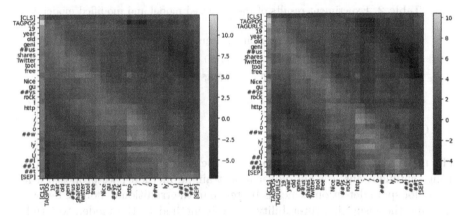

Fig. 7. Attention Head 3–6 of the proposed model and improved model.

URL links, and the rest as *TAGURLL*. We add these labels after the sentiment label. We used the optimized data to retrain the proposed model, resulting in the modified model, as shown in Table 2. We found that after adding the URL tag, the accuracy, F1 score, and precision of the modified model were improved.

Next, we analyze the BERT model using interpretability methods and directly examine the corresponding attention head of the original model. In Fig. 7–8, the left part is the proposed model, and the right is the improved model. In Fig. 7, Head 3–6 is the same as before, with [CLS] paying the most attention to the sentiment tag, but the URL tag has become the second most attention token, which to some extent, proves the validity of the URL tag. In Fig. 8, Head 9–12, the URL part is more prominent than the original model, indicating that the model is paying more attention to the URL part, proving the URL tag's reliability for improving precision and accuracy.

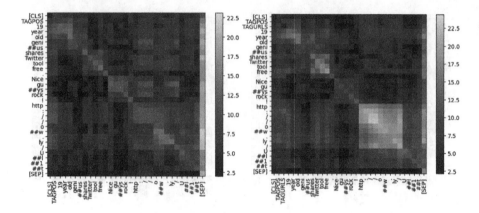

Fig. 8. Attention Head 9–12 of the proposed model and improved model.

Table 2. Experiment results of the proposed model and modified model

	Precision	Accuracy	Recall	F1 Score
Proposed Model	85.97	77.60	74.92	78.49
Improved Model	**87.13**	**77.85**	75.78	78.88

5   Conclusion

The main contributions of this paper can be summarized in two points. First, using sentiment analysis and adversarial training methods, we proposed a new model for spam detection, which is better than the traditional models. Secondly, applying the visual interpretability analysis method to the model, we studied the principle of internal classification of the model, found the reasons for the difference in precision in different models, and proved the effectiveness of the proposed model at the same time, further improving its performance.

The large-scale pre-trained BERT model based on VAT can be extended to other tasks. The attention mechanism can analyze the in-depth features with heavy weights, and these features can effectively improve the accuracy and precision of the task. In future work, we will utilize the VAT and visual interpretation method in other pre-trained language models (e.g., ALBERT, XLNET) to further improve the performance of spam classification.

Acknowledgements. This work is Supported by KAKENHI (20K11830) Japan.

References

1. Ahsan, M., Sharma, T.: Spams classification and their diffusibility prediction on Twitter through sentiment and topic models. Int. J. Comput. Appl. **44**(4), 365–375 (2022)

2. Antonakaki, D., Fragopoulou, P., Ioannidis, S.: A survey of Twitter research: data model, graph structure, sentiment analysis and attacks. Expert Syst. Appl. **164**, 114006 (2021)
3. Bahdanau, D., Cho, K.H., Bengio, Y.: Neural machine translation by jointly learning to align and translate. In: 3rd International Conference on Learning Representations, ICLR 2015 (2015)
4. Clark, K., Khandelwal, U., Levy, O., Manning, C.D.: What does BERT Look at? an analysis of BERT's attention. In: Proceedings of the 2019 ACL Workshop BlackboxNLP: Analyzing and Interpreting Neural Networks for NLP, pp. 276–286 (2019)
5. Dai, A.M., Le, Q.V.: Semi-supervised sequence learning. In: Advances in Neural Information Processing Systems 28 (2015)
6. Dalvi, N., Domingos, P., Mausam, Sanghai, S., Verma, D.: Adversarial classification. In: Proceedings of the Tenth ACM SIGKDD International Conference on Knowledge Discovery and Data Mining, pp. 99–108 (2004)
7. Farías, D.I.H., Patti, V., Rosso, P.: Irony detection in twitter: the role of affective content. ACM Trans. Internet Technol. (TOIT) **16**(3), 1–24 (2016)
8. Floridi, L., Chiriatti, M.: GPT-3: its nature, scope, limits, and consequences. Mind. Mach. **30**, 681–694 (2020)
9. Gao, Y., Gong, M., Xie, Y., Qin, A.K.: An attention-based unsupervised adversarial model for movie review spam detection. IEEE Trans. Multimed. **23**, 784–796 (2021)
10. Go, A., Bhayani, R., Huang, L.: Twitter sentiment classification using distant supervision. CS224N project report, Stanford **1**(12), 2009 (2009)
11. Heidari, M., Jones, J.H.: Using BERT to extract topic-independent sentiment features for social media bot detection. In: 2020 11th IEEE Annual Ubiquitous Computing, Electronics and Mobile Communication Conference (UEMCON), pp. 0542–0547. IEEE (2020)
12. Kabakus, A.T., Kara, R.: A survey of spam detection methods on Twitter. Int. J. Adv. Comput. Sci. Appl. **8**(3) (2017). https://thesai.org/Publications/ViewPaper?Volume=8&Issue=3&Code=IJACSA&SerialNo=5
13. Kenton, J.D.M.W.C., Toutanova, L.K.: BERT: Pre-training of deep bidirectional transformers for language understanding. In: Proceedings of NAACL-HLT, pp. 4171–4186 (2019)
14. Kokhlikyan, N., et al.: Captum: a unified and generic model interpretability library for pytorch. arXiv preprint arXiv:2009.07896 (2020)
15. Li, C., Chen, Z., Zheng, J.: An efficient transformer encoder-based classification of malware using API calls. In: 2022 IEEE 24th International Conference on High Performance Computing and Communications; 8th International Conference on Data Science and Systems; 20th International Conference on Smart City; 8th International Conference on Dependability in Sensor, Cloud and Big Data Systems and Application (HPCC/DSS/SmartCity/DependSys), pp. 839–846. IEEE (2022)
16. Li, C., He, M., Qaosar, M., Ahmed, S., Morimoto, Y.: Capturing temporal dynamics of users' preferences from purchase history big data for recommendation system. In: 2018 IEEE International Conference on Big Data (Big Data), pp. 5372–5374. IEEE (2018)
17. Li, C., Yamanaka, C., Kaitoh, K., Yamanishi, Y.: Transformer-based objective-reinforced generative adversarial network to generate desired molecules. In: Raedt, L.D. (ed.) Proceedings of the Thirty-First International Joint Conference on Artificial Intelligence, IJCAI-22, pp. 3884–3890 (2022)
18. Li, C., Zhang, X., Qaosar, M., Ahmed, S., Alam, K.M.R., Morimoto, Y.: Multi-factor based stock price prediction using hybrid neural networks with atten-

tion mechanism. In: 2019 IEEE International Conference on Dependable, Autonomic and Secure Computing, International Conference on Pervasive Intelligence and Computing, International Conference on Cloud and Big Data Computing, International Conference on Cyber Science and Technology Congress (DASC/PiCom/CBDCom/CyberSciTech), pp. 961–966. IEEE (2019)

19. Li, C., Zheng, J.: API call-based malware classification using recurrent neural networks. J. Cyber Secur. Mobility 617–640 (2021)

20. Lundberg, S.M., Lee, S.I.: A unified approach to interpreting model predictions. In: Advances in Neural Information Processing Systems 30 (2017)

21. Marappan, R., Bhaskaran, S.: Movie recommendation system modeling using machine learning. Int. J. Math. Eng. Biol. Appl. Comput. 1(1), 12–16 (2022). https://www.scipublications.com/journal/index.php/ijmebac/article/view/291

22. Miyato, T., Maeda, S.i., Koyama, M., Ishii, S.: Virtual adversarial training: a regularization method for supervised and semi-supervised learning. IEEE Trans. Pattern Anal. Mach. Intell. 41(8), 1979–1993 (2018)

23. Monica, C., Nagarathna, N.: Detection of fake tweets using sentiment analysis. SN Comput. Sci. 1(2), 1–7 (2020)

24. Perveen, N., Missen, M.M.S., Rasool, Q., Akhtar, N.: Sentiment based Twitter spam detection. Int. J. Adv. Comput. Sci. Appl. 7(7) (2016). https://thesai.org/Publications/ViewPaper?Volume=7&Issue=7&Code=IJACSA&SerialNo=77

25. Radford, A., Narasimhan, K., Salimans, T., Sutskever, I., et al.: Improving language understanding by generative pre-training (2018)

26. Ribeiro, M.T., Singh, S., Guestrin, C.: "Why Should I Trust You?": explaining the predictions of any classifier. In: Proceedings of the 22nd ACM SIGKDD International Conference on Knowledge Discovery and Data Mining, pp. 1135–1144 (2016)

27. Saumya, S., Singh, J.P.: Detection of spam reviews: a sentiment analysis approach. CSI Trans. ICT 6(2), 137–148 (2018)

28. Sundararajan, M., Taly, A., Yan, Q.: Axiomatic attribution for deep networks. In: International Conference on Machine Learning, pp. 3319–3328. PMLR (2017)

29. Vosoughi, S., Roy, D., Aral, S.: The spread of true and false news online. Science 359(6380), 1146–1151 (2018)

30. YASH: Spam detection on Twitter

31. Zhang, X., Li, C., Morimoto, Y.: A multi-factor approach for stock price prediction by using recurrent neural networks. Bull. Network., Comput., Syst. Softw. 8(1), 9–13 (2019)

TSCMR:Two-Stage Cross-Modal Retrieval

Zhihao Chen[1,3] and Hongya Wang[1,2,3(✉)]

[1] School of Computer Science and Technology, Donghua University, Shanghai, China

[2] State Key Laboratory of Computer Architecture, Institute of Computing Technology, CAS, Beijing, China

[3] Shanghai Key Laboratory of Computer Software Evaluating and Testing, Shanghai, China

hywang@dhu.edu.cn

Abstract. Currently, large-scale vision and language models has significantly improved the performances of cross-modal retrieval tasks. However, large-scale models require a substantial amount of computing resources, so the execution of these models on devices with limited resources is challenging. Thus, it is paramount to reduce the model size and minimize computing costs of a model without sacrificing its performance. In this paper, we improved TERAN by dividing cross-modal retrieval into two stages: image-text coarse-grained matching and image-text fine-grained matching. Specifically, we present a novel approach called Two-Stage Cross-Modal Retrieval network(TSCMR). To reduce model size after model training, our approach utilized a new knowledge distillation method for Transformer-based models. Experiments have shown that our approach maintains a performance comparable to TERAN on the MS-COCO 1K test set, while being 2x smaller and 3.1x faster on inference.

Keywords: Cross-modal · Two-Stage Retrieval · Knowledge Distillation

1 Introduction

The rapid development of mobile internet has fueled an explosive growth in the volume of multimodal data comprised of images, text, and videos. Correspondingly, the demands from users with regard to data modalities have become increasingly diversified. Consequently, a significant shift towards cross-modal retrieval from single-modal retrieval has been observed in users' retrieval requests. For instance, corporations like Google have recently attempted to utilize textual descriptions to achieve cross-modal retrieval between text and images. The concept of cross-modal retrieval is aimed at promoting information interaction between different modalities, and as such, is focused on retrieving other modality samples with similar semantics through a modality sample. Given this aim, the presence of semantic relations between modalities becomes pivotal.

© The Author(s), under exclusive license to Springer Nature Switzerland AG 2023
X. Yang et al. (Eds.): ADMA 2023, LNAI 14179, pp. 563–575, 2023.
https://doi.org/10.1007/978-3-031-46674-8_39

In recent years, the mainstream method for cross-modal retrieval has been to train large-scale pre-training models based on the Transformer [1] architecture to learn the semantic relationships between different modalities. These models can be divided into single-stream and dual-stream structures. However, to extract meaningful information from highly redundant datasets, complex models and a large amount of computational resources are required, regardless of the structure used. At present, many models have billions of parameters and demand more than 10GB of GPU memory for deployment, so it is difficult to efficiently execute them on resource-restricted devices. Furthermore, retrieving information using such models takes a long time. In light of these challenges, minimizing the storage and computation costs of the model while ensuring optimal performance is crucial.

TERAN utilizes the cosine similarity to generate the similarity score between each region and word, thus forming a region-word similarity matrix. By applying a pooling technique to the matrix, a global similarity score is obtained for the image and text. Notably, the computational time involved in calculating the similarity between an image and text is significantly higher than that of extracting the features for both. Constructing a matrix for a single image and text pairing is not time-consuming, but for a corpus of a hundred or more, the process becomes protracted.

To optimise the inference speed. This paper proposes a two-stage cross-modal retrieval model. Specifically, the two-stage cross-modal retrieval model divides the retrieval task into coarse-grained and fine-grained matching stages. In the first stage, global features representing images and text are added, and scores are derived from these features to identify top-performing candidates for the second stage. In the second stage, the model uses regional features of the images and word-level features of the text to calculate fine-grained similarity scores, which form the final basis for determining image-text similarity. By selecting top k scoring items from the coarse-grained phase, the model can also attain inference acceleration. Notably, this two-stage process is designed to reduce time and computational resource consumption during the fine-grained matching phase. After training, this paper use a discussion of a newly-developed Transformer distillation method to reduce model size.

2   Related Work

This section provides a comprehensive discussion of prior research on cross-modal retrieval through the use of joint image and text processing. The main architecture of this model, which is the Transformer Encoder architecture, was introduced. Furthermore, we elaborated on knowledge distillation and its implementation in models employing the Transformer Encoder architecture.

2.1   Joint Image and Text Processing for Cross-Modal Retrieval

At present, Transformer-based pretrained models are highly esteemed in both academia and industry for understanding visual and textual information due to their excellent performance in cross-modal retrieval, attracting attention of researchers. These models are classified into two categories: single-stream structure models and dual-stream structure models, based on the current research.

The mainstream method for cross-modal retrieval is to train large-scale pretrained models based on the Transformer architecture to learn the semantic correspondence between different modalities. These models are divided into single-stream [2–4] and dual-stream [5–7] structures. Before inputting the model, image-text pairs require image and text feature extraction. Image features may be region features based on object detection [8], CNN-based global features or patch features like ViT [9] whereas text features usually follow the preprocessing method of BERT [10]. Single-stream structures combine text and image features, inputting them into a single Transformer block, and fusing multiple modality inputs through self-attention mechanisms. The final output value, identified by the cls token, determines the similarity of the inputted image-text pair. Single-stream structures learn cross-modal feature information more effectively, leading to better performance in the final evaluation metrics. Dual-stream structures input text and image features separately into two different Transformer blocks. One block processes image features, the other processes text features, and they each output the cls token representing the global feature for both image and text, respectively. Cosine similarity is then utilized to calculate the similarity between image-text pairs. However, the lack of interaction between image and text features diminishes accuracy. To solve this problem, some models include additional Transformer blocks within the dual-stream structure to achieve interaction between different modality features. Nevertheless, while performance improves, model complexity and parameters increase as well.

The TERAN [11] model proposed by Nicola et al. belongs to a dual-stream architecture that deals with cross-modal retrieval tasks via word-region alignment in image-text matching. The supervision is only employed at a global image-text level in this model. Fine-grained matching is implemented between the low-level components of images and texts, which includes matching of image regions and words to maintain the richness of information in both modalities. TERAN performs as well as single-stream models in image and text retrieval tasks. The fine-grained alignment method from TERAN provides new ideas for large-scale cross-modal information retrieval research.

2.2   Transformer Encoder

The model architecture we propose is mainly composed of Transformer [1] Encoder. Specifically, as shown in Fig. 1, the Transformer Encoder layer mainly includes two sub-layers: multi-head attention(MHA) layer and fully connected feed-forward neural network(FNN) layer.

The Multi-Head Attention (MHA) is constructed by combining multiple self-attention layers altogether. The objective of the attention layer is to gather information on the connection between each token and other tokens to determine their significance in the input sequence. We adopt three input vectors, namely,

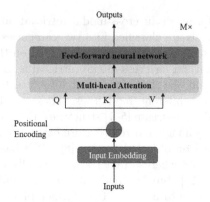

Fig. 1. Overview of Transformer Encoder.

the query(Q) vector, the key(K) vector, and the value(V) vector for our attention layer. The attention function can be expressed as the following formula:

$$\text{Attention}\,(Q, K, V) = \text{soft max}\left(\frac{QK^T}{\sqrt{d_k}}\right)V \tag{1}$$

where d_k is the dimension of keys and acts as a scaling factor, and the factor $\sqrt{d_k}$ is used to mitigate the vanishing gradient problem of the softmax function in case the inner product assumes too large values.In essence, querying is akin to searching for information on a browser. The matching pages returned by the browser are keys, but what we require are the values that carry the desired information. By analyzing specific tokens and other tokens in a given sequence, we can determine their relevance and interdependencies with respect to another token. The self-attention mechanism involves multiple calculations, where different weight matrices are used for Q, K, and V, to facilitate this analysis.

The Transformer encoder incorporates a feedforward neural network layer, comprising of two linear transformation layers and a Rectified Linear Unit (ReLU) activation function, to acquire more comprehensive information.

2.3   Knowledge Distillation

Large-scale models are typically constructed using a single intricate network, or a composite of multiple networks. While these models demonstrate impressive performance and generalizability, small-scale models are often less expressive due to their smaller size. Knowledge distillation involves using knowledge gained from large-scale models to aid training of small-scale models, achieving comparable performance as large-scale models with reduced parameter size, thereby enabling model compression and acceleration.

Hinton et al. introduced the concept of "knowledge distillation" in [12]. The central idea is to improve the training of a small model by utilizing the knowledge learned by a large model. Therefore, the knowledge distillation framework generally comprises a large model (known as the teacher model) and a small model (known as the student model). To enhance the quality of distilled knowledge and improve the performance of the student model, [13] proposed using an ensemble of models as the teacher model. [14] presented a knowledge distillation method based on the Transformer model structure that compresses and accelerates the pre-trained BERT model. Although it introduced a new loss function, [15] conducted experiments on the BERT model. In [16], a task-agnostic model compression method based on the BERT model was proposed.

In the field of natural language processing, the scale of pre-trained language models has been continuously expanding, and model compression has thus become increasingly important. To address this, [17] introduced a structured pruning method specifically designed for certain tasks called CoFi (Coarse and Fine-grained Pruning). The method combines pruning of coarse-grained units, such as self-attention layers and feedforward layers, with that of fine-grained units, such as heads and hidden dimensions. In addition, the authors proposed a hierarchical distillation method to dynamically learn the layer mapping relationship between the teacher and student models, which improves model performance. CoFi-compressed models achieve more than 10 times model acceleration, 95% parameter pruning, and maintain an accuracy rate of over 90% of the original model.

3 Method

In this section, we firstly introduce the model architecture. Then, we delineate the training objectives of the TSCMR. Lastly, we provide a comprehensive description of the knowledge distillation technique that was employed after completing the TSCMR training.

3.1 Model Architecture

Figure 2 displays TSCMR that includes the initial processing of images and text, an image encoder, a text encoder, and a method for calculating image and text similarity. Fast-Rcnn [8] is used for initial image processing and encodes input image I into an embedding sequence: $\{r_1, \cdots, r_n\}$. An image encoder consisting of four transformer encoders and one transformer encoder with two layers is used. The sequence is converted to $\{I_{cls}, r_1, \cdots, r_n\}$, where the token I_{cls} represents the global representation of the image, before inputting it into the image encoder. The text encoder adopts a combination of a 6-layer BERT model and one transformer encoder with two layers, converting input text T into an embedding sequence $\{T_{cls}, w_1, \cdots, w_n\}$. The token T_{cls} signifies the global representation of the text.

3.2 Training Objectives

TSCMR has two training objectives: image-text coarse-grained matching task (ITCG) and image-text fine-grained matching task (ITFG).

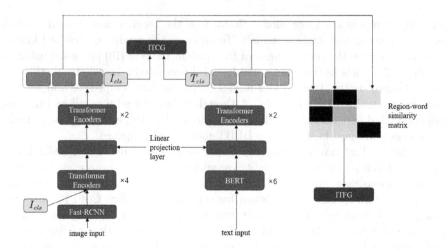

Fig. 2. The proposed TSCMR architecture. ITFG stands for image-text fine-grained matching, ITCG stands for image-text fine-grained matching. The orange boxes represents image region features and the green boxes represents word features. (Color figure online)

Image-Text Coarse-Grained Matching. In contrast to the TERAN, our new model architecture employs I_{cls} and T_{cls} for two-stage retrieval in order to reduce model inference time. After passing through the image and text encoders, we obtain the final image embedded sequence $\{I_{cls}, r_1, \cdots, r_n\}$ and text embedded sequence $\{T_{cls}, w_1, \cdots, w_n\}$. In the image-text coarse-grained matching stage, the S_{IT} similarity score is given by the cosine similarity between I_{cls} and T_{cls}, thus assigning higher scores to matched image and text pairs. The formula is as follows:

$$S_{IT} = \frac{I_{cls}^T T_{cls}}{\|I_{cls}\|\|T_{cls}\|} \tag{2}$$

After computing the coarse-grained similarity between image and text, we can employ the identical approach as described in [18] to compute the loss. This approach involves utilizing the hinge-based triplet ranking loss and directing attention towards hard negatives. The formula for calculating the loss is presented below:

$$L_{ITCG} = \max_{T'}\left[\alpha + S_{IT'} - S_{IT}\right]_+ + \max_{I'}\left[\alpha + S_{I'T} - S_{IT}\right]_+ \tag{3}$$

where $[x]_+ \equiv \max(0, x)$ and α is a margin that defines the minimum separation that should hold between the truly matching image-text pairs and the negative pairs, and calculates the negative examples T' and I' using the following method:

$$T' = arg\max_{z \neq T} S(z, T) \tag{4}$$

$$I' = arg\max_{y \neq I} S(y, I) \tag{5}$$

where (I, T) is a positive pair,z and y is negatives. During training, the dataset is divided into batches, thus negative examples are sampled from each batch.

Image-Text Fine-Grained Matching. At this stage, we drew upon the similarity matrix method employed in the TERAN, albeit abstaining from employing the I_{cls} and T_{cls} used in the previous phase. Cosine similarity is utilized to assess the similarity between the i-th region in I and the j-th word in T. Furthermore, the following approach is taken to compute the similarity matrix A:

$$A_{ij} = \frac{r_i^T w_j}{\|r_i\| \|w_j\|} \quad r_i \in I, w_j \in T \tag{6}$$

To calculate the global similarity between image and text, we used an appropriate pooling function to pool the similarity matrix. Inspired by [19,20], we adopted the max-sum pooling method, which selects the maximum value of each row in the similarity matrix A and sums them up. The specific formula is as follows:

$$S_{IT} = \sum_{w_j \in T} \max_{r_i \in I} A_{ij} \tag{7}$$

During this stage, we drew inspiration from the TopK algorithm. For each image I, we selected the finest K texts from the image-text coarse-grained matching scores to proceed to this stage. We calculated the fine-grained matching scores between I and the selected texts by employing a similarity matrix. Likewise, for each text T, we opt for the top M images with image-text coarse-grained matching scores, enter this stage, and calculate the fine-grained matching scores between T and these M images using similarity matrix. If the matching similarity scores of the text or image that genuinely matches are not in the top K or M sequence, we replace the lowest score with the newly found score. The hinge-based triplet ranking loss method is also implemented in this phase to calculate the loss, while the formula remains identical as follows:

$$L_{I2T-ITFG} = \max_{T'} [\alpha + S_{IT'} - S_{IT}]_+ \quad T' \in K \tag{8}$$

$$L_{T2I-ITFG} = \max_{I'} [\alpha + S_{I'T} - S_{IT}]_+ \quad I' \in M \tag{9}$$

The full training objective of two-stage retrieval model is:

$$L = L_{ITCG} + L_{I2T-ITFG} + L_{T2I-ITFG} \tag{10}$$

3.3 Distilling After Training

To minimize the model size, we utilized a Transformer-based knowledge distillation method to compress TSCMR. Drawing from [14], this work employs a hierarchical distillation technique to distill the multi-head self-attention modules, feedforward neural network modules, and embedding layers of every layer in the model, which is shown in Fig. 3.

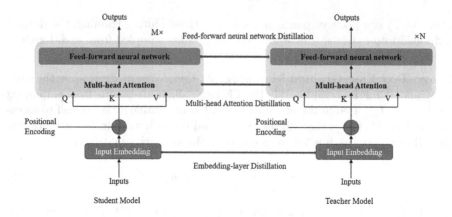

Fig. 3. The details of distillation

Embedding-Layer Distillation. The loss calculation for the embedding layer is as follows:

$$L_{embd} = MSE\left(E^S W_e, E^T\right) \tag{11}$$

where E^S and E^T respectively represent the embeddings of the student network and the teacher network. Since the embedding layer of the teacher network is usually smaller than that of the teacher model to reduce model size, the embedding of the student model is generally linearly transformed to project onto the space where the embedding of the teacher model is located. Finally, the mean squared error method is used to calculate the loss.

Transformer Encoder Distillation. We propose adopting the method of distillation every k layers for the Transformer encoder. Specifically, the loss is calculated every 3 layers when the teacher model consists of 12 layers while the student model has only 4 layers. Correspondingly, the first layer of the student model is aligned with the third layer of the teacher model, the second layer of the student model with the sixth layer of the teacher model, the third layer of the student model with the ninth layer of the teacher model, and the fourth layer of the student model with the twelfth layer of the teacher model. The loss of each Transformer encoder layer includes both the loss of the self-attention layer and the feedforward neural network layer.

The loss calculation of the self-attention layer follows the method below:

$$L_{attn} = \frac{1}{h} \sum_{i=1}^{h} MSE\left(A_i^S, A_i^T\right) \tag{12}$$

where h denotes the number of attention heads, A_i^S represents the attention score matrix of the i-th attention head in the student model, and A_i^T represents the attention score matrix of the i-th attention head in the teacher model.

The loss calculation method for the feedforward neural network layer is as follows:

$$L_{FFN} = MSE\left(H^S W_h, H^T\right) \tag{13}$$

where the matrices H^S and H^T refer to the hidden states of student and teacher networks respectively. Similar to embedding-layer distillation, the output of the student model is mapped to the same space as the output of the teacher network. This mapping enables the student model to learn from the teacher network and improve its performance.

Finally, by implementing the previously stated distillation objectives, we can calculate the overall distillation loss:

$$L = L_{embd} + L_{attn} + L_{FFN} \tag{14}$$

4   Experiments

This section introduces the datasets, evaluation metrics, and training process settings. The efficacy and efficiency of the cross-modal retrieval in TSCMR are evaluated. Moreover, we investigates the performance of TSCMR with the implementation of knowledge distillation in retrieval tasks, and the reduction in model size is also evaluated.

4.1   Datasets and Metric

This work employs two popular datasets, Microsoft COCO (MS-COCO) [21] and Flickr30K (F30K) [22], to train and test cross-modal retrieval tasks and investigate their effectiveness and efficiency. The MS-COCO dataset comprises 123,287 images, and each image has five corresponding texts. We utilize 113,287 images, 5,000 images, and 5,000 images for training, validation, and testing, respectively. The F30K dataset consists of 31,000 images, with five corresponding texts for each image. We select 29,000 images, 1,000 images, and 1,000 images for training, validation, and testing, respectively. For evaluation, this study uses Recall@K, a widely-used metric that precisely assesses the model's performance. The Recall@K value falls between 0 to 1 and indicates the proportion of appropriately identified positive samples in the model.

4.2   Settings

In the training of TSCMR, we use a image encoder consisting of a 4-layer transformer encoder and a 2-layer transformer encoder, and a text encoder consisting of a 6-layer BERT and a 2-layer transformer encoder. Image features and text features are projected into a common space of 1024 dimensions through a linear transformation for the final similarity calculation. In the experiment, we set the dropout rate to 0.1, use the Adam optimizer, set the epoch to 30, set the batch size of the MS-COCO dataset to 40, and set the batch size of F30K to 30. The learning rate is set to $1e-5$ during the first 20 epochs of training and $1e-6$ during the remaining 10 epochs. When selecting the top-k images and texts with high similarity scores before entering the second stage, the k value is set to 15 for the MS-COCO dataset and 10 for the F30K dataset. After completing the training of TSCMR, we performed knowledge distillation. In the image encoder, we use a combination of 2-layer transformer encoder and 1-layer transformer encoder, while in the text encoder, we use a combination of 3-layer BERT and 1-layer transformer encoder. The dimensions and hyperparameters are kept unchanged during model training.

4.3   Results and Analysis

We compare our TSCMR method against the following baselines:VSRN [23],CAMERA [24],PFAN [25],MMCA [26],and TERAN. For the MS-COCO dataset, we present the result on the 1k test set. For 1k images, we computed the result through five-fold cross-validation on the 5k test set while averaging the obtained results.

Table 1. Results on the MS-COCO dataset,on the 1k test set

Model	Image Retrieval			Text Retrieval			
	R@1	R@5	R@10	R@1	R@5	R@10	SpeedUp
VSRN	62.8	89.7	95.1	76.2	94.8	98.2	-
CAMERA	63.4	90.9	95.8	77.5	96.3	98.8	-
PFAN	61.6	89.6	95.2	76.5	96.3	99	-
MMCA	61.6	89.8	95.2	74.8	95.6	97.7	-
TERAN	65	91.2	96.4	77.7	95.9	98.6	1.0x
TSCMR-100	63.6	90.1	95.6	75.2	95.1	98.5	6.7x
TSCMR-300	**64.8**	**91.1**	**96.7**	**77.2**	**95.6**	**98.8**	**3.1x**
TSCMR-500	64.9	91.3	96.8	77.4	95.8	98.9	1.9x

Table 1 reports the results on the MS-COCO dataset. The result reveals that the recall value of our method has experienced a significant downfall particularly in image retrieval with a drop of over a point in Recall@1, and over two points in text processing, when k is fixed to 100. Despite our model furnishing a 6.7 times higher retrieval speed compared to TERAN's method at k=100, our recall value suffered a huge setback. Nevertheless, when k is 300, the recall accuracy closely approximates that TERAN while maintaining a good balance between efficacy and viability. At $k = 500$, there is a minor improvement in recall value, however, the inference speed is only 1.9 times faster than TERAN.

Table 2 demonstrates that selecting at k of 100 results in a significant drop in the recall value, particularly for text retrieval, similar to the MS-COCO dataset. At k of 300 provides a well-balanced performance between recall value and efficiency that is not significantly different from TERAN. Increasing the value of k

Table 2. Results on the F30K dataset

Model	Image Retrieval			Text Retrieval			
	R@1	R@5	R@10	R@1	R@5	R@10	SpeedUp
VSRN	54.7	81.8	88.2	71.3	90.6	96	-
CAMERA	58.9	84.7	90.2	76.5	95.1	97.2	-
PFAN	50.4	78.7	86.1	70	91.8	95	-
MMCA	54.8	81.4	87.8	74.2	92.8	96.4	-
TERAN	59.5	84.9	90.6	75.8	93.2	96.7	1.0x
TSCMR-100	57.6	83.1	90.2	72.7	92.5	96.2	6.7x
TSCMR-300	**59.2**	**84.8**	**90.7**	**74.9**	**93**	**96.6**	**3.2x**
TSCMR-500	59.4	85	90.9	75	93.1	96.8	2x

to 500 does not substantially improve the recall value, but it significantly slows down the inference speed when compared to k set at 300.

During the testing phase, we made multiple selections of the optimal value of K for the MS-COCO and F30k datasets. Ultimately, we found that selecting a K value around 33 % of the size of the test set achieved an optimal balance between effectiveness and efficiency.

Table 3. Results on the MS-COCO dataset,on the 1k test set

Model	Image Retrieval			Text Retrieval			
	R@1	R@5	R@10	R@1	R@5	R@10	model size
TSCMR-300	64.8	91.1	96.7	77.2	95.6	98.8	100%
KL-TSCMR-300	64.1	91	96.5	76.2	94.8	98.3	50%

After the completion of the training phase for the two-stage retrieval model, knowledge distillation was conducted on the MS-COCO dataset. Table 3 of the report indicates that while the recall rate decreased slightly after the application of knowledge distillation, the size of the model reduced by 50%. Overall, this is a commendable achievement, especially for devices with GPU memory limitations.

5   Conclusions and Future Works

This paper proposes a new model architecture TSCMR for cross-modal retrieval, which is different from TERAN. The model consists of two stages: a image-text coarse-grained matching stage, based on global feature extraction, to filter irrelevant content before image-text fine-grained matching between word and image regions. Moreover, knowledge distillation is employed to reduce the model size after the training of the retrieval model. The experimental results demonstrate that our model is capable of achieving outcomes comparable to those of TERAN on the MS-COCO 1K test set, with a 3.1x increase in inference speed and a 50% decrease in model size.

For the future work, the similarity calculation method has space for further improvement, and we plan to optimize it to enhance inference speed. We have currently tested our method on two datasets, and we intend to extend the testing to additional datasets in the future. To reduce model size, we will explore combining knowledge distillation, quantization, and pruning with our method.

Acknowledgments. The work reported in this paper is partially supported by NSF of Shanghai under grant number 22ZR1402000, the Fundamental Research Funds for the Central Universities under grant number 2232021A-08, State Key Laboratory of Computer Architecture (ICT,CAS) under Grant No. CARCHB 202118, Information Development Project of Shanghai Economic and Information Commission (202002009) and National Natural Science Foundation of China (No. 61906035).

References

1. Vaswani, A., Shazeer, N., Parmar, N., et al.: Attention is all you need. In: Advances in Neural Information Processing Systems, 30 (2017)
2. Li, L.H., Yatskar, M., Yin, D., et al.: Visualbert: a simple and performant baseline for vision and language. arXiv preprint arXiv:1908.03557 (2019)
3. Li, G., Duan, N., Fang, Y., et al.: Unicoder-vl: a universal encoder for vision and language by cross-modal pre-training. Proc. AAAI Conf. Artif. Intell. **34**(07), 11336–11344 (2020)
4. Qi, D., Su, L., Song, J., et al.: Imagebert: cross-modal pre-training with large-scale weak-supervised image-text data. arXiv preprint arXiv:2001.07966 (2020)
5. Lu, J., Batra, D., Parikh, D., et al.: Vilbert: pretraining task-agnostic visiolinguistic representations for vision-and-language tasks. In: Advances in Neural Information Processing Systems, 32 (2019)
6. Tan, H., Bansal, M., Lxmert: learning cross-modality encoder representations from transformers. arXiv preprint arXiv:1908.07490 (2019)
7. Huang, Z., Zeng, Z., Liu, B., et al.: Pixel-Bert: aligning image pixels with text by deep multi-modal transformers. arXiv preprint arXiv:2004.00849 (2020)
8. Ren, S., He, K., Girshick, R., et al.: Faster R-CNN: Towards real-time object detection with region proposal networks. In: Advances in Neural Information Processing Systems, 28 (2015)
9. Dosovitskiy, A., Beyer, L., Kolesnikov, A., et al.: An image is worth 16x16 words: Transformers for image recognition at scale. arXiv preprint arXiv:2010.11929 (2020)
10. Devlin, J., Chang, M.W., Lee, K., et al.: Bert: pre-training of deep bidirectional transformers for language understanding. arXiv preprint arXiv:1810.04805 (2018)
11. Messina, N., Amato, G., Esuli, A., et al.: Fine-grained visual textual alignment for cross-modal retrieval using transformer encoders. ACM Trans. Multimed. Comput. Commun. Appl. (TOMM) **17**(4), 1–23 (2021)
12. Hinton, G., Vinyals, O., Dean, J.: Distilling the knowledge in a neural network. arXiv preprint arXiv:1503.02531 (2015)
13. Freitag, M., Al-Onaizan, Y., Sankaran, B.: Ensemble distillation for neural machine translation. arXiv preprint arXiv:1702.01802 (2017)

14. Jiao, X., Yin, Y., Shang, L., et al.: Tinybert: distilling bert for natural language understanding. arXiv preprint arXiv:1909.10351 (2019)
15. Sanh, V., Debut, L., Chaumond, J., et al.: DistilBERT, a distilled version of BERT: smaller, faster, cheaper and lighter. arXiv preprint arXiv:1910.01108 (2019)
16. Sun, Z., Yu, H., Song, X., et al.: Mobilebert: a compact task-agnostic BERT for resource-limited devices. arXiv preprint arXiv:2004.02984 (2020)
17. Xia, M., et al.: Structured pruning learns compact and accurate models. In: Proceedings of the 60th Annual Meeting of the Association for Computational Linguistics (Volume 1: Long Papers), pp. 1513–1528 (2022)
18. Faghri, F., Fleet, D.J., Kiros, J.R., Fidler, S.: VSE++: improving visual-semantic embeddings with hard negatives. In: BMVC. BMV A Press, 12 (2018)
19. Karpathy, A., Fei-Fei, L.: Deep visual-semantic alignments for generating image descriptions. In: Proceedings of the IEEE Conference on Computer Vision and Pattern Recognition, pp. 3128–3137 (2015)
20. Lee, K.H., Chen, X., Hua, G., Hu, H., He, X.: Stacked cross attention for image-text matching. In: Proceedings of the European Conference on Computer Vision (ECCV), pp. 201–216 (2018)
21. Lin, T.-Y., et al.: Microsoft COCO: common objects in context. In: Fleet, D., Pajdla, T., Schiele, B., Tuytelaars, T. (eds.) Computer Vision – ECCV 2014: 13th European Conference, Zurich, Switzerland, September 6-12, 2014, Proceedings, Part V, pp. 740–755. Springer, Cham (2014). https://doi.org/10.1007/978-3-319-10602-1_48
22. Plummer, B.A., Wang, L., Cervantes, C.M., Caicedo, J.C., Hockenmaier, J., Lazebnik, S.: Flickr30k entities: Collecting region-to-phrase correspondences for richer image-to-sentence models. In: ICCV (2015)
23. Li, K., Zhang, Y., Li, K., Li, Y., Fu, Y.: Visual semantic reasoning for image-text matching. In: Proceedings of the IEEE International Conference on Computer Vision, pp. 4654–4662 (2019)
24. Qu, L., Liu, M., Cao, D., Nie, L., Tian, Q.: Context-aware multi-view summarization network for image-text matching. In: Proceedings of the 28th ACM International Conference on Multimedia, pp. 1047–1055 (2020)
25. Wang, Y., et al.: Position focused attention network for image-text matching. arXiv preprint arXiv:1907.09748 (2019)
26. Wei, X., Zhang, T., Li, Y., Zhang, Y., Wu, F.: Multi-modality cross attention network for image and sentence matching. In: Proceedings of the IEEE/CVF Conference on Computer Vision and Pattern Recognition, pp. 10941–10950 (2020)

Effi-Emp: An AI Based Approach Towards Positive Empathic Expressions

Rifat Hossain Rafi[1]([✉])⊙ and Donghai Guan[2]⊙

[1] University of Adelaide, Adelaide, Australia
rifathossain.rafi@student.adelaide.edu.au
[2] Nanjing University of Aeronautics and Astronautics, Nanjing, China
dhguan@nuaa.edu.cn

Abstract. Recently due to the advent of the Web, the ubiquity of search engines, and the widespread use of social media, the modern world has been witnessing technologies that can mimic human behavior. With the rise of Internet usage among youth and the prevalence of cyberbullying, mental health has become a global issue. It is one of the great convictions that newer technologies can also make a valuable contribution to creating an efficient and empathic model which will depend less on computation power and provide the same output as bigger models. This paper presents a novel approach to address a fundamental mental health aspect by leveraging knowledge distillation to generate positive and highly empathic feedback. The proposed method "Effi-Emp", is a hybrid approach that utilizes transformers and attention layers to generate empathic dialogues which closely mimic human dialogues. Sentence coherence and rate of change during sentence generation have been given special attention so that the model generates high-quality empathic sentences. By utilizing the last layer of the model Knowledge Distillation was achieved with about 20% improved run time in comparison with the State-of-Art model. Automatic and baseline evaluation methods demonstrate that the proposed model outperforms all other models by a decent margin.

Keywords: empathy · knowledge distillation · encoder · decoder · Effi-Emp

1 Introduction

1.1 Motivation

In this contemporary time, digitization can be observed with the introduction of the Web, the growing impact of search engines, and the spread of social media. In addition, some of the platforms are trying to mimic human behavior [8]. Nonetheless, social media platforms can be useful in terms of examining mental health and providing mental support as both of these issues are identified as global challenges, especially due to the rise of usage of the internet among

X. Yang et al. (Eds.): ADMA 2023, LNAI 14179, pp. 576–590, 2023.
https://doi.org/10.1007/978-3-031-46674-8_40

teenagers and increased cyberbullying as well. One of the critical components for mental support is "empathy," which is one of the complex concepts that entails caring about others, sharing, comprehending their emotions, and feeling motivated to help them. Evidence refers to the fact that a positive correlation in terms of treating depression in mental health assistance can be found with empathic interactions [17].

1.2 Challenges

Even though it is possible to describe empathy yet, traditional methods of detecting empathy are based on training and through peer review [7]. Most of the methods to detect empathy are generated from scratch. In terms of a few thousand sentences, it scales properly but in terms of millions of sentences these ideas are not feasible [26]. Research on transforming empathy from low to high has been conducted which replaces a certain context of the sentence instead of generating from scratch [18]. However, the number of them is quite a few and proved to be complex [2]. The second challenge is to transform lower empathy into higher empathy with the given expression from the user. The response posts might display warmth and affection, but they still lack the understanding of the feelings that are required to support individuals. Another challenge is to make the model lightweight so that it can work with less power-hungry workstations.

The proposed method Effi-Emp must qualify or accomplish certain specific goals to enhance discourse or relate to empathy in order to develop meaningful empathic phrases.

1.3 Solution and Contribution

To address this existing challenge, a novel approach Effi-Emp is introduced. The proposed model generates a dataset that is more empathic and categorizes the output of the response post. The model is then converted into a lightweight model to improve computational time. Our contribution to this experiment is given below :

1. Generating a cleaner and more empathic dataset which produces better results by using a certain percentage of drop model rate.
2. Relabelling output of generated sentences which are generated by respective models.
3. Implemented knowledge distillation to produce a lightweight version of the complex model to avoid dependencies on hardware and complex neural networks to a certain extent.

To be precise, Sect. 1 provides an overall view of the existing challenge, contributions are portrayed. Section 2 describes the previous works related to empathy and positive emotions. The procedures of the model, functions, and substitutions which were made to cope with hardware are described in Sect. 3. Experiments and results are described in Sect. 4 and Sect. 5, respectively. Finally, Sect. 6 displays an overall review of this experiment and concludes the research work in a future direction.

2 Related Works

This section describes earlier work on Natural Language Processing (NLP) correlation with empathy, creating empathy-based discourse, and knowledge distillation.

2.1 Mental Support and Empathy

Natural Language Processing (NLP) for online mental health support initiatives have mostly concentrated on studying conversational support seeking and giving strategies like context adaptability and answer diversity which is the key factors of this research [1]. Researchers have developed techniques for recognizing therapeutic acts, assessing counsellor's language development, extracting conversational engagement patterns, examining moderation, and spotting cognitive reorganisation in supportive talks [2]. Majumdar et al. stated that most techniques for generating empathetic communication have the propensity to facilitate empathic dialogues by using emotional grounding or emotion mimicry [13]. Castonguay et al. emphasised the conversational skill of empathy, which is essential in counselling and providing mental health care [4]. Hence, this paper heavily revolves around the key concept "empathy" which is one of the foundations of mental support.

This study is built on prior work to understand and create computational methods for identifying empathy in social platforms and text-based peer support. By utilising a hybrid empathic approach, this study has been made even more useful in order to elevate the level of empathy.

2.2 Dialogue Generation Based on Empathy

Vaswani et al. [24] introduced a simple network architecture named transformer which made a huge impact in terms of sentence generation. Model architecture is straightforward. In terms of dataset, they have used WMT 2014 English-German and WMT 2014 English-French datasets. By an estimated sequence length, sentence pairs were grouped together. The usage of self-attention mechanism is displayed.

In order to generate a conversational task, Jacky et al. [3] used two powerful GPUs which trained GPT2 model with 110 million parameters and stack of 12 decoding blocks which employs transformers and attention mechanism to produce semantically sounded text. DailyDialog (DD) dataset was infused with an emotional classifier named DeepMoji which exhibits an unbalanced distribution of emotions. Additionally, they stated that the larger models tend to perform better.

Wenxu et al. [21] used NLPCC2017 dataset which has annotated and labeled dataset with post and response. Wang et al. [25] suggested, combining Bidirectional long-short-term memory(Bi-Lstm) and attention. The emotional expression module will receive these labeled conversation data and respond with the appropriate emotion in accordance with the posting's sentiment [21].

EMPATHETICDIALOGUES (ED), the first empathetic conversation dataset with 25K conversations in 32 emotions, was proposed by Rashkin et al. [15]. Models trained on non-empathetic datasets were less empathic than conversational models, which were trained on the dataset's listener role.

FastText tries to resolve this by considering each word as a collection of subwords. It is more useful for morphologically rich languages where a given word may have a larger number of morphological forms, making it difficult to train good word embeddings [28].

Dataset were gathered from [3,15,21] for this experiment. The usage of word embedding was done to find a correlation between words. Extension of this work was accomplished further by implementing a pre-trained dataset in order to achieve and generate positive empathic sentences and label them into independent categories. Furthermore, label word mapping would be implemented using pre-trained model and a comparison of the legacy model would be displayed.

2.3 Knowledge Distillation

Knowledge distillation uses a certain amount of processing power which can be used on the low-end computational units which turns out to be one of the popular methods in this contemporary time [10]. The idea of teaching cheaper, smaller models (referred to as "students") to imitate more expensive ones (referred to as "teachers") is an old one that was first introduced based on model compression. Nearly right out of the gate, this method can be used with deep neural networks [5]. By distilling knowledge in both directions at each epoch, Zhang et al. trained a pair of models [29]. According to Tarvainen et al., training steps that average multiple student models tend to produce better-performing students [23]. Romero et al. suggested transferring knowledge using both the earlier layers and the logit one. They proposed a regressor to link the intermediate layers of the teacher and student in order to account for the width difference. [16]. Instead of going through the intermediate layer, fetching data from the last k-layer provides better output in terms of training and compressing the model [22].

Yu et al. used a shared representation of layers to address this issue, but it is not simple to select the right layer for matching [27]. Maroto et al. found out that even when both models are on par with each other, a pupil might occasionally outperform a teacher by a slight margin. This result has been linked to the presence of concealed information about the teacher's learned representations in its outputs (such as the similarity between classes), which the student can use more effectively than the original labels [14].

The goal of this research utilizes the usage of Knowledge Distillation(KD) and demonstrates how the student model compares favorably to the original teacher model.

3 Effi-Empathy-Gen

In this section, we first represent the Empathy Generative Architecture in Sect. 3.1 which represents the left section which is used for generating the sentences. In Sect. 3.2, the working mechanism of Rationale Identifier is provided which is used for identifying whether the sentence is rationale or non-rationale.

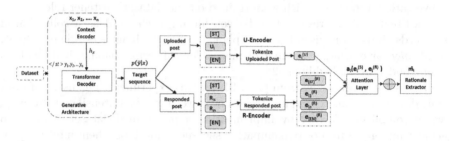

Fig. 1. Empathy Generator and Rationale identifier architecture uses two independently pre-trained RoBERTa$_{BASE}$ encoders for encoding uploaded post and response post respectively. We leverage attention between them for generating uploaded-context aware representation of the response post, used to perform rationale extraction.

3.1 Empathy Generator

Our empathy generation is based on the full transformer model [24], which consists of a single pair of encoder and decoder.

At first, input data will be tokenized and represented using $dx_1 .... dx_n$ based on previous utterances then encoded and stored into h_x vector by the context encoder. h_x will be used as an input to the decoder to generate the start symbol $</st>$ and tokens dy_1, dy_2 till dy_n. The Transformer decoder is trained to reduce the negative log-likelihood of the target sequence and uses the encoder output to predict a sequence of words $\bar{y}$. Through this model prediction of the next sequence of occurrence is possible.

Assuming a given dialogue context d_x of previous occurrences h_i concatenated and tokenized as dx_1 till dx_n followed by target sequence bar y. It will maximize the likelihood $p(\bar{y}|x)$ of producing the target sequence. Afterward, based on the given output path a new file would be generated and it will be used as an input for Rationale Identifier.

3.2 Rationale Identifier

Objective of the Model. Rationale Identifier needs to achieve a specific task. Assuming extracted rationale is a sub-sequence of words of w_i in response posts R_i. This sub-sequence can be represented as mask $mk_i = mk_{i1}, ... , mk_{in}$ over the words in R_i where $m_{ij}\epsilon\{rationale, non-rationale\}$.

Mechanism of the Model. Data generated from the previous model would have uploaded and corresponding response post. We can denote uploaded text U_i and its corresponding responded texts R_i. This model uses two independently pre-trained transformer encoders from RoBERTa$_{BASE}$.

Let $U_i = u_{i1}, ... , u_{in}$ be an uploaded post and $R_i = r_{i1},...,r_{in}$ be the responded post of the corresponded uploaded post. For the pair (U_i, R_i), our experiment determines rationale categorization such as empathy or non-empathy which would be determined by 1 and 0 respectively.

$$e_i^U = U - Encoder([ST], U_i, [EN]) \tag{1}$$

$$e_i^R = R - Encoder([ST], R_i, [EN]) \tag{2}$$

In order to change the model weights during training, cross-entropy loss is used. The goal is to reduce loss; hence, the better the model, the smaller the loss. At first softmax function is implemented to get the probabilities. Logits are transformed into probabilities by softmax. Cross-Entropy is used to calculate the deviation from the truth values by using the output probabilities u_i.

$$D(u, h) = \sum_i h_i \log(f(u_i)) \tag{3}$$

The prediction-making result of our softmax function is s. h_i is factorial. Only one entry in the true class's single hot-encoded label is a one; all the others are zeros.

$$a_i(e_i^R, e_i^U) = softmax(\frac{e_i^R e_i^U}{\sqrt{d}})e_i^U \tag{4}$$

In terms of Attention layer, single-headed attention layer is used. Queries can be denoted with response post encoding e_i^R. Usage of the attention layer takes place to generate uploaded-context aware representations of the response post. Key and values are the uploaded post encoding e_i^U. The hidden size of RoBERTa$_{BASE}$ is denoted as d where $d = 768$.

To achieve the final seeker context-aware representation of the response post, we sum encoded response e_i^R with its representation of attention layer $a_i(e_i^R, e_i^U)$ to generate residual mapping h_i^R.

To generate rationale or non-rationale predictions m_i, usage of final representation from the individual tokens in R_i $[h_i^R(r_{i1}, ...r_{in})]$ and passing it through a linear layer was done.

Using Fastext [27] word embedding, the correlation between the words has been portrayed. It can be observed that some of the words such as "sad, like, happy, angry" are some of the common words which can represent different things based on various scenarios. To automatically mark the emotional categories of posts and responses, which will be labeled as one of the categories of "happy, like, sad, disgusted, angry, and other," a pre-trained sentiment classifier (BiLSTM+Attention classifier) and NLPCC2017 dataset have been utilized with the final representation of the output.

3.3   Teacher Student Model

In terms of Knowledge Distillation, assuming the teacher model Effi-Emp is denoted as $f(x; \theta)$ and student model Effi-Emp(lite) is denoted as $s(x; \theta')$. Since we need to work on the k-layers we denote a model with k-layers which is Effi-Emp$_k$

Let $(kdx_i, kdy_i)^N_i = 1$ refers to training sample of N, where kdx$_i$ is the ith input instance for $f(x, \theta)$ and ground truth label is denoted as kdy_i. Our model starts by computing contextualized embedding $kdh_i = $ Effi-Emp$(kdx_i) \in \mathbb{R}^d$. d indicates the number of hidden layers in the model. Afterward, we will apply a softmax layer $kd\hat{y}$ into the embedding of Effi-Emp output to get classification [22].

$$kd\hat{y} = P(kdy_i|kdx_i) = softmax(\mathbf{W}kdh_i) \tag{5}$$

Over here $\mathbf{W}$ refers to the weight matrix which needs to be learned. Learned parameter θ^t can be calculated as :

$$\theta^t = \arg\min_{\theta} \sum_{i\epsilon[N]} L^t_{CE}(kdx_i, kdy_i; [\theta_{Effi-Emp}, W]) \tag{6}$$

Parameters of the teacher model are denoted as superscript t. N refers to the set of $1, 2, ...N$ and L_{CE} refer to cross-entropy loss function.

Objective of KD Models. The first step of Distilling Knowledge is to train the teacher model. Based on the softmax layer given above to compute any output layer of kdx_i the formula can be calculated as :

$$kd\hat{y} = P^t(kdy_i|kdx_i) = softmax\frac{(W * Effi - Emp(kdx_i; \theta^t)}{T} \tag{7}$$

From the equation above, $P^t(.|.)$ denotes as probability output from the Effi-Emp model. T refers to temperature which controls the reliability of the teacher's soft prediction and $kd\hat{y}_i$ is fixed as soft labels.

Similarly, for the student model, $P^s(.|.)$ can be denoted as the corresponding probability output for the student model and θ^s can be denoted as learned parameters. Let C be denoted as a set of class labels and c as a single class label. Then the formula can be calculated as :

$$L^s{}_{DS} = \sum_{i\epsilon[N]} \sum_{c\epsilon[C]} [P^t(kdy_i = c|kdx_i; \theta^t).\log P^s(kdy_i = c|kdx_i; \theta^s)] \tag{8}$$

In order to fine-tune the target task for the student model, cross-entropy loss is also included. Then the formula stands :

$$L^s{}_{CE} = \sum_{i\epsilon[N]} \sum_{c\epsilon[C]} [1[kdy_i = c].\log P^s(kdy_i = c|kdx_i; \theta^s)] \tag{9}$$

With α denoted as the hyper-parameter which is responsible for balancing cross-entropy and distillation loss, the final objective function can be calculated as :

$$L_{KD} = (1 - \alpha)L^s_{CE} + \alpha L_{DS} \tag{10}$$

Model Compression. The student model is cultivated to imitate the presentations only for the $[ST]$ tokens in the intermediate layers, following the intuition aforementioned that the $[ST]$ token is important in terms of predicting final labels. For an input kdx_i the outputs of the $[ST]$ tokens for all the layers are denoted as :

$$h_i = [h_{i1}, h_{i2}....h_{ik}] = Effi - Emp_k(kdx_i) \epsilon \mathbb{R}^{k \times d} \tag{11}$$

As for the intermediate layer which is to be distilled, it is denoted as I_{pt}. For our model, $I_{pt} = 2, 4, 6, 8, 10$. For this experiment $k = 5$, because the output from the last layer is connected to the softmax layer which is mentioned in Eq. 9. In addition, the teacher model introduces training loss which is defined as the mean-square loss between normalized hidden states :

$$L_T = \sum_{i=1}^{N} \sum_{j=1}^{M} \left\| \frac{h_i, I^t_{ptj}}{\left\| h_i, I^t_{ptj} \right\| 2} \right\| \tag{12}$$

In this equation, M denotes to the number of layers in the student network, and super scripts t in h refers to student and teacher parameters respectively. Like α, β is also used as a hyper-parameter. Combining Eq. 12 to the Eq. 10 we can get :

$$L_{KDN} = (1 - \alpha)L^s_{CE} + \alpha L_{DS} + \beta L_T \tag{13}$$

4   Experimental Setup

4.1   Selection of Dataset

For **Twitter** Dataset Gathering dataset proved to be time-consuming in comparison with other procedures. In order to fetch dataset [3] from Twitter, snscaper package was used. Based on time and given the specific parameter, 1.8 million tweets were gathered. For **Reddit** Dataset, A number of sub-communities, also known as subreddits, are hosted by Reddit (reddit.com), such as r/depression. We make use of threads from 55 Reddit subreddits dedicated to mental health [20]. Approximately 4 million interactions and 1 million threads are in this publicly available dataset. For in-domain pre-training, the entire dataset is utilized, and annotate a subset of 10 thousand interactions on empathy. **Empathy Dialogue (ED) Dataset** [15], which is publicly accessible through the ParlAI framework, consists of 24,850 conversations about a situation description that were collected from 810 different participants. Conversations were roughly divided into three partitions: 80% train, 10% validation, and 10% test.

4.2   Training Details and Experimental Setup

For Effi-Emp, the weights discovered by $RoBERTa_{BASE}$ are used to initialize both the U- and R-Encoders. We also implement a domain-adaptive pre-training method for the two encoders [9] so that they can adjust to conversational and mental health contexts. We employ the datasets of 1M seeker posts and 800K responder posts, respectively, provided from [20], for this additional pre-training of the two encoders. For pre-training, we employ the masked language modeling task (3 epochs, batch size = 8).

For knowledge distillation, the first k layers of parameters are being pre-trained with $RoBERTa_{BASE}$, where k ϵ (3, 6) are transformer layers, are used to initialize $Effi-Emp_k$. We set the number of hidden units in the final softmax layer at 768, the batch size at 32, and the number of epochs at 4 for all trials, with learning rates ranging from $(1e-5, 2e-5, 5e-5)$ in order to minimize the hyper-parameter search space. From all the parameters which need to be learned, we hard-coded the number of α and T to search over β and learning rate. We set $\alpha = 0.2, 0.5, 0.7$, temperature $T = 5, 10, 20$ and $\beta = 10, 100, 500, 100$. A grid search was performed over T, α and learning rate to achieve an accuracy rate.

For the experimental setup, all three of our dataset was divided into train, validation, and test sets 70%, 10%, and 20% respectively. Effi-Emp model was trained based on 4 epochs at a learning rate of $2e-5$, with a batch size of 32. The weight of empathy identification loss is set to 1 and weight of rationale extraction loss is set to 0.5.

5   Result and Discussion

The first subsection provides details about the baseline methods. Later sections provide the comparison data, results from automated, ablation evaluation, and computational results of lightweight models and the state-of-art models.

5.1   Baseline Methods

As the task of empathic rewriting has not been explored before, we compare it against baseline approaches [18]. Our baselines are:

1. PARTNER [18]: Empathic dialogue generation model, usage post-edit method.
2. RoBERTa [12]: RoBERTa has the same architecture as BERT, but uses a byte-level BPE as a tokenizer (same as GPT-2) and uses a different pre-training scheme
3. BERT [6]: Bi-directional transformer for pre-training over a lot of unlabeled textual data to learn a language representation
4. DialoGPT [30]: A large dialogue generation model, based on GPT-2 and pre-trained on Reddit conversations.
5. MIME [13]: An empathic dialogue generation model which exploits emotion mimicking while accounting for emotion polarity (positive or negative).
6. BART [11]: An encoder-decoder model for sequence-to-sequence language generation.

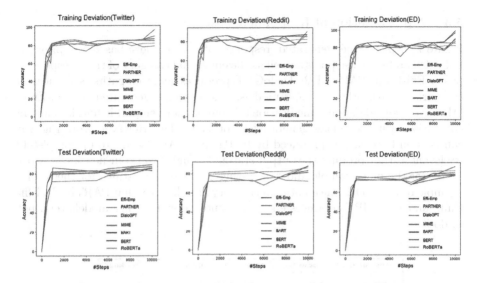

Fig. 2. Deviation of Training and Testing models based on different dataset

5.2 Baseline Comparison

Figure 2 portrays the deviation of training and testing scores for each dataset. Results from Table 1 outperforms other existing State-of-Art model in terms of comparison which was achieved using model dropout. In terms of the Twitter dataset Effi-Emp and PARTNER scored an accuracy rate of 89.04%, 85.81% for accuracy, and 87.34%, 82.98% for the f1 score respectively. Other State-of-Arts achieved over 78% in both f1 and accuracy scores. As for Reddit and ED dataset Effi-Emp scores accuracy rate of 87.51% and 86.06% and f1 score of 87.46% and 80.47% respectively. Other State-of-Art models produced 71% to 86% scores for both Reddit and ED dataset in terms of accuracy and f1 score.

Table 1. Table represents the comparison of accuracy between SOAT models and Effi-Emp for Twitter, Reddit, and ED datasets. The higher accuracy rate refers to the model performing better.

Model	Twitter		Reddit		ED	
	F1	Accuracy	F1	Accuracy	F1	Accuracy
Effi-Emp	**87.3498** ± 0.0248	**89.0381** ± 0.0239	**87.46982** ± 0.0023	**87.5093** ± 0.0192	**80.4765** ± 0.0159	**86.0901** ± 0.3333
PARTNER	82.9876 ± 0.0661	85.8083 ± 0.6389	82.0176 ± 0.2612	86.8575 ± 0.1812	78.8264 ± 0.0206	81.9999 ± 0.2110
DialoGPT	81.9218 ± 0.4129	85.6698 ± 0.3764	72.9042 ± 0.2119	77.4566 ± 0.1573	79.3314 ± 0.5333	80.0777 ± 0.0201
MIME	82.4193 ± 0.2544	87.0976 ± 0.2515	78.7572 ± 0.0073	83.6382 ± 0.0398	72.6551 ± 1.0913	78.2431 ± 0.0067
BART	79.3601 ± 0.0732	85.9751 ± 0.3828	76.8717 ± 0.3485	82.9012 ± 0.5107	69.2286 ± 0.4996	77.0181 ± 0.1762
BERT	78.0285 ± 0.0040	83.5121 ± 0.6127	80.9498 ± 0.0218	81.0986 ± 0.0118	70.5887 ± 0.0484	76.9276 ± 0.0394
RoBERTa	80.9082 ± 0.3185	83.7809 ± 0.2961	74.2186 ± 0.0042	76.8564 ± 0.0026	71.8699 ± 0.5561	77.9832 ± 0.3368

5.3   Representation of Responded Post

Figure 3 displays the generated response of the models based on previous uploaded and response posts as well as the emotional category of those responded posts. Effi-Emp, PARTNER, DialoGPT produces more empathic sentences in comparison with other existing models. The outcome generated by Effi-Emp, "Have you considered any medications?" sounds more empathic and humane in comparison with sentences generated by other models. However, from Fig. 3 it can be seen that output produced by BERT and BART seems to be completely irrelevant as they produce the output stating "Congrats on your new job", "Congrats" for the first example and "disgusting", "it's so disgusting" for the second example respectively. The emotional category also displays that PARTNER, Effi-Emp, and DialoGPT produces empathic sentence while other models tend not to do it.

Uploaded Post	Response Post	Model	Outcome	Emotional Category
I was happy an hour ago, an hour later I am sad, am I getting mad?	It really pains me to hear that you are struggling.	PARTNER	I hope you feel better	Empathic
		MIME	I care	Not Empathic
		BART	Congrats	Not Empathic
		BERT	Congrats on your new job	Not Empathic
		EFFI-EMP	Have you considered any medications ?	Empathic
		DialoGPT	I understand	Empathic
I have sore throat and swollen gums.	I don't want that to happen.	PARTNER	Take medications	Empathic
		MIME	I feel you	Empathic
		BART	Disgusting	Not Empathic
		BERT	It's so disgusting	Not Empathic
		EFFI-EMP	Consider visiting hospital ?	Empathic
		DialoGPT	I feel you	Empathic

Fig. 3. Comparison of responses based on Effi-Emp and State-of-Art models. The comparison shows either the sentence is rationale or non-rationale. Rationale posts are denoted as empathic posts and Non-Rationale posts are denoted as not empathic posts.

5.4   Automated and Ablation Results

A number of automated metrics are used in order to obtain preferred results such as coherence, changes in empathy, and diversity. Calculating empathic change using the classification model for empathy created in [19]. The model generates scores for empathy that range from 0 to 6. Using the scoring system created in [18], a coherence measuring method is implemented. For this experiment, assessing the number of changes between the original response and the rebuilt response using the edit rate was done [20]. Ablation results were evaluated based on the same metrics as automated metrics evaluation. It displays the results without an attention layer, no uploaded post, and no pre-training. In terms of experiments without using the attention layer, we concatenate uploaded and response posts and use concatenated representation as an input for the linear layer.

Table 2. Performance of Effi-emp and comparisons with SOAT baselines on the set of automatic metrics. Eff-Emp outperforms all baselines in terms of empathy change (↑), coherence (↑), and rate of changes(↓). (↑) indicates higher is better, (↓) indicates lower is better.

Model	Change in Empathy(↑)	Coherence (↑)	Rate of change (↓)
Effi-Emp	1.6890	0.7987	0.8381
PARTNER	1.6410	0.7961	0.8654
DialoGPT	1.3992	0.4887	1.8193
BERT	1.1381	0.2556	5.8923
MIME	1.3871	0.3897	1.9821
BART	0.7876	0.1958	2.9849
RoBERTa	0.7812	0.3917	2.0087

Table 2 portrays the outcome of different models and the new approach described in the paper. In terms of sentence coherence, Effi-Emp, and PARTNER performed better scoring 0.7987 and 0.7961 respectively. DialoGPT performed better than other existing models. Other models such as MIME, BART, and BERT scored less than 0.5. From Effi-Emp The rate of change is 0.8981 which is significantly lower than other models. For the change in empathy, both Effi-Emp and PARTNER produced respectable scores which are 1.6980 and 1.6410 respectively.

Ablation results in Table 3 imply the fact that the attention layer has less impact when it comes to sentence generation. However, if the model has to produce results without uploaded post or without prior pre-training, the produced results are significantly lower.

Table 3. Ablation results for empathy change (↑), coherence (↑), rate of changes(↓). Effi-Emp produces decent results without attention layer. However, the results are relatively low if the model is run without uploaded posts and pre-training. (↑) indicates higher is better, (↓) indicates lower is better.

Model	Twitter	Reddit	ED
	Change in Empathy(↑)	Coherence (↑)	Rate of change (↓)
Effi-Emp	1.6890	0.7987	0.8381
no attention layer	1.2833	1.6408	0.9866
no uploaded post	0.8912	3.5671	5.0781
no pre-training	0.6791	3.6512	5.3287

Table 4. Comparison of KD method with other State-of-art methods. We compare the computation time of SOAT baselines with the Lightweight model of Effi-emp. (↓) indicates lower is better for computational time. For accuracy (↑) indicates higher is better.

Model	Computational Time (↓)			Accuracy (↑)		
	Twitter	Reddit	ED	Twitter	Reddit	ED
Eff-Emp(student)	320	409	302	87.0988 ± 0.1836	86.2339 ± 0.1862	84.3411 ± 0.2404
Effi-Emp	380	460	340	89.0381	87.5093	86.0901
PARTNER	410	480	360	85.8083	86.8575	81.9999
BERT	430	500	390	83.5121	81.0986	76.9276
DialoGPT	466	550	410	85.6698	77.4566	80.0777
BART	472	588	418	85.9751	82.9012	77.0181
RoBERTa	486	590	420	83.7809	76.8564	77.9832
MIME	520	600	460	87.0975	83.6382	78.2431

5.5   Results for Knowledge Distillation

The result of Knowledge distillation plays an important role in terms of state-of-art model comparison. As it can be seen from Table 4 among all the models, the Original Effi-Emp model, PARTNER, and Effi-Emp(Student Model) models have the least computation time. Since the unit for computational time is minute in this experiment, Effi-Emp(Student Model) takes the lead with a computational time of 320 min, 409 min, and 302 min while running on Twitter, Reddit, and ED datasets respectively. It proves to be 17.5% to 20% faster in comparison with the State-of-Art models. Apart from that, the Lightweight version of Effi-Emp produces accuracy closer to the teacher model and it is around 1.14 % – 2% efficient in comparison with other State-of-Art Models.

6   Conclusion

Overall, mental health has emerged as one of the serious threats in contemporary society. Existing resources supporting mental health or illness tend to lack the most important aspects of support which is empathy. Without feeling empathic toward the peer solving this emerging problem would be an issue. Effi-Emp solves the problem by decoding the dataset into an empathic dataset that contains uploaded post and their corresponding response. Afterward, provides the required feedback if they are being rationale or non-rationale. Accuracy and f1 score display improvement of 5%–7% over the State-of-Art methods. Automated and ablations results portray the rate of change, coherence, and results without using pre-training which are satisfactory for Effi-Emp model. To run the model with fewer dependencies and fewer network layers, knowledge distillation is implemented. Student model displays improvement of 17%–20% over other existing methods. However, the model is trained with given parameters only. If the model is tested with a different dataset which is not modified specifically for empathy, it does not produce satisfactory scores. The most important aspect of Effi-Emp is that self-harm or insensitive words can be detected. Through extensive automated and baseline evaluations, it can be proven that Effi-emp is able to create and identify higher levels of empathic sentences in comparison with other methods. Nonetheless, with the invention of chatGPT and GPT3 empathy generation is aimed to reach great heights. In the near future, empathy generation might generate sentences that might be on par with human-generated empathic sentences.

References

1. Althoff, T., Clark, K., Leskovec, J.: Large-scale analysis of counseling conversations: an application of natural language processing to mental health. Trans. Assoc. Comput. Linguist. **4**, 463–476 (2016)
2. Buechel, S., Buffone, A., Slaff, B., Ungar, L., Sedoc, J.: Modeling empathy and distress in reaction to news stories. arXiv preprint arXiv:1808.10399 (2018)
3. Casas, J., Spring, T., Daher, K., Mugellini, E., Khaled, O.A., Cudré-Mauroux, P.: Enhancing conversational agents with empathic abilities. In: Proceedings of the 21st ACM International Conference on Intelligent Virtual Agents, pp. 41–47 (2021)
4. Castonguay, L.G., Hill, C.E.: How and why are some therapists better than others?: Understanding therapist effects. In: JSTOR (2017)
5. Cho, J.H., Hariharan, B.: On the efficacy of knowledge distillation. In: Proceedings of the IEEE/CVF International Conference on Computer Vision, pp. 4794–4802 (2019)
6. Devlin, J., Chang, M.W., Lee, K., Toutanova, K.: Bert: Pre-training of deep bidirectional transformers for language understanding. arXiv preprint arXiv:1810.04805 (2018)
7. Gibson, J., et al.: A deep learning approach to modeling empathy in addiction counseling. Commitment **111**(2016), 21 (2016)
8. Grewal, D., Hulland, J., Kopalle, P.K., Karahanna, E.: The future of technology and marketing: a multidisciplinary perspective (2020)
9. Gururangan, S., Marasović, A., Swayamdipta, S., Lo, K., Beltagy, I., Downey, D., Smith, N.A.: Don't stop pretraining: Adapt language models to domains and tasks. arXiv preprint arXiv:2004.10964 (2020)
10. Hinton, G., Vinyals, O., Dean, J.: Distilling the knowledge in a neural network. arXiv preprint arXiv:1503.02531 (2015)
11. Lewis, M.,et al.: Bart: Denoising sequence-to-sequence pre-training for natural language generation, translation, and comprehension. arXiv preprint arXiv:1910.13461 (2019)
12. Liu, Y., et al.: Roberta: A robustly optimized BERT pretraining approach. arXiv preprint arXiv:1907.11692 (2019)
13. Majumder, N., et al.: Mime: Mimicking emotions for empathetic response generation. arXiv preprint arXiv:2010.01454 (2020)
14. Maroto, J., Ortiz-Jiménez, G., Frossard, P.: On the benefits of knowledge distillation for adversarial robustness. arXiv preprint arXiv:2203.07159 (2022)
15. Rashkin, H., Smith, E.M., Li, M., Boureau, Y.L.: Towards empathetic open-domain conversation models: a new benchmark and dataset. arXiv preprint arXiv:1811.00207 (2018)

16. Romero, A., Ballas, N., Kahou, S.E., Chassang, A., Gatta, C., Bengio, Y.: FitNets: Hints for thin deep nets. arXiv preprint arXiv:1412.6550 (2014)
17. Santamaría-García, H., et al.: Empathy for other suffering and its mediators in mental health professionals. Sci. Reports **7**(1), 1–13 (2017)
18. Sharma, A., Lin, I.W., Miner, A.S., Atkins, D.C., Althoff, T.: Towards facilitating empathic conversations in online mental health support: a reinforcement learning approach. In: Proceedings of the Web Conference 2021, pp. 194–205 (2021)
19. Sharma, A., Miner, A.S., Atkins, D.C., Althoff, T.: A computational approach to understanding empathy expressed in text-based mental health support. arXiv preprint arXiv:2009.08441 (2020)
20. Sharma, E., De Choudhury, M.: Mental health support and its relationship to linguistic accommodation in online communities. In: Proceedings of the 2018 CHI Conference on Human Factors in Computing Systems, pp. 1–13 (2018)
21. Shi, W., Shang, Z., Bao, S., Li, G.: Text generation based on empathetic dialogues between nurses and patients. In: 2019 International Conference on Computer, Network, Communication and Information Systems (CNCI 2019), pp. 569–578. Atlantis Press (2019)
22. Sun, S., Cheng, Y., Gan, Z., Liu, J.: Patient knowledge distillation for BERT model compression. arXiv preprint arXiv:1908.09355 (2019)
23. Tarvainen, A., Valpola, H.: Mean teachers are better role models: weight-averaged consistency targets improve semi-supervised deep learning results. In: Advances in Neural Information Processing Systems 30 (2017)
24. Vaswani, A., et al.: Attention is all you need. In: Advances In Neural Information Processing Systems 30 (2017)
25. Wang, Y., Huang, M., Zhu, X., Zhao, L.: Attention-based lstm for aspect-level sentiment classification. In: Proceedings of the 2016 Conference on Empirical Methods in Natural Language Processing, pp. 606–615 (2016)
26. Xiao, B., Can, D., Georgiou, P.G., Atkins, D., Narayanan, S.S.: Analyzing the language of therapist empathy in motivational interview based psychotherapy. In: Proceedings of The 2012 Asia Pacific Signal and Information Processing Association Annual Summit and Conference, pp. 1–4. IEEE (2012)
27. Yu, R., Li, A., Morariu, V.I., Davis, L.S.: Visual relationship detection with internal and external linguistic knowledge distillation. In: Proceedings of the IEEE International Conference on Computer Vision, pp. 1974–1982 (2017)
28. Zeberga, K., Attique, M., Shah, B., Ali, F., Jembre, Y.Z., Chung, T.S.: A novel text mining approach for mental health prediction using Bi-LSTM and BERT model. Comput. Intell. Neurosci. **2022** (2022)
29. Zhang, Y., Xiang, T., Hospedales, T.M., Lu, H.: Deep mutual learning. In: Proceedings of the IEEE Conference on Computer Vision and Pattern Recognition, pp. 4320–4328 (2018)
30. Zhang, Y., et al.: Dialogpt: Large-scale generative pre-training for conversational response generation. arXiv preprint arXiv:1911.00536 (2019)

Industry Track Papers

Research on Image Segmentation Algorithm Based on Level Set

Ping Wu, Mingkun Zhang$^{(\boxtimes)}$, Qing Yue, and Zhimin Gao

AVIC Shenyang Aircraft Design and Research Institute, 40 Tawan Street, Huanggu District, Shenyang City, Liaoning Province, China
zhangmk92@163.com

Abstract. Digital image processing has garnered significant attention and extensive research in both civilian and military domains. Image segmentation serves as a fundamental component of image processing, directly influencing the outcomes of subsequent image processing steps. Hence, it holds immense research value to efficiently and accurately extract target regions from diverse and complex images. In recent years, numerous image segmentation algorithms have been proposed. However, traditional segmentation methods fall short in fully addressing challenges like topological changes in image contours, limiting their applicability across various image types. To enhance the adaptability and universality of image segmentation algorithms, this study employs the level set method to transform the curve evolution problem into a numerical solution involving partial differential equations. This paper initially surveys the research trends in level set-based image segmentation. It then elucidates the characteristics of several classical level set image segmentation algorithms, including CV, RSF, LGIF, LRCV, and RABFLR, while also providing recursive methodologies for the LGIF model, LRCV model, and RABFLR model. Subsequently, an enhanced LGIF model is proposed, which incorporates bilateral filtering for image preprocessing and effectively integrates global and local information. Experimental results demonstrate a significant improvement in segmentation effectiveness compared to existing algorithms.

Keywords: level set · image segmentation · active contour model

1 Introduction

1.1 Overview of Image Segmentation

Image segmentation is an important research direction of computer vision [1], as it significantly impacts the outcomes of subsequent image processing tasks [2]. Therefore, efficiently processing diverse and complex images while accurately extracting the target regions holds immense research significance.

Among the essential image segmentation methods [3], level set has gained widespread adoption in the field of medical image segmentation [4]. The level set

---

P. Wu and M. Zhang—Contributed equally to the paper as co-first authors.

X. Yang et al. (Eds.): ADMA 2023, LNAI 14179, pp. 593–606, 2023.
https://doi.org/10.1007/978-3-031-46674-8_41

image segmentation technique employs evolving closed curves to indirectly segment the images, transforming the intricate curve evolution process into a numerical solution problem involving partial differential equations [5].

1.2 The Classical Model of Level Set

The CV model [6] is a classical level set approach that utilizes energy functional minimization for curve evolution, demonstrating favorable performance in noisy image segmentation. However, the CV model is less suitable for images with complex backgrounds and uneven grayscale. To address this limitation, the RSF model [7] introduces the Gaussian kernel function as the energy term, leveraging scalable regions to process local image information. Nonetheless, the RSF model heavily relies on the initial position of the evolving curve and exhibits suboptimal performance in noise image processing. The LGIF model [8], building upon the CV and RSF models, combines global energy data elements with local energy data elements to exploit both global and local information within an image. This integration yields commendable convergence rates and segmentation accuracy [9]. However, the LGIF model entails more complex calculations and is sensitive to the initial contour of the evolution curve and image noise [10]. Taking inspiration from the CV model, the RABFLR model filters the image using an adaptive bilateral filter before iteratively minimizing the defined energy functional [11]. This approach drives the evolution curve towards approximating the target boundary. The RABFLR model exhibits strong noise reduction capabilities but proves less effective for images with uneven intensity.

To mitigate the sensitivity to image noise and enhance segmentation efficiency [12], this paper proposes an improved LGIF model. Building upon the LGIF model, the image undergoes preprocessing, including noise reduction through the addition of a bilateral filter. This modification ensures the model maintains segmentation speed while reducing its sensitivity to noise-containing images. The proposed model, along with several classical models, is employed to segment real and composite images. The results demonstrate the effectiveness of the proposed model in accurately segmenting images with uneven intensity.

2   Related Work

2.1   LGIF Model

To enhance the image segmentation performance for images characterized by complex backgrounds and uneven grayscale, as well as reduce reliance on the nature of the evolution curve, the LGIF model incorporates a fusion of global energy data elements and local energy data elements. The weights of the global energy fitting items and local energy fitting items are adjusted using a weight coefficient. This adjustment ensures the comprehensive utilization of both global and local image information during the evolution process.

The data fitting energy functional of LGIF model is defined as follows:

$$E(\phi, f_1, f_2, c_1, c_2) = (1 - \omega)\left(\int \left(\int K(x - y)(I(y) - f_1(x))^2 H_\varepsilon(\phi(y)))\right)dy\right)dx$$

$$+ \int \left(\int K(x - y)(I(y) - f_2(x))^2 (1 - H_\varepsilon(\phi(y)))dy\right)dx\right) + \tag{1}$$

$$\omega\left(\lambda_1 \int |I - c_1|^2 H_\varepsilon(\phi)dx + \lambda_2 \int |I - c_2|^2 (1 - H_\varepsilon(\phi))dx\right)$$

In Formula (1), the weight parameter is a constant value employed in the image segmentation process. The formula harnesses both global and local image information to facilitate the evolution of the initial curve towards the target region.

To enhance the accuracy of level set function calculations, Formula (2) introduces the distance penalty function as a regularization term. The formulation is as follows:

$$P(\phi) = \int_\Omega \frac{1}{2}(|\nabla\phi(x)| - 1|)^2 dx \tag{2}$$

In order to maintain smoothness throughout the entire evolution process, an additional length term is incorporated into the energy functional, as depicted in Formula (3):

$$L(\phi) = \int_\Omega |\nabla H(\phi(x))| dx \tag{3}$$

The complete energy functional definition is shown in Eq. (4):

$$F(\phi, f_1, f_2, c_1, c_2) = E(\phi, f_1, f_2, c_1, c_2) + \mu P(\phi) + \nu L(\phi) \tag{4}$$

In Formula (4), the level set function φ is kept unchanged, and f_1, f_2, c_1 and c_2 are minimized, as shown in Formula (5–8):

$$f_1(x) = \frac{\int K(y - x)I(y)H(\phi(y))dy}{\int K(y - x)H(\phi(y))dy} \tag{5}$$

$$f_2(x) = \frac{\int K(y - x)I(y)(1 - H(\phi(y)))dy}{\int K(y - x)(1 - H(\phi(y)))dy} \tag{6}$$

$$c_1(\phi) = \frac{\int_\Omega I(x)H_\varepsilon(\phi)dx}{\int_\Omega H_\varepsilon(\phi)dx} \tag{7}$$

$$c_2(\phi) = \frac{\int_\Omega I(x)(1 - H_\varepsilon(\phi))dx}{\int_\Omega (1 - H_\varepsilon(\phi))dx} \tag{8}$$

The expression of the level set function with respect to time is obtained by using variational method and gradient descending flow, as shown in Formula (9):

$$\frac{\partial\phi}{\partial t} = -\delta_\varepsilon(\phi)(F_1 + F_2) + \mu\left(\nabla^2\phi - div\left(\frac{\nabla\phi}{|\nabla\phi|}\right)\right) + \nu\delta_\varepsilon(\phi)div\left(\frac{\nabla\phi}{|\nabla\phi|}\right) \tag{9}$$

In Formula (9), Fi is defined as shown in Formula (10 and 11):

$$F_1 = (1 - \omega)\left(\int K_\sigma(x - y)(I(y) - f_1(x))^2 - \int K_\sigma(x - y)(I(y) - f_1(x))^2 \right) \quad (10)$$

$$F_2 = \omega\left(\lambda_1(I - c_1)^2 - \lambda_2(I - c_2)^2 \right) \quad (11)$$

In the LGIF model, the local fitting term plays a dominant role near the target boundary, attracting the curve contour around the desired region. On the other hand, the global intensity fitting term incorporates global image information and becomes more influential when the curve is far from the target area, thereby enhancing the model's robustness. Additionally, the LGIF model can effectively handle images with uneven intensity and is not constrained by the initial contour.

However, the LGIF model encounters certain challenges. Firstly, the determination of optimal weight parameters directly impacts the final segmentation results, yet the ideal values may vary for each image, resulting in significant computational overhead. Secondly, the LGIF model employs the fitting energy from the CV model for its global intensity fitting component, leading to suboptimal segmentation outcomes when processing images with similar target areas and backgrounds. The bilateral filter used in this context is defined as follows:

$$h(x) = k_r^{-1}(x) \int_\Omega f(\xi)c(\xi, x)s(f(\xi), f(x))d\xi \quad (12)$$

The regularization term is defined as:

$$k(x) = \int_\Omega c(\xi, x)s(f(\xi), f(x))d\xi \quad (13)$$

where $c(\xi, x)$ represents the geometric proximity between the local region centered on each point x and its nearby point ξ; $s(f(\xi), f(x))$ is an unbiased function representing the luminosity similarity between the central pixel point x and its neighboring point ξ. In the above formula, c(.) And s(.) Are defined as follows:

$$c(\xi, x) = \exp\left(-\frac{1}{2}\left(\frac{\|\xi - x\|}{\sigma_d} \right)^2 \right) \quad (14)$$

$$s(\xi, x) = \exp\left(-\frac{1}{2}\left(\frac{\|f(\xi) - f(x)\|}{\sigma_r} \right)^2 \right) \quad (15)$$

The standard bilateral filter may encounter challenges when accurately estimating edge gradient information in images that contain noise. To address this issue, the RABFLR model employs a range-based adaptive bilateral filter for processing noisy images. The key distinction between this filter and the basic bilateral filter is the definition of the s(.) function, which is as follows:

$$s_r(\xi, x, \zeta) = \alpha \exp\left(-\frac{1}{2}\left(\frac{\|f(\xi) - f(x) - \zeta(x)\|}{\sigma_r} \right)^2 \right) + k \quad (16)$$

$$\zeta(x) = \begin{cases} |f(x) - \text{Mean}(\Omega_y)|, & |x - y| \le \rho \\ = 0, & \text{otherwise} \end{cases} \tag{17}$$

In the above equation, α and k are two positive parameters, and Ω_y represents the pixel set in the $(2N + 1) \times (2N + 1)$ window of the pixel center point x. Mean(Ω_y) represents the average value of the pixels in Ω_y. ρ is a constant that controls the size of the local neighborhood. The function $\zeta(x)$ is range-based and designed to ensure that the adaptive bilateral filter effectively preserves image edges and exhibits robustness when processing noisy images.

The model defines the following data-driven local area energy functional:

$$\mathcal{E}_x(C, f_1, f_2) = \sum_{i=1}^{2} \lambda_i \int_{\Omega_i} K_\rho(x - y)|J(y) - f_i(x)|^2 dy \tag{18}$$

$$K_\rho(x - y) = \begin{cases} = 1, & |x - y| \le \rho \\ = 0, & \text{otherwise} \end{cases} \tag{19}$$

where λ_1 and λ_2 are two constant parameters, respectively. The functions $f_1(x)$ and $f_2(x)$ approximate the image intensity of x processed with a radius ρ in the region centered at point $J(x)$, respectively. $K_\rho(.)$ is an eigenfunction of the region O_x: |x-y| $\le \rho$.

To obtain the target boundary, the objective is to minimize the energy functional E across all points x in the entire image region. To ensure a smooth evolution of the curve, a length penalty term is added to the energy functional of the model. The overall energy functional is defined as follows:

$$E(C, f_1(x), f_2(x)) = \int_\Omega (C, f_1(x), f_2(x)) + v|C| \tag{20}$$

In order to deal with the topological changes of curves in the image, the evolution curve C is implicitly represented by the level set function. Formula (21) can be redefined as:

$$E(\phi, f_1(x), f_2(x)) = \lambda_1 \int \left(\int K_\rho(x, y)|J(y) - f_1(x)|^2 H'(\phi(y)dy) \right) dx$$

$$+ \lambda_2 \int \left(\int K_\rho(x, y)|J(y) - f_2(x)|^2 (1 - H'(\phi(y))dy) \right) dx \tag{21}$$

$$+ \int \delta'|\nabla\phi|dx + \mu \mathcal{R}_p(\phi)$$

where $H'(.)$ is the approximate estimate of *Heaviside* function, and $\delta'(.)$ is the approximate *Dirac* function, which is defined as follows:

$$H'(x) = \frac{1}{2} \cos\left(\frac{\pi}{2} - a \tan\left(\frac{x - \varepsilon}{\varepsilon} \right) \right) + \frac{1}{2} \tag{22}$$

$$\delta'(x) = \frac{1}{2} \sin\left(\frac{\pi}{2} - a \tan \frac{x - \varepsilon}{\varepsilon} \right) \cdot \frac{\varepsilon}{\varepsilon^2 + (x - \varepsilon)^2} \tag{23}$$

A regularization term is constructed to maintain the regularity and smoothness of curve evolution, which is defined as follows:

$$\mathcal{R}_p(\phi) = \int_\Omega p(|\nabla\phi(x)|)dx \tag{24}$$

$$p(s) = \begin{cases} \frac{1}{2}s^2(s^2-1)^2, & 0 \leq s \leq 1 \\ \frac{1}{2}(s-1)^2, & s \geq 1 \end{cases} \tag{25}$$

In the above equation, the regularization term $p(.)$ is denoted as $Rp(.)$ and serves the purpose of not only smoothing the curve but also maintaining it as a signed distance function throughout its evolution. Consequently, the energy functional incorporating the regularization term is expressed as Formula (26):

$$E(\phi, f_1(x), f_2(x)) = \lambda_1 \int \left(\int K_\rho(x,y)|J(y) - f_1(x)|^2 H'(\phi(y)dy) \right)dx$$

$$+\lambda_2 \int \left(\int K_\rho(x,y)|J(y) - f_2(x)|^2(1 - H'(\phi(y))dy) \right)dx \tag{26}$$

$$+ \int \delta'|\nabla\phi|dx + \mu \int_\Omega p(|\nabla\phi(x)|)dx$$

The gradient descent method is employed to solve the minimum energy functional described in Formula (26). Initially, the functionals for $f_1(x)$ and $f_2(x)$ in Formula (26) are derived while keeping the level set function ϕ unchanged. Through the variational method, the following Euler equations for $f_1(x)$ and $f_2(x)$ are obtained:

$$\int K_\rho(x-y)M_i^\varepsilon(\phi(y))(J(y) - f_i(x))^2 dy = 0, i = 1, 2 \tag{27}$$

In the above equation, $M_1^\varepsilon(\cdot) = H'(\cdot)$ and $M_2^\varepsilon(\cdot) = 1 - H'(\cdot)$ represent parameters. Through the derivation of $f_i(x)$, the weighted average strength of the neighborhood around the central point x is obtained, with the size determined by the radius of the area C. This yields the expression for $f_i(x)$:

$$f_i(x) = \frac{K_\rho(x) * [M_i^\varepsilon(\phi(x))J(x)]}{K_\rho(x) * M_i^\varepsilon(\phi(x))}, \quad i = 1, 2 \tag{28}$$

Therefore, the gradient downflow looks like this:

$$\frac{\partial\phi}{\partial t} = -\delta'(\phi)(\lambda_1 d_1 - \lambda_2 d_2) + v\delta'(\phi)\text{div}\left(\frac{\nabla\phi}{|\nabla\phi|}\right) + \mu\text{div}(d_p(|\nabla\phi|)\nabla\phi) \tag{29}$$

In Formula (29), the first term represents the data fitting term, The second term corresponds to the length smoothing term,the last term denotes the regularization term. The variable d_i is defined as follows:

$$d_i(x) = \int K_\rho(y-x)|J(x) - f_i(y)|^2 dy, \quad i = 1, 2 \tag{30}$$

In summary, when applying the RABFLR model for image segmentation, the first step involves using an adaptive bilateral filter to preprocess the image. Subsequently, an iterative process is employed to minimize the defined energy functional and guide the evolution curve towards the desired boundary.

Compared to the RSF model, the RABFLR model demonstrates superior anti-noise capabilities through the utilization of range-based adaptive bilateral filters and the incorporation of local intensity clustering characteristics. Consequently, it successfully segments images containing noise. However, the accuracy of the RABFLR model diminishes when confronted with images exhibiting uneven intensity and complex segmentation structures. Moreover, since the RABFLR model solely relies on local image information during the segmentation process, its segmentation speed is comparatively slower.

3 Improved LGIF Model

The LGIF model combines the global energy data item from the CV model and the local energy data item from the RSF model, thereby incorporating both global and local information of the image. This integration enhances the convergence speed while maintaining a high segmentation accuracy. However, the model exhibits sensitivity to images with noise due to the absence of filtering. In contrast, the RABFLR model improves upon the RSF model by employing an adaptive bilateral filter for image preprocessing. This filter effectively reduces noise while preserving edge information. Nonetheless, the RABFLR model's reliance on local information leads to slower segmentation speeds.

In this study, a preprocessing step that applies a bilateral filter to the input image was proposed, effectively eliminating image noise and reducing the model's sensitivity to noise-containing images without sacrificing segmentation speed. By introducing the global energy fitting item into the local fitting data item and incorporating a weight coefficient to adjust the balance between global and local energy fitting during curve evolution, we ensure that both global and local information are considered in the image processing stage. As a result, the model achieves a higher segmentation accuracy while maintaining a fast convergence speed, ultimately leading to improved segmentation outcomes.

Bilateral filters used in this model are defined as follows:

$$h(x) = k_r^{-1}(x) \int_\Omega f(\xi)c(\xi, x)s(f(\xi), f(x))d\xi \tag{31}$$

where the regularization term is defined as:

$$k(x) = \int_\Omega c(\xi, x)s(f(\xi), f(x))d\xi \tag{32}$$

In the given equation, the term $c(\xi, x)$ represents the geometric proximity between the local region centered on each point x and its neighboring point ξ. Meanwhile, $s(f(\xi), f(x))$ is an unbiased function that quantifies the luminosity similarity between the central pixel point x and its neighboring point ξ. . The definitions of c(.) and s(.) are as follows:

$$c(\xi, x) = \exp\left(-\frac{1}{2}\left(\frac{\|\xi - x\|}{\sigma_d}\right)^2\right) \tag{33}$$

$$s_r(\xi, x, \zeta) = \alpha \exp\left(-\frac{1}{2}\left(\frac{\|f(\xi) - f(x) - \zeta(x)\|}{\sigma_r}\right)^2\right) + k \tag{34}$$

$$\zeta(x) = \begin{cases} |f(x) - \mathrm{Mean}(\Omega_y)|, & |x - y| \le \rho \\ = 0, & \text{otherwise} \end{cases} \tag{35}$$

where α and k are two positive parameters, and Ω_y represents the pixel set in the $(2N + 1) \times (2N + 1)$ window of the pixel center point x. $\mathrm{Mean}(\Omega_y)$ represents the average value of pixels in Ω_y. ρ is a constant that controls the size of the local neighborhood. $\zeta(x)$ is a range-based function.

The energy functional of this model incorporates the global energy fitting term into the local fitting data term. Additionally, a weight coefficient is defined to adjust the relative importance of the global energy fitting term and the local energy fitting term during the curve evolution process.

The data fitting energy functional is defined as follows:

$$E(\phi, f_1, f_2, c_1, c_2) = (1 - \theta)\left(\int \left(\int K_\sigma(x - y)(I(y) - f_1(x))^2 H_\varepsilon(\phi(y))\right)dy\right)dx$$

$$+ \int \left(\int K_\sigma(x - y)(I(y) - f_2(x))^2(1 - H_\varepsilon(\phi(y)))dy\right)dx\bigg)$$

$$+ \theta\left(\lambda_1 \int |I - c_1|^2 H_\varepsilon(\phi)dx + \lambda_2 \int |I - c_2|^2(1 - H_\varepsilon(\phi))dx\right) \tag{36}$$

In the given equation, the weight parameter, denoted by θ, is a constant value. This formula effectively integrates both global and local information from the image to guide the evolution of the curve towards the desired target region.

To ensure that the level set function remains consistent throughout the iterative process and to maintain a smooth and coherent curve evolution, a distance penalty function is introduced as a regularization term in the energy functional.

$$P(\phi) = \int_\Omega \frac{1}{2}(|\nabla\phi(x)| - 1|)^2 dx \tag{37}$$

In order to ensure that the curve is always smooth during the whole evolution process, a length term is added to the energy functional, which is defined as follows:

$$L(\phi) = \int_\Omega |\nabla H(\phi(x))| dx \tag{38}$$

The complete energy functional is defined as follows:

$$F(\phi, f_1, f_2, c_1, c_2) = E(\phi, f_1, f_2, c_1, c_2) + \mu P(\phi) + \nu L(\phi) \tag{39}$$

Keep the level set function φ unchanged in the above formula, and minimize f_1, f_2, c_1 and c_2, then the following expressions can be obtained:

$$f_1(x) = \frac{\int K_\sigma(y - x)I(y)H(\phi(y))dy}{\int K_\sigma(y - x)H(\phi(y))dy} \tag{40}$$

$$f_2(x) = \frac{\int K_\sigma(y-x)I(y)(1-H(\phi(y)))dy}{\int K_\sigma(y-x)(1-H(\phi(y)))dy} \tag{41}$$

$$c_1(\phi) = \frac{\int_\Omega I(x)H_\varepsilon(\phi)dx}{\int_\Omega H_\varepsilon(\phi)dx} \tag{42}$$

$$c_2(\phi) = \frac{\int_\Omega I(x)(1-H_\varepsilon(\phi))dx}{\int_\Omega (1-H_\varepsilon(\phi))dx} \tag{43}$$

Then using variational method and gradient downflow, the level set is obtained as follows:

$$\frac{\partial \phi}{\partial t} = -\delta_\varepsilon(\phi)(F_1 + F_2) + \mu\left(\nabla^2\phi - div\left(\frac{\nabla\phi}{|\nabla\phi|}\right)\right) + v\delta_\varepsilon(\phi)div\left(\frac{\nabla\phi}{|\nabla\phi|}\right) \tag{44}$$

F_i is defined as follows:

$$F_1 = (1-\omega)\left(\int K_\sigma(x-y)(I(y)-f_1(x))^2 - \int K_\sigma(x-y)(I(y)-f_1(x))^2\right) \tag{45}$$

$$F_2 = \theta\left(\lambda_1(I-c_1)^2 - \lambda_2(I-c_2)^2\right) \tag{46}$$

4 Experiments and Results

This section evaluates the reliability and effectiveness of the proposed model using medical images, comparing its performance against the CV model, RSF model, LGIF model, LRCV model, and RABFLR model. The experimental results substantiate the effectiveness of the proposed model.

4.1 Dataset and Evaluation Metrics

To validate the efficacy of the proposed model, this study assesses the segmentation results by comparing them with the standard segmentation results using the difference method. Three skin cancer medical images are employed for testing purposes, enabling an evaluation of the segmentation outcomes produced by each model. This comprehensive analysis aims to identify the strengths and limitations of the proposed model.

This paper employs several evaluation metrics to assess the experimental results, including the Dice coefficient, Jaccard similarity, accuracy, recall, and precision. These indices are defined as follows [13]:

Precision is calculated by dividing the number of pixels that are correctly predicted as positive by the total number of pixels predicted as positive.

$$precision = TP/(TP + FP) \tag{47}$$

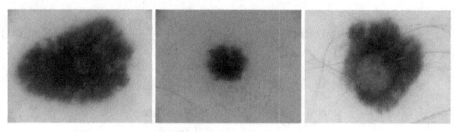

Fig. 1. .   **Fig. 2.** .   **Fig. 3.** .

Recall is the number of pixels that are predicted to be positive and are actually positive divided by the number of pixels that are actually positive

$$recall = TP/(TP + FN) \tag{48}$$

Accuracy is predicting the correct number of pixel points divided by all pixel points:

$$accuracy = (TP + TN)/(TP + TN + FP + FN) \tag{49}$$

Jaccard coefficient is the number of pixels in the intersection of segmentation result and ground truth image set divided by the number of pixels in the union of segmentation result and ground truth image set:

$$Jaccard = TP/(TP + FP + FN) \tag{50}$$

Dice coefficient is the ratio of the intersection of segmentation result and ground truth and the sum of them. That is:

$$Dice = 2 * TP/(2 * TP + FP + FN) \tag{51}$$

4.2 Comparative Experiment

This study employed the CV model, RSF model, LGIF model, LRCV model, RABFLR model, and the proposed model to partition the skin cancer images. Figures 4, 5, 6, 7, 8 and 9 illustrate the visualization of the three medical images through different stages of the segmentation process. Each column in the figures showcases the addition of the initial contour, the intermediate iterative segmentation process, and the final segmentation results achieved by the six models. Each row corresponds to a distinct image undergoing processing.

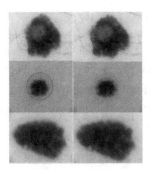

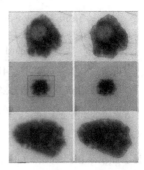

Fig. 4. Segmentation results of improved CV model

Fig. 5. Segmentation results of improved RSF model

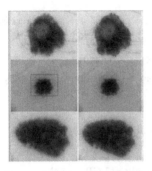

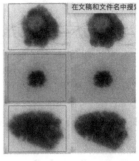

Fig. 6. Segmentation results of improved LRIF model

Fig. 7. Segmentation results of improved LRCV model

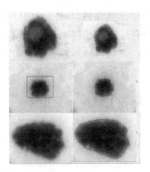

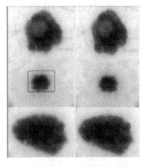

Fig. 8. Segmentation results of improved RABFLR model

Fig. 9. Segmentation results of improved LGIF model

The following is the evaluation index obtained by the segmentation of three images by the above six models (the first five models are iterated 1000 times, and the improved LGIF model is iterated 500 times):

Table 1 shows the segmentation evaluation results of CV model:

Table 1. Segmentation evaluation results of CV model

Image	Dice	Jaccard	accuracy	recall	precession
Fig. 1	0.93134904	0.87824143	0.94589992	0.8739287	0.98371
Fig. 2	0.99224963	0.99732876	0.99924134	0.9999573	0.93182
Fig. 3	0.97986847	0.94946239	0.98403278	0.9497463	0.89983

Table 2 shows the segmentation evaluation indexes of the RSF model:

Table 2. Segmentation evaluation results of the RSF model

Image	Dice	Jaccard	accuracy	recall	precession
Fig. 1	0.78736889	0.5551432	0.80012994	0.55502938	0.67576483
Fig. 2	0.69984761	0.5985431	0.69956481	0.64791243	0.76323532
Fig. 3	0.69865012	0.6978864	0.59985549	0.59709024	0.86783646

Table 3 shows the segmentation evaluation results of LGIF model:

Table 3. Segmentation evaluation results of LGIF model

Image	Dice	Jaccard	accuracy	recall	precession
Fig. 1	0.96393744	0.938472637	0.93832747	0.98647323	0.987492348
Fig. 2	0.93328301	0.921412432	0.91298456	0.92131486	0.913241211
Fig. 3	0.98632432	0.943128431	0.93434121	0.94153879	0.989741313

Table 4 shows the segmentation evaluation results of LRCV model:

Table 4. Segmentation evaluation results of LRCV model

Image	Dice	Jaccard	accuracy	recall	precession
Fig. 1	0.9746975	0.95985385	0.97013578	0.97543705	0.9643367
Fig. 2	0.9597643	0.94987537	0.95884556	0.91986584	0.9403421
Fig. 3	0.8709774	0.93864368	0.94187546	0.94209353	0.9875465

Table 5 shows the segmentation evaluation results of RABFLRmodel:

Table 5. Segmentation evaluation results of RABFLRmodel

Image	Dice	Jaccard	accuracy	recall	precession
Fig. 1	0.6897654	0.54875396	0.80531468	0.540864715	0.7598644
Fig. 2	0.6575379	0.80743681	0.71975378	0.701532686	0.7964327
Fig. 3	0.7176489	0.79245781	0.69853356	0.699217535	0.7601466

Table 6 shows the segmentation evaluation results of the improved LGIF model:

Table 6. Segmentation evaluation results of the improved LGIF model

Image	Dice	Jaccard	accuracy	recall	precession
Fig. 1	0.9685314	0.9843102	0.98147038	0.98535885	0.9927643
Fig. 2	0.9446853	0.9175369	0.91864792	0.91975032	0.9376428
Fig. 3	0.9902181	0.9486535	0.95308657	0.94642894	0.9896531

The evaluation results of the five models in the table indicate notable differences. When comparing the CV model, LGIF model, LRCV model, and the new model, it becomes evident that the RSF model and RABFLR model yield lower evaluation indexes after 1000 iterations. This can be attributed to their reliance solely on local image information, resulting in slower curve evolution processes and longer iterative convergence times. Conversely, it is noteworthy that the improved LGIF model achieves comparable evaluation indexes to the CV model, LGIF model, and LRCV model after 500 iterations. This observation suggests that the improved LGIF model exhibits faster convergence speed and delivers more accurate segmentation results.

Furthermore, the evaluation index of the LRCV model, LGIF model, and new model aligns more closely with a value of 1 compared to the fundamental CV model. The LRCV model combines local image information with the foundations of the CV model, resulting in enhanced segmentation accuracy. Similarly, the LGIF model and new model leverage both global data elements from the CV model and local data elements from the RSF model, thereby achieving superior segmentation effects.

5 Conclusion

This paper introduces a novel image segmentation model that addresses several key challenges. The proposed model incorporates a bilateral filter for image preprocessing and integrates a global energy fitting term into the local fitting data term within the energy functional. This integration effectively considers both global and local information of the image, thereby reducing sensitivity to image noise and improving image segmentation speed. Experimental results demonstrate the model's exceptional performance on medical images, exhibiting high sensitivity and achieving robust segmentation results even in the presence of non-uniform image intensity distributions.

References

1. Shervin, M., Boykov, Y.Y., Porikli, F., Plaza, A.J., Kehtarnavaz, N., Terzopoulos, D.: Image segmentation using deep learning: a survey. IEEE Trans. Pattern Anal. Mach. Intell. (2021)
2. Liu, C.: Research on Segmentation Algorithm of Noisy Image Based on Level Set. Beijing Jiaotong University, Beijing (2020)
3. Qian, Y., Zhang, Y.: A review of image segmentation methods for level sets. J. Image Graph. **1**, 7–13 (2008)
4. Li, J., Yang, F., Song, D., et al.: Liver MRI image segmentation method based on CV model optimization. Mod. Sci. Instrum. (2021)
5. Ming, D.A.I.: Application of Active Contour Model Based on Level Set Method in Image Segmentation. South China University of Technology, Guangzhou (2020)
6. Chan, T.F., Vese, L.A.: Active contours without edges. IEEE Trans. Image Process. **10**(2), 266–277 (2001)
7. Li, C., Kao, C.Y., Gore, J.C., et al.: Minimization of region-scalable fitting energy for image segmentation. IEEE Trans. Image Process. **17**(10), 1940–1949 (2008)
8. Wang, L., Li, C., Sun, Q., et al.: Active contours driven by local and global intensity fitting energy with application to brain MR image segmentation. Comput. Med. Imaging Graph. **33**(7), 520–531 (2009)
9. Bai, P.R., et al.: A Novel Integration Scheme Based on Mean Shift and Region-Scalable Fitting Level Set for Medical Image Segmentation (2015)
10. Han, B., Wu, Y.: Active contours driven by novel LGIF energies for image segmentation. Electron. Lett. **53**(22), 1466–1468 (2017)
11. Yu, H., He, F., Pan, Y.: A scalable region-based level set method using adaptive bilateral filter for noisy image segmentation. Multimedia Tools Appl. **79**(9), 5743–5765 (2020)
12. Fang, J., Lv, Y., Wang, Y., et al.: Medical image segmentation algorithm based on level set. Electron. Sci. Technol. **34**(2), 12–20 (2019)
13. Zhang, Q.M., Li, Y.H., Luo, X.M., et al.: Infrared image segmentation algorithm of power equipment based on improved Chan-Vese model. Infrared Technol. **45**(2), 8 (2023)

Predicting Learners' Performance Using MOOC Clickstream

Kui Xiao[1,2,3] ⓘ, Xueyan Pan[1] ⓘ, Yan Zhang[1,2,3]([✉]) ⓘ, Xiaohui Tao[4] ⓘ,
and Zhifang Huang[5] ⓘ

[1] School of Computer Science and Information Engineering, Hubei University,
Wuhan Hubei 430062, China
[2] Engineering and Technical Research Center of Hubei Province in Educational
Informatization, Wuhan 430062, China
[3] Hubei Province Project of Key Research Institute of Humanities and Social
Sciences at Universities (Research Center of Information Management for
Performance Evaluation), Wuhan, China
zhangyan@hubu.edu.cn
[4] The University of Southern Queensland, Toowoomba Queensland 4072, Australia

[5] Normal School of Hubei University, Wuhan Hubei 435002, China

Abstract. Massive Open Online Courses (MOOCs) have gradually
become a dominant trend in online education. However, due to the large
number of learners participating in MOOCs, teachers usually cannot
accurately know the learning outcomes of each MOOC user. In addition,
many learners did not take the corresponding quiz after watching the
MOOCs' videos, and some MOOC videos even did not contain a quiz,
which makes it difficult to evaluate the learners' performance. In the
absence of learners' test scores, how to evaluate learners' performance
has become a huge challenge. In this paper, we build a MOOC platform
and collect user clickstream data in course videos, and propose a novel
approach for predicting learners' performance based on MOOC clickstream. We use MOOC clickstream data to define handcrafted features
and embedding features of user learning behavior, which are used to infer
learners' performance. Experimental results show that the performance
of the proposed method exceeds that of the state-of-the-art methods.

Keywords: Learners' performance · Learning outcome · Clickstream ·
MOOC · E-learning

1 Introduction

With the popularization of online education, more and more people acquire
knowledge through the Internet. Massive Open Online Courses (MOOCs) [9]
are also gaining popularity as an important online learning resource. MOOC
refers to the establishment of learning communities through unrestricted participation and readily available online courses. In the MOOC platform, students are

X. Yang et al. (Eds.): ADMA 2023, LNAI 14179, pp. 607–619, 2023.
https://doi.org/10.1007/978-3-031-46674-8_42

not limited by time and place, and can flexibly arrange their own study plans. Especially, in traditional classrooms, if students are distracted, they will miss what the teacher taught, and they will have to spend a lot of time reviewing after class. MOOC is different from these traditional courses. If students do not understand the content of the course, they can understand the knowledge by playing the course video repeatedly. In general, in the MOOC learning process, there is no teacher supervision, no entry threshold, and no need to pay expensive fees, which is very convenient for students to carry out personalized learning. Currently, MOOC platforms such as Coursera [13], edX [1], and XuetangX [18] have registered more than 10 million people.

However, the MOOC learning mode also faces some problems. On the one hand, learning in MOOC is mainly based on watching course videos. The number of students participating in a MOOC is much higher than that of a traditional course. It is difficult for teachers to take care of each student. Many students do not continue to participate in the study after enrolling in the course, and the low course completion rate is a common problem faced by MOOC platforms [8,11].

On the other hand, there are many learners who just watch the course video of a MOOC without taking the quiz in the course, and some MOOCs even do not provide a quiz, which made it difficult for MOOC users to know the effect of their learning. Since many MOOCs lack student test scores, it is obviously impossible to directly evaluate their learning outcomes. How to accurately evaluate the learning outcomes of MOOC users has become an urgent problem to be solved in the current MOOC research. Some researchers try to predict the learning outcomes of MOOC users by analyzing the learning history records left by users, and provide early guidance for the users who need help [2,3,5,10].

In fact, it is not easy to obtain the learning history information of students from each MOOC platform, because it is the private data of each platform after all. In this study, we built a MOOC platform using open source code, and collected the learning history information of students, i.e. clickstream data, from the MOOC platform to support the task of prediction of learning outcomes of MOOC users. By analyzing the characteristics of MOOC users' clickstream actions and combining them, we define the handcrafted features of user learning behaviors; in addition, we also input the clickstream data into recurrent neural networks according to time series, and extracted the embedding features of user learning behaviors. Both of the two types of features are used to predict the learning outcomes of MOOC users. Experimental results on four datasets demonstrate effectiveness and robustness of the proposed approach on predicting learning outcomes. Additionally, our approach also significantly outperforms the state-of-the-art methods.

Our main contributions include:

- A novel approach that leverages MOOC users' clickstream data to predict their learning performance. This is useful for assessing student performance in MOOCs that lack student test scores.

- A MOOC platform and four clickstream datasets. The datasets come from four videos of a MOOC and contain all the clickstream actions of MOOC users when watching videos.

The remainder parts of this paper are organized as follows: Sect. 2 discusses related work. In Sect. 3, we introduce the MOOC platform and the process of collecting user clickstream data. We present the architecture of our approach and the definitions of user learning behavior features in Sect. 4. Detailed experimental results and analysis are given in Sect. 5. Section 6 summarizes our work with a brief discussion of future work.

2   Relate Work

In the beginnings of MOOC growth, related researches primarily focused on the quality of MOOC videos such as the length of the video and presentation [7,17]. Later, researches on online learning analytics centered on creating a predictive model to predict dropout rate by examining their participation in MOOC course video events [8,11]. Recently, researchers are beginning to focus on how to evaluate students' performance in MOOCs. Sinha et al. [14] presented the first study that describes usage of detailed clickstream information to form cognitive video watching states that summarize student clickstream.

Brinton et al. [2] studied student behavior and performance in two MOOCs. They presented two frameworks by which video-watching clickstreams can be represented: one based on the sequence of events created, and another on the sequence of positions visited. With the event based framework, they extracted recurring subsequences of student behavior, and they found that some of these behaviors were significantly associated with whether a user would be Correct on First Attempt (CFA).

Chu et al. [3] developed a methodology for predicting student performance on in-video quizzes from their associated video watching behavior. They modelled student video-watching behavior through deep learning operating on raw event data. They developed a clustering guided meta-learning-based training procedure that optimizes the prediction model based on inferred similarities within student behavioral clusters.

Crockett et al. [4] focuses on analysis of clickstream data from the textbook in search of viewing patterns among students. It was found that students typically fit one of three viewing patterns. These patterns can be used in further research to inform creation of new interactive texts for improved student success.

Aouifi et al. [6] analyzed how learners interact with the pedagogical sequences of educational videos, and its effect on their performance. In their study, the video courses were segmented on several pedagogical sequences. They focalized on the interpretation of the path followed by a learner watching an educational video, and the way they navigate the pedagogical sequences of that video, in order to predict whether a learner can pass or fail the video course.

Mubarak et al. [12] exploited a temporal sequential classification problem by analyzing video clickstream data and predicted learner performance by addressing their issues and improving the educational process. They employed a LSTM network on a set of implicit features extracted from video clickstream data to predict learners' weekly performance and enable instructors to set measures for timely intervention.

Yu et al. [15] established a series of learning behaviors using the video clickstream records of students of a MOOC platform to identify seven types of cognitive participation models of learners. They subsequently built practical machine learning models by using KNN, SVM, and ANN algorithms to predict students' learning outcomes via their learning behaviors. Besides, their approach of combining basic clickstream actions into user learning behavioral features has given us great inspiration.

Yurum et al. [16] focuses on the study which is to investigate the predictive relationship between video clickstream behaviors and students' test performance with two consecutive experiments. The first experiment was performed as an exploratory study with 22 university students using a single test performance measure and basic statistical techniques. The second experiment was performed as a conclusive study with 16 students using repeated measures and comprehensive data mining techniques. The findings show that a positive correlation exists between the total number of clicks and students' test performance.

Some of the above methods are similar to our proposed approach, but they only rely solely on handcrafted or embedding features of user learning behavior to predict learners' performance. In this article, we will combine both of the two types of features to predict their learning outcomes.

3 MOOC Platform Construction

We built a MOOC website using the source code provided by the EduSoho[1] platform. Then we upload the video files of courses such as "Software Design and Architecture", "JAVA Application Development", "PHP Technology" of a software engineering major in a Chinese university to the platform. These courses are taught in Chinese, and there are corresponding quizzes in each course video. In this paper, we only choose the data of the MOOC "Software Design and Architecture" for the experiment. Compared with other courses on the MOOC platform, this course has the largest number of students. The number of students currently enrolled in other courses is not large, so the clickstream data of these courses was not used.

In this MOOC platform, various clickstream actions of users watching course videos will be recorded in real time, such as the start play action, the pause action, the forward skipping action, the backward skipping action, accelerating the playrate action, decelerating the playrate action and the ending action. By recording these actions, it will help us analyze the learning behavior of MOOC users, and then predict their performance based on learning behavior.

---

[1] https://www.edusoho.com/open/show.

4   Proposed Approach

The work of this research is to serve the online education platforms such as various MOOC platforms. Generally speaking, students' performance are evaluated through quizzes. This study uses students' learning behaviors to evaluate their learning outcomes. No matter for the students who did not take the quizzes in videos or the students who have not finished watching the videos, our proposed method can be used to evaluate their performance, so that teachers can provide help to the students with poor learning effect as soon as possible. Furthermore, we define both handcrafted features and embedding features of MOOC users' learning behaviors to predict their learning performance. Handcrafted features can better reflect the intention of learners' clickstream actions when watching videos, while embedding features can better reflect the time-series characteristics of clickstream data. The combination of the two can help us more accurately infer learners' performance. The architecture of the proposed approach is shown in Fig. 1.

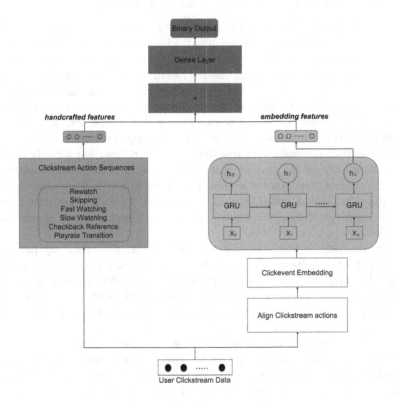

Fig. 1. The architecture of the proposed approach.

4.1   Handcrafted Features

Inspired by [15], our MOOC platform records seven types of clickstream actions: the start play action of the video (Pl), the pause action of the video (Pa), the forward skipping action of the video (Sf), the backward skipping action of the video (Sb), accelerating the playrate action of the video (Rf), decelerating the playrate action of the video (Rs) and the ending action of the video (St).

Most clickstream actions are recorded directly, but Rf and Rs are not. When MOOC users change the playback speed, we only record the action and the changed speed. Then, we know whether the playback speed has increased or decreased by comparing it with the speed before the action occurred.

However, a single clickstream action often cannot reflect the current intention of a MOOC user. Only by observing his or her continuous multiple clickstream actions, will we know what the user wants to do at the moment. Therefore, we define six handcrafted features for user learning behavior, each feature is composed of three clickstream actions. Each feature represents an event that occurs while the user is watching the video, and also reflects the intention of the MOOC user at that time. The six features are as follows.

(1) Rewatch: It is the number of times the user rewatches any part of the video. If a MOOC user wants to replay a part of a video, they usually perform two actions: Sb action and Pl action. In other words, when a user clicks the start play button after performing the backward skipping action, it means that the user wants to rewatch the part of the video at the moment. Here, we define the event rewatch as a clickstream action sequence of Sb and Pl, such as "PlSb*". Besides, "*" can represent any clickstream action.

(2) Skipping: It is the number of times the user skips forward in the video. The event skipping usually indicates that the MOOC user has already mastered the knowledge of the current part of the video, and he or she looks forward to the correct time point of a video they want to watch. Moreover, it usually requires multiple forward skipping actions (Sf) to find an accurate position. Therefore, we define the event skipping as a combination of multiple Sf actions.

(3) Fast watching: It is the number of times the user watches any part of the video at a faster speed. The event fast watching may indicate that a MOOC user has already watched the current part of a video. If the user skips forward directly, it may not be possible to accurately locate the correct position at once. Consequently, we define the event fast watching as a sequence composed of Pl action and Rf action.

(4) Slow watching: It is the number of times the user watches any part of the video at a slower speed. The event slow watching may suggest that the MOOC user wants to watch the current part of the video carefully. This may be because the current content is more important, so users need to watch this part slowly. We define the event slow watching as a sequence of Pl action and Rs action.

(5) Checkback reference: It is the number of times the user skips backward in the video. The event checkback reference is similar to the event skipping.

The MOOC user may want to find an accurate playback point backwards. So, we define this event as a combination of multiple Sb actions.

(6) Playrate transition: It is the number of times the user chooses the playback speed when watching a video. The event playrate transition indicates that a MOOC user is selecting the optimal playback speed for the video. Accordingly, we define this event as a sequence of Rf action and Rs action.

Table 1 shows the details of grouping clickstream action sequences to form MOOC user behavioral features.

Table 1. Grouping clickstream action sequences to form user behavioral features.

Features	Clickstream action sequences
Rewatch	*SbPl*, *SbPl, PlSb*, *PlSb, Sb*Pl, Pl*Sb*
Skipping	*SfSf*, *SfSf, Sf*Sf*
Fast watching	*PlRf*, *PlRf, RfPl*, *RfPl, Pl*Rf, Rf*Pl*
Slow watching	*Pl*Rs, Rs*Pl*
Checkback reference	*SbSb*, *SbSb, Sb*Sb*
Playrate transition	*RfRf*, *RfRf, Rf*Rf, RfRs*, *RfRs, Rf*Rs, RsRs*, *RsRs, Rs*Rs, RsRf*, *RsRf, Rs*Rf*

It should be noted that Pa and St did not appear as key elements in the above-mentioned action sequences. They may appear in the action sequences as non-critical factors denoted by "*".

4.2 Embedding Features

When we use the above handcrafted features, we only record the number of occurrences of corresponding feature events, such as when users watch course videos, their Rewatch times and their Fast Watching times, etc. However, the clickstream data generated by MOOC users while watching videos is a typical time series data. Analyzing the time series data directly can also help predict students' performance. The above mentioned handcrafted features only reflect the number of times of the of certain feature events, and do not reflect the characteristics of this time series.

In order to make full use of the hidden information in MOOC users' clickstream data, we use recurrent neural network (RNN) to extract user learning behavior features from user clickstream data. Derived from feedforward neural networks, RNNs can use their internal state (memory) to process variable length sequences of inputs. However, due to the gradient exploding problem and vanishing gradient problem, RNNs can only learn short-term dependencies, and it is difficult to model long-distance dependencies. In order to solve these problems, a better solution is to introduce a gating mechanism to control the accumulation speed of information, including selectively adding new information and selectively forgetting previously accumulated information. Two typical variants of Gated RNN are Long Short-Term Memory Network (LSTM) and Gated Recurrent Unit (GRU). LSTM uses three gates to control the path of information transmission, namely input gate, forget gate and output gate. Among them, the input gate and the forget gate are complementary and have certain redundancy. The GRU network directly uses a gate to control the balance between input gate and forget gate, i.e. reset gate. In addition, GRU also uses an update gate to control how much information the current state needs to keep from the historical state, and how much new information needs to be accepted from the candidate state. In this paper, we will choose the GRU network to extract user learning behavior features from MOOC clickstream data.

In a course video, the number of clicks performed by different users is usually not the same. Taking the video "Introduction to the Unified Modeling Language" in the course "Software Design and Architecture" as an example, the average number of clicks in this video by users is 31. Among them, 91% of the users have no more than 100 clicks. The user with the most clicks made 494 clicks. In addition, there are very few users who have less than 3 clicks. To solve this problem, we only selected the first 100 clicks of each user in a video for the analysis of user learning behavior. For each click operation, we will use one-hot encoding to generate its vector. If the user clicks less than 100 times in a video, the insufficient part is treated as all-zero. If the user clicks less than 3 times in the video, the learning record of the user will be discarded.

The vector sequence of these clickstream actions will be sequentially input into a GRU network to generate the user's embedding feature vector. The generated embedding features are a 64-dimensional vector. After the handcrafted features and embedding features are concatenated, they will be input into a dense layer to predict the MOOC users' performance. The activation function is sigmoid.

5   Experiments

5.1   Datasets

We selected four course videos from the course "Software Design and Architecture" to build user clickstream datasets, including "Introduction to Unified Modeling Language" (D1), "Object-Oriented Design Principles" (D2), "Factory Patterns" (D3), "Decorator Pattern" (D4). The "Factory Patterns" contains three design patterns: simple factory, factory method and abstract factory. The code and data are available at https://github.com/PxYAN/Clickstream.

We provide ten exercises in each video, all of which are one-choice questions. If a MOOC user answers 9 out of 10 questions correctly, we consider the user to have passed the quiz and mastered the knowledge of the video; otherwise, the user is considered to have failed the quiz and not mastered the knowledge of the video. The test scores of these students will be used as the ground truth for our experiments to evaluate the performance of the proposed approach. Table 2 shows the statistics of the four datasets.

Table 2. Statistics of the four datasets.

	D1	D2	D3	D4
Total number of users	223	194	172	127
Number of users who passed the quiz	111	69	61	85
Number of users who failed the quiz	112	125	111	42
Clicks per capita	31.0	42.5	84.1	48.1
Most clicks	494	779	3138	1637
Number of users with more than 100 clicks	18	20	28	13
Average study time of students (min)	65.27	123.32	169.73	113.07

5.2 Experimental Settings

The evaluation is performed using a 5-fold cross validation. The latent embedding dimension in our model is set to k = 64. During training, the batch size is set to 512, the learning rate to 0.001, and the weight decay to 0.03. We train the proposed model using the Adam optimizer for 100 epochs. And our model is implemented with keras.

To evaluate the performance of the proposed approach, we apply two commonly used classification metrics: prediction accuracy (ACC) and F1 score. ACC measures the average accuracy of students' performance on the test datasets. The F1 score is the harmonic mean of precision and recall.

5.3 Baselines

We compare our approach with the following state-of-the-art methods:

1) kNN model: Aouifi et al. [6] focalized on the interpretation of the path followed by a learner watching an educational video, and the way they navigate the pedagogical sequences of that video, in order to predict whether a learner can pass or fail the video course. They applied educational data mining technique using K-nearest Neighbours and Multilayer Perceptron algorithms to predict learner's performance. In our experiments, we represent the baseline as "kNN".

2) n-gram model: Jeon and Park [8] presented a binary classification model that encodes clicking events as n-gram vectors of event types and uses them as input to GRU networks. In our experiments, we use n = 3 and denote the baseline as "3-gram".

3) HFs model and EFs model: We also use the proposed handcrafted features and embedding features separately to predict the learning outcomes of MOOC users, and name these two models "HFs" and "EFs" respectively.

5.4   Comparison with Baselines

Table 3 lists the performance comparison of different methods on accuracy, precision, recall and F1 score. We can see that the performance of the HFs, EFs, and the proposed method (HFs + EFs) on the four datasets is generally better than that of the 3-gram and kNN models. All three proposed models perform better than 3-gram and kNN except that HFs perform slightly worse than kNN on the dataset "Factory Patterns" (D3) in terms of accuracy and F1. Closer observation shows that the performance of proposed method and EFs is significantly better than that of HFs and kNN, while the performance of HFs and kNN is significantly better than that of 3-gram.

Among all the methods, the proposed method beats others with best accuracy and F1 across all four datasets. Its average accuracy outperforms 3-gram, kNN, HFs and EFs by 21.94%, 20.47%, 16.48% and 0.56%, respectively. Also, its average F1 outperforms the four state-of-the-art methods by 17.16%, 12.24%, 7.92% and 0.81%, respectively. It suggests that using GRU to analyze time-series clickstream data is of great help in predicting the performance of MOOC users.

Table 3. Comparison proposed approach with baseline methods.

		D1	D2	D3	D4
3-gram	Acc	60.21	50.34	48.73	59.82
	P	68.21	60.34	51.42	70.14
	R	74.39	52.98	50.00	71.47
	F1	70.64	54.27	50.31	70.26
KNN	Acc	61.68	58.45	56.45	70.14
	P	**87.21**	76.26	65.00	**91.32**
	R	65.80	62.01	57.88	72.47
	F1	74.84	67.86	61.17	80.72
HFs	Acc	65.67	61.48	53.42	70.94
	P	65.84	62.72	53.94	73.25
	R	**99.26**	85.09	70.00	92.50
	F1	79.16	71.87	60.70	81.44
EFs	Acc	81.59	75.19	74.34	79.49
	P	83.25	77.10	75.26	81.00
	R	90.09	80.64	**77.50**	92.57
	F1	86.27	78.76	75.82	86.32
Proposed method	Acc	**82.15**	**80.74**	**75.08**	**81.27**
	P	82.95	**81.40**	**77.15**	83.27
	R	91.71	**87.13**	76.25	**92.72**
	F1	**87.08**	**83.98**	**76.31**	**87.30**

5.5   Discussion

Let's go back to the question of greatest concern: what kind of learning behaviors do MOOC users have that are more likely to pass the quiz in a course video? From the experimental results, we can see that there are two types of learners who are more likely to pass the quiz in a video. The first type of people are those who learn faster than normal, and the second type of people are those who learn slower than normal.

For the first type of people, during their learning process, there are usually events such as "Skipping" or "Fast watching". In other words, when they watch a MOOC video, they skip forward multiple times, or play fast multiple times. We suppose that these MOOC users have already learned relevant knowledge in traditional classrooms or textbooks before watching the videos. Watching the video is just a review of previously learned knowledge for them, so they watch the video faster than normal.

For the second type of people, when they start watching a course video, there will always be multiple pause actions (Pa), or the decelerating the playrate action (Rs) and then the pause action (Pa). In our opinion, these learners may not have learned the knowledge in the video before, but they feel that the knowledge at the beginning of the video is more important. Once they miss it, the later knowledge will be difficult to understand, so their learning speed at the beginning of the video is very slow. These learners can watch the course videos with a very humble attitude, and the learning outcomes will naturally be very good.

6   Conclusions and Feature Work

Early prediction of the performance of MOOC users will help teachers and educational experts analyze and understand the learning behavior of the users. It is very difficult to identify the learning outcomes of MOOC users without knowing their quiz scores. In this paper, we build a MOOC platform and propose a clickstream based approach that predicts the performance of MOOC users through their learning history records. On the one hand, we combine basic clickstream actions into handcrafted features of users, and on the other hand, we input clickstream data into the GRU network to generate embedding features. Experiments on four datasets show that the performance of the proposed method is significantly better than the state-of-the-art methods.

As for future work, we want to increase the number of courses in the MOOC platform and the number of learners participating in the experiment. In addition, we will also analyze the amount of time learners spend watching each course video and the difficulty of the content. We believe that these factors will help improve the accuracy of MOOC user performance prediction.

Acknowledgement. This work is supported by the Ministry of Education's Youth Fund for Humanities and Social Sciences Project (No.19YJC880036),the National Natural Science Foundation of China (Nos.62102136, 61902114, 61977021), the Key R & D projects in Hubei Province (Nos.2021BAA188, 2021BAA184, 2022BAA044), the Science and Technology Innovation Program of Hubei Province (No.2020AEA008).

References

1. Breslow, L., Pritchard, D.E., DeBoer, J., Stump, G.S., Ho, A.D., Seaton, D.T.: Studying learning in the worldwide classroom research into edX's first MOOC. Res. Pract. Assess. **8**, 13–25 (2013)
2. Brinton, C.G., Buccapatnam, S., Chiang, M., Poor, H.V.: Mining MOOC clickstreams: video-watching behavior vs. in-video quiz performance. IEEE Trans. Sig. Process. **64**(14), 3677–3692 (2016)
3. Chu, Y.W., Tenorio, E., Cruz, L., Douglas, K., Lan, A.S., Brinton, C.G.: Click-based student performance prediction: a clustering guided meta-learning approach. In: 2021 IEEE International Conference on Big Data (Big Data), pp. 1389–1398. IEEE (2021)
4. Crockett, B.: Measuring students' engagement with digital interactive textbooks by analyzing clickstream data. In: Proceedings of the AAAI Conference on Artificial Intelligence, vol. 36, pp. 13132–13133 (2022)
5. El Aouifi, H., El Hajji, M., Es-Saady, Y., Douzi, H.: Predicting learner's performance through video viewing behavior analysis using graph convolutional networks. In: 2020 Fourth International Conference On Intelligent Computing in Data Sciences (ICDS), pp. 1–6. IEEE (2020)
6. El Aouifi, H., El Hajji, M., Es-Saady, Y., Douzi, H.: Predicting learner's performance through video sequences viewing behavior analysis using educational data mining. Educ. Inf. Technol. **26**(5), 5799–5814 (2021)
7. Guo, P.J., Kim, J., Rubin, R.: How video production affects student engagement: an empirical study of MOOC videos. In: Proceedings of the First ACM Conference on Learning@ Scale Conference, pp. 41–50 (2014)
8. Jeon, B., Park, N.: Dropout prediction over weeks in MOOCs by learning representations of clicks and videos. arXiv preprint arXiv:2002.01955 (2020)
9. Kay, J., Reimann, P., Diebold, E., Kummerfeld, B.: MOOCs: so many learners, so much potential... IEEE Intell. Syst. **28**(3), 70–77 (2013)
10. Li, X., Xie, L., Wang, H.: Grade prediction in MOOCs. In: 2016 IEEE Intl Conference on Computational Science and Engineering (CSE) and IEEE Intl Conference on Embedded and Ubiquitous Computing (EUC) and 15th Intl Symposium on Distributed Computing and Applications for Business Engineering (DCABES), pp. 386–392. IEEE (2016)
11. Liang, J., Li, C., Zheng, L.: Machine learning application in MOOCs: dropout prediction. In: 2016 11th International Conference on Computer Science & Education (ICCSE), pp. 52–57. IEEE (2016)
12. Mubarak, A.A., Cao, H., Ahmed, S.A.: Predictive learning analytics using deep learning model in MOOCs' courses videos. Educ. Inf. Technol. **26**(1), 371–392 (2021)
13. Severance, C.: Teaching the world: Daphne Koller and Coursera. Computer **45**(8), 8–9 (2012)
14. Sinha, T., Jermann, P., Li, N., Dillenbourg, P.: Your click decides your fate: Inferring information processing and attrition behavior from MOOC video clickstream interactions. arXiv preprint arXiv:1407.7131 (2014)
15. Yu, C.H., Wu, J., Liu, A.C.: Predicting learning outcomes with MOOC clickstreams. Educ. Sci. **9**(2), 104 (2019)
16. Yürüm, O.R., Taşkaya-Temizel, T., Yıldırım, S.: The use of video clickstream data to predict university students' test performance: a comprehensive educational data mining approach. Educ. Inf. Technol. **28**(5), 5209–5240 (2023)

17. Zhang, G., Patuwo, B.E., Hu, M.Y.: Forecasting with artificial neural networks: the state of the art. Int. J. Forecast. **14**(1), 35–62 (1998)

18. Zhang, T., Yuan, B.: Visualizing MOOC user behaviors: a case study on XuetangX. In: Yin, H., et al. (eds.) IDEAL 2016. LNCS, vol. 9937, pp. 89–98. Springer, Cham (2016). https://doi.org/10.1007/978-3-319-46257-8_10

A Fine-Grained Verification Method for Blockchain Data Based on Merkle Path Sharding

Liang Wen[1], Zhiqiong Wang[2,3]([✉]), Tingyu Cui[1], Caiyun Shi[1], Baoting Li[1], and Zhongming Yao[1]

[1] School of Computer Science and Engineering, Northeastern University, Shenyang 110819, China
yaozming@stumail.neu.edu.cn
[2] College of Medicine and Biological Information Engineering, Northeastern University, Shenyang 110819, China
wangzq@bmie.neu.edu.cn
[3] Key Laboratory of Big Data Management and Analytics (Liaoning Province), Northeastern University, Shenyang 110819, China

Abstract. Blockchain has been widely applied these years. As its volume of data is greatly increasing, it is necessary to shard the data to utilize hardward storage efficiently. After sharding, however, the validated data are stored in different nodes and read cross-shardingly, which is inefficient because of extra communication time. To solve this problem, the authors use the Mp-tree (Merkle Path Tree) structure to replace the Merkle tree in the traditional blockchain, and then shard the blocks according to the Mp-tree architecture. The Mp-tree can validate the accuracy of data to avoid cross-sharding so as to improve the efficiency of validation. It has some advantages: first, it can reduce the volume of storage space by fine-grained slicing; second, it can improve the validation efficiency of blockchain slicing by independent validation. This study demonstrates that our design can optimize the generation efficiency in existing sharding systems, and also exploiting its independent verification feature to improve verification efficiency.

Keywords: Blockchain · Sharding · Merkle Path

1 Introduction

Blockchain is a new type of distributed basic structure. It uses a blockchain data structure to verify and store data, a distributed node consensus algorithm to generate and update data, and a Merkle tree to generate a Hash function

This research was partially supported by the National Key R&D Program of China (No.2022YFB4500800), the National Natural Science Foundation of China (No. 62072089) and the Fundamental Research Funds for the Central Universities of China (Nos.N2116016,N2104001, N2019007).

to secure data transmission and access [1] [2]. This new distributed basic structure and computing paradigm provides a more secure, efficient, transparent, and reliable solution for data management in finance, healthcare, and other fields.

As the volume of data increases, current hardware systems are gradually unable to support independent storage, so blockchain storage expansion has become a popular technology [3]. Among them, sharding is widely used in blockchain systems as it can greatly improve hardware utilization and enable more nodes to participate in the overall blockchain system. In recent years, platforms such as ETH and Hyperledger have also been exposed to and investigated various methods of storage expansion [4,5].ETH shards Ethernet to be multiple concurrent channels to expand the network so as to solving network congestion due to the large volume of data. Another one, Hyperledger uses an endorsement/consensus model where simulation execution and block validation are performed separately in nodes with different roles. Simulation execution is concurrent, which improves expansion and throughput [6].

However, existing sharding architectures ignore the fact that random sharding leads to more communication costs in subsequent verification, to be more specific, inefficient verification, which is a contradiction between speed and expansion in blockchain [7]. In current sharding systems, validation transactions often require cross-shard communication to query the corresponding validation information, a process that is affected by the network environment and physical distance; the larger the data volume, the slower the connecting speed, so the longer the duration of validation process. Accordingly, it is necessary to find a sharding architecture that can meet the storage expansion requirements while avoiding the extra communication cost caused by cross-sharding verification.

In this paper, a Merkle path-based sharding architecture is designed as shown in Fig. 1. This structure is used to replace the original Merkle tree when the blocks are generated and to parallelize the sharding when the transaction information is packaged to generate the blocks so that the sharding time is parallel to the block generation time and the sharding cost is reduced. In addition, common sharding techniques can lead to severe validation delays later on as shown in Fig. 2, requiring reading the data cross-shardingly each time during the validation, which leads to inefficient validation. Based on the features of the Mp-tree, storing the corresponding validation information in each shard separately, so the data can be independently validated at the storage node, avoiding communication delays caused by cross-sharding reading during the validation process, which improves validation efficiency. While, this method faces two major challenges.

Challenges: 1) How to design the Mp-tree sharding architecture to shard it in the block when its generating, so that the operation cost of system sharding is minimized. 2) How to save the verification information after Mp-tree sharding into each shard independently. In the validation process, the Merkle tree is located along the leaf nodes, one by one and finally to the root, and the whole process goes through multiple branches. It is therefore a challenge to independently determine the validation information for the leaf node data.

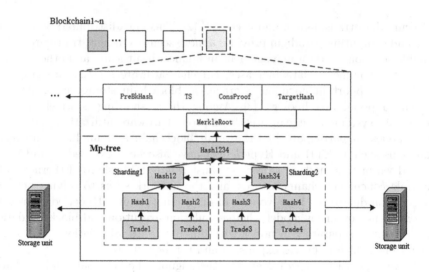

Fig. 1. Structure diagram for fine-grained verification of blockchain data based on Merkle path slicing. Blue block is the Genesis block, and yellow block is a new created block. (Color figure online)

Contributions. 1) To address the problem of large cost of sharding operations, the authors port the Mp-tree inside block generation, so that sharding can be conducted in parallel with block generation. 2) Using Mp-tree to copy the corresponding data validation nodes under the same shard while establishing the tree to ensure independent validation within the shard. (3) Based on the Mp-tree architecture, this study conducts controlled experiments compared to traditional blockchain systems and sharding systems.

Section 2 reviews the related work and provides some necessary backgrounds. Section 3 explains the problem statement. Section 4 presents the design of the Mp-tree. Section 5 proves the correctness of our design and presents a complete set of experiments, and Sect. 6 concludes the paper.

2   Related Work

2.1   Blockchain Sharding

Blockchain sharding research has evolved rapidly over the years, from grouping by features at the beginning gradually to sharding from physical nodes. Many existing blockchain sharding systems have been studied and proposed [8–13]. A new concept called consensus unit (CU) has been proposed by Zihuan Xu [14] et al., which organizes different nodes into a unit and lets them store at least one copy of the blockchain data in the system, but they does not have a complete sharding structure. For the sharding method, Jiaping Wang [15] et al. proposed the Asynchronous BFT Protocols, and Hung Dang [16] et al. used TEE+PBFT to improve the consensus algorithm. They applied inter-block coarse-grained sharding, The former one divides the blockchain into multiple independent and parallel zones, each zone is mining independently with each other in parallel. They designed Relay Transaction, a method for cross-group transactions, which can pack information interaction between zones. The latter one, considering a consortium blockchain platform, proposes that using SGX to additionally deallocate committee members, and by using Rnd to dynamically shuffle each committee to go about solving the problem of throughput degradation due to node allocation during the sharding process. However they are all affected by network latency when communicating across blocks, and the overall experiment is based on a more ideal environment going forward, without considering the cost of chunked communication after latency. Xiaodong Qi [17] et al. proposed using Honey Badger BFT consensus protocol, using RS erasure code check blocks to store the blockchain chain code in shards, and using distributed key generation protocol to ensure communication security, so finally making all nodes reach consensus on the information. However, each time a new node is generated, the entire erasure code system needs to be regenerated, which increases the cost of communication and worsens the impact of network latency on the response speed of the system in a sharding environment.

2.2 Blockchain Verification

Blockchain validation efficiency has been a popular researching direction. Cheng Xu [18] et al. proposed an accumulator-based validation data structure, ADS, which supports the dynamic aggregation of arbitrary query attributes, further developing Vchain using two new indexes to aggregate data records within or between blocks for efficient query validation. Christian ReitwieBner [19] et al. proposed Truebit, a scalable computational framework based on incentive and validation games, which allows arbitrary computational tasks to be conducted under the chain and guarantees the accuracy of the results. Meanwhile, they also designed an adaptable difficulty adjustment mechanism to prevent malicious collaboration between the perpetrators and the verifiers. Schwartz David [20] et al. propose Ripple, a Byzantine fault-tolerant consensus protocol based on groups of trusted subnetworks, which aims at low latency, high throughput, and low-cost transaction processing. By using voting sets and a joint Byzantine protocol, the protocol can reach consensus in the presence of $(n\text{-}1)/\,5$ malicious participants, while also providing a way to dynamically adjust the network topology to accommodate network changes. These studies are all optimized from a verification approach and cannot reduce the cost of cross-sharding communication in sharding verification. As verification efficiency improves, the inefficiency of verification due to communication delays in the sharding system can also be improved. Zhongteng Cai [21] et al. use Benney to decouple functions so that the verification part is chained separately while keeping an additional root part of each verification tree to reduce the verification cost. coSplit [22] et al. maximizes parallelisation. Ethernet proposes a Danksharding scheme based on evidence consistency, where there are no disjoint shards. Validators check the availability of the same block by sampling data availability, so they do not have to download the entire contents of the block to check its availability. Though, there is still inter-block communication latency due to the problem of the sharding structure, which does not inherently address the problems caused by sharding.

3   Problem Statement

3.1   Preliminary Knowledge

Node. A node is a group of devices that store, generate and update blocks in a blockchain system. In this experiment block B=b_0^e,..., b_m^e is saved to node N=n_0^e,..., n_m^e and data is saved to each block according to $S_{data}=data_1$,...,$data_m$ to be saved into each block. Nodes are divided into two types in this experiment, normal nodes and shard nodes for Mp-tree sharding generation. The Mp-tree based shard nodes are Peer nodes and are placed to different positions in the blockchain system, allowing for parallel processing when data is collected, so enhancing the efficiency.

Verification. To ensure system validity and data consistency, a series of computed query costs are required. In traditional sharding systems, validation requires reassembling the corresponding several sharding structures and then traversing them. Frequent cross-sharding requests can significantly increase the cost of validation as the volume of data increases and the bandwidth of the system is delayed. By creating a fine-grained sharding structure C=c_0^e,..., c_m^e within a block via Mp-tree, the Hash verification information $v_j \in \{V\}$ required for each shard part is saved to a separate tree ssharding structure C=c_0^e,..., c_m^e for the corresponding shards. Realizing independent validation data for each shrad during the validation process avoids increasing the cost of validation due to the communication cost of reading across chunks.

3.2   Problem Description

Blockchain sharding technology is a possible system function optimization solution that does not compromise the degree of decentralization and security. The purpose of sharding is to relieve the pressure on nodes to process information, but it may then lead to additional communication costs for major nodes during the broadcast process, further leading to verification delays. The common validation mechanism used in the blockchain is Merkle tree validation, which starts from the bottom data leaf node and cascades upwards. In a sharded system, as multiple messages are scattered across different blocks, this verifying process requires the sharded clusters to be reassembled into complete blocks before validation is performed as shown in Fig. 2. This process of verification under sharding increases the cost of verification, so verification under the Mp-tree structure has to be satisfied without the need to reassemble the blocks after sharding and can be achieved under one shard. However, the coarse-grained sharding system does not support saving verification information under the same shard, so a fine-grained sharding system is proposed here to ensure the efficiency of Mp-tree verification.

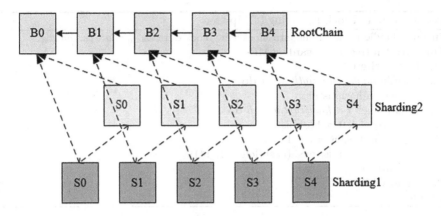

Fig. 2. Inter-block communication in the sharding structure.

4 Systems Models

4.1 Fine-Grained Sharding

A sharding system can theoretically improve the storage function of the whole system, but it may cause some performance loss due to sharding operations. Especially when multiple transactions are concentrated in one node at the same time, if the node then shards, blocking may occur. As the Mp-tree architecture is a fine-grained sharding process included in block generation, it is possible to parallelize the slicing process in the block generation process, thus further reducing the loss caused by the sharding system. The blocks are sharded and saved to physical addresses for subsequent data validation.

The algorithm is shown in the diagram, where the incoming transaction data is packaged into a block and no further data is added to the block when the volume of transaction data reaches a threshold. At the same time, in the process of packing data, Mp-tree is constructed, and the validation information of the data is now saved to the Mp-tree, then pruning and dividing.

The transaction data $S_{data}=data_1,...,data_m$ transmitted by the node is packaged and allocated to each node that performs the Mp-tree construction, which is used to build a dynamic parallel division during block generation. As shown in Fig. 1, the eight transactions are validated and sorted, including checking whether each transaction conforms to the smart contract rules and conditions and whether the digital signature is valid. These transaction records are then packaged and distributed into individual blocks to build the Mp-tree. In this case, there are eight transactions, and the intra-block sharding is tentatively set to two shards. When constructing the left and right subtrees of the Mp-tree, the left subtree is kept at one physical address and the right subtree and its root node are kept at another address. Since the sharding is done in parallel, multiple blocks can be processed at the same time, thus increasing processing efficiency. A cross-copy operation is performed when the left and right subtrees require

---

Algorithm 1. Block Encoding of Mp-tree

Input: Bundle transactions into blocksnew block data $S_{data}=data_1,...,data_m$.
Output: Mp-tree after sharding MP
```
 1: if c > a then
 2:     while hashList.length == 1 do
 3:         return hashList[0];
 4:         for each i < hashList.length do
 5:             const left = hashList[i]
 6:             const right = i + 1;
 7:             newHashList.push(hash(left + right))
 8:         end for
 9:     end while
10: end if
```

---

cross-tree validation information to be saved, and the exact saving process will be described in detail later. Finally, when all nodes have finished processing and verifying their own sharding blocks, they will be merged into the final blockchain. During the merging process, a consensus algorithm operation will also be performed to determine which block will become the latest block on the main chain (Fig. 3).

4.2 Independent Verification

In the traditional block generation process, a Merkle tree is constructed within the block. The Merkle tree itself has a certain segmentation effect on the data. For a transaction, it is possible to separate out the parent nodes associated with that from the leaf node upwards. For intra-block fine-grained sharding, the process of building a Merkle tree is essentially a fine-grained sharding of transactions. However, the traditional Merkle tree requires multiple data from the left and right subtrees to jointly compute the validation in order to verify the son node, and if the sharding is done according to the Merkle tree it will cause a common problem to that of consistent sharding between blocks - the increasong communication cost caused by reading across shards. This is why the Mp-tree architecture is proposed, which has similar validation and generation capabilities as the Merkle tree, but additionally stores the information about the nodes to be validated in a separate subtree, so that intra-chunk independent validation can be achieved after sharding.

Figure 4 shows an example of an Mp-tree, in which each node divides the block into H=H_a,...,H_h in this process. If the sharding is done after the block Mp-tree construction process, it will additionally increase the sharding cost and reduce the system performance. Therefore, during the MP-tree construction process, whenever the parent biparent node is constructed upwards, storing the information of the corresponding transaction data node in a direct shard into the storage node will enable parallel sharding and reduce the post-sharding cost. Figure 4 represents the Mp-tree sharding method within a single block, which

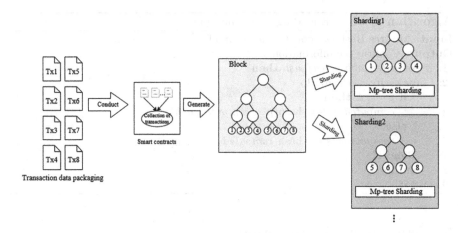

Fig. 3. Fine-grained sharding process.

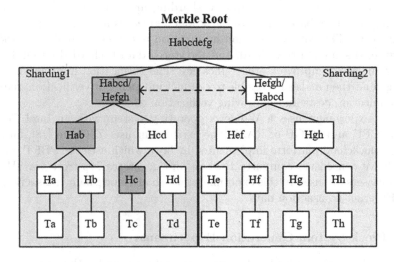

Fig. 4. Mp-tree based independent verification process.

allows parallel sharding when multiple blocks are generated at the same time, improving the sharding efficiency and reducing the cost. In the MP-tree generation process, the root node and another node validation information is saved additionally in the other side, in Fig. 4, Habcd additionally saves the information of Hefgh, and then the Mp-tree is sharded. Each shard corresponds to a sub-tree, and the root node is always saved in the last shardd sub-tree to facilitate data reduction. As in Fig. 4, the example root-based shard is then divided into two parts. When verifying the transaction Td, it can be seen that every parent node storing verification information is in the same shard, avoiding cross-sharding communication.

Algorithm 2. Block Encoding of Verification

Input: Mp-tree Hashlist(root, $leaf_hash$, proof).
Output: Confirm the information.
1: **if** $computed_hash = leaf_hash$ **then**
2:     **for** each $p \in Proof$ **do**
3:         **if** $p == L$ **then**
4:             $computed_hash = $ hash(p $+ computed_hash$);
5:         **else**
6:             $computed_hash = $ hash($computed_hash + $ p[1]);
7:         **end if**
8:     **end for**
9: **end if**

---

5   Experimental Evaluation

This experiment designs an Mp-tree sharding architecture, through which a sharding structure based on intra-block Merkle trees is implemented in this new architecture. The data transaction information is packed within the block in a fine-grained spanning tree shard, and the stored data is sharded into individual storage units to improve storage efficiency. The verification information in the storage unit then makes it possible to conduct independent verification, reducing communication costs and improving verification efficiency.

This experiment uses a Windows operating system with an Intel Core i7-8750H CPU and 16 GB of RAM. The experiment uses ZJChain [18], an open-source blockchain platform implemented in Java, which uses the PBFT protocol to achieve consensus, computes Hashes, and generates Mp-trees via SHA256. Two datasets are used in this experiment, which are blockchain transaction data and Ethernet transaction data.

5.1   Block System Generation Experiments

Figure 5 shows the entire block system generation time for the original system and the slice generation using Mp-tree. The whole experimental environment was first validated by using two data sets for block generation, with groups of 20, 40, 60, 80, and 100 nodes for the experiments, from transaction data packaging validation to block generation, recording the original system time and the Mp-tree system time. As can be seen from the experiments, the split system increases the system generation time by a small amount relative to the original system, but for throughput, it is much more efficient due to the improved storage efficiency, and the experiments demonstrate that the improved system does not cost more for system generation and has system advantages in general.

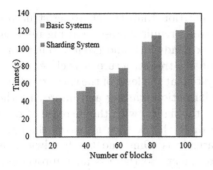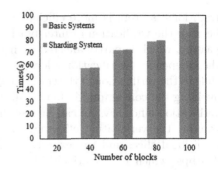

Fig. 5. Validation of the experimental environment.

5.2 Data Processing Time

As shown in Fig. 6, the data processing time within the whole block is experimented. The larger the data volume and the more transactions, the longer the processing time within the block. While the process of slicing increases the data processing time, it can be seen that the Mp-tree structure used in this paper increases the generation time by a small amount compared to the traditional system, but not by much in terms of generation time compared to the traditional system, which is to say that the new construction method we have invoked does not impose a greater cost on the overall transaction processing; and it can be seen that due to the fine-grained slicing, the parallel generation within the block time is less than the time spent on slicing after the entire block has been generated in traditional slicing systems. The experiments show that the Mp-tree architecture-based sharding system outperforms the traditional sharding structure in terms of sharding processing time and does not incur a greater cost than the traditional system.

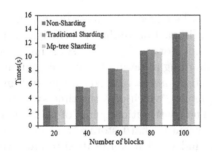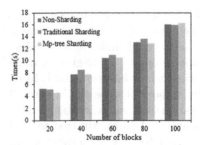

Fig. 6. VIntra-block slice processing time.

5.3 Blockchain Validation Trials

As shown in Fig. 7, the experiment mainly compares the time on re-verification between the Mp-tree structure and the traditional sharding structure. The traditional system is more efficient in verification than the sharding structure because the verification of inter-block sharding requires cross-sharding message passing, which is partly affected by network bandwidth and latency. Therefore, the independent verification adopted by the Mp-tree structure is as close as possible to the verification of the traditional system at the level of the slice structure, avoiding the consumption of cross-slice information transfer and improving the verification efficiency. As can be seen from the figure, with the increase of data volume, the verification efficiency of the direct block slice is better than the traditional slice, and the experimental latency is 49ms and the bandwidth is 30Mbps, which is a relatively ideal environment. So there is an improvement in verification time but not significant. If the environment network is poor, the MP-tree-based independent verification speed is better than traditional coarse-grained inter-block sharding.

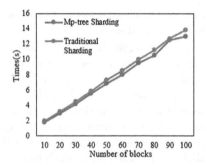 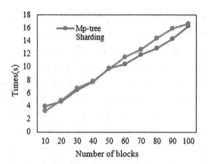

Fig. 7. Verification time.

6   Conclusion and Future Work

Blockchain sharding expansions have always belonged to coarse-grained inter-block sharding, and this structure has a great impact on verification efficiency. In order to enable independent verification within storage units after sharding, this paper proposes a fine-grained intra-block sharding structure to avoid communication delays between shards in the verification process. Based on this system, the authors propose the Mp-tree sharding model, which uses the time synchronization of block generation for sharding. The contribution points of this paper include: (1) using Mp-tree to achieve fine-grained shard storage and improve storage efficiency; (2) proposing intra-block independent verification to improve the efficiency of post-sharding verification. The Mp-tree fine-grained structure designed in this paper has some potentials, and the generated packed data will be further processed in the future to exploit the advantages of fine-grained querying.

References

1. Maiyya, S., Zakhary, V., Agrawal, D., Abbadi, A.E.: Database and distributed computing fundamentals for scalable, fault-tolerant, and consistent maintenance of blockchains. ACM VLDB **11**(12), 2098–2101 (2018)
2. Han, S., Xu, Z., Zeng, Y., Chen, L.: Fluid: A Blockchain Based Framework for Crowdsourcing. In: ACM SIGMOD, pp. 1921–1924 (2019)
3. Ali, M., Nelson, J.C., Shea, R., Freedman, M.J.: Blockstack: a global naming and storage system secured by blockchains. In: USENIX ATC, pp. 181–194 (2016)
4. Nakamoto, S.: Bitcoin: A Peer-to-peer Electronic Cash System (2008)
5. Buterin, V., et al.: Ethereum white paper (2014). https://github.com/ethereum/wiki/White-Paper
6. Zeng, S.: Survey of blockchain: principle, progress and application. J. Commun. **41**(1), 134–151 (2020)
7. Blockchain. Blockchain speeds and the scalability debate. https://blocksplain.com/2018/02/28/transaction-speeds/
8. Hong, Z., Guo, S., Li, P., Wuhui, C.: Pyramid: a layered sharding blockchain system. In: Proceedings of IEEE International Conference on Computer Communications, pp. 1–10 (2021)
9. Al-Bassam, M., Sonnino, A., Bano, S., Hrycyszyn, D., Danezis, G.: Chainspace: a sharded smart contracts platform. In: Proceedings of 25th Annual Network and Distributed System Security Symposium, pp. 1–15 (2018)
10. Tao, Y., Li, B., Jiang, J., Ng, H.C., Wang, C., Li, B.: On sharding open blockchains with smart contracts. In: Proceedings of IEEE 36th International Conference on Data Engineering, pp. 1357–1368 (2020)
11. Li, S., Yu, M., Yang, C.-S., Avestimehr, A.S., Kannan, S., Viswanath, P.: PolyShard: coded sharding achieves linearly scaling efficiency and security simultaneously. In: Proceedings of IEEE International Symposium of Information Theory, pp. 203–208 (2020)
12. Wang, G., Shi, Z.J., Nixon, M., et al.: SoK: sharding on blockchain. In: Proceedings of the 1st ACM Conference on Advances in Financial Technologies, pp. 41–61 (2019)
13. Chen, J., Li, Z.H., Gao, D.X., et al.: A sliced load balancing approach using state imputation. Comput. Sci. **49**(11), 302–308
14. Xu, Z., Han, S., Chen, L.: CUB, a consensus unit-based storage scheme for blockchain system. In: 2018 IEEE 34th International Conference on Data Engineering (ICDE), pp. 173–184. IEEE (2018)
15. Wang, J., Wang, H.: Monoxide: scale out blockchains with asynchronous consensus zones. In: 16th USENIX Symposium on Networked Systems Design and Implementation (NSDI 19), pp. 95–112 (2019)
16. Dang, H., Dinh, T.T.A., Loghin, D., et al.: Towards scaling blockchain systems via sharding. In: Proceedings of the 2019 International Conference on Management of Data, pp. 123–140 (2019)
17. Qi, X., Zhang, Z., Jin, C., et al.: BFT-Store: storage partition for permissioned blockchain via erasure coding. In: 2020 IEEE 36th International Conference on Data Engineering (ICDE), pp. 1926–1929. IEEE (2020)
18. Xu, C., Zhang, C.: vChain+: optimizing verifiable blockchain Boolean range queries. In: Proceedings of the 2022 International Conference on Data Engineering (ICDE) (2022)

19. Jason, T., ReitwieBner, C.: A scalable verification solution for blockchains. arXiv preprint arXiv:1908.04756 (2019)
20. Schwartz, D., Youngs, N., Britto, A.: The ripple protocol consensus algorithm. Ripple Labs Inc White Paper **5**(8), 151 (2014)
21. Cai, Z., Liang, J., Chen, W., et al.: Benzene: scaling blockchain with cooperation-based sharding. IEEE Trans. Parallel Distrib. Syst. **34**(2), 639–654 (2022)
22. Pirlea, G., Kumar, A., Sergey, I.: Practical smart contract sharding with ownership and commutativity analysis. In: Proceedings of 42nd ACM SIGPLAN International Conference on Programming Language Design and Implementation, pp. 1327–1341 (2021)
23. Zestaken, ZJChain. Block https://github.com/zestaken/BlockChainDemo

A Privacy Preserving Method for Trajectory Data Publishing Based on Geo-Indistinguishability

Fengyun Li[1,2(✉)], Jiaxin Dong[2], Mingming Chen[2], and Peng Li[3]

[1] Beijing Key Laboratory of Internet Culture and Digital Dissemination Research,
Beijing 110101, China
lifengyun@mail.neu.edu.cn
[2] School of Computer Science and Engineering, Northeastern University,
Shenyang 110819, China
[3] School of Intelligent and Engineering, Shenyang City University, Shenyang 110112, China

Abstract. Trajectory data contains rich space-time information. Analysis and research on it can provide important information for location-based services. However, these data are usually private to users, and there will be serious privacy leakage if they are directly published to a third party. Therefore, privacy preserving should be carried out before the data is published. Aiming at the problems of existing research in terms of low privacy preserving degree and data utility, a privacy preserving algorithm for trajectory data publishing based on Geo-indistinguishability is proposed. The algorithm mainly includes two parts of constructing a noise trajectory set and selecting the optimal noise trajectory. n constructing a set of noise trajectories, the total noise value of each trajectory is first calculated using a planar Laplace function, so that the algorithm satisfies geographical indistinguishability as a whole. Then, the total noise value is randomly decomposed and assigned to the position points of the trajectory to generate noise trajectories. Then, based on the trajectory uncertainty model, calculate the Manhattan distance between each noise trajectory at each time, obtain the number of indistinguishable trajectories for each noise trajectory, and preferentially select a noise trajectory with the largest number of indistinguishable trajectories to replace the real trajectory for publishing. The experimental results show that compared with other methods, the algorithm proposed in this paper has a higher degree of privacy preserving and better data utility.

Keywords: trajectory privacy preserving · Geo-indistinguishability · differential privacy · trajectory uncertainty

1 Introduction

Nowadays, with the development of information technology, a variety of sensor devices such as smart phones, wearable devices, car navigators, etc. are increasingly pouring into our lives [1]. Especially in the Internet of Vehicles, a large number of population

X. Yang et al. (Eds.): ADMA 2023, LNAI 14179, pp. 633–647, 2023.
https://doi.org/10.1007/978-3-031-46674-8_44

mobility data have been generated. The research on these data can be used for commercial organizations, traffic management departments, some legal information inquiry agencies, and other location-based services to provide important values [2], such as traffic monitoring, route planning, hot spot tracking, targeted advertising and accident warning, etc. Due to the fact that this trajectory information is usually private to the user and sensitive, directly publishing these data will have serious privacy leakage problems. Relevant studies have shown that due to the high recognizability of the original trajectory, four spatiotemporal points are enough to uniquely identify 95% of individuals [3]. When some attackers have already mastered certain background information through some means, analyzing the published trajectory data may reveal more privacy of users. Therefore, privacy preserving before the publishing of trajectory data has become a research hotspot.

Researchers have carried out a lot of research work on the privacy preserving issue of trajectory data publishing, and have also achieved many meaningful research results. Dong et al. [4] proposed an obfuscation method for privacy preserving, by adding different data in the trajectory social graph. The random noise is achieved, but only when the noise level is high enough, they can resist the record link attack. Iwata et al. [5] proposed a segmentation technique, that is, to segment each trajectory and use a different pseudo identifier for each trajectory segment. But this simple technique cannot significantly reduce monotonicity. Based on this, Xu et al. [6] proposed a location privacy preserving method based on the trajectory segmentation of the starting point and ending point. Salas et al. [7] proposed to iteratively exchange partial trajectories between users, and finally output a trajectory composed of multiple user trajectory fragments. This can effectively resist recording link attacks, but greatly reduces the utility of data.

K-anonymity is the first privacy standard proposed by Sweeney [8] in relational databases. It has also been widely used in trajectory databases. Zang and Bolot [9] proposed to realize K-anonymity through spatiotemporal point generalization, which became the most basic method to realize K-anonymity in trajectory database. Mehta et al. [10] proposed l-Diversity, the idea behind it is that the sensitive attributes contained in any query result have at least l different values, so that the attacker can only obtain the user's sensitive information with a probability of $1/l$. Furthermore, Li et al. [11] proposed t-proximity, which imposes statistical constraints on sensitive attributes, rather than numerical attributes defined by l-diversity. Mehta [12] proposed differential privacy, which has become an important method of privacy preserving, and has been well received by everyone. However, due to the sparsity of trajectory data and other characteristics, it cannot be used directly for privacy preserving, so many studies have extended differential privacy. Ye et al. [13] designed differential privacy and sliding window model The combined method achieves privacy preserving for the publishing of real-time trajectory statistical data. Wei et al. [14] successfully extended differential privacy to two-dimensional space and proposed geographic indistinguishability.

Aiming at the existing research issues, in order to improve the degree of data privacy preserving and the effectiveness of data, this paper utilizes geographical indistinguishability to achieve privacy preserving for trajectory data publishing and proposes a trajectory data publishing privacy preserving algorithm.

2 Related Definitions

The trajectory is a spatiotemporal sequence in which the position information of the moving object is arranged in chronological order, which can be expressed as: $tr = \{l_1, l_2, \ldots, l_n\}$. Where $l_i = (x_i, y_i, t_i)$ represents the i-th spatiotemporal point of tr, the sampling time is t_i, , the position is (x_i, y_i), $1 \leq i \leq n$, $t_1 < t_2 < \cdots < t_i < \cdots < t_n$, and n is the length of the trajectory. For the need of calculation accuracy, we discretize the geographic area covered in dataset D into $m \times n$ units. As shown in Fig. 1, the trajectory after the change is still the original representation, but its x_i and y_i are now represented as the discretized x-axis and y-axis coordinates. For example, the trajectory tr can be expressed as $tr = \{<1, 1, t_1>, <2, 2, t_2>, <3, 3, t_3>, <4, 3, t_4>\}$, where t_i is its sampling time, and then convert it to latitude and longitude as needed. The trajectory database D is a collection of trajectories, expressed as: $D = \{tr_1, tr_2, \ldots, tr_m\}$, where m is the number of trajectories.

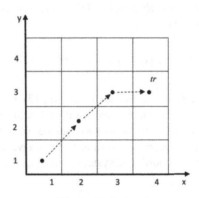

Fig. 1. Trajectory

Definition 1. ε-differential privacy [12]. Given a random algorithm M, for any adjacent dataset D and D', that is, the two are different at most in one record, for all sets $S \subseteq Range(M)$ satisfies the Eq. (1), it is said that the algorithm M satisfies ε-differential privacy.

$$Pr[M(D) \in S] \leq e^{\varepsilon} \cdot Pr[M(D') \in S] \tag{1}$$

Among them, $Range(M)$ is all possible outputs of M, ε is the privacy budget, which represents the degree of privacy preserving provided by algorithm M. The smaller ε, the higher the degree of privacy preserving, and the greater the data disturbance and the worse the data utility.

There are two commonly used implementation mechanisms for differential privacy: Laplace mechanism [12] and Exponential mechanism.

Definition 2. Geographical indistinguishability [14]. For all $x, x' \in X, z \in Z$, where X is the set of real locations and Z is the set of possible disturbed locations. If the mechanism K satisfies ε-geography indistinguishability, for all x, x', that is:

$$K(x)(z) \le e^{\varepsilon d(x,x')} \tag{2}$$

where $d(x, x') \le r$, r is the privacy radius, $K(x)(z)$ represents the probability that the real position point is x but the disturbed position point is z. Equation (2) means if the input is x, the probability of output z is p, then input x', and the probability of output z is in the range of $(p - e^{\varepsilon d(x,x')}, p + e^{\varepsilon d(x,x')})$.

Kacha et al. [15] improved Laplace, and proposed the Planar Laplace function to make it suitable for two-dimensional space and satisfy geographic indistinguishability.

Definition 3. Planar Laplace function [15]. Given $\varepsilon \in R^+$, the original position $x_0 \in R^2$, the perturbation position $x \in R^2$ generated by a certain noise-adding mechanism, its probability density function is:

$$D_\varepsilon(x_0)(x) = \frac{\varepsilon^2}{2\pi} e^{-\varepsilon d(x_0,x)} \tag{3}$$

Among them, $\frac{\varepsilon}{2\pi}$ is the normalization factor, and Eq. (3) is the plane Laplace function centered on x_0. From Eq. (3), it can be seen that the probability density function of the Planar Laplace function is only related to the distance, so it will be more convenient to use the polar coordinate system. According to the conversion equation, the Planar Laplace function of the polar coordinate system with x_0 as the origin is:

$$D_\varepsilon(r, \theta) = \frac{\varepsilon^2}{2\pi} r e^{-\varepsilon r} \tag{4}$$

where, the included angle θ and the radius r are independent of each other. Through derivation, the generation method of the two can be obtained. θ is relatively simple, and the random value is directly taken in the interval $[0, 2\pi]$. The generation of the radius r is more complicated. The calculation equation is:

$$r = C_\varepsilon^{-1}(p) = -\frac{1}{\varepsilon}\left(W_{-1}\left(\frac{p-1}{e}\right) + 1\right) \tag{5}$$

where, p is a uniformly distributed random number in the interval (0,1), and the $W_{-1}(X)$ function is the part of the Lambert function $W(X)$ in the interval $(-\infty, -1)$. This paper will generate the total noise of the trajectory based on this method, perturb it, and generate the noise trajectory.

Definition 3. Manhattan distance. The Manhattan distance between two space-time points at the same time $l_i = (x_i, y_i, t_i)$ and $l_i' = (x_i', y_i', t_i)$ is:

$$d\left(l_i, l_i'\right) = |x_i - x_i'| + |y_i - y_i'| \tag{6}$$

3 Privacy Preserving Algorithm Based on Geographic Indistinguishability

3.1 The Overall Design of the Algorithm

A privacy preserving algorithm for trajectory data publishing based on geographic indistinguishability (GTP for short) is proposed. The algorithm mainly includes two modules: building a set of noisy trajectories and selecting trajectories to be output. The former generates noise for each trajectory, and each trajectory in the set is geographically indistinguishable from the real trajectory. At the same time, in order to preserve the utility of the data, only trajectories with similar trends to the real trajectory are selected. The latter selects the noise trajectory with the highest number of indistinguishable trajectories from the noise trajectory set and publishes it instead, which better protects the privacy of the trajectory. The overall flow of the algorithm is shown in Algorithm 1.

In Algorithm 1, the BNT algorithm (Algorithm 2) is used firstly to construct a set of noise trajectories for each real trajectory, and merge these sets into a set of noise trajectories S. After creating a set of noise trajectories, use the CNT algorithm (Algorithm 3) for each real trajectory to select an optimal noise trajectory for publishing.

Algorithm 1. GTP algorithm

Input: Trajectory dataset T, privacy budget ε, trajectory similarity threshold α
Output: Set of noise trajectories to be published T'
1: $S \leftarrow \varnothing$ /* Total noise trajectory dataset */
2: for $tr \in T$ do
3:   $NS \leftarrow BNT(tr, \varepsilon, \alpha)$ /* Generate noise trajectory set NS for trajectory tr */
4:   $S \leftarrow S \cap NS$
5: end for
6: $T' \leftarrow CNT(S)$ /* Select a noise trajectory to be published for each trajectory*/
7: return T'

3.2 Construction Method of Noise Trajectory Set

The main purpose of this section is to generate a set of noise trajectories that meet the conditions for each real trajectory, and a noise trajectory set building algorithm (Building Noise Trajectory, BNT) is proposed (Algorithm 2). The algorithm is based on the geographic indistinguishability perturbation mechanism to achieve the purpose of adding noise to trajectory data. Among them, the polar coordinate representation method is used to generate noise that satisfies the planar Laplacian random number.

The specific process of constructing the noise trajectory set is as follows:

1) Firstly, define the noise value vector *noise* and the direction vector d of the noise with a length of $2n$, and initialize it to 0 (line 3–4), where n is the length of the trajectory.
2) Calculate the total noise totleNoise (line 5–7). Firstly, randomly select the value p from the uniform distribution of $[0, 1)$, and then obtain a random value r that satisfies the planar Laplace distribution by Eq. (4). The total noise value that can be added to the trajectory tr is $r * 2n$.
3) Randomly select the *totleNoise* number of positions to construct the noise vector, and calculate the number of 1 in the direction vector of the noise (line 8–15).
4) Calculate the set of noise trajectories generated by the noise value vectors and the noise direction vectors. There are q positions in the noise direction vector d that are 1, so there is 2^q combinations for adding noise to these q positions. To construct the $j(1 \leq j \leq 2^q)$ noise vector as an example.

Firstly, make a copy of d to get d_{copy}, convert the value j to q-bit binary b, and traverse each bit in b. If the i-th bit is 1, then the i-th 1 in d_{copy} will be changed to -1. If the i-th bit of b is 0, then the i-th 1 in d_{copy} will not be changed. After the traversal is completed, noise is added to the trajectory tr according to noise and d_{copy} to generate a noise-added trajectory tr_{noise}. In actual implementation, the position point of the noise-added trajectory will be judged whether it exceeds the boundary. If it exceeds the boundary, displacement processing is performed to make the position point just on the boundary. Finally, the correlation between the noise trajectory tr_{noise} and the real trajectory tr is calculated. If $u > \alpha$, it is considered that the trends of the two trajectories are similar. The noise trajectory tr_{noise} can be published instead of tr, and fill it in the noise trajectory set of trajectory tr. On the contrary, it is considered that the trends of these two trajectories are not similar. If it is used instead of tr, it will have poor effectiveness, so it will not be added to the noise trajectory set.

Algorithm 2. BNT algorithm

Input: trajectory tr, privacy budget ε, similarity threshold α
Output: noise trajectory set NS
1: NS $\leftarrow \varnothing$ /* the noise trajectory set of trajectory tr */
2: $n \leftarrow tr$.length
3: noise $\leftarrow R^{2n}$ /* noise value vector, initialized as 2n dimensional Zero vector */
4: $d \leftarrow R^{2n}$
5: $p \leftarrow Uniform(0,1)$
6: $r \leftarrow C_\varepsilon^{-1}(p)$
7: totleNoise $\leftarrow r*2n$ /*total noise*/
8: $q \leftarrow$ totleNoise
9: **while** totleNoise>0 **do**
10:    $i \leftarrow$ Uniform(0,2n)
11:    $noise_i$++ /* The i-th component of the noise value vector increases by 1 */
12:    if (d_i=1) q--
13:    $d_i \leftarrow 1$
14:    totleNoise--
15: end while
16: for j from 0 to 2^q-1 do
17:    $d_{copy} \leftarrow$ copy(d)
18:    $b \leftarrow$ toBinary(j,q)   /*Change j to binary b with q bits */
19:    for k from 1 to q do
20:      if ($b_k = 1$) d_{copy_t} =-1 /*t is the index of the k-th 1 in d$_{copy}$*/
21:    end for
22:    $tr_{noise} \leftarrow tr$+noise$\cdot d_{copy}$
23:    if ($u = r(tr_{noise}, tr)>\alpha$)
          NS $\leftarrow$ NS $\cup$ tr_{noise}
24: end for
25: return NS

---

5) Return the set of noise trajectories.

When the algorithm is implemented, this paper also has a certain limit on the number of noisy trajectories. If the number of noisy trajectories is less than 4, then regenerate the random value p, that is, rerunning the algorithm. If the number of noisy trajectories is greater than 10, it is sorted according to the correlation value, and only the top ten most correlated trajectories are taken to reduce the complexity of the next step.

The above is only the noise trajectory set of a real trajectory, that is, the *NS* in Algorithm 1. The BNT algorithm must be applied to each real trajectory in the dataset *T*, and then adding its noise trajectory set to a set again to form the entire noise trajectory set, that is, *S* in Algorithm 1.

3.3 Optimal Noise Trajectory Selection Algorithm

The uncertainty of the trajectory is objective, and Kacha et al. [15] defined the uncertain trajectory and proposed the (k, δ)-anonymity problem. Based on the inspiration of the (k, δ)-anonymity problem, this paper designs an algorithm on how to select a more private trajectory from a collection of noise trajectories to be published.

The main purpose of this section is to select a noise trajectory from the set of noise trajectories of each real trajectory. For this purpose, an optimal noise trajectory selection (Choosing Noise Trajectory, CNT for short) algorithm is designed, which is based on the idea of (k, δ) to select an optimal trajectory. The specific steps are shown in Algorithm 3.

When the distance between two trajectories at a certain time is less than δ, the two positions are called indistinguishable. When the number of indistinguishable positions of the two trajectories is greater than the product of β and the length of the trajectory, the two trajectories are called indistinguishable.

The algorithm extracts a real trajectory noise trajectory set (denoted as NS) from the noise trajectory set S constructed in the previous step, and extracts a noise trajectory tr_{noise} from NS, and calculates the Manhattan distance between it and all the noise trajectories in S except NS at each time of tr_{noise}. If the distance between a noise trajectory and tr_{noise} exceeds a certain ratio (for example, half), the distance is less than the threshold δ, then this noise trajectory is considered to be similar to tr_{noise}, and the trajectory can be regarded as tr_{noise} indistinguishable trajectories. After that, count the number of indistinguishable trajectories of tr_{noise}, calculate the number of indistinguishable trajectories for each trajectory with added noise in NS, and preferentially select a trajectory with the largest number of indistinguishable trajectories to replace the real trajectory of NS for publication, and delete all trajectories from NS except for this one. Repeat the above operation until the noise trajectory set S is traversed.

Among them, a dictionary type map is defined, where the key is the identifier of the noise trajectory, and the value is the number of positions that satisfy the indistinguishability between the noise trajectory and tr_{noise}. The **map** is calculated by the CNIT (Calculate the Number of Indistinguishable Location of the Trajectory, CNIT) function, which calculates the distance between the trajectory tr_{noise} and each trajectory in the S-NS set at each moment. When the distance between trajectory tr' and tr_{noise} at certain moment is less than the threshold δ, and if the *map* already contains the identifier of this trajectory, add 1 to the value and re-add it to the map, otherwise, set this trajectory identifier as the key, and add the value 1 to the **map**. After tr_{noise} is traversed, then return **map**.

Algorithm 3. CNT algorithm

Input: noise trajectory set S, indistinguishable radius δ, indistinguishable trajectory threshold β
Output: Noise trajectory collection T'
1: $T' \leftarrow \varnothing$
2: **for** i from 1 to S.size **do**
3:    NS $\leftarrow S_i$ /*NS represents the noise trajectory set of the trajectory i */
4:    numoftr $\leftarrow R^{NS.size}$ /* numoftr stores the number of indistinguishable
                                       trajectories for each noise trajectory in NS*/
5:    **for** j from 1 to NS.size **do**
6:      $tr_{noise} \leftarrow NS_j$ /* tr_{noise} represents trajectory i in NS */
7:      map $\leftarrow \varnothing$ /*Map is a dictionary type */
8:      map $\leftarrow$ CNIT (tr_{noise}, S-NS)
9:      $n \leftarrow tr_{noise}$.length
10:     num $\leftarrow$ |map.value $\geq \beta * n$| /*num is the number of values larger than $\beta*n$ in
                                             map */
11:     numoftr$j$ $\leftarrow$ num
12:    **end for**
13:    maxindex $\leftarrow$ IndexOfArr(numoftr) /*The IndexOfArr function finds the index of the
                              element with the highest value in numoftr*/
14:    $T' \leftarrow T' \cup NS_{maxindex}$
15:    $S \leftarrow S$-NS
16:    $S \leftarrow S \cup NS_{maxindex}$
17: **end for**
18: **return T'**
18:    $b \leftarrow$ toBinary(j,q)   /*Change j to binary b with q bits */
19:    for k from 1 to q do
20:      if ($b_k = 1$)
       $d_{copy_t} = -1$ /*t is the index of the kth 1 in d_{copy}*/
21:    end for
22:    $tr_{noise} \leftarrow tr$+noise$\cdot d_{copy}$
23:    if ($u = r(tr_{noise}, tr) > \alpha$)
       NS $\leftarrow$ NS $\cup tr_{noise}$
24: **end for**
25: **return NS**

---

4 Experimental Analysis

4.1 Experimental Data and Environment

This experiment is based on an environment with a processor of 3.20 GHz and a memory of 8 GB in Java language. The experiment is based on the classic algorithms GNoise [16] and PNoise [16] in the same application scenario, and use three real datasets T-Drive and Geolife for comparative analysis. Among them, the T-Drive dataset describes the GPS trajectories of 10,357 taxis in Beijing within a week, including about 15 million location points, and the total distance of the trajectory is about 9 million kilometers. Geolife dataset is the trajectory of 182 users collected by Microsoft Research Asia in five years, the sampling interval is about 5s, and the total distance is 1,292951 km. Data preprocessing steps mainly includes: Keep the time to minutes, and keep the latitude, longitude, and precision to a few decimal places according to the data situation. Remove the repeated time and the data whose latitude and longitude are not much different from the previous moment. According to the length of the user's trajectory, the dataset is divided into three parts with lengths between 5–10, 10–15 and 15–20.

The number of trajectories in the dataset after processing is shown in Table 1.

Table 1. Number of trajectories in datasets

datasets	trajectories length		
	5–10	10–15	15–20
T-Drive	379	34	80
Geolife	20	20	39

4.2 Security Analysis

In this paper, the security of the algorithm is evaluated by the probability of sub-trajectory link attack. The probability of sub-trajectory link attack means that the attacker knows the location of the user at a certain moment, and the dataset published by the trajectory collection server contains the user. The attacker passes through these time and space. The points infer the probability of the user corresponding to the trajectory in the dataset.

Given stator trajectory $tr' = \{l'_1, l'_2, \ldots, l'_m\}$, trajectory tr, positioning error δ. If tr contains all the sampling moments in tr'. And for each spatiotemporal point l'_i ($i \in [1, m]$) in tr', it satisfies $dis(tr(t_i), l_i) \leq \delta$, it is said that tr contains tr'. Otherwise, it is said that tr does not contain tr'. Based on the above definition, given a sub-trajectory, the number of sub-trajectories contained in the trajectory dataset T can be calculated (see Eq. (7) and (8)).

$$num(tr, tr') = \begin{cases} 1 & \forall l \in tr' \rightarrow dis(tr(t_i), l_i) \leq \delta \\ 0 & otherwise \end{cases} \tag{7}$$

$$num\left(T, tr'\right) = \sum_{tr \in T} num(tr, tr') \tag{8}$$

The sub-trajectory link attack is carried out from this point in time and space, and the probability of success is shown in Eq. (9).

$$p\left(T, tr'\right) = 1/nmu(T, tr') \tag{9}$$

In this experiment, the lengths of three sub-trajectories are set to 2, 5, and 7. Among them, the sub-trajectories with lengths 2 and 5 run on all datasets, and the sub-trajectories with length 7 are only tested in the range of 10–15 and 15–20. The privacy budget parameters $\varepsilon = 0.1$, trajectory similarity threshold $\alpha = 0.7$, positioning error $\delta = 1$, which is much larger than the noise added by the position point, ensuring that the sub-trajectory will be linked to the real trajectory. But this will also make the number of trajectories linked to the sub-trajectory increase, and the probability of success of the attack will decrease. Figure 2 and Fig. 3 show the results of our algorithm and GNoise, PNoise running on datasets T-Drive and Geolife, when the sub trajectory lengths are 2, 5, and 7. a, b, and c indicate that the trajectory length is between 5–10, 10–15 and 15–20, respectively.

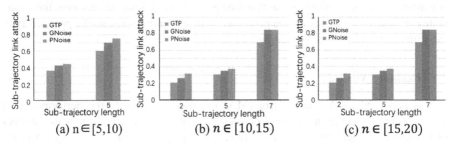

(a) n ∈ [5,10)   (b) $n \in [10,15)$ (c) $n \in [15,20)$

Fig. 2. Experiments based on the T-Drive dataset

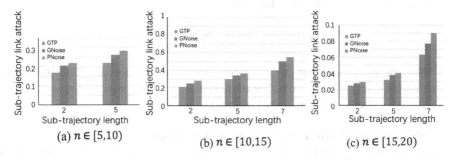

(a) $n \in [5,10)$ (b) $n \in [10,15)$ (c) $n \in [15,20)$

Fig. 3. Experiments based on Geolife dataset

From the experimental results of the two datasets, it can be found that the algorithm GTP in this paper is better than the GNoise and PNoise algorithms in resisting sub-trajectory link attacks, that is, the probability of success of the attack is lower. This is because the algorithm in this paper randomly decomposes the total noise into a few copies, and then randomly select a few spatiotemporal points to add noise, and the position where the noise is added is more disturbed. And when selecting and outputting a noise trajectory from the set of noise trajectories, the distance between the noise trajectory and other noise trajectories will be calculated. Distinguish the noise trajectory with the most trajectories, so that more trajectories are close in distance, so there will be more trajectories when using sub-trajectories to link attacks, the probability of successful attack is lower.

Secondly, in the same dataset, the greater the length of the sub-trajectory, the higher the probability of successful attack. Since the longer the length of the sub-trajectory, the fewer trajectories can satisfy the condition, so the probability of linking to the real trajectory is higher. Taking the length range of 5 to 10 as an example, the T-Drive dataset has a sampling time interval of one week, and the time span is not very large. Moreover, this length interval contains 379 trajectories, with a large number. Therefore, several spatiotemporal points may be linked to multiple trajectories, and the probability of sub trajectory linking attacks is not very high or particularly small. The sampling interval of the Geolife dataset is four years, with a large span, and only 20 trajectories are included in the 5 to 10 length range. Therefore, many trajectory time points may not intersect. Therefore, when conducting sub trajectory linking attacks, there will be fewer trajectories containing the time requirements at that time point. In these fewer trajectories, there may be fewer trajectories that meet the position distance requirements, so the probability of sub trajectory linking attacks is higher.

4.3   Usability Analysis

In this paper, Mean Absolute Error (MAE) is used to evaluate the performance of the algorithm. Assume the original dataset $T = \{tr_1, tr_2, \ldots, tr_m\}$, m is the number of trajectories, where $tr_i = \{ <x_1, y_1, t_1>, <x_2, y_2, t_2>, \ldots, <x_{in}, y_{in}, t_{in}> \}$, in is the length of the trajectory tr_i. The dataset published after running the algorithm is $T' = \{tr'_1, tr'_2, \ldots, tr'_m\}$, $tr'_i = \{ <x'_1, y'_1, t'_1>, <x'_2, y'_2, t'_2>, \ldots, <x'_m, y'_m, t'_m> \}$, then the MAE of these two datasets is shown in Eq. (10).

$$MAE\left(T, T'\right) = \frac{1}{n}\sum\nolimits_{i=1}^{m} \frac{1}{in} \cdot \sum\nolimits_{j=1}^{in} (\left|x_j - x'_j\right| \left|y_i - y'_i\right|) \tag{10}$$

Next, the error evaluation and comparison analysis of the algorithm GTP (this paper proposed), GNoise and PNoise will be carried out from the two aspects of privacy budget parameters and trajectory similarity threshold.

(1) The influence of privacy budget parameter ε on error
In this experiment, the trajectory similarity threshold $\alpha = 0.7$, and the value of ε is from 0.1 to 1, increasing by 0.1. Figure 4 shows the results of our algorithm, GNoise and PNoise on the datasets T-Drive and Geolife, respectively. Since the total noise is calculated by multiplying the Lambeau value of a random number by the trajectory length, and the error calculation is divided by the trajectory length, the trajectory length does not have much influence on the average absolute error of the entire dataset. In this experiment, these two datasets are no longer divided into three parts according to the length of the trajectory.

From the experimental results, it can be seen that the MAE of our algorithm is much lower than the other two algorithms. This is because GNoise and PNoise randomly allocate noise according to the privacy budget parameters, and then directly output the trajectory of the added noise. After generating noise based on privacy budget parameters, our algorithm GTP decomposes the noise and constructs multiple noise trajectories that meet the parameters, forming a set of noise trajectories for the real trajectory. The algorithm also sets a trajectory similarity threshold, retaining only trajectories that have a similar trend to the real trajectory, further preserving the utility of the data. In addition, the Mean absolute error MAE varies with the privacy budget parameter ε. This is because

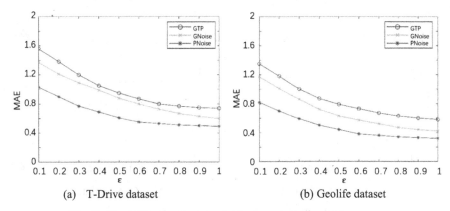

(a)   T-Drive dataset   (b) Geolife dataset

Fig. 4. The influence of privacy budget parameter ε on data error

the availability of data is directly proportional to the privacy budget, thus, the lager ε, the larger the noise added, the smaller the data error.

The algorithm in this paper is generating according to the privacy budget parameters. After the noise, the noise is decomposed, and a number of noise trajectories meeting the parameters are constructed to form a set of noise trajectories of the real trajectory, and the trajectory similarity threshold is set, and only trajectories that are not much different from the real trajectory are retained. The utility of the data is preserved. In addition, the average absolute error MAE decreases as the privacy budget parameter ε increases. This is because the availability of data is directly proportional to the privacy budget, so the larger ε, the smaller the added noise, and the smaller the data error.

(2) The influence of trajectory similarity α on error
In the experiment, set the privacy budget ε = 0.1, and the similarity threshold α is 0.4 to 1, increasing by 0.1. GNoise and PNoise have no correlation with α, so they are always fixed values. Figure 5 respectively show the running results of our algorithm and GNoise, PNoise based on datasets T-Drive and Geolife.

From Fig. 5, it can be seen that no matter what the value of α is, the MAE of our algorithm is smaller than the other two algorithms, and as α increases, the MAE of the proposed algorithm gradually becomes smaller. This is because when generating a set of noise trajectories, the greater the trajectory similarity α, the more stringent requirements on the similarity between the noise-added trajectory and the real trajectory. When α increases to a certain value, the change in MAE is no longer obvious. This is because as α increases, the number of trajectories in the noise trajectory set decreases. In this way, when selecting the noise trajectory to be output from the noise trajectory set, there will be few trajectories selected, which makes the probability of successful sub-trajectory link attacks increase. When α reaches a certain level, the MAE will no longer decrease significantly.

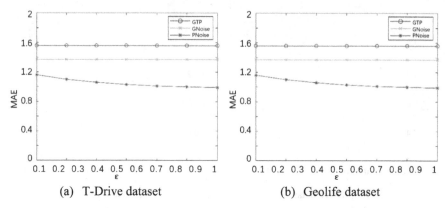

(a) T-Drive dataset   (b) Geolife dataset

Fig. 5. The influence of trajectory similarity α on error

5   Conclusion

Aiming at the privacy preserving problem of trajectory data publishing, a privacy preserving algorithm for trajectory data publishing based on geographic indistinguishability is proposed. The algorithm mainly includes two parts of constructing a set of noise trajectories and optimal noise trajectory selection. In the first part, the planar Laplace function is used to calculate the total noise value, randomly decompose it to the trajectory position points to generate a noise trajectory, which effectively guarantees the data utility. In the selection of noise trajectories to be published, based on the (k, δ)-anonymous model, the noise trajectory with the largest number of indistinguishable trajectories is selected for each trajectory to replace it for publication. Finally, the algorithm is compared with the classic algorithms GNoise and PNoise. Analysis and experimental results verify that the proposed algorithm has good performance in terms of data utility and privacy preserving.

Acknowledgement. This work has been supported the National Key Research and Development Program of China (No. 2022YFE0114200), the Open Research Fund from Beijing Key Laboratory of Internet Culture and Digital Dissemination Research (No. 2021021900051).

References

1. Cheng, W., Wen, R., Huang, H., Miao, W., Wang, C.: OPTDP: towards optimal personalized trajectory differential privacy for trajectory data publishing. Neurocomputing **472**, 201–211 (2022)
2. Xiao, Y., Jia, Y., Cheng, X., Wang, S., Mao, J., Liang, Z.: I know your social network accounts: a novel attack architecture for device-identity association. IEEE Trans. Dependable Secure Comput. **20**(2), 1017–1030 (2023)
3. Saito, S., Matsushima, T.: Upper bound on privacy-utility tradeoff allowing positive excess distortion probability. IEICE Trans. Fundam. Electron. Commun. Comput. Sci. **E105.A**(3), 425–427 (2022)
4. Dong, X., Zhang, W., Zhang, Y., You, Z., Gao, S., Shen, Y., et al.: Optimizing task location privacy in mobile crowdsensing systems. IEEE Trans. Ind. Inf. **18**(4), 2762–2772 (2022)

5. Iwata, N., Kamei, S., Alam, K.M.R., Morimoto, Y.: Location data anonymization retaining data mining utilization. In: Chen, W., Yao, L., Cai, T., Pan, S., Shen, T., Li, X. (eds.) ADMA 2022. LNCS, vol. 13726, pp. 407–418. Springer, Cham (2022). https://doi.org/10.1007/978-3-031-22137-8_30

6. Xu, J., Liu, L., Zhang, R., Xie, J., Duan, Q., Shi, L.: IFTS: a location privacy protection method based on initial and final trajectory segments. IEEE Access **9**, 18112–18122 (2021)

7. Salas, J., Megías, D., Torra, V.: SwapMob: swapping trajectories for mobility anonymization. In: Domingo-Ferrer, J., Montes, F. (eds.) PSD 2018. LNCS, vol. 11126, pp. 331–346. Springer, Cham (2018). https://doi.org/10.1007/978-3-319-99771-1_22

8. Sweeney, L.: K-anonymity: a model for protecting privacy. Int. J. Uncertain. Fuzziness Knowl. Based Syst. **10**(5), 557–570 (2002)

9. Zang, H., Bolot, J.: Anonymization of location data does not work: a large-scale measurement study. In: 17th Annual International Conference on Mobile Computing and Networking, pp. 145–156. Springer, Cham (2011)

10. Mehta, B.B., Rao, U.P.: Improved l-diversity: scalable anonymization approach for privacy preserving big data publishing. J. King Saud Univ. Comput. Inf. Sci. **34**(4), 1423–1430 (2022)

11. Li, N., Li, T., Venkatasubramanian, S.: t-Closeness: privacy beyond k-anonymity and l-diversity. In: 23rd IEEE International Conference on Data Engineering, pp. 106–115 (2007)

12. Luo, H., Zhang, H., Long, S., et al.: Enhancing frequent location privacy-preserving strategy based on geo-Indistinguishability. Multimedia Tools Appl. **80**, 21823–21841 (2021)

13. Ye, F., El Rouayheb, S.: Intermittent private information retrieval with application to location privacy. IEEE J. Sel. Areas Commun. **40**(3), 927–939 (2022)

14. Wei, J., Lin, Y., Yao, X., Zhang, J.: Differential privacy-based location protection in spatial crowdsourcing. IEEE Trans. Serv. Comput. **15**(1), 45–58 (2022)

15. Kacha, L., Zitouni, A., Djoudi, M.: KAB: a new k-anonymity approach based on black hole algorithm. J. King Saud Univ. Comput. Inf. Sci. **34**(7), 4075–4088 (2022)

16. Zhao, C., Zhu, Y., Du, Y., Liao, F., Chan, C.-Y.: A novel direct trajectory planning approach based on generative adversarial networks and rapidly-exploring random tree. IEEE Trans. Intell. Transp. Syst. **23**(10), 17910–17921 (2022)

HA-CMNet: A Driver CTR Model for Vehicle-Cargo Matching in O2O Platform

Zilong Jiang[1,3], Xiang Zuo[1], Kaifu Yuan[2], Lin Li[3(✉)], Dali Wang[1], and Xiaohui Tao[4]

[1] Department of Computer Science and Technology, Guizhou University of Finance and Economics, Guiyang, China
[2] School of Business Administration, Guizhou University of Finance and Economics, Guiyang, China
[3] School of Computer Science and Artificial Intelligence, Wuhan University of Technology, Wuhan, China
cathylilin@whut.edu.cn
[4] University of Southern Queensland, Toowoomba, Australia

Abstract. Vehicle-cargo matching is a key task in freight O2O platform, which involves the complex interactions of drivers, vehicles, cargos, cargo owners and environmental context. Many existing works mainly study the matching of vehicle routing problems, the matching based on the credit evaluation of both drivers and cargo owners, and the matching based on game theory from the perspective of management. Since the freight O2O platform is also the producer of big data, this study proposes a driver CTR prediction model for vehicle-cargo matching task from the perspective of data mining. Specifically, we first use the bottom-level attention network to model fine-grained preferences in driver historical behaviors, such as route interest and search interest, as well as fine-grained preferences such as vehicle type and vehicle length interest, and route interest in cargo owner historical behaviors. Then, the driver basic profile vector, cargo owner basic profile vector, cargo description vector, driver preferences vector and cargo owner preferences vector are feeded into the neural network composed of two deep components for feature interactions learning, and then a top-level attention network is used to learn the influencing factors of different information on the vehicle-cargo matching task. Finally, a multi-classifier is used for matching prediction. We conduct comprehensive experiments on real dataset, and the results show that, compared with the existing solutions, considering user preferences and adopting deep components collaborative modeling can improve the performance of vehicle-cargo matching to a certain extent, which verifies the effectiveness and superiority of the proposed model.

Keywords: Recommendation system · Online To Offline platform · Vehicle-cargo matching · User preferences modeling · Hierarchical attention network

© The Author(s), under exclusive license to Springer Nature Switzerland AG 2023
X. Yang et al. (Eds.): ADMA 2023, LNAI 14179, pp. 648–664, 2023.
https://doi.org/10.1007/978-3-031-46674-8_45

1   Introduction

In China, individual truck drivers undertake a significant portion of the road freight volume, but these truck drivers have low loading efficiency due to asymmetric logistics and distribution information, resulting in waste of social resources such as road resources and energy consumption. In order to eliminate the pain point of information asymmetry in the industry, freight O2O, a "Internet +logistics" model, has emerged. Freight O2O is a vehicle and cargo information matching platform for road logistics, in which cargo owners release vehicle demand (including route, time, cargo type, weight and volume, etc.) on the Internet and vehicle owners choose whether or not to receive orders. This platform makes use of huge industry information resources on the Internet and integrates supply chains by connecting "empty vehicles" and "cargo sources" based on big data technology. The intermediate link is omitted and the transportation cost and logistics management cost are reduced. Typical examples include intercity freight platforms such as YMM and Huochebang, and intra-city freight platforms such as Huolala.

As a key module of freight O2O platform, vehicle-cargo matching determines its stowing efficiency. Accurate and efficient vehicle-cargo matching can quickly find the most suitable vehicle owner to serve the cargo owner, which not only improves the operation efficiency of the vehicle owner, but also saves the time and cost of the cargo owner, effectively achieving the optimal allocation of resources. However, due to the complexity of the matching process involving multiple factors such as drivers, cargos, cargo owners, road transportation, and freight rates, existing researchers in the logistics field mainly focus on route optimization [1,2], multi-objective programming [3,4], credit evaluation [5–7], and freight rate prediction. The industry is paying more attention to the architecture and support technology of the freight O2O platform, their research on vehicle-cargo matching mainly focuses on mechanical matching using first-level indicators such as "location+route+freight". Therefore, the existing matching mode cannot fully meet the needs of both sides of the stow, and the matching accuracy and efficiency still need to be improved.

Even a small improvement in the accuracy of vehicle-cargo matching can effectively improve the experience of drivers and cargo owners, better connect "empty vehicles" with "cargo sources", and improve logistics costs and transportation efficiency to a certain extent. Based on the large-scale data accumulated by freight O2O platform, this study design an intelligent matching model considering user preferences to mine the potential rules contained in the historical information of freight data. Specifically, on the basis of traditional selection of first-level indicators such as "location+route+freight", we focus on exploring the potential interests of both drivers and cargo owners, and then adopt multi-component collaborative feature interactions learning module to obtain high-order abstract feature vectors containing complex correlation relationships, and finally feed them into the prediction module to calculate the final predicted value.

The main contributions of this paper are summarized as follows:

- In order to improve the matching efficiency between drivers and cargos on the freight O2O platform, we propose a personalized recommendation algorithm for cargos to drivers, named HA-CMNet. This model can not only mine the potential preference information of drivers and cargo owners, but also learn the complex correlation relationships between multiple features by using the fitting ability of deep components, which will help improve the accuracy of vehicle-cargo matching.
- We design a special multi-party information fusion module considering user preferences. Specifically, the bottom-level attention network is first used to model various fine-grained preferences in drivers' historical behaviors and cargo owners' historical behaviors, and then the top-level attention network is used to learn the influence factors of different information on vehicle -cargo matching task.
- We conduct extensive experiments on real world dataset, the results show that the proposed model achieves better results on the current task compared to the advanced and most relevant mainstream baselines. In addition, in order to verify the effectiveness of the proposed model, the influence of key parameters and model structure on performance is studied extensively and deeply.

The rest of this paper is organized as follows. In Sect. 2, the problem studied are briefly stated and the proposed model and its architecture are described in detail. In Sect. 3, the comprehensive experiments are conducted to verify the performance of the proposed model. In Sect. 4, the related work is introduced. In Sect. 5, the study work is summarized and the future work is prospected.

2   Our Approach

2.1   Problem Statement

On the premise that both driver and cargo owner are registered as members of the freight O2O platform, our study takes the driver as the reference point, and the matching module in the system pushes appropriate cargos information to the driver. This process involves the following steps:

Step 1: The driver logs in to the platform and release the information related to vehicle source. The system selects cargos from the original database that meet the first level indicators such as "origin-destination", "vehicle type-vehicle length", and "transportation cost" based on the driver's target route;

Step 2: If the cargos information obtained in Step 1 is too much, the intelligent matching model will further select the accurate target cargos from the sources set satisfying Step 1 by combining multiple second-level indicators of driver, cargos, cargo owner and context, and then push them to the driver after sorting.

Step 3: If the cargos information obtained in Step 1 is too little, the intelligent matching model will select relevant target cargos from the original source database by combining the driver's location information and a variety of second-level indicators such as driver, cargos, cargo owner, and context, and push them to the driver after sorting.

This study mainly focuses on Step 2 and Step 3, and the second-level indicators will vary according to specific scenarios.

2.2 Architecture of the Proposed HA-CMNet Model

In this section, we describe the proposed HA-CMNet model, which consists of four functional modules, namely input data representation module, multi-party information fusion module considering user preferences, feature interactions learning module and matching calculation and loss function module. The complete model is a hybrid structure composed of these four modules, as shown in Fig. 1.

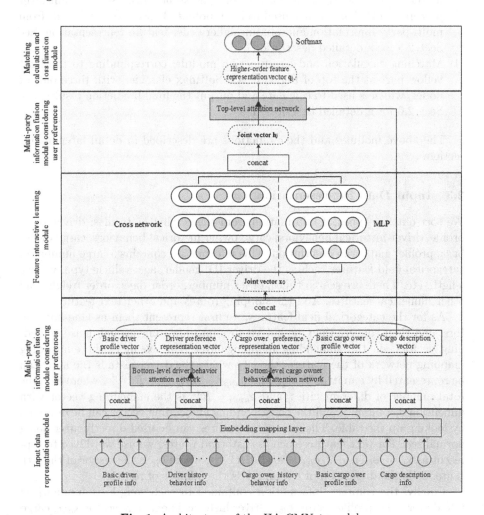

Fig. 1. Architecture of the HA-CMNet model.

The role of each module is as follows:

(1) Input data representation module: corresponding to the light red part at the bottom of Fig. 1, it is a embedding mapping layer, which is used to convert sparse features of input data into low-dimensional dense embedding vectors. See Sect. 2.3 for a detailed description.

(2) Multi-party information fusion module considering user preferences: corresponding to the light gray part in the middle of Fig. 1, it is a two-layer attention network, which is used to obtain the user preferences representation and the influence factors of different information on the vehicle -cargo matching task. See Sect. 2.4 for a detailed description.

(3) Feature interactions learning module: corresponding to the light blue part in the middle of Fig. 1, it consists of a cross network [8] and multilayer perceptron(MLP), which is used to learn potential correlation relations from multi-party information and obtain higher-order feature representation. See Sect. 2.5 for a detailed description.

(4) Matching calculation and loss function module: corresponding to the light yellow part at the top of Fig. 1, it is a Softmax classifier with three output nodes, which is used to calculate and output the final prediction results. See Sect. 2.6 for a detailed description.

The above modules and their functions are described in detail later in this section.

2.3   Input Data Representation

We sort out five types of information from the original data files: driver basic profile, driver historical behaviors, cargo owner historical behaviors, cargo owner basic profile, and cargo description. This information contains a large number of categorical field features such as the driver ID, gender, age, vehicle type, vehicle length, etc. There are search times, order number, order days, order freight and other numerical features. And the shipping routes and other text features.

As for the categorical field features, we first represent them as One-hot vectors, and then convert the One-hot vectors into low-dimensional dense embedding vectors via the embedding mapping layer. For example, the embedding mapping network of the feature "the city where the driver installs the app" can be represented by a matrix $\mathbf{V} = [\mathbf{e}_{app1}, \mathbf{e}_{app2}, ..., \mathbf{e}_{appK}] \in R^{K \times d_v}$, where K is the total number of different cities, and $\mathbf{e}_{appj} \in R^{d_v}$ is the embedding vector with dimension d_v of item j, and the corresponding embedding vector can be obtained by looking up the table. The numerical features can be used directly after being normalized. For text features such as historical routes, we use word2vec to represent as $\mathbf{x}_b = [\mathbf{e}_{l1}, \mathbf{e}_{l2}, ..., \mathbf{e}_{lt}, ..., \mathbf{e}_{lT}] \in R^{T \times d_e}$, where $\mathbf{e}_{lt}$ is the vector of the t-th route, d_e is the dimension of $\mathbf{e}_{lt}$, and T is the number of routes.

Finally, the complex information contained in each sample is summarized into driver basic profile vector $\mathbf{x}_{dbp}$, driver historical behaviors vector $\mathbf{x}_{dhb}$, cargo owner historical behaviors vector $\mathbf{x}_{cohb}$, cargo owner basic profile vector $\mathbf{x}_{cobp}$, and cargo description vector $\mathbf{x}_{cd}$ after primary key association. These vectors are concatenated into the joint input vector $\mathbf{x}_{input} = [\mathbf{x}_{dbp}, \mathbf{x}_{dhb}, \mathbf{x}_{cohb}, \mathbf{x}_{cobp}, \mathbf{x}_{cd}]$ of the sample . In addition, the class label of each sample is a One-hot vector represented by three kinds of tags: browse cargos, click cargos and make phone calls.

2.4 Multi-party Information Fusion Considering User Preferences

For driver users, whether the transaction of the target cargo is finally concluded is not only related to the first-level index of "location + route + freight rate", but also involves the actual willingness of both the driver and the cargo owner. Understanding and restoring the potential intentions of both the driver and the cargo owner is the key issue to be addressed in this section. The attention mechanism can consider the weight relationship between different elements on the basis of a global perspective. Taking advantage of this advantage, we adopt hierarchical attention network [9] to first model the fine-grained preferences of driver and cargo owner, and then learn the influence factors of different information on the vehicle-cargo matching task, and finally incorporate them into the subsequent matching prediction calculation.

2.4.1 The Bottom-Level Attention Network Based User Preferences Modeling

• *Driver Preferences Modeling*

We design a specialized driver behavior attention network to model driver preferences, in order to obtain driver preferences representation vector. The input of this part comes from the joint input vector $\mathbf{x}_{input}$. First, we extract each behavior $\mathbf{x}_{dhbi}$ from the driver historical behaviors vector $\mathbf{x}_{dhb} = [\mathbf{x}_{dhb1}, ..., \mathbf{x}_{dhbM}]$, and then concatenate it with the driver basic profile vector $\mathbf{x}_{dbp}$, the cargo owner basic profile vector $\mathbf{x}_{cobp}$, and the cargo description vector $\mathbf{x}_{cd}$ to obtain the input vector $[\mathbf{x}_{dbp}, \mathbf{x}_{cobp}, \mathbf{x}_{cd}, \mathbf{x}_{dhbi}]$ for this part. The output is the learned driver preferences representation vector $\mathbf{x}_{dhb}'$.

Firstly, we model the attention score α_m for each driver behavior, as shown in Eq. (1),

$$\alpha_m = \mathbf{w}_1^T \cdot s\left(\mathbf{W}_1[\mathbf{x}_{dbp}, \mathbf{x}_{cobp}, \mathbf{x}_{cd}, \mathbf{x}_{dhbi}]\right), \tag{1}$$

where $\mathbf{w}_1 \in R^{d_2}$ and $\mathbf{W}_1 \in R^{d_2 \times d_1}$ are parameters of the driver behavior attention network, d_1 is the dimension of the input vector of the attention network, d_2 is the dimension of the output vector of the attention network, and $s(x)$ is the nonlinear activation function.

Then, by normalizing the attention score α_m in Eq. (1), we obtain the final attention score α_m' as follows,

$$\alpha_m' = \frac{exp(\alpha_m)}{\sum_{l=1}^{M} exp(\alpha_l)}, \tag{2}$$

Finally, we concatenate the driver behavior vectors weighted by the final attention scores into a driver preferences representation vector as follows,

$$\mathbf{x}'_{dhb} = \sum_{m=1}^{M} \alpha'_m \mathbf{x}'_{dhbm}. \tag{3}$$

• *Cargo Owner Preferences Modeling*

Similarly, we design a specialized cargo owner behavior attention network to model cargo owner preferences, in order to obtain cargo owner preferences representation vector. The input of this part comes from the joint input vector $\mathbf{x}_{input}$. First, we extract each behavior $\mathbf{x}_{cohbi}$ from the cargo owner historical behaviors vector $\mathbf{x}_{cohb} = [\mathbf{x}_{cohb1}, ..., \mathbf{x}_{cohbN}]$, and then concatenate it with the driver basic profile vector $\mathbf{x}_{dbp}$, the cargo owner basic profile vector $\mathbf{x}_{cobp}$, and the cargo description vector $\mathbf{x}_{cd}$ to obtain the input vector $[\mathbf{x}_{dbp}, \mathbf{x}_{cobp}, \mathbf{x}_{cd}, \mathbf{x}_{cohbi}]$ for this part. The output is the learned cargo owner preferences representation vector $\mathbf{x}'_{cohb}$.

Firstly, we model the attention score β_n for each cargo owner behavior, as shown in Eq. (4),

$$\beta_n = \mathbf{w}_2^T \cdot s\left(\mathbf{W}_2 [\mathbf{x}_{dbp}, \mathbf{x}_{cobp}, \mathbf{x}_{cd}, \mathbf{x}_{cohbi}]\right), \tag{4}$$

where $\mathbf{w}_2 \in R^{d_3}$ and $\mathbf{W}_2 \in R^{d_4 \times d_3}$ are parameters of the cargo owner behavior attention network, d_3 is the dimension of the input vector of the attention network, d_4 is the dimension of the output vector of the attention network, and $s(x)$ is the nonlinear activation function.

Then, by normalizing the attention score β_n in Eq. (4), we obtain the final attention score β'_n as follows,

$$\beta'_n = \frac{exp(\beta_n)}{\sum_{l=1}^{M} exp(\beta_l)}, \tag{5}$$

Finally, we concatenate the cargo owner behavior vectors weighted by the final attention scores into a cargo owner preferences representation vector as follows,

$$\mathbf{x}'_{cohb} = \sum_{n=1}^{N} \beta'_n \mathbf{x}_{cohbn}. \tag{6}$$

2.4.2   The Top-Level Attention Network Based Influence Modeling of Different Information

Information from different sources has different activity distribution and is of different importance to vehicle-cargo matching. We adopt the top-level attention network to model the influence of different information on the final matching task. The input of this part is six kinds of information, including driver basic profile vector $\mathbf{x}_{dbp}$, cargo owner basic profile vector $\mathbf{x}_{cobp}$, cargo description vector $\mathbf{x}_{cd}$, driver preferences representation vector $\mathbf{x}'_{dhb}$ and cargo owner preferences representation vector $\mathbf{x}'_{cohb}$ obtained in Sect. 2.4.1, and higher-order abstract feature representation $\mathbf{h}_j$ obtained in the deep components of subsequent Sect. 2.5. The output is the corresponding influence factor of various information and the higher-order feature representation $\mathbf{q}_j$.

Firstly, we model the attention score $\delta_{jj'}$ for each kind of information, as shown in Eq. (7),

$$\delta_{jj'} = \mathbf{w}^T \cdot tanh\,(\mathbf{V}\mathbf{s}_{j'} + \mathbf{W}\mathbf{h}_j)\,, \tag{7}$$

where $\mathbf{w}$, $\mathbf{V}$ and $\mathbf{W}$ are parameters of the top-level attention network, $\mathbf{s}_{j'}$ $(j' = 1, 2, 3, 4, 5)$ is the vector representations of five kinds of information obtained from the bottom-level attention network. Specifically, $\mathbf{s}_1 = \mathbf{x}_{dbp}, \mathbf{s}_2 = \mathbf{x}_{cobp}, \mathbf{s}_3 = \mathbf{x}_{cd}, \mathbf{s}_4 = \mathbf{x}'_{dhb}, \mathbf{s}_5 = \mathbf{x}'_{cohb}$. And $\mathbf{h}_j$ is the high-level feature representation obtained by subsequent feature interactions learning module when predicting the j-th classification.

Then, by normalizing the attention score $\delta_{jj'}$ in Eq. (7), we can obtain the final attention score $\delta'_{jj'}$ for predicting the above five aspects of information in multiple classifications, which is the impact factor of different information on the vehicle-cargo matching task. The calculation formula is as follows,

$$\delta'_{jj'} = \frac{exp(\delta_{jj'})}{\sum_{l=5}^{M} exp(\delta_{jl})}, \tag{8}$$

Finally, when predicting the j-th class tag, the higher-order feature representation vector $\mathbf{q}_j$ obtained from the top-level attention network can be calculated as:

$$\mathbf{q}_j = \sum_{j'=1}^{5} \delta'_{jj'} \cdot \mathbf{s}_{j'}. \tag{9}$$

2.5   Feature Interactions Learning

In order to fully explore the potential correlation relationships between multiple features, we use cross network and MLP for collaborative modeling, and learn higher-order abstract feature representation in both explicit and implicit ways. The input of this part is a joint vector $\mathbf{x}_0 = [\mathbf{x}_{dbp}, \mathbf{x}_{cobp}, \mathbf{x}_{cd}, \mathbf{x}'_{dhb}, \mathbf{x}'_{cohb}]$ concatenated by the driver basic profile vector $\mathbf{x}_{dbp}$, the cargo basic profile vector $\mathbf{x}_{cobp}$, the cargo description vector $\mathbf{x}_{cd}$ obtained in Sect. 2.3, and the driver preferences representation vector $\mathbf{x}'_{dhb}$ and the cargo owner preferences representation vector $\mathbf{x}'_{cohb}$ obtained in Sect. 2.4.1. The output is a joint vector $\mathbf{h}_j$ concatenated by the higher-order feature representations obtained from the cross network and MLP.

• *Cross Network based Explicit Feature Interactions Learning*

We use a cross network [8] to explicitly learn vector-level higher-order cross features, and the input of cross network is the joint vector $\mathbf{x}_0$ mentioned above. The cross network is composed of cross layers, with each layer having the following Eq. (10):

$$\mathbf{x}_{l+1} = \mathbf{x}_0 \mathbf{x}_l^T \mathbf{w}_l + \mathbf{b}_l + \mathbf{x}_l = f(\mathbf{x}_l, \mathbf{w}_l, \mathbf{b}_l) + \mathbf{x}_l, \tag{10}$$

where $\mathbf{x}_l, \mathbf{x}_{l+1} \in R^d$ are column vectors denoting the outputs from the l-th and $(l+1)$-th cross layers, respectively, $\mathbf{w}_l, \mathbf{b}_l \in R^d$ are the weight and bias parameters of the l-th layer, and $l \in [1, .., L_1]$. Each cross layer adds back its input after a feature crossing f, and the mapping function $f : R^d \mapsto R^d$ fits the residual of $\mathbf{x}_{l+1} - \mathbf{x}_l$. A visualization of one cross layer is shown in Fig. 2.

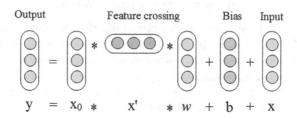

$$y \quad = \quad \mathbf{x}_0 \quad * \quad \mathbf{x}' \quad * \quad w \quad + \quad b \quad + \quad \mathbf{x}$$

Fig. 2. Visialization of a cross layer.

The output $\mathbf{x}_{L_1}$ of the cross network is the feature representation vector obtained from its last layer, which will be concatenated with the higher-order feature representation vector obtained from the last layer of the MLP network to form a joint vector $\mathbf{h}_j$.

• *MLP based Implicit Feature Interactions Learning*

We use MLP network to implicitly learn bit-level high-order cross features. The input of MLP is also the joint vector $\mathbf{x}_0$ mentioned above.

The forward propagation process of the network can be formally described as:

$$\mathbf{x}^l = \sigma \left(\mathbf{W}^l \mathbf{x}_0 + \mathbf{b}^l \right), \tag{11}$$

$$\mathbf{x}^{l+1} = \sigma \left(\mathbf{W}^{l+1} \mathbf{x}^l + \mathbf{b}^{l+1} \right), \tag{12}$$

where $l \in [1, .., L_2]$ is the number of hidden layer, $\mathbf{W}^l$ is the connection weight matrix of the $(l-1)$-th and l-th layers of MLP network, $\mathbf{b}^l$ is the offset of the l-th layer of the MLP network, σ is the nonlinear activation function, and $\mathbf{x}^l$ is the output vector of the l-th layer.

The output $\mathbf{x}_{L_2}$ of the MLP network is the feature representation vector obtained from its last layer, which will be concatenated with the higher-order feature representation vector obtained from the last layer of the crossover network to form a joint vector $\mathbf{h}_j$.

2.6 Matching Calculation and Loss Function

The task studied in this paper belongs to the multi-classification problem. We use the Softmax classifier for matching prediction calculation. The input of this part is the higher-order feature representation vector $\mathbf{q}_j$ obtained from the top-level attention network in Sect. 2.4.2, and the output is a predicted value calculated by the following Eq. (13),

$$\hat{y} = P(y = j) = Softmax(\mathbf{q}_j) = \frac{exp(\mathbf{q}_j)}{\sum_{j'=0}^{2} exp(\mathbf{q}_{j'})}, j \in \{0, 1, 2\}, \quad (13)$$

In addition, we use the cross entropy loss function with regularization term to optimize the model parameters,

$$L = P(y = j) = -\frac{1}{N} \sum_{i=1}^{N} \left(\sum_{j=0}^{2} y_i log(\hat{y}_i) \right) + \lambda \sum_{l} \|w_l\|^2, \quad (14)$$

where $\hat{y}_i$ is the predicted value calculated according to Eq. (13), y_i is the label value, N is the total number of samples, and λ is the penalty factor for the regularization term.

3 Experiments

In this section, we introduce the experiments in detail, including the dataset, evaluation metrics, baselines, comparative study, hyperparameters setting and sensitivity analysis, and ablation study.

3.1 Dataset

This study use a dataset provided by China's largest intercity freight O2O platform, Full Truck Alliance Co. Ltd [10]. We sort out five types of information from the original file of the dataset by means of primary key association: driver basic profile including driver ID, gender, age, registration time, vehicle type, vehicle length, APP platform, device brand, etc.; driver historical behaviors including the cargos sent by the driver, recent search routes in different time periods, number of browsing sources, number of clicks on sources, and number of phone calls made; cargo owner historical behaviors including their historical shipping routes, categories, total recent shipments in different time periods, number of days shipped, the quantity of goods complained, etc.; cargo owner basic profile including ID, gender, age, authentication time, activation date, APP platform, device brand, package type, and other features; cargo description including features such as cargo ID, cargo owner ID, origin, destination, weight, volume, type, freight rate, required vehicle type, required vehicle length, etc.

3.2 Evaluation Metrics

The prediction task of this study belongs to the multi-classification problem, and we use three evaluation metrics, macro-precision, macro-recall and macro-F1 score, for evaluation [11]. Among them, macro F1-score is a balanced comprehensive index. The larger the value of these metrics, the better the classification effect.

3.3   Baselines

In order to fully verify the effectiveness of the proposed model, we compare it with several state of the art intelligent vehicle-cargo matching methods. The details of the baselines are as follows:

LightGBM [12]: This method can well model the feature interactions and is widely used in industry.

XGBoost [13]: This method can also better model the potential interactions between different features in vehicle and cargo information, and obtain valuable and interpretable high-order features, and has a good effect in industry.

DNN based competition plan [14]: This method can fit the potential law among the features, and currently some leading enterprises begin to exploit the deep learning technology for intelligent matching of vehicle, cargo and context information.

A-SENet [15]: This method introduces SENet and attention on the basis of the DNN based competition plan, which is currently a relatively novel solution.

3.4   Comparative Study

For HA-CMNet trained from scratch, we use Gaussian distribution (mean value is 0, standard deviation is 0.01) to randomly initialize model parameters. The training loss of the proposed model on the dataset is shown in Fig. 3, and the hyperparameters setting is shown in Sect. 3.5. In this section, we compare HA-CMNet with the baselines on the same task.

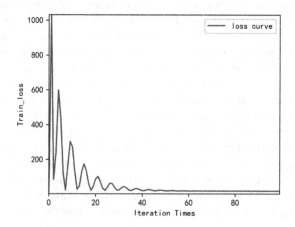

Fig. 3. Training loss of HA-CMNet on this dataset.

- *Parameters Setting of Baselines*

In a consistent experimental environment, in order to make baselines reach their best states, we also optimize them through many experiments. Finally, for Light-GBM, we set learning_rate to 0.05, min_child_sample to 18, and max_depth to 5. For XGBoost, we set booster to gbtree, max_depth to 3, learning rate to 0.001, objective to multi:softprob, eval_metric to mlogloss, early_stop is 15. For the DNN based competition plan and A-SENet, we adopt an MLP structure of 100-100-100, a learning rate of 0.0001, and an optimizer of Adam. To avoid overfitting, L2 regularization is adopted and the dropout is set at 0.5.

• *Results and Comparative Analysis*

Table 1. Performance comparison of all models.

Model	Macro-precision	Macro-recall	Macro-F1 score
LightGBM	0. 8104	0.7712	0.7834
XGBoost	0.7936	0.8074	0.8011
DNN based competition plan	0.8272	0.8322	0.8273
A-SENet	0.8372	0.8410	0.8351
HA-CMNet	0.8593	0.8465	0.8502

Table 1 shows the experimental results of all models on the current prediction task. LightGBM and XGBoost are traditional methods used in the industry, which train multiple decision trees to fit the association relationships in multiparty information, and mining the potential association relationships between features is proven to be helpful for prediction. Compared with XGBoost, LightGBM has faster training speed and lower computational overhead, but we find that its accuracy is slightly lower than XGBoost, which may be due to its decision tree node splitting method and tree growth method.

DNN based competition plan achieve better results than LightGBM and XGBoost, demonstrating that the DNN based solution outperforms the traditional tree based model due to its superior extraction ability of complex association relationships compared to the tree based model. Compared with DNN based competition plan, A-SENet adds SENet modules to the left and right towers, which can dynamically represent the importance of different features and improve the prediction accuracy to a certain extent.

And our proposed HA-CMNet achieve better results than DNN based competition plan and A-SENet, mainly due to the following two advantages: (i)By introducing the hierarchical attention network, selecting significant latent information from different levels is helpful to improve the modeling ability of the model; (ii) By combining cross network and MLP to learn multi-party feature interactions, prediction accuracy can be improved more effectively than separately learning the driver latent vector and the cargo owner latent vector.

In conclusion, by sufficient comparison with the mainstream baselines, the proposed model can achieve better results on the experimental dataset, which proves the validity of the HA-CMNet model design.

3.5   Hyperparameters Setting and Sensitivity Analysis

In this section, we mainly study the influence of hyperparameters setting on the performance of HA-CMNet, including: (i) MLP network structure; (ii) Number of cross network layers; (iii) Learning rate; (iv) Activation function; (v) Penalty factor; (vi) Optimizer. We adopt control variable method to conduct experiments, that is, first fix the values of other hyperparameters, and then change the value of a hyperparameter to be understood within a certain range to observe its influence on the prediction results.

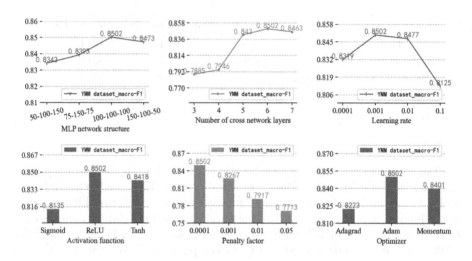

Fig. 4. Hyperparameters setting and sensitivity analysis.

As shown in Fig. 4, we test the influence of four network structures [Diamond, Constant, Incrementing, and Decreasing] on the final performance. It can be observed that when the MLP network structure is Constant, the performance of HA-CMNet is relatively best, which may be because this structure is more suitable for the current prediction task. We also find that if the number of hidden layers continues to increase, the model performance will not be significantly improved, and this phenomenon is caused by excessive parameters leading to overfitting.

For the cross network, we find that the effect is relatively stable when the number of cross layers is 6, and increasing the number of cross layers no longer improves the effect. This indicates that introducing a higher degree of feature interactions is not helpful. Finally, we choose the MLP network structure as 100-100-100, and the number of layers of the cross network is 6.

As for the learning rate, we test in four values [0.0001, 0.001, 0.01, 0.1], and find that 0.001 is better in convergence speed and accuracy. By comparing three activation functions: Sigmoid, ReLU and Tanh, it is found that ReLU is slightly better than the other two. This may be because ReLU is a left saturation function, which alleviates the problem of gradient disappearance of neural network to a certain extent, and can accelerate the convergence rate of gradient decline.

Based on experience, the penalty factor for the regularization term is tested among four values: [0.0001, 0.001, 0.01, 0.05], and it is found that 0.0001 is slightly better. The effect is relatively good when the optimizer is Adam. The dropout rate is set to 0.5, and we find that this value matches the ReLU activation function in the experiment.

3.6 Ablation Study

In the ablation test, in order to further evaluate the design rationality and validity of the proposed model, we will analyze the influence of different components in HA-CMNet in detail.

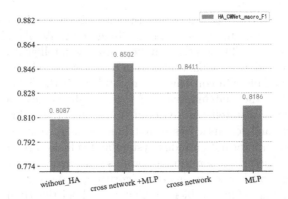

Fig. 5. The effect of the proposed models with different components.

• *The Influence of Hierarchical Attention Network on Prediction*
We test the effect of models with and without hierarchical attention networks, respectively. From Fig. 5, we can see that the former achieve significantly better results, which can be explained by the fact that it contains two levels of improvement: first, the low-level attention network focuses on fine-grained user preferences; second, the top-level attention network strengthens the influence of different information on matching prediction. This is attributed to the fact that the attention network can effectively select significant latent information from the feature level, which helps improve the proposed model's expressive ability.
• *The Influence of Cross Network and MLP on Prediction*
We test the effect of feature interactions learning modules using three different methods: cross network+MLP, cross network, and MLP. It can be observed from Fig. 5 that the effect of MLP is relatively low, while that of cross network is significantly higher, and cross network +MLP achieves the best effect. We also find that combinations including cross network can achieve better results,

which is an interesting phenomenon that explicit modeling of high-order feature interactions has a better effect on matching prediction, and can be regarded as an experience or rule.

In conclusion, theoretically and practically, the proposed model can effectively select significant latent information by using hierarchical attention network, and can simultaneously learn feature interactions from different perspectives by using cross network and MLP, which is a reasonable and effective design mode for vehicle-cargo matching task.

4 Related Work

As a new supply and demand stowing mode, vehicle-cargo matching in the freight O2O platform attracts researchers and engineers at home and abroad to study from different perspectives due to its complexity and importance in the current logistics transportation.

Li et al. studied the one vehicle multi-point delivery service [1] in the freight O2O platform and the supply-demand matching and route planning of the zero-carload cargo business [2]. Some researchers also use intelligent algorithms such as semantic web and evolutionary algorithm to study the vehicle-cargo matching problem. For example, Gu constructed an ontology in the field of vehicle-cargo matching, studied the matching rate of road freight vehicles and cargos, and implemented the basic intelligent reasoning function of the highway freight information platform and semantic based matching between vehicles and cargos [3]. Liu constructed an information index system and vehicle-cargo matching model based on multi-objective programming, proposed a solving algorithm for vehicle-cargo matching model based on genetic algorithm, and optimized and analyzed the vehicle-cargo matching problem of Chuanhua Logistics Highway Port [4].

Some researchers also use game theory and credit evaluation methods to study problems related to vehicle-cargo matching. For example, Jia et al. constructed a bilateral user transaction game model for the vehicle-cargo matching platform, exploring the control problem of platform users evolving from multi attribution to single attribution [5]. Shao et al. selected four platforms such as huochebang, YMM as the analysis object, first constructed the competitiveness evaluation index system for the vehicle-cargo matching platform, and then used analytic hierarchy process to determine the weights of each evaluation index, finally, used the fuzzy comprehensive evaluation method to evaluate the selected platform and obtained its competitiveness level [6]. Bing used analytic hierarchy process and fuzzy comprehensive evaluation to construct a credit evaluation system for vehicle owners and cargo owners, and established a one-to-one vehicle-cargo matching scheduling model and a one-to-many vehicle-cargo matching scheduling model with the goal of minimizing matching costs [7].

The industry also begins to explore solutions based on big data technology. They adopted the embedding mapping layer and DNN to model driver latent vector and cargo latent vector, and then adopted softmaxclassifier for classification [14]. On this basis, Fang et al. proposed a driver CTR model A-SENet, which differed in that SENet was used to calculate the cargo latent vector, while attention and SENet were used to calculate the driver latent vector [15].

5 Conclusion and Future Work

This paper proposes a driver CTR model named HA-CMNet, which aims to improve the efficiency and accuracy of vehicle-cargo matching task in freight O2O platform. The model has the following advantages: (i)It can effectively mine the preferences information of drivers and cargo owners; (ii) It can learn the influence factors of different information on the vehicle-cargo matching task; (iii) It can learn the higher-order feature interactions at both the bit-level and the vector-level, and obtain the cross features containing rich semantic information. Detailed and comprehensive experiments are conducted on the real dataset provided by Full Truck Alliance Co. Ltd., and the results show that the proposed model can achieve a certain degree of better performance than the baselines.

There are two directions for future work. Firstly, we explore the introduction of graph comparison learning [16], graph convolutional networks [17], and other graph methods [18,19] to improve item representation accuracy; Secondly, we consider extending our solution to the order-rider matching problem in O2O platforms for life services.

Acknowledgement. This work is supported by National Natural Science Foundation of China (NO.72261003, 62276196, 62106070), Guizhou Provincial Science and Technology Project (NO.Qiankehejichu-ZK[2022]yiban019), and Key Research and Development Project of Hubei Province (NO.2021BAA030).

References

1. Li, J., Zheng, Y., Dai, B., Yu, J.: Implications of matching and pricing strategies for multiple delivery-points service in a freight O2O platform. Transp. Res. Part E Logistics Transp. Rev. **136**, 101871 (2020)
2. Li, J., Zhou, T., Xu, L., Dai, B.: Matching optimization decision of city LTL carpool based on time windows on the freight O2O platform. Syst. Eng. – Theor. Pract. **40**(4), 979–988 (2020)
3. Gu, J.: Vehicle-Cargo Matching System Based on Semantic Web Technology. Tsinghua University, Beijing (2013)
4. Liu, X.: Research on the Optimization of the Matching of Truck and Cargo in the Highway Port of Transfar. Tsinghua University, Beijing (2013)
5. Jia, X., Hai, F., Dong, R.: Control design for promoting single-homing user ratio of vehicles and cargos matching two-sided platform. Comput. Integr. Manuf. Syst. **23**(4), 903–912 (2017)
6. Shao, Y., Yang, M., Wang, Y.-J., Wei, Y.-Z.: Research on competitiveness of vehicle-freight matching platform for carrier. In: 15th International Conference on Service Systems and Service Management. Hangzhou, China (2018)

7. Bing, C.: Research on Vehicle-Cargo Matching in Highway Trunk Freight Transportation - Taking T Platform as an Example. Zhejiang Sci-Tech University, Hangzhou (2018)
8. Wang, R., Fu, B., Fu, G., Wang, M.: Deep & Cross network for ad click predictions. In: 23th ACM SIGKDD Conference on Knowledge Discovery and Data Mining Workshop-ADKDD&TargetAD. Halifax, NS, Canada (2017)
9. Sun, J., Zhu, M., Jiang, Y., Liu, Y.-Z., Wu, L.: Hierarchical attention model for personalized tag recommendation. J. Am. Soc. Inf. Sci. **72**(2), 173–189 (2021)
10. YMM-TECH algorithm contest dataset. https://ymmtech.ymm56.com. Accessed 6 Feb 2020
11. Zhou, Z.-H.: Machine Learning, 2nd edn. Tsinghua University Press, Beijing (2021)
12. Ke, G., et al.: LightGBM: a highly efficient gradient boosting decision tree. In: The Thirty-First Annual Conference on Neural Information Processing Systems. Long Beach, CA, USA, pp. 3146–3154 (2017)
13. Chen, T., Guestrin, C.: XGBoost: a scalable tree boosting system. In: Proceedings of the 22nd ACM SIGKDD International Conference on Knowledge Discovery and Data Mining. San Francisco, CA, USA, pp. 785–794 (2016)
14. DNN based competition plan. http://api.ymm56.com/article4153.html. Accessed 12 Oct 2022
15. Fang, F., Wang, C.: Driver click-trough rate prediction model based on SENet two-tower model considering attention mechanism in vehicle-cargo matching. Logistics Sci. Technol. **2**, 91–97 (2022)
16. Song, X., Li, J., Lei, Q., Zhao, W., Chen, Y., Mian, A.: Bi-CLKT: Bi-graph contrastive learning based knowledge tracing. Knowl.-Based Syst. **241**, 108274 (2022)
17. Song, X., Li, J., Tang, Y., Zhao, T., Chen, Y., Guan, Z.: A joint graph convolutional network based deep knowledge tracing. Inf. Sci. **580**, 510–523 (2021)
18. Xue, G., Zhong, M., Li, J., Chen, J., Zhai, C., Kong, R.: Dynamic network embedding survey. Neurocomputing **472**, 212–223 (2022)
19. Cai, T., Li, J., Mian, A., Li, R.-H., Sellis, T., Yu, J.X.: Target-aware holistic influence maximization in spatial social networks. IEEE Trans. Knowl. Data Eng. **34**(4), 1993–2007 (2022)

A Hybrid Intelligent Model SFAHP-ANFIS-PSO for Technical Capability Evaluation of Manufacturing Enterprises

Tingting Liu[1,2], Xuefeng Ding[1,2], Yuming Jiang[1,2], and Dasha Hu[1,2(✉)]

[1] College of Computer Science, Sichuan University, Chengdu 610065, China
hudasha@scu.edu.cn
[2] Big Data Analysis and Fusion Application Technology Engineering Laboratory of Sichuan Province, Chengdu 610065, China

Abstract. In the collaborative production environment of manufacturing tasks, the evaluation of enterprise technical capability in advance has a direct impact on the high-performance collaboration between the supplier and the demander of production tasks. In order to solve the problem of efficiency and accuracy in the past evaluation methods, this paper applies the adaptive neuro fuzzy inference system (ANFIS) to the field of enterprise technical capability evaluation for the first time, and proposes an improved hybrid intelligent model SFAHP-ANFIS-PSO based on fuzzy theory. The model takes the four key indexes which have the deepest impact on enterprise technical capability evaluation as the input variables of the model, and the input data are evaluated by ANFIS with high accuracy. Spherical fuzzy analytic hierarchy process (SFAHP) calculated the weight of each evaluation index to preprocess the ANFIS model. Particle swarm optimization algorithm (PSO) continuously optimized the model parameters in the training process, shortened the model training time, improved the evaluation efficiency, and further improved the accuracy of the evaluation results. The experimental results show that the SFAHP-ANFIS-PSO model has high convergence speed and accuracy of results, and can be used to evaluate the technical capability of enterprises with high efficiency and high accuracy.

Keywords: Technical Capability Evaluation · ANFIS · Spherical Fuzzy Analytic Hierarchy Process · Particle Swarm Optimization

1 Introduction

As knowledge becomes more and more specialized and decentralized, more enterprises choose to achieve complementary advantages and resource sharing through good cooperation with other enterprises [1]. High-performance collaboration maximizes the benefits of all stakeholders [2]. In collaborative networks, the in-depth study of the relevant factors affecting collaboration performance is of great significance for optimizing the selection of collaboration partners and improving the collaboration efficiency between enterprises. In recent years, domestic and foreign researchers have widely mentioned the importance of enterprise technical capability on collaboration performance in the fields like collaborative supplier selection and evaluation, collaborative innovation performance and collaborative result performance evaluation [3–5]. Therefore, how to evaluate the technical capability of enterprises is an extremely important problem. Enterprise technical capability evaluation is a decision problem with great uncertainty or fuzziness. In previous research on capability evaluation, a two-stage approach is often adopted. Firstly, key indicators for the evaluation problem are determined, followed by the use of different evaluation methods to obtain evaluation results. Commonly used evaluation methods include questionnaire survey, multi-criteria decision method, fuzzy comprehensive evaluation method, Back Propagation (BP) Neural Network, etc. Most of these evaluation methods are characterized by strong subjectivity, low accuracy of evaluation results, complicated calculation and low efficiency of evaluation process. BP neural network introduces intelligent algorithms into the evaluation process, but BP neural network has high requirements for data samples and parameter Settings. From a mathematical perspective, it is highly prone to getting trapped in local extrema and suffers from slow convergence speed. In addition, the ability of BP neural networks to handle uncertainty problems is inadequate.

In this paper, an improved hybrid intelligent model SFAHP-ANFIS-PSO based on fuzzy theory is proposed to evaluate the technical capability of enterprises accurately and efficiently. The SFAHP-ANFIS-PSO model combines neural network and fuzzy reasoning, quantifies the technical capability of enterprises based on fuzzy rules, and effectively solves the evaluation of uncertainty problems. To further enhance the efficiency and accuracy of the assessment results, the SFAHP algorithm [6] is used to calculate the weights of evaluation indicators. Preprocessing is performed on the model before training, and the PSO algorithm [7] is utilized to optimize the model parameters during the training process.

2 Related Work

In the field of capability evaluation, reference [8] aims at the maturity evaluation of intelligent manufacturing capability in the chair industry, proposes a multi-level evaluation index system, and uses the analytic network process (ANP) and SD software to complete the maturity evaluation of industrial intelligent manufacturing capability. Reference [9] addresses the evaluation of technological innovation capabilities in Chinese prefabricated construction companies. They propose an evaluation indicator system and utilize information entropy to determine the weights of the evaluation indicators. Finally, they quantitatively assess the technological innovation capabilities of the companies using Gray Relational Analysis (GRA). Reference [10] combines BP neural networks with the Firefly Algorithm and Sparrow Search Algorithm in the field of intelligent manufacturing capability maturity assessment. Aiming at improving enterprise performance and venture capital management, reference [11] studied the positive impact of optimizing supply chain technology innovation and venture capital on enterprise performance improvement, and proposed an enterprise performance evaluation index system. Reference [12] proposed a supplier evaluation method, which takes factors such as cost and enterprise technical capability as evaluation criteria, and uses expert scoring method to quantify qualitative indicators. Previous research methods rely on strong subjectivity in the evaluation process, the calculation process is complex, and the ability to deal with uncertain problems is insufficient. These methods have resulted in lower accuracy of assessment results and low efficiency in the evaluation process. ANFIS [13] has been widely applied to various evaluation and prediction problems and has demonstrated strong advantages in handling uncertainty. Reference [14] proposed four hybrid neural network models for spatial prediction of wildfire probability. The four models combined ANFIS with metaheuristic optimization algorithms GA, PSO, SFLA and ICA respectively. Reference [15] proposed an ANFIS-PNN-GA model to predict the air overpressure caused by blasting, in which the GA algorithm was utilized to optimize the ANFIS-PNN structure and improve the model efficiency. Reference [16] developed prediction models for compressive strength (CS) of ordinary concrete (NC) and high-performance concrete (HPC) using a combination of ANFIS and Grey Wolf Optimization (GWO) algorithm. The results demonstrated that the hybrid approach of ANFIS and GWO effectively improved the performance of the models and enhanced the accuracy of predicting CS.

In summary, the evaluation of technological capabilities in enterprises requires an evaluation method that is efficient, accurate, and capable of handling uncertainty or ambiguity. The combination of ANFIS and intelligent optimization algorithm can solve a variety of evaluation and prediction problems. Therefore, this paper proposes a SFAHP-ANFIS-PSO model for high-efficiency and high-accuracy evaluation of enterprise technical capability.

3 Implementation of the Enterprise Technological Capability Evaluation Model

3.1 Enterprise Technical Capability Evaluation Problem

The evaluation of technological capabilities in enterprises is a systematic evaluation process that assesses their strength, potential, and other aspects in the field of technology. When evaluating the technological capabilities of manufacturing enterprises, factors such as technical personnel, research and development capabilities, and technological infrastructure need to be considered to determine key evaluation indicators. In the actual production environment, each indicator may have a different impact on the evaluation results of technological capabilities. Based on this, scientific methods are used to analyze and calculate the data of each indicator, leading to the final evaluation results. In the context of collaborative networks and supply-demand relationships, the evaluation results of technological capabilities in enterprises can help demand-side enterprises understand the technical strength of the supply-side enterprises before cooperation. It facilitates the evaluation of the feasibility and reliability of collaboration, thereby ensuring the successful completion of high-performance collaborative processes.

3.2 Determine the Evaluation Index and Index Weight

Evaluation Index. Based on the characteristics of manufacturing enterprises, expert experience and an analysis and summary of previous relevant literatures, this paper has determined the main factors for considering enterprise technical capability, including the number of senior technical engineers, the number of product patents, the proportion of R&D investment, and the degree of technology intensity.

Determine Index Weights Based on SFAHP. The impact of different indexes on the evaluation of enterprise technical capability varies. This paper combines expert opinions with spherical fuzzy analytic hierarchy process to determine index weights. An index weight evaluation model based on SFAHP [10] is designed. The combination of spherical fuzzy set and traditional AHP in the model allows decision makers to express their uncertainty in the decision-making process through a nine-level spherical fuzzy scale, which effectively improves the scientific and accuracy of the evaluation results.

3.3 Construct the Improved Hybrid Intelligent Model SFAHP-ANFIS-PSO

Feasibility Analysis of ANFIS for Technical Capability Assessment. ANFIS is a hybrid fuzzy inference system that combines neural networks with fuzzy logic. FIS transforms the fuzzy human experience and knowledge into the rule base of the system, which maps the nonlinear relationship between a set of inputs and outputs by using the past experience. Neural network makes ANFIS have the ability of self-learning, which determines that even without enough prior experience to construct the rule base of ANFIS, based on reasonable membership function setting, a set of rules that approximate the provided data set can be generated by learning. Based on the above advantages, ANFIS can effectively simulate the reasoning process of experts when dealing with imprecise problems related to assessing technological capabilities in enterprises. By learning from historical data of technological capability assessments, ANFIS is capable of capturing the expertise and knowledge inherent in the data. Using evaluation indicator data as input to the model, ANFIS transforms the input data into membership degrees of different fuzzy sets. Through rule operations, it ultimately maps to a specific value representing the technological capability of the enterprise.

Preprocess the ANFIS Model. The ANFIS model has a five-layer network hierarchy, in which the adjustable parameters are in the first layer and the fourth layer. The parameters of Layer1 are all the membership function parameters, called the premise parameters, and the consequence parameters of the model are included in Layer4. This paper uses the evaluation index weights obtained by SFAHP method to preprocess all consequence parameters of the model. The index weights obtained by SFAHP combined with expert experience are much closer to the optimal solution of consequence parameters compared to random numbers. Preprocessing can effectively improve the evaluation accuracy and convergence speed of the model.

Construct the SFAHP-ANFIS-PSO Model. Particle Swarm Optimization (PSO) algorithm is an intelligent optimization algorithm that finds the optimal solution through the cooperation and information sharing between individuals in the swarm. The algorithm is simple and easy to understand and implement, which has strong global search ability to avoid falling into local optimal solution.

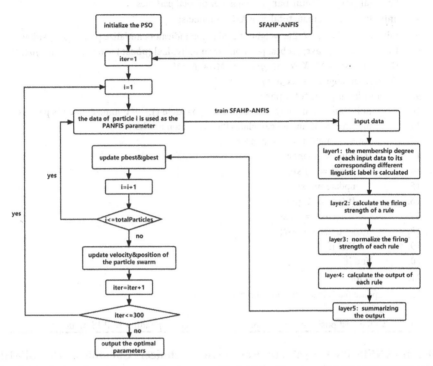

Fig. 1. Work flowchart of constructing improved hybrid intelligent model SFAHP-ANFIS-PSO

In this paper, PSO algorithm is applied to ANFIS model optimization, and an improved hybrid intelligence model, SFAHP-ANFIS-PSO, was constructed (Fig. 1) to assess the technical capabilities of enterprises quantitatively. In the construction process of the SFAHP-ANFIS-PSO model, the model parameters are preprocessed using the SFAHP algorithm to calculate the weights of the indicators. Then, leveraging the advantages of the PSO algorithm, the model continuously searches for the global optimal solution for all adjustable parameters, leading to model updates. Ultimately, the model achieves optimality in terms of convergence speed and assessment result accuracy. The pseudo-code of the SFAHP-ANFIS-PSO model construction algorithm is shown in Algorithm 1.

Algorithm 1: SFAHP-ANFIS-PSO Model Construction Algorithm
1. Calculates the weight of evaluation indicators by SFAHP;
2. Set max_iter;   // max_iter is max Number of iterations
3. Set total_par;   // total_par is the number of total particles
4. anfis_learners = create total_par ANFIS models;
5. Initialize the *PSO*;   // the weight of evaluation indicators initialize particles' position
6. Define *pbest, gbest*;   // best position for an individual (*pbest*) and population (*gbest*)
7. Define $pcost = MAX\_NUM, gcost = MAX\_NUM$;
8. For *iter* in range (1, max_iter):
9.   For *i* in range (1, total_par):
10.   build_anfis_param (anfis_learners *i*, particle *i*);   // optimize model's params
11.   the *i*th ANFIS model calculates the evaluation results;
12.   rmse = calculation result error;
13.   If (rmse < p*cost*):
14.   pcost = *rmse*;
15.   update *pbest*;
16.   If (rmse < g*cost*):
17.   g*cost* = rmse;
18.   update *pbest*;
19.   End if
20.   End if
21.   End for
22.   update particle swarm position and velocity;
23. End for
24. Save the best position of the population and its corresponding ANFIS model.

SFAHP-ANFIS-PSO Evaluate Enterprise Technical Capability. The SFAHP-ANFIS-PSO model takes four indicators as input data: the number of senior technical engineers in the enterprise, the number of product patents, the proportion of research and development investment, and the technological intensity. The overall evaluation process of the model consists of five hierarchical steps. Firstly, fuzzy membership assignment: For each input variable in the model, a corresponding number of fuzzy sets and their optimized membership functions are defined. This step determines the degree of membership for the input data in the fuzzy sets. Then, rule triggering strength calculation: Based on the obtained membership degrees from the previous step, the triggering strength of each evaluation rule is calculated. The number of rules is determined by multiplying the number of fuzzy sets for each input variable. Then, normalization of rule triggering strengths: The triggering strengths of each rule are normalized to ensure their consistency and comparability. completed optimal linear function. The output of each rule is then calculated by optimizing the completed optimal linear function. Finally, the final enterprise technical capability score is obtained by summing the output of each rule.

4   Experimental Results

To verify the superiority of the proposed model in this paper, MAE and RMSE were selected as performance parameters to quantitatively evaluate the model performance. The SFAHP-ANFIS-PSO model and the ANFIS-PSO model were compared from two aspects: model convergence speed and evaluation result accuracy. The program for the proposed model is implemented using the Python programming language, specifically version 3.8.0. The execution environment is set as follows: 16 GB of memory and the Windows 10 operating system. For the experimental setup, the parameters are as follows: the population size is set to 50, the number of iterations is set to 300, and each input variable has 4 fuzzy sets. The experimental comparison results are shown in Table 1 and Fig. 2.

Table 1. Comparison between the performances of the assessment models.

Assessment model	Train dataset		Test dataset	
	RMSE	MAE	RMSE	MAE
ANFIS-PSO	0.0493	0.1948	0.0954	0.2471
SFAHP-ANFIS-PSO	0.0199	0.1260	0.0381	0.1744

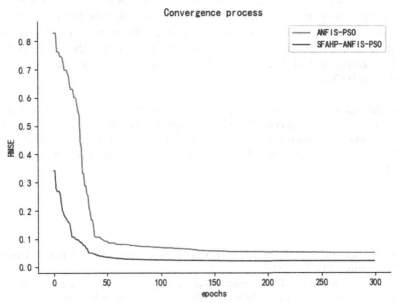

Fig. 2. Model convergence process

As shown in Table 1, both SFAHP-ANFIS-PSO and ANFIS-PSO models can achieve high accuracy evaluation of enterprise technical capabilities. Among them, the error of the evaluation results of SFAHP-ANFIS-PSO is significantly smaller. The highest evaluation accuracy on the training and test datasets is (RMSE $= 0.0199$ and MAE $= 0.1260$) and (RMSE $= 0.0381$ and MAE $= 0.1744$). In terms of convergence speed, Fig. 2 reflects that the technical capability evaluation models based on ANFIS all have extremely fast convergence speed. The ANFIS-PSO model converges after about 160 iterations, and the SFAHP-ANFIS-PSO model converges after about 60 iterations thanks to the preprocessing of SFAHP.

5 Conclusion

In order to achieve high efficiency and precision evaluation of enterprise technical capability, this paper proposes a SFAHP-ANFIS-PSO model for enterprise technical capability evaluation. Firstly, four key evaluation indexes of enterprise technical capability were determined as the input parameters for the model. Then, the SFAHP-ANFIS-PSO model is constructed to complete the evaluation of technical ability. By utilizing ANFIS's ability to convert human experiential knowledge into fuzzy logic and its efficient self-learning capability, input data can be transformed into membership degrees of different fuzzy sets that closely resemble human experience. This enables efficient rule learning to take place. The SFAHP is used to calculate the weight of the evaluation indexes for model preprocessing, and the intelligent optimization algorithm PSO is used to optimize the model. The optimal solution of the model parameters is constantly searched in the training process to further improve the convergence efficiency of the model and the accuracy of the evaluation results. The experimental results show that the SFAHP-ANFIS-PSO model converges after about 60 iterations, which greatly improves the evaluation efficiency. After the validation data set test, the results show that the model evaluation results have extremely high accuracy, and the evaluation accuracy is (RMSE $= 0.0381$ and MAE $= 0.1744$).

Acknowledgements. This work was supported in part by the National Key R&D Program of China under Grant No. 2020YFB1707900 and 2020YFB1711800; the National Natural Science Foundation of China under Grant No. 62262074, U2268204 and 62172061; the Science and Technology Project of Sichuan Province under Grant No. 2022YFG0159, 2022YFG0155, 2022YFG0157.

References

1. Ansell, C., Gash, A.: Collaborative governance in theory and practice. J. Public Admin. Res. Theory **18**(4), 543–571 (2007)
2. Choi, I., Moynihan, D.: How to foster collaborative performance management? Key factors in the US federal agencies. Public Manag. Rev. **21**(10), 1538–1559 (2019)
3. Prashanth, K.D., Parthiban, P., Dhanalakshmi, R.: Evaluation of the performance and ranking of suppliers of a heavy industry by TOPSIS method. J. Sci. Ind. Res. **79**(2), 144–147 (2020)
4. Haleem, A., Khan, S., Luthra, S., Varshney, H., Alam, M., Khan, M.I.: Supplier evaluation in the context of circular economy: a forward step for resilient business and environment concern. Bus. Strateg. Environ. **30**(4), 2119–2146 (2021)

5. Zhou, J., Liu, Z., Li, J., Zhang, G.: The effects of collaboration with different partners: a contingency model. IEEE Trans. Eng. Manage. **68**(6), 1546–1557 (2021)
6. Kutlu, G.F., Kahraman, C.: A novel spherical fuzzy analytic hierarchy process and its renewable energy application. Soft. Comput. **24**(6), 4607–4621 (2020)
7. Kennedy, J., Eberhart, R.: Particle swarm optimization. In: IEEE International Conference on Neural Networks, pp. 1942–1948 (1995)
8. Wang, W., Wang, J., Chen, C., Su, S., Chu, C., Chen, G.: A capability maturity model for intelligent manufacturing in chair industry enterprises. Processes **10**(6) (2022)
9. Dou, Y., Xue, X., Wang, Y., Xue, W., Huangfu, W.: Evaluation of enterprise technology innovation capability in prefabricated construction in China. Constr. Innov. **22**(4), 1059–1084 (2021)
10. Shi, L., Ding, X., Li, M., Liu, Y., Ahmad, M.: Research on the capability maturity evaluation of intelligent manufacturing based on firefly algorithm, sparrow search algorithm, and BP neural network. Complexity, 1–26 (2021)
11. Wang Z., Lu, J., Li, M., Yang, S., Wang, Y., Cheng, X.: Edge computing and blockchain in enterprise performance and venture capital management. Comput. Intell. Neurosci. 2914936 (2022)
12. Tuzkaya, G.: An intuitionistic fuzzy Choquet integral operator based methodology for environmental criteria integrated supplier evaluation process. Int. J. Environ. Sci. Technol. **10**(3), 423–432 (2013)
13. Jang, J.S.R.: ANFIS: adaptive-network-based fuzzy inference system. IEEE Trans. Syst. Man Cybern. **23**(3), 665–685 (1993)
14. Jaafari, A., Zenner, E.K., Panahi, M., Shahabi, H.: Hybrid artificial intelligence models based on a neuro-fuzzy system and metaheuristic optimization algorithms for spatial prediction of wildfire probability. Agric. For. Meteorol. **266–267**, 198–207 (2019)
15. Harandizadeh, H., Armaghani, D.J.: Prediction of air-overpressure induced by blasting using an ANFIS-PNN model optimized by GA. Appl. Soft Comput. **99**(1), 106904 (2021)
16. Golafshani, E.M., Behnood, A., Arashpour, M.: Predicting the compressive strength of normal and high-performance concretes using ANN and ANFIS hybridized with Grey Wolf Optimizer. Constr. Build. Mater. **232**, 117266 (2020)

A Method for Data Exchange and Management in the Military Industry Field

Ping Wu, Xingqiao Wang[✉], Xin Zhang, and Zhimin Gao

AVIC Shenyang Aircraft Design and Research Institute, 40 Tawan Street, Huanggu, Shenyang, Liaoning, China
912392877@qq.com

Abstract. With the ongoing integration of industrialization and informatization, enterprise information infrastructure has been steadily advancing, leading to a substantial surge in data volume generated within organizations. Military enterprises, in particular, face unique challenges such as distributed data sources, data confidentiality concerns, and limited data sharing capabilities. The traditional approach of manually transferring data using physical media yields low transmission efficiency and requires significant human resources. Furthermore, the lack of comprehensive planning and standardized frameworks during the initial stages of enterprise information system development has resulted in data silos throughout the organization. Consequently, the seamless integration of data links and efficient data management has emerged as a critical priority for enterprises. This research paper presents a comprehensive methodology for data exchange and management in the military industry sector. It encompasses key aspects such as establishing data links, designing top-level architectural plans, constructing and implementing robust data models, implementing effective data warehouse management, and ultimately achieving a unified and centralized data display process. Through the implementation of this methodology, the aim is to facilitate efficient data flow, provide users with clear visualizations of data processing outcomes, enable streamlined data management, and enhance the value of data assets within the enterprise.

Keywords: Data Architecture · Data Visualization · Data Transmission · Data Processing

1 Introduction

1.1 A Subsection Sample

As the enterprise's digital transformation efforts gradually unfold, the advancement of information construction has become a key focus. In response to diverse business requirements, numerous information systems have been established, resulting in substantial data accumulation. While these information systems have brought operational

---

P. Wu and X. Wang—Contributed equally to the paper as co-first authors.

© The Author(s), under exclusive license to Springer Nature Switzerland AG 2023
X. Yang et al. (Eds.): ADMA 2023, LNAI 14179, pp. 674–680, 2023.
https://doi.org/10.1007/978-3-031-46674-8_47

convenience, they have also uncovered underlying challenges. Military enterprises, in particular, face the need to maintain data confidentiality, leading to data barriers among different units. As information is transmitted across systems, data redundancy becomes evident. The lack of correlation and traceability between systems hampers database access and increases the risk of crashes. Furthermore, the absence of top-level architecture and institutional processes within the enterprise contributes to fragmented data management and a lack of centralized control, resulting in duplicated data definitions and difficulties in realizing its value.

To address these issues, this research paper proposes a comprehensive method for data exchange and management in the military industry domain. This method offers robust support for enterprises in designing top-level data architectures, establishing related management systems, facilitating data interface connectivity between systems and different locations, building data warehouses, and creating data visualization platforms. By implementing this approach, enterprises can achieve enhanced data management capabilities, improved data correlation, and more effective data utilization, ultimately driving the success of digital transformation initiatives.

2 Key Technology Research

2.1 Data Architecture

Data architecture plays a crucial role within the broader context of enterprise architecture. It encompasses the business architecture and serves as the foundation for designing application architecture [1]. The diagram provided illustrates the comprehensive structure of the data architecture [2].

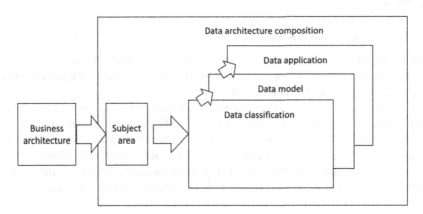

Fig. 1. Data Architecture

The diagram illustrates the data architecture's comprehensive structure, which can be divided into four main components: subject domain, data classification, data model, and data application (Fig. 1).

a) Subject Domain

The subject domain serves as the foundational structure of the data classification system within the enterprise architecture. It establishes a fundamental alignment with the business domain and provides the basis for defining the data architecture.

b) Data Classification

Data classification entails the systematic categorization of all essential data objects. It involves dividing the data objects into distinct groups based on their respective topic domains and conducting analysis and classification within each domain.

c) Data Model

The data model encompasses the definition of conceptual and logical models for data. These models represent the content and relationships of data objects, providing a framework for understanding and organizing the data within the architecture.

d) Data Application

Data application involves the strategic planning of the top-level architecture and implementation approach for data application systems. It is driven by the enterprise's vision and guides the construction and implementation of data management application systems throughout the organization.

2.2 Data Warehouse Construction

A data warehouse [3, 4] is a strategic repository that facilitates comprehensive data support for decision-making processes across all levels of an enterprise. It serves as a centralized data store specifically designed for analytical reporting and decision support, offering valuable insights to enhance business process optimization and enabling enterprises to monitor aspects such as time, cost, quality, and control. This dedicated data resource plays a crucial role in empowering organizations with business intelligence capabilities.

2.3 Data Visualization Technology

Data visualization [5, 6] refers to the process of visually representing location-based information within a vast dataset using graphics or images, leveraging data analysis and development tools to their full potential.

Finereport, a leading provider of big data BI and analysis platforms in China, specializes in the fields of business intelligence and data analysis. Offering a user-friendly interface and robust features, Finereport caters to various requirements such as diverse report display, interactive analysis, data entry, permission management, scheduling, print output, portal management, and large screen display. Its extensive range of features reduces implementation costs while meeting the daily management needs of enterprises.

2.4 Data Exchange

Data exchange [7, 8] is the procedure of establishing a transient interconnection pathway for data communication between multiple data terminal devices. It enables the seamless transfer of data between any two terminal devices. Data exchange encompasses various modes, including circuit exchange, message switching, packet exchange, and hybrid exchange.

3 Design and Implementation

3.1 Enterprise Model Management Platform

After investigating and analyzing the current basic situation of the enterprise, the proposed approach is the "1 + 2 + N" framework. This framework focuses on developing a robust data management system as the core, while simultaneously nurturing the capabilities to construct a comprehensive data architecture and data platform. The data is then empowered through the implementation of various value-added initiatives.

a) Building Data Management System

By establishing high-level data management entities such as data management committees and data management offices. Defining top-level systems and departmental standards and developing data management processes, the enterprise can effectively govern the operations of data management.

b) Data Architecture and Data Management Platform

Utilizing the new era quality management system, the business processes within the enterprise are thoroughly examined. This includes organizing business objects, mapping them to corresponding data entities, constructing conceptual and logical data models, and establishing a comprehensive, distinct, and authoritative definition of business objects in the enterprise's data landscape. This standardized foundation serves as a reference for subsequent information system development, database design, data integration, and data sharing initiatives.

3.2 Data Warehouse Construction

In order to store and use date more efficiently, the enterprises build their own data warehouse, which contains stage layer, CDM (Common Dimension Model) layer and ADS (Application Data Service) layer. The CDM layer includes DWD (Data Warehouse Detail) layer and DWS (Data Warehouse Service) layer. The model hierarchy is shown in the figure.

The different layers of the system serve distinct functions as outlined below:

1) Stage

The primary role of the stage layer is to facilitate data synchronization by directly storing data from various information systems within the enterprise into a centralized data warehouse. Additionally, it involves the archival or cleansing of historical data based on the specific requirements of data operations (Fig. 2).

2) CDM

Within the CDM layer, the data from the stage layer undergoes processing to generate detailed fact data and dimension table data. Subsequently, this processed data is further transformed to generate summarized public indicator data.

3) ADS

The application data layer (ADS) functions as a repository for personalized statistical indicator data derived from processing activities based on the stage layer and the common dimension model (CDM) layer (Fig. 3).

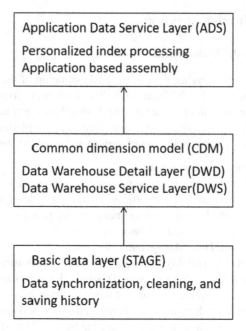

Fig. 2. Model Hierarchy

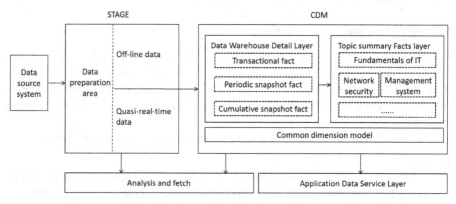

Fig. 3. Circulation relationship

When constructing a data warehouse, it adheres to the principles of data architecture design and aligns with the data management standards established within the enterprise. This approach aims to manage the growth trajectory of data volume, thereby enhancing the efficiency of data exploration and development while optimizing data storage space and reducing associated costs.

3.3 Big Data Center

In order to unlock the inherent value of data, transform it into tangible insights, and enhance the enterprise's data-driven capabilities, an enterprise big data center is established utilizing the Finereport tool. Tailored to cater to diverse user groups, the big data center offers varying types of foundational data information. Recognizing the confidentiality requirements of military enterprises and the need for access control, the big data center implements a dual-layer authorization mechanism. This approach grants authorization to data visualization pages within Finereport, while also managing user visibility through the data center management system's menu access privileges. The dual-layer control ensures robust data security measures are in place.

3.4 Remote Data Transmission

During the initial construction phase, the decentralized generation of data across multiple locations resulted in a lack of unified and standardized data. This led to a complex network of data transmission, as illustrated in the diagram. The intricate mesh of data transmission links gave rise to issues such as redundant transmissions, duplicated requirements, missing attributes, and unclear process records. To address these challenges, our enterprise will establish dedicated data centers in various regions. These centers will serve as centralized hubs, consolidating and harmonizing the data requirements. Consequently, the complex mesh of data transmission links will be transformed into a more efficient "dual star data link" configuration (Fig. 4).

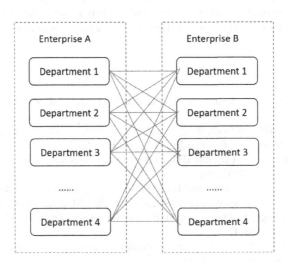

Fig. 4. Old Data Link

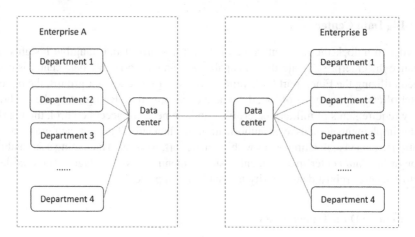

Fig. 5. New Data Link

4 Prospect

This article presents a comprehensive approach to data exchange and management in the military industry (Fig. 5). It outlines a systematic method that encompasses various components such as data management system planning, data model management platform development, data warehouse establishment, big data center construction, and remote data transmission connectivity. By implementing this approach, it enables standardized definition management, efficient storage, unified display, and seamless data exchange and sharing among individual information systems within enterprises.

References

1. Wang, S., Wang, H.: Architecture big data: challenges. Curr. Situation Prospects **34**(10), 1741–1752 (2011)
2. McKendrick, J.: Data-to-value: designing aiviodern enterprise data architecture that delivers more %T for the business. Database Trends Appl. **36**(6) (2022)
3. Yulin, Y., Xianmin, Z.: A reliable ETL strategy and architecture design for data warehouse. Journal **10**, 172–174+229 (2005)
4. Chen, Y., et al.: Chinese intracranial hemorrhage imaging database: constructing a structured multimodal intracranial hemorrhage data warehouse. Chin. Med. J. **136**(13), 1632–1634 (2022)
5. Xueqi, C., Xiaolong, J.: Overview of big data systems and analysis technologies. Journal **25**(09), 1889–1908 (2014)
6. Ma, R., Sun, E.D., Zou, J.: A spectral method for assessing and combining multiple data visualizations. Nat. commun. **14**(1) (2023)
7. Yanchen, X.: Data Communication and Computer Networks. 2nd edn. People's Post and Telecommunications Publishing House Co. Ltd. (2015)
8. Daniška, D., Vrban, B., Nečas, V.: Development of database structures and data exchange principles for nuclear decommissioning planning. Radiat. Prot. Dosim. **198**(9–11), 740–746 (2022)

Multi-region Quality Assessment Based on Spatial-Temporal Community Detection from Computed Tomography Images

Yiwen Liu[1]($\boxtimes$), Tao Wen[1,2], Tongze Xu[1], Baoting Li[1], Wei Sun[3], and Zhenning Wu[1]

[1] Northeastern University, Shenyang, China
liuyw@mail.neu.edu.cn
[2] Dalian Neusoft University of Information, Dalian, China
[3] Neusoft Institute Guangdong, Foshan, China

Abstract. Computed Tomography (CT) images are widely used due to their low cost and high effectiveness. However, artifacts caused by human motion lead to a decline in image quality, which affects diagnostic accuracy and prognosis. Recently, significant progress has been made in motion blur detection using Convolutional Neural Networks (CNNs). However, these CNN-based methods still fall short of meeting the requirements of the medical field. Furthermore, CNN-based artifacts can only handle the regular node, but do not suitable for the irregular node distribution scenario, which result in ignorance of the relationship between CT images. In this paper, a novel construction method for head CT images based on complex networks theory has been proposed. Firstly, the spatial-temporal information is utilized to construct the graph of head CT images. The relationship between different head CT images is depicted from a comprehensive perspective. The head CT images are mapped to a topology of CT image network. Secondly, structural differences are reflected by comparing topological characteristics between graph construction based on spatial-temporal domain and spatial information. Finally, multi-region image quality is classified using spatial-temporal community detection. Experimental results demonstrate that the spatial-temporal community detection method significantly improves the performance of multi-region quality assessment, achieving an accuracy of up to 99.79%. Moreover, it better satisfies the clinical requirement for the interpretability.

Keywords: Head Computed Tomography Images · Spatial-Temporal Domain · Multi-region · Graph Theory · Community Detection · Quality Assessment

1 Introduction

Computed Tomography (CT) image is drawing increasing attention and playing a crucial role in diagnosis and prognosis, due to its low cost and non-invasive nature. To achieve high diagnostic accuracy, the prerequisite is the high quality of CT images. However, CT image may be affected by individual motion, such as shaking, breathing, and other factors, leading to a decline in image quality that may not meet the requirements of clinical scenes, ultimately resulting in decreased diagnostic accuracy. Therefore, effective detection of motion artifacts which may affect the diagnosis is of great necessity [1]. It is challenging that it requires extensive clinical experience and comprehensive knowledge to evaluate the quality of CT images, which is time-consuming, subjective and laborious. It is rather difficult to automatically quantify the decline in image quality caused by motion artifacts for the complexity of the CT image quality declination. Consequently, the academia and industry have been in a race to automatically classify the quality of CT images and determine whether a CT image has motion artifacts to increase the diagnosis efficiency and relieve repetitive workloads.

As an important task in medical imaging classification task, researchers have conducted a significant amount of valuable works for accurate quality assessment. Several research mainly focus on handcrafted features designation. Although substantial progresses have been made to extract manual features more exactly, the selected features may result in poor generality as they are lack of robustness due to the complexity and uncertainty of formation mechanism, especially for deterioration situations. On the contrary, deep learning methods [2–5], particularly, Convolution Neural Networks (CNNs), have achieved unprecedented superior performance. However, CNN-based methods need a great deal of annotated samples, which is not applicable in the medical domain where the cost of manual annotation is high. CNNs are effectively opaque black boxes and there is no way to understand how they generate decisions, which makes it challenging to troubleshoot them when necessary. Additionally, CNNs do not take the relationship between the CT images into account, and may not comprehensively depict the topological structure characteristics of the CT image from a global perspective. It is usually hard to embed priors such as physical and topological properties into CNN-based methods, which may neglect useful relationship for classification. Furthermore, CNNs lack generality and interpretability, failing to meet the practical clinical requirement as they do not provide any hint when used to classify medical images. Features are extracted in an implicit manner, few hints about how to make the decision have been demonstrated, resulting in a wonder why the algorithms predict a label for domain experts who care the interpretability most. Moreover, CNNs can only handle regular nodes, and are not suitable for complex scenes with irregular node distributions. Another challenge in head CT image quality classification is that decrease in image quality at different regions vary substantially.

The motivation of this paper is that, from the above analysis, the core issues of medical image classification are how to model the relationship of the head CT images, and how to efficiently fuse domain knowledge to improve the classification accuracy. To achieve these goals, in this paper, as it is essential to take region heterogeneity into account, we introduce complex network theory to model the relationship between head CT images based on anatomical prior knowledge. Great efforts have been dedicated to facilitate the understanding the relationship between topology and function and the network characteristics in a systematic manner. The relationship may help to understand the images of different qualities, thereby providing better explainability. It is intuitively showed that by examining the content of each slice and analyzing the relationship among slices comprehensively, informative representations could be learned, the quality of CT images could be predicted in a more interpretable fashion, and since then diagnosis and treatment plans may be better formulated (Fig. 1).

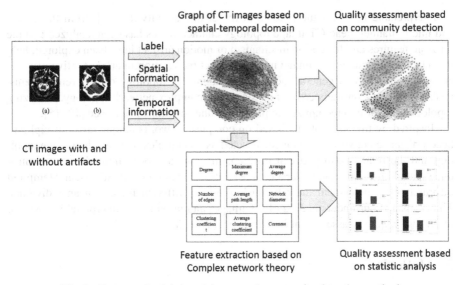

Fig. 1. Framework of the spatial-temporal community detection method.

The main contributions are as follows.

Firstly, in order to comprehensively characterize the relationship between CT images, a novel construction method of CT image networks based on spatial-temporal domain is proposed.

Secondly, network features of hybrid quality of CT images have been explored by analyzing the complex network topology. The effectiveness of temporal information has been validated by comparing the spatial-temporal domain construction method and spatial information method.

Third, by utilizing community detection, multi-region quality assessment is implemented based on the spatial-temporal domain.

2   Related Works

To better present and understand the relationship, complex networks have attracted increasing attention due to the powerful capability, great flexibility and generality, many cross-disciplinary applications based on complex networks have been made, including transportation networks, social networks, biology, earthquake networks and so on.

Graph theory [6, 7] has demonstrated effective in understanding and recognizing the complexity of real systems, which represents the relationship of topology and function of complex systems from an overall perspective and investigates the internal mechanism of the system in a new way. In recent years, the application of graph theory has been extensively explored and has made significant progress. Researchers have applied graph theory to study the characteristics of the macro-topological dynamic nodes of the Internet, and reveals the internal evolution law of the Internet [8, 9]. In [10], spatio-temporal influence domain has been utilized to construct the seismic network, which provides an interpretable basis for the prediction of seismic activity. In [11], from the macro and micro manner, head CT image topology characteristics has been analyzed and the intrinsic features and formation mechanism of motion artifact have been explored, help us to better understand CT image motion artifact recognition in an innovative way. As complex network is a general method of recognizing topology of different systems, researchers focus on transferring real system to complex network topology, analyzing topological feature to recognize the properties and functions in essence [12–15].

Inspired by the applications based on complex networks among cross-disciplines, head CT image network has been constructed using complex network theory, the topology and the relationship between different images have been figured out, we introduce spatial information to understand the relationship among individuals, and temporal information to understand the relationship among different images of an individual. This deepened our understanding of CT images and provided an interpretable basis for multi-region quality assessment in the medical domain.

3 Methods

3.1 Problem Description and Basic Assumptions

Artifacts may lead to a sharp decline in image quality, compromise the post-processing and dramatically affect the diagnosis accuracy, which brings significant confusion in the diagnostic process, and results in the bias of diagnosis, treatment planing and prognosis. CT image artifacts, including motion artifacts, metal artifacts, stripe artifacts, and so on, have degraded the quality of CT image severely.

In this paper, we make several assumptions to simplify the investigation.

Firstly, the quality is evaluated based on whether motion artifacts has affected the diagnosis process.

Additionally, each single ST scan consists of multiple slices, with a maximum of 32 slices. Each slice is composed of multiple pixels, and all the slices have a consistent resolution.

Furthermore, it is assumed that the qualities and anatomical prior knowledge are correctly labeled.

Figure 2 Shows typical artifacts-free CT images and artifacts-affected CT images.

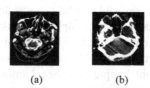

(a)   (b)

Fig. 2. Typical CT images of different qualities. (a) artifacts-free CT image (b) artifacts-affected CT image.

The Image acquisition log provided below includes information such as acquisition time, slice location information, and image quality annotations. The image quality is divided into two categories, one is artifact-free image with image quality marked "high" and the other is with image quality marked "low". It is assumed that the data collected at the same time belong to the same individual. There are differences in the number of images at the same acquisition time, resulting in the absence of some region belonging to a certain individual. Thus, the acquired images are heterogeneous. Table 1 depicts the information of CT images.

Table 1. Acquisition information.

Image ID	Acquisition time	Image quality	Region ID
1	T1	high	S5
2	T2	low	S2
3	T2	high	S5
4	T3	low	S2
5	T3	low	S7
6	T4	low	S2
7	T4	high	S5
8	T5	low	S5
9	T5	low	S6
10	T5	high	S2
11	T5	low	S4
12	T6	low	S2
13	T6	low	S3
14	T6	low	S4
15	T6	low	S5
16	T6	low	S6
17	T6	low	S7
18	T7	high	S2
19	T7	high	S3
20	T7	high	S6

3.2 CT Image Graph Construction Based on Spatial-Temporal Domain

Suppose G (V, E), where G denotes the undirected unweighted graph, V denotes node, which represent head CT image, and E denotes edge between head CT images. When the artifact identification problem of CT images is transformed into a complex network node classification problem, [11] constructed the relationship between images based on region information, which provides a topological basis for artifacts identification. In order to characterize the relationship between CT images more comprehensively, this paper integrates temporal information, which reflects individual information, and then displays the relationship between images more comprehensively.

Each CT image can be seen as a node, and the relationship between the CT images can be seen as the edges. For simplicity, in this paper, an undirected unweighted graph has been constructed.

If two CT images are acquired at the same time and share the same label or the two CT images have the same spatial information and share the same label, an edge is generated between two CT images.

The network topology constructed from the log information in Table 1 has been shown in Fig. 3 below.

Fig. 3. Network topology of spatial-temporal domain.

By taking temporal information into consideration, image 2 and image 3 are connected, however, the edge does not exist in the construction of network topology based on merely spatial information. Graph construction based on spatial information, image 13,14,19,20 are isolated.

3.3 Topological Features of Complex Networks

Complex network topological features [16–18] contain node-level features and graph-level features. By investigating network topology characteristics, network structure can be recognized and the function of network can be understood in a topological way. The implications of the topological features are as follows.

Degree [19]: degree belongs to the node-level network characteristics, which depicts the properties of a single node.

Average degree: average degree belongs to the graph-level characteristics, which can be used to describe the overall characteristics of the network.

Number of edges: graph-level features which can be used to characterize network size.

Average path length [20]: the average distance between any two nodes.

Network diameter: the maximum distance between any two nodes in the network.

Clustering Coefficient [20]: the degree of clustering of the network. Average clustering coefficient is a measure of how closely a node connects to its neighbors.

Coreness: the hierarchy can be recognized by k-core decomposition.

3.4 Construction of CT Image Network Topology Based on Spatial-Temporal Domain

The annotation of head CT images used in this paper includes both spatial and temporal information, and the number of images collected at the same time varies. An undirected and unweighted graph has been constructed, as depicted in Fig. 4.

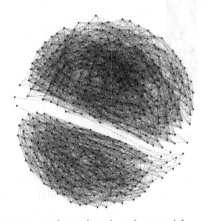

Fig. 4. CT image network topology based on spatial-temporal domain.

The topological characteristics were investigated after constructing the graph based on the spatial-temporal domain, as shown in Table 2 below.

Table 2. The network topological characteristics of spatial-temporal graph.

Dataset	Average clustering coefficient	Average Path length	Average Degree
Hybrid CT images	0.588	2.159	23.238
CT images without artifacts alone	0.567	2.072	19.765
CT images with artifacts alone	0.601	2.195	25.454

4 Experiments and Results

4.1 Effectiveness of Temporal Information

This paper compares the network topological characteristics based on spatial-temporal information with those based solely on spatial information.

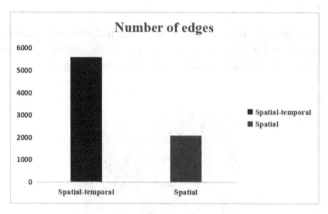

Fig. 5. Number of edges of spatial-temporal and spatial information construction.

As can be seen from Fig. 5, when considering temporal information, and the network size has been enlarged. That is to say, isolated nodes may decrease, and the influence of node may increase.

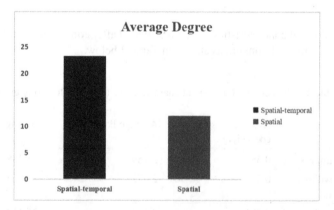

Fig. 6. Average degree of spatial-temporal and spatial information construction.

As can be seen from Fig. 6, average degree has increased when taking temporal information into consideration, indicating that node becomes more important.

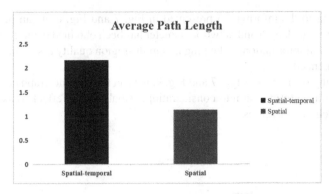

Fig. 7. Average path length of spatial-temporal and spatial information construction.

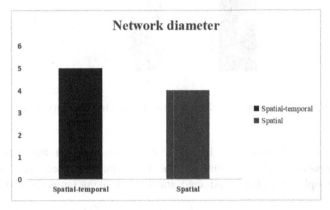

Fig. 8. Average path length of spatial-temporal and spatial information construction.

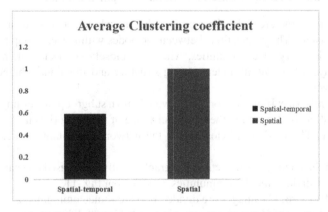

Fig. 9. Average clustering coefficient of spatial-temporal and spatial information construction.

According to the information presented in Fig. 7 and Fig. 8, it can be seen that a larger average path length and network diameter has been obtained in the graph based on spatial-temporal information, indicating that multi-region quality assessment capability has been enhanced.

Combining the results of Fig. 7 and Fig. 9, we can see that the graph constructed by taking temporal information into consideration, "small world" effect weakens and the size of the "circle" increases.

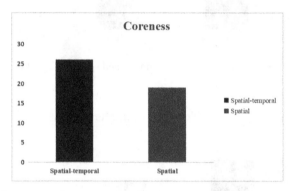

Fig. 10. Coreness of spatial-temporal and spatial information construction.

As can be seen from Fig. 10, the coreness is larger in the spatial-temporal information-based graph than that of spatial information-based one, indicating an increased complexity of the constructed graph when taking temporal information into account.

4.2   Community Detection Based on Spatial-Temporal Domain

The community structure is the characteristic that the whole network consists of several groups or clusters. The connections between the nodes within each group are relatively close. The discovery of communities, which are closely connected elements in real systems, plays a key role in understanding structure and functional characteristics of various networks.

Community size: If the network G is divided into n subgroups according to a certain community division algorithm, the community size of the network is n.

Modular: The modular is closer to 1, the network community structure is more obvious.

Through the community detection, the graph construction based on spatial-temporal information is divided into 9 communities, as shown in Fig. 11.

The artifacts-free CT images are divided into the same community, as shown in the purple part of Fig. 11. The artifacts ones are divided into 8 different communities.

As can be seen from Fig. 11-Fig. 13, when taking temporal information into consideration, smaller modular has occurred, indicating that cliques are no longer obvious and information may flowed across different regions by studying other images of a

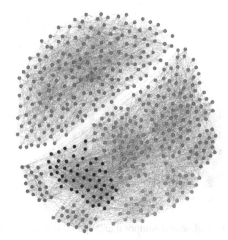

Fig. 11. Community detection in CT network.

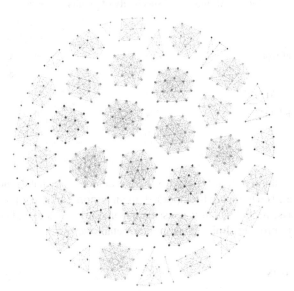

Fig. 12. Community detection with spatial information.

certain individual. In contrast, the spatial information-based graph (Fig. 12) shows 44 communities, suggesting a more refined and less generalized community division. The graph constructed based on the spatial-temporal information may be more suitable for multi-region quality assessment in the clinical practices.

Table 3 gives comparisons of accuracy and specificity among different methods, indicating that the spatial-temporal domain community detection achieves the best performance in terms of accuracy and specificity.

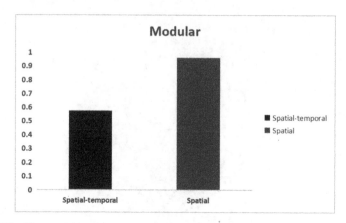

Fig. 13. Modular of spatial-temporal and spatial information construction.

Table 3. Accuracy and specificity of quality assessment

Method	Accuracy	Specificity
Spatial-Temporal Domain	99.79%	99.59%
Spatial Domain	96.67%	96.77%
MADM-CN + SVM [11]	98%	96%
MADM-CN + RF [11]	97%	98%
CNN [4]	76.67%	66.67%

Furthermore, the community detection results demonstrate that the network topology construction based on the spatial-temporal domain is more suitable for the study of multi-region quality assessment. Quality assessment from the perspective of complex network can be helpful to provide clinicians with interpretable results.

5 Conclusions

Head CT image quality assessment is a critical part of image classification. In the process of head CT image graph construction, combining domain knowledge with complex networks theory, the graph constructed based on the spatial-temporal domain depicts the relationship between CT images from a comprehensive perspective. Network features of hybrid qualities of CT images have been explored by analyzing the complex network topology. The influence and closeness between nodes are represented by the edges. Multi-region quality classification problem has been solved by spatial-temporal community detection when taken the individual differences into consideration, which validates the effectiveness of the temporal information. Understanding these relationships between nodes and the influence on each other help us to recognize the intrinsic mechanism of image quality assessment. Experimental results on a real-world dataset demonstrate the effectiveness of the spatial-temporal community detection method. This paper provides scientific guidance for the study of motion artifacts identification and CT image quality assessment.

References

1. Boas, F., Fleischmann, D.: CT artifacts: causes and reduction techniques. Imaging Med. **4**(2), 229–240 (2012)
2. Litjens, G., et al.: A survey on deep learning in medical image analysis. Med. Image Anal. **42**, 60–88 (2017)
3. Chen, X., et al.: Recent advances and clinical applications of deep learning in medical image analysis. Med. Image Anal. **79**, 102444 (2022)
4. Krizhevsky, A., Sutskever, I., Hinton, G.: ImageNet classification with deep convolutional neural networks. Adv. Neural Inf. Process. Sys., t1097–1105 (2012)
5. Guan, Q., Huang, Y., Luo, Y., Liu, P., Xu, M.: Yang, Y: Discriminative feature learning for thorax disease classification in chest X-ray images. IEEE Trans. Image Process. **30**, 2476–2487 (2021)
6. Strogatz, S.: Exploring complex networks. Nature **410**(6835), 268–276 (2001)
7. He, X., Wang, L., Liu, Z., Liu, Y.: Similar seismic activities analysis by using complex networks approach. Symmetry **12**(5), 778 (2020)
8. Liu, Y., Zhao, H., Ai, J., Jia, S.: Characteristics of birth and death nodes with IP-level topology. J. Northeastern Univ. (Nat. Sci.) **34**(9), 1232–1235 (2013)
9. Liu, Y., Zhao, H., Ai, J., Wang, J.: Research on correlation between internet measurement levels and dynamic nodes characteristics. J. Northeastern Univ. (Nat. Sci.) **35**(2), 195–198 (2014)
10. He, X., Wang, L., Zhu, H., Liu, Z.: Statistical properties of complex network for seismicity using depth-incorporated influence radius. Acta Geophys. **67**(6), 1515–1523 (2019)
11. Liu, Y., et al.: Graph-based motion artifacts detection method from head computed tomography images. Sensors **22**(15) (2022). 5666
12. Zhou, T., Lü, L., Zhang, Y.C.: Predicting missing links via local information. Eur. Phys. J. B. Condens. Matter Phys. **71**(4), 623–630 (2009)
13. Wu, T., Huang, Q., Liu, Z., Wang, Y., Lin, D.: Distribution-balanced loss for multi-label classification in long-tailed datasets. In: Vedaldi, A., Bischof, H., Brox, T., Frahm, J.-M. (eds.) ECCV 2020. LNCS, vol. 12349, pp. 162–178. Springer, Cham (2020). https://doi.org/10.1007/978-3-030-58548-8_10
14. Lewis, T.G.: Network Science: Theory and Applications. John Wiley & Sons, Hoboken, New Jersey (2009)
15. Senior, A.W., et al.: Improved protein structure prediction using potentials from deep learning. Nature **577**(7792), 706–710 (2020)
16. Barabási, A.L., Albert, R.: Emergence of scaling in random networks. Science **286**, 509–512 (1999)
17. da Fontoura Costa, L., Rodrigues, F.A., Travieso, G., Villas Boas, P.R.: Characterization of complex networks: a survey of measurements. Adv. Phys. **56**(1), 167–242 (2007)
18. Dorogovtsev, S.N., Mendes, J.F.F.: Evolution of networks. Adv. Phys. **51**, 1079–1187 (2002)
19. Albert, R., Barabási, A.L.: Statistical mechanics of complex networks. Rev. Mod. Phys. **74**(1), 47–97 (2002)
20. Watts, D.J., Strogatz, S.H.: Collective dynamics of 'small-world' networks. Nature **393**(6684), 440–442 (1998)

Author Index

X. Yang et al. (Eds.): ADMA 2023, LNAI 14179, pp. 695–697, 2023.
https://doi.org/10.1007/978-3-031-46674-8

Printed in the United States
by Baker & Taylor Publisher Services